Lecture Notes in Artificial Intelligence   5179

Edited by R. Goebel, J. Siekmann, and W. Wahlster

Subseries of Lecture Notes in Computer Science

Ignac Lovrek   Robert J. Howlett
Lakhmi C. Jain (Eds.)

Knowledge-Based Intelligent Information and Engineering Systems

12th International Conference, KES 2008
Zagreb, Croatia, September 3-5, 2008
Proceedings, Part III

Springer

Series Editors

Randy Goebel, University of Alberta, Edmonton, Canada
Jörg Siekmann, University of Saarland, Saarbrücken, Germany
Wolfgang Wahlster, DFKI and University of Saarland, Saarbrücken, Germany

Volume Editors

Ignac Lovrek
University of Zagreb, Faculty of Electrical Engineering and Computing
Unska 3, 10000 Zagreb, Croatia
E-mail: ignac.lovrek@fer.hr

Robert J. Howlett
KES International, 2nd Floor
145-157 St. John Street, London EC1V 4PY, UK
E-mail: rjhowlett@kesinternational.org

Lakhmi C. Jain
University of South Australia, School of Electrical and Information Engineering
Mawson Lakes Campus, Adelaide, SA 5095, Australia
E-mail: lakhmi.jain@unisa.edu.au

Library of Congress Control Number: 2008933454

CR Subject Classification (1998): I.2, H.4, H.3, J.1, H.5, K.6, K.4

LNCS Sublibrary: SL 7 – Artificial Intelligence

ISSN   0302-9743
ISBN-10   3-540-85566-1 Springer Berlin Heidelberg New York
ISBN-13   978-3-540-85566-8 Springer Berlin Heidelberg New York

Springer is a part of Springer Science+Business Media

springer.com

© Springer-Verlag Berlin Heidelberg 2008
Printed in Germany

Typesetting: Camera-ready by author, data conversion by Scientific Publishing Services, Chennai, India
Printed on acid-free paper   SPIN: 12511365   06/3180   5 4 3 2 1 0

Preface

Delegates and friends, we are very pleased to extend to you a warm welcome to this, the 12th International Conference on Knowledge-Based and Intelligent Information and Engineering Systems organised by the Faculty of Electrical Engineering and Computing at the University of Zagreb, in association with KES International.

For over a decade, KES International has provided an annual wide-spectrum intelligent systems conference for the applied artificial intelligence research community. Having originated in Australia and been held there during 1997–99, the conference visited the UK in 2000, Japan in 2001, Italy in 2002, the UK in 2003, New Zealand in 2004, Australia in 2005, the UK in 2006, Italy in 2007, and now in Zagreb, Croatia in 2008. It is planned that KES 2009 will be held in Santiago, Chile before returning to the UK in 2010. The KES conference is mature and regularly attracts several hundred delegates. As it encompasses a broad range of intelligent systems topics, it provides delegates with an opportunity to mix with researchers from other groups and learn from them. The conference is linked to the International Journal of Intelligent and Knowledge-Based Systems, published by IOS Press under KES editorship. Extended and enhanced versions of the best papers presented at the KES conference may be published in the Journal.

In addition to the annual wide-range intelligent systems conference, KES has run successful symposia in several specific areas of the discipline. Agents and Multi-Agent Systems is a popular area of research. The first KES symposium on Agents and Multi-Agent Systems took place in 2007 in Wroclaw, Poland (KES-AMSTA 2007) followed in 2008 by a second in Incheon, Korea (KES-AMSTA 2008). The third in the series is planned to be held in the historic city of Uppsala in Sweden (June 3–5, 2009). Intelligent Multi-Media is a second area of focus for KES. The first KES symposium on Intelligent and Interactive Multi-Media Systems and Services (KES IIMSS 2008) will be held in Athens, Greece, in 2008, followed by a second in Venice, Italy, in 2009 (dates to be notified). A third area of interest supported by KES is Intelligent Decision Support Technologies, and the first KES symposium on this subject (KES IDT 2009) is planned for Hyogo in Japan for next year (April 23–25, 2009). Over time, each of these areas will be supported by a KES focus group of researchers interested in the topic, and if appropriate, by a journal. To this end, the International Journal of Intelligent Decision Support Technologies is published by IOS Press under the editorship of a developing KES IDT focus group.

For the future we have plans and a vision for KES. Firstly, we describe the plans.

We plan to maintain and increase the quality of KES publications. The KES quality principle is that we do not seek to expand KES activities by publishing inferior papers. However, equally, we do not believe it serves authors or the research community to reject good papers on the basis of an arbitrary acceptance / rejection ratio. Hence papers for KES conferences and symposia, and the KES Journal, will be rigorously reviewed by experts in the field, and published only if they are of a sufficiently high level, judged by international research standards.

We will further develop KES focus groups, and where appropriate, we will adopt journals and symposia, where this supports and helps us maintain our quality principle.

We introduced the concept of KES membership several years ago to provide returning KES delegates with discounted conference fees. We plan to supplement the benefits of KES membership by launching a profile page system, such that every KES member will have their own profile page on the KES web site, and be able to upload a description of themselves, their research interests and activities. The member site will act as a contact point for KES members with common interests and a potential channel to companies interested in members' research.

Printing technology is changing and this will have an effect on publishers and publications. We are conducting trials with rapid publication technology that makes it possible to print individual copies of a book on demand. The KES Rapid Research Results book series will make it convenient to publish books appealing to niche markets, for example specialised areas of research, in a way that would not have been economic before.

Many KES members and supporters have research interests outside intelligent systems. In fact, intelligent systems may just be a tool used for an application which is the main interest. A significant number of those involved in KES have interests in environmental matters. In 2009, KES will address the issue of sustainability and renewable energy through its first conference in this area, Sustainability in Energy and Buildings (SEB 2009), which will be held in Brighton, UK, April 30–May 1, 2009. SEB 2009 will address a broad spectrum of sustainability issues relating to renewable energy and the efficient use of energy in domestic and commercial buildings. Papers on the application of intelligent systems to sustainability issues will be welcome. However, it will not be compulsory that papers for SEB 2009 have significant intelligent systems content (as is a criterion for other KES conferences and symposia).

In addition to the firm plans for KES there is a longer term vision. In the time that it has been in existence, KES International has evolved from being the organiser of just a single annual conference, to a provider of an expanding portfolio of support functions for the research community. Undoubtedly, KES will continue to develop and enhance its knowledge transfer activities. The KES community consists of several thousand members, and potentially it could play a significant role in generating synergy and facilitating international research co-operation. A long term vision for KES is for it to evolve into an international academy providing the means for its members to perform international collaborative research projects. At the moment we do not have the means to turn this vision into a reality, but we will work towards this aim.

The annual KES conference continues to be a major feature of the KES organisation, and KES 2008 will continue the tradition of excellence in KES conferences.

The papers for KES 2008 were submitted either to Invited Sessions, chaired and organised by respected experts in their fields, or to General Sessions, managed by Track Chairs. Each paper was thoroughly reviewed by two members of the International Review Committee, and also inspected by a Track Chair or Invited Session Chair. A decision about whether to publish the paper was made, based on the KES quality principle described above. If the paper was judged to be of high enough quality to be accepted, the Programme Committee then decided on oral or poster presentation, based on the subject and content of the paper. All papers at KES 2008 are considered to be of equal weight and importance, no matter whether they were oral or poster presentations. This has resulted in the 316 high-quality papers included in these proceedings.

Thanks are due to the very many people who have given their time and goodwill freely to make the conference a success.

We would like to thank the KES 2008 International Programme Committee for their help and advice, and also the International Review Committee, who were essential in providing their reviews of the papers. We are very grateful for this service, without which the conference would not have been possible. We thank the high-profile keynote speakers for providing interesting expert talks to inform and inspire subsequent discussions.

An important feature of KES conferences is the Invited Session Programme. Invited Sessions give both young and established researchers an opportunity to organise and chair a set of papers on a specific topic, presented as a themed session. In this way, new topics at the leading edge of intelligent systems can be presented to interested delegates. This mechanism for feeding new ideas into the research community is very valuable. We thank the Invited Session Chairs who have contributed in this way.

The conference administrators, and the local organising committee, have all worked extremely hard to bring the conference to a high level of organisation. In this context, we would like to thank Mario Kusek, Kresimir Jurasovic, Igor Ljubi, Ana Petric, Vedran Podobnik and Jasna Slavinic (University of Zagreb, Croatia); Peter Cushion, Nicola Pinkney and Antony Wood (KES Operations, UK).

A vital contribution was made by the authors, presenters and delegates without whom the conference could not have taken place. So finally, but by no means least, we thank them for their involvement.

June 2008   Bob Howlett
                                                                 Ignac Lovrek
                                                                 Lakhmi Jain

Organisation

KES 2008 Conference Organisation

KES 2008 was organised by KES International, Innovation in Knowledge-Based and Intelligent Engineering Systems, and the University of Zagreb, Faculty of Electrical Engineering and Computing.

KES 2008 Conference Chairs

General Chair	Ignac Lovrek, University of Zagreb, Croatia
Executive Chair	Robert J. Howlett, University of Brighton, UK
Invited Sessions Chair	Lakhmi C. Jain, University of South Australia, Australia
Local Chair	Mario Kusek, University of Zagreb, Croatia
Award Chair	Bogdan Gabrys, University of Bournemouth, UK

KES Conference Series

KES 2008 is a part of the KES Conference Series.

Conference Series Chairs	Lakhmi C. Jain and Robert J. Howlett
KES Executive Chair	Robert J. Howlett, University of Brighton, UK
KES Founder	Lakhmi C. Jain, University of South Australia, Australia

Local Organising Committee

Kresimir Jurasovic, Igor Ljubi, Ana Petric, Vedran Podobnik, Jasna Slavinic (University of Zagreb, Croatia), Peter Cushion (KES Operations, UK)

International Programme Committee

Abe, Akinori	ATR Knowledge Science Laboratories, Japan
Adachi, Yoshinori	Chubu University, Japan
Angulo, Cecilio	Technical University of Catalonia, Spain
Apolloni, Bruno	University of Milan, Italy
Baba, Norio	Osaka Kyoiku University, Japan
Balachandran, Bala M.	University of Canberra, Australia
Dalbelo Basic, Bojana	University of Zagreb, Croatia
Beristain, Andoni	Universidad del Pais Vasco, Spain

Bianchini, Monica	Universita degli Studi di Siena, Italy
Castellano, Giovanna	University of Bari, Italy
Chen, Yen-Wei	Ritsumeikan University, Japan
Cheng, Jingde	Saitama University, Japan
Corchado, Emilio	University of Burgos, Spain
Cuzzocrea, Alfredo	University of Calabria, Italy
Damiani, Ernesto	University of Milan, Italy
Di Noia, Tommaso	Technical University of Bari, Italy
Esposito, Floriana	University of Bari, Italy
Gabrys, Bogdan	University of Bournemouth, UK
Gao, Kun	Zhejiang Wanli University, China
Hartung, Ronald L.	Franklyn University, USA
Hakansson, Anne	Uppsala University, Sweden
Holmes, Dawn	University of California Santa Barbara, USA
Howlett, Robert J.	University of Brighton, UK
Ishida, Yoshiteru	Toyohashi University of Technology, Japan
Ishii, Naohiro	Aichi Institute of Technology, Japan
Jain, Lakhmi C.	University of South Australia, Australia
Jevtic, Dragan	University of Zagreb, Croatia
Karny, Miroslav	Academy of Sciences of the Czech Republic, Czech Republic
Kunifuji, Susumu	Japan Advanced Institute of Science and Technology, Japan
Lee, Hsuan-Shih	National Taiwan Ocean University, Taiwan
Lim, Chee-Peng	University of Science, Malaysia
Liu, Honghai	University of Portsmouth, UK
Lovrek, Ignac	University of Zagreb, Croatia
Maojo, Victor	Universidad Politecnica de Madrid, Spain
Markey, Mia	University of Texas at Austin, USA
Mumford, Christine	Cardiff University, UK
Munemori, Jun	Wakayama University, Japan
Nakamatsu, Kazumi	University of Hyogo, Japan
Nakano, Ryohei	Nagoya Institute of Technology, Japan
Nakao, Zensho	University of the Ryukyus, Japan
Nauck, Detlef	BT Intelligent Systems Research Centre, UK
Negoita, Mircea Gh.	KES International
Nguyen, Ngoc Thanh	Wroclaw University of Technology, Poland
Nicoletti, Maria do Carmo	Federal University of Sao Carlos, Brazil
Nikolos, Ioannis K.	Technical University of Crete, Greece
Nishida, Toyoaki	Kyoto University, Japan
Nuernberger, Andreas	University of Magdeburg, Germany
Palade, Vasile	University of Oxford, UK
Park, Gwi-Tae	Korea University, Seoul, Korea
Pham, Tuan	James Cook University, Australia
Phillips-Wren, Gloria	Loyola College in Maryland, USA

Sharma, Dharmendra   University of Canberra, Australia
Shkodirev, Viacheslaw   St Petersburg State Polytechnic University,
                                Russia
Sik Lanyi, Cecilia   University of Pannonia, Hungary
Sunde, Jadranka   Defence Science and Technology Organisation,
                                Australia
Tagliaferri, Roberto   University of Salerno, Italy
Taki, Hirokazu   Wakayama University, Japan
Tsuda, Kazuhiko   University of Tsukuba, Japan
Turchetti, Claudio   Università Politecnica delle Marche, Italy
Vassanyi, Istvan   University of Pannonia, Hungary
Veganzones, Miguel A.   Universidad del Pais Vasco, Spain
Vellido, Alfredo   Technical University of Catalonia, Spain
Watada, Junzo   Waseda University, Japan
Watanabe, Toyohide   Nagoya University, Japan
Yamashita, Yoshiyuki   Tokio University of Agriculture and
                                Technology, Japan

International Review Committee

Abe, Akinori   ATR Knowledge Science Laboratories, Japan
Abe, Jair   University of Sao Paulo, Brazil
Abu Bakar, Rohani   Waseda University, Japan
Abulaish, Muhammad   Jamia Millia Islamia, India
Adachi, Yoshinori   Chubu University, Japan
Adli, Alexander   University of the Ryukyus, Japan
Akama, Seiki   C-Republics, Japan
Al-Hashel, Ebrahim   University of Canberra, Australia
Alquezar, Ren   Technical University of Catalonia, Spain
Angelov, Plamen   Lancaster University, UK
Anguita, Davide   University of Genoa, Italy
Angulo, Cecilio   Technical University of Catalonia, Spain
Anisetti, Marco   University of Milan, Italy
Aoyama, Kouji   Fujitsu Laboratories Limited, Japan
Apolloni, Bruno   University of Milan, Italy
Appice, Annalisa   University of Bari, Italy
Aritsugi, Masayoshi   Kumamoto University, Japan
Arroyo-Figueroa, Gustavo   Instituto de Investigaciones Electricas, Mexico
Azzini, Antonia   University of Milan, Italy
Baba, Norio   Osaka Kyoiku University, Japan
Balachandran, Bala M.   University of Canberra, Australia
Balas, Marius   Aurel Vlaicu University of Arad, Romania
Balas, Valentina E.   Aurel Vlaicu University of Arad, Romania
Balic, Joze   University of Maribor, Slovenia
Bandini, Stefania   University of Milan Bicocca, Italy
Baruque, Bruno   University of Burgos, Spain

Basili, Roberto   University of Rome, Italy
Bassis, Simone   University of Milan, Italy
Bayarri, Vicente   GIM Geomatics S.L., Spain
Belanche, Lluis   Technical University of Catalonia, Spain
Bellandi, Valerio   University of Milan, Italy
Berendt, Bettina   Katholieke Universiteit Leuven, Belgium
Bianchini, Monica   Università degli Studi di Siena, Italy
Bidlo, Michal   Technical University of Brno, Czech Republic
Bielikov, Mria   Slovak University of Technology, Slovakia
Billhardt, Holger   Universidad Rey Juan Carlos I, Spain
Bingul, Zafer   Kocaeli University, Turkey
Bioucas, Jose   Instituto Superior Tecnico, Portugal
Bogdan, Stjepan   University of Zagreb, Croatia
Borghese, Alberto   University of Milan, Italy
Borzemski, Leszek   Wroclaw University of Technology, Poland
Bouchachia, Abdelhamid   University of Klagenfurt, Austria
Bouquet, Paolo   Università degli Studi di Trento, Italy
Bouridane, Ahmed   Queen's University, Belfast, UK
Bruzzone, Lorenzo   Università degli Studi di Trento, Italy
Buciu, Ioan   University of Oradea, Romania
Cabestany, Joan   Technical University of Catalonia, Spain
Calpe-Maravilla, Javier   University of Valencia, Spain
Camino-Gonzalez, Carlos Luis   Forestry and Technology Center of Catalonia,
                                        Spain
Camps-Valls, Gustavo   University of Valencia, Spain
Cao, Jiangtao   University of Portsmouth, UK
Capkovic, Frantisek   Slovak Academy of Sciences, Slovakia
Carpintero-Salvo, Irene Rosa   Universidad de Granada and Consejeria de
                                        Medio Ambiente, Spain
Castellano, Giovanna   University of Bari, Italy
Castiello, Ciro   University of Bari, Italy
Castillo, Elena   University of Cantabria, Spain
Cavar, Damir   Indiana University, USA
Ceccarelli, Michele   University of Sannio, Italy
Ceravolo, Paolo   University of Milan, Italy
Chan, Chee Seng   University of Portsmouth, UK
Chang, Chan-Chih   Industrial Technology Research Institute,
                                        Taiwan
Chang, Chuan-Yu   National Yunlin University of Science &
                                        Technology, Taiwan
Chen, Mu-Yen   National Changhua University of Education,
                                        Taiwan
Chen, Yen-Wei   Ritsumeikan University, Japan
Cheng, Jingde   Saitama University, Japan
Chetty, Girija   University of Canberra, Australia

Chi Thanh, Hoang	Hanoi University of Science, Vietnam
Ciaramella, Angelo	University of Naples Parthenope, Italy
Cicin Sain, Marina	University of Rijeka, Croatia
Claveau, Vincent	IRISA-CNRS, France
Coghill, George	University of Auckland, New Zealand
Colucci, Simona	Technical University of Bari, Italy
Corbella, Ignasi	Technical University of Catalonia, Spain
Corchado Rodriguez, Juan M.	University of Salamanca, Spain
Corchado, Emilio	University of Burgos, Spain
Costin, Mihaela	Institute for Computer Science, Romanian Academy, Romania
Cox, Robert	University of Canberra, Australia
Crippa, Paolo	Università Politecnica delle Marche, Italy
Cruz, Antonio	Federal University of Sao Carlos, Brazil
Csipkes, Gabor	Technical University of Cluj-Napoca, Romania
Curzi, Alessandro	Università Politecnica delle Marche, Italy
Cuzzocrea, Alfredo	University of Calabria, Italy
Czarnowski, Ireneusz	Gdynia Maritime University, Poland
D'Amato, Claudia	University of Bari, Italy
D'Apuzzo, Livia	University of Naples Federico II, Italy
Dalbelo Basic, Bojana	University of Zagreb, Croatia
Damiani, Ernesto	University of Milan, Italy
Davidsson, Paul	Blekinge Institute of Technology, Sweden
de Campos, Cassio Polpo	Rensselaer Polytechnic Institute, USA
De Gemmis, Marco	University of Bari, Italy
De Santis, Angela	University of Alcala, Spain
Deguchi, Toshinori	Gifu National College of Technology, Japan
Dell'Endice, Francesco	University of Zurich, Switzerland
Di Noia, Tommaso	Technical University of Bari, Italy
Di Sciascio, Eugenio	Technical University of Bari, Italy
Diani, Marco	University of Pisa, Italy
Diaz-Delgado, Ricardo	Estacion Biologica de Donana-CSIC, Spain
Dobsa, Jasminka	University of Zagreb, Croatia
Dorado, Julian	Universidad de la Coruna, Spain
Dujmic, Hrvoje	University of Split, Croatia
Dumitriu, Luminita	Dunarea de Jos University, Romania
Duro, Richard	Universidade da Coruna, Spain
Edman, Anneli	Uppsala University, Sweden
Erjavec, Tomaz	Josef Stefan Institute, Slovenia
Esposito, Anna	University of Naples Federico II and IIASS, Italy
Esposito, Floriana	University of Bari, Italy
Etchells, Terence	Liverpool John Moores University, UK
Lee, Eun-Ser	Soong Sil University, Korea
Fang, H.H.	Taipei College of Maritime Technology, Taiwan

Fasano, Giovanni	University of Venice, Italy
Feng, Jun	Hohai University, China
Fernandez-Caballero, Antonio	Universidad de Castilla-La Mancha, Spain
Fras, Mariusz	Wroclaw University of Technology, Poland
Frati, Fulvio	University of Milan, Italy
Fuchino, Tetsuo	Tokyo Institute of Technology, Japan
Fujimoto, Taro	Fujitsu Laboratories Limited, Japan
Fujinami, Tsutomu	Japan Advanced Institute of Science and Technology, Japan
Fukue, Yoshinori	Fujitsu Laboratories Limited, Japan
Gallego-Merino, Miren Josune	Universidad del Pais Vasco, Spain
Gamberger, Dragan	Rudjer Boskovic Institute, Croatia
Gao, Kun	Zhejiang Wanli University, China
Gao, Ying	Saitama University, Japan
Garcia-Sebastian, Maite	Universidad del Pais Vasco, Spain
Gastaldo, Paolo	University of Genoa, Italy
Gendarmi, Domenico	University of Bari, Italy
Georgieva, Petia	University of Aveiro, Portugal
Gianfelici, Francesco	Università Politecnica delle Marche, Italy
Gianini, Gabriele	University of Milan, Italy
Giordano, Roberto	Federal University of Sao Carlos, Brazil
Giorgini, Paolo	University of Trento, Italy
Gledec, Gordan	University of Zagreb, Croatia
Gold, Hrvoje	University of Zagreb, Croatia
Goldstein, Pavle	University of Zagreb, Croatia
Gomez-Dans, Jose Luis	University College London, UK
Goto, Yuichi	Saitama University, Japan
Grana, Manuel	Universidad del Pais Vasco, Spain
Greenwood, Garrison	Portland State University, USA
Gu, Dongbing	University of Essex, UK
Guo, Huawei	Shanghai Jiao Tong University, China
Hakansson, Anne	Uppsala University, Sweden
Halilcevic, Suad	University of Tuzla, Bosnia & Herzegovina
Hammer, Barbara	Clausthal University of Technology, Germany
Hanachi, Chihab	University Toulouse 1 and IRIT Laboratory, France
Hara, Akira	Hiroshima City University, Japan
Harada, Koji	Toyohashi University of Technology, Japan
Harris, Irina	Cardiff University, UK
Harris, Richard	University of Lancaster, UK
Hartung, Ronald L.	Franklin University, USA
Hasegawa, Shinobu	Japan Advanced Institute of Science and Technology, Japan
Hayashi, Hidehiko	Naruto University of Education, Japan
Hernandez, Carmen	Universidad del Pais Vasco, Spain

Herrero, Alvaro   University of Burgos, Spain
Hildebrand, Lars   Technical University of Dortmund, Germany
Hiroshi, Mineno   Shizuoka University, Japan
Handa, Hisashi   Okayama University, Japan
Hocenski, Zeljko   University of Osijek, Croatia
Hori, Satoshi   Monotsukuri Institute of Technologists, Japan
Howlett, Robert J.   University of Brighton, UK
Hruschka, Eduardo   University of Sao Paulo, Brazil
Huang, Xu   University of Canberra, Australia
Huljenic, Darko   Ericsson Nikola Tesla, Croatia
Ichimura, Takumi   Hiroshima City University, Japan
Inuzuka, Nobuhiro   Nagoya Institute of Technology, Japan
Ioannidis, Stratos   University of the Aegean, Greece
Ishibuchi, Hisao   Osaka Prefecture University, Japan
Ishida, Yoshiteru   Toyohashi University of Technology, Japan
Ishii, Naohiro   Aichi Institute of Technology, Japan
Ito, Hideaki   Chukyo University, Japan
Ito, Kazunari   Aoyama University, Japan
Ito, Sadanori   Tokio University of Agriculture and
                                         Technology, Japan
Itou, Junko   Wakayama University, Japan
Iwahori, Yuji   Chubu University, Japan
Jacquenet, Francois   University of Saint-Etienne, France
Jain, Lakhmi C.   University of South Australia, Australia
Jarman, Ian   Liverpool John Moores University, UK
Jatowt, Adam   Kyoto University, Japan
Jevtic, Dragan   University of Zagreb, Croatia
Jezic, Gordan   University of Zagreb, Croatia
Jiang, Jianmin   University of Bradford, UK
Jimenez-Berni, Jose Antonio   IAS-CSIC, Spain
Johnson, Ray   Defence Science and Technology Organisation,
                                         Australia
Ju, Zhaojie   University of Portsmouth, UK
Jung, Jason   Yeungnam University, Korea
Juszczyszyn, Krzysztof   Wroclaw University of Technology, Poland
Kaczmarek, Tomasz   Poznan University of Economics, Poland
Karny, Miroslav   Academy of Sciences of the Czech Republic,
                                         Czech Republic
Karwowski, Waldemar   University of Central Florida, USA
Katarzyniak, Radoslaw   Wrocaw University of Technology, Poland
Kato, Shohei   Nagoya Institute of Technology, Japan
Katsifarakis, Konstantinos   Aristotelian University of Thessaloniki, Greece
Kazienko, Przemyslaw   Wroclaw University of Technology, Poland
Kecman, Vojislav   University of Auckland, New Zealand
Keysers, Daniel   Google Switzerland, Switzerland

Kim, Dongwon	Korea University, Korea
Kimura, Masahiro	Ryukoku University, Japan
Kojiri, Tomoko	Nagoya University, Japan
Kolaczek, Grzegorz	Wroclaw University of Technology, Poland
Koukam, Abder	Université de Technologie de Belfort Montbeliard, France
Kovacic, Zdenko	University of Zagreb, Croatia
Krajnovic, Sinisa	Nippon Ericsson K.K., Japan
Krol, Dariusz	Wroclaw University of Technology, Poland
Kubota, Naoyuki	Tokyo Metropolitan University, Japan
Kucheryavskiy, Sergey	Aalborg University Esbjerg, Denmark
Kume, Terunobu	Fujitsu Laboratories Limited, Japan
Kunifuji, Susumu	Japan Advanced Institute of Science and Technology, Japan
Kuroda, Chiaki	Tokyo Institute of Technology, Japan
Kusek, Mario	University of Zagreb, Croatia
Lanjeri, Siham	University of Cordoba, Spain
Le, Kim	University of Canberra, Australia
Lee, Hsuan-Shih	National Taiwan Ocean University, Taiwan
Lee, Huey-Ming	Chinese Culture University, Taiwan
Lee, Malrey	ChonBuk National University, Korea
Lennox, Barry	University of Manchester, UK
Leray, Philippe	University of Nantes, France
Lhotska, Lenka	Czech Technical University, Czech Republic
Lisboa, Paulo	Liverpool John Moores University, UK
Liu, Honghai	University of Portsmouth, UK
Liu, Yonghuai	Aberystwyth University, UK
Loia, Vincenzo	University of Salerno, Italy
Lonetti, Francesca	University of Pisa, Italy
Loo, Chu-Kiong	Multimedia University, Malaysia
López, Beatriz	Universitat de Girona, Spain
Lops, Pasquale	University of Bari, Italy
Lovrek, Ignac	University of Zagreb, Croatia
Ma, Wanli	University of Canberra, Australia
MacDonald, Bruce	University of Auckland, New Zealand
Magnani, Lorenzo	University of Pavia, Italy
Malchiodi, Dario	University of Milan, Italy
Maldonado-Bautista, Jose O.	Universidad del Pais Vasco, Spain
Maojo, Victor	Universidad Politecnica de Madrid, Spain
Maratea, Antonio	University of Naples Parthenope, Italy
Margaris, Dionissios	University of Patras, Greece
Marinai, Simone	University of Florence, Italy
Markovic, Hrvoje	Tokyo Institute of Technology, Japan
Marrara, Stefania	University of Milan, Italy
Martin, Luis	Universidad Politecnica de Madrid, Spain

Martin-Sanchez, Fernando	Institute of Health Carlos III, Spain
Masulli, Francesco	University of Genoa, Italy
Matijasevic, Maja	University of Zagreb, Croatia
Matsuda, Noriyuki	Wakayama University, Japan
Matsudaira, Kazuya	Shizuoka University, Japan
Matsui, Nobuyuki	University of Hyogo, Japan
Matsumoto, Hideyuki	Tokyo Institute of Technology, Japan
Matsushita, Mitsunori	NTT Communication Science Labs, Japan
McCormac, Andrew	Alpha Data Ltd, UK
Mencar, Corrado	University of Bari, Italy
Meng, Qinggang	University of Loughborough, UK
Menolascina, Filippo	Technical University of Bari, Italy
Mera, Kazuya	Hiroshima City University, Japan
Minazuki, Akinori	Kushiro Public University of Economics, Japan
Mineno, Hiroshi	Shizuoka University, Japan
Minoru, Fukumi	Tokushima University, Japan
Misue, Kazuo	University of Tsukuba, Japan
Mitsukura, Yasue	Tokyo University of Agriculture and Technology, Japan
Mituhara, Hiroyuki	Tokushima University, Japan
Miura, Hirokazu	Wakayama University, Japan
Miura, Motoki	Japan Advanced Institute of Science and Technology, Japan
Miyadera, Youzou	Tokyo Gakugei University, Japan
Mizuno, Tadanori	Shizuoka University, Japan
Mohammadian, Masoud	University of Canberra, Australia
Moraga, Claudio	University of Dortmund, Germany
Mukai, Naoto	Tokyo University of Science, Japan
Mumford, Christine	Cardiff University, UK
Munemori, Jun	Wakayama University, Japan
Nachtegael, Mike	Ghent University, Belgium
Nakada, Toyohisa	Japan Advanced Institute of Science and Technology, Japan
Nakamatsu, Kazumi	University of Hyogo, Japan
Nakamura, Tsuyoshi	Nagoya Institute of Technology, Japan
Nakano, Ryohei	Nagoya Institute of Technology, Japan
Nakao, Zensho	University of the Ryukyus, Japan
Napolitano, Francesco	University of Salerno, Italy
Nara, Shinsuke	Saitama University, Japan
Nara, Yumiko	The Open University of Japan, Japan
Nascimiento, Jose	Instituto Superior de Engenharia de Lisboa, Portugal
Nebot, Angela	Technical University of Catalonia, Spain
Negoita, Mircea Gh.	KES International

Ng, Wilfred	Hong Kong University of Science and Technology, China
Nguyen, Ngoc Thanh	Wroclaw University of Technology, Poland
Nicolau, Viorel	Dunarea de Jos University of Galati, Romania
Nicoletti, Maria do Carmo	Federal University of Sao Carlos, Brazil
Nicosia, Giuseppe	University of Catania, Italy
Nijholt, Anton	University of Twente, The Netherlands
Nishida, Toyoaki	Kyoto University, Japan
Nishimoto, Kazushi	Japan Advanced Institute of Science and Technology
Nobuhara, Hajime	University of Tsukuba, Japan
Nowe, Ann	VUB, Belgium
Nowostawski, Mariusz	University of Otago, New Zealand
O'Grady, Michael	University College Dublin, Ireland
Okamoto, Takeshi	Kanagawa Institute of Technology, Japan
Oltean, Gabriel	Technical University of Cluj-Napoca, Romania
Ortega, Juan	Universidad de Sevilla, Spain
Ozawa, Seiichi	Kobe University, Japan
Palade, Vasile	University of Oxford, UK
Pan, Dan China	Mobile Group Guangdong Branch, China
Pan, Jeng-Shyang	National Kaohsiung University of Applied Sciences, Taiwan
Pandzic, Igor S.	University of Zagreb, Croatia
Papathanassiou, Stavros	National Technical University of Athens, Greece
Park, Gwi-Tae	Korea University, Korea
Parra-Llanas, Xavier	Technical University of Catalonia, Spain
Pasero, Eros	Politecnico di Torino, Italy
Paz, Abel Francisco	University of Extremadura, Spain
Pedrycz, Witold	University of Alberta, Canada
Pehcevski, Jovan	MIT Faculty of Information Technologies, Macedonia
Perez del Rey, David	Universidad Politecnica de Madrid, Spain
Perez, Rosa M.	University of Extremadura, Spain
Perez-Lopez, Carlos	Technical University of Catalonia, Spain
Pessa, Eliano	University of Pavia, Italy
Pham, Tuan	James Cook University, Australia
Phillips, Phil	Office for National Statistics, UK
Phillips-Wren, Gloria	Loyola College in Maryland, USA
Picasso, Francesco	University of Genoa, Italy
Pieczynska, Agnieszka	Instytut Informatyki Technicznej, Poland
Pirrone, Roberto	University of Palermo, Italy
Plaza, Antonio	University of Extremadura, Spain
Popa, Rustem	Dunarea de Jos University of Galati, Romania
Popescu, Daniela	University of Oradea, Romania

Ragone, Azzurra	Technical University of Bari, Italy
Raiconi, Giancarlo	University of Salerno, Italy
Raimondo, Giovanni	Politecnico di Torino, Italy
Ramon, Jan	Katholieke Universiteit Leuven, Belgium
Ranawana, Romesh	ClearView Scientific, UK
Razmerita, Liana	Copenhagen Business School, Denmark
Reghunadhan, Rajesh	Bharathiar University, India
Remagnino, Paolo	Kingston University, UK
Resta, Marina	University of Genoa, Italy
Reusch, Bernd	Technical University of Dortmund, Germany
Rizzo, Donna	University of Vermont, USA
Rohani, Bakar	Waseda University, Japan
Romero, Enrique	Technical University of Catalonia, Spain
Rosic, Marko	University of Split, Croatia
Rovas, Dimitrios	Technical University of Crete, Greece
Rozic, Nikola	University of Split, Croatia
Saito, Kazumi	University of Shizuoka, Japan
Sakai, Hiroshi	Kyushu Institute of Technology, Japan
Sakamoto, Ryuuki	Wakayama University, Japan
Sanchez, Eduardo	Logic Systems Laboratory IN-Ecublens, Switzerland
Sassi, Roberto	University of Milan, Italy
Sato-Ilic, Mika	University of Tsukuba, Japan
Scarselli, Franco	University of Siena, Italy
Schanda, Janos	University of Veszprem, Hungary
Schwenker, Friedhelm	University of Ulm, Germany
Sebillot, Pascale	IRISA/INSA de Rennes, France
Semeraro, Giovanni	University of Bari, Italy
Sergiadis, Georgios	Aristotelian University of Thessaloniki, Greece
Serra-Sagrista, Joan	Universitat Autonoma Barcelona, Spain
Sharma, Dharmendra	University of Canberra, Australia
Shiau, Yea-Jou	China University of Technology, Taiwan
Shin, Jungpil	University of Aizu, Japan
Shinagawa, Norihide	Tokyo University of Agriculture and Technology, Japan
Shizuki, Buntarou	University of Tsukuba, Japan
Shkodirev, Viacheslaw	St. Petersburg State Polytechnic University, Russia
Sidhu, Amandeep	Curtin University of Technology, Australia
Signore, Oreste	Istituto di Scienza e Tecnologie dell' Informazione A. Faedo, Italy
Sikic, Mile	University of Zagreb, Croatia
Sinkovic, Vjekoslav	University of Zagreb, Croatia
Smuc, Tomislav	Rudjer Boskovic Institute, Croatia
Snajder, Jan	University of Zagreb, Croatia

Sobecki, Janusz	Wroclaw University of Technology, Poland
Sohn, Surgwon	Hoseo University, Korea
Somol, Petr	Institute of Information Theory and Automation, Czech Republic
Staiano, Antonino	University of Naples Federico II, Italy
Stamou, Giorgos	National Technical University of Athens, Greece
Stankov, Slavomir	University of Split, Croatia
Stecher, Rodolfo	L3S Research Center, Germany
Stellato, Armando	University of Rome, Italy
Stoermer, Heiko	University of Trento, Italy
Strahil, Ristov	Rudjer Boskovic Institute, Croatia
Sugihara, Taro	Japan Advanced Institute of Science and Technology, Japan
Sugiyama, Kozo	Japan Advanced Institute of Science and Technology, Japan
Sulaiman, Ross	University of Canberra, Australia
Supek, Fran	Rudjer Boskovic Institute, Croatia
Surjan, Gyoergy	National Institute for Strategic Health Research, Hungary
Suzuki, Nobuo	KDDI Corporation, Japan
Tabakow, Iwan	Wroclaw University of Technology, Poland
Tadanori, Mizuno	Shizuoka University, Japan
Tadic, Marko	University of Zagreb, Croatia
Tagawa, Takahiro	Kyushu University, Japan
Tagliaferri, Roberto	University of Salerno, Italy
Takahashi, Osamu	Future University-Hakodate, Japan
Takahashi, Shin	University of Tsukuba, Japan
Takeda, Kazuhiro	Shizuoka University, Japan
Takenaka, Tomoya	Shizuoka University, Japan
Taki, Hirokazu	Wakayama University, Japan
Tamani, Karim	University of Savoie, France
Tanahashi, Yusuke	Nagoya Institute of Technology, Japan
Tateiwa, Yuichiro	Nagoya University, Japan
Tesar, Ludvik	Academy of Sciences of the Czech Republic, Czech Republic
Thai, Hien	University of the Ryukyus, Japan
Ting, Hua Nong	University of Malaya, Malaysia
Ting, I-Hsien	National University of Kaohsiung, Taiwan
Tohru, Matsuodani	Debag Engineering Ltd, Japan
Tonazzini, Anna	Istituto di Scienza e Tecnologie dell' Informazione A. Faedo, Italy
Torsello, Maria Alessandra	University of Bari, Italy
Tran, Dat	University of Canberra, Australia
Trentin, Edmondo	University of Siena, Italy

Trzec, Krunoslav — Ericsson Nikola Tesla, Croatia
Tsourveloudis, Nikos — Technical University of Crete, Greece
Tsumoto, Shusaku — Shimane University, Japan
Turchetti, Claudio — Università Politecnica delle Marche, Italy
Tweedale, Jeffrey — Defence Science and Technology Organisation, Australia

Uchino, Eiji — Yamaguchi University, Japan
Ugai, Takanori — Fujitsu Laboratories Limited, Japan
Ushiama, Teketoshi — Kyushu University, Japan
Vaklieva-Bancheva, Natasha — Bulgarian Academy of Sciences, Bulgaria
Vassanyi, Istvan — University of Pannonia, Hungary
Vega, Miguel — University of Granada, Spain
Veganzones, Miguel Angel — Universidad del Pais Vasco, Spain
Vellido, Alfredo — Technical University of Catalonia, Spain
Vitabile, Salvatore — University of Palermo, Italy
Vohland, Michael — Trier University, Germany
Wang, Jin-Long — Ming Chuan University, Taiwan
Wang, Yang — University of Portsmouth, UK
Watada, Junzo — Waseda University, Japan
Watanabe, Toyohide — Nagoya University, Japan
Watanabe, Yuji — Nagoya City University, Japan
Weber, Cornelius — Frankfurt Institute for Advanced Studies, Germany

Whitaker, Roger — Cardiff University, UK
Xia, Feng — Queensland University of Technology, Australia
Yamada, Kunihiro — Tokai University, Japan
Yamashita, Yoshiyuki — Tokyo University of Agriculture and Technology, Japan

Yasuda, Takami — Nagoya University, Japan
Yip, Chi Lap — University of Hong Kong, China
Yi-Sheng, Huang — Chung Cheng Institute of Technology, Taiwan
Yoshida, Kenichi — University of Tsukuba, Japan
Yoshida, Kouji — Shonan Institute of Technology, Japan
Yoshiura, Noriaki — Saitama University, Japan
Younan, Nick — Mississippi State University, USA
Yu, Donggang — James Cook University, Australia
Yu, Zhiwen — Kyoto University, Japan
Yuizono, Takaya — Japan Advanced Institute of Science and Technology, Japan

Yukawa, Takashi — Nagaoka University of Technology, Japan
Zalili, Musa — Waseda University, Japan
Zare, Alina — University of Florida, USA
Zebulum, Ricardo — NASA Jet Propulsion Laboratory, USA
Zeng, An — Guangdong University of Technology, China
Zeng, Xiangyan — University of California Davis, USA

Zhang, Bailing	Xi'an Jiaotong-Liverpool University, China
Zhu, Goupu	Sun Yat-Sen University, China
Zippo, Antonio	University of Milan, Italy
Zunino, Rodolfo	University of Genoa, Italy

General Track Chairs

Artificial Neural Networks and Connectionist Systems
Bruno Apolloni, University of Milan, Italy

Fuzzy and Neuro–Fuzzy Systems
Bernd Reusch, Technical University of Dortmund, Germany

Evolutionary Computation
Zensho Nakao, University of the Ryukyus, Japan

Machine Learning and Classical AI
Floriana Esposito, University of Bari, Italy

Agent Systems
Ngoc Thanh Nguyen, Wroclaw University of Technology, Poland

Knowledge Based and Expert Systems
Anne Håkansson, Uppsala University, Sweden

Hybrid Intelligent Systems
Vasile Palade, University of Oxford, UK

Intelligent Vision and Image Processing
Tuan Pham, James Cook University, Australia

Knowledge Management, Ontologies and Data Mining
Bojana Dalbelo Basic, University of Zagreb, Croatia

Web Intelligence, Text and Multimedia Mining and Retrieval
Andreas Nuernberger, University of Magdeburg, Germany

Intelligent Signal Processing
Miroslav Karny, Academy of Sciences of the Czech Republic, Czech Republic

Intelligent Robotics and Control
Honghai Liu, University of Portsmouth, UK

Invited Session Chairs

Advanced Groupware
Jun Munemori (Wakayama University, Japan), Hiroshi Mineno (Shizuoka
    University, Japan)

Advanced Knowledge-Based Systems
Alfredo Cuzzocrea (University of Calabria, Italy)

Advanced Neural Processing Systems
Monica Bianchini, Marco Maggini, Franco Scarselli (Università degli Studi di
    Siena, Italy)

Agent and Multi-Agent Systems: Technologies and Applications
Bala M. Balachandran, Dharmendra Sharma (University of Canberra, Australia)

Ambient Intelligence
Cecilio Angulo (Technical University of Catalonia, Spain), Honghai Liu
    (University of Portsmouth, UK)

Application of Knowledge Models in Healthcare
István Vassányi (University of Pannonia, Hungary), Gyoergy Surjan (National
    Institute for Strategic Health Research, Hungary)

Artificial Intelligence Driven Engineering Design Optimization
Ioannis K. Nikolos (Technical University of Crete, Greece)

Biomedical Informatics: Intelligent Information Management from
    Nanomedicine to Public Health
Victor Maojo (Universidad Politecnica de Madrid, Spain)

Chance Discovery
Akinori Abe (ATR Knowledge Science Laboratories, Japan), Yukio Ohsawa
    (University of Tokyo, Japan)

Communicative Intelligence
Ngoc Thanh Nguyen (Wroclaw University of Technology, Poland), Toyoaki
    Nishida (Kyoto University, Japan)

Computational Intelligence for Image Processing and Pattern Recognition
Yen-Wei Chen (Ritsumeikan University, Japan)

Computational Intelligence in Human Cancer Research
Alfredo Vellido (Technical University of Catalonia, Spain), Paulo J.G. Lisboa
    (Liverpool John Moores University, UK)

Computational Intelligence Techniques for Knowledge Discovery
Claudio Turchetti, Paolo Crippa, Francesco Gianfelici (Università Politecnica
    delle Marche, Italy)

Computational Intelligence Techniques for Web Personalization
Giovanna Castellano, Alessandra Torsello (University of Bari, Italy)

Computational Intelligent Techniques for Bioprocess Modelling, Monitoring and
    Control
Maria do Carmo Nicoletti, Teresa Cristina Zangirolami, Estevam Rafael
    Hruschka Jr. (Federal University of Sao Carlos, Brazil)

Contributions of Intelligent Decision Technologies (IDT)
Gloria Phillips-Wren (Loyola College in Maryland, USA), Lakhmi C. Jain
    (University of South Australia, Australia)

Engineered Applications of Semantic Web – SWEA
Tommaso Di Noia, Marco Degemmis, Giovanni Semeraro, Eugenio Di Sciascio
    (University of Bari, Italy)

Enhance Secure User Authentication Through Intelligent and Strong Techniques
Ernesto Damiani, Antonia Azzini (University of Milan, Italy)

Evolutionary Multiobjective Optimization
Christine Mumford (Cardiff University, UK)

Evolvable Hardware and Adaptive Systems – Advanced Engineering Design
    Methodologies and Applications
Mircea Gh. Negoita (KES International), Sorin Hintea (Technical University of
    Cluj-Napoca, Romania)

Evolvable Hardware Applications in the Area of Electronic Circuits Design
Mircea Gh. Negoita (KES International), Sorin Hintea (Technical University of
    Cluj-Napoca, Romania)

Hyperspectral Imagery for Remote Sensing: Intelligent Analysis and
    Applications
Miguel A. Veganzones, Manuel Grana (Universidad del Pais Vasco, Spain)

Immunity-Based Systems
Yoshiteru Ishida (Toyohashi University of Technology, Japan)

Innovation-oriented Knowledge Management Platform
Toyohide Watanabe (Nagoya University, Japan), Taketoshi Ushiama (Kyushu
    University, Japan)

Innovations in Intelligent Multimedia Systems
Cecilia Sik Lanyi (University of Pannonia, Hungary), Lakhmi C. Jain
    (University of South Australia, Australia)

Innovations in Virtual Reality
Cecilia Sik Lanyi (University of Pannonia, Hungary), Lakhmi C. Jain
    (University of South Australia, Australia)

Intelligent Computing for Grid
Kun Gao (Zhejiang Wanli University, China)

Intelligent Data Processing in Process Systems and Plants
Yoshiyuki Yamashita (Tokyo University of Agriculture and Technology, Japan),
    Tetsuo Fuchino (Tokyo Institute of Technology, Japan)

Intelligent Environment Support for Collaborative Learning
Toyohide Watanabe, Tomoko Kojiri (Nagoya University, Japan)

Intelligent Systems in Medicine and Healthcare
Chee-Peng Lim, (University of Science Malaysia, Malaysia), Lakhmi C. Jain
    (University of South Australia, Australia), Robert F. Harrison (University of
    Sheffield, UK)

Intelligent Systems in Medicine: Innovations in Computer–Aided Diagnosis and
Treatment
Mia Markey (University of Texas at Austin, USA), Lakhmi C. Jain (University
of South Australia, Australia)

Intelligent Utilization of Soft Computing Techniques
Norio Baba (Osaka Kyoiku University, Japan)

Knowledge-Based Interface Systems [I]
Naohiro Ishii (Aichi Institute of Technology, Japan), Yuji Iwahori (Chubu
University, Japan)

Knowledge-Based Interface Systems [II]
Yoshinori Adachi (Chubu University, Japan), Nobuhiro Inuzuka (Nagoya
Institute of Technology, Japan)

Knowledge Interaction for Creative Learning
Toyohide Watanabe, Tomoko Kojiri (Nagoya University, Japan)

Knowledge-Based Creativity Support Systems
Susumu Kunifuji (Japan Advanced Institute of Science and Technology, Japan),
Kazuo Misue (University of Tsukuba, Japan), Motoki Miura (Japan Advanced
Institute of Science and Technology, Japan), Takanori Ugai (Fujitsu
Laboratories Limited, Japan)

Knowledge-Based Multi-criteria Decision Support
Hsuan-Shih Lee (National Taiwan Ocean University, Taiwan)

Knowledge-Based Systems for e-Business
Kazuhiko Tsuda (University of Tsukuba, Japan)

Neural Information Processing for Data Mining
Ryohei Nakano (Nagoya Institute of Technology, Japan), Kazumi Saito
(University of Shizuoka, Japan)

Neural Networks in Image Processing
Monica Bianchini, Marco Maggini, Franco Scarselli (Università degli Studi di
Siena, Italy)

New Advances in Defence and Security Systems in Intelligent Environments
Jadranka Sunde (Defence Science and Technology Organisation, Australia),
Lakhmi C. Jain (University of South Australia, Australia)

Novel Foundation and Applications of Intelligent Systems
Valentina E. Balas (Aurel Vlaicu University of Arad, Romania), Chee-Peng
Lim, (University of Science Malaysia, Malaysia), Lakhmi C. Jain (University
of South Australia, Australia)

Reasoning-Based Intelligent Systems
Kazumi Nakamatsu (University of Hyogo, Japan)

Relevant Reasoning for Discovery and Prediction
Jingde Cheng, Yuichi Goto (Saitama University, Japan)

Skill Acquisition and Ubiquitous Human Computer Interaction
Hirokazu Taki (Wakayama University, Japan), Satoshi Hori (Monotsukuri
    Institute of Technologists, Japan)

Soft Computing Approach to Management Engineering
Junzo Watada (Waseda University, Japan), Huey-Ming Lee (Chinese Cultural
    University, Taiwan), Taki Kanda (Bunri University of Hospitality, Japan)

Smart Sustainability
Robert J. Howlett (University of Brighton, UK)

Spatio-temporal Database Concept Support for Organizing Virtual Earth
Toyohide Watanabe (Nagoya University, Japan), Jun Feng (Hohai University,
    Japan), Naoto Mukai (Tokyo University of Science, Japan)

Unsupervised Clustering for Exploratory Data Analysis
Roberto Tagliaferri (University of Salerno, Italy), Michele Ceccarelli (University
    of Sannio, Italy)

Use of AI Techniques to Build Enterprise Systems
Ronald L. Hartung (Franklin University, USA)

XML Security
Ernesto Damiani, Stefania Marrara (University of Milan, Italy)

3D Approaches for Visual Facial Expression and Emotion Dynamics Recognition
    in a Real Time Context
Andoni Beristain, Manuel Grana (Universidad del Pais Vasco, Spain)

Sponsoring Institutions

Ministry of Science, Education and Sports of the Republic of Croatia
University of Zagreb, Faculty of Electrical Engineering and Computing
Ericsson Nikola Tesla, Zagreb, Croatia
Croatian National Tourist Board
Zagreb Tourist Board

Table of Contents – Part III

IV Intelligent Processing

Intelligent Data Processing in Process Systems and Plants

Neural Information Processing for Data Mining

Soft Computing Approach to Management Engineering

V Intelligent Systems

Advanced Groupware

Agent and Multi-agent Systems: Technologies and Applications

Engineered Applications of Semantic Web – SWEA

Evolvable Hardware and Adaptive Systems – Advanced Engineering Design Methodologies and Applications

Evolvable Hardware Applications in the Area of Electronic Circuits Design

Hyperspectral Imagery for Remote Sensing: Intelligent Analysis and Applications

Immunity-Based Systems

Innovations in Intelligent Multimedia Systems and Virtual Reality

Intelligent Environment Support for Collaborative Learning

Intelligent Systems in Medicine and Healthcare

Knowledge Interaction for Creative Learning

Novel Foundation and Applications of Intelligent Systems

Skill Acquisition and Ubiquitous Human Computer Interaction

Smart Sustainability

Unsupervised Clustering for Exploratory Data Analysis

Use of AI Techniques to Build Enterprise System

Development of Activity Models of Integrated Safety and Disaster Management for Industrial Complex Areas

Yukiyasu Shimada[1] and Hossam A. Gabbar[2]

[1] National Institute of Occupational Safety and Health, 1-4-6, Umezono, Kiyose, Tokyo,
204-0024 Japan
shimada@s.jniosh.go.jp
[2] Okayama University, 3-1-1, Tsushima-Naka, Okayama, Okayama, 700-8530 Japan
gabbar@cc.okayama-u.ac.jp

Abstract. There are increasing challenges that face industrial organizations to manage safety in normal and to minimize risks when disaster occurs. Such challenges increase when many production plants are placed in industrial complex area. This paper describes activity models to manage plant safety and disasters of production plants in industrial complex areas. The proposed activity models are used to design and develop integrated decision support and management system for safety and disaster management of industrial complex areas, which is validated using case study scenarios.

Keywords: Safety and Disaster Management, Activity Model, Agent System.

1 Introduction

In industrial complex area, there are many interconnected and organizationally linked factories such as petrochemical plant, chemical process plant, ironworks, automobile factory etc. These factories are normally integrated with various utilities and service entities such as supply-demand relationships of raw material, water, power, and products. Such integration represents a threat of accidents or incidents to occur as a result of fault propagation among upstream and downstream facilities. Industrial plants might face faults in their production process, which is a deviation from normal condition in one or more process equipment. This is considered as an abnormal situation where corrective or preventive actions are required. Safety is defined, as per IEC61508, as the freedom from unacceptable risks. The assurance of process safety will reduce the possibility of accidents / disasters. Not only abnormal situation management, but also re-planning of production and transportation and disaster prevention activities as a whole industrial complex area is required. Disaster prevention will provide means to prevent causes for any disaster.

Many engineering support systems (e.g. agent systems) are proposed to support safety assessment, constrained and safe control, fault diagnosis, safety operation management, etc. of chemical process plant. However, these systems have limited capabilities and can't be considered a complete safety management solution for the whole plant.

I. Lovrek, R.J. Howlett, and L.C. Jain (Eds.): KES 2008, Part III, LNAI 5179, pp. 1–8, 2008.

Similarly, disaster management and prevention systems are typically implemented separately in individual companies without integrated views among interconnected production plants. So far, the research and development activities about disaster management systems are performed separately from safety management activities. This represents limitation to the proposed integrated safety solutions.

In this paper, activity models using IDEF0 standard format are proposed with the aim of developing integrated environment to support safety and disaster management activities for industrial complex areas. Practical safety and disaster management support systems (agent systems) are proposed and validated using case study scenarios. The proposed decision support system includes fault diagnosis, abnormal operation support, production and transportation planning, and other related production and engineering systems. The proposed support systems include process monitoring (read and analysis of sensor data), safety assessment (evaluate associated risks), safety management (manage the overall safety life cycle, counteractions, etc.), and disaster prevention (preparedness, loss prevention activities, etc.) functionalities. IDEF0-based activity models make it possible to summarize the information related to each activity needed to develop each agent system and to describe information flow among them. The proposed activity models facilitated the design of agents to manage the overall integrated production systems.

2   Activity Models Using IDEF0 Standards

Fig. 1 shows the standard activity models using IDEF0. The rectangle represents each activity, and the arrows describe information [1]. The information is classified into four categories (ICOM: Input, Control, Output and Mechanism); i.e. 'Input' to be changed by the activity, 'Control' to constraint the activity, 'Output' to be results of the activity and 'Mechanism' to be resources for the activity. The 'Input', 'Control' and 'Mechanism' are fed to left side, top and bottom of the activity respectively. The 'Output' is to go out from right side of the activity. Each activity is developed to sub-activities hierarchically.

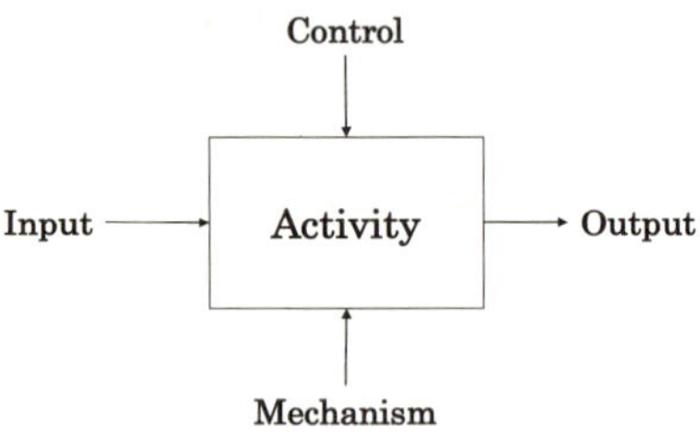

Fig. 1. IDEF0 standard format

Fuchino et al. proposed some IDEF0 activity models for the safety management of chemical process plant at previous KES international conferences. One is for chemical process design to reuse the process design rationale information effectively to other new process design [2], the other is for performing effective process hazard analysis like a HAZOP [3]. Furthermore, this team developed the activity models for performing effective maintenance activities and defined the specification of knowledge and information management environment needed to perform each activity [4]. Sugiyama showed activity modeling approach and design framework to obtain optimized process flowsheet during the design of chemical process [5].

3   Activity Models of Integrated Safety and Disaster Management

As normal activities, safety of process plant is evaluated and safety countermeasures are discussed. In this process, first, some process malfunctions are assumed, then the causes and effects of them are inferred and safety countermeasures are considered. Through this process safety design activity, cause-and-effect models, emergency operation procedures, etc. to be able to support fault diagnosis and safety operation management can be prepared. Various computer-aided support systems have been proposed and developed for process safety assessment, fault diagnosis, safety operation support in abnormal situations during plant operation. In addition, the proposed tools support the re-planning of production and transportation processes, more specifically, in abnormal situations. The support of disaster-prevention activities will enable production plants in complex areas to increase their profit with minimized risks by minimizing the interruption to their production schedules. However, conventional activity support systems for process safety management have been discussed and developed independently. In other words, they cannot support the activity in case of emergency situations in an integrated manner.

The following are the purposes of developing activities for safety and disaster management in abnormal situations (which are developed as part of the underlying project of Grant-in-Aid Scientific Research in Okayama Univ.).

(1)   Plant data processing: To analyze data from process plant and transform it into a form that each agent system can treat
(2)   Safety monitoring and control: To monitor plant condition and control its states via safety operation when abnormal situations occur
(3)   Plant fault diagnosis: To detect plant abnormal situation and to infer the cause and effect of it
(4)   Plant operation support: To indicate the next operation to plant operator based on data and information related with plant condition
(5)   Production and transportation planning: To reconsider the plan of production and transportation in disaster situations and to plan the maintenance and recovery efforts of production function as one industrial complex area
(6)   Disaster prevention activities: To recognize the condition of whole plant and industrial complex area in times of disaster and to support initial response activities for disaster prevention, while ensuring consistence with plant operation support

In case of plant abnormal situation and/or accident, it is required to develop the integrated environment to support safety and disaster management activity and to minimize the damages as whole industrial complex area from the viewpoints of safety, productivity, environmental impact, product quality, etc. In this paper, IDEF0-based activity models are developed to summarize the information related to each activity and the information flow to link them sequentially and to support the development of this integrated environment for the safety and disaster management above described. These activity models provide appropriate links among agents.

Hereinafter, only higher-level models to clarify the relationships between each safety and disaster management activity are described.

3.1 Top Level Activity Model

The top level activity reflects the ultimate goal of supporting safety and managing disasters of production plants in industrial complex areas, as shown in Fig.2. The inputs include plant information such as structure, behavior, and operation. Plant structural elements include process equipment, control devices, and instruments. Plant behavior includes the changes of process states. And plant operation shows the steps and procedures to produce required products. Plant operational data is another input to the top level activity, which include running values of process variables. The controls of the top activity include policies and regulations of risk management for individual organizations and the whole industrial complex area. In addition, social requirements should be also considered to control the whole activity.

In order to achieve such target activity, set of methods and mechanisms are adopted such as safety assessment methods, agent system, simulator, and process and equipment control methods. The target activity will produce set of deliverables or outputs, which include safety design documentation, fault diagnosis results, operation support instructions, plans of production and transportation, and disaster prevention support information.

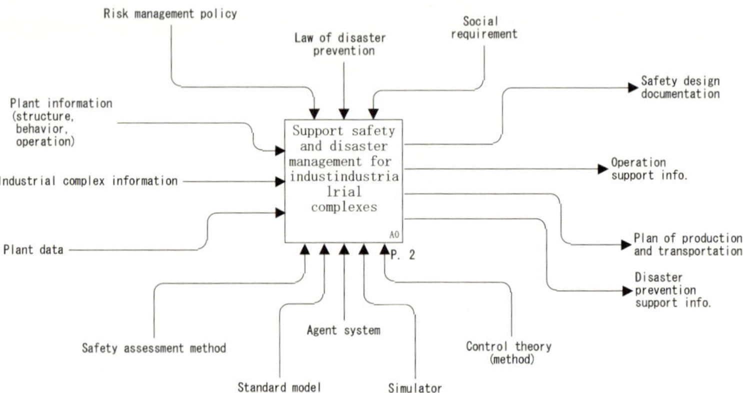

Fig. 2. Integrated safety and disaster management for industrial complex areas

3.2 Second Level Activity Model

In the second level activity model, detailed activities are described, as shown in Fig.3. There are three major activities that are stated. The first, detailed, activity (A1) is the management of the whole process where risk management of individual companies and industrial complex area are used as a control to produce management policy for safety and disaster prevention. Second activity (A2) is the safety assessment and response planning of disaster prevention, which is meant to be in the offline stage. Such activity requires information on plants and industrial complex areas, and produce results of safety assessment and response plans for disaster prevention. The third activity (A3) is the support of disaster prevention in industrial complex, which is meant to be in the online mode. Such activity will produce results of fault diagnosis, operation support information, plan of production and transportation and disaster prevention support information by using the result of activity (A2) and company's management policy and social requirement, etc. On the other hand, the results of activity (A3) are fed back to activity (A2) through the management activity (A1) to improve operation support and their emergency response plans, etc.

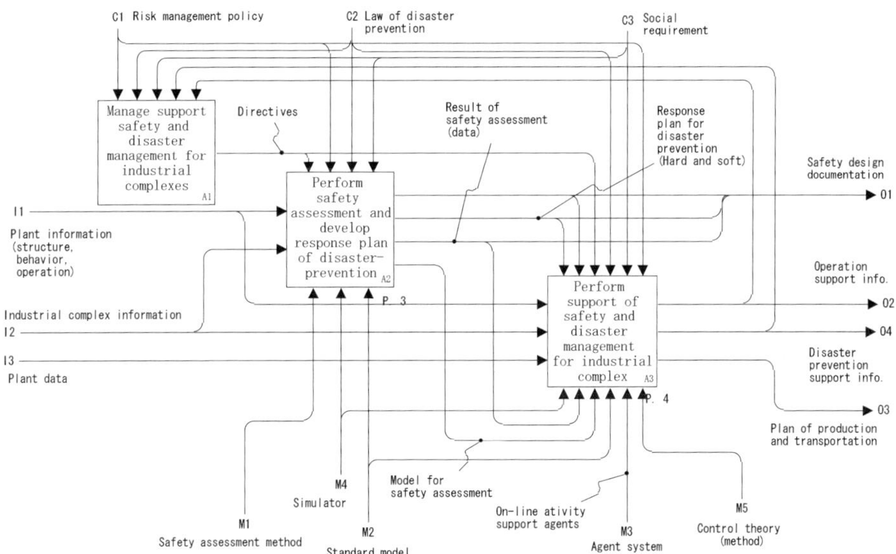

Fig. 3. Activity model for "Support safety and disaster management for industrial complexes"

3.3 Third Level Activity Model

3.3.1 Support Safety Assessment – Offline Mode

Each activity in the second level will be further explored to explain detailed activities underneath. The activity of safety assessment and response planning of disaster prevention (A2) is further expanded into five further activities as shown in Fig.4. Activity (A21) shows the management of the whole activity of safety assessment and countermeasure

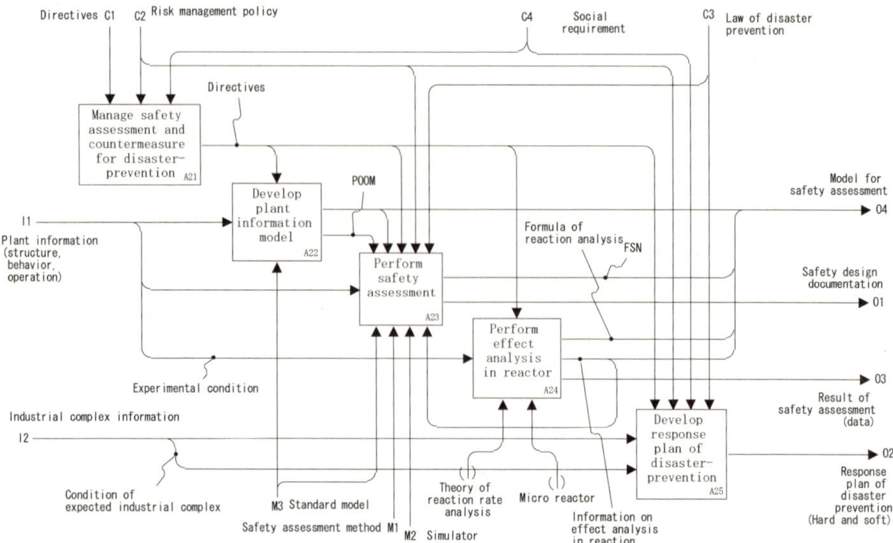

Fig. 4. Activity model for "Perform safety assessment and develop response plan of disaster prevention"

planning of disaster prevention. Activity (A22) is the development of plant information model as well as for the whole industrial complex areas. The generated information models are constructed in the form of POOM, plant/process object oriented methodology [6]. Activity (A23) is the safety assessment to identify all possible hazards, causes, and consequences, along with the associated risks. Activity (A24) is the effect analysis of critical hazards, including reaction processes. Activity (A25) is the response planning for disaster prevention. The outputs from each activity are used to perform online activity (A3).

3.3.2 Support Disaster Prevention in Industrial Complex – Online Mode

Activity (A3) is the support for minimization of the damages as whole industrial complex area from the viewpoints of safety, productivity, environmental impact, product quality, etc. This is further expanded as shown in Fig.5. Activity (A31) directs each support agent to execute the activity. Activity (A32) is process monitoring, safety control and fault diagnosis. This includes the collection of real time process data, which are used for fault diagnosis. Plant condition includes causes and consequences of identified faults, as well as the monitored and controlled safety information, in case of abnormal situation. This is followed by operation support as shown in activity (A33). Activity (A34) represents the planning of production and transportation using plant and industrial complex area information based on the analysis of disaster prevention and consequence analysis. Finally, activity (A35) will generate consolidated information for the disaster prevention based on information from activities (A32, A33 and A34) and inform it to headquarters, site workers etc.

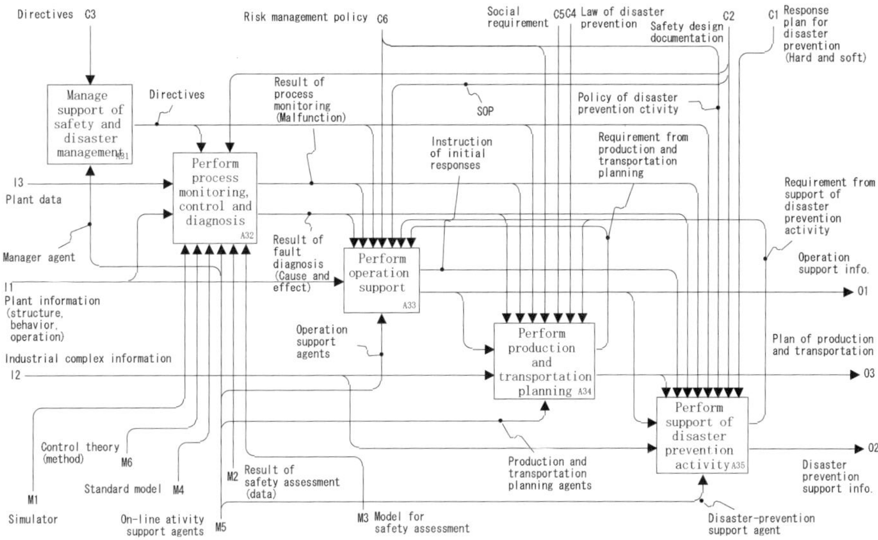

Fig. 5. Activity model for "Perform support of safety and disaster management for industrial complex"

3.4 Discussions

In order to develop the integrated safety and disaster management in industrial complex area, activity models have been developed by using IDEF0 format. Activity model in Figs 2-5 shows the information needed for developing the each agents and information flow to link with each agent system.

As results of this activity modeling, following some advantages are clarified.

(1) Activity models can distinguish offline activities which should discuss in the planning stage with online activities which support the plant safety management at plant abnormal situation and/or the disaster prevention.

(2) Input and output information and constraint condition needed for performing these activities such as fault diagnosis, safety operation support can be clarified.

(3) The position and the purpose of tools and methods to support each activity execution can be clarified. And the specification of agent system to support each activity can be defined.

(4) The purpose of 'manager agent' which supports each activity execution can be defined. In order to manage activity execution, manager agent should be developed for management of activities within each level. This manager agent requires for performing activity based on requirement from each activity and constraint condition such as company's management policy and social requirement. If each agent system outputs multiple solutions, the manager agent has to judge which is better and to manage their activities.

(5) Framework to develop the integrated environment for safety and disaster management in the industrial complex area can developed.

4 Conclusions

Due to the complex operation of interlinked production plants within industrial complex areas, it is essential to provide an integrated environment to ensure safety and to support the disaster prevention activities. To achieve such target, IDEF0-based activity models have been proposed. Safety and control activities are carried out during the design and operation of production plants to assure safe operation and to prevent disasters to fail-safe states. The proposed activity models provide input-output analysis for decision support for safety management, production and transportation planning, disaster prevention, and fault diagnosis of individual production plants as well as industrial complex areas. The proposed system design is based on integrated and distributed multi-agent structure where each of these activities can be carried out independently. Each activity shows input and output information, constraint condition, and tools and methods related to perform the underlying activity. The purpose of agent systems (activity support tools) and the specifications required to develop them are explained.

In future, the environment to develop the integrated information system will be improved and each agent system will be developed based on the proposed activity models. Furthermore, manager agents will be developed to convey the required information between agents and to manage and maintain the performance of each agent.

Acknowledgements

The authors would like to gratefully acknowledge the support of the Grant-in-Aid Scientific Research (16101005) from Japan Society for the Promotion of Science.

References

1. IDEF0 Overviews, http://www.idef.com.idef0.html
2. Fuchino, T., Shimada, Y.: IDEF0 Activity Model Based Design Rationale Supporting Environment for Lifecycle Safety. In: Palade, V., Howlett, R.J., Jain, L. (eds.) KES 2003. LNCS, vol. 2773, pp. 1281–1288. Springer, Heidelberg (2003)
3. Fuchino, T., Batres, R., Shimada, Y.: A Knowledge-Based Approach for the Analysis of Abnormal Situations. In: Apolloni, B., Howlett, R.J., Jain, L.C. (eds.) KES2007/WIRN2007. LNAI, vol. 4693, pp. 712–719. Springer, Heidelberg (2007)
4. Fuchino, T., Miyazawa, M., Naka, Y.: Business Model of Plant Maintenance for Lifecycle Safety. In: 17th European Symp. on Computer-Aided Process Engineering. Elsevier, Amsterdam (2007)
5. Sugiyama, H.: Decision-making Framework for Chemical Process Design Including Different Stages of Environmental Health and Safety (EHS) Assessment, Diss. ETH No.17186 (2007)
6. Gabbar, A., Chung, P.W.H., Suzuki, K., Shimada, Y.: Utilization of Unified Modeling Language (UML) to Represent the Artifacts of the Plant Design Model. In: Int. Symp. on Design, Operation, and Control of Next Generation Chemical Plants (PSE Asia 2000), pp. 387–392 (2000)

LCA of the Various Vehicles in Environment and Safety Aspect

Kazuhiro Takeda[1], Shingo Sugioka[1], Yukiyasu Shimada[2], Takashi Hamaguchi[3], Teiji Kitajima[4], and Tetsuo Fuchio[5]

[1] Shizuoka University, 3-5-1, Johoku, Naka-ku, Hamamatsu, Shizuoka 432-8561, Japan
tktaked@ipc.shizuoka.ac.jp
[2] National Institute of Occupational Safety and Health, 1-4-6, Umezono, Kiyose, Tokyo 204-0024, Japan
[3] Nagoya Institute of Technology, Gokiso-cho, Showa-ku, Nagoya 466-8555, Japan
[4] Tokyo University of Agriculture and Technology, 2-24-16, Naka-cho, Koganei, Tokyo, 184-8588, Japan
[5] Tokyo Institute of Technology, 2-12-1, Oookayama, Meguro, Tokyo 152-8552, Japan

Abstract. LCA (Life Cycle Assessment) about a FCV (Fuel Cell Vehicle) is performed regularly, but these studies have not considered closely enough the safety issues that are important for practical, everyday use. In this study, we considered the safety of the FCV, GV (Gasoline Vehicle), HEV (Hybrid Electric Vehicle) and EV (Electric Vehicle), and estimated the vehicles in relation to environment and safety. As a result of LCA, FCV may not have a higher environmental performance than the other vehicles in the whole life cycle. So, it is important for FCV to reduce platinum catalyst and select more cleanly hydrogen production processes. They were effective against improving an environmental performance of FCV, but the improved environmental performance was inferior to that of the other vehicles.

Keywords: Life Cycle Assessment, Fuel Cell Vehicle, Gasoline Vehicle, Hybrid Electric Vehicle, Electric Vehicle, Safety Issues.

1 Introduction

From a surge of consciousness for the environment, a clean energy source for fossil fuels has attracted attention recently. In this regard, hydrogen is a good energy medium, because its combustion effluent gas is only water, and its application reduces greenhouse gas emissions and can prevent global warming which is an important subject of concern.

Fuel cells are developed to use hydrogen energy, because they are a popular, clean and more highly efficient energy medium than the internal combustion engine [1-3]. It has been tried as a source of power for a vehicle. However, it is necessary to investigate in an economic, a technical, and an environmental aspect, before an innovative system is introduced. The environmental investigation should be considered as a total life cycle of the whole system.

I. Lovrek, R.J. Howlett, and L.C. Jain (Eds.): KES 2008, Part III, LNAI 5179, pp. 9–16, 2008.

So, LCA (Life Cycle Assessment) about a FCV (Fuel Cell Vehicle) is performed regularly [4, 6, 7], but these studies have not considered closely enough the safety issues that are important for practical, everyday use.

In this study, we considered the safety of the FCV, GV (Gasoline Vehicle), HEV (Hybrid Electric Vehicle) and EV (Electric Vehicle), and estimated the vehicles in relation to energy and environment.

2 LCA (Life Cycle Assessment)

LCA [7, 8] is the technical research of environmental influence through the whole life cycle of the product such as resource, production, distribution, sales, use, recycling and disposal quantitatively, and analyzing objectively. It consists of four steps as follows.

1. Goal and scope definition: In this step, the purpose of LCA becomes clear and sets a border of the process and function unit that fitted in the purpose.
2. Inventory analysis: In this step, based on the range to investigate, input or output data such as resources and energy into the process, the product and the waste from the process is collected and then we make the inventory of emission of each material as environmental load.
3. Impact assessment: In this step, environmental impact from a product is estimated from the data which is provided by inventory analysis.
4. Interpretation: From the result of the inventory analysis and impact assessment, the refinement and the characteristics of the products are expressed.

3 Practice of LCA

We used the LCA software tool JEMAI-LCApro [5]. The data was collected related to the FCV, GV, HEV and EV [5, 6, 9-12], and laws and issues were investigated about the safe use of the hydrogen [6, 13, 14].

3.1 Goal and Scope Definition

LCA was carried out for the purpose of comparing the GV, HEV, EV and FCV in environment and safety aspect.

The range of the investigation was the whole life cycle of the vehicle and fuel, such as production, running, parts exchange and disposal. It was assumed that the transportation of the product was not evaluated. The setting conditions are shown in Table 1 [6, 15, 16].

3.2 Impact Assessment

In this study, LIME (Life-cycle Impact assessment Method based on Endpoint modeling) [5] has been used as evaluation method to assess the economic cost. LIME has been developed as an end-point assessment method based on damage calculation particularly for Japanese industries and society. The characterization factors of LIME are used to calculate the amount of damage to safeguard subjects (global warming, resource depletion, urban air pollution, and so on) by using causal relation modeling.

Table 1. The setting conditions for LCA

	GV	HEV	EV	FCV
Fuel	gasorine	gasorine + electoricity	electoricity	hydrogen
Fuel consumption	16.9 km/L	30.67 km/L	10 km/kWh	7.81 km/Nm3
Mileage (km)	100,000			
Type	ordinary vehicle			
Common parts	body, door, etc. (1,068 kg)			
Original parts	engine, etc. (191 kg)	GV parts, battery, motor (353 kg)	battery, motor (322 kg)	FC stack, motor, etc. (469 kg)
Weight (kg)	1,157	1,421	1,390	1,537

The damages are integrated as the economic loss by using conjoint analysis to determine weighting factors.

As a result of impact assessment, economic loss by the environmental impact for the various vehicles is shown in Fig.1. EV has the least impact, and FCV has the worst impact. Notably, impact of global warming and urban air pollution for FCV is bigger than the other vehicles.

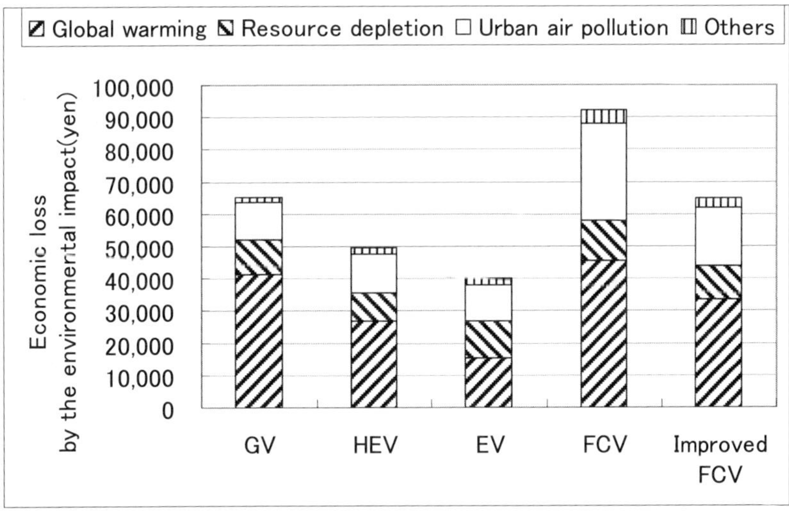

Fig. 1. Economic loss by the environmental impact for the various vehicles

To improve FCV, proportion of emission was investigated. These results are calculated by use of materials used in life cycle and database in JAMAI-LCApro. In Fig. 2, proportion of emission gases for FCV, such as CO_2 for global warming and NOx and SOx for urban air pollution, are displayed. It is revealed that CO_2 emission is mainly caused by production process of hydrogen as fuel and NOx and SOx emission are mainly caused by platinum included as catalyst in FC.

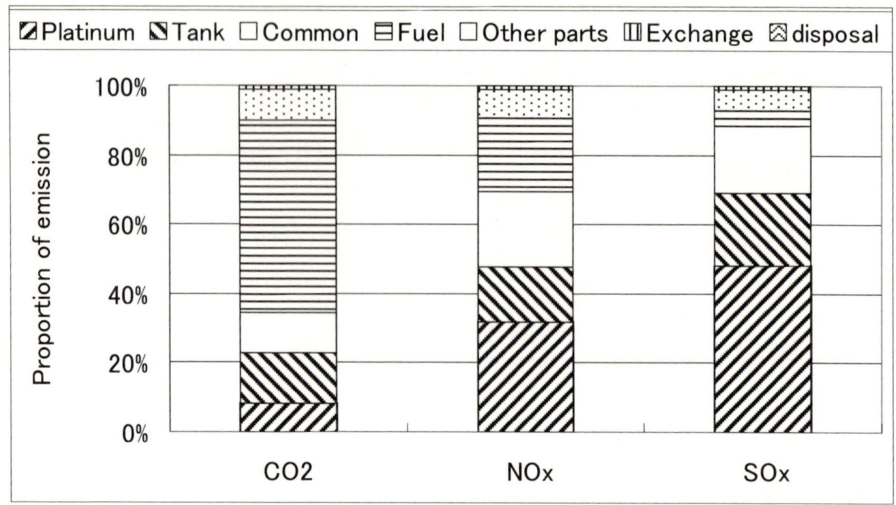

Fig. 2. Proportion of emission for FCV

3.3 Review of FCV Production Process

To Revise FCV production process, following four plans were examined. Plan A to C are plans for production process of hydrogen. In inventory analysis, hydrogen production stage was exchanged from Steam reforming of city gas to each plan. Plan D is reduction of Platinum.

- Plan A: Steam reforming of city gas and by-product hydrogen from COG (Coke Oven Gas).
- Plan B: Steam reforming of naphtha and by-product hydrogen from oil factory.
- Plan C: Electrolysis of water.
- Plan D: Reduction Platinum consumption to one-tenth of present quantity.

The results are shown in Fig. 3. Using plan A or B, CO2 emission is able to reduce about 20 %. To product hydrogen by electrolysis of water is attractive in hydrogen recycle aspect, but is worse to environment. Using plan D, NOx and SOx (especially SOx) emission are able to reduce. As shown in Fig. 2, CO2 emission is mainly caused by fuel and NOx and SOx emission are mainly caused by platinum. So, plan A for production process of hydrogen is employed to reduce CO2 emission, and plan D for Platinum reduction is employed to reduce NOx and SOx emission.

Economic loss for improved FCV is lower than that for FCV and as much as that for GV as shown in Fig. 1.

3.4 LCA for Fuel Stations

When an operation aspect of FCV is thought about, it is important to consider the hydrogen station. Therefore, LCA was carried out for comparing the hydrogen station (HS) with the gas station (GS). For HS, hydrogen production stage was out of count,

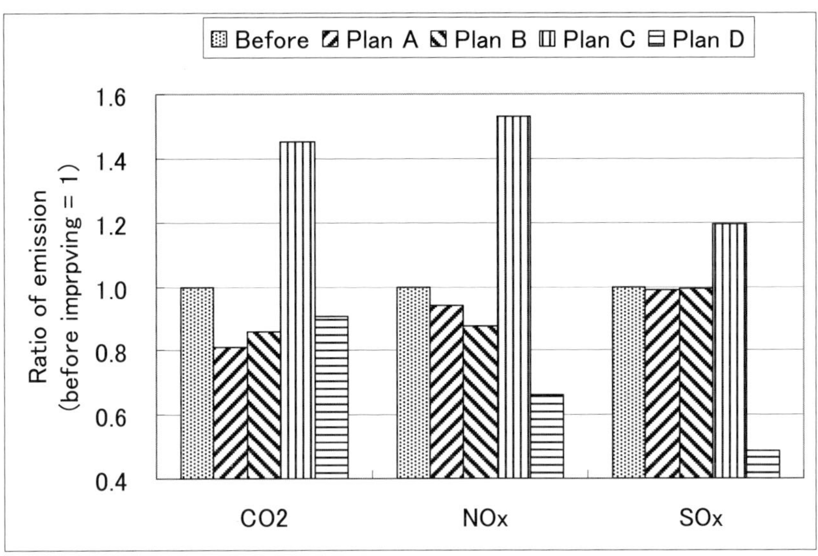

Fig. 3. Ratio of emission for improved FCV

because the stage was assessed as vehicle's LCA. For GS, ground pollution by gasoline was counted. Some setting conditions are shown in Table 2.

Table 2. The setting conditions for LCA of stations

	GS	HS
Area (m2)	2788.99	3238.99
Scope (year)	10	
Common parts	office building, main canopy	
Original parts	gasoline tank, gasoline dispensers	hydrogen production plant, hydrogen dispensers

Economic loss by the environmental impact for GS and HS is shown in Fig. 4. That for HS is slightly worse than that for GS. The main difference of that for GS and HS are caused by urban air pollution, which are NOx and SOx emission.

3.5 Safety Assessment

For the integration of an environmental score and a safety score, the safety score should take a common unit with environmental score (yen). So, the safety score was calculated by the following equation.

$$ss = p \ s = (f \ d) \ s \tag{1}$$

Where ss is safety score (yen), p is potential loss (yen/year), s is score (year), f is frequency of accident (1/year), and d is damage of accident (yen).

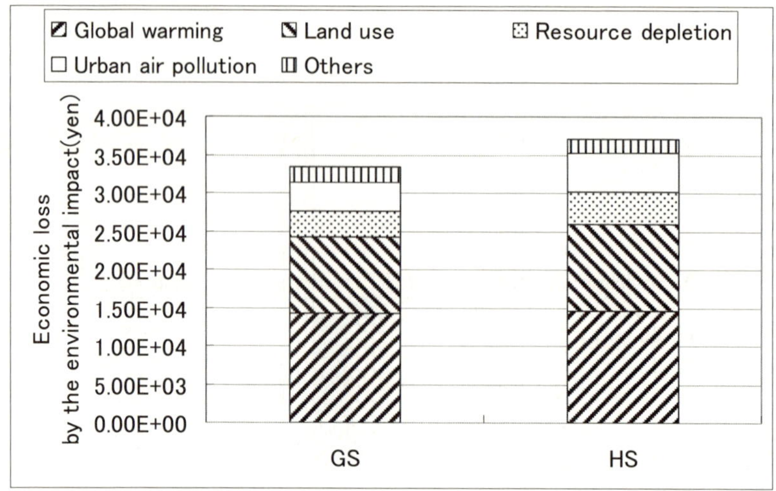

Fig. 4. Economic loss by the environmental impact for GS and HS

The scope of assessment is 10 years. The Frequency and damage of accident of GS are estimated by actual performance. On the other hand, those of HS are estimated by HAZOP report [6]. The result of calculation was shown in Fig. 5. Economic loss for FCV and improved FCV are the same. That for FCV or improved FCV is mainly by loss for station. FCV and improved FCV have the same score. This study let station use frequency for HEV be 16.9 / 30.67 (ratio of fuel cost of HEV and GV), and EV may not use a station.

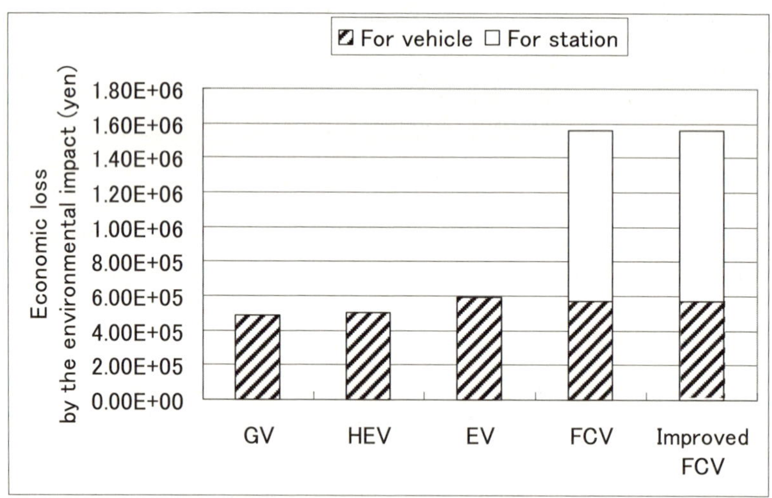

Fig. 5. Economic loss by the environmental impact for various vehicles and stations in safety aspect

3.6   Total Score for Environment and Safety

The environmental score and the safety score for various vehicles are integrated in
Fig. 6. Because of the safety score for station, FCV and improved FCV are the worst
score. Although, even if all of the safety scores for station are the same, total score for
FCV and improved FCV have the worst score.

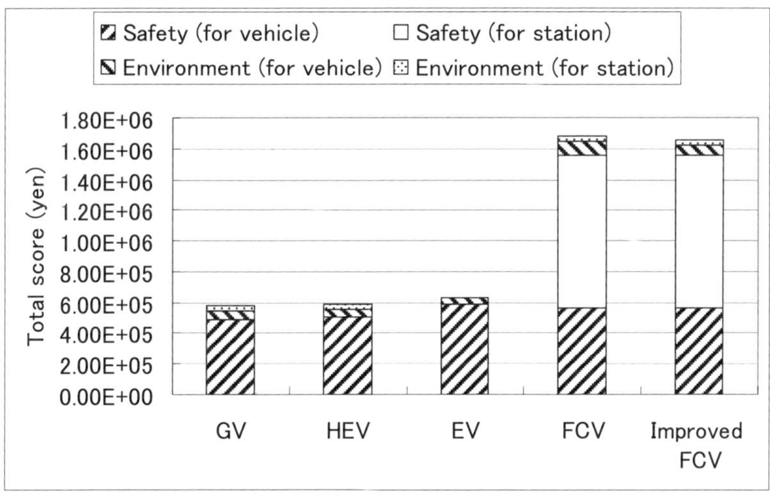

Fig. 6. Total score about environment and safety for various vehicles and stations

4   Conclusion

In this study, LCA was performed for various vehicles in environment and safety
aspect. There were some improved points which were reducing Platinum consumption
and changing the production process of hydrogen in the FCV process. Improved FCV
had as high environmental performance as the GV during the whole life cycle. Even
though, improved FCV had a lower environmental performance than the other clean
energy vehicles such as HEV and EV during the whole life cycle. It followed that the
FCV was not so attractive environmentally. For the safety side, FCV was evaluated
lower than the other vehicles.

References

1. http://www.jhfc.jp/data/reports/data/fcv2005report_01.pdf
2. Marubayashi, K.: Japanese action for realization of hydrogen society. J. Jap. Soc. for
   Safety Eng. 44, 364–372 (2005)
3. Sakaguchi, J., Kokubun, N.: Hydrogen society-hydrogen supply chain management.
   JHPI 42, 121–129 (2004)
4. Nada, Z., Li, X.: Life cycle analysis of vehicles powered by a fuel cell and by internal
   combustion engine for Canada. J. power source 155, 297–310 (2006)

5. http://www.jemai.or.jp/CACHE/lca_index.cfm
6. http://www.tech.nedo.go.jp/index.htm
7. Pehnt, M.: Life-cycle assessment of fuel cell stacks. Int. J. of Hydrogen Energy 26, 91–101 (2001)
8. Lunghi, P., Bove, R., Desideri, U.: Life-cycle-assessment of fuel-cells-based landfill-gas energy conversion technologies. J. Power Sources 131, 120–126 (2004)
9. Funazaki, A., Taneda, K.: A study of inventories for Automobile LCA (4). JARI Res. J. 23, 548–555 (2001)
10. Examination and investigation about the calculation of the basic unit and total amount of the vehicle exhaust, SUURI-KEIKAKU Co., Ltd (2004)
11. http://www.jhfc.jp/data/reports/data/h17/h17_kekka_main.pdf
12. http://www.jhfc.jp/data/seminar_report/04/pdf/04_h17seminar.pdf
13. The question and answer about laws and ordinances about the service station 4th edition, Sankyo Hoki Publishing Co., Ltd (1999)
14. http://www.jhfc.jp/data/reports/data/fcv2005report_02.pdf
15. http://www.jhfc.jp/data/seminar_report/03/pdf/5_H16JHFC.pdf
16. Ministry of Land, Infrastructure and Transportation: The list of the fuel consumption of the vehicle (2007)
17. http://www.tepco.co.jp/cc/press/05090203-j.html

Plant Model Generation for Countermeasures Planning

Takashi Hamaguchi[1], Kazuhiro Takeda[2], Yukiyasu Shimada[3],
and Yoshihiro Hashimoto[1]

[1] Nagoya Institute of Technology, Gokiso-cho, Showa-ku, Nagoya 466-8555, Japan
hamachan@nitech.ac.jp
[2] Shizuoka University, 3-5-1, Johoku, Hamamatsu, Shizuoka 432-8561, Japan
[3] National Institute of Occupational Safety and Health, 1-4-6, Umezono, Kiyose,
Tokyo 204-0024, Japan

Abstract. It is desirable to support operator activities in abnormal situations in chemical plants. Alarm systems have consequently become an important tool used by operators to maintain safe operation. An effective alarm must be connected to adequate countermeasures. Although, fault diagnosis systems have very fertile grounds for theoretical and industrial development, there are few countermeasures planning systems. For adequate countermeasures planning, plant situations must be reflected dynamically on the system. In this paper, we modified countermeasure planning system based on CE-matrices to change of plant situations. This approach is useful for designing effective alarm management system.

Keywords: Alarm Management, Countermeasures Planning, CE-matrices.

1 Introduction

In most chemical plants, a DCS (Distributed Control System) is installed to keep the process variables stable. The introduction of the DCS with easily programmable alarm settings has resulted in large number of alarm signals being configured on many sites in order to provide plant information to human operators. Alarm systems have consequently become an important tool used by operators to maintain safe operation. For each alarm, the operator has to comprehend the information provided by the alarm, decide if it requires any action and take necessary actions to maintain the safe operation of the plant.

There are a number of fundamental guidelines such as the EMMUA (Engineering Equipment and Material Users Association) No.191 for alarm management [1]. Alarm management is systems thinking to manage the design of an alarm system to increase usability. Most often the major usability problem is that there are too many alarms annunciated in a plant upset, commonly referred to as alarm flood. There are also other problems with alarm systems such as poorly designed alarms, improperly set alarm points, ineffective annunciation, and unclear alarm messages. Without alarm rationalization efforts, they become a serious problem and increase the risk of safety and environmental incidents.

I. Lovrek, R.J. Howlett, and L.C. Jain (Eds.): KES 2008, Part III, LNAI 5179, pp. 17–24, 2008.

The role of operator on a plant generally encompasses a range of different activities including plant operation, optimization of production, fault identification, countermeasures planning, and action. The tasks involved change depending on pant state, e.g. whether it is in normal operation, upset operation, emergency shut-down, planned shut-down, start up, or operating mode change. The key to an effective alarm is that it should mark the point at which the operator has to take action. For example, alarms with variables reflecting operation limits should be set on the boundary between the normal and the upset state of the plant.

To solve this problem, the border of normal and upset must be considered based on the operator intervention such as fault identification, countermeasures planning, and action. The combination of fault trees and cause-effect model are useful for the problem. The roots of the fault trees are "safety", "quality of product", and "quantity of product" as shown in Fig.1. If countermeasures for different troubles are common, the alarms can be unified and the number of alarm can be decreased. Therefore, countermeasures planning system is important for the abnormal situation management.

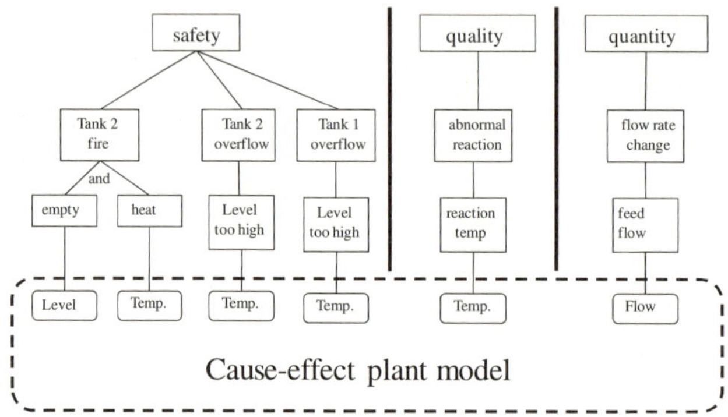

Fig. 1. Relationships between fault trees and cause-effect model

This paper addresses the importance of effective alarm systems and countermeasures planning. It also describes a countermeasure planning system by using CE-matrices for providing valuable advisory information to help the operator plan and take correct and quick countermeasures. The key is how to generate a dynamic suitable plant model for countermeasure planning based on plant situation. In this paper, we propose a unit modules composition approach.

This paper is organized as follows. In Section 2, the problem of a countermeasure planning system and an alarm management system are outlined an example scenario. A detailed description of the countermeasures planning system by using CE-matrices is provided in Section 3. A plant model generation by unit modules composition is provided in Section 4. Conclusions are given in Section 5.

2 Example Scenario for Countermeasures Planning and Alarm Management

In this section, the problem of a countermeasure planning system and an alarm management system are outlined by an example scenario. A controller configuration

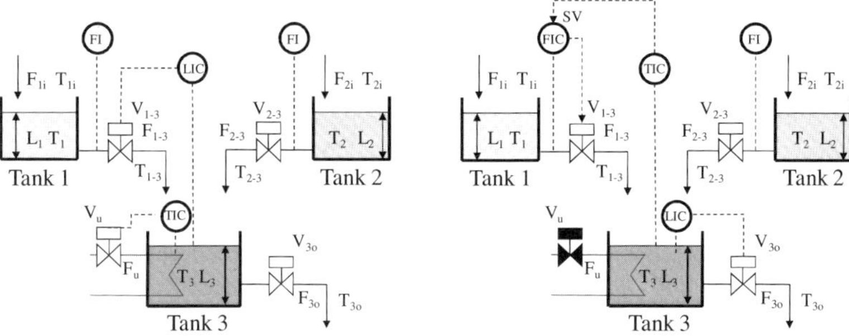

Fig. 2. Example plant   **Fig. 3.** Countermeasure 1

setting of an example plant is shown in Fig.2. This plant had two raw material flows, which were $F_{1\text{-}3}$ and $F_{2\text{-}3}$, and one product flow rate, which was F_{3o}. Controlled variables are temperature T_3 and level L_3 in the tank 3. T_3 is controlled by using heat medium flow rate F_u. L_3 is controlled by using $F_{1\text{-}3}$.

First in this example scenario, the valve V_u becomes stuck and the opening of the valve is smaller than the normal value. Therefore, T_3 cannot be controlled by F_u. The alarm which shows that "T_3 is low" is activated. The operator recognizes that there is a decrease of T_3 and suspects that the cause is V_u's sticking. It is assumed that the following process variables are conditions in this trouble situation.

Temperature T_1 > Temperature T_3 > Temperature T_2

The temperature $T_{1\text{-}3}$ of flow rate $F_{1\text{-}3}$ equals T_1. Hence, the operator can control T_3 by using $F_{1\text{-}3}$ and L_3 by using F_{3o}. The countermeasure 1 for this trouble is shown in Fig.4. In such a situation, the operator switches the part of the controller from auto to manual.

For countermeasures planning, the human does not deal with much information at once. In the first stage of the countermeasure planning, the operator manipulates something by focusing on some observed variables as in the example scenario. This is similar to controller configuration of single loop controllers. It is too difficult to solve this configuration's solution for the whole plant. Hence, there is little systematic approach that can be taken. The Relative Gain Array (RGA) approach is well known for paring problem considering interaction [2]. Since the approach is required to identify steady-state gain matrix, it is improper to apply the approach upon plant wide-problem. Moreover, it is not practical for the design of the controller configuration to assume the existence of dynamic simulator to the whole plant. On the other hand, the qualitative model can easily deal with the information of the whole plant. We already proposed qualitative models and automatic controller configuration algorithms based on them [4]. The combination of the classification of change of operation conditions based on the sets of controlled variables and the automatic controller configuration algorithm enable discovery of the achievable operation states and the set of operator actions [5]. A method to select operating modes, in which the abnormal states could be settled down safely, is based on three fault trees for safety, quality, and quantity [6]. Controlled variables are abandoned in the order of quantity, quality, and safety. The operator can know the reason why the operating mode is selected and the feasibility of operating modes is judged by using CE-matrices. The controller configuration algorithm is expanded for cascade control or ratio control [7]. The applied controller expression is needed for effective countermeasures.

However, our proposed method is based on a static qualitative plant model. If the difference between T_1 and T_3 is decreasing in the countermeasure 1 as shown in Fig.4, the flow rate $F_{1\text{-}3}$ cannot use the manipulated variable for T_3. To deal with the change of the plant situation, the qualitative plant model with fault trees must be reflected automatically.

3 The Method for Controller Configuration

In this section, the method for controller configuration is explained. This method is based on the qualitative model expression using matrices by cause-effect. We call them CE-matrices. These CE-matrices have elements 0 or 1. Each column corresponds to a cause and each row to an effect. When the (i, j) element of the CE-matrices is "1", the j-th column variable affects the i-th row variable.

Consider the example plant shown in Fig.2 and Fig.3. The CE-matrix G in Fig.4 explains the model of the example plant. Its elements are "0" or "1". The CE-matrix G has rows and columns corresponding to the explanatory variables, controlled variables, and manipulated variables. Note that (F_{3o}, L_3) is "1". The "1" means that if the liquid level L_3 is changed from steady-state value, then flow rate F_{3o} will be affected from L_3.

G	F_{1i}	T_{1i}	L_1	T_1	$F_{1\text{-}3}$	$T_{1\text{-}3}$	F_{2i}	T_{2i}	L_2	T_2	$F_{2\text{-}3}$	$T_{2\text{-}3}$	F_u	L_3	T_3	F_{3o}	T_{3o}	$V_{1\text{-}3}$	$V_{2\text{-}3}$	V_u	V_{3o}
F_{1i}	0	0	0	0	0	0	0	0	0	0	0	0	0	0	0	0	0	0	0	0	0
T_{1i}	0	0	0	0	0	0	0	0	0	0	0	0	0	0	0	0	0	0	0	0	0
L_1	1	0	0	0	1	0	0	0	0	0	0	0	0	0	0	0	0	0	0	0	0
T_1	0	1	0	0	0	0	0	0	0	0	0	0	0	0	0	0	0	0	0	0	0
$F_{1\text{-}3}$	0	0	1	0	0	0	0	0	0	0	0	0	0	0	0	0	0	1	0	0	0
$T_{1\text{-}3}$	0	0	0	1	0	0	0	0	0	0	0	0	0	0	0	0	0	0	0	0	0
F_{2i}	0	0	0	0	0	0	0	0	0	0	0	0	0	0	0	0	0	0	0	0	0
T_{2i}	0	0	0	0	0	0	0	0	0	0	0	0	0	0	0	0	0	0	0	0	0
L_2	0	0	0	0	0	0	1	0	0	0	1	0	0	0	0	0	0	0	0	0	0
T_2	0	0	0	0	0	0	0	1	0	0	0	0	0	0	0	0	0	0	0	0	0
$F_{2\text{-}3}$	0	0	0	0	0	0	0	0	1	0	0	0	0	0	0	0	0	0	1	0	0
$T_{2\text{-}3}$	0	0	0	0	0	0	0	0	0	1	0	0	0	0	0	0	0	0	0	0	0
F_u	0	0	0	0	0	0	0	0	0	0	0	0	0	0	0	0	0	0	0	1	0
L_3	0	0	0	0	1	0	0	0	0	0	1	0	0	0	0	1	0	0	0	0	0
T_3	0	0	0	0	1	1	0	0	0	0	1	1	1	0	0	0	0	0	0	0	0
F_{3o}	0	0	0	0	0	0	0	0	0	0	0	0	0	1	0	0	0	0	0	0	1
T_{3o}	0	0	0	0	0	0	0	0	0	0	0	0	0	0	1	0	0	0	0	0	0

Fig. 4. CE-matrix of plant G

The CE-matrix C in Fig.5 explains the controller configuration shown in Fig.2. Note that $(V_{1\text{-}3}, L_3)$ is "1". The "1" means that L_3's controller LIC is connected to the actuator $V_{1\text{-}3}$. This CE-matrix C can be generated automatically from unit matrix. Because each column of controller matrix C always has only "1" and each row of it has at most one "1", all single loop combinations can be generated easily to apply substitutions of unit matrix. The CE matrix for countermeasure 1 can also generated by the cascade controller generation algorithm. Therefore, controller configurations can be designed by computers.

C	F_{1i}	T_{1i}	L_1	T_1	F_{1-3}	T_{1-3}	F_{2i}	T_{2i}	L_2	T_2	F_{2-3}	T_{2-3}	F_u	L_3	T_3	F_{3o}	T_{3o}
F_{1i}	1	0	0	0	0	0	0	0	0	0	0	0	0	0	0	0	0
T_{1i}	0	1	0	0	0	0	0	0	0	0	0	0	0	0	0	0	0
L_1	0	0	1	0	0	0	0	0	0	0	0	0	0	0	0	0	0
T_1	0	0	0	1	0	0	0	0	0	0	0	0	0	0	0	0	0
F_{1-3}	0	0	0	0	1	0	0	0	0	0	0	0	0	0	0	0	0
T_{1-3}	0	0	0	0	0	1	0	0	0	0	0	0	0	0	0	0	0
F_{2i}	0	0	0	0	0	0	1	0	0	0	0	0	0	0	0	0	0
T_{2i}	0	0	0	0	0	0	0	1	0	0	0	0	0	0	0	0	0
L_2	0	0	0	0	0	0	0	0	1	0	0	0	0	0	0	0	0
T_2	0	0	0	0	0	0	0	0	0	1	0	0	0	0	0	0	0
F_{2-3}	0	0	0	0	0	0	0	0	0	0	1	0	0	0	0	0	0
T_{2-3}	0	0	0	0	0	0	0	0	0	0	0	1	0	0	0	0	0
F_u	0	0	0	0	0	0	0	0	0	0	0	0	1	0	0	0	0
L_3	0	0	0	0	0	0	0	0	0	0	0	0	0	0	0	0	0
T_3	0	0	0	0	0	0	0	0	0	0	0	0	0	0	0	0	0
F_{3o}	0	0	0	0	0	0	0	0	0	0	0	0	0	0	0	1	0
T_{3o}	0	0	0	0	0	0	0	0	0	0	0	0	0	0	0	0	1
V_{1-3}	0	0	0	0	0	0	0	0	0	0	0	0	0	1	0	0	0
V_{2-3}	0	0	0	0	0	0	0	0	0	0	0	0	0	0	0	0	0
V_u	0	0	0	0	0	0	0	0	0	0	0	0	0	0	1	0	0
V_{3o}	0	0	0	0	0	0	0	0	0	0	0	0	0	0	0	0	0

Fig. 5. CE-matrix of Controller C

In Fig.6, the control loop model GC, which explains the effect of the set-point changes on the plant, is presented by the Boolean multiplication of two matrices G and C. Note that (F_{3o}, L_3) is changed "0" from CE-matrix G in Fig.4. This change means that L_3 is kept by V_{1-3}'s operation.

GC	F_{1i}	T_{1i}	L_1	T_1	F_{1-3}	T_{1-3}	F_{2i}	T_{2i}	L_2	T_2	F_{2-3}	T_{2-3}	F_u	L_3	T_3	F_{3o}	T_{3o}
F_{1i}	0	0	0	0	0	0	0	0	0	0	0	0	0	0	0	0	0
T_{1i}	0	0	0	0	0	0	0	0	0	0	0	0	0	0	0	0	0
L_1	1	0	0	0	1	0	0	0	0	0	0	0	0	0	0	0	0
T_1	0	1	0	0	0	0	0	0	0	0	0	0	0	0	0	0	0
F_{1-3}	0	0	1	0	0	0	0	0	0	0	0	0	0	1	0	0	0
T_{1-3}	0	0	0	1	0	0	0	0	0	0	0	0	0	0	0	0	0
F_{2i}	0	0	0	0	0	0	0	0	0	0	0	0	0	0	0	0	0
T_{2i}	0	0	0	0	0	0	0	0	0	0	0	0	0	0	0	0	0
L_2	0	0	0	0	0	0	1	0	0	0	1	0	0	0	0	0	0
T_2	0	0	0	0	0	0	0	1	0	0	0	0	0	0	0	0	0
F_{2-3}	0	0	0	0	0	0	0	0	1	0	0	0	0	0	0	0	0
T_{2-3}	0	0	0	0	0	0	0	0	0	1	0	0	0	0	0	0	0
F_u	0	0	0	0	0	0	0	0	0	0	0	0	0	0	1	0	0
L_3	0	0	0	0	1	0	0	0	0	0	1	0	0	0	0	1	0
T_3	0	0	0	0	1	1	0	0	0	0	1	1	1	0	0	0	0
F_{3o}	0	0	0	0	0	0	0	0	0	0	0	0	0	0	0	0	0
T_{3o}	0	0	0	0	0	0	0	0	0	0	0	0	0	0	0	0	0

Fig. 6. CE-matrix of control loop GC

The reachability matrix R is defined by using GC as shown in Eq.(1);

$$R = \sum_{k=1}^{\infty} (\mathbf{GC})^k \tag{1}$$

If the dimension of GC is n, the reachability matrix R is defined as shown in Eq.(2);

$$R = \sum_{k=1}^{n} (\mathbf{GC})^k \tag{2}$$

By using this method, results of the reachability matrix R indicate the effect between variables in the steady state of the controlled plant. The reachability matrix R is shown in Fig.7. If all of the diagonal elements for controlled variables (L_3 and T_3) are "1", then all control loops can be judged effective. The signal of L_3 set-points arrives at the L_3 and the feedback control of L_3 can work. The signal of T_3 set-points arrives at the T_3 and the feedback control of T_3 can work, too. So the feasibility of the controller configuration C in Fig.5 can be judged from the reachability matrix R in Fig.7. The countermeasure 1 can also be judged from the reachability matrix.

Manual intervention by the operator is discussed as the selection problem of the observed variables and manipulated variables, because the action of the operator can be regarded as a manipulation of observed variables by a controller. The elements of the CE-matrix C show the set of countermeasures. Effective controller configurations are automatically generated, so automatic countermeasures planning is possible by using C that is regarded as the countermeasure.

R	F_{1i}	T_{1i}	L_1	T_1	F_{1-3}	T_{1-3}	F_{2i}	T_{2i}	L_2	T_2	F_{2-3}	T_{2-3}	F_u	L_3	T_3	F_{3o}	T_{3o}
F_{1i}	0	0	0	0	0	0	0	0	0	0	0	0	0	0	0	0	0
T_{1i}	0	0	0	0	0	0	0	0	0	0	0	0	0	0	0	0	0
L_1	1	0	1	0	1	0	1	0	1	0	1	0	0	1	0	1	0
T_1	0	1	0	0	0	0	0	0	0	0	0	0	0	0	0	0	0
F_{1-3}	1	0	1	0	1	0	1	0	1	0	1	0	0	1	0	1	0
T_{1-3}	0	1	0	1	0	0	0	0	0	0	0	0	0	0	0	0	0
F_{2i}	0	0	0	0	0	0	0	0	0	0	0	0	0	0	0	0	0
T_{2i}	0	0	0	0	0	0	0	0	0	0	0	0	0	0	0	0	0
L_2	0	0	0	0	0	0	1	0	1	0	1	0	0	0	0	0	0
T_2	0	0	0	0	0	0	0	1	0	0	0	0	0	0	0	0	0
F_{2-3}	0	0	0	0	0	0	1	0	1	0	1	0	0	0	0	0	0
T_{2-3}	0	0	0	0	0	0	0	1	0	1	0	0	0	0	0	0	0
F_u	1	1	1	1	1	1	1	1	1	1	1	1	1	1	1	1	0
L_3	1	0	1	0	1	0	1	0	1	0	1	0	0	1	0	1	0
T_3	1	1	1	1	1	1	1	1	1	1	1	1	1	1	1	1	0
F_{3o}	0	0	0	0	0	0	0	0	0	0	0	0	0	0	0	0	0
T_{3o}	0	0	0	0	0	0	0	0	0	0	0	0	0	0	0	0	0

Fig. 7. Reachability matrix R

4 Plant Model Generation by Unit Modules Composition

In this section, we propose a method to synthesize plant model G by using unit modules composition. This method can deal with the change of the cause-effect in the process variables as the change of the plant situation.

At first, CE matrices of plant module are prepared for each unit. Their process variables are modeled by inner variables like T and L, and interface variables like F_{in}, F_{out}, T_{in}, and T_{out}. Each module has their fault tree for safety, quality, and quantity. The tag names of process variables in Tank 1 module $G1$, tank 2 module $G2$, and tank 3 module $G3$ are modified based on the units connection as shown in Fig.8.

The plant model G' in the example scenario situation, is shown in Fig.9. It can be synthesized from CE matrix $G1$, $G2$, and $G3$ with following procedures.

Prodesure1: To combine unit modules, Boolean addition is applied for same elements.
Procedure 2: In combination plant matrix, it is assigned to "0" for null elements.

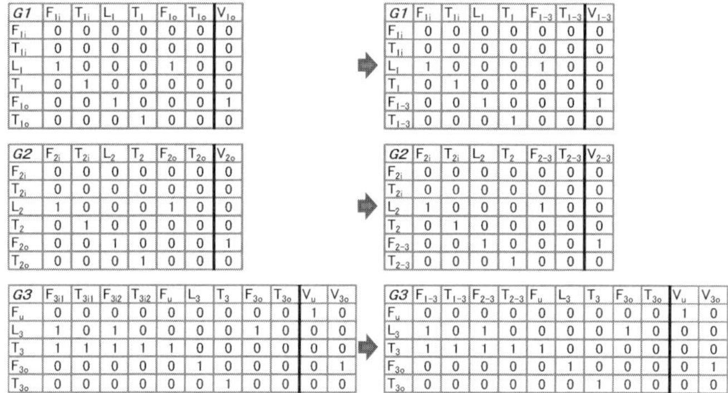

$G1$

$G1$	F_{1i}	T_{1i}	L_1	T_1	F_{1o}	T_{1o}	V_{1o}
F_{1i}	0	0	0	0	0	0	0
T_{1i}	0	0	0	0	0	0	0
L_1	1	0	0	0	1	0	0
T_1	0	1	0	0	0	0	0
F_{1o}	0	0	1	0	0	0	1
T_{1o}	0	0	0	1	0	0	0

$G1$	F_{1i}	T_{1i}	L_1	T_1	F_{1-3}	T_{1-3}	V_{1-3}
F_{1i}	0	0	0	0	0	0	0
T_{1i}	0	0	0	0	0	0	0
L_1	1	0	0	0	1	0	0
T_1	0	1	0	0	0	0	0
F_{1-3}	0	0	1	0	0	0	1
T_{1-3}	0	0	0	1	0	0	0

$G2$	F_{2i}	T_{2i}	L_2	T_2	F_{2o}	T_{2o}	V_{2o}
F_{2i}	0	0	0	0	0	0	0
T_{2i}	0	0	0	0	0	0	0
L_2	1	0	0	0	1	0	0
T_2	0	1	0	0	0	0	0
F_{2o}	0	0	1	0	0	0	1
T_{2o}	0	0	0	1	0	0	0

$G2$	F_{2i}	T_{2i}	L_2	T_2	F_{2-3}	T_{2-3}	V_{2-3}
F_{2i}	0	0	0	0	0	0	0
T_{2i}	0	0	0	0	0	0	0
L_2	1	0	0	0	1	0	0
T_2	0	1	0	0	0	0	0
F_{2-3}	0	0	1	0	0	0	1
T_{2-3}	0	0	0	1	0	0	0

$G3$	F_{3i1}	T_{3i1}	F_{3i2}	T_{3i2}	F_u	L_3	T_3	F_{3o}	T_{3o}	V_u	V_{3o}
F_u	0	0	0	0	0	0	0	0	0	1	0
L_3	1	0	1	0	0	0	0	1	0	0	0
T_3	1	1	1	1	1	0	0	0	0	0	0
F_{3o}	0	0	0	0	0	1	0	0	0	0	1
T_{3o}	0	0	0	0	0	0	1	0	0	0	0

$G3$	F_{1-3}	T_{1-3}	F_{2-3}	T_{2-3}	F_u	L_3	T_3	F_{3o}	T_{3o}	V_u	V_{3o}
F_u	0	0	0	0	0	0	0	0	0	1	0
L_3	1	0	1	0	0	0	0	1	0	0	0
T_3	1	1	1	1	1	0	0	0	0	0	0
F_{3o}	0	0	0	0	0	1	0	0	0	0	1
T_{3o}	0	0	0	0	0	0	1	0	0	0	0

Fig. 8. CE matrix $G1$, $G2$, and $G3$

Procedure 3: The cause-effect in combination plant model is modified by using the relationship rule of interface variables and inner variables in each unit module, as "If $T_{in} \neq T$, then $(F_{in} \to T)$".

Procedure 1 and 2 are normal connection rules. The arrangement of the relationship rule for procedure 3 in this module composition approach is needed for designing alarm management system.

G'	F_{1i}	T_{1i}	L_1	T_1	F_{1-3}	T_{1-3}	F_{2i}	T_{2i}	L_2	T_2	F_{2-3}	T_{2-3}	F_u	L_3	T_3	F_{3o}	T_{3o}	V_{1-3}	V_{2-3}	V_u	V_{3o}
F_{1i}	0	0	0	0	0	0	0	0	0	0	0	0	0	0	0	0	0	0	0	0	0
T_{1i}	0	0	0	0	0	0	0	0	0	0	0	0	0	0	0	0	0	0	0	0	0
L_1	1	0	0	0	1	0	0	0	0	0	0	0	0	0	0	0	0	0	0	0	0
T_1	0	1	0	0	0	0	0	0	0	0	0	0	0	0	0	0	0	0	0	0	0
F_{1-3}	0	0	1	0	0	0	0	0	0	0	0	0	0	0	0	0	0	1	0	0	0
T_{1-3}	0	0	0	1	0	0	0	0	0	0	0	0	0	0	0	0	0	0	0	0	0
F_{2i}	0	0	0	0	0	0	0	0	0	0	0	0	0	0	0	0	0	0	0	0	0
T_{2i}	0	0	0	0	0	0	0	0	0	0	0	0	0	0	0	0	0	0	0	0	0
L_2	0	0	0	0	0	0	1	0	0	0	1	0	0	0	0	0	0	0	0	0	0
T_2	0	0	0	0	0	0	0	1	0	0	0	0	0	0	0	0	0	0	0	0	0
F_{2-3}	0	0	0	0	0	0	0	0	1	0	0	0	0	0	0	0	0	0	1	0	0
T_{2-3}	0	0	0	0	0	0	0	0	0	1	0	0	0	0	0	0	0	0	0	0	0
F_u	0	0	0	0	0	0	0	0	0	0	0	0	0	0	0	0	0	0	0	1	0
L_3	0	0	0	0	1	0	0	0	0	0	1	0	0	0	0	1	0	0	0	0	0
T_3	0	0	0	0	0	1	0	0	0	0	1	1	1	0	0	0	0	0	0	0	0
F_{3o}	0	0	0	0	0	0	0	0	0	0	0	0	0	1	0	0	0	0	0	0	1
T_{3o}	0	0	0	0	0	0	0	0	0	0	0	0	0	0	1	0	0	0	0	0	0

Fig. 9. Plant model G' for example scenario

Consequently, the (T_3, F_{1-3}) is changed "0" from CE matrix G in Fig.4. This change of element describes that the operator cannot use flow date F_{1-3} for T_3 control. The reachability matrix based on G' can judge that the countermeasure 1 becomes an inadequate countermeasure in the new situation.

This unit modules composition approach has following merits for countermeasures planning and alarm management system. The plant modeling and maintenance can be easy by unit modules composition. This approach can be adapted to changes in plant situations.

To apply the proposed method to countermeasures design and/or alarm management, time to root cause detection or time to action must be included in the system.

The fault detection problems are discussed by another model of CE-matrices [3]. The later latter problem will be proposed in the next paper.

5 Conclusion

In this paper, we discussed that plant models must be modified based on the plant situation. The unit modules composition for countermeasure planning is proposed to deal with the problem. This approach is useful for the designing of effective alarm management systems. This method can be considered as the first step to bridge the discontinuity between operator support systems for abnormal situations.

References

[1] EEMUA 191, Alarm Systems: A Guide to Design Management and Procurement, EE-MUA, 2nd Edition (2007)

[2] Luyben, W.L., Luyben, M.L.: Essential of Process Control. The McGraw-Hill, New York (1997)

[3] Hamaguchi, T., Miura, H., Yoneya, A., Hashimoto, Y., Togari, Y.: Faults Diagnosis Utilizing CE-Matrices. Kagaku Kogaku Ronbunshu 24(4), 615–619 (1998)

[4] Hamaguchi, T., Kamiya, T., Yoneya, A., Hashimoto, Y., Togari, Y.: Controller Configuration Design and Troubleshooting Planning Using CE-Matices. Kagaku Kogaku Ronbunshu 25(3), 384–388 (1999)

[5] Hamaguchi, T., Hashimoto, Y., Yoneya, A., Togari, Y.: Countermeasures Planning based on Controller Configuration. In: Proceedings of the IASTED International Conferences Intelligent Systems and Control, Hawaii, USA, pp. 205–209 (2000)

[6] Hamaguchi, T., Hashimoto, Y., Itoh, T., Yoneya, A., Togari, Y.: Selection of Operating Modes in Abnormal Situations. Journal of Chemical Engineering of JAPAN 35(12), 1282–1289 (2002)

[7] Hamaguchi, T., Hashimoto, Y., Itoh, T., Yoneya, A., Togari, Y.: Abnormal Situation Correction Based on Controller Reconfiguration. Journal of Process Control 13, 16–175 (2003)

Trend Analysis for Decision Support in Control Actions of Suspension Polymerization

Hideyuki Matsumoto, Masaya Honda, and Chiaki Kuroda

Department of Chemical Engineering, Tokyo Institute of Technology,
Tokyo 152-8550, Japan
hmatsumo@chemeng.titech.ac.jp

Abstract. This paper presents an application method of trend analysis for decision support in the control action of divided addition of stabilizer in suspension polymerization. The proposed method can extract segments online from time-series temperature data, and it is shown that segments extracted by appropriate setting of thresholds can give us information useful for determination of timing of the second addition of suspension stabilizer. Moreover, it is seen that dynamic analysis of slope of extracted segments can provide us with information relevant to unfavorable behavior of droplets in a plain form. It is demonstrated that the dynamic analysis of segments is also useful for prevision of reactor fouling.

1 Introduction

In monitoring time-series process data acquired by conventional measurements, e.g. temperature, flow rate and pressure, operator's reasoning is qualitative rather than based on precise numbers. As considering that operators process symbolic data, an advanced monitoring and supervision system must process large amount of numerical data into information that is comprehensible to the operators. Trend analysis is a useful approach to extract information from numerical process data and represent it symbolically, in a qualitative or semi-qualitative way [1]. Its objective is to convert online numerical data into knowledge usable for operator support. A qualitative process trend is a description of the evolution of the qualitative state of a process variable, in a time interval, using a set of symbols called primitives.

Various representations of process trends have been reported in the previous studies. Charbonnier *et al.* proposed a methodology to extract on-line temporal trends from a time series [1]. The proposed method for trend analysis can deal with noisy data online. In representation of trends extracted online, three qualitative primitives are used: Steady, Increasing, Decreasing. Then analysis of process trends using the simple representation has demonstrated to be useful in process engineering applications; food process monitoring and abnormal situation management in the three-tanks system. Yamashita presented an effective online method to extract qualitative representation of movements in a

I. Lovrek, R.J. Howlett, and L.C. Jain (Eds.): KES 2008, Part III, LNAI 5179, pp. 25–32, 2008.

two variables plane [2]. The method was motivated by the idea of the Charbonnier's approach for an univariate time series, and usefulness of the method was demonstrated on an application to a diagnosis of valve stiction in an industrial plant.

This paper presents an application method of trend analysis for decision support in control actions of suspension polymerization. In suspension polymerization that is sometimes called peal polymerization or bead polymerization, droplets of monomer into which initiator is dissolved are generally distributed in water. The polymerization proceeds in the droplets, and generated polymer particles precipitate in water. One of problems for the suspension polymerization is to generate polymers with broad particle size distribution. Moreover, the agglomeration between droplets and sticking of droplets to the reactor wall lead to serious troubles from heat build-up and the formation of large masses, making it difficult to carry out prolong batch operation [3]. Divided addition of suspension stabilizer is well known as a useful action of controlling dynamic behavior of droplets in polymerization and final particle size distribution [4]. However, it seems difficult to explain clearly how to determine timing of the second addition of suspension stabilizer.

According to Tanaka *et al.* [4], conversion is considered to be reference for determining the second addition of suspension stabilizer because rates of coalescence and breakup of droplets change with conversion. As based on heat flow model for reaction calorimetry [5], it is considered that trajectory of temperature data inside reactor includes information about variations of the conversion and the overall heat-transfer coefficient. For example, a high-gain nonlinear cascade state estimator BenAmor *et al.* proposed can be used to estimate conversion online from temperature data without samples [6]. Since relationships among conversion, physical properties and dynamic behaviors of droplets are complicated, it is considered that advance of the above-mentioned estimator based on theoretical models will enhance necessity of the trend analysis that can convert online numerical data into knowledge for control actions.

Hence, purpose of this paper is to investigate on correlation of primitives extracted from temperature data with knowledge for timing of the stabilizer addition. The Charbonnier's approach [1] is used in splitting the time-series data into line segments. Furthermore, applicability of this method for trend analysis will be discussed from the viewpoint of prevision of reactor fouling.

2   Method of Trend Extraction in Suspension Polymerization Process

2.1   Online Segmentation of Time-Series Process Data

A segmentation algorithm consists of splitting data into successive line segments of the form:

$$\hat{y}(t) = (t - t_0)p + y_0 \tag{1}$$

where t_0 is the time when the segment beings, p is its slope and y_0 is the ordinate at time t_0. Parameter identification is based on the least-square method. The segmentation algorithm determines online the moment when the current linear approximation is no longer acceptable and when the new linear function that now best fit the data should be calculated. Criterion of this determination is based on the cumulative sum (cusum).

$$cusum(t_1 + k\Delta t) = cusum(t_1 + (k - 1)\Delta t) + e(t_1 + k\Delta t) \qquad (2)$$

where

$$e(t_1 + k\Delta t) = \hat{y}(t_1 + k\Delta t) - (t_1 + k\Delta t - t_{01})p_1 - y_{01} \qquad (3)$$

As shown in Fig.1, the absolute value of the cusum is compared to two thresholds $th1$ and $th2$ at each sampling time $(t1 + k\Delta t)$.

- If $|\ cusum(t_1 + k\Delta t)\ | \ < \ th1$: the linear current model is acceptable.
- If $|\ cusum(t_1 + k\Delta t)\ | \ \geq \ th1$: the signal value and corresponding time are stored.
- If $|\ cusum(t_1 + k\Delta t)\ | \ > \ th2$: the linear model is no longer acceptable and a new linear model is calculated on the stored values during the period.

Once the new linear model has been calculated, the cusum is reset to zero.

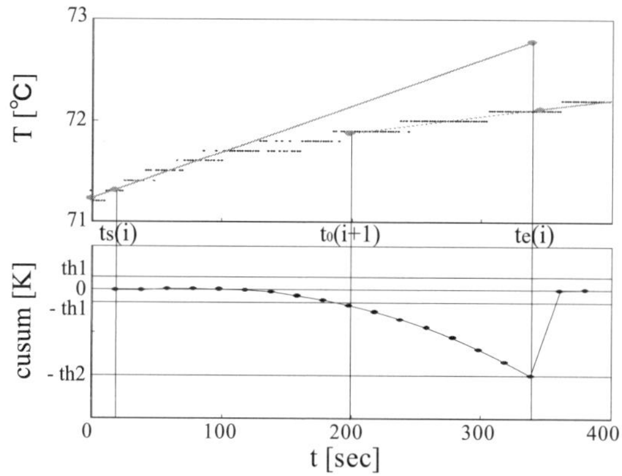

Fig. 1. Schematic diagram of online segmentation based on the cusum

2.2   Effects of Two Thresholds on Trend Extraction

In this paper, process data of suspension polymerization were collected in a laboratory-scale. In experiments for data acquisition, a reactor with a jacket was prepared, and water and styrene monomer were fed into the reactor. Volume fraction of styrene was 5 %. Azobisisobutyronitrile (AIBN) was used as an initiator and its concentration was 0.05 mol/l. As a suspension stabilizer, polyvinyl alcohol (PVA) was prepared at 0.02 wt% concentration. Rotational speed of a flat-bladed turbine impeller was set to 320 rpm, where complete dispersion of styrene monomer droplets was confirmed. Then temperature of a coolant flowing into the jacket was kept 75 °C.

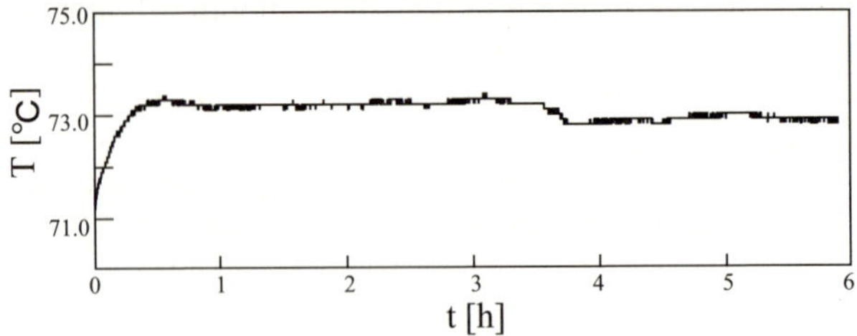

Fig. 2. Variation of temperature inside the reactor

Figure 2 shows a trajectory of temperature inside the reactor, which is recorded at a time interval of two seconds. Polymerization of styrene is an exothermic reaction, whereas change of temperature is not volatile in the suspension polymerization. It is because continuous phase of water can remove heat of reaction generated in dispersed droplets of styrene. As seeing carefully slow and small change of the temperature, its decrease from t=3.5 h to t=4 h was considered to show termination of reaction. Moreover, tendency in that the temperature rose slightly and temporarily was seen at about 3 h. We considered that the tendency was caused by increase of reaction rate in the last stage of polymerization, as based on numerical simulation of radical polymerization process.

In order to relate some segments, which are extracted online from temperature data, to knowledge about process dynamics as mentioned above, determination of thresholds ($th1$ and $th2$) is significant in the procedure explained in the previous subsection. Figure 3 shows effect of change of th2 on segmentation of the temperature data plotted in Fig. 2. In all these segmentations, sampling time (Δt) and number of sampling data for the initial linear model (d) were set at 4.0 s and 50. And the value of $th1$ was set at 2.5. As comparing Fig. 3 (a) with (b), it was seen that lower value of $th2$ could cause extraction of many short segments, which was more sensitive to tiny fluctuation of data. On the other hand, Fig. 3 (c) showed that higher value of $th2$ could slow down response to steep change of the data.

Thus, in this paper, the case (b) where $th2 = 20$ was supposed to be preferable one for setting of parameters. Moreover, effect of $th1$ on the segmentation was investigated for a case when value of $th2$ was set at 20. In a case when th1 was zero, it was seen that many short segments were extracted over a range of 4 to 6 h. It, however, was considered that the problem of generating many short segments could be solved by application of a method for classification and aggregation of segments that Charbonnier et al. [1] proposed.

3   Results and Discussions

3.1   Extraction of Timing of Adding Suspension Stabilizer

First, an effect of divided addition of PVA on formation of polymer particles was investigated empirically. PVA solution at a 0.02 wt% was fed into the reactor

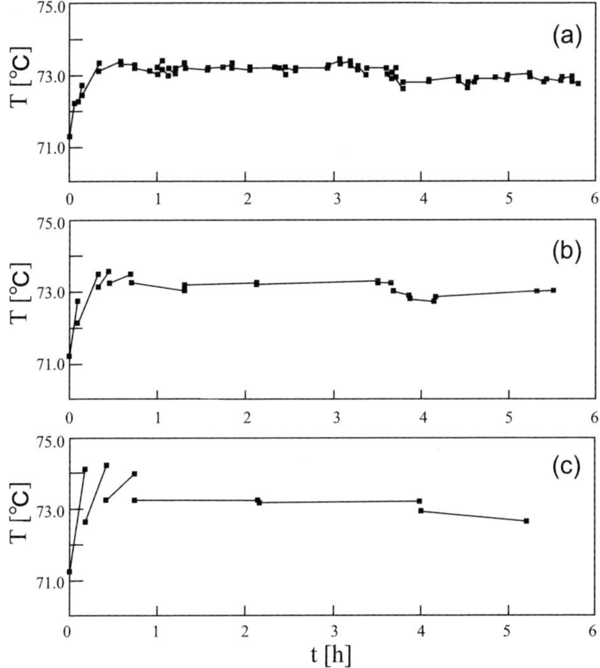

Fig. 3. Effect of $th2$ on segmentation: (a) $th2$=5.0, (b) $th2$=20, (c) $th2$=100

before initiating polymerization. Then PVA solution at a 0.01 wt% was fed again between 0.5 and 4.5 h. As a result of the experiments, Fig. 4 showed two images of surface for a polymer particle that were shot by the scanning electron microscope (SEM). As comparing Fig. 4 (b) with Fig. 4 (a) where PVA solution at a 0.03 wt% was fed before start of polymerization, it was considered that the second addition of the stabilizer at 3 h brought about reduction of adhesion of minute particles to surface of polymer particle. Moreover, it was seen that minute particles could be easily separated from the polymer particles, only when the second addition time was about 3 h.

Next, we investigated whether extracted trend of temperature could be related to the above-mentioned experimental results or not. It was seen in Fig. 3 (a) that new segment was extracted at about 3 h, whereas Figs. 3 (b) and (c) did not show the desired segmentation. So value of $th2$ was changed over a range of 5.0 to 20, as setting it at multiples of $th1$. After much trial and error for determining $th1$, we thought that trend as illustrated in Fig. 5 (a) facilitated inferring of process dynamic behavior. In Fig. 5 (a) where $th1$ and $th2$ were 3.5 and 10.5, a plain and short segment of "Increasing" over a time st_1 was considered to sign increase of reaction rate as described in subsection 2.2. Then, in try and error for determination of two thresholds, it was seen that the segmentation was sensitive to change of $th1$ rather than $th2$. So it seemed difficult to determine $th1$ theoretically. On the other hand, we thought that a method of setting

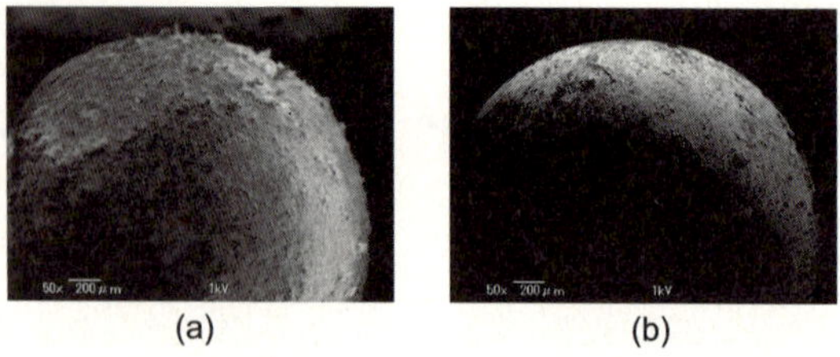

Fig. 4. Effect of divided addition of PVA: (a) no divided addition, (b) the second addition time was 3 h

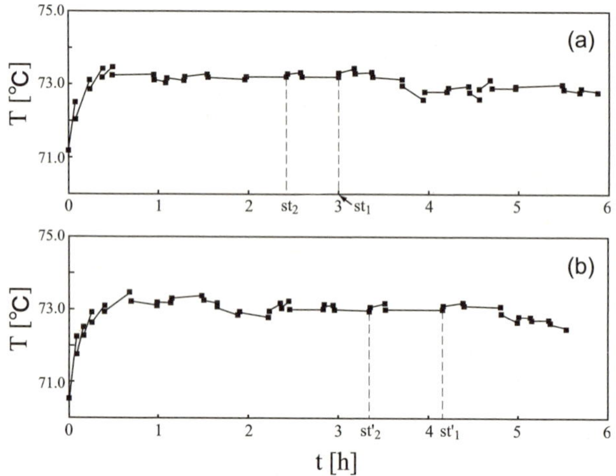

Fig. 5. A comparison between two runs in behaviors of segments: (a) run no. 1, (b) run no. 2

$th2$ at three times value of $th1$ was appropriate, as based on the three-sigma method.

Moreover, time-series data of temperature that were acquired in another run (run no.2) were analyzed. It was observed in this run that more minute particles were adhered to surface of polymer particles than the previous run (run no.1), although both of two runs were made under the same operational condition. Figure 5 (b) shows a result of segmentation for run no.2 by setting $th1$ and $th2$ at 3.0 and 9.0. Since difference of trend between Figs. 5 (a) and (b) was seen over a range of 0 to 2 h, time transition of slope p of extracted segments (see Eq. 1) was analyzed in each run. It was estimated that value of p for run no.2 did not approach zero stably, which was supposed to indicate process dynamic behavior involved with the adhesion of minute particles. On the other hand, it was thought that behaviors of segments at about st_1 and st_2 in Fig. 5 (a) were like to ones at about st'_1 and st'_2 in Fig. 5 (b). Since the likeness of trend between two runs was assumed to time delay of polymerization, it was considered that a

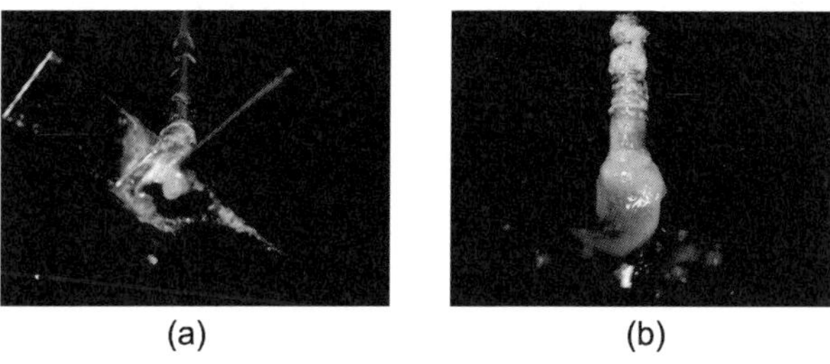

Fig. 6. Images of impellers after a lapse of two hours: (a) a normal case, (b) a case when the valve was clogged

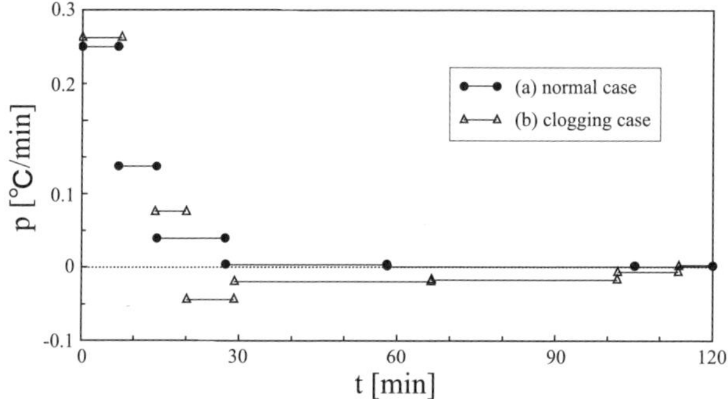

Fig. 7. Time transition of slope of extracted segment: (a) a normal case, (b) a case when the valve was clogged

time st_2' in Fig. 5 (b) (st_2 in Fig. 5 (a)) would be a candidate for determination of timing of stabilizer addition.

3.2 Trend Analysis for Prevision of Reactor Fouling

We carried out processing of releasing a valve at bottom of the reactor when two hours elapsed from start of polymerization, on the assumption that the above-mentioned batch process would be extended to multistage semibatch process. In the experiments, a pitched paddle impeller was used. As a result of the experiments, it was observed that the valve was clogged with sticky small particles. As to trajectory of temperature data for the case when the clogging occurred, distinction between this case and the normal case seemed to be difficult on the surface. On the other hand, adhesion of much white substance to the impeller shaft was seen in Fig. 6 (b).

Thus extraction of segments was implemented for temperature data in both the normal case and the case when the clogging occurred. In the trend analysis, number of sampling data for the initial linear model was set at 50. Values of $th1$ and $th2$ were set at 2.5 and 20. In this subsection, changes of slope p of extracted segments were analyzed. Two analytical results plotted by circle and triangle in Fig. 7 correspond to two images in Fig. 6 (a) and (b). As a result of comparison between two cases, it was seen that value of p for the clogging case did not approach zero stably. It was considered that this trend of p in the case of fouling was similar to a result of trend analysis for adhesion of minute particles that was described in the previous subsection. Therefore we thought that the trend analysis applied in this paper might provide operators with information about dynamic behavior of droplets in a plain form.

4 Conclusion

This paper has demonstrated that the trend analysis based on the Charbonnier's approach was applicable to online monitoring of temperature data in suspension polymerization process. It was seen that an appropriate setting of thresholds ($th1$ and $th2$) could extract segments by which process dynamic behavior was inferred easily. In the control action of divided addition of suspension stabilizer, the segmentation showed extraction of some characteristic times for the second addition of stabilizer. Moreover, it was seen that dynamic analysis of slope p of extracted segments could provide us with information about agglomeration between droplets and sticking of droplets to the the valve.

References

1. Charbonnier, S., Garcia-Beltan, C., Cadet, C., Gentil, S.: Trend extraction and analysis for complex system monitoring and decision support. Eng. App. Artif. Intel. 18, 21–36 (2005)
2. Yamashita, Y.: On-line extraction of qualitative movements for monitoring process plants. In: Gabrys, B., Howlett, R.J., Jain, L.C. (eds.) KES 2006. LNCS (LNAI), vol. 4252, pp. 595–602. Springer, Heidelberg (2006)
3. Hatate, Y., Ikeura, T., Shinonome, M., Kondo, K., Nakashio, F.: Suspension polymerization of styrene under ultrasonic irradiation. J. Chem. Eng. Japan 14, 38–43 (1981)
4. Tanaka, M., Tanaka, H., Kimura, I., Saito, N.: Effect of divided addition of suspension stabilizer on particle size distribution in suspension polymerization of styrene. Kagaku Kogaku Ronbunshu 18, 528–534 (1992)
5. Hergeth, W.-D., Jaeckle, C., Krell, M.: Industrial process monitoring of polymerization and spray drying processes. Polym. React. Eng. 11, 663–714 (2003)
6. BenAmor, S., Colombie, D., McKenna, T.: Online reaction calorimetry. Application to the monitoring of emulsion polymerization without samples or models of the heat-transfer coefficient. Ind. Eng. Chem. Res. 41, 4233–4241 (2002)

An Operational Model and a Computer Support Environment for Batch Plants Based on Adaptive Scheduling

Hisaaki Yamaba, Makoto Hirosaki, and Shigeyuki Tomita

University of Miyazaki, 1-1 Gakuen Kibanadai Nishi, Miyazaki, Japan

Abstract. Scheduling is one of the most important feature in production management. Production systems are operated on schedules in order to minimize costs of production and complete production of given demands by their due dates. However, production cannot be often carried out as scheduled because of troubles or accidents such as malfunction of reactors, tardiness of chemical reactions, and so on. Even in such situations, production activities have to be continued with modifying original schedules. As matters stand, such modifications of schedules are carried out depending on human experiences. In this work, an operational model of a production system was considered which is used to continue production under production environments with several uncertainties. First, a computer aided environment of which simulators and schedulers form the core is developed based on the Object-Oriented approach. Next, a series of experiments was carried out using this environment in order to evaluate performances of various parameters used in proposed methods.

1 Introduction

In order to minimize costs of productions and complete production of given production demands by their due dates, production systems are operated as scheduled. However, production cannot be carried out as given schedules because of troubles or accidents such as malfunction of reactors or tardiness of chemical reactions. Even in such situations, production activities have to be continued with modifying original schedules. As matters stand, such modifications of schedules are carried out depending on human experiences. Therefore, some rational manner for dealing with troubles and accidents are strongly expected.

In this work, an operational model of a production system was considered which is used to continue production activities under production environments with various uncertainties. Concretely, following features were concerned.

1. How to determine whether to carry out re-scheduling or not?
2. How to divide jobs into targets of re-scheduling and jobs which are continued to be processed under an original schedule?
3. How to process jobs which were decided to be processed on an original schedule which no longer work out well ?

I. Lovrek, R.J. Howlett, and L.C. Jain (Eds.): KES 2008, Part III, LNAI 5179, pp. 33–40, 2008.

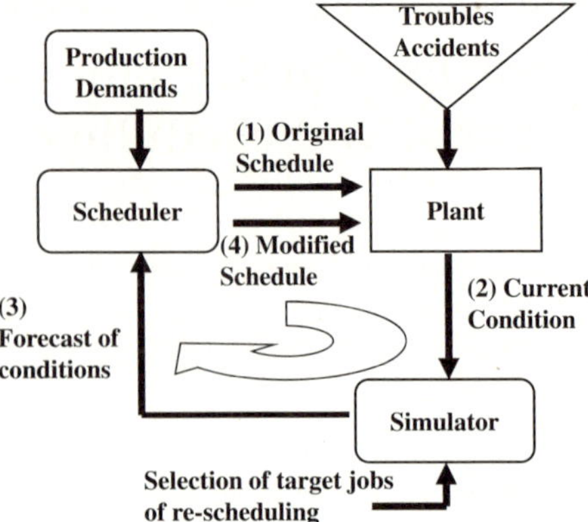

Fig. 1. Concept of Proposed Support Method of Production

These factors should be fixed depending on configurations of target plants. In order to determine the three factors above, a simulation based approach was adopted. A computer aided environment which consists of simulators and schedulers is developed based on the Object-Oriented approach. Software components are defined as classes of Java language. A computer aided environment for each plant with various configurations can be build easily using the components.

In this work, a series of experiments was carried out using this environment in order to evaluate performances of various parameters used in proposed methods.

2   Basic Concept of Operational Model

Fig.1 shows the basic concept of operational model proposed here. The model makes good use of schedulers and simulators.

(1) First, a scheduler of a plant makes original schedules for given demands. The plant starts production as scheduled. However, when it becomes hard to continue productions under the current schedule because of troubles or accidents such as malfunctions of machines or delay in processes, it is decided that re-scheduling is carried out and the production is shift from the original schedule to the new one after a given "switching time".

(2) For making out a modified schedule, the progress of production in the target time window is needed as the initial condition for the re-scheduling. In order to obtain the information, a simulation system of the plant is introduced. The simulator works following the condition of the plant with the operational policy of the plant.

(3) The results of the simulation (working conditions of the plant, etc) is given to the scheduler of the support environment. Using the information as the initial condition, the scheduler makes out a modified schedule which is start from the switching time point.

(4) By switching to the new modified schedule at the switching time, the target plant can continue a seamless production.

3   Class of Target Production System

It is assumed that the target plants discussed here have the following aspect.

Multi-purpose production with different combination of machines is intended. Parallel operations for multi-products are most likely required. A sequence of processes which is required for producing a final product forms a tree structure.

In this study, a "job" means 1 process in a sequence of processes required for each final product. Use of different combinations of machines which can carry out the same job is allowed together with alternation of the "recipes." Several operations such as "feed of a material", "heating", "cleaning of the reactor", etc, are listed in recipes in the required order.

Plants of the target class consist of some chemical reactors, raw material tanks, product tanks and pipelines for transferring materials.

Production demands are given to the plant in a periodic fashion. A schedule of demands given in the $i-1$th period are made out in the ith period and the demands are processed in the $i+1$th period. It is assumed that a productive capacity of a target plant is balanced with a load of production of give demands. Therefore, a schedule of the ith period has to be created under the condition that some jobs in the schedule of $i-1$th period are pushed out to the ith period.

The concepot of "Adaptive Scheduling" [1] was proposed in order to cope with such class of production systems by iterative scheduling.

4   Operational Method

In this paper, tardiness of operation time is focused on as the concrete troubles.

The basic concept of the proposed method is shown in Fig. 2. Lower rows of each machine in the Gantt charts represent schedules, and upper rows represent their progress.

1. First, the original schedule is given (Fig. 2 (a)).
2. When the progress of production comes to be disagreement with the original schedule because of troubles or accidents, the operator of the plant has to decide whether he should modify current schedule or not. In this work, the determination is carried out when any job do not end although a fixed time has passed after the scheduled ending time of the job. The fixed length of time is called the "determination time" in this work (Fig. 2 (b)).
3. In case that it is decided that the re-scheduling is carried out, the "switching time" is decidedat first. The "switching time" means the start time of the schedule window of the re-scheduling. Several indices can be adopted as the criterion for deciding the switching time. In this work, the modified schedule will start at n unit time after the "determination time" (Fig. 2 (b)).

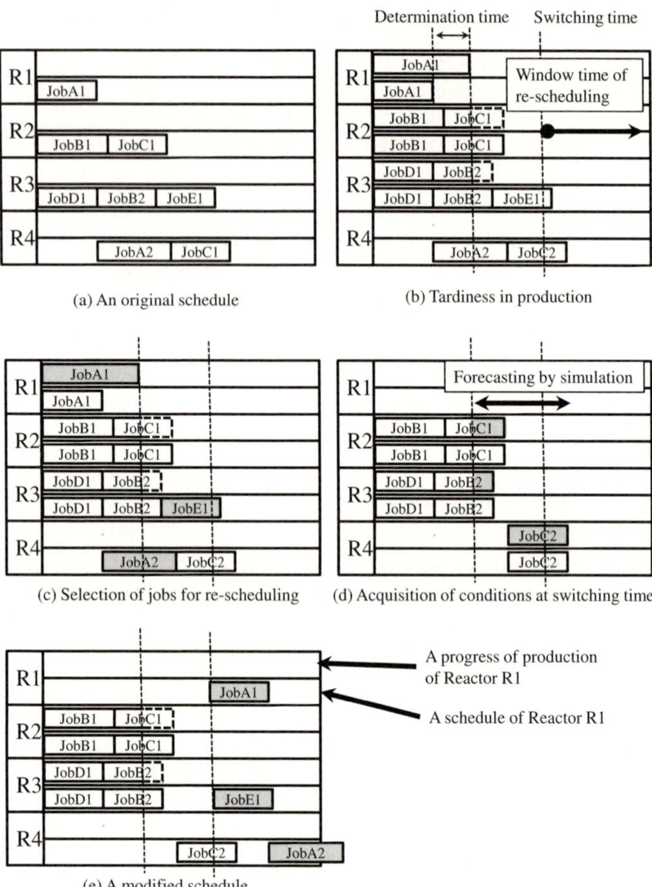

Fig. 2. Proposed Operation Method

4. Next, jobs which are not completed divided into two groups, jobs which are the target of re-scheduling and jobs which are carried out as the current schedule. In this work, jobs which do not start at the "determination time" are selected as targets of re-scheduling. (Fig. 2 (c)).

5. The jobs in the former group are carried out under the same policy mentioned in 8.

6. As for the jobs in the latter group, a modified schedule is made out. The target scheduling span is from the switching time to the end time of the current production period. In order to create a feasible schedule, the condition of the plant in the target scheduling span has to be given. The simulator of the plant is used to forecast the condition. (Fig. 2 (d)).

7. After the "switching time", the plant is operated to continue production on the modified schedule. (Fig. 2 (e)).

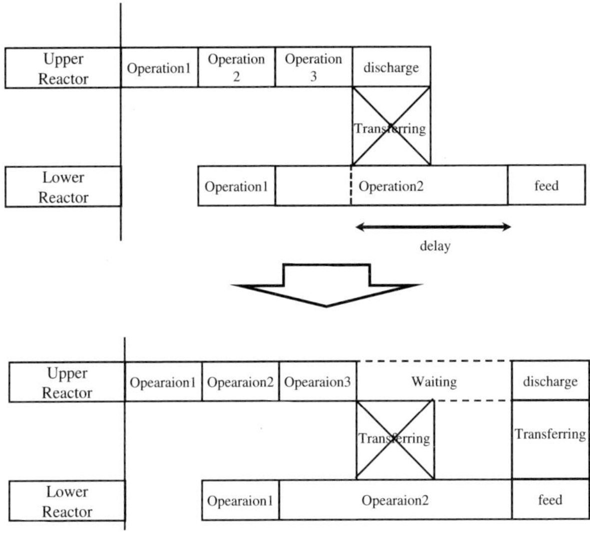

Fig. 3. Inserting a waiting operation to synchronize transferring operations

8. During the period from the "decision time" to the "switching time", production
   of a plant is executed basically on the original schedule. However, a start time
   each job is delayed according to tardiness of the schedule. Especially, in order
   to synchronize a time of transfering materials between an uppper reactor and
   lower one, an additional "waiting operation" has to be inserted (Fig. 3).

It is considered that an appropriate length of the "determination time" and
a length of a time span from the "determination time" to the "switching time"
depend on features of each production environment and each plant. In this work,
they are decided through simulation experiments. In the section of experiments,
performances of several candidates of the "determination time" under some production environments are investigated.

5 Development of Operation Support Environment

The operation support environment for production systems which consists of
simulators and schedulers was developed based on the operation model described
above.

Schedulers and simulators are the core of the operation support environment
developed here. As for the scheduling system, "Bpos" [2] is adopted as the scheduler of the environment. Some heuristic rules for scheduling obtained through preliminary experiments are mebedded in Bpos. Scheduling by Bpos is carried out
through placing segments corresponding to jobs on a gantt-chart using the rules.

On the other hand, simulators of plants are developed newly by means of
Object-Oriented approach. In order to deal with various configurations of plants

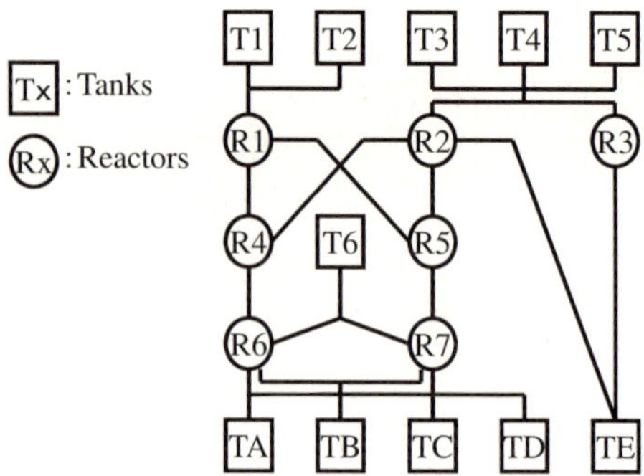

Fig. 4. Model plant used in experiments

and various batch processes, many Java classes were defined with their methods such as Reactor, Tank, Recipe, PipeLine, Batch, Process, Operation, Terminal-Condition, and so on.

Objects of Class Reactor are instantiated corresponding to each of reactors in target plants. Productive abilities of each reactor are represented by giving appropriate values to their variables.

Batch processes executed in reactors are abstracted and described by 4 classes: Operatoin, Process, Batch and TerminalCondition. Class Batch represents a set of materials in each reactor. Concretely, volumes and temperatures of materials are managed by an object of Class Batch.

Through simulations, an object of Class Operation is activated on a given schedule. The "operation" starts its corresponding process such as "transferring of a material", "heating ", "waiting", and so on. Each type of processes is represented by subclasses of Class Process. In order to realize parallel production in a plant, each process has its own thread. An object of Class TerminalCondition keeps watching on a designated batch. As soon as its conditions specified by the concrete subclass (Class OperationTimeCondition, TemperatureToHeatCondition, etc) and a parameter value (a time length of operation, a goal temperature to heat, etc) are satisfied, the object of Class TerminalCondition stop the thread of the corresponding process.

6   Experiments

A series of experiments was carried out in order to evaluate the influence of operational parameters mentioned above to the performance of a model plant. Another simulator was also developed and used as the role of the model plant.

6.1   Conditions

Fig. 4 shows the model batch plant used in the experiments. The plant consists of 6 raw material tanks, 7 batch reactors and 5 products tanks. A production

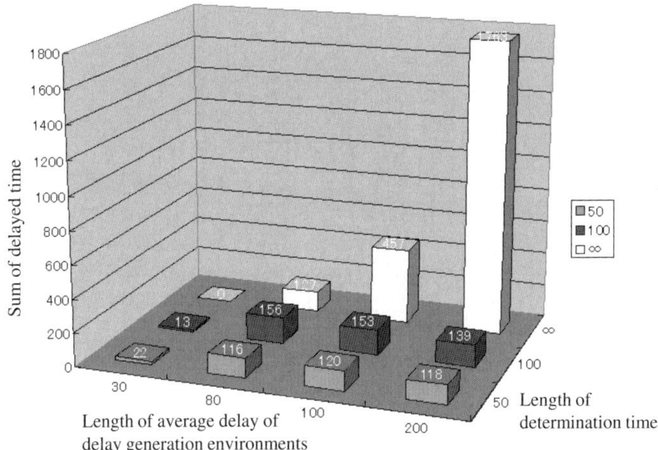

Fig. 5. Results of Experiments

period was 1 day. 1 experiment was a series of 7 days simulation. The "switching time" was set to 30 unit time after the time when re-scheduling was decided to be executed.

The purpose of the experiments is investigating an influence of the length of "determination time " on a productive performance of a plant. Four delay generation environment were introduced against the three candidates of "determination time" . The results were evaluated by the sum of delayed time of jobs which are not completed in the scheduled day.

One delay was generated in one day. A delayed operation was selected at random. The length of delayed time was calculated according to some regular distributions shown in table 1. The three candidates of "determination time " were 50 unit time, 100 unit time and "never re-schedule".

6.2   Results of Experiments

The experiments were carried out on the 15 combinations of the 3 "determination time" and 4 delay generation environments. 7 days simulations were repeated 5 times for each of 15 experiments.

Fig. 5 shows the results of the experiments. The averages of the sums of delayed time were plotted in the chart.

In the experiments using the delay generation environment 3 and 4 (the averages of delay time were 100 and 200), the shortest delay time was obtained when "decision time" was 50. In condition that re-scheduling was not executed, the sum of delay time was the longest. This means that the re-scheduling was effective in the the delay generation environments which generate longer delays. On the other hand, in the experiments 1 (the average of delay time was 30), no

Table 1. Delay generation environments

	average	distribution
1	30	50
2	80	50
3	100	50
4	200	50

job was delayed in the condition that re-scheduling was not executed. Besides, some jobs were delayed when 'decision time" was set to 50 or 100.

In general, some jobs selected for re-scheduling have been planned to be processed between "determination time" and "switching time" in the original schedule. Since new schedules of such jobs are postponed after the "switching time" in the modified schedule, the modified schedule tends to be a tight one. This means that re-scheduling results in lower performance when a length of expected delay time is shorter. However, in case that a length of expected delay time is longer, earlier determinations of re-scheduling save more time. The boundary of expected delay time whether re-scheduling works effectively or not was seemed to be 80 in the experiments.

Since the validity of knowledge obtained form these results are limited to the conditions of these experiments, it was confirmed that the support environment was useful to obtain parameters of the management method proposed in this work.

7   Conclusions

In order to manage a batch type chemical plant under uncertain conditions, a management support method was introduced together with the computer based support system which consists of simulators and schedulers. A series of experiments was carried out using this environment in order to evaluate performances of various parameters used in proposed methods. The results of the experiments show the effectiveness of this environment.

References

1. Tomita, S., Yamaba, H., O'shima, E.: On an Intelligent Scheduling System for a Parallel Distributed Process of Multi-Product – An attempt of developing an 'Adaptive scheduling' system. In: Proceeding of 4th International Symposium on Process Systems Engineering (PSE 1991), vol. II, pp. 19.1–19.15 (1991)
2. Tomita, S., Yamaba, H., O'shima, E.: Development of an intelligent Scheduling System for Managing Multipurpose Chemical Batch Plant – An attempt to exploit heuristic rules for developing an 'Adaptive' scheduling system. Kagaku Kogaku Ronbunshu 17, 740–749 (1991) (in Japanese)

Improving Search Efficiency of Incremental Variable Selection by Using Second-Order Optimal Criterion

Kazumi Saito[1], Nobuaki Mutoh[1], Tetsuo Ikeda[1],
Toshinao Goda[2], and Kazuki Mochizuki[2]

[1] School of Administration and Informatics, University of Shizuoka
[2] Department of Food and Nutritional Sciences, University of Shizuoka
52-1 Yada, Suruga-ku, Shizuoka 422-8526 Japan
{k-saito,muto,t-ikeda,gouda,kmochi}@u-shizuoka-ken.ac.jp

Abstract. We address the problem of improving search efficiency of incremental variable selection. As one application, we focus on generalized linear models that are linear with respect to their parameters, but their objective functions are not restricted to a standard sum of squared error. In this paper, we present a method for incrementally selecting a set of relevant variables together with a newly proposing criterion based on second-order optimality for our models. In our experiments using a synthetic dataset with tens of thousands of variables, we show that the proposed method was able to completely restore the relevant variables. Moreover, the method substantially improved the search efficiency in comparison to a conventional calculation method. Furthermore, it is shown that we obtained promising initial results using a real dataset in health-checkup.

1 Introduction

With an unprecedented spread of computers and Internet, we can conveniently obtain more and more data in various areas of application. Such areas typically include text processing of Web documents, gene expression array analysis and user profile utilization. Obviously, without some suitable processing of data, its mere existence would not become a useful asset that can benefits our purpose, and many other matters. In this paper, as primitive data processing, we focus on selecting a relatively small number of relevant variables from datasets with tens or hundreds of thousands of variables.

There exist a large amount of work focusing on variable selection, as presented in a survey paper [7] and a recent book [11]. Among several problem settings, one primitive task is to select relevant variables by assuming the linear least-square model provided with supervised information [18,23]. However, this simple model is likely to have an intrinsic limitation in the case that the Gaussian noise assumption is completely inadequate. To cope with such situations, as a straightforward extension, we can consider adopting suitable generalized linear models, which are typically formalized as logistic regression [21] or ordinal regression [2].

I. Lovrek, R.J. Howlett, and L.C. Jain (Eds.): KES 2008, Part III, LNAI 5179, pp. 41–49, 2008.

In this paper, we present a method for incrementally selecting a relatively small number of relevant variables with a newly proposing criterion based on second-order optimality. In what follows, we describe our problem formalization, and then explain our proposing method in Section 2. In section 3, we report our initial experimental results. We finally discuss related work and future directions in Section 4.

2   Proposing Method

Let $D = \{(\mathbf{x}_n, y_n) : 1 \leq n \leq N\}$ be a set of samples, where $\mathbf{x}_n = (x_{n,1}, \cdots, x_{n,M})^T$ denotes an M-dimensional input vector and y_n a target value corresponding to $\mathbf{x}_n$. Note that $\mathbf{a}^T$ denotes the transposed vector of $\mathbf{a}$. Now we define a partial list of distinct K variables by $R_K = (r(k) : 1 \leq k \leq K)$, where $1 \leq r(k) \leq M$, and $r(k) \neq r(k')$ if $k \neq k'$. We also introduce a $(K+1)$-dimensional parameter vector as $\mathbf{w}_K = (w_0, w_1, \cdots, w_K)^T$, and we focus on generalized linear models [21] consisting of a linear function with respect to $\mathbf{x}_n$, i.e., $z_n = w_0 + \sum_{k=1}^{K} w_k x_{n,r(k)}$. In this paper, for a given variable list size K, we address a problem of finding a variable list R_K and its corresponding optimal parameter vector $\hat{\mathbf{w}}_K$, which minimize a certain objective function as described below.

A standard objective function is the following sum of squared error.

$$\mathcal{E}_1(\mathbf{w}_K; R_K) = \frac{1}{2} \sum_{n=1}^{N} (y_n - z_n)^2. \tag{1}$$

This formalization is referred to as linear regression with variable selection. In the case that $0 \leq y_n \leq 1$, or typically $y_n \in \{0, 1\}$, another standard objective function is the following sum of cross-entropy terms.

$$\mathcal{E}_2(\mathbf{w}_K; R_K) = \sum_{n=1}^{N} -(y_n \log \sigma(z_n) - (1 - y_n) \log(1 - \sigma(z_n))), \tag{2}$$

where $\sigma(z_n) = 1/(1 + \exp(-z_n))$ stands for a sigmoidal (logistic) function [21]. This formalization is referred to as logistic regression with variable selection. In the case that y_n is a ranking score for the n-th sample, we want to preserve the ranking order of the samples by using the predicted values $\{z_n\}$. Then its reasonable objective function is the following pair-wise sum of cross-entropy terms.

$$\mathcal{E}_3(\mathbf{w}_K; R_K) = \sum_{n=1}^{N-1} \sum_{n'=n+1}^{N} -(\delta(y_n, y_{n'}) \log \sigma(z_n - z_{n'}) + (1 - \delta(y_n, y_{n'})) \log(1 - \sigma(z_n - z_{n'}))),$$

$$\tag{3}$$

where $\delta(y_n, y_{n'}) = 1$ if $y_n > y_{n'}$; $\delta(y_n, y_{n'}) = 0$ if $y_n < y_{n'}$; otherwise $\delta(y_n, y_{n'}) = 0.5$. This formalization is referred to as ordinal regression (or rank learning [2]) with variable selection.

As a basic approach to our problem, we focus on an incremental algorithm, i.e., the algorithm performs variable selection by adding at each step the variable that most decreases the objective function. More specifically, for a given variable list size K, we can summarize the incremental algorithm using $\mathcal{E}_i$ ($i \in \{1, 2, 3\}$) as follows.

step 1. Initialize $k = 1$ and $R_0 = \emptyset$;
step 2. compute $\hat{s} = \arg\min_s \mathcal{E}_i(\hat{\mathbf{w}}_k; (R_{k-1}, s))$, and set $R_k = (R_{k-1}, \hat{s})$;
step 3. if $k = K$, then output R_K and $\hat{\mathbf{w}}_K$, and terminate the iteration;
step 4. set $k = k + 1$, and return to **step 2.**;

Here (R_{k-1}, s) means we append the variable number s behind the last element of R_{k-1}.

In what follows, we derive a second-order optimal criterion for variable selection, in order to improve the computational efficiency at **step 3.** described above. One may think that this could be implemented by directly minimizing the objective function for each variable. Obviously, such an approach would be computationally demanding. As another idea, one may focus on some standard criteria like correlation coefficients between $(y_1, \cdots, y_N)$ and $(x_{1,s}, \cdots, x_{N,s})$. However, this approach does not directly address minimizing the objective function.

2.1   Second-Order Optimal Criterion

We consider a parameter vector set by $\bar{\mathbf{w}}_k = (\hat{\mathbf{w}}_{k-1}^T, 0)^T$, and define a modification vector calculated as $\Delta\mathbf{w}_k = \mathbf{w}_k - \bar{\mathbf{w}}_k$. For arbitrary variable number s and $R_k = (R_{k-1}, s)$, the second order Taylor expansion of $\mathcal{E}(\mathbf{w}_k; R_k)$ around $\bar{\mathbf{w}}_k$ gives the following:

$$\mathcal{E}(\mathbf{w}_k; R_k) - \mathcal{E}(\bar{\mathbf{w}}_k; R_k) \approx \frac{\partial\mathcal{E}(\bar{\mathbf{w}}_k; R_k)}{\partial\mathbf{w}_k^T}\Delta\mathbf{w}_k + \frac{1}{2}\Delta\mathbf{w}_k^T\frac{\partial^2\mathcal{E}(\bar{\mathbf{w}}_k; R_k)}{\partial\mathbf{w}_k\partial\mathbf{w}_k^T}\Delta\mathbf{w}_k. \quad (4)$$

In most cases, the following modification vector $\widehat{\Delta\mathbf{w}}_k$ minimizes the r.h.s. of Eq. (4).

$$\widehat{\Delta\mathbf{w}}_k = -\left(\frac{\partial^2\mathcal{E}(\bar{\mathbf{w}}_k; R_k)}{\partial\mathbf{w}_k\partial\mathbf{w}_k^T}\right)^{-1}\frac{\partial\mathcal{E}(\bar{\mathbf{w}}_k; R_k)}{\partial\mathbf{w}_k}. \quad (5)$$

Thus by substituting Eq. (5) into Eq. (4), we can obtain the second-order optimal improvement as follows.

$$\mathcal{E}(\mathbf{w}_k; R_k) - \mathcal{E}(\bar{\mathbf{w}}_k; R_k) \approx -\frac{1}{2}\frac{\partial\mathcal{E}(\bar{\mathbf{w}}_k; R_k)}{\partial\mathbf{w}_k^T}\left(\frac{\partial^2\mathcal{E}(\bar{\mathbf{w}}_k; R_k)}{\partial\mathbf{w}_k\partial\mathbf{w}_k^T}\right)^{-1}\frac{\partial\mathcal{E}(\bar{\mathbf{w}}_k; R_k)}{\partial\mathbf{w}_k}. \quad (6)$$

Here note that in the case of linear regression using $\mathcal{E}_1(\mathbf{w}_k; R_k)$, Eq. (6) gives the exact optimal improvement.

In the case of our problem settings, we can efficiently calculate Eq. (6). More specifically, due to the optimality condition at the $(k-1)$-th iteration, the gradient vector becomes as follows.

$$\frac{\partial\mathcal{E}(\bar{\mathbf{w}}_k; R_k)}{\partial\mathbf{w}_k^T} = \left(\frac{\partial\mathcal{E}(\hat{\mathbf{w}}_{k-1}; R_{k-1})}{\partial\mathbf{w}_{k-1}^T}, \frac{\partial\mathcal{E}(\bar{\mathbf{w}}_k; R_k)}{\partial w_{k,k}}\right) = \left(\mathbf{0}_{k-1}^T, \frac{\partial\mathcal{E}(\bar{\mathbf{w}}_k; R_k)}{\partial w_{k,k}}\right), \quad (7)$$

where $\mathbf{0}_{k-1}$ indicates the $(k-1)$-dimensional vector whose elements are zeros. Therefore, we can calculate Eq. (6) as follows.

$$\mathcal{E}(\mathbf{w}_k; R_k) - \mathcal{E}(\bar{\mathbf{w}}_k; R_k) \approx \left[\left(\frac{\partial^2 \mathcal{E}(\bar{\mathbf{w}}_k; R_k)}{\partial \mathbf{w}_k \partial \mathbf{w}_k^T} \right)^{-1} \right]_{k,k} \left(\frac{\partial \mathcal{E}(\bar{\mathbf{w}}_k; R_k)}{\partial w_{k,k}} \right)^2. \tag{8}$$

Here, by using a well-known matrix inversion formula [20], we can obtain the element of the inverted matrix as follows.

$$\left[\left(\frac{\partial^2 \mathcal{E}(\bar{\mathbf{w}}_k; R_k)}{\partial \mathbf{w}_k \partial \mathbf{w}_k^T} \right)^{-1} \right]_{k,k}^{-1} = \frac{\partial^2 \mathcal{E}(\bar{\mathbf{w}}_k; R_k)}{\partial w_{k,k} \partial w_{k,k}} - \frac{\partial^2 \mathcal{E}(\bar{\mathbf{w}}_k; R_k)}{\partial w_{k,k} \mathbf{w}_{k-1}^T} \left(\frac{\partial^2 \mathcal{E}(\hat{\mathbf{w}}_{k-1}; R_{k-1})}{\partial \mathbf{w}_{k-1} \partial \mathbf{w}_{k-1}^T} \right)^{-1} \frac{\partial^2 \mathcal{E}(\bar{\mathbf{w}}_k; R_k)}{\partial w_{k,k} \mathbf{w}_{k-1}}.$$

$$\tag{9}$$

Consequently, we propose a method that recursively selects new variables by using Eq. (8) combined with Eq. (9).

2.2 Computational Complexity

We consider computational complexity of our proposing method at the $(k+1)$-th iteration for convenience. Obviously, we need to prepare the matrix $\partial^2 \mathcal{E}(\hat{\mathbf{w}}_k; R_k) / \partial \mathbf{w}_k \partial \mathbf{w}_k^T$, and calculate its inversion appearing in Eq. (9). These computational complexities are $O(Nk^2)$ and $O(k^3)$, respectively. Here note that we can utilize the same inverted matrix for arbitrary variable number s. For each variable number s, we need to calculate the gradient vector, the $(k+1)$-th element vector of the Hessian matrix, and the quadratic form appearing in Eq. (9). These total computational complexities are $O((M-k)N)$, $O((M-k)Nk)$ and $O((M-k)k^2)$, respectively. Thus, by adopting a natural setting that $k \ll M$, $k \ll N$, we can obtain the dominant computational complexity as $O(MNk)$.

On the other hand, we consider a conventional calculation method which directly solves Eq. (6) for each variable number s. Then the method additionally requires the computational complexity of $O(Mk^3)$. Compared with our proposing method, the conventional calculation method works with comparable efficiency until $k^2 \approx N$, but requires a substantial amount of computational load when $k^2 \gg N$. In our experiments, we confirm this fact by using a synthetic data set.

3 Evaluation by Experiments

3.1 Synthetic Data

First, we applied our method to a synthetic data set. Our regression problem is to find a linear function described as $y = w_0 + \sum_{k=1}^{K^*} w_k x_k$. Here we assigned each of $\{w_0, \cdots, w_{K^*}\}$ to a value generated uniformly at random in the range of $[-2, -1]$ or $[1, 2]$. Obviously, in the case that w_k is substantially small, it is almost impossible to select the corresponding variable x_k. In our experiments, we set $K^* = 30$ and show the actually generated values for $\{w_0, \cdots, w_{K^*}\}$

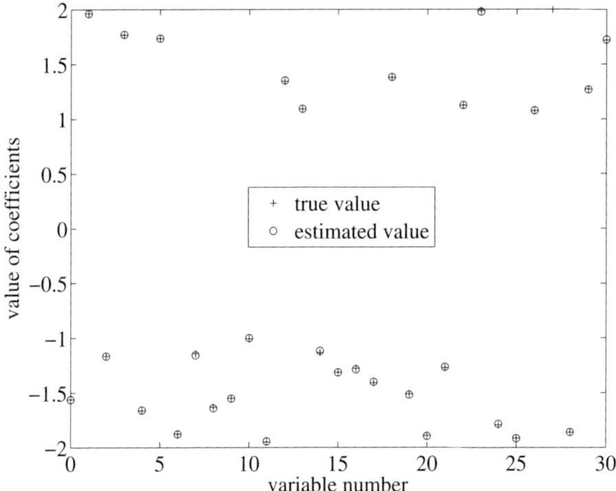

Fig. 1. Evaluation of learning quality

later in Figure 1. In addition to the above-mentioned variables $\{x_1, \cdots, x_{K*}\}$, we introduced lots of irrelevant variables: $M - K^*$ irrelevant ones. Here we set $M = 10,000$. As for each sample, we generated each variable value uniformly at random in the range of $[-1, 1]$. Then we calculated the corresponding value of y by following the above linear function, and also added Gaussian noise with a mean of 0 and a standard deviation of 0.1. We prepared $1,000$ samples for our problem ($N = 1,000$).

In our experiments, after setting that $w_0 = \sum_{n=1}^{N} y_n/N$, we applied our method form $k = 1$ to $k = 100$. Figure 1 compares the parameter values obtained by our method with the actually generated ones. Here note that at the first 30 iterations, our proposed method completely restored all of the relevant variables. Moreover, this figure shows that each difference between the generated and obtained parameter values are quite small. This experimental result indicates that our approach is vital.

Figure 2 shows the learning efficiency of our method with respect to k, in comparison to a conventional calculation method that requires computational complexity with $\max(O(MNk), O(Mk^3))$. This experimental result indicates that our method substantially improved the search efficiency compared with the conventional calculation method. Especially, we can see that the efficiency of the conventional method becomes much worse from the point around $k^2 \approx N$, i.e., $k \approx 30$. This suggests the validity of our arguments about computational complexity described in Section 2.2.

3.2   Real Data

Next, we applied our method to a real dataset in health-checkup. Recently, many Japanese people is becoming the life-style related diseases such as diabetes, lipid abnormality and hypertension, probably caused by life style changes

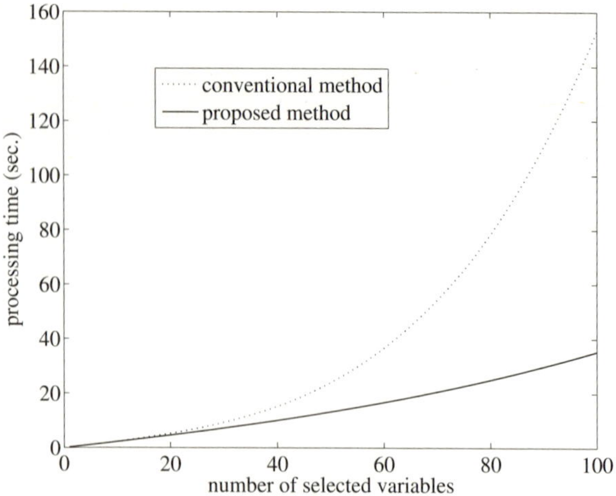

Fig. 2. Evaluation of learning efficiency

from Japanese traditional diet to westernized one. Particularly, it is known that the mortality and incidence of cardiovascular disease are elevated pronouncedly when such symptoms occurred simultaneously with obesity. Thus, predicting and preventing such diseases are important.

Recently, one circulating protein in plasma called adiponectin, is focused as a molecule for predicting and preventing the risk of such diseases. Adiponectin, an adipose tissue-secreted cytokine, is known to enhance insulin sensitivity by enhancing glucose incorporation in skeletal muscle and adipocyte, and by activating β-oxidation of lipid in liver [5,14]. It has been already demonstrated that higher adiponectin concentrations in plasma in both healthy and type 2 diabetic subjects in Western countries as well as in Asian countries including Japan are negatively associated with parameters indicating hyperglycemia, lipid abnormality and hypertension such as fasting glucose, triacylglycerol, and total/LDL cholesterol concentrations and systolic/diastolic blood pressure. Also, it is reported that hemoglobin concentrations are negatively associated with adiponectin concentrations. Additionally, adiponectin has higher association with concentrations of HDL-cholesterol, which has positive effects for improving lipid abnormality [22,6,12,3,19,9,13]. Therefore, it is believed that adiponectin in plasma is important for preventing diabetes and cardiovascular disease [24,26,15,4]. However, adiponectin concentrations are speculatively affected by many factors related to life style such as diet and exercise, as well as clinical parameters in health checkup.

In our experiments, as the top three variables highly associated with adiponectin concentrations in plasma, our method extracted the list, $R_3 =$ (hemoglobin concentrations, tomato intake, body mass index(BMI)), from a large number of variables in the real dataset. Previous many studies have already shown that addiponectin concentrations are strongly and negatively associated with each of hemoglobin concentrations and BMI. Our current data are consistent with the data and supports strongly that BMI is one of the factors for predicting the plasma adiponectin concentrations. Surprisingly, we found strong association between adiponectin concentrations and tomato intake. The studies examining the relation between adiponectin concentrations and dietary factors are few, although a few study reported that adiponectin concentrations are strongly associated with intake of fiber in Western countries [16,17]. In Japanese population, one study reported that healthy men consuming deep-yellow vegetables every day tended to have higher concentrations of adiponectin in plasma [25]. Our experimental results strongly support their result because tomato is also categorized as a deep-yellow vegetable.

4   Related Work and Discussion

As described earlier, there exist a large amount of work focusing on variable selection [7,11], and one primitive task is to select relevant variables by assuming the linear least-square model provided with supervised information. For instance, the Gram-Schmidt orthogonalization procedure permits the performance of forward variable selection by adding at each step the variable that most decreases the mean-squared error [18,23]. However, in the case of some generalized linear model typically formalized as logistic regression [21] or ordinal regression [2], it is not guaranteed that this procedure minimizes the objective function by adding the variable. In contrast, our method adds at each step the variable that most decreases the objective function in terms of second-order optimality.

The second-order optimal criterion introduced in the variable selection can be considered as an extension of a measure used in the Hessian-based weight pruning for neural networks [1], such as the optimal brain damage (OBD) [10] and the optimal brain surgeon (OBS) [8]. The OBS computes the full Hessian matrix, while the OBD makes an assumption that the matrix is diagonal. These methods prune a network one by one by finding a weight which minimizes the increase in the objective function. They show that even small weights may have a substantial effect on the objective function. However, they didn't suggest any variable selection as proposed in the present paper.

Although we have been encouraged by our results to date, there remain several issues we must solve before our method can become a useful tool for incremental variable selection. Clearly, we need to evaluate our method by applying a wider variety of generalized linear models. We also want to extend our method to be applicable to a wider variety of problems. Some problems may require neural networks of complex structures. Although we believe that the method is potentially applicable to complex ones, this claim must be confirmed by our further experiments.

5   Conclusion

In this paper, we addressed the problem of improving search efficiency of incremental variable selection. As one application, we focused on generalized linear models that are linear with respect to their parameters, but their objective functions are not restricted to a standard sum of squared error. For this problem, we presented a method for incrementally selecting a set of relevant variables together with a newly proposing criterion based on second-order optimality. In our experiments using a synthetic dataset and a real dataset in health checkup, we obtained promising initial results. In future, we plan to evaluate the proposed method by performing further experiments.

Acknowledgment

This work was partly supported by Special Grant from University of Shizuoka.

References

1. Bishop, C.M.: Neural networks for pattern recognition. Clarendon Press (1995)
2. Burges, C., Shaked, T., Renshaw, E., Lazier, A., Deeds, M., Hamilton, N., Hullender, G.: Learning to rank using gradient descent. In: Proc. Int. Conf. on Machine Learning, pp. 89–96 (2005)
3. Chandran, M., Phillips, S.A., Ciaraldi, T., Henry, R.R.: Adiponectin: more than just another fat cell hormone? Diabetes Care 26, 2442–2450 (2003)
4. Duncan, B.B., Schmidt, M.I., Pankow, J.S., Bang, H., Couper, D., Ballantyne, C.M., Hoogeveen, R.C., Heiss, G.: Adiponectin and the development of type 2 diabetes: the atherosclerosis risk in communities study. Diabetes 53, 2473–2478 (2004)
5. Englund Ogge, L., Brohall, G., Behre, C.J., Schmidt, C., Fagerberg, B.: Alcohol consumption in relation to metabolic regulation, inflammation, and adiponectin in 64-year-old Caucasian women: a population-based study with a focus on impaired glucose regulation. Diabetes Care 29, 908–913 (2006)
6. Gottsater, A., Szelag, B., Kangro, M., Wroblewski, M., Sundkvist, G.: Plasma adiponectin and serum advanced glycated end-products increase and plasma lipid concentrations decrease with increasing duration of type 2 diabetes. Eur. J. Endocrinol. 151, 361–366 (2004)
7. Guyon, I., Elisseeff, A.: An introduction to variable and feature selection. Journal of Machine Learning Research 3, 1157–1182 (2003)
8. Hassibi, B., Stork, D.G., Wolf, G.: Optimal brain surgeon and general network pruning. In: Proc. IEEE Int. Conf. on Neural Networks, pp. 293–299 (1992)
9. Jaleel, F., Jaleel, A., Aftab, J., Rahman, M.A.: Relationship between adiponectin, glycemic control and blood lipids in diabetic type 2 postmenopausal women with and without complication of ischemic heart disease. Clin. Chim. Acta. 370, 76–81 (2006)
10. LeCun, Y., Denker, J.S., Solla, S.A.: Optimal brain damage. Advances in Neural Information Processing Systems 2, 598–605 (1990)
11. Liu, H., Motoda, H.: Computational methods of feature selection. Chapman & Hall/CRC, Boca Raton (2007)
12. Mannucci, E., Ognibene, A., Cremasco, F., Dicembrini, I., Bardini, G., Brogi, M., Terreni, A., Caldini, A., Messeri, G., Rotella, C.M.: Plasma adiponectin and hyperglycaemia in diabetic patients. Clin. Chem. Lab. Med. 41, 1131–1135 (2003)

13. Matsubara, M., Namioka, K., Katayose, S.: Relationships between plasma adiponectin and blood cells, hepatopancreatic enzymes in women. Thromb Haemost, 360–366 (2004)
14. Nishida, M., Moriyama, T., Sugita, Y., Yamauchi-Takihara, K.: Abdominal obesity exhibits distinct effect on inflammatory and anti-inflammatory proteins in apparently healthy Japanese men. Cardiovasc Diabetol. 6, 27 (2007)
15. Pischon, T., Girman, C.J., Rifai, N., Hotamisligil, G.S., Rimm, E.B.: Association between dietary factors and plasma adiponectin concentrations in men. Am. J. Clin. Nutr. 81, 780–786 (2005)
16. Qi, L., Meigs, J.B., Liu, S., Manson, J.E., Mantzoros, C., Hu, F.B.: Dietary fibers and glycemic load, obesity, and plasma adiponectin levels in women with type 2 diabetes. Diabetes Care 29, 1501–1505 (2006)
17. Qi, L., Rimm, E.B., Liu, S., Rifai, N., Hu, F.B.: Dietary glycemic index, glycemic load, cereal fiber, and plasma adiponectin concentration in diabetic men. Diabetes Care 28, 1022–1028 (2005)
18. Rivals, I., Personnaz, L.: MLPs (Mono-Layer Polynomials and Multi-Layer Perceptrons) for Nonlinear Modeling. Journal of Machine Learning Research 3, 1383–1398 (2003)
19. Sakuta, H., Suzuki, T., Yasuda, H., Ito, T.: Adiponectin levels and cardiovascular risk factors in Japanese men with type 2 diabetes. Endocr. J. 52, 241–244 (2005)
20. Scharf, L.L.: Statistical signal processing. Addison-Wesley, Reading (1991)
21. Seber, G.A.F., Wild, C.J.: Nonlinear Regression. John Wiley & Sons, Chichester (1989)
22. Shetty, G.K., Economides, P.A., Horton, E.S., Mantzoros, C.S., Veves, A.: Circulating adiponectin and resistin levels in relation to metabolic factors, inflammatory markers, and vascular reactivity in diabetic patients and subjects at risk for diabetes. Diabetes Care 27, 2450–2457 (2004)
23. Stoppiglia, H., Dreyfus, G., Dubois, R., Oussar, Y.: Ranking a random feature for variable and feature selection. Journal of Machine Learning Research 3, 1399–1414 (2003)
24. Takefuji, S., Yatsuya, H., Tamakoshi, K., Otsuka, R., Wada, K., Matsushita, K., Sugiura, K., Hotta, Y., Mitsuhashi, H., et al.: Smoking status and adiponectin in healthy Japanese men and women. Prev Med. (2007)
25. Tsukinoki, R., Morimoto, K., Nakayama, K.: Association between lifestyle factors and plasma adiponectin levels in Japanese men. Lipids Health Dis. 4, 27 (2005)
26. Vozarova, B., Weyer, C., Hanson, K., Tataranni, P.A., Bogardus, C., Pratley, R.E.: Circulating interleukin-6 in relation to adiposity, insulin action, and insulin secretion. Obes. Res. 9, 414–417 (2001)

A Biphase-Bayesian-Based Method of Emotion Detection from Talking Voice

Jangsik Cho, Shohei Kato, and Hidenori Itoh

Dept. of Computer Science and Engineering, Graduate School of Engineering,
Nagoya Institute of Technology,
Gokiso-cho Showa-ku Nagoya 466-8555 Japan
{cho,shohey,itoh}@juno.ics.nitech.ac.jp

Abstract. This paper propose a Bayesian-based method of emotion detection from talking voice. Development of a entertainment robot and joyful communication between human and robot have given us the motivation for a computational method for robot to detect its dialogist's emotion from his talking voice. The method is based on the Bayesian networks which represent the dependence and its strength between dialogist's utterance and his emotion, by using a Bayesian modeling for prosodic feature quantities extracted from emotionally expressive voice data. In this paper, we propose a biphase inference method using the Bayesian networks. This inference method has two steps: to reduce the choice of emotion at the first step and to infer a certain emotion reliably from little choice at the second step. The paper also reports some empirical reasoning performance of this method.

1 Introduction

For several past decades, robot technology has been done with distinct success in industrial and manufacturing. Nowadays, robotics research has been shifting from industrial to domestic application, and several domestic robots, aimed at communicating expressively with human, have been developed (e.g., [11,5,6,7]). To live with people, robots need to understand people's instruction through communication with them. Communication, even if it is between robots and human, should involve not only conveying messages or instructions but also psychological interaction, such as comprehending mutual sentiments, sympathizing with the other person, and enjoying conversation itself. To communicate this way, a robot requires several mechanisms that enable the robot to recognize human emotions, to have emotions, and to express emotions. In this research, as a first step of the target, we focus on recognition of dialogist's emotion.

On the other hand, Bayesian networks, one of the eminently practical probabilistic reasoning techniques for reasoning under uncertainty, are becoming an increasingly important area for research and application in the entire field of Artificial Intelligence (e.g., [16,13,2,1]). In this paper, we propose a learning approach to Bayesian network modeling of the knowledge to detect emotional contents from a talking voice, and a biphase inference method using the Bayesian networks.

I. Lovrek, R.J. Howlett, and L.C. Jain (Eds.): KES 2008, Part III, LNAI 5179, pp. 50–57, 2008.
© Springer-Verlag Berlin Heidelberg 2008

2   Bayesian Networks

A Bayesian network (BN) is a graphical structure that allows us to represent and reason about an uncertain domain [13]. The graph structure is constrained to be a directed acyclic graph (or simply dag). The node in a Bayesian network represent a set of random variables from the domain. A set of directed arcs (or links) connects pairs of nodes, representing the direct dependencies between variables. Assuming discrete variables, the strength of the relationship between variables is quantified by conditional probability distributions associated with each node.

Most commonly, BNs are considered to be representations of joint probability distributions. Consider a BN containing the n nodes, X_1 to X_n, taken in that order. A particular value in the joint distribution $P(X_1 = x_1, \cdots, X_n = x_n)$ is calculated as follows:

$$P(X_1 = x_1, ..., X_n = x_n) = \prod_{i=n}^{n} P(x_i | Parents(X_i)), \tag{1}$$

where $Parents(X_i) \subseteq \{x_1, \cdots, x_{i-1}\}$ is a set of parent nodes of X_i. This equation means that node X_i is dependent on only $Parents(X_i)$ and is conditionally independent of nodes except all nodes preceding X_i.

Once the topology of the BN is specified, the next step is to quantify the relationships between connected nodes. Assuming discrete variables, this is done by specifying a conditional probability table (CPT). Consider that node X_i has n possible values $y_1, \cdots, y_m$ and its parent nodes $Parents(X_i)$ have m possible combinations of values $\boldsymbol{x}_1, \cdots, \boldsymbol{x}_m$.

Once the topology of the BN and the CPT are given, we can do the probabilistic inference in the BN by computing the posterior probability for a set of query nodes, given values for some evidence nodes. Belief propagation (BP) proposed in [16] is a well-known inference algorithm for singly-connected BNs, which has simple network structure called a polytree. Assume X is a query node, and there is some set of evidence nodes $\boldsymbol{E}$ (not including X). The task of BP is to update the posterior probability of X by computing $P(X|\boldsymbol{E})$.

In the most general case, the BN structure is a multiply-connected network, where at least two nodes are connected by more than one path in the underlying undirected graph, rather than simply a tree. In such networks, BP algorithm does not work, and then, several enhanced algorithms, such as junction tree [9], logic sampling [8] and loopy BP [15], are proposed as an exact or approximate inference method in multiply-connected networks.

In this research, we intend to reduce the scale of knowledge-base and computational cost drastically, by utilizing Bayesian networks for knowledge representation for emotion detection. Reasoning from partial evidence is forte of Bayesian networks. This is a great advantage to a reasoning mechanism for human-robot interaction.

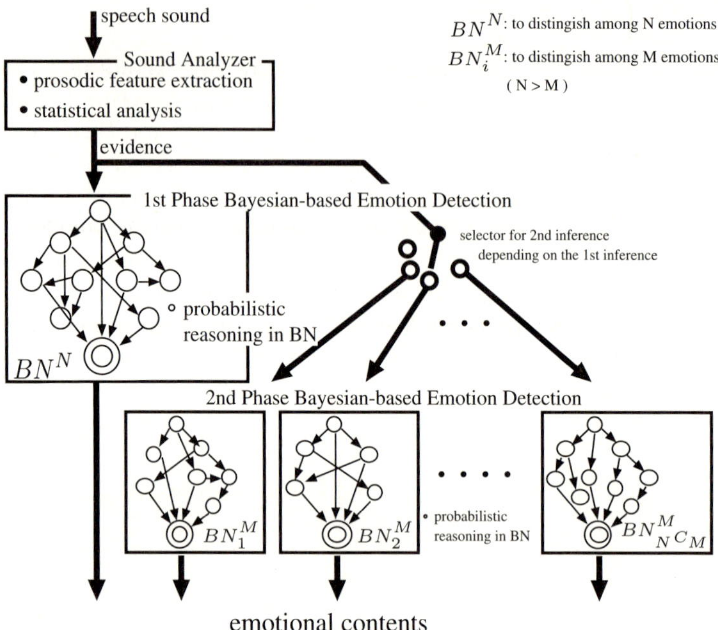

Fig. 1. Biphase Bayesian-based emotion detection from talking voice

3   Biphase Bayesian Approach

At the beginning of this research, we have developed a Bayesian approach to emotion detection from talking voice [12]. The paper provided a Bayesian modeling of prosodic features of dialogist's voice for emotion detection by building Bayesian network from numbers of segments of emotive, expressive voice samples. And the practical usefulness of probabilistic reasoning in the BN was reported. However, the reasoning method was quite simple: to output a certain emotion that has the highest probability in the BN. There is no consideration of the probability distribution among emotions. In this paper, we propose a biphase inference method using the Bayesian networks.

Considering practical application for human-robot communication, a sophisticated inference mechanism using BNs should be invent for emotion detection in the case of subtle difference in probability values. BN can reduce the choice, even though it does not infer a certain emotion with height probability. In this case, our method can re-infer using a BN modelling with the reduced emotions at the second phase. The inference method is the following procedure.

Preparation. In advance of emotion detection inference, one BN and a set of BNs are built up from training voice samples for the first and the second phase, respectively. The BN at the first phase distinguishes N kinds of emotions, and a BN at the second phase distinguishes $M(< N)$ kinds of emotions. The set of BNs for the second phase is a M-combination from a set with N emotions. The BN at the first phase is denoted by BN^N and a BN at the second phase is denoted by BN_i^M ($i = 1, \cdots, _N C_M$).

First phase. Probabilistic reasoning in BN^N with some evidences of prosodic features of speech sound from sound analyzer makes conditional probability of emotion, given the evidences. The method outputs a certain emotional label and terminates if the probability satisfies the following condition:

$$\max_{e \in Emotions} (P(e)) > \alpha \quad \text{and} \quad \underset{e \in Emotions}{\operatorname{argmax}} (P(e)) \text{ is unique,}$$

where α means the threshold of emotion determination at the first phase BN^N; otherwise the method selects BN_i^M according to the emotional labels within the M-th highest probability and then shifts to the second phase.

Second phase. Probabilistic reasoning in BN_i^M makes conditional probability of emotion, given the evidences. The method determines an emotional label with the highest probability.

The system flow of the above mentioned method of biphase Bayesian-based emotion detection is illustrated in Figure 1.

4   Learning of Emotion Detection Engine

In this paper, we focus on the prosodic features of dialogist's voice as a clue to what emotion the dialogist expresses. The section describes a BN modeling for this problem.

4.1   Training Data

Speech data for learning should be expressive of emotions. In this research, we used numbers of segments of voice samples, which are spoken emotionally by actors and actresses, from films, TV dramas, and so on. In segmentation, transcriber selects six emotional labels (anger, disgust, sadness, fear, surprise or happiness). This label is a goal attribute for emotion detection. It should be noticed that all voice samples in the training data are labeled with any of six emotions, that is, the training data has no voice with neutral emotion. We think that a BN learned from this training data can detect neutral emotion by giving nearly flat probabilities to all emotions.

4.2   Feature Extraction for BN Modeling

Voice has three components (prosody, tone and phoneme). It becomes obvious from past several researches that prosodic component is the most relative to emotional voice expressions (e.g., [17,3]). In this research, as attributes of training data, we adopt three prosodic events: energy, fundamental frequency and duration as acoustic parameters for BN modeling. Prosodic analysis is done for 11 ms frames passed through Hamming extracted from voice waveforms sampled at 22.05 kHz.

The attributes concerning energy, maximum energy (PW_{MAX}), minimum energy (PW_{MIN}), mean energy (PW_{MEAN}) and its standard deviation (PW_S) are determined by the energy contours for the frames in a voice waveform. The attributes concerning fundamental frequency, maximum pitch ($F0_{MAX}$), minimum pitch ($F0_{MIN}$), mean pitch ($F0_{MEAN}$) and its standard deviation ($F0_S$) are determined by short time Fourier transforms for the frames in a voice waveform. As an attribute concerning duration, we measure the duration per a single mora (Tm). Then we added an attribute of dialogist's sexuality (SE).

In this paper, the goal attribute ($EMOT$) and the above nine prosodic feature values and dialogist's sexuality (total eleven attributes) are assigned to the nodes of a BN model. For learning the discrete causal structure of a BN model, all prosodic features are converted to discrete values.

4.3   Learning Causal Structure of the BN Model

The section describes how to specify the topology of the BN model for emotion detection and to parametrize CPT for connected nodes. The emotion detection BN modeling is to determine the qualitative and quantitative relationships between the output node containing goal attribute (emotions) and nodes containing prosodic features. In this paper, we adopt a model selection method based on the Bayesian information criterion (BIC), which has information theoretical validity and is able to learn a high prediction accuracy model by avoidance of over-fitting to training data. Let $\mathcal{M}$ be a BN model, $\hat{\theta}_{\mathcal{M}}$ be the maximum likelihood (ML) estimate of the parameter representing $\mathcal{M}$, and d be the number of parameters. A BN model $\mathcal{M}$ is evaluated by BIC of $\mathcal{M}$ defined as

$$BIC(\hat{\theta}_{\mathcal{M}}, d) = \log_{\hat{\theta}_{\mathcal{M}}}^{\mathcal{N}} P(D) - \frac{d \log \mathcal{N}}{2}, \tag{2}$$

where D is training samples and $\mathcal{N}$ is the number of the samples. In case that D is partially observed, expectation maximization (EM) algorithm [4] is utilized for estimating θ asymptotically with incomplete data in the training samples.

We adopt K2 [2] as the search algorithm for a BN model which maximizes BIC. K2 needs the pre-provided variable order. We, thus, consider every possible permutation of three node groups: PW, $F0$ and Tm.

5   Empirical Results

The section describes an Experimentation of emotion inference of our biphase-based Bayesian approach. In advance of emotion detection inference, firstly, we collected 1600 segments of voice waveform and labeled any of six emotional contents, anger, disgust, sadness, fear, surprise or happiness, as mentioned in Section 4.1, and then extracted nine prosodic features and one dialogist's sexuality from each of the segments and assigned them to the attributes as mentioned in Section 4.2. In the prosodic analysis, we used the Snack sound toolkit [10]. And then we randomly selected 1400 samples as training data and discretized

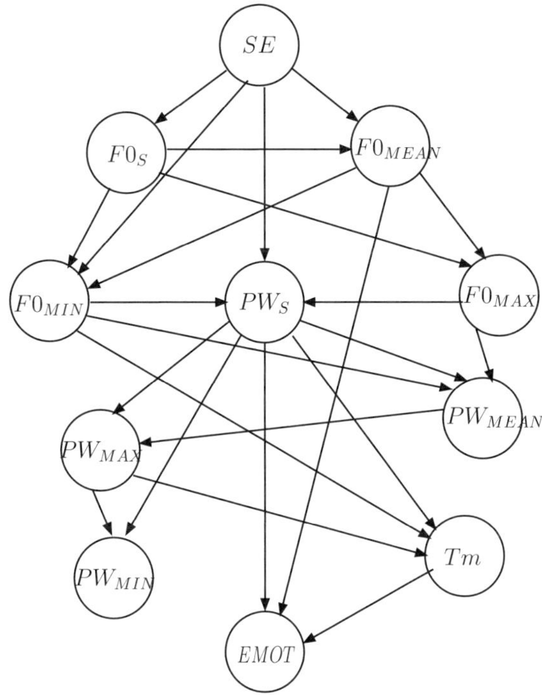

Fig. 2. A Bayesian network for emotion detection (at the first phase BN^6)

their attribute values with five values. In this experiment, we determined the threshold for the discretization on the basis of the idea of even-sized chunk, that is, each of the discrete values covers 20% of the training data. Then we modeled one Bayesian network BN^6 which distinguishes six kinds of emotions and fifteen Bayesian networks BN_i^2 $(i = 1, \cdots, 15)$ which distinguish two kinds of emotions for the first and the second phase, respectively (i.e., $N = 6$ and $M = 2$ for the biphase-based method in Section 3). BNs were built from the above mentioned training voice samples with changing six variable orders by Bayes Net Toolbox [14]. Fig. 2 shows one of the results with the variable order $SE \prec F0 \prec PW \prec Tm \prec EMOT$ for the first phase BN^6.

5.1   Inference Performance of Biphase-Bayesian-Based Method

We, then, converted prosodic features of the rest 200 samples into discrete value by the same thresholds for the training data, and then we examined the inference performance of the biphase-based BN model by the emotion detection test for the rest 200 samples. In reasoning with the BN, we used junction tree [9] as the inference algorithm with BNs, which is well-known as an exact inference algorithm in multiply connected BNs. The examination was done by changing the threshold α.

Table 1. The Accuracy Rates

Emotion	Singleton BN	Accuracy Rates (%)					
		Biphase BNs					
		$\alpha \leq 0.2$	$\alpha = 0.3$	$\boldsymbol{\alpha = 0.4}$	$\alpha = 0.5$	$\alpha = 0.6$	$\alpha \geq 0.7$
Anger	55.8	65.1	65.1	67.4	**69.8**	**69.8**	65.1
Sadness	32.4	56.8	59.5	**62.2**	56.8	56.8	56.8
Disgust	28.2	**53.8**	**53.8**	51.3	43.6	43.6	43.6
Fear	22.2	48.1	48.1	**63.0**	59.3	59.3	59.3
Surprise	33.3	**48.1**	**48.1**	44.4	40.7	40.7	40.7
Happiness	55.6	**59.3**	**59.3**	51.9	33.3	33.3	33.3

Table 1 shows the results. We examined the inference performance using singleton BN for comparison. Reasoner using singleton BN cannot determine a certain emotional label if there is more than one emotional label with the highest probability. This case is a frequent occurrence, and it makes a lower accuracy rate. On the other hand, the results indicate that biphase-based method has largely acceptable accuracy rates for all emotions in comparison with singleton BN. In this particular examples, the accuracy rates were the best average using biphase-based method with $\alpha = 0.4$; In the method with $\alpha \geq 0.7$ BN^6, BN at the first phase, could not determinate and any of BN_i^2s at the second phase was used for all of 200 samples.

6   Conclusion

This paper proposed a biphase-based Bayesian method of detecting emotion from talking voice. The method provided a Bayesian modeling of prosodic features of talking voice for emotion detection, and a probabilistic reasoning technique using double-layered BNs. Considering practical human robot communication, unfortunately, accurate realtime speech recognition and voice analysis is not fully guaranteed. In such case, Bayesian approach proposed in this paper gives much benefit to emotion detection by probabilistic inference from partial evidence and reasoning under uncertainty.

We consider that emotional contents are conveyed not only by voice but also by verbal, facial expression, gesture and so on. In future works, we will dedicate to a mixture of BN modeling of these various information for emotion detection during conversation, and describe a multi-phase Bayesian approach in the forthcoming papers.

Acknowledgment

Ifbot was developed as part of an industry-university joint research project among the Business Design Laboratory Co,. Ltd., Brother Industries, Ltd., A.G.I. Inc, ROBOS Co., and the Nagoya Institute of Technology. We are grateful to all of them for their input. This work was supported in part by the Tateishi Science and Technology Foundation, and by the Hori Information Science Promotion Foundation, and by a Ministry of Education, Science, Sports and Culture, Grant–in–Aid for Scientific Research under grant #17500143 and #20700199.

References

1. Akiba, T., Tanaka, H.: A Bayesian approach for user modelling in dialog systems. In: 15th International Conference of Computational Linguistics, pp. 1212–1218 (1994)
2. Cooper, G.F., Herskovits, E.: A Bayesian method for the induction of probabilistic networks from data. Machine Learning 9, 309–347 (1992)
3. Cowie, R., Douglas-Cowie, E., Tsapatsoulis, N., Votsis, G., Kollias, S., Fellenz, W., Taylor, J.G.: Emotion recognition in human-computer interaction. IEEE Signal Processing Magazine 18(1), 32–80 (2001)
4. Dempster, A., Laird, N., Rubin, D.: Maximum likelihood from incomplete data via the EM algorithm. Journal of the Royals Statistical Society B 39, 1–38 (1977)
5. Endo, G., Nakanishi, J., Morimoto, J., Cheng, G.: Experimental studies of a neural oscillator for biped locomotion with QRIO. In: IEEE International Conference on Robotics and Automation (ICRA 2005), pp. 598–604 (2005)
6. Fujita, M.: Development of an Autonomous Quadruped Robot for Robot Entertainment. Autonomous Robots 5, 7–18 (1998)
7. Fujita, M., Kitano, H., Doi, T.: Robot Entertainment. In: Druin, A., Hendler, J. (eds.) Robots for kids: exploring new technologies for learning, ch. 2, pp. 37–70. Morgan Kaufmann, San Francisco (2000)
8. Henrion, M.: Propagating uncertainty in Bayesian networks by logic sampling. Uncertainty in Artificial Intelligence 2, 149–163 (1988)
9. Jensen, F.V.: Bayesian Networks and Decision Graphs. Springer, Heidelberg (2001)
10. Sjölander, K.: The Snack Sound Toolkit, `http://www.speech.kth.se/snack/`
11. Kato, S., Ohshiro, S., Itoh, H., Kimura, K.: Development of a communication robot ifbot. In: IEEE International Conference on Robotics and Automation (ICRA 2004), pp. 697–702 (2004)
12. Kato, S., Sugino, Y., Itoh, H.: A Bayesian Approach to Emotion Detection in Dialogist's Voice for Human Robot Interaction. In: Gabrys, B., Howlett, R.J., Jain, L.C. (eds.) KES 2006. LNCS (LNAI), vol. 4252, pp. 961–968. Springer, Heidelberg (2006)
13. Korb, K.B., Nicholson, A.E.: Bayesian Artificial Intelligence. Chapman & Hall/CRC, Boca Raton (2003)
14. Murphy, K.P.: Bayes Net Toolbox, `http://www.cs.ubc.ca/~murphyk/Software/BNT/bnt.html`
15. Murphy, K.P., Weiss, Y., Jordan, M.I.: Loopy belief propagation for approximate inference: an empirical study, pp. 467–475 (1999)
16. Pearl, J.: Probabilistic Reasoning in Intelligent Systems. Morgan Kaufmann, San Francisco (1988)
17. Scherer, K.R., Johnstone, T., Klasmeyer, G.: Vocal expression of emotion. In: Davidson, R.J., Goldsmith, H., Scherer, K.R. (eds.) Handbook of the Affective Sciences, pp. 433–456. Oxford University Press, Oxford (2003)

EM Algorithm with PIP Initialization and Temperature-Based Selection

Yuta Ishikawa and Ryohei Nakano

Nagoya Institute of Technology
Gokiso-cho, Showa-ku, Nagoya 466-8555 Japan
{ishikawa,nakano}@ics.nitech.ac.jp

Abstract. The EM algorithm is an efficient algorithm to obtain the ML estimate for incomplete data, but has the local optimality problem. The deterministic annealing EM (DAEM) algorithm was once proposed to solve this problem, which begins a search from the *primitive initial point (PIP)*. Then, proposed was the mes-EM algorithm, which runs the EM repeatedly in many various directions from the PIP, and achieves good solution quality with high computing cost. This paper proposes a variant of the mes-EM, called mes-EM(β), which uses the temperature to select a small promising portion of the mes-EM runs. Our experiments for the Gaussian mixture estimation showed the proposed algorithm was much faster than the original mes-EM without degrading its solution quality.

Keywords: EM Algorithm, Gaussian mixture model, Primitive initial point.

1   Introduction

The Expectation-Maximization (EM) algorithm [1] is a general-purpose iterative algorithm for maximum likelihood estimation for incomplete data. The EM and its variants [2] have been applied in a large number of applications. The EM is not without its limitations; one of the most serious may be the local optimality problem. The problem makes the EM performance dependent on its starting points. Thus, obtaining good starting points has been long addressed [3].

To solve this problem fundamentally, the deterministic annealing EM (DAEM) algorithm [4] was once proposed by introducing an inverse temperature β. At the lowest $\beta = 1$, the DAEM is reduced to the EM, and at the highest $\beta = 0$ the DAEM converges to a singular point, called the *primitive initial point (PIP)*, where the posterior becomes uniform and any component has the same parameter values.

However, the DAEM will not always achieve the global optimality mainly because it employs a single token search. Then, the mes-EM algorithm [5] was proposed, incorporating a multiple token search into the EM, employing *the PIP* as its initial point. The mes-EM finds good solutions, however its computing cost is high because the algorithm generates many search tokens near *the PIP* to cover a high-dimensional search space.

I. Lovrek, R.J. Howlett, and L.C. Jain (Eds.): KES 2008, Part III, LNAI 5179, pp. 58–66, 2008.

This paper proposes a variant of the mes-EM, called the mes-EM(β), which uses the temperature to select a small promising portion of the mes-EM runs. First, we consider predicting the final solution quality from halfway quality, and investigate correlations between halfway and the final qualities. We could not find any correlation in a likelihood space, but found meaningful one in a high temperature space. Then, based on the results, we propose the mes-EM(β) and evaluate its performance for the Gaussian mixture estimation.

2   Background

2.1   EM Algorithm

Suppose that a sample $(\boldsymbol{x}, \boldsymbol{z})$ is generated with the density $p(\boldsymbol{x}, \boldsymbol{z}|\boldsymbol{\theta})$, where only $\boldsymbol{x}$ is observable and $\boldsymbol{z}$ is hidden or missing. Here $\boldsymbol{\theta}$ denotes a parameter vector , and let $p(\boldsymbol{x}|\boldsymbol{\theta})$ be the density generating $\boldsymbol{x}$. In the EM context, $\{(\boldsymbol{x}^{\mu}, \boldsymbol{z}^{\mu}), \mu = 1, \ldots, N\}$ is called *complete data*, and $\{\boldsymbol{x}^{\mu}, \mu = 1, \ldots, N\}$ is called *incomplete data*.

The purpose of the ML (maximum likelihood) estimation is to maximize the following log-likelihood from incomplete data.

$$\mathcal{L}(\boldsymbol{\theta}) = \sum_{\mu} \log p(\boldsymbol{x}^{\mu}|\boldsymbol{\theta}). \tag{1}$$

The EM algorithm performs the ML estimation by iteratively maximizing the following *Q-function*, the expectation of $\mathcal{L}_{cmp}(\boldsymbol{\theta})$ over the posterior $P(\boldsymbol{z}^{\mu}|\boldsymbol{x}^{\mu}, \boldsymbol{\theta}^{(t)})$. Here $\boldsymbol{\theta}^{(t)}$ denotes the estimate obtained after the t-th iteration.

$$Q(\boldsymbol{\theta}^{(t)}, \boldsymbol{\theta}) = \sum_{\mu} \sum_{\boldsymbol{z}^{\mu}} P(\boldsymbol{z}^{\mu}|\boldsymbol{x}^{\mu}, \boldsymbol{\theta}^{(t)}) \log p(\boldsymbol{x}^{\mu}, \boldsymbol{z}^{\mu}|\boldsymbol{\theta}). \tag{2}$$

$$P(\boldsymbol{z}^{\mu}|\boldsymbol{x}^{\mu}, \boldsymbol{\theta}^{(t)}) = \frac{p(\boldsymbol{x}^{\mu}, \boldsymbol{z}^{\mu}|\boldsymbol{\theta}^{(t)})}{\sum_{\boldsymbol{z}^{\mu}} p(\boldsymbol{x}^{\mu}, \boldsymbol{z}^{\mu}|\boldsymbol{\theta}^{(t)})}. \tag{3}$$

A general flow of the EM algorithm is described below.

1. Initialize $\boldsymbol{\theta}^{(0)}$ and $t \leftarrow 0$.
2. Iterate the following EM-step until convergence.
       E-step: Compute $Q(\boldsymbol{\theta}^{(t)}, \boldsymbol{\theta})$ by computing the posterior $P(\boldsymbol{z}^{\mu}|\boldsymbol{x}^{\mu}, \boldsymbol{\theta}^{(t)})$.
       M-step: $\boldsymbol{\theta}^{(t+1)} = argmax_{\boldsymbol{\theta}} Q(\boldsymbol{\theta}^{(t)}, \boldsymbol{\theta})$, and $t \leftarrow t + 1$.

2.2   mes-EM Algorithm

The mes-EM algorithm [5] was proposed to improve the solution quality of the m-EM algorithm[6]. The mes-EM makes use of the characteristics of the landscape of a likelihood surface around the primitive initial point (PIP).

Let the Hessian of the target function have R-fold smallest negative eigen values at the PIP. Let $\boldsymbol{U} = \{\boldsymbol{u}_r\}$, $r = 1, \ldots, R$ be the orthonormal set of the corresponding eigen vector. Search tokens are generated along the following

60   Y. Ishikawa and R. Nakano

directions in addition to the orthonormal set U. The number of vectors included in V amounts to 2^R. The rationale for generating V is given in [5].

$$V = \sum_{r=1}^{R}(\pm u_r)$$
$$= \{(u_1 + \cdots + u_R), \ldots, (-u_1 - \cdots - u_R)\} \tag{4}$$

A method of running the EM $2R + 2^R$ times starting from the same initial point PIP with $\pm U \cup V$ as their search directions is called *mes (multi-directional in eigen space)-EM algorithm*. A general flow of the mes-EM is described below.

1. Calculate all eigen values of the Hessian of $F_1(\boldsymbol{\theta})$.
2. Generate search directions $\pm U \cup V$ by using R-fold smallest negative eigen value.
3. For each direction repeat the EM algorithm with the PIP as its starting point.

The following free energy $F_\beta(\boldsymbol{\theta})$ is defined in the DAEM framework [4]:

$$F_\beta(\boldsymbol{\theta}) = -\frac{1}{\beta}\log Z(\boldsymbol{\theta}), \tag{5}$$
$$Z(\boldsymbol{\theta}) = \prod_\mu \sum_{z^\mu} p(x^\mu, z^\mu|\boldsymbol{\theta})^\beta. \tag{6}$$

The free energy $F_1(\boldsymbol{\theta})$ appeared above is given as follows:

$$F_1(\boldsymbol{\theta}) = -\sum_\mu \log p(x^\mu|\boldsymbol{\theta}) = -\mathcal{L}(\boldsymbol{\theta}) \tag{7}$$

3   Proposed Method

3.1   Gaussian Mixture Estimation

Consider a K-dimensional Gaussian mixture model with C components. Here a hidden variable z^μ is the component number $c(= 1, \ldots, C)$ to which a sample x^μ belongs. The density for the c-th component is defined as follows.

$$p(x^\mu, c|\boldsymbol{\theta}_c) = \pi_c g_c(x^\mu|m_c, \Sigma_c), \tag{8}$$
$$g_c(x^\mu|m_c, \Sigma_c) = \frac{1}{(2\pi)^{K/2}|\Sigma_c|^{1/2}} \exp\left(-\frac{1}{2}(x^\mu - m_c)^T \Sigma_c^{-1}(x^\mu - m_c)\right). \tag{9}$$

Here π_c, m_c, and Σ_c are a mixing ratio, a mean vector, and a covariance matrix for the component respectively. A covariance matrix is set to be diagonal. Then, the density of a Gaussian mixture is defined as follows.

$$p(x^\mu|\boldsymbol{\theta}) = \sum_{c=1}^{C} p(x^\mu, c|\boldsymbol{\theta}). \tag{10}$$

The *primitive initial point (PIP)* $\boldsymbol{\theta}_{pip} = \{\boldsymbol{\theta}_c^{pip}\}$, $\boldsymbol{\theta}_c^{pip} = (\pi_c^{pip}, \boldsymbol{m}_c^{pip}, \boldsymbol{\Sigma}_c^{pip})$ for a Gaussian mixture is given below. Note that each component has the same parameters.

$$\pi_c^{pip} = \frac{1}{C} \tag{11}$$

$$\boldsymbol{m}_c^{pip} = \frac{1}{N} \sum_{\mu=1}^{N} \boldsymbol{x}^{\mu} \tag{12}$$

$$\boldsymbol{\Sigma}_c^{pip} = \frac{1}{N} \sum_{\mu=1}^{N} (\boldsymbol{x}^{\mu} - \boldsymbol{m}_c^{pip})(\boldsymbol{x}^{\mu} - \boldsymbol{m}_c^{pip})^T. \tag{13}$$

Here we generate data sets for a Gaussian mixture as follows; i.e., the number of dimensions K is set to be 5, the number of components C is either 9 or 12, and for each C, three different data sets are generated. For each component, we denote 3 data sets by #1, #2, and #3. The number of samples N for each data set is set as dependent on the number of parameters $M = C(2K + 1) - 1$; that is, $N = 20M$. For $C = 9$, we have $M = 98$ and $N = 1,960$, and for $C = 12$, we have $M = 131$ and $N = 2,620$.

3.2   Predicting Final Quality from Halfway Quality

Usually the multiplicity R for a Gaussian mixture is not small, and the number of search tokens for the mes-EM increases exponentially with the order of 2^R. Thus, the mes-EM generates a large number of search tokens by using Eq.(4). To reduce computing cost of the mes-EM, we consider selecting a small promising portion of mes-EM runs.

We know excellent solutions including the global optimum can be reached from the PIP [5], which means there is a path to an excellent solution from the PIP. Moreover, we know solutions are located around a constant distance from the PIP [5].

On the other hand, we know the EM is very slow at the final stage of an EM path because the algorithm converges linearly. In fact, in our experiments of EM runs starting from the PIP, the CPU time needed to reach a 90 % point is on average around only 10 % of the total CPU time needed until convergence. Here a 90 % point $\boldsymbol{\theta}_{0.9}$ denotes a point on the EM path, which satisfies $\|\boldsymbol{\theta}_{0.9} - \boldsymbol{\theta}_{pip}\| \approx 0.9 \times \|\boldsymbol{\theta}_{final} - \boldsymbol{\theta}_{pip}\|$, where $\boldsymbol{\theta}_{final}$ is the final point of the EM path.

Based on the above facts, we consider predicting the final solution quality from the quality at a 90 % point. If it works, all we have to do is just to select good EM runs and continue such EM runs until convergence.

With these in mind, we investigate correlation between the quality at a 90% point and the final quality. Three examples of the results obtained using artificial data sets and real data sets are shown in Figs. 1, 2, and 3.

Each value is measured by negative log-likelihood $-\mathcal{L}(\boldsymbol{\theta})(= F_1(\boldsymbol{\theta}))$. In each figure, a left-hand graph shows the correlation in a likelihood space ($\beta = 1$), while a right-hand graph shows the correlation in a high temperature space ($\beta = 0.3$).

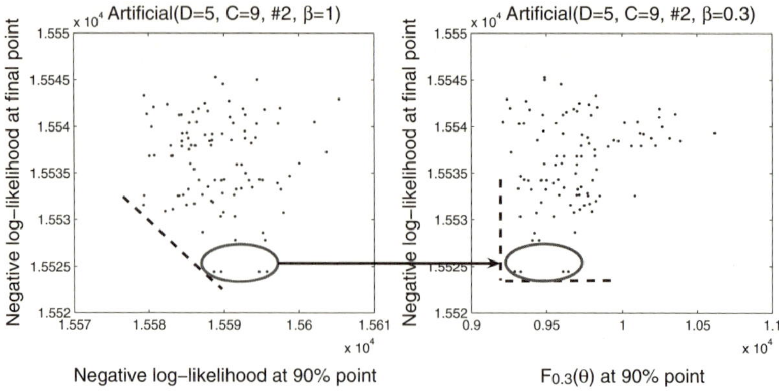

Fig. 1. Correlation between 90% point quality and the final quality (Artificial: C=9)

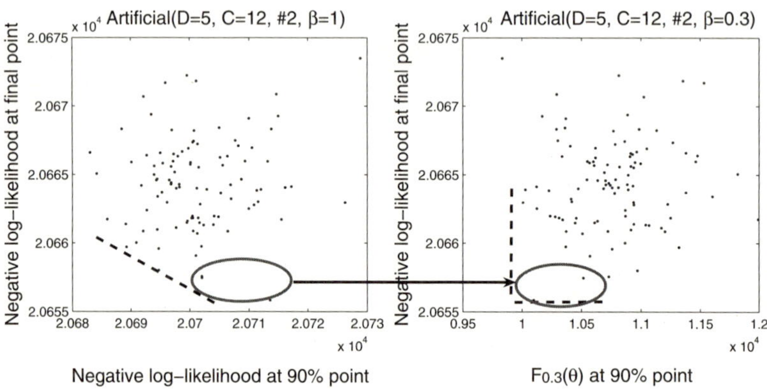

Fig. 2. Correlation between 90% point quality and the final quality (Artificial: C=12)

We could not find any preferable correlation in a likelihood space, which means we cannot predict the final excellence based on 90% qualities in this space. Then, we investigated the correlation under different temperatures ($\beta = 0.1$, 0.3, 0.6), and happened to find meaningful correlation in a high temperature $\beta = 0.3$. The results are shown in right-hand graphs, which show the preferable tendency that excellent final runs were also excellent or good at a 90 % point. Such tendency allows us to preform good prediction.

3.3   mes-EM(β) Algorithm

Making good use of the above results, we propose a lighter variant of the mes-EM, called *the mes-EM(β) algorithm*. To get the average distance between final solutions and the PIP, we run the EM I times and obtain I solutions: $\{\boldsymbol{\theta}_1, \ldots, \boldsymbol{\theta}_I\}$. Then, we calculate the average distance as below:

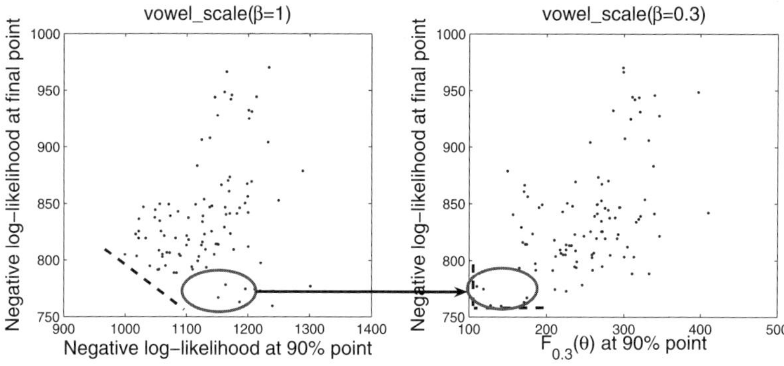

Fig. 3. Correlation between 90% point quality and the final quality (Real data: Vowel_scale)

$$\overline{D} = \frac{1}{I} \sum_{i=1}^{I} \|\boldsymbol{\theta}_i - \boldsymbol{\theta}_{pip}\| \tag{14}$$

where the PIP $\boldsymbol{\theta}_{pip}$ is given by Eqs. (11), (12) and (13). Then we run the mes-EM until each search token reaches a 90% point which is located at a distance $0.9 \times \overline{D}$ from the PIP. When a token reaches the 90 % point, we calculate the free energy $F_\beta(\boldsymbol{\theta})$ with $\beta = 0.3$. Then, we select M tokens which have the lowest M free energies and continue the EM estimation until convergence.

Naturally, by selecting a good small portion of mes-EM runs, the mes-EM(β) is expected to reduce high computing cost of the mes-EM. A general flow of the mes-EM(β) is described below.

Step 1: Run the EM estimation I times starting from the PIP.
Step 2: Calculate the average distance $\overline{D}$ using I solutions obtained at Step 1.
Step 3: Run the mes-EM until each token reaches a 90% point.
Step 4: Compute $F_\beta(\boldsymbol{\theta})$ for each token.
Step 5: Select the best M tokens according to $F_\beta(\boldsymbol{\theta})$ obtained above.
Step 6: Continue the EM until convergence for each token selected at Step 5.

4   Performance Evaluation

4.1   Artificial Data Sets

We evaluated the performance of the mes-EM(β) algorithm by comparing with the EM and the mes-EM using six artificial data sets generated in Section 3.1.

The EM are repeated 100 times, where initial points are generated by using the K-means algorithm. The mes-EM is performed by using Eq.(4). For the mes-EM(β), β is set to be 0.3, while system parameters I and M are set to be 5 and 50 respectively.

The results for artificial data sets are shown in Table 1. In the table, "Best" and "Ave" show the best and the average obtained by each algorithm, while

Table 1. Performance Comparison (Artificial data sets)

Data	Algorithm	Best	Ave	Std	Total Time (sec)
C9_1	EM	15529	15544	5.59	565
	mes-EM	**15524**	15536	**5.02**	758
	mes-EM(β)	**15524**	**15535**	6.09	**434**
C9_2	EM	15681	15692	6.67	514
	mes-EM	**15680**	**15687**	3.43	770
	mes-EM(β)	15682	**15687**	**3.17**	**452**
C9_3	EM	16161	16176	9.62	436
	mes-EM	**16160**	16176	8.05	731
	mes-EM(β)	**16160**	**16173**	**6.80**	**369**
C12_1	EM	21175	21195	7.45	1,329
	mes-EM	**21174**	21190	5.96	12,267
	mes-EM(β)	21176	**21189**	**5.54**	**1,178**
C12_2	EM	20656	20667	5.25	**1,425**
	mes-EM	**20653**	**20664**	3.60	17,071
	mes-EM(β)	20658	**20664**	**3.41**	3,174
C12_3	EM	**20864**	20881	6.93	**1,386**
	mes-EM	20867	20884	5.53	17,685
	mes-EM(β)	20867	**20878**	**3.72**	2,412

"Std" denotes the standard deviation of solutions. These values are measured in negative log-likelihood. Although the bests of the mes-EM(β) are much the same or slightly worse than the mes-EM, the averages or standard deviations of the mes-EM(β) are much the same or better than the mes-EM. Compared with the EM, the mes-EM(β) shows much the same quality in the bests and the averages, but its standard deviations are surely smaller than those of the EM, which means the mes-EM(β) produces excellent solutions more stably than the EM. From these results, the solution quality of the mes-EM(β) was much the same as the mes-EM. As for computing cost, the mes-EM(β) achieves significant reduction compared with the mes-EM. The tendency was remarkable for $C = 12$.

4.2   Real Data Sets

We evaluated the proposed algorithm by using five real data sets: "abalone" (UCI), "mg_scale" (GWF01a), "letter_scale" (Statlog), "vehicle_scale" (Statlog) and "vowel_ scale" (UCI). The set-up of the algorithms was the same as the previous experiments. The number of components C for each data set was set to be 10, the number of dimensions and a data size for each data set are shown in Table 2.

The results for real data sets are shown in Table 3. As for the bests, the three algorithms showed much the same performance. As for the averages and standard deviations, the mes-EM(β) outperformed the other two, which means the mes-EM(β) more stably produced excellent or good solutions than the other two. As for computing cost, the mes-EM(β) was more than three times as fast as the mes-EM. These results showed that the mes-EM(β) worked well for these real data sets.

Table 2. Real Data Sets

Name	abalone	mg_scale	letter_scale	vehicle_scale	vowel_scale
Dimensions (D)	8	6	16	18	10
Components (C)	10	10	10	10	10
Data Size (N)	4,177	1,385	15,000	596	528

Table 3. Performance Comparison (Real data sets)

Data	Algorithm	Best	Ave	Std	Total Time (sec)
abalone	EM	-39592	-38398	1091	1,432
	mes-EM	**-39597**	-39158	327	4,035
	mes-EM(β)	-39595	**-39206**	**90**	**964**
mg_scale	EM	**906**	**1064**	89	157
	mes-EM	911	1083	103	458
	mes-EM(β)	911	1065	**84**	**134**
letter_scale	EM	-30393	-25229	2447	9,072
	mes-EM	**-31015**	-27698	2029	27,317
	mes-EM(β)	**-31015**	**-28170**	**1647**	**7,298**
vehicle_scale	EM	-5175	-4687	285	144
	mes-EM	**-5179**	-4531	312	393
	mes-EM(β)	**-5179**	**-4849**	**225**	**110**
vowel_scale	EM	762	863	59	79
	mes-EM	**760**	839	47	232
	mes-EM(β)	**760**	**814**	**35**	**76**

5 Conclusion

In this paper we proposed a variant of the mes-EM, called the mes-EM(β) algorithm. The proposed algorithm uses the temperature to effectively select a small promising portion of mes-EM runs. The effective selection can be done in a high temperature space ($\beta = 0.3$). Our experiments for the Gaussian mixture estimation using artificial and real data sets showed the mes-EM(β) showed much the same solution quality as the mes-EM, but was more than three times as fast as the mes-EM for real data sets. In the future, we plan to investigate a free energy surface to find the theoretical rationale for the proposed algorithm.

References

1. Dempster, A.P., Laird, N.M., Rubin, D.B.: Maximum-likelihood from incomplete data via the EM algorithm. J. Royal Statist. Soc. Ser. B 39, 1–38 (1977)
2. McLachlan, G.J., Krishnan, T.: The EM Algorithm and Extensions. John Wiley & Sons, Chichester (1997)
3. McLachlan, G.J., Peel, D.: Finite Mixture Models. John Wiley & Sons, Chichester (2000)

4. Ueda, N., Nakano, R.: Deterministic annealing EM algorithm. Neural Networks 11, 271 (1998)
5. Ishikawa, Y., Nakano, R.: Landscape of a Likelihood Surface for a Gaussian Mixture and its Use for the EM Algorithm. In: Proc. Int. Joint Conf. on Neural Networks (IJCNN 2006), pp. 2413–2419 (2006)
6. Takada, M., Nakano, R.: Threshold-based dynamic annealing for multi-thread DAEM and its extreme. In: Proc. Int. Joint Conf. on Neural Networks (IJCNN 2003), pp. 501–506 (2003)

Prediction of Information Diffusion Probabilities for Independent Cascade Model

Kazumi Saito[1], Ryohei Nakano[2], and Masahiro Kimura[3]

[1] School of Administration and Informatics, University of Shizuoka
52-1 Yada, Suruga-ku, Shizuoka 422-8526, Japan
k-saito@u-shizuoka-ken.ac.jp
[2] Department of Computer Science and Engineering, Nagoya Institute of Technology
Gokiso-cho, Showa-ku, Nagoya 466–8555 Japan
nakano@ics.nitech.ac.jp
[3] Department of Electronics and Informatics, Ryukoku University
Otsu, Shiga 520-2194, Japan
kimura@rins.ryukoku.ac.jp

Abstract. We address a problem of predicting diffusion probabilities in complex networks. As one approach to this problem, we focus on the independent cascade (IC) model, and define the likelihood for information diffusion episodes, where an episode means a sequence of newly active nodes. Then, we present a method for predicting diffusion probabilities by using the EM algorithm. Our experiments using a real network data set show the proposed method works well.

1   Introduction

Recently, attention has been devoted to investigating complex networks such as social, computer and biochemical networks [3,18,15,13,16]. A network can play an important role as a medium for the spread of various information [20,15]. For example, innovation, hot topics and even malicious rumors can propagate through social networks among individuals, and computer viruses can diffuse through email networks. Widely-used fundamental probabilistic models of information diffusion through networks are the *independent cascade (IC) model* [8,10,9]. The IC models can also be identified with the so-called *susceptible/infective/recoverd (SIR) model* for the spread of disease in a network [15]. Here we consider information diffusion phenomena in a given network on basis of the IC model.

The IC model needs to be provided with some adequate parameter values in advance. More specifically, the *diffusion probability* through each link in the network must be specified for the IC model in advance. However, it is usually difficult to know the diffusion probabilities through links for any real network in advance. Therefore, it is an important research issue to infer the diffusion probabilities through links from an observed data set of information diffusion.

I. Lovrek, R.J. Howlett, and L.C. Jain (Eds.): KES 2008, Part III, LNAI 5179, pp. 67–75, 2008.

2   Proposed Method

In this section, after explaining some preliminaries, we formalize a problem for estimating probabilities of information diffusion using a data set obtained from a directed network. Then we propose a method for estimating information diffusion probabilities.

2.1   Preliminaries

For a given directed network (or equivalently graph) $G = (V, E)$, let V be a set of nodes (or vertices) and E a set of links (or edges), where we denote each link by $e = (v, w) \in E$ and $v \neq w$, meaning there exists a directed link from a node v to a node w. For each node v in the network G, we define $F(v)$ as a set of child nodes of v as follows:

$$F(v) = \{w : (v, w) \in E\}. \tag{1}$$

Similarly, we define $B(v)$ as a set of parent nodes of v as follows:

$$B(v) = \{u : (u, v) \in E\}. \tag{2}$$

We introduce the IC model [8,10,9]. In this model, for each directed link $e = (v, w)$, we specify a real value $\kappa_{v,w}$ with $0 < \kappa_{v,w} < 1$ in advance. Here $\kappa_{v,w}$ is referred to as the *diffusion probability* through link (v, w). The diffusion process proceeds from a given initial active set $D(0)$ in the following way. When a node $v(\in D(t))$ first becomes active at time-step t, it is given a single chance to activate each currently inactive child node w, and the attempt succeeds with probability $\kappa_{v,w}$. If v succeeds, then w becomes active at time-step $t + 1$, i.e., $w(\in D(t + 1))$. If multiple parent nodes of w first become active at time-step t, then their activation attempts are sequenced in an arbitrary order, but all the attempts are performed at time-step t. Whether or not v succeeds, v will not make any further attempts to activate w in subsequent rounds. The process terminates if no more activations are possible.

2.2   Problem Formulation

Let an information diffusion episode $D = D(0) \cup D(1) \cup \cdots \cup D(T)$ be the union of ordered sets, where a set $D(t)$ consists of nodes newly made active at time-step t. For some link $e = (v, w)$, where $v \in D(t)$ and $w \in D(t + 1)$, it is probable that the node v succeeded in activating the node w at time-step t through the link e. Since we should consider other possibilities that some other nodes $v' \in D(t) \cap B(w)$ also succeeded in activating the node w, we calculate the probability that the node w becomes active at time-step $t + 1$ as follows:

$$P_w(t + 1) = 1 - \prod_{v \in B(w) \cap D(t)} (1 - \kappa_{v,w}). \tag{3}$$

Here note that if $w \in D(t + 1)$, it is guaranteed that $D(t) \cap B(w) \neq \emptyset$.

Next, let $C(t)$ denote a set of nodes made active by time-step t, i.e., $C(t) = \cup_{\tau \leq t} D(\tau)$. In the case that $v \in D(t)$ and $w \notin C(t+1)$, we know that the node v definitely failed in activating the node w through the link e. Clearly, in the case that (a) $v \in D(t)$ and $w \in C(t)$, or (b) $v \notin D$, we cannot obtain any information about the attempt with respect to the link $e = (v, w)$. Therefore, for an episode D we can define the following likelihood function with respect to $\boldsymbol{\theta} = \{\kappa_{v,w}\}$:

$$L(\boldsymbol{\theta}; D) = \left(\prod_{t=0}^{T-1} \prod_{w \in D(t+1)} P_w(t+1) \right) \left(\prod_{t=0}^{T-1} \prod_{v \in D(t)} \prod_{w \in F(v) \setminus C(t+1)} (1 - \kappa_{v,w}) \right) \tag{4}$$

Let $\{D_s : s = 1, \cdots, S\}$ be a set of S independent information diffusion episodes. Then we can define the following objective function with respect to $\boldsymbol{\theta}$.

$$L(\boldsymbol{\theta}) = \sum_{s=1}^{S} \log L(\boldsymbol{\theta}; D_s)$$

$$= \sum_{s=1}^{S} \sum_{t=0}^{T_s-1} \left(\sum_{w \in D_s(t+1)} \log P_w^{(s)} + \sum_{v \in D_s(t)} \sum_{w \in F(v) \setminus C_s(t+1)} \log(1 - \kappa_{v,w}) \right) \tag{5}$$

$$P_w^{(s)} = 1 - \prod_{v \in B(w) \cap D_s(t_w^{(s)}-1)} (1 - \kappa_{v,w}) \tag{6}$$

Here $P_w^{(s)}$ stands for the probability that a node w becomes active at time-step $t_w^{(s)}(= t + 1)$ in an episode D_s. Given D_s, the time-step $t_w^{(s)}(= t + 1)$ when w becomes active is known, and is omitted in representing $P_w^{(s)}$. Then, our problem is to obtain the set of information diffusion probabilities, $\boldsymbol{\theta} = \{\kappa_{v,w}\}$, which maximizes Eq. (5).

2.3   Learning Method

Since it is rather hard to analytically maximize Eq. (5), we apply the Expectation-Maximization (EM) algorithm in order to obtain its solution.

For a link (v, w) in an episode D_s where $v \in D_s(t)$, we know an activation attempt through the link (v, w) was surely performed. If $w \in D_s(t + 1)$, then the attempt through (v, w) succeeded with the probability $\widehat{\kappa}_{v,w}/\widehat{P}_w^{(s)}$, and failed with the probability $1 - (\widehat{\kappa}_{v,w}/\widehat{P}_w^{(s)})$. Here $\widehat{\kappa}_{v,w}$ stands for a current estimate of $\kappa_{v,w}$, and the value $\widehat{P}_w^{(s)}$ is calculated by using Eq. (6) and current estimates $\widehat{\boldsymbol{\theta}} = \{\widehat{\kappa}_{v,w}\}$. On the other hand, if $w \notin C_s(t+1)$, then the attempt through (v, w) failed with no doubt. Considering these cases, we have the following Q-function for S episodes.

$$Q(\boldsymbol{\theta}|\widehat{\boldsymbol{\theta}}) = \sum_{s=1}^{S} \sum_{t=0}^{T_s-1} \sum_{v \in D_s(t)} \left(\sum_{w \in F(v) \cap D_s(t+1)} \left(\frac{\widehat{\kappa}_{v,w}}{\widehat{P}_w^{(s)}} \log \kappa_{v,w} + (1 - \frac{\widehat{\kappa}_{v,w}}{\widehat{P}_w^{(s)}}) \log(1 - \kappa_{v,w}) \right) \right.$$

$$\left. + \sum_{w \in F(v) \setminus C_s(t+1)} \log(1 - \kappa_{v,w}) \right) \tag{7}$$

By solving the optimality condition $\partial Q / \partial \kappa_{v,w} = 0$, we have the following new estimate of $\kappa_{v,w}$.

$$\kappa_{v,w} = \frac{1}{|S_{v,w}^+| + |S_{v,w}^-|} \sum_{s \in S_{v,w}^+} \frac{\widehat{\kappa}_{v,w}}{\widehat{P}_w^{(s)}} \tag{8}$$

Here $S_{v,w}^+$ indicates a set of episodes, each of which satisfies both $v \in D_s(t)$ and $w \in D_s(t+1)$ for the link (v, w), while $S_{v,w}^-$ denotes a set of episodes, each of which satisfies both $v \in D_s(t)$ and $w \notin D_s(t+1)$ for the link (v, w). Moreover, $|S|$ indicates the number of elements of a set S.

We can summarize our learning method based on the EM algorithm. The method repeats the following two EM steps until convergence.

step 1. Estimate each success probability $\widehat{P}_w^{(s)}$ by using Eq. (6).
step 2. Update each diffusion probability $\kappa_{v,w}$ by using Eq. (8).

We consider that such probabilities estimated by the above method can be used to a wider variety of applications.

3   Evaluation by Experiments

3.1   Network Data

We describe the details of the network data used in our experiments.

Blogs are personal on-line diaries managed by easy-to-use software packages, and they have spread rapidly through the World Wide Web [9]. They are equipped with a link creation function called a *trackback*, and so *bloggers* (i.e., blog authors) discuss various topics and establish mutual communications by putting trackbacks on each other's blogs. We treated a link created by a trackback as bidirectional for simplicity, and employed a trackback network of blogs as an example of information propagation network.

We exploited the blog "Theme salon of blogs" in the site "goo"[1], where a blogger can recruit trackbacks of other bloggers by registering an interesting theme. By tracing up to ten steps back in the trackbacks from the blog of the theme "JR Fukuchiyama Line Derailment Collision", we collected a large connected trackback network in May, 2005. This network was a directed graph of $12,047$ nodes and $79,920$ links.

3.2   Experimental Settings

We generated a value uniformly at random in some range $[\alpha, \beta]$ ($0 < \alpha < \beta < 1$), then assigned this value to the diffusion probability $\kappa_{u,v}$ for any directed link (u, v) of a network; that is, $\kappa_{u,v} \in [\alpha, \beta]$. We determine the typical values of α, β for the blog network, and use them in the experiments. It is known that the IC model is equivalent to the bond percolation process that independently declares

---

[1] `http://blog.goo.ne.jp/usertheme/`

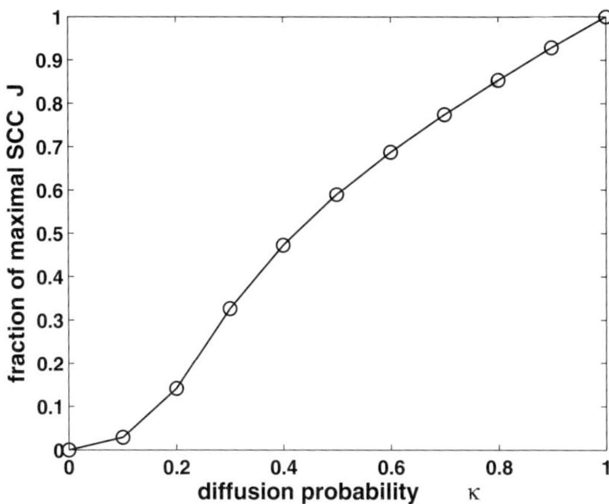

Fig. 1. The fraction J of the maximal SCC as a function of the diffusion probability κ

every link of the network to be "occupied" with probability κ [15]. Let J denote the expected fraction of the maximal strongly connected component (SCC) in the network constructed by occupied links. Note that J is an increasing function of κ. We focus on the point κ_* at which the average rate of the change of J, $dJ/d\kappa$, attains the maximum, and regard it as the typical value of κ for the network. Note that κ_* is a critical point of $dJ/d\kappa$, and defines one of the features intrinsic to the network. Figure 1 shows the values of J as a function of κ. Here, we estimated J using the bond percolation method with the same parameter value as [11]. From this figure we experimentally estimated κ_* to be $\kappa_* = 0.2$ for the blog network. Thus, we set the range such that $\alpha = 0.1$ and $\beta = 0.3$.

3.3   Experimental Results

Figure 2 shows how the learning performance changed with respect to S, where we changed S from 1 to 100 at intervals of 10 episodes. In our preliminary experiment, we set the size of the initial active set to be just one, and an initial active node is selected at random. We evaluated the learning performance by using the average absolute errors between true values and the corresponding estimated ones of diffusion probabilities, and standard deviations of the errors. As expected, the learning performance improved as the number S of episodes increased.

Figure 3 shows the distribution of T_s for the case $S = 100$. Since an initial node is selected at random, T_s is widely distributed ranging from one to 53. Here recall that there exist $79,920$ links in the network. These experimental results indicate that our approach worked well for this size of network.

4   Related Work and Discussion

There exist a large amount of work for information diffusion through networks.

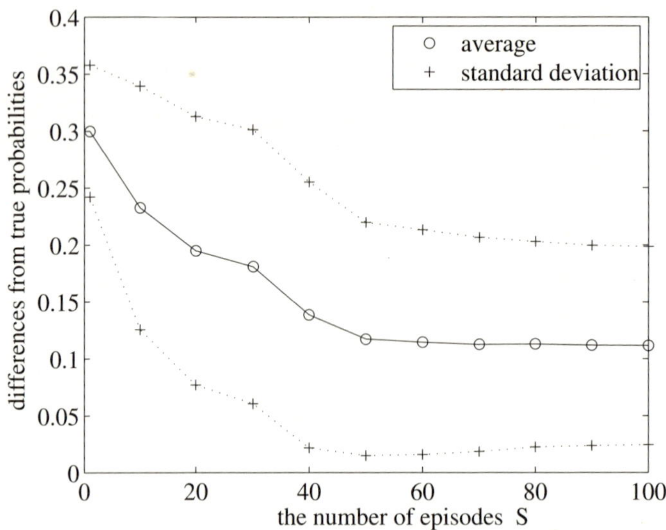

Fig. 2. How learning performance changes when the number of episodes S increases

Gruhl *et al.* [9], Adar and Adamic [1], and Leskovec, Adamic and Huberman [12] addressed the problem of tracking the propagation patterns of topics and influence through network spaces. Albert, Jeong and Barabási [2], Broder *et al.* [5], Callaway *et al.* [6], and Newman, Forrest and Balthrop [14] showed that the strategy of removing nodes in decreasing order of out-degree is effective for preventing the spread of undesirable things through networks. Balthrop *et al.* [4] studied effective "vaccination" strategies for preventing the spread of computer viruses through networks.

In contrast, a piece of information can diffuse through a social network in the form of "word-of-mouth" communication. Thus, it is important to find a limited number of influential nodes that are effective for the spread of information through a social network in terms of sociology and "viral marketing". Domingos and Richardson [7], Richardson and Domingos [17], Kempe, Kleinberg and Tardos [10], and Kimura, Saito and Nakano [11] studied a combinatorial optimization problem called the *influence maximization problem*. The problem is to extract a set of k nodes to target for initial activation such that it yields the largest expected spread of information, where k is a given positive integer. In particular, Kempe, Kleinberg and Tardos [10], and Kimura, Saito and Nakano [11] investigated the influence maximization problem for the IC model on which we focus in this paper.

Unlike discrete-time diffusion models in networks as above, Song *et al.* [19] proposed a continuous-time information diffusion model for a set of users. More specifically, they incorporated the diffusion rate that captures how efficiently the information can diffuse among the users. On the basis of their model, they predicted how likely the information will propagate from a specific sender to a specific receiver within a limited time, and applied it to a recommendation

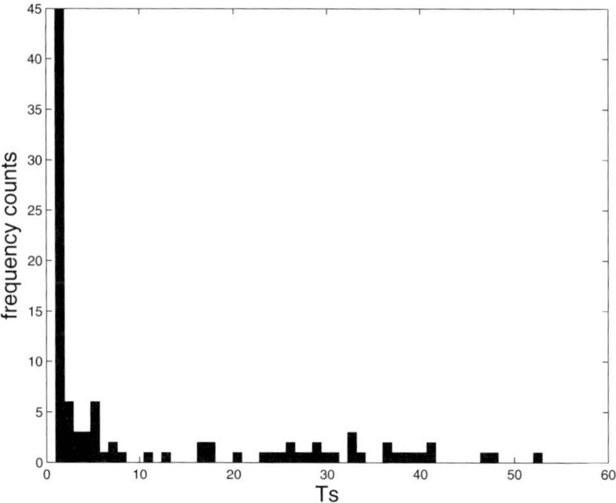

Fig. 3. Distribution of Ts for the case $S = 100$

system. They also estimated the expected time for information diffusion to a particular user, and applied it to a user ranking system. Our future work also includes applying our proposed method for the IC model to recommendation and ranking systems.

For the discrete-time IC-like model, Gruhl *et al.* [9] have already addressed a learning problem of diffusion parameters, and intuitively derived an EM-like algorithm for this purpose. Here when each parameter of geometric distribution is set such that $r_{v,w} = 1$, their model substantially reduces to our model. In this setting and our formalization, their posterior probability becomes as follows.

$$q(w|v) = \kappa_{v,w} \Big/ \sum_{u \in B(w) \cap D(t)} \kappa_{u,w}. \tag{9}$$

On the other hand, the posterior probability of our algorithm gives the following.

$$q(w|v) = \kappa_{v,w} \Big/ \left(1 - \prod_{u \in B(w) \cap D(t)} (1 - \kappa_{u,w}) \right). \tag{10}$$

We can easily see that the denominator of Eq. (9) is an approximation to that of Eq. (10) by ignoring cross terms of κ's.

5   Conclusion

We addressed the problem of predicting diffusion probabilities in complex networks. As one approach to this problem, we focused on the IC model, and defined the likelihood for multiple episodes. Then, we presented a method for predicting diffusion probabilities by using the EM algorithm. Our experiments using a real blogroll network showed that the proposed method improved the prediction performance with the increase of episodes. In future, we plan to introduce the delay of propagation into the IC model, and extend our learning method to cope with the model extension.

Acknowledgment

This work was partly supported by the Grant-in-Aid for Scientific Research (C) (No. 20500147) from the Ministry of Education, Culture, Sports, Science and Technology, Japan.

References

1. Adar, E., Adamic, L.: Tracking information epidemics in blogspace. In: Proc. of the 2005 IEEE/WIC/ACM International Conference on Web Intelligence (WI 2005), pp. 207–214 (2005)
2. Albert, R., Jeong, H., Barabási, A.L.: Error and attack tolerance of complex networks. Nature 406, 378–382 (2000)
3. Barabási, A.L., Albert, R.: Emergence of scaling in random networks. Science 286, 509–512 (1999)
4. Balthrop, J., Forrest, S., Newman, M.E.J., Williampson, M.W.: Technological networks and the spread of computer viruses. Science 304, 527–529 (2004)
5. Broder, A., Kumar, R., Maghoul, F., Raghavan, P., Rajagopalan, S., Stata, R., Tomkins, A., Wiener, J.: Graph structure in the Web. In: Proc. of the 9th International World Wide Web Conference (WWW 2000), pp. 309–320 (2000)
6. Callaway, D.S., Newman, M.E.J., Strogatz, S.H., Watts, D.J.: Network robustness and fragility: Percolation on random graphs. Physical Review Letters 85, 5468–5471 (2000)
7. Domingos, P., Richardson, M.: Mining the network value of customers. In: Proc. of the 7th ACM SIGKDD International Conference on Knowledge Discovery and Data Mining (KDD-2001), pp. 57–66 (2001)
8. Goldenberg, J., Libai, B., Muller, E.: Talk of the network: A complex systems look at the underlying process of word-of-mouth. Marketing Letters 12, 211–223 (2001)
9. Gruhl, D., Guha, R., Liben-Nowell, D., Tomkins, A.: Information diffusion through blogspace. In: Proc. of the 13th International World Wide Web Conference (WWW 2004), pp. 107–117 (2004)
10. Kempe, D., Kleinberg, J., Tardos, E.: Maximizing the spread of influence through a social network. In: Proc. of the 9th ACM SIGKDD International Conference on Knowledge Discovery and Data Mining (KDD-2003), pp. 137–146 (2003)
11. Kimura, M., Saito, K., Nakano, R.: Extracting influential nodes for information diffusion on a social network. In: Proc. of the 22nd AAAI Conference on Artificial Intelligence (AAAI 2007), pp. 1371–1376 (2007)
12. Leskovec, J., Adamic, L., Huberman, B.A.: The dynamics of viral marketing. In: Proc. of the 7th ACM Conference on Electronic Commerce (EC-2006), pp. 228–237 (2006)
13. McCallum, A., Corrada-Emmanuel, A., Wang, X.: Topic and role discovery in social networks. In: Proc. of the 19th International Joint Conference on Artificial Intelligence (IJCAI-2005), pp. 786–791 (2005)

14. Newman, M.E.J., Forrest, S., Balthrop, J.: Email networks and the spread of computer viruses. Physical Review E 66, 035101 (2002)
15. Newman, M.E.J.: The structure and function of complex networks. SIAM Review 45, 167–256 (2003)
16. Palla, G., Derényi, I., Farkas, I., Vicsek, T.: Uncovering the overlapping community structure of complex networks in nature and society. Nature 435, 814–818 (2005)
17. Richardson, M., Domingos, P.: Mining knowledge-sharing sites for viral marketing. In: Proc. of the 8th ACM SIGKDD International Conference on Knowledge Discovery and Data Mining (KDD-2002), pp. 61–70 (2002)
18. Strogatz, S.H.: Exploring complex networks. Nature 410, 268–276 (2001)
19. Song, X., Chi, Y., Hino, K., Tseng, B.L.: Information flow modeling based on diffusion rate for prediction and ranking. In: Proc. of the 16th International World Wide Web Conference (WWW-2007), pp. 191–200 (2007)
20. Watts, D.J.: A simple model of global cascades on random networks. In: Proc. of the National Academy of Sciences of the United States of America, vol. 99, pp. 5766–5771 (2002)

Reducing SVR Support Vectors by Using Backward Deletion

Masayuki Karasuyama, Ichiro Takeuchi, and Ryohei Nakano

Nagoya Institute of Technology
Gokiso-cho, Showa-ku, Nagoya 466-8555 Japan
{krsym,nakano}@ics.nitech.ac.jp, takeuchi.ichiro@nitech.ac.jp

Abstract. Support Vector Regression (SVR) is one of the most famous sparse kernel machines which inherits many advantages of Support Vector Machines (SVM). However, since the number of support vectors grows rapidly with the increase of training samples, sparseness of the SVR is sometimes insufficient. In this paper, we propose two methods which reduce the SVR support vectors using backward deletion. Experiments show our method can dramatically reduce the number of support vectors without sacrificing the generalization performance.

Keywords: Support Vector Regression, Support Vectors, Backward Deletion.

1 Introduction

Support Vector Machines (SVM) [1,2] have attracted wide interest as a way of pattern classification and regression estimation. Support Vector Regression (SVR) is expected to inherit the good SV characteristics that it generalizes well to unseen data.

One of the important property of the SVR is their sparseness. By using ε-insensitive loss function, the SVR can provide sparse representation of regression function. However, since the number of support vectors grows rapidly with the increase of training samples, sparseness of the SVR is sometimes insufficient.

In this paper, we propose two methods which reduce the SVR support vectors by using backward deletion. The basic idea is iteratively removing the support vectors which have the smallest effect on the objective function. Experiments show that our method can dramatically reduce the number of support vectors without sacrificing the generalization performance.

2 Support Vector Regression

Suppose we are given multivariate data D with a target variable $y : \{(\boldsymbol{x}^{\mu}, y^{\mu}), \mu = 1, \cdots, N\}$. In non-linear SVR, $\boldsymbol{x}$ is mapped to $\Phi(\boldsymbol{x})$ in some feature space, and represented as

$$f(\boldsymbol{x}) = \langle \boldsymbol{w}, \Phi(\boldsymbol{x}) \rangle + b. \tag{1}$$

I. Lovrek, R.J. Howlett, and L.C. Jain (Eds.): KES 2008, Part III, LNAI 5179, pp. 76–83, 2008.

Here $\langle \cdot, \cdot \rangle$ denotes the dot product. The target function to be minimized can be defined as follows, where the coefficient C denotes a penalty factor.

$$\frac{1}{2}\|\boldsymbol{w}\|^2 + \frac{C}{2}\sum_{\mu=1}^{N}\ell_2^{\varepsilon}(f^{\mu}, y^{\mu}), \qquad f^{\mu} \equiv f(\boldsymbol{x}^{\mu}). \tag{2}$$

The first term seeks a smaller $\|\boldsymbol{w}\|^2$, which means seeking as flat a line as possible. The second term is the ε-insensitive penalty. Here, we employ the ε-insensitive ℓ_2 loss function $\ell_2^{\varepsilon}(f^{\mu}, y^{\mu})$.

$$\begin{aligned} \ell_2^{\varepsilon}(f^{\mu}, y^{\mu}) &= |f^{\mu} - y^{\mu}|_{\varepsilon}^2 \\ &= \Big(\max(0, |f^{\mu} - y^{\mu}| - \varepsilon)\Big)^2. \end{aligned} \tag{3}$$

Then our optimization problem can be formalized as follows.

$$\text{minimize} \quad \tfrac{1}{2}\|\boldsymbol{w}\|^2 + \tfrac{C}{2}\sum_{\mu=1}^{N}((\xi^{\mu})^2 + (\xi^{*\mu})^2), \tag{4}$$

$$\text{subject to} \quad \begin{cases} y^{\mu} - (\langle \boldsymbol{w}, \Phi(\boldsymbol{x}^{\mu})\rangle + b) \leq \varepsilon + \xi^{\mu}, \\ (\langle \boldsymbol{w}, \Phi(\boldsymbol{x}^{\mu})\rangle + b) - y^{\mu} \leq \varepsilon + \xi^{*\mu}, \\ \xi^{\mu}, \xi^{*\mu} \qquad\qquad\qquad\quad \geq \quad 0. \end{cases}$$

Slack variables $\xi^{\mu}, \xi^{*\mu}$ denote deviations from an ε-zone. The Lagrangian function of the above optimization problem can be reduced to as follows.

$$\begin{aligned} \max_{\boldsymbol{\alpha}}\min_{b} J(\boldsymbol{\alpha}, b) &= -\frac{1}{2}\sum_{\mu=1}^{N}\sum_{\mu'=1}^{N}\alpha^{\mu}\alpha^{\mu'}K(\boldsymbol{x}^{\mu}, \boldsymbol{x}^{\mu'}) - b\sum_{\mu=1}^{N}\alpha^{\mu} \\ &\quad + \sum_{\mu=1}^{N}y^{\mu}\alpha^{\mu} - \varepsilon\sum_{\mu=1}^{N}|\alpha^{\mu}| - \frac{1}{2C}\sum_{\mu=1}^{N}(\alpha^{\mu})^2. \end{aligned} \tag{5}$$

Here, K denotes a kernel function such that $K(\boldsymbol{x}^{\mu}, \boldsymbol{x}^{\mu'}) = \langle \Phi(\boldsymbol{x}^{\mu}), \Phi(\boldsymbol{x}^{\mu'})\rangle$, and α^{μ} is a parameter which forms the following regression function.

$$f(\boldsymbol{x}; \boldsymbol{\theta}) = \sum_{\mu \in SV} \alpha^{\mu}K(\boldsymbol{x}^{\mu}, \boldsymbol{x}) + b. \tag{6}$$

To obtain the optimal α^{μ}, usually we solve the constrained optimization problem by taking the derivative of eq. (5) with respect to b. There are many implementation to solve this optimization problem such as LIBSVM [3]. The samples which has non-zero α^{μ} are called the support vectors ($SV \equiv \{\mu | \alpha^{\mu} \neq 0\}$). The support vectors lie in the outside of the ε-tube (Fig. 1).

3   Reducing Support Vectors

By using ε-insensitive loss function, the SVR can provide a sparse representation of regression function. However, sparseness of the SVR is often insufficient. Here,

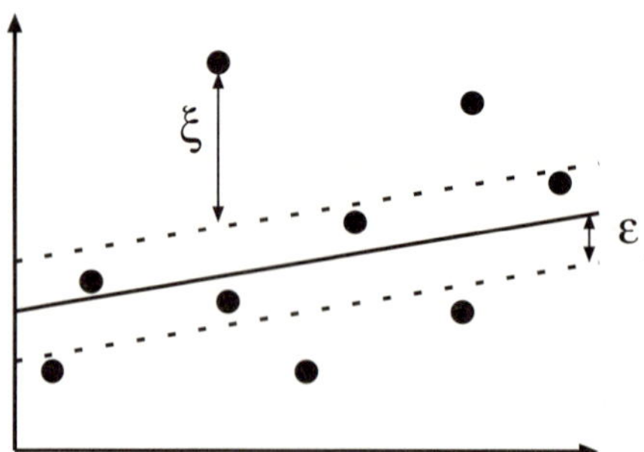

Fig. 1. In SVR, the regression function has an insensitive tube whose radius is ε. The training error of the samples which are in that tube is ignored.

we consider to derive the backward deletion of the SVR support vectors to obtain more sparseness. The basic idea is iteratively removing the support vectors which have the smallest effect for the objective function.

We can rewrite eq. (5) by defining a parameter vector $\boldsymbol{\theta} \equiv [\boldsymbol{\alpha}^T \ b]^T$ and $\boldsymbol{K}$ which denotes a kernel matrix whose (i, j) element is $K(\boldsymbol{x}^i, \boldsymbol{x}^j)$.

$$J(\boldsymbol{\alpha}, b) = -\frac{1}{2}\boldsymbol{\theta}^T \begin{bmatrix} \boldsymbol{K} + \frac{1}{C}\boldsymbol{I} & \mathbf{1} \\ \mathbf{1}^T & 0 \end{bmatrix} \boldsymbol{\theta} + \begin{bmatrix} \boldsymbol{y} - \mathrm{sgn}(\boldsymbol{\alpha})\varepsilon \\ 0 \end{bmatrix}^T \boldsymbol{\theta}. \tag{7}$$

$\mathrm{sgn}(a)$ denotes the sign function which takes the value of $-1(a < 0)$ or $0(a = 0)$ or $1(a > 0)$, and $\mathrm{sgn}(\boldsymbol{\alpha})$ is a vector of the $\mathrm{sgn}(\alpha^\mu)$. Here, we define the vectors $\boldsymbol{\alpha}_{SV} = [\alpha^\mu](\mu \in SV), \boldsymbol{\theta}_{SV} = [\boldsymbol{\alpha}_{SV}^T \ b \ ]^T, \boldsymbol{y}_{SV} = [y^\mu](\mu \in SV)$. Setting the derivative of eq. (5) with respect to $\boldsymbol{\theta}_{SV}$ as $\mathbf{0}$, we obtain

$$-\begin{bmatrix} \boldsymbol{K}_{SV} + \frac{1}{C}\boldsymbol{I} & \mathbf{1} \\ \mathbf{1}^T & 0 \end{bmatrix} \boldsymbol{\theta}_{SV} + \begin{bmatrix} \boldsymbol{y}_{SV} - \mathrm{sgn}(\boldsymbol{\alpha}_{SV}) \ \varepsilon \\ 0 \end{bmatrix} = \mathbf{0}, \tag{8}$$

where $\boldsymbol{K}_{SV} \equiv \boldsymbol{K}(SV, SV)$. Here $\boldsymbol{K}(S_1, S_2)$ denotes a submatrix of $\boldsymbol{K}$, whose rows indexed by a set S_1 and whose columns are indexed by a set S_2. Henceforth we use the following matrix and vector.

$$\boldsymbol{M} \equiv \begin{bmatrix} \boldsymbol{K}_{SV} + \frac{1}{C}\boldsymbol{I} & \mathbf{1} \\ \mathbf{1}^T & 0 \end{bmatrix}, \quad \boldsymbol{p} \equiv \begin{bmatrix} \boldsymbol{y}_{SV} - \mathrm{sgn}(\boldsymbol{\alpha}_{SV}) \ \varepsilon \\ 0 \end{bmatrix}. \tag{9}$$

By using this, we can see the optimal $\boldsymbol{\theta}_{SV}$ has the following relation.

$$\boldsymbol{\theta}_{SV} = \boldsymbol{M}^{-1}\boldsymbol{p}. \tag{10}$$

Substituting eq. (10) into objective function eq. (7), we have

$$J_{opt} = -\frac{1}{2}\boldsymbol{\theta}_{SV}^T \ \boldsymbol{M} \ \boldsymbol{\theta}_{SV} + \boldsymbol{p}^T \boldsymbol{\theta}_{SV}$$

$$= \frac{1}{2} \ \boldsymbol{p}^T \boldsymbol{M}^{-1}\boldsymbol{p}. \tag{11}$$

Here, we consider to remove a support vector $s_\nu \in SV = \{s_1, \cdots, s_l\}$. We iteratively select the ν which has the minimum decrease of J_{opt} that follows the approach of [4]. To simplify the notation, we drive the decrease of J_{opt} when we remove s_1. Let matrix $\boldsymbol{M}$ be partitioned into a block form.

$$\boldsymbol{M} = \begin{bmatrix} m_{11} & \boldsymbol{m}_1^T \\ \boldsymbol{m}_1^T & \boldsymbol{M}_1 \end{bmatrix}.$$

By using the Block Matrix Inversion, $\boldsymbol{M}^{-1}$ can be expressed as follows.

$$\boldsymbol{M}^{-1} = \begin{bmatrix} 0 & \boldsymbol{0}^T \\ \boldsymbol{0} & \boldsymbol{M}_1^{-1} \end{bmatrix} + \frac{1}{\kappa} \begin{bmatrix} -1 \\ \boldsymbol{M}_1^{-1} \boldsymbol{m}_1 \end{bmatrix} \begin{bmatrix} -1 & \boldsymbol{m}_1^T \boldsymbol{M}_1^{-1} \end{bmatrix}. \tag{12}$$

where $\kappa = m_{11} - \boldsymbol{m}_1^T \boldsymbol{M}_1^{-1} \boldsymbol{m}_1$. Substituting eq. (12) into eq. (11) and using eq. (10), we can obtain

$$J_{opt} = \frac{1}{2} \boldsymbol{p}^T \begin{bmatrix} 0 & \boldsymbol{0}^T \\ \boldsymbol{0} & \boldsymbol{M}_1^{-1} \end{bmatrix} \boldsymbol{p} + \kappa (\alpha^{s_1})^2. \tag{13}$$

The first term is a optimal value of the objective function after s_1 is removed from SV. Therefore, we can select a removing support vector s_ν by the following.

$$\nu = \underset{\nu}{\operatorname{argmin}} \frac{(\alpha^{s_\nu})^2}{\boldsymbol{M}^{-1}(\nu, \nu)}, \tag{14}$$

where $\boldsymbol{M}^{-1}(\nu, \nu)$ is a (ν, ν) element of $\boldsymbol{M}^{-1}$. Here, we use $\boldsymbol{M}^{-1}(\nu, \nu) = \kappa^{-1}$. We remove the selected ν from the support vectors $SV \leftarrow \{SV \setminus \nu\}$, and update $\boldsymbol{M}^{-1}, \boldsymbol{p}, \boldsymbol{\theta}_{SV}$. Then, we assume the sign of parameters $\operatorname{sgn}(\boldsymbol{\alpha}_{SV})$ don't change. If the $\operatorname{sgn}(\alpha^\mu)$ changes by this update, we remove this parameter, too.

We can iterate this procedure until some convergence criterion is met. We refer to this method as RSVR1 (Reduced SVR). However, this method completely ignores the removed support vectors. We derive another method for reducing the support vectors which tries to duplicate original regression function. Here, we define two type of support vectors: complete support vectors SVc and active support vectors SVa. SVc is the support vectors which is obtained by usual training of the SVR without removing any support vector. SVa is a remaining support vectors after removing some support vectors $R \subset SV : SVa \equiv \{SVc \setminus R\}$. We seek the value of parameter $\boldsymbol{\theta}_{SVa}$ which satisfies following relation.

$$\begin{bmatrix} \boldsymbol{K}(SVc, SVa) & \boldsymbol{1} \end{bmatrix} \boldsymbol{\theta}_{SVa} = \boldsymbol{f}_{SVc}. \tag{15}$$

Here, $\boldsymbol{f}_{SVc} = \begin{bmatrix} \boldsymbol{K}(SVc, SVc) & \boldsymbol{1} \end{bmatrix} \boldsymbol{\theta}_{SVc}$ is a vector of output for SVc which is obtained by the usual SVR. Since the left hand side of eq. (15) is output for SVc which is obtained by using SVa, this equation means that the output for SVc is invariant for removing the support vectors. By defining $\boldsymbol{\Phi} \equiv \begin{bmatrix} \boldsymbol{K}(SVc, SVa) & \boldsymbol{1} \end{bmatrix}$, $\boldsymbol{q} \equiv \boldsymbol{\Phi}^T \boldsymbol{f}_{SVc}$, we can obtain the following.

$$\boldsymbol{\theta}_{SVa} = (\boldsymbol{\Phi}^T \boldsymbol{\Phi})^{-1} \boldsymbol{q}. \tag{16}$$

However, since the matrix $\boldsymbol{\Phi}^T\boldsymbol{\Phi}$ sometimes close to singular, practically we replace this by $\boldsymbol{\Phi}^T\boldsymbol{\Phi} + \eta\boldsymbol{I}$, where η is a small constant, and we set $\eta = 10^{-8}$. As a objective function, we consider a squared error.

$$
\begin{aligned}
E &= \frac{1}{2}\left(\boldsymbol{\Phi}\,\boldsymbol{\theta}_{SVa} - \boldsymbol{f}_{SVc}\right)^T\left(\boldsymbol{\Phi}\,\boldsymbol{\theta}_{SVa} - \boldsymbol{f}_{SVc}\right) \\
&= \frac{1}{2}\boldsymbol{\theta}_{SVa}^T\,\boldsymbol{\Phi}^T\boldsymbol{\Phi}\,\boldsymbol{\theta}_{SVa} - \boldsymbol{f}_{SVc}^T\,\boldsymbol{\Phi}\,\boldsymbol{\theta}_{SVa} + const.
\end{aligned}
\tag{17}
$$

Removing constant term and substituting eq. (16), we have

$$
E_{opt} = -\frac{1}{2}\,\boldsymbol{q}^T(\boldsymbol{\Phi}^T\boldsymbol{\Phi})^{-1}\boldsymbol{q}.
\tag{18}
$$

As is the case with J_{opt}, we can derive the change of E_{opt} when we remove the ν-th active support vector from $SVa = \{a_1, \cdots, a_k\}$. We can select a removing support vector a_ν by the following.

$$
\nu = \underset{\nu}{\operatorname{argmin}} \frac{(\alpha^{a_\nu})^2}{(\boldsymbol{\Phi}^T\boldsymbol{\Phi})^{-1}(\nu,\nu)},
\tag{19}
$$

where $(\boldsymbol{\Phi}^T\boldsymbol{\Phi})^{-1}(\nu,\nu)$ is a (ν,ν) element of $(\boldsymbol{\Phi}^T\boldsymbol{\Phi})^{-1}$. We refer to this method as RSVR2.

4   Experiments

We evaluate the performance of our method by using some data sets. Our implementation is in MATLAB, and we employ the LIBSVM [3] for SVM solver. We use the following RBF kernel.

$$
K(\boldsymbol{x}, \boldsymbol{x}') = \exp\left(-\frac{\|\boldsymbol{x} - \boldsymbol{x}'\|^2}{2\sigma^2}\right).
\tag{20}
$$

For all experiments, we set three hyperparameters as follows: $\varepsilon = 0.05, C = 1, \sigma = 1$.

4.1   Artificial Data Set

We consider the following target function.

$$
y = \frac{\sin(x)}{x}, \qquad x \in [-10, 10].
\tag{21}
$$

The size of data D is $N = 200$; with additive Gaussian noise (mean 0 and standard deviation 0.1). The x values are sampled on a uniformly spaced grid in the input space. To evaluate generalization performance, we generate 1000 noiseless samples for test, independent of D. Experiments are conducted 10 times by changing D.

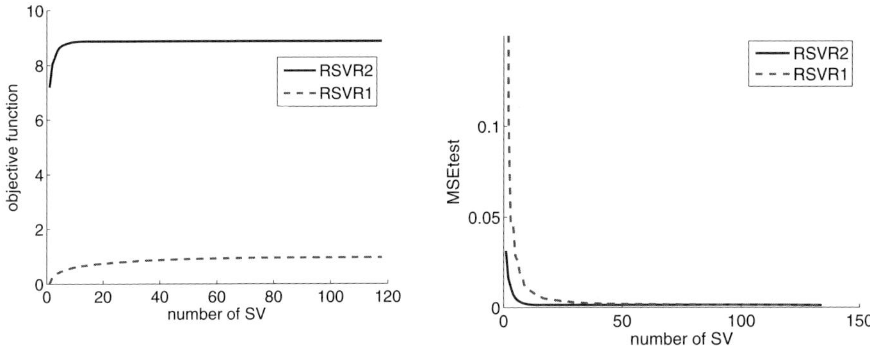

Fig. 2. The value of objective functions for artificial data set

Fig. 3. MSEtest for artificial data set

Table 1. The result of RSVR1 for artificial data set

p	time(sec)	J_{opt}	#SV	MSEtest(e-03)
1	0.01	0.96	126.20 ± 6.01	1.44 ± 0.32
0.995	0.09	0.96	89.40 ± 4.65	1.47 ± 0.34
0.95	0.11	0.91	52.50 ± 4.55	1.88 ± 0.46
0.80	0.11	0.77	22.70 ± 2.95	3.83 ± 1.48

Table 2. The result of RSVR2 for artificial data set

p	time(sec)	$-E_{opt}$	#SV	MSEtest(e-03)
1	0.01	8.87	126.20 ± 6.01	1.44 ± 0.32
0.9999	0.13	8.87	12.80 ± 0.63	1.47 ± 0.34
0.999	0.13	8.86	9.90 ± 0.99	1.74 ± 0.53
0.99	0.13	8.72	6.10 ± 1.52	4.44 ± 2.09

Figs.2 and 3 show how the value of objective function and Mean Squared Error for test set changed during decreasing the number of support vectors. We can see the value of objective function of the RSVR2 $(-E_{opt})$ suddenly decrease certain point, and corresponding test error increases. On the other hand, the value of RSVR1 (J_{opt}) changes more slowly. Tables 1 and 2 show the result of some point whose objective function reaches $p * \mathrm{Obj}_1$. Here, Obj_1 is the first value of the objective function which is obtained by usual SVR. We can see that the RSVR1 and the RSVR2 has almost the same generalization performance as usual SVR when $p = 0.995, p = 0.9999$, respectively. However, the number of support vectors of the RSVR2 much smaller than the RSVR1.

4.2 Housing Data Set

The Housing data set from the UCI Repository of MLDB [5] contains data on the housing and environmental conditions related to housing values in suburbs

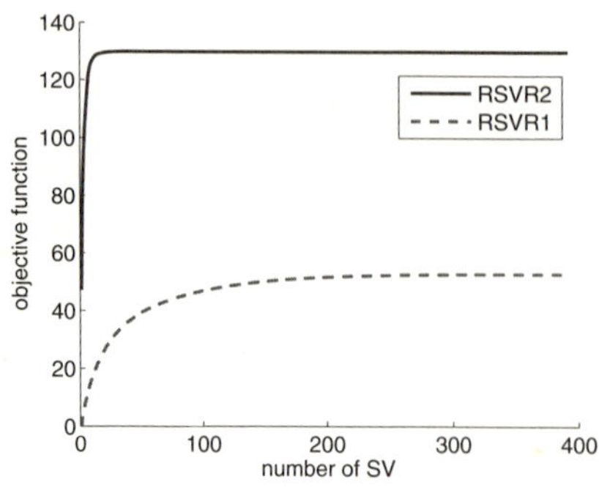

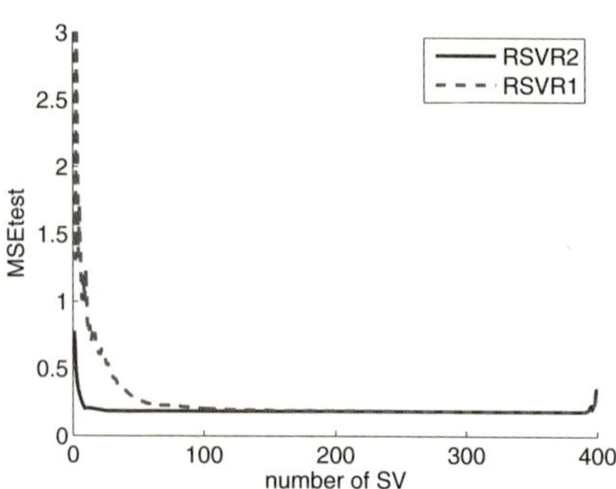

Fig. 4. The value of objective functions for housing data set

Fig. 5. MSEtest for housing data set

Table 3. The result of RSVR1 for housing data set

p	time(sec)	J_{opt}	#SV	MSEtest(e+01)
1	0.07	52.90	394.90 ± 2.60	1.49 ± 0.70
0.995	0.77	52.63	268.80 ± 2.15	1.49 ± 0.69
0.95	0.99	50.23	147.90 ± 2.28	1.54 ± 0.67
0.80	1.03	42.23	63.50 ± 1.43	1.88 ± 0.61

Table 4. The result of RSVR2 for housing data set

p	time(sec)	$-E_{opt}$	#SV	MSEtest(e+01)
1	0.07	129.77	394.90 ± 2.60	1.49 ± 0.70
0.9999	1.46	129.75	38.10 ± 1.85	1.49 ± 0.69
0.999	1.47	129.62	24.80 ± 1.03	1.51 ± 0.66
0.99	1.47	128.23	12.40 ± 1.07	1.66 ± 0.68

of Boston. The data set consists of 13 numeric explanatory variables and one target variable. The set has 506 samples with no missing value. We randomly divided the data into 455 training samples and 51 test samples 10 times. Thus, we have 10 training sets and 10 associated test sets.

Tables 3 and 4 show the results. Here again the RSVR1 and the RSVR2 has almost the same generalization performance as the usual SVR when $p = 0.995, p = 0.9999$, respectively. Then the RSVR2 has much smaller support vectors than the RSVR1 and the usual SVR.

By using this data set we also compare the performance of the RSVR2 and the BDFS [4]. BDFS is feature selection method for Sparse Kernel Ridge Regression (SKRR). Since the Kernel Ridge Regression (KRR) [6] is equivalent to the SVR when the value of ε is equal to 0, here we set $\varepsilon = 0$ in RSVR2. Other hyperparameters are set to be $C, \sigma = 1$ for both method. Fig. (6) shows comparison of MSEtest with decreasing support vectors. We can see the similar result for two

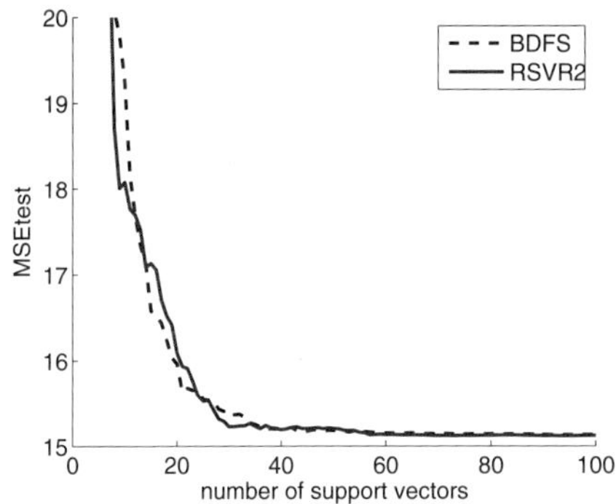

Fig. 6. MSEtest for housing data set
(elapsed CPU times are BDFS: 1.33, RSVR2: 2.06 (sec))

methods. However, the BDFS for the SKRR must start from N support vectors which means to calculate $(N+1) \times (N+1)$ matrix inverse. On the other hand, when we set $\varepsilon > 0$, the RSVR2 can start from $(|SVc|+1) \times (|SVc|+1)$ matrix.

5   Conclusion

In this paper, we proposed a backward deletion of the SVR support vectors. Experiments show that our method can produce more sparse model than the usual SVR without sacrificing generalization performance. A future study is to optimize hyperparameters of the SVR while pursuing a sparse model.

References

1. Vapnik, V.N.: The Nature of Statistical Learning Theory. Springer, Heidelberg (1995)
2. Cristianini, N., Shawe-Taylor, J.: An Introduction to Support Vector Machines and other kernel-based learning methods. Cambridge University Press, Cambridge (2000)
3. Chang, C.C., Lin, C.J.: Libsvm - a library for support vector machines, http://www.csie.ntu.edu.tw/~cjlin/libsvm/
4. Wang, L., Bo, L., Jiao, L.: Sparse kernel ridge regression using backward deletion. In: Yang, Q., Webb, G. (eds.) PRICAI 2006. LNCS (LNAI), vol. 4099, pp. 365–374. Springer, Heidelberg (2006)
5. Asuncion, A., Newman, D.: UCI machine learning repository (2007), http://www.ics.uci.edu/~mlearn/MLRepository.html
6. Saunders, G., Gammerman, A., Vovk, V.: Ridge regression learning algorithm in dual variables. In: Proc. 15th International Conf. on Machine Learning, pp. 515–521. Morgan Kaufmann, San Francisco (1998)

Structure Analysis of Fuzzy Node Fuzzy Graph and Its Application to Sociometry Analysis

Hiroaki Uesu

1-104, Totsuka-machi, Shinjuku-ku, Tokyo 169-8050, Japan, Waseda University
uesu@suou.waseda.jp

Abstract. We could generally analyze the inexact information efficiently and investigate the fuzzy relation by applying the fuzzy graph theory. We would extend the fuzzy graph theory, and propose a fuzzy node fuzzy graph. And we transform it to a fuzzy graph by using T-norm family. In this paper, we would discuss about four subjects, (1) fuzzy node fuzzy graph, (2) new T-norm ``quasi logical product'', (3) decision analysis of the optimal fuzzy graph G_{λ_0} in the fuzzy graph sequence $\{G_\lambda\}$. By using the fuzzy node fuzzy graph theory and this new T-norm, we could clarify the relational structure of fuzzy information, and by using the decision of an optimal level on a partition tree, we could analyze the clustering relation among nodes. Moreover, we would illustrate its practical effectiveness with the case study concerning sociometry analysis.

Keywords: fuzzy node fuzzy graph, sociometry analysis, T-norm.

1 Fuzzy Node Fuzzy Graph and T-norm

1.1 Fuzzy Node Fuzzy Graph

Definition 1. Crisp Node Fuzzy Graph[6]

A crisp node fuzzy graph G is defined by

$$G = (V, F) : V = \{v_i\}, F = (f_{ij}), 0 \le f_{ij} \le 1 \tag{1}$$

where V is the set of the nodes and F is an $n \times n$ matrix whose (i, j) component f_{ij} is a fuzziness of the arc from the node v_i to the node v_j.

Definition 2. Fuzzy Node Fuzzy Graph[14]

A fuzzy node fuzzy graph G is defined by

$$G = (V, F) : V = \{v_i / u_i\}, Y = (y_{ij}), 0 \le u_i \le 1,\ 0 \le y_{ij} \le 1 \tag{2}$$

where V is the set of the nodes and the fuzziness u_i is a fuzziness of the node v_i. Y is an $n \times n$ matrix whose (i, j) component y_{ij} is a fuzziness of the arc from the node v_i to the node v_j.

A fuzzy node fuzzy graph is characterized by the fuzziness of the nodes and the fuzziness of the arcs. Therefore, the structure of a fuzzy node fuzzy graph is usually

I. Lovrek, R.J. Howlett, and L.C. Jain (Eds.): KES 2008, Part III, LNAI 5179, pp. 84–91, 2008.

very complicated. Then, it should be interesting to transform a fuzzy node fuzzy graph to a crisp node fuzzy graph and we present a method to transform a fuzzy node fuzzy graph to a crisp node fuzzy graph.

Definition 3. Transformation from Fuzzy Node Fuzzy Graph to Crisp Node Fuzzy Graph

A fuzzy node fuzzy graph $G=(V, Y)$ can be transformed to a crisp node fuzzy graph $G=(V,F)$ by the following method: Let

$$G = (V, F) : V = \{v_i\}, F = (f_{ij}), f_{ij} = T(u_i, y_{ij}) \tag{3}$$

where the fuzziness f_{ij} of the arc from the node v_i to the node v_j could be defined by applying T-norms.

This fuzzy node fuzzy graph analysis would be applied to the sociometry analysis, the instruction structure analysis and so on.

1.2 T-Norm and Its Family

T-norm (Triangular Norm) is a binary operation that is defined by the following:

Definition 4. T-norm is a binary operation

$$p, q \in [0,1] \rightarrow T(p,q) \in [0,1] \tag{4}$$

satisfying the following properties:

(1) Commutativity : $T(p,q) = T(q,p)$ $\tag{5}$

(2) Associativity : $T(p,T(q,r)) = T(T(p,q),r)$ $\tag{6}$

(3) Monotonicity : $p \le q, \quad r \le s \Rightarrow T(p,r) \le T(q,s)$ $\tag{7}$

(4) Boundary conditions : $T(p,0) = 0, \quad T(p,1) = p$ $\tag{8}$

The typical T-norms are the followings:

(1) Logical product : $T_L(p,q) = p \wedge q$ $\tag{9}$

(2) Algebraic product : $T_A(p,q) = pq$ $\tag{10}$

(3) Multi-valued product (Lukaszewicz product) : $T_M(p,q) = (p+q-1) \vee 0$ $\tag{11}$

(4) Drastic product : $T_D(p,q) = \begin{cases} 0 & , p \vee q < 1 \\ p \wedge q, & p \vee q = 1 \end{cases}$ $\tag{12}$

Here, "$a \wedge b$" means "$\min(a,b)$", "$a \vee b$" means "$\max(a,b)$".

Definition 5. Order of T-norms

For two T-norms T_α and T_β, if a relation

$$T_\alpha(p,q) \le T_\beta(p,q), (p,q) \in [0,1]^2$$

holds, we denote it by $T_\alpha \le T_\beta$.

For any T-norm T, a relation $T_D \le T \le T_L$ always holds.

Definition 6. T-norm Family

For any $\lambda \in [a,b]$, when T_λ is T-norm, then we say that $\{T_\lambda\}$ is T-norm family that connects T_α with T_α.

The typical T-norm families are

$$\text{(1) Dubois product}: T_\lambda(p,q) = \frac{pq}{p \vee q \vee \lambda}, \lambda \in [0,1] \tag{13}$$

$$\text{(2) Weber product}: T_\lambda(p,q) = 0 \vee \{(1+\lambda)(p+q-1) - \lambda pq\}, \lambda \ge -1 \tag{14}$$

$$\text{(3) Schweizer product}: T_\lambda(p,q) = \sqrt[\lambda]{0 \vee (p^\lambda + q^\lambda - 1)}, \lambda > 0 \tag{15}$$

(4) Quasi-logical product (Uesu product) :

$$T_\lambda(p,q) = \begin{cases} 0 & , p \vee q < 1-\lambda \\ p \wedge q, & p \vee q \ge 1-\lambda \end{cases}, \lambda \in [0,1] \tag{16}$$

and so on.

The relation of typical T-norm families and quasi-logical product could be illustrated in Fig.3.

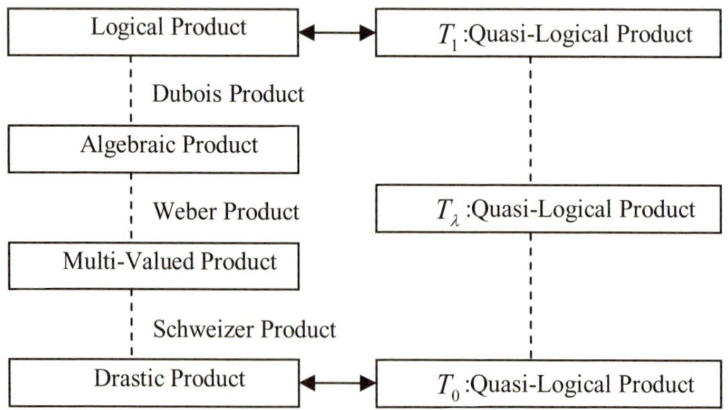

Fig. 1. Relation of Typical T-norm Families

2 Fuzzy Graph and Its Analysis

2.1 Cluster Analysis of Fuzzy Graph

A fuzzy graph G is defined by

$$G = (V,Y), V = \{v_i\}, F = (f_{ij}), 0 \le f_{ij} \le 1, f_{ii} = 1, 1 \le i \le n \tag{17}$$

where f_{ij} is a fuzziness of the arc from the node v_i to the node v_j.

In order to analyze the similarity structure of nodes for a fuzzy graph, we use the symmetric relation matrix $S = (s_{ij})$.

A symmetric relation matrix S could be defined by using the arithmetic mean, the geometric mean, the harmonic mean and so on. Here, we define the symmetric relation matrix S by using the harmonic mean.

A symmetric relation matrix S is defined by

$$S = (s_{ij}), \frac{2}{s_{ij}} = \frac{1}{f_{ij}} + \frac{1}{f_{ji}} \tag{18}$$

where $s_{ij}=0$ if $f_{ij} \cdot f_{ji}=0$.

In order to analyze the clustering structure among nodes, we have its max-min transitive closure $\hat{S} = (\hat{s}_{ij})$ which is computed by $\hat{S} = S^n$.

After that, we define the c-cut matrix S_c of $\hat{S} = (\hat{s}_{ij})$ as follows:

$$S_c = (s_{ij}^c), \quad s_{ij}^c = \begin{cases} 1 & (\hat{s}_{ij} \ge c) \\ 0 & (\hat{s}_{ij} < c) \end{cases}, 0 \le c \le 1 \tag{19}$$

From the matrix S_c, we define the cluster $CL_{Sc}(i)$.

$$J_c(i) = \{ j \mid s_{ij}^c = 1, 1 \le j \le n \}, \ CL_{Sc}(i) = \{ v_j \mid j \in J_c(i) \} \tag{20}$$

The cluster $CL_{Sc(i)}$ gives an equivalence relation among nodes.

Thence, we can construct the partition tree by changing the level c of the c-cut matrix which represents the clustering situation of nodes in a fuzzy graph.

2.2 Decision Analysis of Optimal Value λ_0

Concerning the transformation from a fuzzy node fuzzy graph $G = (V,Y)$ to a fuzzy graph $G_\lambda = (V, F_\lambda)$, we could analyze a fuzzy node fuzzy graph by using quasi-logical product T_λ.

By changing the parameter λ, a sequence $\{G_\lambda\}$ of fuzzy graphs is composed.

Here, we would choose the optimal fuzzy graph G_{λ_0} from this fuzzy graph sequence $\{G_\lambda\}$.

Then, we would discuss the decision method of the optimal parameter λ_0.

In order to decide the optimal fuzzy graph G_{λ_0}, we would define two functions $d(\lambda)$ and $e(\lambda)$ as follows:

Definition 7. Distance Function $d(\lambda)$ and Connectivity Function $e(\lambda)$

$$d(\lambda) = d(F_\lambda, S_{c_0}) = \frac{1}{n^2 - n} \sum_{i=1}^{n} \sum_{j=1}^{n} \left| f_{ij} - s_{ij}^{c_0} \right| \tag{21}$$

$$e(\lambda) = e(F_\lambda) = \frac{\gamma(F_\lambda)}{n^2 - n} \tag{22}$$

where, $\gamma(F_\lambda) = \#(\Gamma_\lambda), \Gamma_\lambda = \{ f_{ij}^\lambda \in F_\lambda | f_{ij}^\lambda > 0 \}$.

And F_λ, S_{c_0} are given as follows:

$$F_\lambda = (f_{ij}^\lambda), \quad f_{ij}^\lambda = \begin{cases} 0 & , u_i \vee y_{ij} < 1 - \lambda \\ u_i \wedge y_{ij} & , u_i \vee y_{ij} \geq 1 - \lambda \end{cases} \tag{23}$$

$$S_{c_0} = (s_{ij}^{c_0}), \quad s_{ij}^{c_0} = \begin{cases} 1 & (\hat{s}_{ij} \geq c_0) \\ 0 & (\hat{s}_{ij} < c_0) \end{cases} \tag{24}$$

Here, $d(\lambda)$ evaluates the feature between the fuzzy node fuzzy graph $G_\lambda = (V, F_\lambda)$ and the optimal clustering cut level c_0. If the value of $d(\lambda)$ is large, then G_λ reasonably shows the feature of the clustering level c_0.

On the other hand, $e(\lambda)$ evaluates the connectivity information of G_λ. If the value of $e(\lambda)$ is small, then G_λ reasonably shows the feature of the connectivity information.

Here, we would normalize the values of $d(\lambda)$ and $e(\lambda)$ respectively, and define $f_d(\lambda)$ and $f_e(\lambda)$ as follows:

Definition 8. Fuzzy Distance Function $f_d(\lambda)$ and Fuzzy Connectivity Function $f_e(\lambda)$

$$f_d(\lambda) = \frac{d_M - d(\lambda)}{d_M - d_m} \tag{25}$$

$$f_e(\lambda) = \frac{e(\lambda) - e_m}{e_M - e_m} \tag{26}$$

where,

$$d_M = \bigvee_{\lambda \in [0,1]} \{d(\lambda)\} \qquad d_m = \bigwedge_{\lambda \in [0,1]} \{d(\lambda)\} \qquad e_M = \bigvee_{\lambda \in [0,1]} \{e(\lambda)\} \quad \text{and}$$

$$e_m = \bigwedge_{\lambda \in [0,1]} \{e(\lambda)\} .$$

By applying the maximal decision of the fuzzy decision, we could reasonably find the optimal value λ_0 concerning the sequence $\{G_\lambda\}$.

Definition 9. Decision of Optimal Value λ_0

$$f_m(\lambda) = f_d(\lambda) \wedge f_e(\lambda)$$
$$\lambda_0 = \wedge \{\lambda : f_m(\lambda) = \bigvee_{x \in [0,1]} f_m(x)\} \tag{27}$$

3   Application to Sociogram Analysis

We would present a practical case study of the fuzzy sociogram analysis that has been experimented in high school.

If we executed the questionnaires 10 students, we could obtain the response table L.

From the table L, we obtained the response matrix $K^{(1)}, K^{(2)}, K^{(3)}$ and the evaluation matrix R.

By analyzing the matrix R, we have the preferring matrix $Y=(y_{ij})$ and the amicable matrix $S=(s_{ij})$. From the amicable matrix S, we obtained the fuzziness u_i of member v_i. We illustrated the fuzzy node fuzzy graph $G=(V, Y)$.

We could transform the fuzzy node fuzzy graph $G=(V, Y)$ to the fuzzy graph by applying quasi-logical product . By changing parameter λ, we obtained fuzzy graph sequence $\{G_\lambda\}$. Here, the fuzzy graph $G_\lambda = (V, F_\lambda)$ corresponds to the fuzzy graph transformed by the logical product, and the fuzzy graph $G_\lambda = (V, F_\lambda)$ corresponds to the fuzzy graph transformed by the drastic product.

Next, we could obtain the partition tree P if we execute the cluster analysis concerning the amicability matrix S.

Here, we could decide the optimal cut level $c_0 =0.58$.

By using c_0-cut matrix Sc_0 and the fuzzy graph sequence $\{G_\lambda\}$, we could decide the optimal fuzzy graph G_{λ_0}. Here, by calculating the distance function $d(\lambda)$ and the connectivity function $e(\lambda)$.

Then, we have normalized the value of $d(\lambda)$ and $e(\lambda)$, so we have calculated $f_d(\lambda)$ and $f_e(\lambda)$.

By applying the fuzzy decision, we could obtain the optimal value $\lambda_0 = 0.309$.

We would illustrate the fuzzy sociogram U^z, $z =c_0=0.58$ in Fig.2 which is based on the fuzzy graph $G_{\lambda_0} = (V, F_{\lambda_0})$, $\lambda_0 = 0.309$. In this graph, we indicated the students in

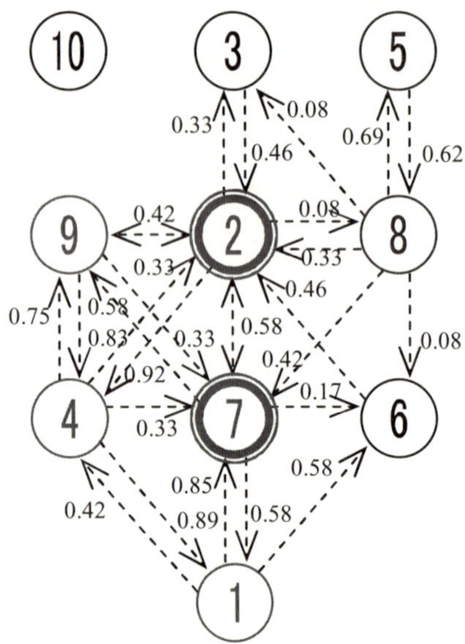

Fig. 2. Fuzzy Sociogram U^z, $\lambda_0 = 0.309$, $z = 0.58$

the same cluster at the optimal cut level $z = c_0 = 0.58$ with the same color and the opinion leaders are boldly marked.

This fuzzy sociogram U^z effectively represents the following results:

(1) The opinion leaders are 2 and 7 because these nodes is connected with many other nodes.
(2) The relatively isolated member is 10 because this node has not been connected with the others.
(3) The subgroups are {5, 8} and {1,2,4,7, 9}.

In this way, we could effectively analyze and clarify the interpersonal relation among members by applying the fuzzy node fuzzy graph and our analysis method.

References

1. Zadeh, L.: Fuzzy Sets. Information and Control VIII 353 (1965)
2. Rosenfeld, A.: Fuzzy Graphs. In: Zadeh, L. (ed.) Fuzzy Sets and Their Applications to Cognitive and Decision Processes, pp. 77–95. Academic Press, Inc., London (1974)
3. Kaufmann, A.: Introduction to the Theory of Fuzzy Subsets. Academic Press, London (1975)
4. Negoita, C.: Application of Fuzzy Sets to System Analysis. Birkhauser Verlag, Basel (1975)
5. Moreno, J.: The Sociometry Readers. Free Press (1960)

6. Nishida, T., Takeda, E.: Fuzzy Sets and its Applications. Morikita Shuppan (1978) (Japanese)
7. Romesburg, H.: Cluster Analysis for Researchers. Lifetime Learning Publications (1984)
8. Aumonn, R.: Lecture on Game Theory. Westview Press (1989)
9. Uesu, H.: Sociometry Analysis Applying Fuzzy Node Fuzzy Graph. Journal of Japan Society for Fuzzy Theory and Systems 14(3) (2002) (Japanese)
10. Uesu, H., Yamashita, H.: Connectivity Properties of T-Norm Families and its Application. In: Int'l Conference on Computer, Communication and Control Technologies (2003)
11. Uesu, H., Tsuda, E.: Clustering Level Analysis Applying Fuzzy Theory and its Application. In: Conference of Japan Society for Educational Technology (2003) (Japanese)
12. Uesu, H.: Fuzzy Node Fuzzy Graph, T-Norm and its Application. Journal of Japan Society for Fuzzy Theory and Intelligent Informatics 16(1), 188–195 (2004) (Japanese)
13. Uesu, H.: Level analysis of Fuzzy Partition Tree and its Application. In: IEEE International Workshop on Soft Computing Applications (2005)
14. Uesu, H.: Optimal Fuzzy Graph Based on Fuzzy Node Fuzzy Graph Analysis. In: The 3rd International Conference on Soft Computing and Intelligent Systems (2006)
15. Analysis of Fuzzy Node Fuzzy Graph and its Application. Int'l Conference on Soft Computing and Human Science (2007)

Consumer Behaviors in Taiwan Online Shopping — Case Study of A Company

Lily Lin[1], Huey-Ming Lee[2], and Li-Hsueh Lin[2]

[1] Department of International Business
China University of Technology
56, Sec. 3, Hsing-Lung Road Taipei (116), Taiwan
lily@cute.edu.tw
[2] Department of Information Management
Chinese Culture University
55, Hwa-Kung Road, Yang-Ming-San, Taipei (11114), Taiwan
hmlee@faculty.pccu.edu.tw, linlicherg@gmail.com

Abstract. This paper focuses on the customer behaviors of Taiwan "Internet Commerce/e-shopping". The member consumption records of A company in 2006 were employed and cross-analyzed to find out the distribution of member variables. We applied RFM model to cluster customers and understand the contribution of each group, and used the decision tree technique, CHAID (Chi-squared Automatic Interaction Detection) to explore member variables and product classifications groups.

Keywords: Online Shopping, Consumer Behaviors, RFM Analysis, CHAID Analysis.

1   Introduction

In the previous study [4], a survey was conducted to compare consumers' characteristics, motivations and differentiated between the stages of the buying process and behavior among those three non-store channels. In this study, we aim to investigate consumer behaviors in Taiwan online shopping and customer development strategies with a focus on only Internet channel using the transaction records of A company. The member consumption records of A company in 2006 were employed and cross-analyzed to find out the distribution of member variables. We applied RFM model (Recency, Frequency, and Monetary value) to cluster customers and understand the contribution of each group, and used the decision tree technique, CHAID (Chi-squared Automatic Interaction Detection) to explore member variables and product classifications groups.

In this study, consumer behaviors in Taiwan Internet Commerce/E-Shopping were analyzed. It was expected that the research results could be beneficial to the understanding of the association between customer groups and product portfolios and make substantial contribution to enterprises' planning of marketing and customer development strategies.

I. Lovrek, R.J. Howlett, and L.C. Jain (Eds.): KES 2008, Part III, LNAI 5179, pp. 92–97, 2008.

2 Cross Analyses of Data

The member consumption records of A company in 2006 were employed and cross-analyzed to find out the relation among variables such as the gender, age, amount of order, region and product category of online shopping group as follows.

2.1 Cross Analyses of Data of Members of Online Shopping Group

The cross analyses and statistics of gender, age and region of online shopping group of A company was described as follows.

(1) Female, ages 30 -39 and living in the Northern Taiwan had highest percentage (17.89%), followed by female, ages 20-29 and live in the Northern Taiwan (17.6%).

(2) For the percentage of male members, most of them are 30 to 39 year old, living in Northern Taiwan (9.66%), followed by those ages 20-29, living in the Northern Taiwan (6.50%).

(3) In the subjects who were over 60 years old and lived in surrounding islands, only two were online shoppers; the subjects, who were over 80 years old, living in the Southern, Eastern and surrounding islands, had no experiences in online shopping. But 0.02% of the subjects who are over 80 years old and living in Central Taiwan had online shopping experiences, the highest percentage in the group of over 80 years old.

Furthering to analyze the distributions in the counties and cities in the Northern Taiwan, we found that female, ages 20-29 and living in Taipei County had highest number of times (5185 times), followed by female, ages 30-39 and living in Taipei County (5135 times) and the third place was female, ages 30-39 and living in Taipei City (4212 times).

2.2 The Relationship between Gender, Age and Amount of Order in Online Shopping Group

The cross analyses and statistics of gender, age and amount of order of online shopping group of A company was described as follows.

(1) The amount of order NT$1000-2999 had the highest percentage spent by female, ages 30-39 had the highest percentage (16.75%), followed by NT$1000-2999 spent by female, ages 20-29 (16.46%).

(2) For the percentage of male, the amount of order NT$1000-2999 spent by whom were 30-39 years old (7.30%), followed by NT$1000-2000 spent by whom were ages 20-29 (4.93%).

(3) In the group with amount of over NT$15,000, male had higher percentage than female at the same age level. For example, male, ages 30-39 (0.34%) was 1% higher than female (0.24%), ages 30-39 in the amount NT$20,000-29999 that means male had more contributions to the amount of order.

Table 1. The relationship between gender, age and product category among online shopping group

Variables	Gender	Age	Product Category
Gender	❶female (68.02) ❷male (31.98)	❶ age: 30~39/female (28.85) ❷ age: 20~29/female (28.82) ❸ age: 30~39/male (15.62)	❶ personal fashion products/female (32.63) ❷ products for daily use/female (13.86) ❸ leisure and health products /female (13.59)
Age	❶ age: 30~39/female (28.85) ❷ age: 20~29/female (28.82) ❸ age: 30~39/male (15.62)	❶ age: 30~39(44.47) ❷ age: 20~29 (39.48) ❸ age: 40~49 (13.00)	❶ personal fashion products/age:30~39 (18.97) ❷ personal fashion products/age:20~29 (17.97) ❸ products for daily use / age: 30~39 (9.85)
Product Category	❶ personal fashion products/female (32.63) ❷ products for daily use/female (13.86) ❸ leisure and health products /female (13.59)	❶ personal fashion products/age:30~39 (18.97) ❷ personal fashion products/age:20~29 (17.97) ❸ products for daily use/age:30~39 (9.85)	❶ personal fashion products (43.08) ❷ products for daily use (21.27) ❸ leisure and health products (20.31)

the highest percentage three groups by cross analyses among 3 variables：
❶female／age: 20~29／personal fashion products (14.15)
❷female／age: 30~39／personal fashion products (14.03)
❸female／age: 30~39／products for daily use (6.14)

P-value=0.000, df=30, statistically significant

Note:
1. represent the highest, the second highest and the third highest respectively
2. In parentheses ( ) is percentage   3. Source: compiled by authors

2.3 The Relationship between Gender, Age and Product Category among Online Shopping Group

The cross analyses and statistics of gender, age and product category of online shopping group of A company was described as follows.

(1) In product category, "personal fashion products" had the highest percentage, followed by "products for daily use" and "leisure and health products" shown in Table 1.

(2) In gender and product category, "personal fashion products" bought by female had the highest percentage, followed by "articles for daily use". "personal fashion product" had the highest percentage in the product categories bought by male shown in Table 1.

3 The RFM Analysis of Online Shopping Group

Total number of customers of online shopping group was 39611. We selected the consumption data of the first 20,000 customers to conduct RFM analyses. With January

Table 2. The RFM reports of 12 equal segments of online shopping group of A company

Ranking	Sample numbers	Latest consumption within last one month (months)	Group's total consumption amount (NT$)	Consumption percentage (%)	Accumulated consumption percentage (%)	Group's total consumption times (times)	Average consumption times (times)	Average unit price of 'time (NT$)	Average unit price of 'each customer' (NT$)
1	4329	1	62303303	25.54%	25.54%	18686	4.32	3334	14392
2	2573	2	31428598	12.88%	38.42%	5408	2.10	5812	12215
3	2085	3	26183361	10.73%	49.15%	3868	1.86	6769	12558
4	1828	4	20620469	8.45%	57.61%	3273	1.79	6300	11280
5	1692	5	19006253	7.79%	65.40%	2921	1.73	6507	11233
6	1573	6	18217431	7.47%	72.87%	2743	1.74	6641	11581
7	1527	7	17408679	7.14%	80.00%	2448	1.60	7111	11401
8	1217	8	13089831	5.37%	85.37%	1978	1.63	6618	10756
9	1527	9	9674944	3.97%	89.33%	1278	1.51	7570	11477
10	1217	10	11091852	4.55%	93.88%	1365	1.42	8126	11518
11	1806	11	8427080	3.45%	97.34%	1014	1.27	8311	10573
12	1370	12	6501805	2.67%	100.00%	731	1.28	8894	11347
Total	20000		243953606	100.00%	100.00%	45713			

Note:

1. "Latest consumption within last one month" means the consumption month was December 2006; "latest consumption within last two month" means the consumption month was November 2006, and so forth.

2. "Group's total consumption amount" means total accumulated consumption amount of consumers in each equal segment in 2006.

3. "Consumption percentage" = total consumption amount of consumers in each equal segment/group's total consumption amount

4. "Group's total consumption times" means accumulated order times of consumers in each equal segment in 2006.

5. "Average consumption times" = total consumption times of each equal segment / number of customers in each equal segment

6. "Average unit price of 'time'" = total consumption amount of each equal segment / total consumption times of each equal segment.

7. "Average unit price of 'each customer'" / number of customers in each equal segment

2007 as base month, we also conducted the arrangement of rankings of "latest consumption (R)" and produced reports of 12 equal segments. The results of analyses were as follows:

(1) The percentage of the consumption amount by consumers within last one month and at least two times more than the amount of the next lower rank was more than 25% of total consumption amount in the analysis data. The average consumption times in 2006 was 4.32, average NT$14,392 each customer and average NT$3,334 each time shown in Table 2.

(2) 30% of customers contributed 50% of total consumption amount in the data m while 70% of customers contributed 80% of total consumption amount shown in Table 2.

(3) The average consumption times of customers in the 11th equal segments was 1.27 times, average NT$10573 each customer and average NT$8,311 each time shown in Table 2.

4   CHAID Analysis – Group Background Decision Tree of Product Category in Online Shopping

In the samples of the research, with the categories of product bought by online shopping groups as the criterion variables and the gender, age in the data of online shopping members as predictor variables, we tried to understand the interaction between information and products bought by online shopping group.

Most of online shopping members were female of ages 30-39 and the percentage of buying "gift and present" was 28.9%.

The percentage of female of ages20-29, and over 80 years old buying "women underwear" was 28.8%.

Most of male online shopping members were at their ages 30-39. The percentage of buying "gift and present" was 15.6%. Analyzing through Odds, we found that the percentage of buying "general food" among male of ages 50-59 was 11.7%, 2.02 times of the percentage of all the subjects in online shopping group who bought "general food".

5   Conclusions

Young, preferring low price products and the trading time section focusing on 09:00-1159 and 21:00-2359 are the characteristics of online shopping group. According the report of RFM 12 equal segments, the number of customer and consumption amount had significant growth. That means A Company's corporate image had won the confidence of customers who became loyal customers. But the average unit price of "time" was in inverse ratio to the ranking of "latest consumption (R)" that indicated the probability for customers who purchased high unit price products to repurchase within one year is not high. 30% of customers in online shopping group contributed 50% of total consumption amount. With the classification of product category in the research as criterion variables and gender, age of the information of customers as predictor variables, we used CHAID algorism of decision tree to group that we could define target groups for product categories and lock in consumption model and target market. "Gift and present" had the highest percentage in the categories of products bought by online shopping groups.

References

1. Boxersox, D.J., Cooper, M.B.: Strategic Marketing Channel Management. Mcgraw-Hill International Edition (1992)
2. Engel, J.F., Blackwell, R.D., Kollat, D.T.: Consumer Behavior, 7th edn. Fort Worth Dryden Press (1993)
3. Huang, W.-C.: A Study on Consumer Behavior of Travel Products by Multilevel Marketing. Graduate Thesis, Nanhua University (2003)
4. Lin, L., Lee, H.-M., Lin, L.-H.: Comparison of Consumer Behaviors in Taiwan Non-Store Channels. In: Apolloni, B., Howlett, R.J., Jain, L. (eds.) KES 2007, Part III. LNCS (LNAI), vol. 4694, pp. 363–369. Springer, Heidelberg (2007)
5. Narayana, C.L., Markin, R.J.: Consumer Behavior and Product Performance : An Alternative Conceptualization. Journal of Marketing 39, 1–6 (1975)
6. Tsai, Y.-F.: The Influence of Purchase Involvement, Purchase Motivation and Website Environment Characteristics on Internet Purchase Intention for Fresh Vegetables. Graduate Thesis, National Chung Hsing University (2001)
7. Wu, M.-C.: A Study for the Purchasing Intention of Consumer in PDA. Graduate Thesis, National Taipei University (2001)

A New Fuzzy Risk Assessment Approach

Huey-Ming Lee[1] and Lily Lin[2]

[1] Department of Information Management, Chinese Culture University
55, Hwa-Kung Road, Yang-Ming-San, Taipei (11114), Taiwan
[2] Department of International Business, China University of Technology
56, Sec. 3, Hsing-Lung Road, Taipei (116), Taiwan
`hmlee@faculty.pccu.edu.tw, lily@cute.edu.tw`

Abstract. In this paper, we present computational rule inferences to tackle the rate of aggregative risk in fuzzy circumstances. Because the proposed assessment method directly uses the fuzzy numbers rather than the linguistic values to evaluate, it can be executed faster than before. The proposed fuzzy assessment method is easier, closer to evaluator real thinking and more useful than the ones they have presented before.

Keywords: Risk assessment; fuzzy risk assessment.

1   Introduction

Generally, risk is the traditional manner of expressing uncertainty in the systems life cycle. Risk assessment is a common first step and also the most important step in a risk management process. Risk assessment is the determination of quantitative or qualitative value of risk related to a concrete situation and a recognized threat. In a quantitative sense, it is the probability at such a given point in a system's life cycle that predicted goals can not be achieved with the available resources. Due to the complexity of risk factors and the compounding uncertainty associated with future sources of risk, risk is normally not treated with mathematical rigor during the early life cycle phases [1]. Risks result in project problems such as schedule and cost overrun, so risk minimization is a very important project management activity [11]. Up to now, there are many papers investigating risk identification, risk analysis, risk priority, and risk management planning [1-4, 6-7].

In evaluating the rate of risk factors, most decision-makers or project-managers, in fact, viewed those factors as linguistic values (terms), e.g., very high, high, middle, low, very low and etc. After fuzzy sets theory was introduced by Zadeh [12] to deal with problem in which vagueness is present, linguistic value can be used for approximate reasoning within the framework of fuzzy sets theory [13] to effectively handle the ambiguity involved in the data evaluation and the vague property of linguistic expression, and normal triangular fuzzy numbers are used to characterize the fuzzy values of quantitative data and linguistic terms used in approximate reasoning. Based on [2-4, 6-7], Lee [9] classified the risk factors into six attributes, divided each attribute into some risk items, and built up the hierarchical structured model of aggregative risk and the evaluating procedure of structured model, ranged the grade of risk for each risk item into eleven ranks, and proposed the procedure to evaluate the rate of

I. Lovrek, R.J. Howlett, and L.C. Jain (Eds.): KES 2008, Part III, LNAI 5179, pp. 98–105, 2008.

aggregative risk using two stages fuzzy assessment method. Chen [5] ranged the grade of risk for each risk item into thirteen ranks, proposed the other arithmetic operations instead of the two stages fuzzy assessment method, and defuzzified the trapezoid or triangular fuzzy numbers by the median.

In [5, 9], they used eleven or thirteen linguistic values for ranking the grades of risk to each risk item, where the linguistic values were represented by the triangular fuzzy numbers. But, it is very complicated to compute. Also, the evaluator only chooses one grade from grades of risk for each risk item. It has difficulty in reflecting the evaluator's incomplete and uncertain thought. Therefore, if we can use fuzzy sense of assessment to express the degree of evaluator's feelings based on his/her own concepts, the results will be closer to the evaluator's real thought. Therefore, Lin and Lee [10] proposed a new fuzzy assessment method to tackle the rate of aggregative risk in fuzzy circumstances. This method directly uses the fuzzy numbers rather than the linguistic values to evaluate, it can be easier, and meet the evaluator real thinking. Based on Lin and Lee [10], we apply computational rule inference to evaluate the aggregative risk in this study. The proposed method is easier than they presented before.

2 The Proposed Fuzzy Risk Assessment Method

We present the fuzzy assessment method as follows;

Step 1: Assessment form for the risk items:

The criteria ratings of risk are linguistic variables with linguistic values V_1, V_2, ..., V_7, where V_1 = extra low, V_2 = very low, V_3 = low, V_4 = middle, V_5 = high, V_6 = very high, V_7 = extra high. These linguistic values are treated as fuzzy numbers with triangular membership functions as follows:

$$\tilde{V}_1 = (0, 0, \frac{1}{6}),$$

$$\tilde{V}_k = (\frac{k-2}{6}, \frac{k-1}{6}, \frac{k}{6}), \ for \ k = 2, 3,..., 6 \tag{1}$$

$$\tilde{V}_7 = (\frac{5}{6}, 1, 1)$$

In previous studies [5, 9], the evaluator only chooses one grade from grade of risk for each risk item, it ignores the evaluator's incomplete and uncertain thinking. Therefore, if we use fuzzy numbers of assessment in fuzzy sense to express the degree of evaluator's feelings based on his own concepts, the computing results will be closer to the evaluator's real thought.

The assessment for each risk item with fuzzy number can reduce the degree of subjectivity of the evaluator, express the degree of evaluator's feelings based on his own concepts. The results will be closer to the evaluator's real thought. Based on the structured model of aggregative risk proposed by Lin and Lee [10] and evaluating form of

Table 1. Contents of the hierarchical structure model [10]

Attribute	Risk item	Weight-2	Weight-1	Linguistic variables						
				V_1	V_2	V_3	V_4	V_5	V_6	V_7
X_1: Personal		$W_2(1)$								
	X_{11}: Personal shortfalls, key person(s) quit		$W_1(1,1)$	$m_{11}^{(1)}$	$m_{11}^{(2)}$	$m_{11}^{(3)}$	$m_{11}^{(4)}$	$m_{11}^{(5)}$	$m_{11}^{(6)}$	$m_{11}^{(7)}$
X_2: System requirement		$W_2(2)$								
	X_{21}: Requirement ambiguity		$W_1(2,1)$	$m_{21}^{(1)}$	$m_{21}^{(2)}$	$m_{21}^{(3)}$	$m_{21}^{(4)}$	$m_{21}^{(5)}$	$m_{21}^{(6)}$	$m_{21}^{(7)}$
	X_{22}: Developing the wrong software function		$W_1(2,2)$	$m_{22}^{(1)}$	$m_{22}^{(2)}$	$m_{22}^{(3)}$	$m_{22}^{(4)}$	$m_{22}^{(5)}$	$m_{22}^{(6)}$	$m_{22}^{(7)}$
	X_{23}: Developing the wrong user interface		$W_1(2,3)$	$m_{23}^{(1)}$	$m_{23}^{(2)}$	$m_{23}^{(3)}$	$m_{23}^{(4)}$	$m_{23}^{(5)}$	$m_{23}^{(6)}$	$m_{23}^{(7)}$
	X_{24}: Continuing stream requirement changes		$W_1(2,4)$	$m_{24}^{(1)}$	$m_{24}^{(2)}$	$m_{24}^{(3)}$	$m_{24}^{(4)}$	$m_{24}^{(5)}$	$m_{24}^{(6)}$	$m_{24}^{(7)}$
X_3: Schedules and budgets		$W_2(3)$								
	X_{31}: Schedule not accurate		$W_1(3,1)$	$m_{31}^{(1)}$	$m_{31}^{(2)}$	$m_{31}^{(3)}$	$m_{31}^{(4)}$	$m_{31}^{(5)}$	$m_{31}^{(6)}$	$m_{31}^{(7)}$
	X_{32}: Budget not sufficient		$W_1(3,2)$	$m_{32}^{(1)}$	$m_{32}^{(2)}$	$m_{32}^{(3)}$	$m_{32}^{(4)}$	$m_{32}^{(5)}$	$m_{32}^{(6)}$	$m_{32}^{(7)}$
X_4: Developing technology		$W_2(4)$								
	X_{41}: Gold-plating		$W_1(4,1)$	$m_{41}^{(1)}$	$m_{41}^{(2)}$	$m_{41}^{(3)}$	$m_{41}^{(4)}$	$m_{41}^{(5)}$	$m_{41}^{(6)}$	$m_{41}^{(7)}$
	X_{42}: Skill levels inadequate		$W_1(4,2)$	$m_{42}^{(1)}$	$m_{42}^{(2)}$	$m_{42}^{(3)}$	$m_{42}^{(4)}$	$m_{42}^{(5)}$	$m_{42}^{(6)}$	$m_{42}^{(7)}$
	X_{43}: Straining hardware		$W_1(4,3)$	$m_{43}^{(1)}$	$m_{43}^{(2)}$	$m_{43}^{(3)}$	$m_{43}^{(4)}$	$m_{43}^{(5)}$	$m_{43}^{(6)}$	$m_{43}^{(7)}$
	X_{44}: Straining software		$W_1(4,4)$	$m_{44}^{(1)}$	$m_{44}^{(2)}$	$m_{44}^{(3)}$	$m_{44}^{(4)}$	$m_{44}^{(5)}$	$m_{44}^{(6)}$	$m_{44}^{(7)}$
X_5: External resource		$W_2(5)$								
	X_{51}: Shortfalls in externally furnished components		$W_1(5,1)$	$m_{51}^{(1)}$	$m_{51}^{(2)}$	$m_{51}^{(3)}$	$m_{51}^{(4)}$	$m_{51}^{(5)}$	$m_{51}^{(6)}$	$m_{51}^{(7)}$
	X_{52}: Shortfalls in externally performed tasks		$W_1(5,2)$	$m_{52}^{(1)}$	$m_{52}^{(2)}$	$m_{52}^{(3)}$	$m_{52}^{(4)}$	$m_{52}^{(5)}$	$m_{52}^{(6)}$	$m_{52}^{(7)}$
X_6.: Performance		$W_2(6)$								
	X_{61}: Real-time performance shortfalls		$W_1(6,1)$	$m_{61}^{(1)}$	$m_{61}^{(2)}$	$m_{61}^{(3)}$	$m_{61}^{(4)}$	$m_{61}^{(5)}$	$m_{61}^{(6)}$	$m_{61}^{(7)}$

structured model proposed by Lee [10], we propose the assessment form of the structured model as shown in Table 1 and propose a new assessment method using computational rule inference to tackle the rate of aggregative risk in software development.

In Table 1,

$$\sum_{i=1}^{6} W_2(i) = 1, \quad 0 \le W_2(i) \le 1 \tag{2}$$

for each $i = 1, 2, \ldots 6$.

$$\sum_{i=1}^{n_k} W_1(k,i) = 1, \ 0 \le W_1(k,i) \le 1 \tag{3}$$

for $k=1, 2,\ldots, 6$; $n_1 = 1, n_2 = 4, n_3 = 2, n_4 = 4, n_5 = 2, n_6 = 1.$; $i = 1, 2,\ldots, n_k$.

$$\sum_{l=1}^{7} m_{ki}^{(l)} = 1, 0 \le m_{ki}^{(l)} \le 1 \tag{4}$$

for l=1, 2, ..., 7; k=1, 2, ..., 6; i=1, 2, ..., n_k.

From Table 1, we directly use the fuzzy numbers ($m_{ki}^{(l)}$) rather than the linguistic values to evaluate. Also, we may express the risk item X_{ki} as fuzzy discrete type

$$X_{ki} = \frac{m_{ki}^{(1)}}{V_1} \oplus \frac{m_{ki}^{(2)}}{V_2} \oplus \frac{m_{ki}^{(3)}}{V_3} \oplus \frac{m_{ki}^{(4)}}{V_4} \oplus \frac{m_{ki}^{(5)}}{V_5} \oplus \frac{m_{ki}^{(6)}}{V_6} \oplus \frac{m_{ki}^{(7)}}{V_7} \tag{5}$$

Step 2: Weighted triangular fuzzy numbers

For easy to express, we take some one attribute, saying X_j, and the items, saying X_{j1}, X_{j2}, ..., X_{jn_j} in Table 1, and introduce the weighted triangular fuzzy numbers as shown in Table 2, for j=1, 2, ..., 6, and $n_1 = 1, n_2 = 4, n_3 = 2, n_4 = 4, n_5 = 2, n_6 = 1$.

Let $B = \{\tilde{V}_1, \tilde{V}_2, \ldots, \tilde{V}_7\}$. From Table 2, we can form the fuzzy relation on X_j and B with weighted triangular fuzzy number elements as follows:

Table 2. Contents of the weighted triangular fuzzy number for item X_{jk}

Attribute	Risk item	Linguistic variables						
		V_1	V_2	V_3	V_4	V_5	V_6	V_7
X_j								
	X_{j1}	$m_{j1}^{(1)}$	$m_{j1}^{(2)}$	$m_{j1}^{(3)}$	$m_{j1}^{(4)}$	$m_{j1}^{(5)}$	$m_{j1}^{(6)}$	$m_{j1}^{(7)}$
Weighted triangular fuzzy number		$m_{j1}^{(1)}\tilde{V}_1$	$m_{j1}^{(2)}\tilde{V}_2$	$m_{j1}^{(3)}\tilde{V}_3$	$m_{j1}^{(4)}\tilde{V}_4$	$m_{j1}^{(5)}\tilde{V}_5$	$m_{j1}^{(6)}\tilde{V}_6$	$m_{j1}^{(7)}\tilde{V}_7$
	X_{j2}	$m_{j2}^{(1)}$	$m_{j2}^{(2)}$	$m_{j2}^{(3)}$	$m_{j2}^{(4)}$	$m_{j2}^{(5)}$	$m_{j2}^{(6)}$	$m_{j2}^{(7)}$
Weighted triangular fuzzy number		$m_{j2}^{(1)}\tilde{V}_1$	$m_{j2}^{(2)}\tilde{V}_2$	$m_{j2}^{(3)}\tilde{V}_3$	$m_{j2}^{(4)}\tilde{V}_4$	$m_{j2}^{(5)}\tilde{V}_5$	$m_{j2}^{(6)}\tilde{V}_6$	$m_{j2}^{(7)}\tilde{V}_7$
	$\cdot$	$\cdot$	$\cdot$	$\cdot$	$\cdot$	$\cdot$	$\cdot$	$\cdot$
	$\cdot$	$\cdot$	$\cdot$	$\cdot$	$\cdot$	$\cdot$	$\cdot$	$\cdot$
	X_{jm_j}	$m_{jm_j}^{(1)}$	$m_{jm_j}^{(2)}$	$m_{jm_j}^{(3)}$	$m_{jm_j}^{(4)}$	$m_{jm_j}^{(5)}$	$m_{jm_j}^{(6)}$	$m_{jm_j}^{(7)}$
Weighted triangular fuzzy number		$m_{jm_j}^{(1)}\tilde{V}_1$	$m_{jm_j}^{(2)}\tilde{V}_2$	$m_{jm_j}^{(3)}\tilde{V}_3$	$m_{jm_j}^{(4)}\tilde{V}_4$	$m_{jm_j}^{(5)}\tilde{V}_5$	$m_{jm_j}^{(6)}\tilde{V}_6$	$m_{jm_j}^{(7)}\tilde{V}_7$

102   H.-M. Lee and L. Lin

$$\tilde{R}_j = \begin{bmatrix} m_{j1}^{(1)}\tilde{V}_1 & m_{j1}^{(2)}\tilde{V}_2 & \cdots & m_{j1}^{(7)}\tilde{V}_7 \\ m_{j2}^{(1)}\tilde{V}_1 & m_{j2}^{(2)}\tilde{V}_2 & \cdots & m_{j2}^{(7)}\tilde{V}_7 \\ \cdot \\ \cdot \\ \cdot \\ m_{jn_j}^{(1)}\tilde{V}_1 & m_{jn_j}^{(2)}\tilde{V}_2 & \cdots & m_{jn_j}^{(7)}\tilde{V}_7 \end{bmatrix} \qquad (6)$$

Step 3: The first stage computational rule of inference

We let

$$(\tilde{T}_{j1}, \tilde{T}_{j2}, \ldots, \tilde{T}_{j7}) = (w_1(j,1), w_1(j,2), \ldots, w_1(j,n_j)) \bullet \tilde{R}_j \qquad (7)$$

where

$$\tilde{T}_{jq} = w_1(j,1)m_{j1}^{(q)}\tilde{V}_q \oplus w_1(j,2)m_{j2}^{(q)}\tilde{V}_q \oplus \ldots \oplus w_1(j,n_j)m_{jn_j}^{(q)}\tilde{V}_q \qquad (8)$$

for $j=1, 2, \ldots, 6$; $q=1, 2, \ldots, 7$. We have that $\tilde{T}_{jq}$ is a triangular fuzzy number.

Step 4: The second stage computational rule of inference

We let

$$(\tilde{S}_1, \tilde{S}_2, \ldots, \tilde{S}_7)$$

$$= (w_2(1), w_2(2), \ldots, w_2(6)) \bullet \begin{bmatrix} \tilde{T}_{11} & \tilde{T}_{12} & \ldots & \tilde{T}_{17} \\ \tilde{T}_{21} & \tilde{T}_{22} & \ldots & \tilde{T}_{27} \\ \cdot \\ \cdot \\ \cdot \\ \tilde{T}_{61} & \tilde{T}_{62} & \ldots & \tilde{T}_{67} \end{bmatrix} \qquad (9)$$

where

$$\tilde{S}_q = w_2(1)\tilde{T}_{1q} \oplus w_2(2)\tilde{T}_{2q} \oplus \ldots \oplus w_2(6)\tilde{T}_{6q} \qquad (10)$$

for $q=1, 2, \ldots, 7$.

Then we have the following Proposition 1.

Proposition 1 Let S_{jq} be the centroid of $\tilde{T}_{jq}$, then, we have

(1) for the attribute X_j, and rating risk V_q, the rate of risk is S_{jq}

(2) the rate of risk for the attribute X_j is $Q^{(j)} = \sum_{q=1}^{7} S_{jq}$.

(3) let T_q be the centroid of $\tilde{S}_q$, then, the aggregative rate of risk is

$$Rate\_Aggregative\_Risk = \sum_{j=1}^{6} w_2(j)Q^{(j)}$$

4 Numerical Example

Example: Assume that we have the following attributes, weights, grade of risk for each risk item as shown in Table 3 [10].

Table 3. Contents of the example [10]

Attribute	Risk item	Weight-2	Weight-1	Linguistic variables						
				V1	V2	V3	V4	V5	V6	V7
X_1		0.3								
	X_{11}		1	0	0.17	0.83	0	0	0	0
X_2		0.3								
	X_{21}		0.4	0	0.53	0.47	0	0	0	0
	X_{22}		0.4	0	0.89	0.11	0	0	0	0
	X_{23}		0.1	0.25	0.75	0	0	0	0	0
	X_{24}		0.1	0.61	0.39	0	0	0	0	0
X_3		0.1								
	X_{31}		0.5	0	0.17	0.83	0	0	0	0
	X_{32}		0.5	0	0.53	0.47	0	0	0	0
X_4		0.1								
	X_{41}		0.3	0	0.89	0.11	0	0	0	0
	X_{42}		0.1	0	0.17	0.83	0	0	0	0
	X_{43}		0.3	0	0.17	0.83	0	0	0	0
	X_{44}		0.3	0	0.53	0.47	0	0	0	0
X_5		0.1								
	X_{51}		0.5	0	0	0.81	0.19	0	0	0
	X_{52}		0.5	0	0	0.81	0.19	0	0	0
X_6		0.1								
	X_{61}		1	0	0.17	0.83	0	0	0	0

(1) By the Proposition 1 shown in Section 2, we have

$Q^{(1)} = 0.304729$ is the rate of risk of the attribute X_1;

$Q^{(2)} = 0.195727$ is the rate of risk of the attribute X_2

$Q^{(3)} = 0.274795$ is the rate of risk of the attribute X_3;

$Q^{(4)} = 0.250848$ is the rate of risk of the attribute X_4

$Q^{(5)} = 0.36473$ is the rate of risk of the attribute X_5;

$Q^{(6)} = 0.304729$ is the rate of risk of the attribute X_6

Rate_Aggregative_Risk =0.269647 is the rate of aggregative risk.

(2) Comparison with Lin and Lee [10]

a) In [10], the rate of aggregative risk is 0.26983. By the proposed method in this study, the computed result is 0.269647. The relative error is (0.269647-0.26983)/0.26983= -0.00068. It is very small. But, the proposed method is easier than in [10].

b) We can tackle the risk rate of each attribute by the proposed method in this study.

5 Conclusion

In general survey forces evaluator to assess one grade from the grade of risk to each risk item, but it ignores the uncertainty of human thought. For instance, when the evaluator need to choose the assessment from the survey which lists eleven choices including "definitely unimportant", "extra unimportant", "very unimportant", "unimportant", "slightly unimportant", "middle", "slightly important", "important", "very important", "extra important", and "definitely important", the general survey becomes quiet exclusive. The assessment of evaluation with fuzzy numbers can reduce the degree of subjectivity of the evaluator. In this paper, we propose a new assessment method to evaluate the rate of aggregative risk. Because the proposed method directly uses the fuzzy numbers rather than the linguistic values to evaluate, it can be executed much fast. Therefore, the evaluator can assess the risk grade by fuzzy numbers to each risk item, which making evaluation process is also easier than the ones presented before [6, 10-11].

Acknowledgments. The authors would like to express their gratitude to Professor Jing-Shing Yao, an Emeritus Professor at the National Taiwan University, for his helpful suggestions.

References

1. AFSC: Software Risk Abatement, U. S. Air Force Systems Command, AFSC/AFLC pamphlet 800-45, Andrews AFB, MD, September, pp. 1–28 (1988)
2. Boehm, B.W.: Software Risk Management. CS Press, Los Alamitos (1989)
3. Boehm, B.W.: Software Risk Management: Principles and Practices. IEEE Software 8, 32–41 (1991)
4. Charette, R.N.: Software Engineering Risk Analysis and Management. Mc Graw-Hill, New York (1989)
5. Chen, S.M.: Evaluating the Rate of Aggregative Risk in Software Development Using Fuzzy Set Theory. Cybernetics and Systems: International Journal 30, 57–75 (1999)

6. Conger, S.A.: The New Software Engineering. Wadsworth Publishing Co., Belmont (1994)
7. Gilb, T.: Principles of Software Engineering Management. Addison-Wesley Publishing Co., New York (1988)
8. Kaufmann, A., Gupta, M.M.: Introduction to Fuzzy Arithmetic Theory and Applications Van Nostrand Reinhold, New York (1991)
9. Lee, H.-M.: Applying fuzzy set theory to evaluate the rate of aggregative risk in software development. Fuzzy sets and Systems 79, 323–336 (1996)
10. Lin, L., Lee, H.-M.: A Fuzzy Assessment Model for Evaluating the Rate of Aggregative Risk in Software Development. International Journal of Computers 1(3), 43–47 (2007)
11. Sommerville, I.: Software Engineering, 6th edn. Pearson Education Limited, England (2001)
12. Zadeh, L.A.: Fuzzy Sets. Information and Control 8, 338–353 (1965)
13. Zadeh, L.A.: The Concept of a Linguistic Variable and it's Application to Approximate Reasoning. Information Sciences 8, 199–249 (I) (1975); 9, pp. 301–357 (II) (1976); pp. 43–58 (III)
14. Zimmermann, H.-J.: Fuzzy Set Theory and Its Applications, Second Revised Edition. Kluwer Academic Publishers, Dordrecht (1991)

Making Scale with Units by Using Order Statistics from the Normal Distribution in AHP

Taki Kanda

Bunri University of Hospitality, Department of Service Management,
311-1 Kashiwabarashinden Shinogawara, Sayama,
Saitama 350-1336, Japan

Abstract. AHP (Analytic Hierarchy Process) was created as a method to help us make decision by Tomas Saaty in 1970s and has been widely adopted in various fields such as economic problems, management problems, economic problems, energy problems or city planning etc. in many areas – U.S.A., Europe and other areas. In AHP paired comparison is used and the values to answers in paired comparison are ratio-based scale. It is generally said that AHP has a problem such that the values for evaluation do not have units. In order to let the values for evaluation have units, order statistics from the normal distribution are used for the values to answers in paired comparison. By using order statistics for the values to answers in paired comparison, the values for evaluation are obtained, the units of which are the standard deviations of the standard normal distribution($\sigma = 1$) on the basis of 0. Here taking a problem of selecting meals based upon human feeling, AHP is applied so as to obtain the values for evaluation of meal products which make us possible to easily compare our meal preference.

1 Introduction

Recently people tend to attach importance to environment or human lives and the need to develop goods perceiving consumer's feelings is increasing, and the studies to deal with human feelings such as sensory test or *Kansei* Engineering (*Kansei* means human feelings in Japanese.) are in very active especially in Japan. Human feelings are very complicated and it is not easy to evaluate them. It is generally said that human feelings are nonlinear and it is also considered that human feelings have hierarchical structure while AHP [1] uses paired comparison which could be useful to deal with nonlinear problems and also clarify hierarchy structure of the problem. AHP might be therefore considered useful to evaluate human feelings. But actually AHP has been mainly used for decision making and not a typical method for evaluating human feelings. Here taking a problem of selecting meals it is discussed how to apply AHP for evaluation of human feelings.

2 Paired Comparison Matrix in AHP

Table 1 shows nine answers to be chosen for the comparison of stimuli A and B and values to answers commonly used in paired comparison of AHP. Based upon answers of a subject the following paired comparison matrix is made.

I. Lovrek, R.J. Howlett, and L.C. Jain (Eds.): KES 2008, Part III, LNAI 5179, pp. 106–110, 2008.

$$\mathbf{M} = \begin{bmatrix} m_{11} & m_{12} & m_{13} \\ m_{21} & m_{22} & m_{23} \\ m_{31} & m_{32} & m_{33} \end{bmatrix} \tag{1}$$

Where

$$m_{11}=m_{22}=m_{33}=1, m_{21}=1/m_{12}, m_{31}=1/m_{13}, m_{32}=1/m_{23}.$$

Table 1. Values to answers in paired comparison

Answers	A	B
A is extremely preferred to B.	9	1/9
A is much preferred to B.	7	1/7
A is considerably preferred to B.	5	1/5
A is a little preferred to B.	3	1/3
A and B are equally preferred.	1	1
B is a little preferred to A.	1/3	3
B is considerably preferred to A.	1/5	5
B is much preferred to A.	1/7	7
B is extremely preferred to A.	1/9	9

3 Evaluation in AHP

In AHP evaluation of weights to criteria and alternatives under each criterion is made. In this study for evaluation of alternatives under each criterion paired comparison is not used and 5 steps evaluation method is used. A method to give measures for evaluation in the case of 3 criteria has been proposed [2]. In the method the measurements for evaluation are given so that the units of them can be considered to be the standard deviations of the standard normal distribution ($\sigma = 1$) and the total of the measurements of three criteria to each alternative can be statistically considered to be 0. The measurements obtained by this method are very useful to express subjective characteristics of alternatives.

4 Values to Answers in Paired Comparison

Paired comparisons are used to obtain weights to criteria. There are many methods of paired comparisons which have been proposed by Thurston, Scheffe etc. since Psychophysics had been proposed in nineteenth century, and AHP is one of the methods the essence of which is the paired comparison. The values to give to the answers of paired comparisons vary with method and should be considered according to the purpose to use paired comparisons. For example, in Guttman's method which quantifies paired comparisons by eigenvalue problems as well as AHP proposed by Saaty the values to the answers are 1 or 0 and in Fuzzy AHP they are fuzzy numbers. In this study order statistics defied based upon the standard normal distribution are used to make the units of the scales for meal intentions equivalent to the standard deviations of the standard normal deviation. For the comparison of stimuli A and B, the scales by order statistics are given in accordance with subjects' answers as shown in Table 2.

Table 2. Scales for answers by order statistics

Answers	A	B
A is extremely preferred to B.	1.485	-1.485
A is much preferred to B.	0.932	-0.932
A is considerably preferred to B.	0.572	-0.572
A is a little preferred to B.	0.275	-0.275
A and B are equally preferred.	0.000	0.000
B is a little preferred to A.	-0.275	0.275
B is considerably preferred to A.	-0.572	0.572
B is much preferred to A.	-0.932	0.932
B is extremely preferred to A.	-1.485	1.485

5 Calculation of Weights to Criteria

Now consider the case where there are n criteria and order statistics are used for the values of paired comparison. In this case the paired comparison matrix to obtain the weights to each criterion, the elements of which are order statistics, comes out as follows.

$$M = \begin{bmatrix} m_{11} m_{12}, \cdots, m_{1n} \\ m_{21} m_{22}, \cdots, m_{2n} \\ \vdots \quad \vdots \quad \ddots \quad \vdots \\ m_{nn} m_{n2}, \cdots, m_{nn} \end{bmatrix} \tag{2}$$

Where

$$m_{11} = m_{22} =, \cdots, = m_{nn} = 0 .$$
$$m_{21} = -m_{12}, m_{23} = -m_{32}, \cdots, m_{n-1n} = -m_{nn-1}$$

Since the matrix (1) is not the positive reciprocal matrix, the geometric mean method which is considered a convenient method of the eigenvvalue method is not able to be used. Therefore let $w_1, w_2, \cdots, w_n$ be the weights to each criterion and they are given as the arithmetic average of each row of (1), that is

$$w_i = \frac{\sum_{k=1}^{n} m_{ki}}{n} , \quad k = 1, 2, \cdots, n \tag{3}$$

Since $w_1, w_2, \cdots, w_n$ are the average of the values of order statistics, the units of them are able to be considered the standard deviations of the standard normal distribution.

6 Evaluation of Alternatives

In this section we consider how to calculate the importance of alternatives in the case of using order statistics for the values of paired comparison in AHP. Now suppose there are n criteria and m alternatives, and let w_i ($i = 1, 2, \cdots, n$) be the weights to criterion i and x_{ij} ($i = 1, 2, \cdots, n$, $j = 1, 2, \cdots, m$) be the values for evaluation of alternative j with respect to criterion i. Then the importance of alternative j ($j = 1, 2, \cdots, m$) is calculated by

$$(\,\mathrm{i}\,) \ \ \text{if} \ \sum_{i=1}^{n} w_i x_{ij} \geq 0, \quad I_j = \sqrt{\sum_{i=1}^{n} w_i x_{ij}} \tag{4}$$

$$(\,\mathrm{ii}\,) \ \ \text{if} \ \sum_{i=1}^{n} w_i x_{ij} < 0, \quad I_j = -\sqrt{-\sum_{i=1}^{n} w_i x_{ij}} \tag{5}$$

7 Evaluation of Human Meal Preference by AHP

As an example of application of AHP to evaluate human feeling using order statistics for the values to answers in paired comparison, a hierarchy for preference of menus on home dining tables was build as shown in Figure 1. In this diagram first hierarchy level (Level 1) is "selection of meals" (goal), second hierarchy level (Level 2) is "subjective characteristics of food" (criteria), third hierarchy level (Level 3) is "menus on home dining tables (alternatives).

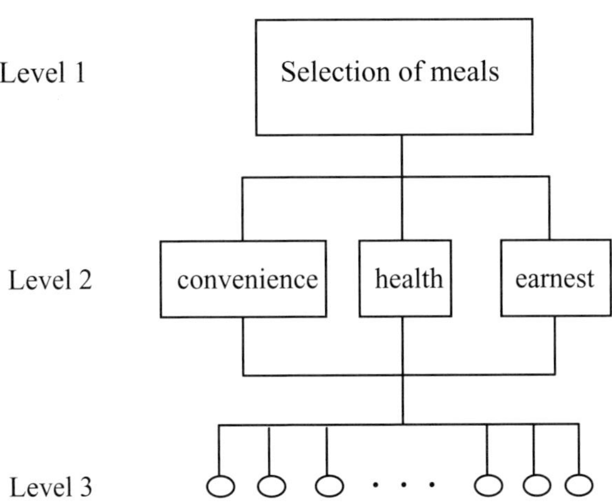

Fig. 1. Hierarchical diagram for selecting menus on home dining table

Based upon the hierarchical diagram, measurements for evaluation of alternatives (menus on home dining tables) are obtained, the units of which are the standard deviations of the standard normal distribution ($\sigma = 1$) on the basis of 0 . The measurement is useful to compare subjective characteristic of food among many alternatives.

8 Final Remarks

Here it was discussed how to obtain measurement for evaluation of alternatives with units in AHP and a method was proposed to obtain the measurement the units of which are the standard normal distributions ($\sigma = 1$) on the basis of 0 .using order statistics from the normal distribution.

References

[1] Saaty, T.L.: A Scaling Method for Priorities in Hierarchical Structures. Journal of Mathematical Psychology 15, 234–281 (1977)
[2] Kanda, T.: A Method to Evaluate Human Meal Kansei. International Journal of Kansei: Engineering (Kansei: Engineering International) 3(3), 13–20 (2002)

Placement Problem in an Industrial Environment

Shamshul Bahar Yaakob and Junzo Watada

Graduate School of Information, Production and Systems, Waseda University, 2-7 Hibikino,
Wakamatsu-ku, Kitakyushu-shi, Fukuoka 808-0135 Japan
shamshul@fuji.waseda.jp, junzow@osb.att.ne.jp

Abstract. A problem of workers' evaluation and placement in an industrial environment is studied in this paper; an effect of workers' relationship on their placement is newly included in this paper. Evaluating workers' suitability is important when decision makers select proper candidates under various evaluation criteria among available human resources and jobs. For problems of this type, an analysis using a fuzzy number approach promises to be potentially effective. In order to make a more convincing and accurate decision, the relationship between jobs is included in the workers' assignment in an industrial environment. The fuzzy suitability evaluation is performed by means of aggregating the decision makers' fuzzy assessment. Examples of typical application are also presented: the results demonstrate that the workers' relationship is an important factor and our method is effective for the decision making process.

Keywords: Fuzzy sets, Decision making, Workers' relationship, Workers' placement.

1 Introduction

The evaluation of workers is important for decision makers (DMs) to select proper workers under various evaluation criteria in an industrial environment [12], [3]. The aim of this research is to help the DMs make more effective selections from optional candidates [10]. The workers' placement is concerned with seeking the optimal matching between the workers and jobs within the constraints of available human resources and jobs [5], [4]. In non-'fuzzy' conventional approaches for workers' placement approaches, the evaluation of workers' suitability tends to use exact values. Kim et al discussed it from personal network [14], [15]. However, due to the vagueness of job demands as well as the complexity of human attributes, the exact evaluation of workers' suitability is quite difficult. The fuzzy theory developed by Zadeh [7], [8] and the concept of fuzzy numbers for example Dubois and Prade [1] can be applied to improve the assessments and the expressions for the assessment results in an industrial environment. Liang and Wang [3] and Kim et al [16] applied the concepts of combining the fuzzy set theory and weighted complete bipartite graphs to develop a polynomial time algorithm for solving personnel placement in a fuzzy environment.

In an industrial environment, an evaluation of workers' relationship, i.e., a group evaluation is also important as well as individual evaluation. In this paper, we develop a new method in which the workers' relationship is included to determine the optimal

I. Lovrek, R.J. Howlett, and L.C. Jain (Eds.): KES 2008, Part III, LNAI 5179, pp. 111–118, 2008.

workers' placement. Triangular fuzzy numbers [3] are used to describe the suitability of workers and the approximate reasoning of linguistic values [8], [9]. The operations of fuzzy addition, subtraction and multiplication derived based on the extension principle [11] are used to implement our algorithm.

In the following sections, triangular fuzzy numbers are briefly reintroduced, and the inclusion of the relationship among the workers is proposed and discussed. Typical examples are also presented in order to demonstrate the effectiveness of our proposal.

2 Fuzzy Numbers and Linguistic Variables

The aim of the fuzzy set theory is to deal with problems which have a source of vagueness. The membership function $f_z(x)$ of fuzzy set Z represents the degree of membership or the grade of x in the fuzzy set Z. The larger $f_z(x)$ is, the stronger the belonging degree of x in Z. Under a fuzzy environment fuzzy numbers are useful in promoting the representation and the information processing. A fuzzy number z in $\mathbb{R}$(real line) is triangular, if its membership function $fz : \mathbb{R} \rightarrow [0, 1]$ is defined as follows:

$$fz(x) = \begin{cases} (x - a)/(b - a), & a < x < b, \\ (x - c)/(b - c), & b < x < c, \\ 0, & \text{otherwise,} \end{cases}$$

where $-\infty < a \leq b \leq c < \infty$. Let us denote a triangular fuzzy number by z = <a, b, c>. Using the extension principle [8], the fuzzy sum, $\oplus$ and the fuzzy subtraction $\ominus$ of any two triangular fuzzy numbers is a triangular fuzzy number. The product $\otimes$ of any two triangular fuzzy numbers can be denoted approximately by a triangular fuzzy number. For example, let $zl = (al, bl, cl)$ and $z2 = (a2, b2, c2)$. Then, $zl \oplus z2 = (al + a2, bl + b2, cl + c2)$, $zl \ominus z2 = (al - c2, b2 - bl, cl - a2)$ and $g \otimes z = (ga, gb, gc)$. Here g is a real number. If $al \geqslant 0$ and $a2 \geqslant 0$, then $zl \otimes z2 \cong (al a2, bl b2, cl c2)$. The triangular fuzzy numbers are used to denote the fuzzy suitability of workers and the approximate reasoning of linguistic values. The center value of "b" presents the maximal grade of $f_z(x)$ and is the most possible value of the workers' suitability. The "a" and "c" are the upper and the lower bounds of available area of the workers' suitability. Linguistic descriptions of complex situations or strategies generally include fuzzy denotations [6]. In this paper, the triangular fuzzy number is employed and assigned to a linguistic variable. For example, a set of {very slow, slow, normal, fast, very fast} is described by a set of triangular fuzzy numbers {(2, 3, 4), (4, 6, 9), (9, 11, 13), (13, 16, 18), (18, 19, 20)}. The mutual compatibility functions of these linguistic values are subjectively defined by the DMs. The linguistic values are used to characterize the DMs' linguistic assessments about criteria weightings and workers' suitability relative to various evaluation criteria.

The proposed evaluation method represents final suitability scores using fuzzy numbers. During ranking process, fuzzy numbers are defuzzified to obtain their best non-fuzzy performance values (BNP) [6]. There are various defuzzification approaches proposed [2]. In this study, in order to rank fuzzy numbers, the center of area (COA) approach has been selected because this method is simple, practical and does not involve evaluator preference. The COA method generates the center of gravity of the possibility distribution of a fuzzy number. Meanwhile, the BNP value of a triangular fuzzy number $Z=(a,b,c)$ can be obtained by equation (1)

$$BNP = a + [(c-a)+(b-a)]/3 \cdot \qquad (1)$$

Therefore, the workers are ranked according to the BNP values of their suitability score.

3 Workers' Placement Problem

In this section, the evaluation including the relationship among the workers is developed in order to tackle the workers' placement problems efficiently. Using the concepts of triangular fuzzy numbers and linguistic variables, the workers' suitability evaluation is performed. The evaluation criteria may be classified into three factors:

 a. Social factors include communication skill, professional knowledge, cooperation, leadership, sense of responsibility, relationship to other members, etc.
 b. Performance factors include speed, quality, attendance condition, late coming, overtime, experience, etc.
 c. Mental factors include intelligence, problem solving ability, creativity, self-confidence, etc.

In this paper the relationship is evaluated by that between two workers. The relationship evaluation is performed via the sum of all the evaluation results between any couple of workers. This prescription can be generally applied to any size of the worker group, and is appropriate for our purpose and for computations. The DMs may also choose a linguistic weighting set $W = \{$not important, not so important, normal, important, very important$\}$ to evaluate the importance of each criterion. In general each criterion has its importance weight depending on the nature of jobs. Therefore in the following computation method the weighted sum is performed (see Eq. (2)). Suppose the following situation: the DMs are responsible for assessing the suitability of m workers *(Pi, i = 1,…, m)* under each of the k criteria *(Ct, t = 1,… k)*. Let $e(J, i, Ct) =$ *(a, b, c)* be a triangular fuzzy number, which is a rating assigned to a worker *Pi* by the DMs for a criterion *(Ct)* for a job (J). Let *W(J, Ct)* be the importance weight of the criterion *Ct* for the job J. The DMs can fix the total worker number assigned to each job depending on the job feature, if required. If not, the total number of workers is also determined in our algorithm. When the ranking order is determined, the center values of fuzzy triangle numbers are primarily used. If there is a tie on the grade value *(a, b, c)*, then the subtraction *((c - b) - (b - a))* between (the upper bound - the center value) and (the center value - the lower bound) of the triangle number is employed to fix the ranking order. This prescription is based on the following: a worker having a larger value of *(c - b) - (b - a)* may have relatively a high ability. The computation flow of our method is as follows:

Step **1:** Determine the evaluation criteria. Select the appropriate rating scale to assess the importance weights of the criteria and the suitability of the workers to the criteria. Assign the linguistic variables to the triangular fuzzy numbers. Tabulate suitability ratings (S) assigned to each worker (P) for each criterion *(Ct)* by each DM. Tabulate importance weightings *(W(J, Ct))* assigned to each criterion *(Ct)* for each job (J) by the DMs.

Step 2: A fuzzy suitability ranking of each worker Pi for the job J can be obtained by standard fuzzy arithmetic operations:

$$E_{eval}(J,i) = \frac{1}{k}\sum_{t=1}^{k} e(J,i,Ct) \otimes W(J,Ct). \tag{2}$$

In Eq. (2) the summation result is divided by the total number k of criteria employed so that $E_{ev,i}(J, i)$ does not depend on k. The ranking order is determined by the total grade value $E_{eval}(J, i)$ for each job J.

Step 3: In order to find possible combinations *PCs*, the DMs assign a fuzzy triangle number to the minimum grade value required for each job.

Step 4: Based on the workers' suitability evaluation result, the possible combinations *PCs* are obtained in order of the ranking each worker having a larger grade value is selected and assigned to the *PC*. The total grade value for the possible combination $E_{pc}(J)$ for the job J is as follows:

$$E_{PC}(J) = \sum_{i}^{PC} E_{eval}(J,i), \tag{3}$$

where the summation is performed over all members of the *PC*. The results are listed in the ranking order based on the total grade value $E_{PC}(J)$ for the possible combinations for each job. If any *PC* does not satisfy the minimum grade value, return to Step 3.

Step 5: Evaluate the relationship among the workers in a combination for a job: the relationship among the workers is computed as follows:

$$E_{RL}(J) = W_{RL}(J) \otimes \left[\frac{1}{{}_f C_2}\sum_{(i,j)} e_{RL}(i, j) \right], \tag{4}$$

where ${}_f C_2 = f(f-1)/2$, $e_{RL}(i, j)$ is a value assigned to the relationship between two workers (Pi and Pj) in the combination, the summation is taken over all couples of workers in the job J, $W_{RL}(J)$ is the importance weight of the relationship for the job J, and $E_{RL}(J)$ is the total relationship-evaluation value.

The summation is normalized by ${}_f C_2$ so that $E_{RL}(J)$ does not depend on the number (${}_f C_2$) of couples of workers. The relationship among the workers is evaluated only for possible combinations in order to save computation time. The final evaluation is computed as follows:

$$E_{comb}(J) = E_{PC}(J) + E_{RL}(J). \tag{5}$$

The result for final evaluation is listed in the ranking order based on the grade value $E_{comb}(J)$.

Step 6: If the total job number TJ is one, the highest- grade combination is the best one. When the total job number TJ is more than one, the DMs specify if one worker can be assigned to plural jobs or not, depending on the job nature. If one worker is not assigned to plural jobs, an overlapped assignment of one worker is checked and avoided in the total combination construction. Based on this information, the total final combination evaluation TE_{comb} is as follows:

$$TE_{comb} = \sum_{J=1}^{TJ} E_{comb}(J). \qquad (6)$$

The result for the total final evaluation TE_{comb} is listed in the ranking order. The combination that has the highest grade value is the result for the workers' placement problem.

4 Illustrative Examples

In this section, a typical example problem of workers' placement is designed to demonstrate the effectiveness of the method that has been proposed in this paper. An example is focused on a production line in an industrial environment.

Suppose that the DMs want to find the better workers' placement for a production line. The information for the problem is as follows:

(a) the workers are 20 persons that are identified by ID number from 1 to 20,
(b) there are 5 evaluation criteria.
(c) each worker should be assigned to only one job,
(d) the number of jobs is 5 and
(e) the number of grouping workers in each job are not fixed.

A stepwise description of workers' evaluation and placement is as follows:

Step 1: The DMs input the grade of the linguistic values of workers for related criteria. To evaluate the relative importance of the five criteria, the DMs fix the linguistic weighting scales.

Step 2: By using Eq. (2), a fuzzy suitability ranking order of each worker is obtained. The result of workers' evaluation is listed in the ranking order as shown in Table 1 for the job 1.

Step 3: The DMs input the information about the minimum grade value required.

Step 4: By using Eq. (3), the possible combinations are computed as shown in Table 2. The best combination is (13, 14, 8, 16), (1, 11, 6, 20), (4, 9, 10, 19), (17, 12, 3, 15) and (5, 18, 2, 7) for the three jobs with the center value 549.0 of the total grade value.

Step 5: By using Eq. (4), the relationship grade among the workers is computed. By using Eq. (5), the combination of the workers is evaluated.

Step 6: By using Eq. (6), the total grade value of the final combination is computed. The result for the final combination is listed in Table 3. The best combination is ((13, 14, 8, 16), (1, 11, 6, 20), (4, 9, 10, 19), (17, 12, 3, 15), (5, 18, 2, 7)) for the three jobs with the center value 791.9 of the total grade value,

Table 1. Result for workers' evaluation in ranking order with grade values $E_{eval}(J, i)$

JOB 1	
Total Grade = 95.8	Workers' ID = 13
Total Grade = 90.7,	Workers' ID = 14
Total Grade = 88.3,	Workers' ID = 8
Total Grade = 86.6,	Workers' ID = 16
Total Grade = 84.5,	Workers' ID = 4
Total Grade = 81.2,	Workers' ID = 19
Total Grade = 76.9,	Workers' ID = 9
Total Grade = 75.9,	Workers' ID = 10
Total Grade = 73.7,	Workers' ID = 15
Total Grade = 69.9,	Workers' ID = 5
Total Grade = 64.4,	Workers' ID = 7
Total Grade = 60.8,	Workers' ID = 18
Total Grade = 54.0,	Workers' ID = 6
Total Grade = 53.3,	Workers' ID = 11
Total Grade = 50.9,	Workers' ID = 17
Total Grade = 47.2,	Workers' ID = 1
Total Grade = 45.7,	Workers' ID = 12
Total Grade = 42.2,	Workers' ID = 2
Total Grade = 42.2,	Workers' ID = 20
Total Grade = 26.0,	Workers' ID = 3

Table 2. Result for workers' combination via $E_{PC}(J)$ without the evaluation for relationship among the workers (The number of the job is 5 and each worker must be assigned to one job)

JOB 1
Ranking top 10 before the evaluation for relationships as follows :
(13, 14, 8, 16) grade value is 90.4
(13, 14, 8, 4) grade value is 89.8
(13, 14, 16, 4) grade value is 89.4
....

JOB 2
Raking top 10 before the evaluation for relationships as follow :
(1, 11, 6, 20) grade value is 92.4
(1, 11, 6, 16) grade value is 92.1
(1, 11, 20, 16) grade value is 91.5
....

JOB 3
Ranking top 10 before the evaluation for relationships as follows :
(4, 9, 10, 19) grade value is 88.3
(4, 9, 10, 3) grade value is 87.6
(4, 9, 19, 3) grade value is 86.9
....

JOB 4
Ranking top 10 before the evaluation for relationships as follows :
(17, 12, 3, 15) grade value is 91.7
(17, 12, 3, 6) grade value is 90.7
(17, 12, 15, 6) grade value is 90.1
....

JOB 5
Ranking top 10 before the evaluation for relationships as follows :
(5, 18, 2, 7) grade value is 91.2
(5, 18, 2, 14) grade value is 90.5
(5, 18, 7, 14) grade value is 89.8
....

The best combination is (13, 14, 8, 16), (1, 11, 6, 20), (4, 9, 10, 19), (17, 12, 3, 15) and the total grade value is 549.0

In Tables 2 and 3 only the center values of the triangle numbers are presented for clarity. Table 2 shows the grade value of the workers' suitability indices by using the conventional method. Table 1 shows the result of the individual evaluation. By using the method presented in this paper, both the individual evaluation and the group one are performed, and the result is presented in Table 3.

Table 3. Result for workers' combination via E_{comb} $(\mathcal{J})$ with the evaluation for relationship among the workers (The number of job is 5 and each worker must be assigned to one job)

Result after the evaluation for relationships		Workers' Relationship Results
JOB 1	$E_{comb}(\mathcal{J})$	$E_{RL}(\mathcal{J})$
(13, 14, 8, 4) grade value is	174.8	85.0
(13, 14, 16, 4) grade value is	169.4	80.0
(13, 8, 16, 4) grade value is	163.8	75.0
JOB 2		
(1, 11, 20, 16) grade value is	163.5	72.0
(11, 6, 20, 16) grade value is	161.7	72.0
(1, 11, 6, 16) grade value is	148.1	56.0
JOB 3		
(9, 10, 19, 3) grade value is	179.3	95.0
(4, 10, 19, 3) grade value is	160.0	75.0
(4, 9, 10, 3) grade value is	157.6	70.0
JOB 4		
(17, 12, 15, 6) grade value is	147.1	57.0
(12, 3, 15, 6) grade value is	145.7	57.0
(17, 3, 15, 6) grade value is	139.8	51.0
JOB 5		
(5, 18, 2, 7) grade value is	127.2	91.2
(5, 18, 7, 14) grade value is	121.8	90.5
(18, 2, 7, 14) grade value is	121.3	89.8

The best combination is (13, 14, 8, 4), (1, 11, 20, 16), (9, 10, 19, 3), (17, 12, 15, 6), (5, 18, 2, 7) and the total grade value TE_{comb} is 791.9

Without the group evaluation, the center value of the fuzzy grade for the best combination ((13, 14, 8, 16), (1, 11, 6, 20), (4, 9, 10, 19), (17, 12, 3, 15), (5, 18, 2, 7)) in Table 2 is 549.0. Even for this fixed combination, we can compute and include the relationship-evaluation grade values, and it is 659.1. After the evaluation for relationship among the workers, the center grade value 791.9 for the best combination in Table 3 is higher than the total grade value 659.1 for the best combination in Table 2 by 20.1%. It is concluded that by using our method a more effective solution is obtained for the workers' placement in an industrial environment.

5   Conclusions

A new proposal was presented to solve the problem of workers' placement in an industrial environment. The relationship among the workers in the group (relationship) is one of the important factors, In order to make a more convincing and accurate decision, the group evaluation is required. In this paper, not only the individual evaluation but also the group evaluation are performed and included to find a better combination. We applied the proposed method to typical application examples. The results demonstrate that the workers' relationship is one of the key issues and our proposal is effective for the decision making process. The domain of workers' placement illustrated in this paper is focused on a production line in an industrial environment. It may be also applied to other types of placement problems. In this paper, five levels of linguistic values are designated in the rating and weighting scales. However, the number rating and weighting scales. However, the number of levels can be adjusted correspondingly based on the needs of detailed evaluation and the available data characteristics.

This paper was focused on the group evaluation among the workers assigned to a group via the fuzzy approach. In general the workers' placement problem consists of several key issues including the selection of evaluation criteria, the evaluation methods, an optimization scheme and so on. These issues, except for those solved in the former papers [3], [4], [10], [12] and this paper, should also be studied in the future.

Acknowledgments. The first author would like to thank Universiti Malaysia Perlis and the Ministry of Higher Education Malaysia for a study leave in Waseda University under the SLAI-KPT scholarship. The first author also would like to thank to Global Center of Excellence (GCOE) for KES 2008 participation financial support.

References

1. Dubois, D., Prade, H.: Operations on fuzzy numbers. Int. J. Syst. Sci. 9, 613–626 (1978)
2. Bortolan, G., Degani, R.: A review of some methods for ranking fuzzy subsets. Fuzzy Sets and Systems 15(1), 1–19 (1985)
3. Liang, G.S., Wang, M.J.: Personnal placement in a fuzzy environment. Comput. Oper. Res. 19, 107–121 (1992)
4. Ishii, H.: Fuzzy combinatorial optimization. J. Jpn. Soc. Fuzzy Theory Systems 4, 31–40 (1992)
5. Cadenas, J.M., Verdegay, J.L.: PROBO: an interactive system in fuzzy linear programming. Fuzzy Sets and Systems 76, 319–332 (1995)
6. Teng, J.Y., Tzeng, G.H.: Fuzzy multicriteria ranking of urban transportation investment alternatives. Transportation Planning and Technology 20(1), 15–31 (1996)
7. Zadeh, L.A.: Fuzzy sets. Inform. and Control 8, 338–353 (1965)
8. Zadeh, L.A.: The concept of a linguistic variable and its application to approximate reasoning, Part 1 and Part 2. Inform. Sci. 8, 199–249 (1975)
9. Biswas, R.: An application of fuzzy set in students' evaluation. Fuzzy Sets and Systems 74, 187–194 (1995)
10. Bellman, R.E., Zadeh, L.A.: Decision-making in fuzzy environment. Management Sci. 17, 141–164 (1970)
11. Yaakob, S.B., Kawata, S.: Workers' placement in an industrial environment. Fuzzy Sets and Systems 106, 289–297 (1999)
12. Duffuaa, S.O., Raouf, A.: A simulation model determining maintenance staffing in an industrial environment. Simulation 59, 93–99 (1992)
13. Nawijn, W.M., Dorhout, B.: On the expected number of assignment in reduces matrices for the linear assignment problem. Oper. Res. Lett. 8, 329–335 (1989)
14. Kim, I., Jeng, D.J.-F., Watada, J.: Redesigning Subgroups in a Personnel Network Based on DNA. IJICIC 2(4), 885–896 (2006)
15. Kim, I., Watada, J., Jeng, D.J.-F.: Building a Managerial Support System for Work Rotation Based on Bio-Soft Computing. In: Proceedings, NAFIPS 2006. IEEE, Los Alamitos (2006)
16. Kim, I., Jeng, D.J.-F., Watada, J.: Fuzzy DNA-Based Computing for Analyzing Complicated Personnel Networks. In: Proceedings, ISME2008, International Symposium of Management Engineering 2008, March 15-17, 2008 (2008)

Motion Tracking Using Particle Filter

Zalili Binti Musa and Junzo Watada

Graduate School of Information, Production and System
Waseda University
2-7 Hibikino, Wakamatsu, Kitakyushu, Fukuoka 808-0135 Japan
zalili@ump.edu.my, junzow@osb.att.ne.jp

Abstract. Video tracking system raises a wide possibility in today's society. This system is used in various applications such as security, monitoring, robotic, and nowadays in day-to-day applications. However the video tracking systems still have many open problems and various research activities in a video tracking system are explored. In this study, we have developed a prototype to track a bounce ball movement. Generally, the movement of a bounce ball is fast and the sizes of objects are different regarding on camera view. Therefore, the aim of this study is to construct a motion tracking for an object movement using a particle filter formula. Where, at the end of this paper, the detail outcome and result are discussed using experiments of this method.

Keywords: Tracking system, Particle Filter, Kalman Filter, Condesation, Object Movement.

1 Introduction

Various techniques for object tracking systems have been developed by researchers. Researchers have used Kalman filter technique to track an object movement. For example, in real-time tracking of multiple people using continuous detection, David Beymer and Kurt Konilige used the Kalman filtering technique to track people movement [1]. However, based on Isard and Blake, this technique has limited use because they are based on unimodal Gaussian densities that cannot support simultaneous alternative motion hypotheses [2].

In this article, the aim is to construct a motion tracking for an object movement. In our research, we have developed a motion tracking prototype using MatLab 7 where we used a bounce ball movement as our data. Based on our observation during the testing, a bounce ball has a fast movement and the sizes of an object regarding on camera view are different. Therefore, we need an available technique to ensure the bounce ball can be tracked as well. In this paper, we used a Particle Filter formula to track the bounce ball. At the end of this paper, the detail outcome and result are discussed using experiments of this method.

In the remain of this paper, we will explain our literature review in Section 2. Then, in Section 3, we describe the architecture of the system including formula and

I. Lovrek, R.J. Howlett, and L.C. Jain (Eds.): KES 2008, Part III, LNAI 5179, pp. 119–126, 2008.

algorithm that we use in our method. In Section 4, we discuss about our result and conclusion.

2 Tracking System and Particle Filter

In the last few years, visual tracking has been an active research area in computer vision. This research is applied to security system, monitoring system, robotic and others. The objective of tracking method is to generate the path of an object by locating an object position in every frame from a video stream. Basically, two main purposes of tracking method are to determine when a new object enters the system and secondly, to estimate the position of an object over time.

Generally tracking object movements is a challenging task because a target changes dynamically. In a traditional way, an object motion can be measured by two models: stochastic models with deterministic and random components and secondly stochastic model by classical deterministic mechanics [3]. Fig. 1 shows such examples as a model includes constant position, constant velocity and constant acceleration. In tracking systems, it is difficult to track an object motion based on constant velocity and constant acceleration model because this model is not a linear motion and also will affect the size of an object during movement. Therefore we need a compatible method to overcome this problem.

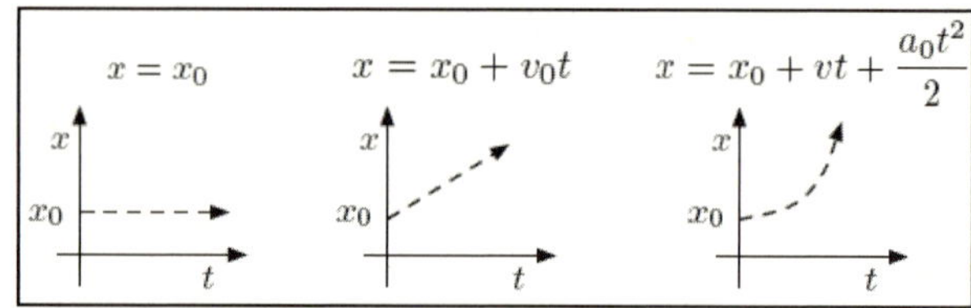

Fig. 1. Traditional motion models[3]

Currently, the popular approach to analyze object motion is used a particle filter and its extension which mainly based on the Bayes' rule. Starting from Blake and Isard's condensation algorithm, particle filter has become a very popular method for visual tracking in computer vision application such as for security system that is used to track a human body, face, hand gesture and so on; a monitoring system for example object motion such as bounce ball, vehicles in a highway and leaf motion.

Based on [4][5], particle filter has been shown to offer improvements in performance over some conventional methods such as Kalman filter, especially in non-linear or non-Gaussian environment. Where, these methods provide a robust tracking of moving objects in a cluttered environment. Related to visual tracking, a particle filter uses multiple discrete "particle" to represent the belief distribution over the location of a tracked object.

Generally, the particle filters, also known as Sequential Monte Carlo methods (SMC), where the probability density is represented by a set of weighted samples (called particle) that implemented by a recursive Bayesian filter. In the Bayes filtering [5][6], the recursively update the posterior distribution over the current state of the system at time t is represented by a random variable x_t. Assuming that there are T frame of data to be processed, and at time t only data from $1 \ldots t-1$ are available. The measurements at time t are labeled z_t and will contain a list of feature measurements. The measurement are denoted $Z^t = z_1, \ldots, z_t$ be a sequence of states up to time t. Therefore using Bayes formula, we denote

$$p(x_t|Z^t) = p(x_t|z_t, Z^{t-1})$$

$$= \frac{p(x_t|z_t, Z^{t-1})p(x_t|Z^{t-1})}{p(z_t|Z^{t-1})} \tag{1}$$

If $p(x_t|x_{t-1})$ represents the dynamic model of probability distribution, Eq.(1) can be expressed as:

$$\frac{p(z_t|x_t, Z^{t-1}) \int_{x_{t-1}} p(x_t|x_{t-1})p(x_{t-1}|Z^{t-1})dx_{t-1}}{p(z_t|Z^{t-1})} \tag{2}$$

Using observation independence and Markov assumptions, Eq.(2) can be expressed as:

$$\frac{p(z_t|x_t) \int_{x_{t-1}} p(x_t|x_{t-1})p(x_{t-1}|Z^{t-1})dx_{t-1}}{p(z_t)} \tag{3}$$

where the likelihood $p(z_t|x_t)$ expresses the measurement model and $p(x_t|x_{t-1})$ is the motion model. In a particle filter, the posterior probability $p(x_{t-1}|Z^{t-1})$ show a set of N weighted samples $\{x_{t-1}^i, W_{t-1}^i\}_{i=1}^N$, where W_{t-1}^i is the weight for particle x_{t-1}^i.

Regarding on our study, we identified that there exist many methods to track an object motion using a particle filter. Based on condensation algorithm[7], the process has classified into three categories: first Select a sample, second prediction and third measure the new position by the weight samples. The Condensation algorithm is shown in Fig. 2.

Meanwhile based on [8][9], the principal steps in the particle filter algorithm as shown in Fig. 3. According to Fig. 3, the detail algorithm can be referred at [9].

Iterate

From the "old" sample-set $\{s_{t-1}^{(n)}\pi_{t-1}^{(n)}, c_{t-1}^{(n)}, n=1,...,N\}$ at time step $t-1$, construct a "new" sample-set $\{s_t^{(n)}\pi_t^{(n)}, c_t^{(n)}\}n=1,...,N$ for time t.

Construct the nth of N new samples as follow:

1. Select a sample $s{'}_t^{(n)}$ as follows:

    (a) generate a random number $r \in [0,1]$, uniformly distributed.

    (b) Find, by binary subdivision, the smallest j for which $c_{t-1}^{(j)} \geq r$

    (c) Set $s{'}_t^{(n)} = s_{t-1}^{(j)}$

2. Predict by sampling form

$$p(x_t | x_{t-1} = s{'}_t^{(n)})$$

To choose each $S_t^{(n)}$. For instance, in the case that the dynamics are governed by a linear stochastic differential equation, the new sample value may be generated as: $s_t^{(n)} = As{'}_t^{(n)} + Bw_t^{(n)}$ where $w_t^{(n)}$ is a vector of standard normal random varieties, and BB^T is the process noise covariance.

3. Measure and weight the new position in terms of the measured features Z_t :

$$\pi_t^{(n)} = p(z_t | x_t = s_t^{(n)})$$

then normalize so that $\sum_n \pi_t^{(n)} = 1$ and store together with cumulative probability as $(s_t^{(n)}\pi_t^{(n)}, c_t^{(n)})$ where

$$c_t^{(0)} = 0,$$
$$c_t^{(n)} = c_t^{(n-1)} + \pi_t^{(n)} (n=1,...,N)$$

Once the N samples have been constructed: estimate, if desired, moments of the tracked position at time-step t as

$$\varepsilon[f(x_t)] = \sum_{n=1}^{N} \pi_t^{(n)} f(s_t^{(n)})$$

Obtaining, for instance, a mean position using $f(x) = x$.

Fig. 2. Condensation Algorithm[8]

3 Overview of the Proposed Method

Generally, this method consists of four main modules: 1) data acquisition, 2) target detection and 3) tracking target.

3.1 Data Acquisition

The objective of an image acquisition module is to capture object motions using a video digital camera. In this research, an image video has been captured about 20-30 second for one scene and the size of each video frame is 320 x 240 pixels in avi format.

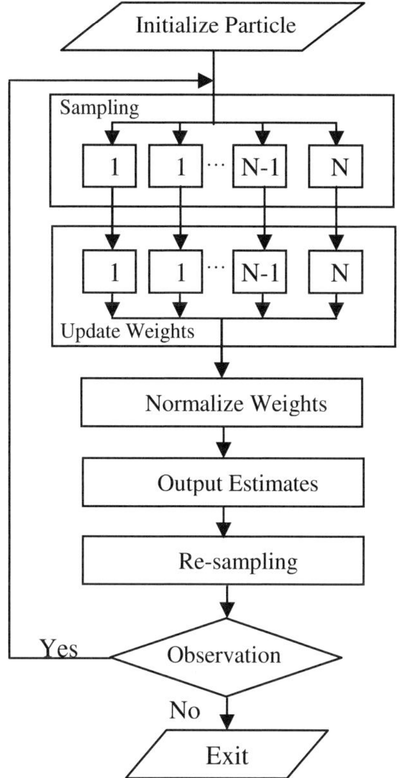

Fig. 3. Flowchart of the particle filter[9]

3.2 Target Detection

In this stage we divided a process into two sub processes which are blurring process and subtraction.

In blurring process, we used Prewitt operator measure to remove some feature of an image. The Prewitt operator component is calculated with kernel Kx and the horizontal edge component is calculated with kernel Ky. |Kx|+|Ky| gives an indication of the gradient in the current pixel [10]. Fig. 4 shows illustration of Prewitt horizontal and vertical operators.

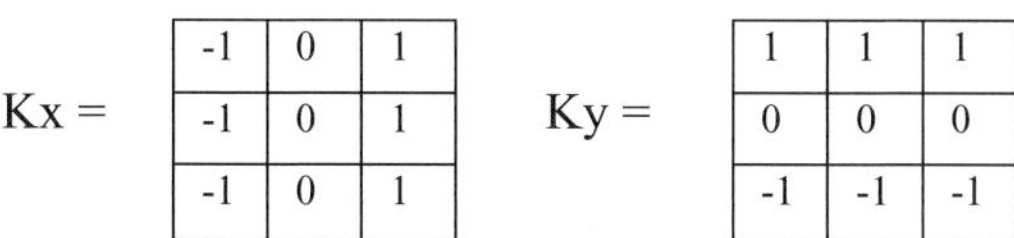

$$Kx = \begin{array}{|c|c|c|} \hline -1 & 0 & 1 \\ \hline -1 & 0 & 1 \\ \hline -1 & 0 & 1 \\ \hline \end{array} \qquad Ky = \begin{array}{|c|c|c|} \hline 1 & 1 & 1 \\ \hline 0 & 0 & 0 \\ \hline -1 & -1 & -1 \\ \hline \end{array}$$

Fig. 4. Illustration of Prewitt horizontal and vertical operators for filtering process

The detail explanation of filtering algorithm can be refer to[11].

As we mentioned above, the second process is subtraction process. After the blurring process, an image subtraction is done as the formula below:

$$P(i) = \begin{cases} 1 : u(i) - v(i) > 10 \\ 0 : otherwise \end{cases} \qquad (6)$$

where $P(i)$ is a new value pixel, $u(i)$ is a value of fixed background pixel and $v(i)$ is a value of frame pixel. i is a number of pixel.

3.3 Tracking Target

The objective of this module is to track a location of several object motions beyond video scene. In this process we used a particle filter algorithm to track an object movement as shown as below:

1.   Initialize
        Generate a set a particle for the target pixel i at time 0:

$$S_i^0 = [x_j, w_j : j = 1...N] \qquad (7)$$

2.   Observation
        The observation model is:

$$p(z_t | x_t) = C + r \qquad (8)$$

   where C and r is a centroid and radius of the target ball respectively . In this case we identify that r as:

$$-1 \times r : r / n : r \qquad (9)$$

   and n is a number of pixel of ball perimeter.

3.   Prediction
        In order to predict the probability distribution of the target (state), we need to have a model of the effect of noise on the resulting from Eq.(8). Therefore the formula for prediction is:

$$p(x_t | x_{t-1}) = C + r + u \qquad (10)$$

   where u is a normal random matrix with normally or Gaussian distributed elements

4.   Output Estimate
        To obtain the estimation value, we employed the formula as shown below:

$$\overline{\overline{x_t}} = \max p(x_t | x_{t-1}) \qquad (11)$$

5.   Re-sampling
        The detail explanation of this formula can be founded in[4].
6.   Normalize the weight

For $i = 1,..., N$, normalize the importance weights as a followed formula:

$$\bar{w}_t^{(i)} = w_t^{(i)} \left[\sum_{j=1}^{N} w_t^{(j)} \right]^{-1} \tag{12}$$

4 Result and Discussion

The proposed method and algorithm were developed, tested and applied. In this case, we have developed our proposed method by using Matlab 7 and this method is executed on Dell compatible Pentium D 2.80GHz. In order to evaluate the proposed method, the 60 frame of data videos have been tested.

As we mentioned before, a bounce ball has fast movement. In a real-time, about 22.33 second is taken to complete bounce ball movement until it stop. Therefore, Table 1 shows a processing time versus number of particle that we have tested in this study.

Table 1. Processing times for different numbers of particle

Numbers of particle(N)	Time Taken (second)
200	34.625
150	33.344
100	31.391
50	29.813
25	28.110
15	27.781

Regarding on results on Table 1, shows that the processing time to track a ball has increase significantly. Even though, time processing are increase, the tracking result is better when we compare with Kalman filter. Table 2 show the different performance of tracking bounce ball between these two techniques.

Table 2. Processing times for comparison method

Methods	Percentage
Particle Filter	96.67
Kalman Filter	36.67

Acknowledgments. The first author would like to thank University Malaysia Pahang, Kementerian Pengajian Tinggi Malaysia for supporting study leave and GCOE program of Waseda University for supporting to attend KES2008.

References

1. Beymer, D., Konolige, K.: Real-time tracking of multiple people using continous detection. In: InternationalConference on Computer Vision (ICCV 1999) (1999)
2. Isard, M., Blake, A.: Contour tracking by stochastic propagation of conditional density. In: European Conference on Computer Vision, pp. 343–356 (1996)
3. Greg, W., Danette Allen, B., Adrian, I., Gary, B.: Measurement Sample Time Optimization for Human Motion Tracking/Capture Systems. In: Proceedings of Trends and Issues in Tracking for Virtual Environments, Workshop at the IEEE Virtual Reality 2007 Conference (2007)
4. Xu, X., Li, B.: Rao-Blackwellised particle filter for tracking with application in visual surveillance. In: IEEE International Workshop on Visual Surveillance and Performance Evaluation of Tracking and Surveillance, pp. 17–24 (2005)
5. Ye, Z., Liu, Z.-Q.: Tracking Human Hand Motion Using Genetic Particle Filter. In: IEEE International Conference on Systems, Man and Cybernetics SMC 2006, pp. 4942–4947 (2006)
6. Ryu, H.R., Huber, M.: A particle filter approach for multi-target tracking. In: IEEE/RSJ International Conference on Intelligent Robots and Systems (IROS 2007), pp. 2753–2760 (2007)
7. Isard, M., Blake, A.: Condensation - condition density propagation for visual tracking. In: International Journal of Computer Vision, pp. 5–28 (1998)
8. Li, P., Zhang, T.: Visual contour tracking based on particle filter. In: Proceedings of Generative-Model-Based Vision, pp. 61–70 (2002)
9. Jung, U.C., Seung, H.J., Xuan, D.P., Jae, W.J., Jong, E.B., Hoon, K.: A real-time object tracking system using a particle filter. International Conference on Intelligent Robots and Systems, 2822–2827 (2006)
10. Neoh, H.S. and Hazanchuk, A.: Adaptive Edge Detection for Real-Time Video Processing using FPGAs (2007), http://www.altera.com/literature/cp/gspx/edge-detection.pdf
11. Musa, Z., Watada, J.: Dynamic Tracking System for Object Recognition. In: International Conference on Knowledge-Based and Intelligent Information & Engineering Systems, pp. 378–385 (2007)

Regression Model
Based on Fuzzy Random Variables

Shinya Imai, Shuming Wang, and Junzo Watada*

Graduate School of Information, Production and Systems, Waseda University,
2-7 Hibikino, Wakamatsu, Kitakyushu 808-0135, Fukuoka, Japan
Mobile: +81-90-3464-4929; Fax: +81-93-692-5179
`shinimai@violin.ocn.ne.jp`,
`smwangips@gmail.com`,
`junzow@osb.att.ne.jp`

Abstract. In real-world regression problems, various statistical data may be linguistically imprecise or vague. Because of such co-existence of random and fuzzy information, we can not characterize the data only by random variables. Therefore, one can consider the use of fuzzy random variables as an integral component of regression problems.

The objective of this paper is to build a regression model based on fuzzy random variables. First, a general regression model for fuzzy random data is proposed. After that, using expected value operators of fuzzy random variables, an expected regression model is established. The expected regression model can be developed by converting the original problem to a task of a linear programming problem. Finally, an explanatory example is provided.

Keywords: Fuzzy random variable; Expected value; Fuzzy regression model.

1   Introduction

Classical regression analysis leads to a very effective statistical model to deal with statistical data. In the past two decades, to cope with the fuzzy environment where human subjective estimation becomes available various fuzzy regression models were presented for fuzzy input-output data and dealt with the theory of fuzzy sets and possibility. For instance, Tanaka *et al.* [20] presented linear regression analysis to cope with fuzzy data in stead of statistical data. Watada and Tanaka [23] [24] [27] presented possibilistic regression analysis based on the concept of possibility in stead of fuzziness. Watada *et al.* built fuzzy time-series model using intersection of fuzzy numbers [30] [32]. Also Watada tried to solve fuzzy regression model for fuzzy data [31] and employed heuristic methods to solve production between fuzzy numbers. Watada and Mizunuma [33] and Yabuuchi and Watada [26] built switching fuzzy regression model to analyze mixed data obtained from various sources. Linguistic regression model is proposed by

---

* Corresponding author.

I. Lovrek, R.J. Howlett, and L.C. Jain (Eds.): KES 2008, Part III, LNAI 5179, pp. 127–135, 2008.

Toyoura and Watada [25]. On the other hand, the concept of fuzzy statistics plays a central role in building a fuzzy regression model [28] as well as the concept of fuzzy numbers.

In practical applications, statistical data may include stochastic information and fuzzy information at the same time. For example, in a factory, the lifetime of some kinds of elements may be described like this: "about 5 months" with probability 0.2, "about 3 months" with probability 0.4, and "about 2 months" with probability 0.4, where "about 5 months", "about 3 months" and "about 2 months" are all linguistic which can be characterized by fuzzy numbers or fuzzy variables. In such a case, the lifetime of the element has the distribution as below:

$$X \sim \begin{pmatrix} \tilde{5} & \tilde{3} & \tilde{2} \\ 0.2 & 0.4 & 0.4 \end{pmatrix}$$

which can not be described by only one of the random variables or fuzzy variables. Therefore, we have to combine the two and turn to a new tool so as to study such two-fold uncertain data. Fuzzy random variable was introduced by Kwakernaak [8,9] in 1978 to study randomness and fuzziness simultaneously. It was defined as a measurable function from a probability space to a collection of fuzzy numbers. Since then, its variants as well as extensions were developed by other researchers for different purposes, e.g., Kruse and Meyer [7], Liu and Liu [11], López-Diaz and Gil [13], Luhandjula [14], and Puri and Ralescu [18]. Wang and Watada discuss T-independence condition for fuzzy random vectors [34] and discuss the fuzzy renewal process with a queueing application [35].

Based on fuzzy random variables and the expected value operators, this paper aims to build a regression model for fuzzy random values. The remainder of this paper is organized as follows. In Section 2, we recall some preliminaries on fuzzy random variables. Section 3 discusses regression model based on fuzzy random variables. In Section 4, an explanatory example is provided to illustrate the proposed fuzzy random regression analysis model. Finally, concluding remarks are given in Section 5.

2   Fuzzy Random Variables

Given a universe Γ, let Pos be a possibility measure defined on the power set $\mathcal{P}(\Gamma)$ of Γ. Let $\Re$ be the set of real numbers. A function $Y : \Gamma \to \Re$ is said to be a fuzzy variable defined on Γ, and the possibility distribution μ_Y of Y is defined by $\mu_Y(t) = \text{Pos}\{Y = t\}$, $t \in \Re$, which is the possibility of event $\{Y = t\}$. For fuzzy variable Y with possibility distribution μ_Y, the possibility and credibility of event $\{Y \leq r\}$ can be given respectively by

$$\text{Pos}\{Y \leq r\} = \sup_{t \leq r} \mu_Y(t),$$

$$\text{Cr}\{Y \leq r\} = \frac{1}{2} \left[1 + \sup_{t \leq r} \mu_Y(t) - \sup_{t > r} \mu_Y(t) \right]. \tag{1}$$

Definition 1 ([10]). *Let Y be a fuzzy variable. The expected value of Y is defined as*

$$E[Y] = \int_0^\infty \mathrm{Cr}\{Y \geq r\}\mathrm{d}r - \int_{-\infty}^0 \mathrm{Cr}\{Y \leq r\}\mathrm{d}r \qquad (2)$$

provided that one of the two integrals is finite.

Example 1. Suppose $Y = (a, b, c)$ is a triangular fuzzy variable, whose possibility distribution is

$$\mu_Y(x) = \begin{cases} (x-a)/(b-a), & \text{if } a \leq x \leq b \\ (c-x)/(c-b), & \text{if } b \leq x \leq c \\ 0, & \text{otherwise.} \end{cases}$$

By (2), we can compute the expected value of Y as

$$E[Y] = \frac{a + 2b + c}{4}.$$

In the following, we will introduce the definition of fuzzy random variables and the expected value operators .

Definition 2 ([11]). *Suppose that $(\Omega, \Sigma, \mathrm{Pr})$ is a probability space, $\mathcal{F}_v$ is a collection of fuzzy variables defined on possibility space $(\Gamma, \mathcal{P}(\Gamma), \mathrm{Pos})$. A fuzzy random variable is a map $X : \Omega \to \mathcal{F}_v$ such that for any Borel subset B of $\Re$, $\mathrm{Pos}\{X(\omega) \in B\}$ is a measurable function of ω.*

Suppose X is a fuzzy random variable on Ω, from the above definition, we know for each $\omega \in \Omega$, $X(\omega)$ is a fuzzy variable. Further, a fuzzy random variable X is said to be positive if for almost every ω, fuzzy variable $X(\omega)$ is positive almost surely.

Example 2. Let V be a random variable defined on probability space $(\Omega, \Sigma, \mathrm{Pr})$. Define that for every $\omega \in \Omega$,

$$X(\omega) = (V(\omega) - 2, V(\omega) + 2, V(\omega) + 6)$$

which is a triangular fuzzy variable defined on some possibility space $(\Gamma, \mathcal{P}(\Gamma), \mathrm{Pos})$. Then, X is a (triangular) fuzzy random variable.

To a fuzzy random variable X on Ω, for each $\omega \in \Omega$, the expected value of the fuzzy variable $X(\omega)$, denoted by $E[X(\omega)]$, has been proved to be a measurable function of ω (see [11, Theorem 2]), i.e., it is a random variable. Based on such fact, the expected value of the fuzzy random variable X is defined as the mathematical expectation of the random variable $E[X(\omega)]$.

Definition 3 ([11]). *Let X be a fuzzy random variable defined on a probability space $(\Omega, \Sigma, \mathrm{Pr})$. The expected value of X is defined as*

$$E[X] = \int_\Omega \left[\int_0^\infty \mathrm{Cr}\{X(\omega) \geq r\}\,\mathrm{d}r - \int_{-\infty}^0 \mathrm{Cr}\{X(\omega) \leq r\}\,\mathrm{d}r \right] \mathrm{Pr}(\mathrm{d}\omega). \qquad (3)$$

Example 3. Consider the triangular fuzzy random variable X defined in Example 2. Suppose the V is a discrete random variable, which takes values $V_1 = 3$ with probability 0.2, and $V_2 = 6$ with probability 0.8. Try to calculate the expected value of X.

From the distribution of random variable V, we know the fuzzy random variable X takes fuzzy variables $X(V_1) = (1,5,9)$ with probability 0.2, and $X(V_2) = (4,8,12)$ with probability 0.8. Further, we need to compute the expected values of fuzzy variables $X(V_1)$ and $X(V_2)$, respectively. That is

$$E[X(V_1)] = \frac{1 + 2 \times 5 + 9}{4} = 5,$$

and

$$E[X(V_2)] = \frac{4 + 2 \times 8 + 12}{4} = 8.$$

Finally, by Definition 3, the expected value of X is

$$E[X] = E[X(V_1)] \times 0.2 + E[X(V_2)] \times 0.8 = 7.4.$$

3   Regression Model Based on Fuzzy Random Variables

Fuzzy Arithmetic or fuzzy Arithmetic operations with fuzzy numbers by the extension principle [16], [17], [36] have been studied in [1]-[15]. These studies have been done through the concept of possibility. In 1984, Sanchez [19] discussed the solution of fuzzy equations in the same way as described in the fuzzy relational equations. Tanaka and Watada [24] pointed out that Fuzzy equations discribed by Sanchez can be regarded as possibilistic equations from our viewpoint.

A possibilistic system has been applied to the linear regression analysis [20],[21]. In this paper our main concerns are on properties of possibilistic linear model and a new formulation of fuzzy linear regression model in the case of fuzzy random variables. A possibilistic linear system can be used as a model for interval analysis whose examples are possibilistic linear regression discussed here [22].

Regression Model of Fuzzy Random Data. Table 1 illustrates data dealt here. Y_i, X_{ik} for all $i = 1. \cdots, N$ and $k = 1, \cdots, K$ are fuzzy random data defined probabilistically as

$$Y_i = \bigcup_{t=1}^{M_{Y_i}} \{(Y_i^t, Y_i^{t,l}, Y_i^{t,r})_{LR}, p_i^t\}, \quad X_{ik} = \bigcup_{t=1}^{M_{X_{ik}}} \{(X_{ik}^t, X_{ik}^{t,l}, X_{ik}^{t,r})_{LR}, q_{ik}^t\}, \quad (4)$$

respectively. That means all values are given fuzzy numbers with its probability, where fuzzy variables $(Y_i^t, Y_i^{t,l}, Y_i^{t,r})_{LR}$ and $(X_{ik}^t, X_{ik}^{t,l}, X_{ik}^{t,r})_{LR}$ are obtained with probability p_i^t and q_{ik}^t for $i = 1, 2, \cdots, N$, $k = 1, 2, \cdots, K$ and $t = 1, 2, \cdots, M_{Y_i}$ or $t = 1, 2, \cdots, M_{X_{ik}}$, respectively.

Table 1. Input - Output Data

No.	Output	Inputs				
i	Y	X_1	X_2	$\cdots$	X_i	$\cdots,\ X_K$
1	Y_1	X_{11}	X_{12}	$\cdots$	X_{1i}	$\cdots,\ X_{1K}$
2	Y_2	X_{21}	X_{22}	$\cdots$	X_{2i}	$\cdots,\ X_{2K}$
$\vdots$	$\vdots$	$\vdots$	$\vdots$		$\vdots$	$\vdots$
j	Y_j	X_{j1}	X_{j2}	$\cdots$	X_{ji}	$\cdots,\ X_{jK}$
$\vdots$	$\vdots$	$\vdots$	$\vdots$		$\vdots$	$\vdots$
N	Y_N	X_{N1}	X_{N2}	$\cdots$	X_{Ni}	$\cdots\ X_{NK}$

Table 2. Expectation of Input - Output Data

No.	Output	Inputs				
i	$E(Y)$	$E(X_1)$	$E(X_2)$	$\cdots$	$E(X_i)$	$\cdots,\ E(X_K)$
1	$E(Y_1)$	$E(X_{11})$	$E(X_{12})$	$\cdots$	$E(X_{1i})$	$\cdots,\ E(X_{1K})$
2	$E(Y_2)$	$E(X_{21})$	$E(X_{22})$	$\cdots$	$E(X_{2i})$	$\cdots,\ E(X_{2K})$
$\vdots$	$\vdots$	$\vdots$	$\vdots$		$\vdots$	$\vdots$
j	$E(Y_j)$	$E(X_{j1})$	$E(X_{j2})$	$\cdots$	$E(X_{ji})$	$\cdots,\ E(X_{jK})$
$\vdots$	$\vdots$	$\vdots$	$\vdots$		$\vdots$	$\vdots$
N	$E(Y_N)$	$E(X_{N1})$	$E(X_{N2})$	$\cdots$	$E(X_{Ni})$	$\cdots\ E(X_{NK})$

Let us denote fuzzy linear model using fuzzy coefficients $\bar{A}_1{}^*, \cdots, \bar{A}_K{}^*$ as follows:

$$\bar{Y}^* = \bar{A}_1{}^* X_1 + \cdots + \bar{A}_N{}^* X_K, \tag{5}$$

where $\bar{Y}_i^*$ denotes estimation and $\bar{A}_i^* = (\{\bar{A}_i^{*l} + \bar{A}_i^{*r}\}/2, \bar{A}_i^{*l}, \bar{A}_i^{*r})_{LR}$ symmetric triangular fuzzy coefficient when triangular fuzzy random data X_{ik} are given for $i = 1, \cdots, N$ and $k = 1, \cdots, K$ as shown in Table 1.

When we know fuzzy random output $Y_i = \bigcup_{t=1}^{M_{Y_i}} \{(Y_i^t, Y_i^{t,l}, Y_i^{t,r})_{LR}, p_i^t\}$ are given at the same time, we can decide fuzzy random linear model so that the estimation of the model includes all given fuzzy random outputs. Therefore, the following relation should hold:

$$\bar{Y}_i^* = \bar{A}_1{}^* X_{i1} + \cdots + \bar{A}_N{}^* X_{iK} \underset{FR}{\supset} Y_i, \quad i = 1, \dots N \tag{6}$$

where $\underset{FR}{\supset}$ is a fuzzy random inclusion relation. The fuzzy random inclusion relation $\underset{FR}{\supset}$ can be defined in various ways, for instance, the chance based inclusion, the expected value based inclusion, and so on. In this paper, we employ the expected value based inclusion as illustrated in Equation (8), which combines the fuzzy inclusion relation at grade h with expected values of fuzzy random vairbles.

Under the fuzzy Arithmetic calculations, the problem to obtain a fuzzy linear regression model results in the following mathematical programming problem:

[Regression model of fuzzy random data]

$$\left.\begin{aligned} &\min_{\bar{A}}\ J(\bar{A}) = \sum_{k=1}^{K} (\bar{A}_k^r - \bar{A}_k^l) \\ &\text{subject to} \\ &\quad \bar{A}_i^r \geq \bar{A}_i^l, \\ &\quad \bar{Y}_i^* = \bar{A}_1{}^* X_{i1} + \cdots + \bar{A}_n{}^* X_{iK} \underset{FR}{\supset} Y_i, \\ &\quad \text{for}\ \ i = 1, \dots N. \end{aligned}\right\} \tag{7}$$

This formulation is defined using fuzzy random variables. But when the expectation of these data [34] are taken as shown in Table 2, we can formulate the new model. This model is corresponding to a conventional regression model but we can consider confidence interval when we take variance of fuzzy random variable into consideration. The discussion of variance of fuzzy random variable will be left for a subsequent paper.

[Regression model of expected fuzzy random data] Let us consider the expectation of fuzzy random variable, which are given in Tables 2 and 1, respectively. Then it will be a conventional fuzzy regression model as in the followings:

$$\left.\begin{array}{l} \min_{\bar{A}} \ J(\bar{A}) = \sum_{k=1}^{K} (\bar{A}_k^r - \bar{A}_k^l) \\[2mm] \text{subject to} \\[1mm] \bar{A}_i^r \geq \bar{A}_i^l, \\ \bar{Y}_i^* = \bar{A}_1^* E(X_{i1}) + \cdots + \bar{A}_n^* E(X_{iK}) \underset{\sim}{\underset{h}{\supset}} E(Y_i), \\[2mm] \text{for} \quad i = 1, \ldots N \end{array}\right\} \tag{8}$$

where $\underset{\sim}{\underset{h}{\supset}}$ denotes fuzzy inclusion relation at grade h.

Although the mathematical programming (7) is obtained to solve the fuzzy regression model, the problem is not easy to solve, because the product between a fuzzy parameter and a fuzzy value distorts the shape of a triangular fuzzy number. The problem results in heuristic algorithm as mentioned in Watada *et al.* [31]. On the other hand, the mathematical programming (8) is a conventional fuzzy regression model. It is easily solved. This model can be corresponding to a conventional regression model. When we consider the variance of fuzzy random

Table 3. Expectation of Input - Output Fuzzy Random Data

No.	Output	Inputs		
i	$E(Y)$	$E(X_1)$	$\cdots$	$E(X_K)$
1	$\sum\limits_{t=1}^{M_{Y_1}} \dfrac{2\bar{Y}_1^t + \bar{Y}_1^{t,r} + \bar{Y}_1^{t,l}}{4} p_1^t$	$\sum\limits_{t=1}^{M_{X_{11}}} \dfrac{2\bar{X}_{11}^t + \bar{X}_{11}^{t,r} + \bar{X}_{11}^{t,l}}{4} q_{11}^t$	$\cdots$	$\sum\limits_{t=1}^{M_{X_{1K}}} \dfrac{2\bar{X}_{1K}^t + \bar{X}_{1K}^{t,r} + \bar{X}_{1K}^{t,l}}{4} q_{1K}^t$
2	$\sum\limits_{t=1}^{M_{Y_2}} \dfrac{2\bar{Y}_2^t + \bar{Y}_2^{t,r} + \bar{Y}_2^{t,l}}{4} p_2^t$	$\sum\limits_{t=1}^{M_{X_{21}}} \dfrac{2\bar{X}_{21}^t + \bar{X}_{21}^{t,r} + \bar{X}_{21}^{t,l}}{4} q_{21}^t$	$\cdots$	$\sum\limits_{t=1}^{M_{X_{2K}}} \dfrac{2\bar{X}_{2K}^t + \bar{X}_{2K}^{t,r} + \bar{X}_{2K}^{t,l}}{4} q_{2K}^t$
$\vdots$	$\vdots$	$\vdots$		$\vdots$
j	$\sum\limits_{t=1}^{M_{Y_j}} \dfrac{2\bar{Y}_j^t + \bar{Y}_j^{t,r} + \bar{Y}_j^{t,r}}{4} p_j^t$	$\sum\limits_{t=1}^{M_{X_{j1}}} \dfrac{2\bar{X}_{j1}^t + \bar{X}_{j1}^{t,r} + \bar{X}_{j1}^{t,l}}{4} q_{j1}^t$	$\cdots$	$\sum\limits_{t=1}^{M_{X_{jK}}} \dfrac{2\bar{X}_{jK}^t + \bar{X}_{jK}^{t,r} + \bar{X}_{jK}^{t,l}}{4} q_{jK}^t$
$\vdots$	$\vdots$	$\vdots$		$\vdots$
N	$\sum\limits_{t=1}^{M_{Y_N}} \dfrac{2\bar{Y}_N^t + \bar{Y}_N^{t,r} + \bar{Y}_N^{t,r}}{4} p_N^t$	$\sum\limits_{t=1}^{M_{X_{N1}}} \dfrac{2\bar{X}_{N1}^t + \bar{X}_{N1}^{t,r} + \bar{X}_{N1}^{t,l}}{4} q_{N1}^t$	$\cdots$	$\sum\limits_{t=1}^{M_{X_{NK}}} \dfrac{2\bar{X}_{NK}^t + \bar{X}_{NK}^{t,r} + \bar{X}_{NK}^{t,l}}{4} q_{NK}^t$

variable, it is possible to build a confidence interval for the regression model of expected fuzzy random variables.

4   An Explanatory Example

Next, as an explanatory example for the usage of the model, we will discuss the fuzzy regression model based on the expectation of fuzzy random variables using triangular fuzzy numbers as shown in Table 3. Let $h = 0$. That is, we take the expectation of all fuzzy random data as defined in Definition 3.

[Fuzzy regression model of expected fuzzy random data]

$$\min_{\bar{A}} \; J(\bar{A}) = \sum_k (\bar{A}_k^r - \bar{A}_k^l)$$

subject to

$$\bar{A}_k^r \geq \bar{A}_k^l,$$

$$\sum_{t=1}^{M_{Y_i}} \frac{|2\bar{Y}_i + \bar{Y}_i^r + \bar{Y}_i^l|}{4} p_i^t \leq \sum_{k=1}^{K} \bar{A}_k^r \left(\sum_{t=1}^{M_{X_{ik}}} \frac{|2X_{ik} + X_{ik}^r + X_{ik}^l|}{4} q_{ik}^t \right),$$

$$\sum_{t=1}^{M_{Y_i}} \frac{|2\bar{Y}_i + \bar{Y}_i^r + \bar{Y}_i^l|}{4} p_i^t \geq \sum_{k=1}^{K} \bar{A}_k^l \left(\sum_{t=1}^{M_{X_{ik}}} \frac{|2X_{ik} + X_{ik}^r + X_{ik}^l|}{4} q_{ik}^t \right)$$

$$\text{for} \quad i = 1, \ldots, N. \tag{9}$$

This calculation is obtained directly from Examples 1 and 3. The model (9) is a simple LP problem. The model (9) can be obtained the solusion so that the regression model includes all the expectations.

5   Concluding Remarks

In this paper, employing fuzzy random variables, we built a regression model for fuzzy random data. By taking their expectation we can simplify the model as shown in Equation (9). This model illustrates that we can take the expectation of fuzzy random data without considering the fuzziness of fuzzy random variables. Therefore, the width of fuzzy regression model will be narrowed because we do not consider the fuzziness of all fuzzy random data. But this result is more significant in real applications. On the other hand, we can discuss confidence interval considering the variance based on fuzzy random variables in the future work.

References

1. Bass, S.M., Kwakernaak, H.: Rating and ranking of multiple aspect alternatives using fuzzy sets. Automatica 13, 47–58 (1977)
2. Cooman, G.D.: Possibility theory III. International Journal of General Systems 25, 352–371 (1997)

3. Deschrijver, G.: Arithmetic operators in interval-valued fuzzy set theory. Information Sciences 177, 2906–2924 (2007)
4. Dubois, D., Prade, H.: Additions of interactive fuzzy numbers. IEEE Trans. AC-26(4), 926–936 (1981)
5. Dubois, D., Prade, H.: Ranking fuzzy numbers in the seting of possibility theory. Information Sciences 30, 183–224 (1983)
6. Dubois, D.: Linear programming with fuzzy data. FIP-84 at Kauai, Hawaii (July 22-26, 1984)
7. Kruse, R., Meyer, K.D.: Statistics with Vague Data. D. Reidel Publishing Company, Dordrecht (1987)
8. Kwakernaak, H.: Fuzzy random variables–I. Definitions and theorems, Information Sciences 15, 1–29 (1978)
9. Kwakernaak, H.: Fuzzy random variables–II. Algorithm and examples, Information Sciences 17, 253–278 (1979)
10. Liu, B., Liu, Y.-K.: Expected value of fuzzy variable and fuzzy expected value models. IEEE Transactions on Fuzzy Systems 10, 445–450 (2002)
11. Liu, Y.-K., Liu, B.: Fuzzy random variable: A scalar expected value operator. Fuzzy Optimization and Decision Making 2, 143–160 (2003)
12. Liu, Y.-K., Liu, B.: On minimum-risk problems in fuzzy random decision systems. Computers & Operations Research 32, 257–283 (2005)
13. López-Diaz, M., Gil, M.A.: Constructive definitions of fuzzy random variables. Statistics and Probability Letters 36, 135–143 (1997)
14. Luhandjula, M.K.: Fuzziness and randomness in an optimization framework. Fuzzy Sets and Systems 77, 291–297 (1996)
15. Nahmias, S.: Fuzzy variables. Fuzzy Sets and Systems 1(2), 97–111 (1978)
16. Negoita, C.V., Ralescu, D.A.: Application of Fuzzy Sets to Systems Analsyis, pp. 12–24. Birkhauser Verlag, Basel (1975)
17. Nguyen, H.T.: A note on the extension principle for fuzzy sets. J. of Math. Appli. 64, 369–380 (1978)
18. Puri, M.L., Ralescu, D.A.: Fuzzy random variables. J. Math. Anal. Appl. 114, 409–422 (1986)
19. Sanchez, E.: Solution of fuzzy equations with extended operations. Fuzzy Sets and Systems 12, 237–248 (1984)
20. Tanaka, H., Uejima, S., Asai, K.: Linear regression analysis with fuzzy model. IEEE Trans. SMC-12, 903–907 (1982)
21. Tanaka, H., Shimomura, T., Watada, J., Asai, K.: Fuzzy linear regression analysis of the number of staff in local goverment, FIP-84 at kauai, Hawaii (July 22-26, 1984)
22. Tanaka, H., Asai, K.: Fuzzy linear programming problems with fuzzy number. Fuzzy Sets and Systems 13(1), 1–10 (1984)
23. Tanaka, H., Hayashi, I., Watada, J.: Possibilistic linear regression for fuzzy data. European J. of Operational Research 40(3), 389–396 (1989)
24. Tanaka, H., Watada, J.: Possibilistic linear systems and their application. Int. J. Fuzzy Sets and Systems 27, 275–289 (1988)
25. Toyoura, Y., Watada, J., Khalid, M., Yusof, R.: Formulation of linguistic regression model based on natural words. Soft Computing Journal 8(10), 681–688 (2004)
26. Yabuuchi, Y., Watada, J.: Formulation of fuzzy switching regression model. Official Journal of Japan Association of Intteligent Information and Fuzzy Systems 16(1), 53–59 (2004)

27. Watada, J., Tanaka, H.: The perspective of possibility theory in decision making. In: Sawaragi, Y., Inoue, K., Nakayama, H. (eds.) Post Conference Book, Multiple Criteria Decision Making - Toward Interactive Intelligent Decision Support Systems, VII-th Int. Conf., pp. 328–337. Springer, Heidelberg (1986)

28. Watada, J., Tanaka, H.: Fuzzy quantification methods. In: Proceedings, the 2nd IFSA Congress, Tokyo, pp. 66–69 (1987)

29. Watada, J., Tanaka, H., Asai, K.: Analysis of time-series data by posibilistic model. In: Proceedings, Int.Workshop on Fuzzy System Applications, Fukuoka, pp. 228–233 (1988)

30. Watada, J.: Fuzzy time-series analysis and forecasting of sales volume. In: Kacprzyk, J., Fedrizzi, M. (eds.) Fuzzy Regression Analysis, pp. 211–227. Omnitech Press, Warsaw (1992)

31. Watada, J.: 5. 5 Multi-attribute decision-making, applied fuzzy system. In: Terano, T., Asai, K., Sugeno, M. (eds.) Applications in Business, AP Professional, ch. 5, pp. 231–268 (1994)

32. Watada, J.: Possibilistic time-series analysis and its analysis of consumption. In: Duboir, D., Yager, M.M. (eds.) Fuzzy Information Engineering, pp. 187–200. John Wiley Sons, Inc., Chichester (1996)

33. Watada, J., Mizunuma, H.: Fuzzy switching regression model based on genetic algorithm. In: Proceedings, The 7th International Fuzzy Systems Association World Congress(IFSA 1997), Prague, Czeck Republic, pp. 113–118 (1997)

34. Wang, S., Watada, J.: T-independence condition for fuzzy random vector based on continuous triangular norms. Journal of Uncertain Systems 2(2), 155–160 (2008)

35. Wang, S., Watada, J.: L-R style fuzzy random renewal process with a queueing application. In: Proc. of the 5th International Symposium of Management Engineering, Kitakyushu, Japan, pp. 312–319 (2008)

36. Zadeh, L.A.: The concept of a linguistic variable and its application to approximate reasoning-I. Inform. Sci. 8, 199–249 (1975)

Biological Clustering Method for
Logistic Place Decision Making

Rohani Binti Abu Bakar and Junzo Watada

Graduate School of Information, Production and Systems, Waseda University
2-7, Hibikino, Wakamatsu-Ku, Kitakyushu-Shi, Fukuoka-Ken, 808-0135, Japan
Tel.: +81-93-692-5179; Fax: +81-93-692-5179
rohani@ump.edu.my, junzow@osb.att.ne.jp

Abstract. One of the main tasks in supply chain network is to identify the determination of logistic location. The main factors could influence the selections are costs and profits for the company itself. Most appropriate place is urgently essentials in today business world to ensure the company could be more competitive then other competitors in the industry. A lot of considerations should be taken during selecting a location to build a logistic place to serve other retailers city effectively. Currently, there are so many algorithms based on different approaches are proposed by other researchers. Thus, this paper intends to propose DNA computing approach to solve the problem. In this study, a cluster-based approach is employ when all cities are grouped before we choose a right city as distribution center. A case study is presented at the end of this paper to illustrate how the proposed technique works.

Keywords: Biologically inspired computing, DNA computing, Logistic problem, Logistic location, cluster-based, Determination.

1 Introduction

It is critical to that products are moved efficiency and effectively from raw material sites to processing facilities, component fabrication plants and finished products assembly plants at distribution centers then to retailers and last to customers. It is reported that approximately 10% of the gross domestic product is devoted to supply-related activities [1].

Generally, three main factors will influence in selecting the location of DCs location. There are; (a) delivery time, (b) transportation cost [2] and (c) demands frequency [3] should be considered with a decision. Among these three factors, the cost of transportation is one of the most important factors to be considered. Transportation cost from DCs to retailer cities will be given impact on the total operational costs of a company especially when the company deals with consumer products such as food and etc. Maranzana points out that the location of factories, warehouses and supply points in general is often influenced by transportation costs [4].

In order to select the most appropriate place to build warehouses or logistic centers as a DCs, it is a difficult and complicated task. We need, supposedly to consider all

I. Lovrek, R.J. Howlett, and L.C. Jain (Eds.): KES 2008, Part III, LNAI 5179, pp. 136–143, 2008.

feasible combinatorial solutions among retailer cities in calculating optimal combinatorial solution. This process can be classified as a NP-hard problem. However, to calculate all possible combinatorial by using a conventional computer is a very difficult task, may be almost impossible, especially when we deal with a huge number of subjects or variables. It is because of limitation that a machine itself such as data storage, processing capabilities and so on.

Thus, this paper intends to purpose an algorithm to solve a DCLP problem by using a DNA computing approach. As a reminder, the organization of this paper is as follows. First, we introduce our study background before we will briefly discuss current researches and trend in the determination distribution center community in Section 2. In Section 3, we discuss our proposed solution. Concluding remarks will be provided in Section 4.

2 Determination of Distribution Location Center

There are a lot of researches regarding the determination of the most appropriate location for DCs. This problem is one of the most important topics to discuss among operations research and transportation research communities. Distribution centers should act as a center for a company to facilitate the flow of its raw material and finished goods [5]. Dixon [6] stated that companies generally have two options in order to select a DC. The first option is to purchase or lease a center and operate it with internal resources and the second option is to use a center owned or leased by an outside provider who also serves as an operator. But the question is where the location for this DC should be.

Cui and Li have proposed genetic algorithm combining with simulated annealing to give an optimal solution [7]. The algorithm sets the initial values and the parameters of the genetic algorithm. The population size of this algorithm is 60 with the maximum number of generations is 300. At the end of the algorithm, the proposed algorithm can search the optimal solution in the analytical space.

On the other hand, Chen has proposed a new fuzzy multiple criteria group decision-making method to solve a selection problem of distribution center location [8]. Chen claims that his proposed method can help a decision maker make a suitable decision under a fuzzy environment and also be performing easily with other aggregation functions to accumulate the fuzzy ratings of experts or decision-makers.

Barreto et al. [9] have proposed a clustering analysis approach in order to group retailer cities before selecting the appropriate one to serve as a DC. Specifically, Baretto et al. have introduced a technique called the sequential heuristic type of a distribution-first, location-second approach. In this technique, Baretto et al. construct a group of customers first based on their capacity limits before determining the distribution route in each customer group. The process will become an iteration one in order to improve the route itself before they can locate the DCs and assign the routes to them.

For comprehensive study related to issues of model and method in finding optimum distribution center in supply chain network, Nagy and Sahli[10] offer a good review and updates current stages of researches for this topic. However, in this paper we employ a DNA computing capability to identify the most appropriate location for DCs and the optimal assignment of each local city to some DC.

As mentioned above, for our study, we only considered to find out the most effective cost for local transportation. The local transportation means transportation from DCs to their retailers. We do not consider the cost of transportation from a plant or manufacturer to DCs. It is because we consider that the goods which distribute to retailers will be stored at DCs that we selected.

To determine the effective costs, we consider the distances between a local city or a retailer and another local city. So that, in this study, we intend to identify which retailer city could serve as a DC when it has the shortest distance with the other local cities in their region.

3   Biological Inspired Computing to Solve Distribution Center Location Problem (DCLP)

In this section, we will discuss details regarding our proposed algorithm to solve DCLP in molecular computing points of view. This work is extension work from [13]. In solving this problem, we apply clustering whereby all retailers are clustered into several cluster groups before we assign one retailer in every group as a DC. In order to select the appropriate retailer that will serve as a DC, we calculate and consider all possible total distances from a retailer point and the shortest one should be selected as a DC.

Before we discuss details of our proposed algorithm, let us consider the number of retailers and a potential DC will be selected among them. We consider a map illustrated in Fig. 1 as a location where the retailer cities are located. We intend to select an optimal place from 22 cities that will serve as DCs in this supply chain management. In our study, we used ordered value to represent real value of distances.

Even though values will be changed, but the end result should show the same patterns. Fig. 2 shows a general algorithm that has been used to solve a proposed problem. Based on purposed algorithm, a number of steps are performed. The computation processes are discussed as below.

Step 1. All retailer cities are denoted as $C_A, C_B....., C_V$. In the same time, all retailers are considered too as a candidate for a distribution center, so we present it as $D_A, D_B...., D_V$. On the other hand, $D_1, D_2,...., D_{23}$ represent cost or distance in ordered values between two cities.

Step 2. The distances between retailers are presented in matrix form as shown as *matrix* Δ. As we mentioned before, we mapped from these values by using ordered values and perform a new matrix by using these ordered values. A new matrix is shown as *matrix O*. Table 1 shows the ordered value that we changed real values.

Step 3. Assign a unique DNA sequence for each retailer city. Table 2 shows an example of DNA sequences that represent a part of the retailer cities and distances. Distances between retailer cities are presented by constant length random sequences that increase by a constant factor as proposed by Narayanan and Zorbalas in [11] to solve traveling salesman problem (TSP).

To generate DNA sequences that represent DCs and retailer cities, we used software named as DNASequenceGenerator developed by [12] where we select a melting

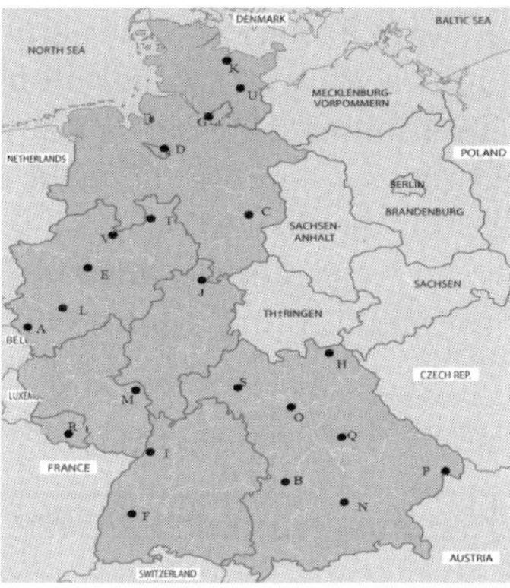

Fig. 1. The location where retailers cities are located for purposes of study

```
START
1.     Synthesized strands (DCi, Ci) into T0
2.     In T0
3.        for i ← 1 to N do
4.             Merge +( DCi, Ci) and -( DCi)
5.        end for
6.        if T0 is not empty

                              n
7.                if strands  ∀ { ∑ C and mark > 0}
                             i=0

8.                   Move to T1
9.                 else Reject !
10.              end if
11.           end if
12.    Sort T1 based on strand length
13.    Pick-up an appropriate strand as a final result
            (e.g a shortest one, or strand that fulfill initial stated
       requirement)
14.    Read out detail strands
END
```

Fig. 2. A proposed algorithm based on biological computing to solve determination problem logistic location using clustering approach

temperature as a constraint. On the other hand, we generate another special sequence named as *mark*. *Mark* is a special oligos that generated to include in DCs encoding scheme. The objective of including *mark* sequence in DCs encoding scheme is because we need to differentiate sequences that represent DCs (Dn) and retailer cities (Cn).

Step 4. Synthesize the oligos for every retailer and distances as shown in Tab.2 based on the proposed encoding scheme and the distance. Fig.3 and Fig.4 show the encoding design scheme to represent a candidate of DCs and retailer cities. In Fig.3, the encoding scheme to represent DCs is started with second part (10-mer) of oligos that represent distribution center i (e.g : D_A) followed by special oligos that we created and named as *mark* for 20-mer and another first half from 20-mer of distribution center $i+1$ (e.g : D_B).

On the other hands, Fig.4 shows another encoding scheme that represents retailer cities. We started our encoding scheme with second half from 20-mer of candidate of distribution center i (e.g : D_B) followed by oligos that represent distance between distribution center i to retailer city i (e.g : D2). Then, it will be followed by retailer city I (e.g : C_A) and followed by the first half of candidate distribution center i . So that, the oligos length will directly represent the different distances from retailer cities to their candidate distribution center.

Table 1. Ordered value to represent real distance between two cities in the case study

Real Distances	Ordered Value
0 – 14	0
15 – 25	1
26 – 35	2
⋮	⋮
236 – 245	23

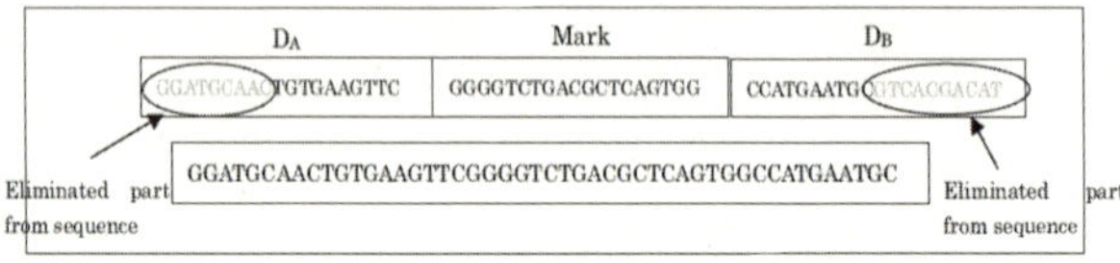

Fig. 3. An example of DNA sequence to represent candidate of DCs for retailer A and B

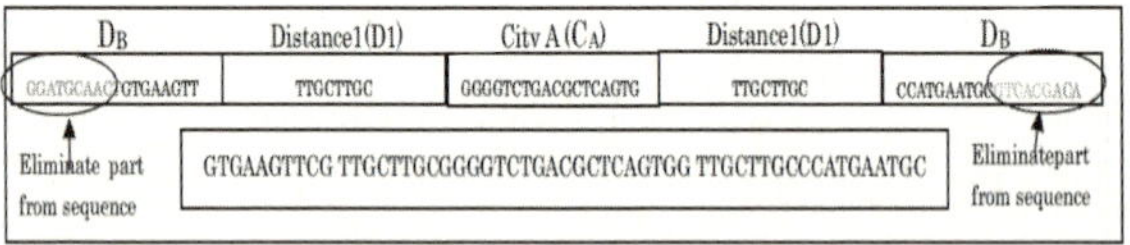

Fig. 4. An example of DNA sequence to represent retailer city A and city B as a candidate of DCs for city A

Table 2. Sample of DNA sequences that represent a part of retailer cities and ordered distances

City / Distance	DNA Sequences
C_A	GGATGCAACTGTGAAGTTCG
C_B	CCATGAATGCGTCACGACAT
⋮	⋮
C_U	CGGTTGTGTTCAACGACACA
C_V	ATCCTGTACGTGCTGTTCGT
D_1	GGT
D_2	TGAGTG
D_3	CCGCAACGA
D_4	GTTTTTCGTGCG

Step 5. All the synthesized oligos are then poured into a test tube (denoted as T_0) for ligation and hybridization processes. All the sequences react to each other to perform possible solutions. In this process, we used strand complementary of each candidate distribution center (DC) sequence to allow two different sequences bind together. Fig. 5 illustrates how synthesized sequences bind together (in encoding scheme design mode) in *T0* during ligation and hybridization process to perform all feasible combinatorial solutions.

Step 6. In this stage, we need to examine the selection that only strands (that we got from the previous stage) include all retailer city and at least one candidate of DC to be considered as a final solution. We used a complementary strand of each retailer cities and complementary of special *mark* that we included before for examine process. In this stage, we used magnetic beads separation procedure where we attached complementary strands of *mark oligos,* that we insert in candidates of DCs sequences, to tiny magnetic beads.

When the complementary oligos anneal to the target strands, that means the strand include at least one candidates of DCs, a magnet is used to pull out the solution with the target strands has been attached to magnetic beads will be extracted and put it in another new test tube (denoted as *T1*). After that, in order to ensure all retailer cities will be included in the feasible strand, we need to run another magnetic bead separation process on strands that available in *T1*. In this time, complementary strands of each retailer city will be bind into tiny magnetic beads. Only strands that included all retailer cities and at least one *mark* will be consider to the next processes.

Step 7. In this stage, we examine strands that only available in *T1*. Gel electrophoresis process is used as a separation process of the most economical path from other possible combinations. Most economical path in this study are reflex to shortest strand that separated by gel electrophoresis process. When we find the most economical one, we examine the strand by using special microscope (e.g atomic force microscope (AFM)) to find out and calculate how many *marks* (that we inserted before in encoding oligos, for candidates of DCs, stage) manually. Each *mark* represents one cluster point, meaning all retailers cities within special marks should gather in one DC group and the city that tabs with special mark should serve as DCs. Table 3 shows total distances from DCs to their retailers in ordered value and real value. Meanwhile Fig.7 shows a location for each cluster group with selected city (circled with small circle) served as a distribution center for their cluster.

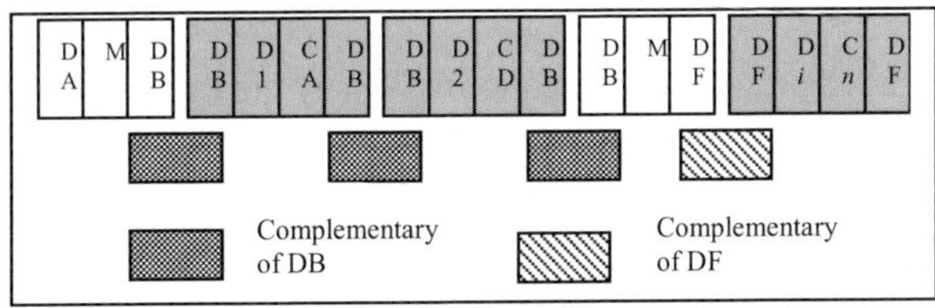

Fig. 5. The illustration on how sequences bind together to present all feasible combinatorial solutions with complementary of candidates of DCs (e.g: DB and DF) react as a binder or linker to bind two different strands

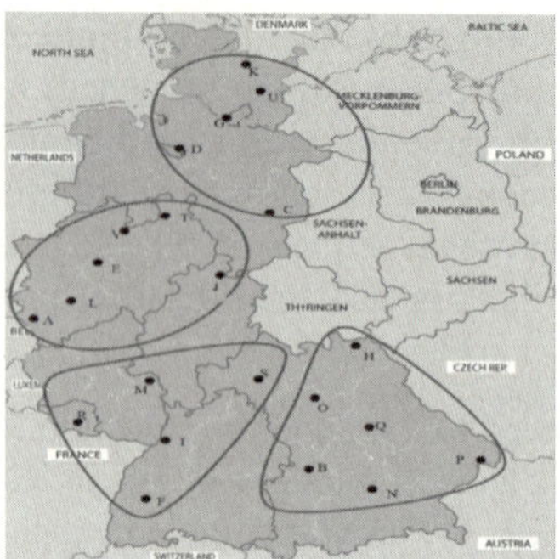

Fig. 6. Results from step 7, where 22 cities are divided into 4 clusters and each cluster group has their own distribution center as indicates in a small circle

Table 3. Total distance from DCs to their retailers in each cluster group

DCs	Cities	Distances	Total (In ordered value)	Real Distance
G	C	4(51)	9	127
	D	2(31)		
	K	2(26)		
	U	1(19)		
E	A	2 (31)	15	198
	J	8 (86)		
	L	1 (17)		
	T	3(40)		
	V	1(24)		
I	F	3(38)	10	130
	M	1 (17)		
	R	3 (36)		
	S	3 (39)		
Q	B	3 (37)	13	154
	H	4 (47)		
	N	2(35)		
	O	2 (35)		
	P	2 (34)		

4   Concluding Remarks

In this paper, we have presented an idea to implement clustering method in molecular algorithm by clustering to identify an appropriate allocation of a distribution center of products for end customers in a supply chain network. The proposed algorithm is able to find the economical place for setting up a warehouse as a DC and to identify which retailer should connect to which DC under the consideration of the shortest distance between a retailer and a DC. We considered a total distance between DCs and retailers in assigning them because of the frequency of travels in a certain time period to deliver products for example; weekly or monthly will be given directly impacts on the cost of transportation itself.

In this study, we manage to solve the proposed problem without considering any prior information regarding the expected end results. However, all results that we obtained from this study at this stage is through a simulation program, not from wet experiments yet. Our current stage is still in theoretical work. So, many works should be done in ahead in order to prove the algorithm which is workable in a real life situation.

One of the major problems in this study is we cannot represent directly a real distance value into DNA sequence because it will be used a huge number of sequences. So, an efficient method to represent a real value in DNA computing is really urgent needed in order to make DNA computing can solve easily real world problem.

Acknowledgments. The first author would like to thank University Malaysia Pahang, Kementerian Pengajian Tinggi Malaysia for supporting study leave and GCOE program of Waseda University for supporting to attend KES2008.

References

[1] El-Houssaine, A., Birger, R., Landeghem, H.V.: Modelling inventory routing problems in supply chains of high comsumption products. European Journal of Operational Research 69(3), 1048–1063 (2006)

[2] Nozick, K.L., Turnquist, M.A.: A two-echelon inventory allocation and distribution center location analysis. Transportation Research Part E 37, 425–441 (2001)

[3] Groothedde, B., Ruijgrok, C., Tavasszy, L.: Towards collaborative, intermodal hub networks: A case study in the fast moving consumer goods market. Transportation Research Part E 41, 567–583 (2005)

[4] Maranzana, F.E.: On the location of supply points to minimize transport costs. Operational Research Quarterly 15, 261–270 (1964)

[5] White Paper, European transport policy for 2010 : Time to decide, The European Commission, COM (2001) 370

[6] Dixon, R., The important of distribution center (last accessed on July 20, 2006), http://www.devicelink.com/mddi/achive/98/03/006.html

[7] Cui, G.B., Li, Y.J.: Study on the location of distribution center: A genetic algorithm combining mechanism of simulated annealing. In: Proceedings of The Third International Conference on Machine Learning and Cybernetics, Shanghai, China, August 26 - 29, 2004 (2004)

[8] Chen-Tung, C.: A fuzzy approach to select the location of the distribution center. Journal of Fuzzy Sets and Systems 118(1), 65–73 (2001)

[9] Baretto, S., Ferreira, C., Paixao, J., Santos, B.S.: Using clustering analysis in capacitated location-routing problem. European Journal of Operation Research 179(3), 968–977 (2006)

[10] Nagy, G., Sahli, S.: Location-Routing, Issues, models and methods. European Journal of Operational Research 177, 649–672 (2007)

[11] Narayanan, A., Zorbalas, S.: DNA computing for efficient encoding of paths. In: Proceedings of Genetic Algorithms, pp. 718–723 (1998)

[12] Udo, F., Sam, S., Wolfgang, B., Hilmar, R.: DNA sequence generator: A program for the construction of DNA sequences. In: Seventh International Workshop on DNA Based Computer, pp. 23–32 (2001)

[13] Bakar, R.A.B., Watada, J.: A Biologically Inspired Computing Approach to Solve Cluster-Based Determination of Logistic Problem, will be appear at Special Issue on The 20th Anniversary Special Issue of International Journal of Biomedical Soft Computing and Human Sciences

Web Application Construction by Group Work and Practice

Kouji Yoshida[1], Isao Miyaji[2], Hiroshi Ichimura[3], and Kunihiro Yamada[4]

[1] Shonan Institute of Technology
1-1-25, Tsujido Nishikaigan Fujisawa Kanagawa 251-8511, Japan
[2] Okayama University of Science, 1-1 Ridaicho Okayama Okayama 700-0005, Japan
[3] Salesian Polytechnic, 8-6-4 Oyamagaoka Machida Tokyo 194-0215, Japan
[4] Tokai University, 1117 Kitakinme Hiratsuka Kanagawa 259-1292, Japan
`yoshidak@info.shonan-it.ac.jp, miyaji@mis.ous.ac.jp,`
`ichimura@salesio-sp.ac.jp, yamadaku@tokai.ac.jp`

Abstract. In recent years, it is important that information technology (IT) students have the experience of developing software development in a group similar to that of large-scale software development. Many students are not only weak in communication, but also are not good at cooperation and collaboration when designing a program. Therefore, we had IT students practice group work in a seminar course and analyzed its effects and problems.

Generally, students are good at the acquisition of knowledge that is constant in distance learning, but find that it is difficult to finish a program by oneself. In distance learning, it is very important to employ mutual teaching and regular communication between students. When there is active discussion, cooperation, and differing opinions among students, new software ideas can more effectively take shape. On the other hand, group collaboration can also limit the freedom of a student's expression of individual ideas. We report that it was very important to effectively balance student collaboration with the individual's freedom to make their own decisions.

Keywords: Group work, e-Learning, Programming, Module combination, e-Collaboration, Information-technology.

1 Introduction

In recent years, there is growing interest in the progress of information and communication systems, high-speed networking, and multimedia environments. Software development becomes large scale and complicated. Therefore we don't almost make an independent system. It is increasing that students are not good at system design and communication. Therefore we must consider about the learning of good communication and practicing a connection of a program by the group work.

When we learn program development jointly, we need to be taught the same thing repeatedly in order for it to become practical knowledge. We have produced a distance education system, which is able to teach with repetition. This is a good example of innovative software development.

I. Lovrek, R.J. Howlett, and L.C. Jain (Eds.): KES 2008, Part III, LNAI 5179, pp. 144–151, 2008.

But, it is difficult to completion of a program only by using this distance learning system. When we use group work technique together and this distance learning system, we considered that it will be able to upbringing effectively.

2 Problems and Present Conditions

We display the ability, technique and knowledge where the individual stands as displayed below. (See figure 1).

After we teach the student through lectures and seminars, we noticed that there was a problem, in particular to the category next to the student. Primary, there are many students whom cannot continue to concentrate. Concentration does not continue throughout lecturing and seminars. There are students surfing the net, and games that grasp their interest. Secondly, there are many students who have no self-expression and self-assertion. In addition, when they argue with another person, they have no patience. It is an inferior ability to not comprehend another's opinion. Thirdly, they do not practice independence or self-restraint. They cannot make a plan while also needing help to decide the heading of their graduation thesis. Their tolerance for stress is poor. They are not able to bear with problems and excessive pressure. They also give up immediately and do not attend the university. To deal with these problems, it becomes important for the consciousness of a learning person to make positive behaviors an everyday habit.

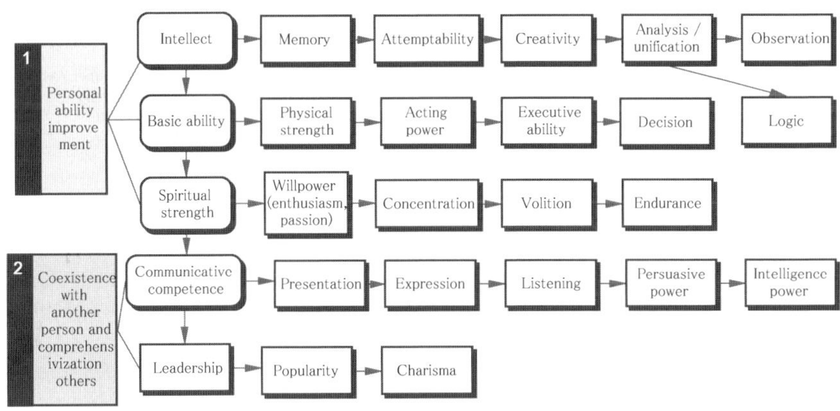

Fig. 1. Ability /technology / knowledge

To deal with these problems, we think about the system which considered using group work technique. There are three points of next superiority that group learning compare with individual learning.

(1) Students raise an intellectual power through an argument among groups and learn deepened understanding.
(2) When students have an interaction with other groups and communication, there is an effect in reinforcing it with confirmation of mutual understanding and understanding of self.

(3) When students listen to announcement of other groups, they improve of own understanding and evaluate deeply through the process.

3 System Overview

This system consists from basic software ideas from a design to programming through making learning of a program. Also students learn application from basics of programming by perl language. They learn an argument and communication through combination and a process of a program among groups. This system has a characteristic that they can experience to acquire the important points when they join a program together.

This system is supported by both the Linux and Windows operating systems. The student accesses the web browser via personal computer. Students start forward learning and group work by learning support, communication support. They start personal learning and group work by learning support, communication support. They check understanding degree by solving the issue of confirmation. They can communicate between a learner and lecturers. They can examine a function of a language with a database. This System of Group Work is shown in Figure 2.

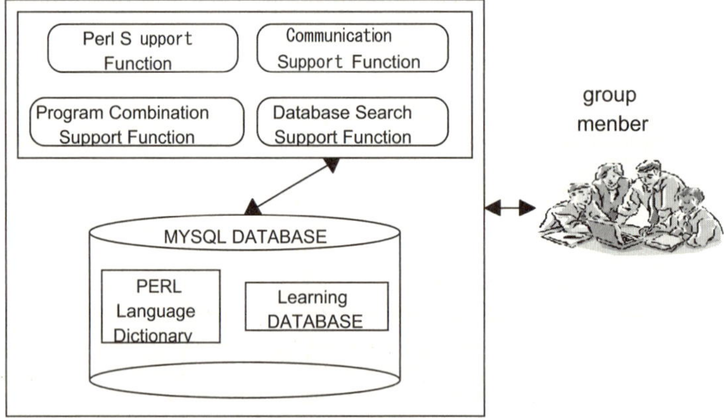

Fig. 2. The System of Group Work

3.1 A Support Function and the Usage

(1) Perl Language Support Function
When students resolve any problems, a problem of a fundamental program, the system then checks their result. It is possible to confirm a program when completed. When there is an error, an error message is sent over to the system. Students arrive at a right answer when they revise a program according to it.
 · Usage of The Perl language
    The program classifies it cording to the item and displays a step-by-step process on how to use the Perl language functions. Each function is then entirely displayed

making it easier to examine the parameter. The main functions are printf, scanf, if-else, array, for and while.

· Exercise to understand The Perl Program

The purpose of this exercise is to confirm what content was understood by the student from using the Perl explanations given. The exercise is selective and provides a percentage representing the degree of comprehension.

· The Perl Program Exercise

This displays both the Perl program's mock validation exercise and the implementation section. Step 1 consists of the validation, and it displays the content (e.g. the parameters and results of the program). After the students input their functions into the text field, they can confirm the entry by running the program. These ideas are displayed in Figure 3.

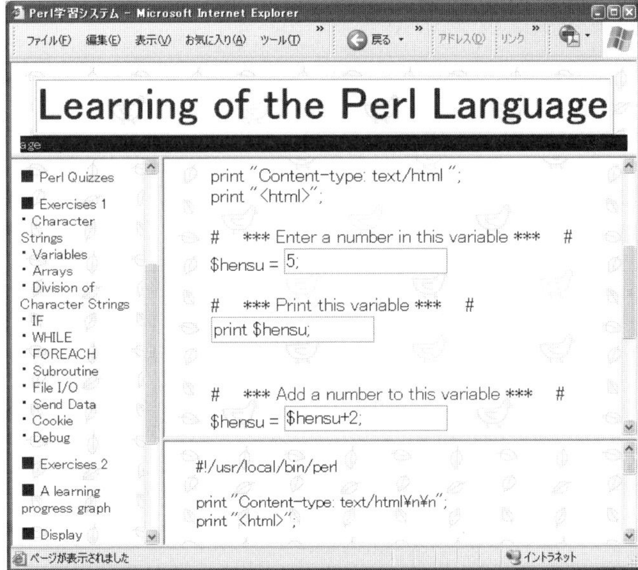

Fig. 3. The programming image

(2)Communication support Function

· Discussion Board Q&A

This section is used for questions and an information exchange between the teacher and students. This section also allows the teacher to respond or create messages. The page's background color is changed whenever a new message is posted alerting students of the message.

· Message Exchange

This section carries sent questions and other information exchanges on a peer-to-peer level. Even with encrypted mail addresses, messages can be sent with a user defined nickname.

· Mail transmission function and data exchange
· Mail receiver function
· Mail transmission and history function
· Learning schedule setting and communication function

(3) Database search support Function
We support database functions about registration, update, deletion and search. It is illustrate on the screen by using GUI showing by figure 4. We display and scroll with another frame screen, because there are many information which is detailed explanation of a function. This function supports a function of a function of PERL, JAVA, C language. In addition, we can use the contents which exchanged information as a learning database to have registration and searches.

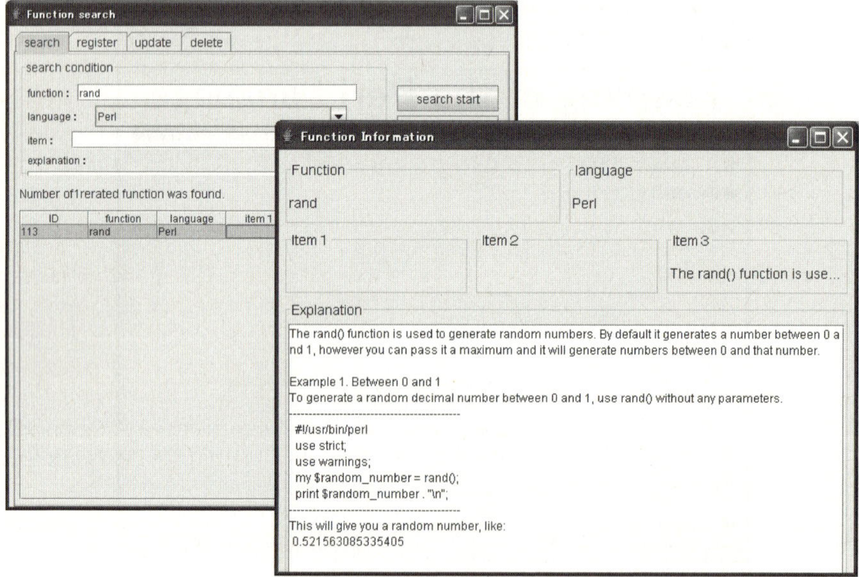

Fig. 4. The database image

3.2   Program Combination Support Function

Practice combination of a basic program to do module combination to a student, and it is support function programming to experience module combination artificially afterwards.

We enumerate contents to explain at each level next.

· An exercise of data consistency
· A revision exercise of program combination
· Para-combination practice of plural programs

4 Group Work Practice

4.1 Period and Method

We carried it out with a seminar method and distance learning system for the third grader of our department. This period was passed four months. It is described below the content that we share it to three groups and carried out. We divide 11 students with 4,4,3 members at the three groups.

On the premise of understanding about C language, students learn about the relation between HTML and CGI. Students learn oneself self-study about a PERL language function which was supported using PERL language. For next one month, students learn from basics of a PERL language to file access, and they shall understand a total PERL language. They learn a basic matter about connection of a program by a program combination support function.

Students started group work in the third month seriously. First of all, they choose a leader among the members. Nextly, they argue among groups, and decide the subject of the program which they want to make. In this circumstances, they learn using by this support system and communication by the mail or chat, and the seminar was held by 2 times a week.

4.2 Contents and Result

After they argue by a program to make among groups, they decide a title and perform basic specifications design. They decide the charge part of a program by the member after having decided external design and user interface. Students must design the data structure of the program of the charge program, and each member write the documentations. After each program that they debugged was completed, they combine programs and reach completion.

They made the problem as "(1) The bulletin board which recorded an access history (2) An electronic shopping lacing braid purchasing system (3) The Web page that combined a typing touch with a game ".

All groups had a hard time about input and output of a file by CGI and combination of a program. It is result that a program of an aim finished by the group work. And, they were improved from a combination of a program to arriving at a high technology through a generation process. It was improved about standardization and documentation at the certain level by this practice, too.

5 Questionnaire Result and Evaluation

·Questionnaire contents
We had a questionnaire for ability of basics and group learning and communication in an item of ability / knowledge of figure 1.

The learning support ability of PERL was evaluated and differed by the programming level of the students. Students still had trouble understanding some of the complex examples. It is a reference with documentation and standardization. Communication via the Internet is effective in the transmission of knowledge, but it is difficult to explain complex contents concretely. Therefore, it is necessary to explain complex ideas using not only writings, but also animations. Additionally, it is hard to convey one's feelings in words.

Students are reluctant to ask another class member to go over course material. However, students exceeded when practicing consultation and communication. Total communication improved by having performed group work. PERL's structures and mechanisms helped to activate group cooperation. Because students learned about documentation before and practiced it, the consciousness was high. Necessity of documentation seems to have been really realized again at the time of group work.

We showed the questionnaire items and their respective count results in table 1. The questionnaire items were rated on a scale of 1 to 5 and mean ratings were calculated.

Table 1. The questionnaire item and the result

	Item	Intermediate result	Last result
Group	Helping each other ?	3.5	4
	Member's good output?	3.1	3.3
	Communication between members	4	4
	Cooperation between members	3.3	4
	Problems about members	3	2.2
Program	Programming	2.6	2.6
	Debugging	2.6	2.6
Ability for basics	Ability for completion	2.5	2.8
	Responsibility	2.5	3.1
	Concentration	2.6	3.1
	Continuation power	2.5	2.8
Documentations Other	Documentation	2.6	3.5
	Understanding	3.1	3.1
	Planning	2.3	2.6
Communicative competence	Did the communication improve?	3.5	4.3
	Did the presentation improve?	2.6	3.3
	Did the persuasive power improve?	2.8	2.8
	Did the listening closely improve?	3.1	3.8
free opinion			

•Evaluation and consideration

These results should raise the awareness that there is a remarkable room for improvement in peer-to-peer communication. In addition, they collaborated to make a program through an exercise, they learned cooperation with a program of a person and difficulty when they joined. Besides, they understood that a document was reflected greatly by this communication and connection of a program. They noticed importance of documentation of student themselves and helped that they improved motivation. It was in particular remarkable about the item of document, communication and listening. This time, because we used both electronic media and communication by a meeting, then they reached to the completion which was supplemented each other in a good meaning.

6 Conclusion and Potential

Students learned at the distance learning system and support function, then it was effective in acquirement of knowledge but it was not completed in this entirely. It was effective in usefully that they exceed a group at the communication. On the other hand, electronic communications are not enough for mutual understanding on some feelings side. However, the problem was solved so that communication by meetings such as seminars had degree twice a week.

As we follow up systematically, we want to support a communications gap by an internet media and a frame of a group. In the future with construction of a system prescribing an effective frame, we proceed with construction of a system taking in a support method of the advice that guidance of a teacher is effective.

Acknowledgement

The authors wish to acknowledge those who have been of assistance to the research in way or another. This study has received assistance from the scientific research costs subsidy "18500737", in addition to the above.

References

1. Neal, L.: Virtual Classrooms and Communities. In: Proceedings of ACM GROUP 1997 Conference, Phoenix, AZ, November 16-19, 1997 (1997)
2. Dumont, R.A.: Teaching and Learning in Cyberspace. IEEE Transactions on Professional Communication 39(4), 192–204 (1996)
3. Lopez, N., Nunez, M., Rodriguez, I., Rubio, F.: Including Malicious Agents into a Collaborative Learning Envionment. In: Cerri, S.A., Gouardéres, G., Paraguaçu, F. (eds.) ITS 2002. LNCS, vol. 2363, pp. 51–60. Springer, Heidelberg (2002)
4. NSW Department of Education and Training, "Blended Learning",
   `http://www.schools.nsw.edu.au/learning/yrk12focusareas/`
   `learntech/blended/index.php`
5. Yoshida, K., Miyaji, I., Yamada, K., Ichimura, H.: Distance Learning System for Programming and Software Engineering. In: Knowledge-Based Intelligent Information and Engineering Systems: KES 2007-WIRN 2007, pp. 460–468 (2007)

Effect Analysis in Implementing IRM (Information Resources Management) to Enterprise

Kazuya Matsudaira[1], Teruhisa Ichikawa[2], and Tadanori Mizuno[3]

[1] Graduate school of Science and Technology, Shizuoka University
3-5-1 Jyohoku, Hamamatsu City, 432-8011 Japan
matsudaira@mizulab.net
[2] Faculty of Informatics, Shizuoka University
3-5-1 Jyohoku, Hamamatsu City, 432-8011 Japan
ichikawa@ia.inf.shizuoka.ac.jp
[3] Faculty of Informatics, Shizuoka University:Graduate School of Science & Technology,
Shizuoka University
3-5-1 Jyohoku, Hamamatsu City, 432-8011 Japan
mizuno@inf.shizuoka.ac.jp

Abstract. In global circumstances, speedy and high quality of information required by CEO of Enterprise for quicker and better decision-making. IRM is the new concept and methods for improving the quality of CEO's decision. However, IRM is still understood by CEO as a bit/byte tool. These phenomena is caused by CEO's ignorance of importance of Information Resource. CEO must understand IRM as new role for enterprise. This paper reports variety of tangible analysis views of IRM merits and demerits. CEO may have a vision of IRM as a powerful management approach. This analysis is based on about 100 enterprises' experiences and one third of those were successful examples of IRM implementations in Japan. KSF is CEO's endeavor who deeply knew of tangible merits. Any of them requested the quantified data of direct and measurable savings and investments. They could understand how investments into IRM is justified by savings through actual figures shown by analysis. By this analysis, CEO could imagine both the best-case scenario and worst-case scenario. Once CEO decided to move into IRM implementation after he had gotten belief of medium-case scenario that the enterprise is keeping smoothly-up annual sales volume and continues to improve revenues in step by step. Information enabled enterprise will suffer from rapid cost-up of information related expenses. Vincent [1] proposed "Information-Investment Accounting" and "Reporting in Financial Statements". Japanese Enterprises are to succeed in controlling Information related costs, after grasping those costs, then in the next step, the Accountant should restate current PL statement. Vincent's idea is that instead of not replacing the regular P&L, but providing further information with CEO. If one column is added, the figures of Information expenses and investments become very apparent. Vincent moreover, requested Restating the Balance Sheet.

Keywords: IRM, BEP, ROI, COST/BENEFIT, CEO.

1 Introduction

Kaizen is the world wide accepted Japanese word. Most CEO loves concrete results of Kaizen, that is "PROFIT" numerically expressed on the bottom line of P&L.

I. Lovrek, R.J. Howlett, and L.C. Jain (Eds.): KES 2008, Part III, LNAI 5179, pp. 152–158, 2008.

Basic functions of Kaizen is Plan-Do-Check-Action, PDCA. Any CEO knows PDCA cycle. But, Most CEO doesn't understand what IRM is. IRM has same PDCA cycle.

Through Kaizen process, CEO, generally, expect a lot of savings. If Kaizen could justify the costs with increase in sales, or profit after tax where lies Profit/Loss statements. However, in the past, most enterprises had experienced only cost up after IT projects. Generally speaking, many of projects got doubled in cost, and savings decreased to half of estimate. IT projects must mean a Value-Added investments at the enterprise. IRM is the strategy that has been one of the most pervasive intra-enterprise planning efforts to date. IRM concept is working in Japan.

2 Views of Tangible Analysis for IRM Implementation

To implement IRM, it needs variety of expenses [2]. Main items are IT related costs like hardware of PCs and servers and communication lines. Also, software related costs should be counted. Depends upon the project size, sometimes, it may needs investments to building and other facilities. Costs or expenditures are relatively easy to pick up and count, but not easy to make clear on tangible merits or effects. By implementing new management concepts, it would affect on profit and result change of PL/BS/Cash-flow. But it is very difficult to separate only influences by IRM. Same is as profits. Increased figures of sales, or more orders received, win of bids of engineering projects, so on. Those merits may be driven by mixed operations. Reduction manpower of operators/clerical employees or even managers/supervisors may be counted as savings. Cut down of outsourcing agents bring direct cost saves to enterprise. Change of local vendors to overseas vendors may get drastic cost down of manufacturing or purchasing cost of parts or materials or subcontracting costs. Decrease of number of vendors may happen to cut cost of inventory costs. These merits should be clearly counted and list up. Of course, demerits are also to be counted. Organization-restructuring may cause labor cuts. Program like recommendation of early retirements and lay-off of employees impact at the levels of managers/executives. Tentative stop of new faces employment is also tactically deployed.

Regarding of labor cut, once such a actions announced, these actions are reported as bad rumors by journalism. Also, Japanese traditional culture dislikes of lay-off. CEO must take it into consideration. By moving to the new IRM-based enterprise, you can reduce overhead costs by simplifying management structures.

3 Tangible Effect and Intangible Effect

There are variety of ways for analysis of effect that stand for costs/benefits [3]. Empirically, we list up several points of views as below.

3.1 IT Investment Deemed as Enterprise Strategy to be Competitive

Since IT is considered as Enterprise's business infrastructure, in such a case, IT investment is strategically needed. At giant organization, a Big IRM project might be commenced. Completion of the project is absolutely vital. If the Project failed, the enterprise would be suffered from survival risk. In this situation, CEO must prepare fund for IT even if the investment is too big and difficult. If project failed, CEO would resign because of heavy responsibility [4].

Recently, two Japanese big banks decided to become one integrated banking institution, so the their two on-line banking systems must be integrated into new one as soon as possible, but could not finish within time and budget limit. CEO resigned from the institution naturally. The bank has lost 10 % of customers of this system fail.

3.2 Total IT Cost Criteria to Keep Enterprise Health

The enterprise is keeping at the averaged rate of industry criteria which is derived from the total IT related costs versus total sales. In Japan, enterprise who performs smartly, 1.4-1.5% is the averaged number. My surveys from sample enterprises (12 enterprises) that function well revealed 1.6% [5]. If we divide in to two smaller samples, one group is production industry that shows 1.8%, the other sample group that belongs to retail industry shows 0.8% in terms of rate (total IT cost/total sales volume).

3.3 Project Cost/Saving/Benefits Analysis

Before start of IT Project, proposed cost and savings would be evaluated and used for project initiation decision. BEP/ROI is necessary to be calculated. After project finished, actual number is to be counted based on same criteria of calculation of estimates. Instead of converting man-hours, computer time, so on to money, key performance indicator like produced lines of COBOL program per man-month also calculated as productivity statistics (Table 1).

In the case, project of Marketing Support MIS was carefully planned before starting Phase 1.Because the market of electric home-appliances is very active and tough competition of pricing is being operated. Top heavy project style was deployed to keep severe delivery time of information system and new Price-Product Data-base. Then at Phase-1, 20% of man-hour input was planned (actual: 18%), alsoP-2 , 15%(11%), P-3,15%(18%). This means 47% of total man-hour was used for design work of MS-MIS system which is dramatically improving productivity of sales man and cut of clerical jobs.

Recent study shows there is few case of high Rate of ROI among IT projects. It means Break Even Point of months is getting longer. Our study from these samples, however, the enterprises shows 30 % as ROI and 36 months as BEP that is significant performance.

Not only Tangible, but also Intangible benefits are to be reported as below.

- New system provides significantly more information than current system, particularly in the area of managerial policy information in the new marketing division.
- New system readily expandable without additional investments if new products appeared, plants modifications happened, segment of markets added, etc. can be easily incorporated.
- Each operating units which are newly organized have complete control over its areas of interest.

Table 1. Project Data of Productivity at Japanese Electric Company

Phase by Phase	Period	% of periods		Man-Month		% of Man-Month	
Phase1	6 Months	25%		84	20%	18	
Phase2	4 Months	17%	63%	54	15%	11	47%
Phase3	5 Months	21%		84	15%	18	
Phase4 - I	5 Months	Parallel With P-4-II		15	8%	3	
Phase4 - II	5 Months		37%	123	20%	25	
Phase5 ~ 7	4 Months			120	22%	25	

· Total Man-Month ;480

· Period 24 Months (2years ;2001 ~ 2002)

· Total Steps Yielded ;375Ksteps

· Project Productivity ;781Steps/Man-Month

From view point of efficiency, productivity of software development is increased 50 % than before. This case shows 781 steps per man-month. in the past, normal standard criteria number is around 500 steps per man-month.

Delivery date of information system was just same date of the planned date of Phase-1 report.

3.4 Information System Department Productivity

To appeal IS department performances to users departments of systems, it is important to measure productivity shown in following 10 figures, and should report to CEO/CFO/CIO, etc.

(1) Actual performance effect contributed by IS Department.

Usually, effect is indicated in Million-yen per man of ISD in a year. This data is very difficult to collect. We need efforts of ISD people and prepared special reporting system automatically getting data of effect or improvements.

We also set the standard time for Systems analysts and Programmers and Data-administrator and Operators and Librarian and Clerks in department.

(2) Produced lines in K steps per man per year.

(3) Maintained lines in K steps per man per year

(4) Produced lines in K steps per K hours per year

(5) Produced lines in K steps per man per year

(6) Actively operated programs in K steps per year

(7) Outsourced % of development stuff per year

(8) Projects number in working per year

(9) The project size and length of project completed and its' productivity in the year

(10) Utilization rate of new Data-base

These figures should be collected before implementation of IRM. The steel company has such historical data, we are lucky when we compare before/after. Before IRM implemented, any project was relatively small, and scope of system was narrow. They experienced several projects were out of control ("RUNAWAY")

System life was diminished because end-users complained of inconveniences of administrative procedures and ineffectiveness of outputs reports. After new methodology, IRM was installed, size of project got bigger and compacted duration of project period was kept. Delivery time was never over 3 years.

Fig. 1 shows 1985 year is the beginning of evolution at this Information Systems Department. IRM as new management concept started at the enterprise.

Reg of Information Retrieval, 10K times is counted in one month after newly designed Data-base opened to users. End users retrieved to Data-base by themselves for getting their needed information. New organization actively access to Big DB for unique analysis and process data resulted in information with upon-request.

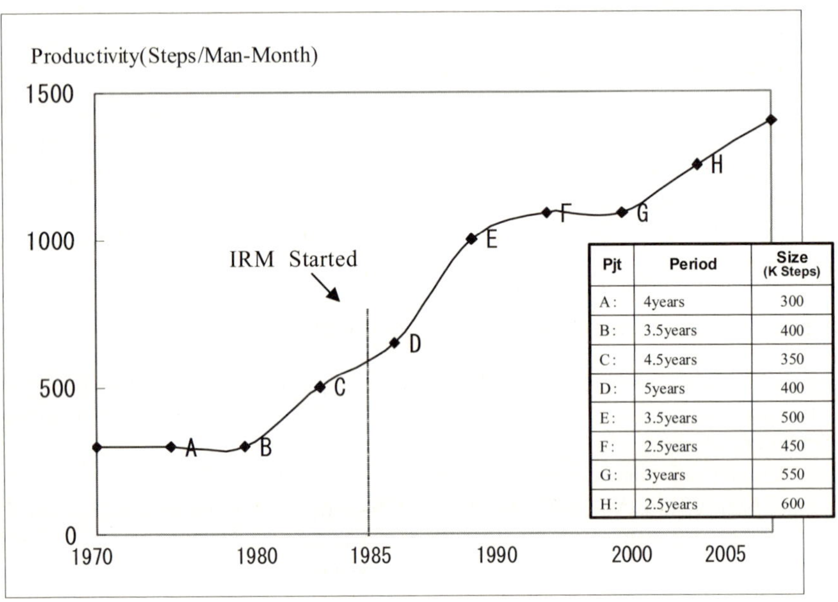

Pjt	Period	Size (K Steps)
A :	4years	300
B :	3.5years	400
C :	4.5years	350
D :	5years	400
E :	3.5years	500
F :	2.5years	450
G :	3years	550
H :	2.5years	600

Fig. 1. History of productivity at Steel company

3.5 Life-Cycle Cost

Most Japanese enterprises spent twenty years since started of computer use in business. Now, they have maintenance problems that more than 80 % of their IT budget spent into modification and improvements of existing systems. Those complicated systems increased that money. CIO is thinking of moderate system life-cycle management. It is important that yearly averaged cost of the System life (years in use) is minimized. IRM is targeting to maximize length of system in use and minimize life-cycle-cost (Fig.2). Above data is to be collected every month by weekly work reports of system engineers/programmers. Data must be analyzed by program number/project code/personnel name/time-basis. To prove effectiveness of IRM, we must show how life-cycle cost could down and keep system in longer use.

T: Averaged Life cycle cost of the Year
T0: Averaged Development cost
T1:Averaged Maintenance Cost in i th year(i=1~N)
Mi: Maintenance cost occurred in i-th year

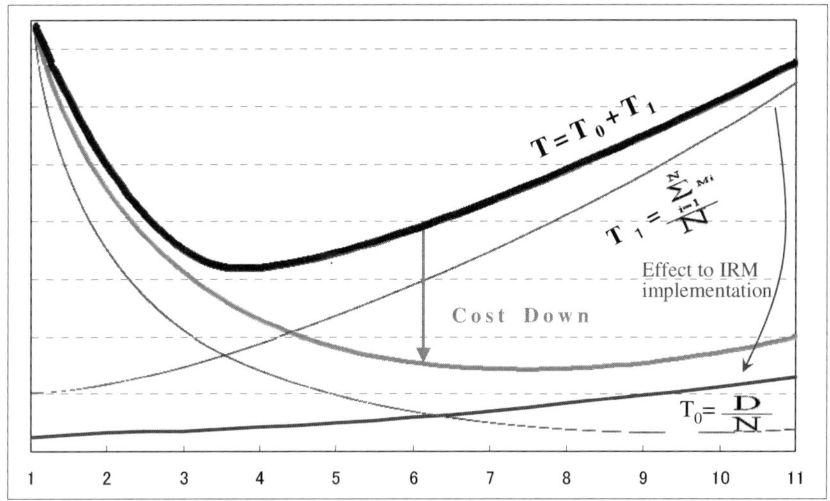

Fig. 2. Life-cycle cost of System

Not only savings and benefits, but also wrong affects by IT project must be surveyed. Following are some report from IRM implementations.

It is true that a kind of craftsman-ship is lost in the enterprise. Skills of work like Estimating cost, Drawing of mechanical in the design department, Tooling's and Jigs in the Industrial Engineering department is never thought for use to improve human sensitivity. Some of those are built in computer procedures as automatic procedure.

Young engineers never learn of those human techniques.

Regarding of Intangible effects, we learnt and pick up following items to be generally stated [6] [7].

1) Top managements recognized IRM is a management function, not merely IT techniques.
2) Through internal IRM education class at enterprise, one third of managers and managements are trained what is IRM. They participate diligently with the IRM projects, resulted in level up of their literacy of information.
3) Systems professional are very active in consulting with end users on improving productivity of their operations.

4) Standardized documents visualized how systems work. Systems department has been able to respond for any organizational changes, system maintenance requests.
5) Information empowerment has become penetrated to all levels of organization structures. Middle class managers deem information is important company asset as same as money.
6) Enterprise has been competitive, since earlier than competitors information like product price policy/new products/marketing channels are able to be caught.
7) Customers satisfaction rated higher and higher after IRM implemented.
8) Important stakeholders like stockholders recognize visibility up of their invested enterprises. Quicker responds to stockholders changes on performances of enterprises.
9) Government is certified with complete and adequate timing of reports on all duty information.
10) Supply Chain Management has promoted chained control of current customers to get more tightened fixed customers.

4   Conclusions

Practical measurement of tangible effect after quantitative analysis forced ISD to insist existing values of the department. Without ISD, end users cannot maintain stable operations in modern age. Specially, COO has become more systems oriented.

CEO believes IRM is a practical solution based on common-sense. His trial to integrate IRM plans with business strategy plan, then this affects evolutional change of enterprise. Now, CEO may be suggested to start IRM revolution on total group enterprises that have over multiple divisions. Also, he must believe Qualitative analysis is always important to make decisions on IT project.

IRM needs in a prerequisite enterprise executive management commitment in long time. Information resource is consist of Information, Data, System, and Organization. To control it as resources is the mission of IS Department [8]. To use information is responsibility of each employee. Empowerment of employee is vital to modern enterprise.

IRM Advisory Committee is to be organized and would oversee the process of Plan-Do-Check-Action of IRM. High quality of information is utilized by all levels organization structure, then might be produced better decisions that IRM targets.

References

1. Vincent, D.R.: The information-based corporation. Dow Johns, Irwin (1990)
2. Matsudaira, K., et al.: New Methodology for Information Resource Management by 4 Aspects Parallel Approach. In: Proceeding of KES 2008 (2008)
3. Senjyu, S., Fushimi, T.: Engineering Economy. Japan Management Association (1967)
4. Strassman, P.A.: Information Payoff. Macmillian Publishing Company (1985)
5. Fukuzawa, H.: Tangible Analysis. First Press Inc. (2008)
6. Barny, J.B.: Gaining and Sustaining Competitive Advantages, 2nd edn. Pearson Education, London (2002)
7. Duman, M., et al.: Process-Aware information systems: Bridging people and software through Process Technology. John Willey & Sons, Inc. (2005)
8. Matsudaira, K., et al.: MUD approach to Organization design in the IRM environment. In: Proceeding of IWIN (2008)

Automatic GUI Generation for Meta-data Based PUCC Sensor Gateway

Masatoshi Ogura[1], Hiroshi Mineno[1], Norihiro Ishikawa[2], Tomoyuki Osano[2], and Tadanori Mizuno[1]

[1] Shizuoka University, 3-5-1 Johoku,Naka-ku, Hamamatsu-shi, Shizuoka 432-8011, Japan
Ogura@mizulab.net
[2] NTT DoCoMo,Inc, 2-11-1 Nagatatyou, Tiyoda-ku, tokyo, 100-8011, Japan

Abstract. Many kinds of communication protocols exist for sensor networks. Therefore, it is difficult to use the sensing data from different kinds of sensor networks. The Peer-to-Peer Universal Computin a Consortium (PUCC) is investigating studies on a seamless peer-to-peer communications technology platform that enables the creation of ubiquitous high level services between connected network devices as well as sensor networks. We developed an automatic graphical user interface (GUI) generation module for a meta-data based PUCC sensor gateway (GW). This module makes it easy to configure a personal service that enables the combination of different kinds of events in heterogeneous sensor networks.

1   Introduction

Recently, various kinds of sensor networks have been developed, and each network is formed by sensor nodes that support its specific communication protocols. For example, with the use of equipments, such as cameras, a sensor network can obtain optical information on animal ecology. On the other hand, ecological sensors could obtain be used toinformation on changes inside an animals body. By combining and analyzing all of the data from many kinds of sensors, we can obtain much information including that necessary for life-sustaining activities. Although a sensor network could obtain a lot of data, it is possible to obtain more information and develop new applications by combining many different kinds of sensor networks.

Exploiting the sensor networks to their full potential, however, raises several hard data management and user interface challenges. For example, since many sensor nodes developed today adopt individual control protocols for communication performance and power saving, different sensor networks have difficulty communicating with each other. Therefore, much research has been done on developing a system that integrates and manages different sensor networks [1].

In this paper, we aim to improve usability by generating a graphical user interface (GUI) automatically to make it easy to configure a personal service that enables different kinds of events in heterogeneous sensor networks. This generation is done based on the meta-data that is provided by the Peer-toPeer Universal Computing Consortium (PUCC) sensor gateway (GW).

I. Lovrek, R.J. Howlett, and L.C. Jain (Eds.): KES 2008, Part III, LNAI 5179, pp. 159–166, 2008.

2 Framework

We propse a system using a sensor GW in ofder to obtain information from many different types of sensor networks. The sensor GW is an equipment that works between user and sensor devices.

2.1 PUCC (Peer to Peer Universal Computing Consortium)

In this system the PUCC protocol is utilized in the communication part. PUCC is a standardization organization involved in researching is performing various apparatus for the purpose of interconnection and technical development to employ using a P2P network. Devices that are dispersed in various network environments are connected in a seamless P2P network. In this way, we can execute various applications making use of the portable terminal. However, because the sensor node throughput is low, mounting the P2P platform is difficult, PUCC has tackled the problem by making use of a sensor gateway [2].

2.2 System Components

The sensor network monitoring framework of our system is composed of the following four functional parts. Figure 1 shows how these parts are interconnected.

Sensor device. This small compornent stores information from various devices and has communication equipment that quantizes data. A sensor device, also known as a sensor node, is a node in a wireless sensor network that is capable of performing some processing, gathering sensory information, and communicating with other connected nodes in the network. The main components of a sensor node are the microcontroller, transceiver, external memory, power source, and one or more sensors.

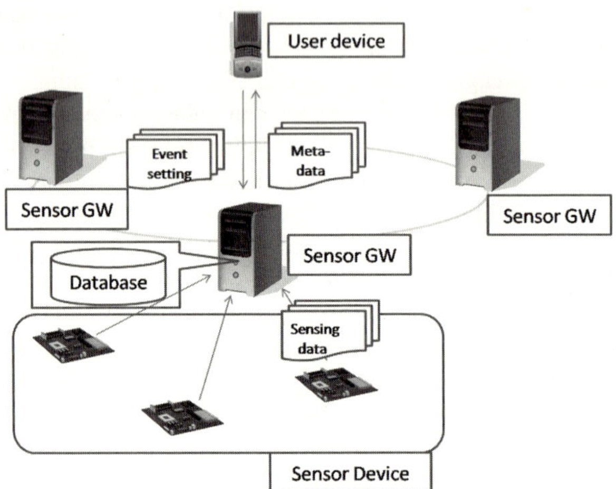

Fig. 1. System overview

Sensor Gateway (GW). Because sensor devices may be built using many different platforms that vary widely in processing power, energy, and bandwidth capabilities, they may have different interfaces to access them. To get around these differences, the sensors are connected to a sensor GW that provides a uniform interface to all components connected to it. The sensor GW implements sensor specific methods to communicate with the sensor. The sensor GW may also implement sharing policies defined by the developer. For instance, the sensor GW may maintain all raw data in its local database.

Database. The database is a data storage system that saves sensor data. We use PostgreSQL as the database management software in this paper.

User Device. The user device is assumed to be a portable terminal. It communicates by using the communication protocol developed by PUCC, and information on an arbitrary sensor device is obtained via the sensor GW [3].

2.3   Event

Only a few kinds of sensor network monitoring tools are available that provide a sensor network data display capability. Many monitoring tools are systems to obseve information on the limited kinds of sensors [4] . To use information on a variety of sensor networks, we added an event the function. When the device returns data that are applicable to a condition the user has indicated with the sensor GW it is referred to as an event. Conditions are set vis-a-vis the multiple devices, and an event that occurs when the device satisfies various conditions simultaneously is designated as a compound event.

3   Automatic GUI Generation

3.1   Sensor GW

The sensor GW has the protocol of the managed various sensors and can communicate with sensor devices. The user can obtain data from the sensor devices because the user can access the sensor GW by using the communication protocol developed by PUCC. The meta-data of the managed sensor devices are sent to the user side when the sensor GW connects the user with the subscription relationship. The service the user wants to use is selected based on the meta-data and indicated.

Sensor devices can compose a vareity of services. First, as a basic service, there is one application directly provides the sensed information to the user. This idicates the actual data the sensor detected, such as surface temperature and sound volume, as numerical values or analog data. Most normal sensors, such as temperature sensors and acceleration sensors, could be utilized this way. Another application provides information through the cooperation of multiple sensors. This is sent to the user or is used to start some kind of event when multiple sensors reach the threshold value set by the user. There is also a service that can activate household electrical equipment by communication with the equipment and using the sensor data as a trigger.

3.2 Meta-data

Meta-data are particularly useful, since they are information on each device. However, when unnecessary data are placed, they become less versatile, and the settings become more difficult.

For a home appliance product to be recognized as an individual service, service contents, when you recognize, as type of device it freezes the function which the refrigerator and the air conditioner and the device offer, video recording and playback etc defines. These devices because this information is stored and selected when needed. It therefore becomes possible to control [5]. In the same way, an event can be formed on the basis of the data the sensor device acquires by making use of the sensor GW. The system classifies sensors according to specifications and functions making it easy to search and control with the manuscript sensor device.

Meta-data are described as two separate data using an XML format. The first is the sensor network information, for example :

- Number of sensor devices
- Sensor node information
- Type and specification of loaded sensors

The second is the service information that a sensor GW offers, for example :

- Raw data showing the measured value of the sensor
- Event data format
- Event data list

The event data list is a collection of events that are judged on the basis of multiple raw data and state variables. The device name and the service name described in the meta-data indicate the contents of the service.

There are only a few kinds of information that a single sensor can transmit, and the transmission method is fixed. Therefore, using this information, it is possible to generate a GUI automatically.

Specifically, there are two methods to form a GUI dynamically. The first method is to construct the GUI by combining each service of each sensor as a single module. The second method is to prepare the frame, which is the basis of the application, beforehand, and change the structure inside on the basis of available sensor devices. The two methods are depicted Figure 2.

When using the sensor information as a module, every service that is read from the meta-data is displayed or an input column is made, and it is indicated in the order it is programmed in the meta-data. In the case of a GUI that browses the sensor information, the category of inspected information obtained by the service name of the meta-data is shown on the display so that it can be selected as desired information. Since the detected information from each sensor is transmitted to the sensor GW, information of the sensor the user chose could be obtained from the sensor GW. This is the flow of browsing the detected information, as in the upper part of Fig 2.

When services are operated through the cooperation of multiple sensor devices, the GW reports to the user which devices are usable by scanning the meta-data transmitted

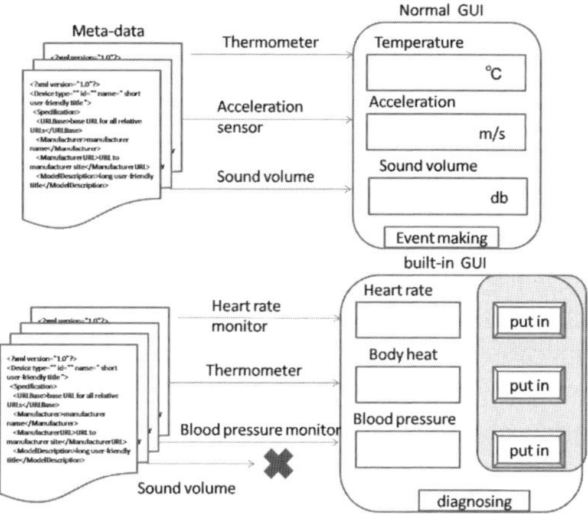

Fig. 2. GUI generation from meta-data

from the multiple devices, then indicates the data that the user can access, and lets them make a choice. The dynamic GUI scans the information of the upper and lower limits and the unit of the data value. Then it indicates the range of the value that the user can input. By selection of conditions, such as turning lights on and off, or indicating when the range of the value is narrow, it presents a list, and lets the user choose appropriate information.

Another case is when the frame of the application is prepared beforehand. A frame is a GUI framework that assumes a special service. Basically, this is almost the same as when using the GUI as a module. For example, when a sensor device saves a detected data, a save button is added to the GUI, or a send button is added when sending information to another place. We hope to develop many functions to extend the field of services using sensors, and will investigate necessary functions in the future. An application that utilizes the GUI with a specific purpose, such as building a network of home appliance products and a cooperative service of home appliance products and sensors, is assumed. When using the GUI of an existing application, the sensor part is usually added to an existing GUI. The existing application side is able to obtain information from the sensors just by defining the category of the necessary data, and information from unnecessary sensor device will not be displayed.This is illustrated in the lower part of Fig 2.

Conditional expression. With the present sensor GW, an event that occurs from a device happens with a change in the state variable value, which the device manages. However, the state variable value of the sensor device often changes. If the change in the state variable value is the only condition to cause on event occurrence, many events will occur, and it will be difficult to achieve inclear detailed control with just the condition of a change in the state variable value in one sensor device. Therefore, it is unsuitable for setting a complicated event occurrence condition. In the future, it will be necessary to improve the occurrence condition of events by subdividing the definition of the conditional expression.

In order to reflect the variation of a variable state event occurrence condition in detail, it is necessary to use a conditional expression of the event occurrence condition and to combine conditional expressions. This way it is possible to create detailed event occurrence conditions.

Various values are measured by a sensor, but even the same temperature sensor has different in measurement methods, and the objects it measures arealso different. To define the sensor clearly, there needs to be an accurate description of what it measures.

3.3   Dynamic Generation of GUI

The method to generate a dynamic GUI is based on the device information used for the meta-data data, such as the device name, kind of value, etc.. The meta-data are classified in a part showing a tag and the contents of the tag retrieved by a parser. By reading the indication content beforehand, the length of the indication department in consideration of the number of devices is decided.

The merit of utilizing the meta-data for the automatic generation of GUI is that the meta-data is described in XML. XML can cooperate with many tools; it does not depend on the OS, nor does it depend on the programming language. Accordingly, XML is widely utilized and it is easy to use with other applications even with large expansions. In the future, the number of movile phones operating Java environment is expected to increase and to respond to this demand, the GUI is programmed with Java.

3.4   User Customization

For user customization of the GUI, the user individually sets the details of the presentation method by selecting the background of the GUI. The user can to search for devices from many sensor devices that correspond to the service by selecting the ID or type of sensor device to be used. An application that extracts only the condition, such as information from patrol sensors inside the home, is one possible case. Since PUCC carries out cooperation of various types of equipment (home appliances, etc.) today, the situations in which compound events can be utilized are expanded by using sensor information as the trigger to start them automatically. The future objective is to put this to practical use, and to investigate the use of cooperative equipment other than sensors.

4   Prototype Development

We have currently developed a prototype system by using Java (J2SE 1.5.0) on Microsoft WindowsXP. This system consists of the user device, the GW device, and sensor devices. The generated GUI consists of two parts, a service selection part and an event-setting part.

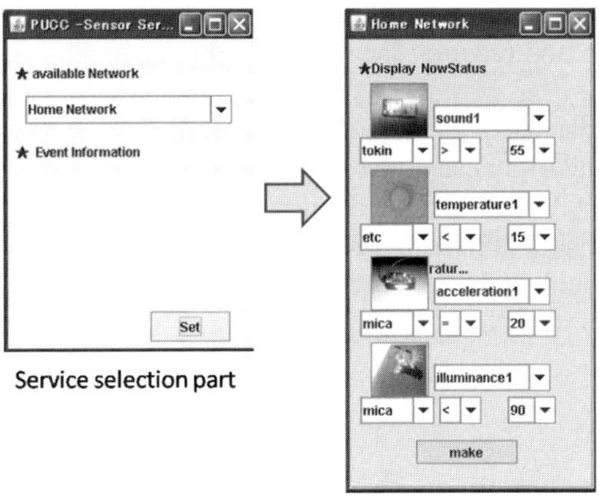

Service selection part

Event-Setting part

Fig. 3. Sample of generated GUI

4.1   Service Selection Part

There are two methods for setting the service, selecting sensors by client setting, and recording on the meta-data that the sensor GW offers. An abridged general view of the prototype is shown in Figure 3. The service part is the part in which the user chooses available services indicated by the sensor GW. This part of the GUI is shown in the left of the Figure. First, the service provider chooses as many sensor devices as needed, gives a service name, and registers it. The registered service name is registered in the service choice department. The user can choose a necessary sensor simply by choosing a service name displayed by the GUI.

4.2   Event-Setting Part

The operation of the sensor to be used is carried out in the event-setting part. The event-setting department is shown in the right side of Figure 3. On the basis of the sensor device data that are obtained from the meta-data, the name of the sensor device, conditional expression, value, and a picture of the sensor is displayed. The data that are used in this prototype are the five items shown below.

- Device ID: ID of sensor device that is managed by the sensor GW
- Device Name : Type of sensor
- Minimum/Maximum: Minimum and maximum available value of device
- Step: The minimum increase and decrease in the value of a device
- Icon: An image to depict a sensor device

These are utilized mainly to inform the user of the sensor specifications. The location is required in order to pinpoint the position of service from the position of the sensor GW, but it was not indicated this time.

When a user selects a service in the service-setting part, a sensor for that service is displayed. When a user does not select a service, the user judges the sensor device from the sensor name and picture. Then the user sets a conditional expression which generates an event.

In this prototype, the system automatically generates a GUI to allow selection of a service part and setting of an event that using a sensor device. Based on the elements of the meta-data, the GUI can change the number of sensors and the range of values to display. Even if the user communicates with a different sensor GW, they can use the service of the sensor in the GUI. As a result, the system reduces the download time of the service that the sensor GW provides.

5   Conclusion

In this paper, we described a system we cpnstructed that automatically generates a GUI for users of the system that manages different kinds of sensor networks that are integrated by using a sensor GW. In the automatic generation of a GUI, we succeeded in to setting a threshold for the data that the sensor detects by connecting to a sensor GW. The user can easily access service to sensor devices by using this system. In the future, we will examine ways of incorporateing this system in the services that other networks offer to household electrical appliance networks.

References

1. Balazinska, M., Deshpande, A., Franklin, M., Gibbons, P., Gray, J., Nath, S., Hansen, M., Liebhold, M., Szalay, A., Tao, V.: Data Management in the Worldwide Sensor Web. IEEE Pervasive Computing 6, 30–40 (2007)
2. Ishikawa, N., Kato, T., Sumino, H., Murakami, S., Hjelm, J.: PUCC Architecture, Protocols and Applicationsh. In: IEEE Consumer Communications and Networking Conference (CCNC 2007), pp. 788–792 (2007)
3. Sumino, H., Uchida, Y., Ishikawa, N., Tsutsui, H., Ochi, H., Nakamura, Y.: Home Appliance Control from Mobile Phonesh. In: IEEE Consumer Communications and Networking Conference (CCNC 2007), pp. 793–797 (2007)
4. Yu, M., Song, J., Kim, J., Shin, K., Mah, P.: NanoMon: A Flexible Sensor Network Monitoring Software. In: The 9th International Conference on Advanced Communication Technology (ICAST2007), vol. 2, pp. 1423–1426 (2007)
5. Alesheikh, A., Ghorbani, M., Mohammadi, H.: Design and Implementation of Sensor Metadata on Internet. XXth ISPRS Congress, Commission 4, 1148–1152 (2004)

The Pictograph Chat Communicator II

Jun Munemori, Taro Fukuda, Moonyati Binti Mohd Yatid, and Junko Itou

Wakayama University, Faculty of Systems Engineering, 930 Sakaedani,
Wakayama, Japan
`{munemori,s095045,s105054,itou}@sys.wakayama-u.ac.jp`

Abstract. In Japan and other countries, pictographs have widely spread to add nuance to mails in mobile phones. We have developed a pictograph chat system, which enables communication using nothing but pictographs. We have prepared 550 pictograph symbols. The system has a history tab, which can save and reuse all pictographs used in the previous chat lines. We applied the system for communication to 2 groups of foreign students and Japanese students. We have carried out experiments 4 times. We report the results of the experiments as below. (1) The average number of chat lines was 29 lines in 30 minutes. (2) The subjects understood 83% of the chat contents. (3) The history tab is used about 24% of all selected pictographs. (4) History tab was used frequently in relation to the number of chat lines.

Keywords: Groupware, Pictograph, Chat, History tab, Cross-cultural communication, PC.

1 Introduction

The communication of exchanging information by E-mail, chatting, and electronic bulletin board has widespread by the spread of networks. Moreover, communication can be easily done using MSN Messenger [1] etc. with text base. In addition, face marks and pictographs have appeared, used to convey feelings [2].

Language becomes a barrier if we think about communication between people from different countries, language and background, and if a common language is not understood, it is difficult to communicate through a text base conversation. Moreover, to learn a foreign language requires considerable time. Therefore, attention was given to pictographs that are used to convey feelings and slight nuances based on the idea that communication is possible if pictographs are used, even if a common language like English is not understood.

To add pictographs to chatting when experimenting on a teleconference between Japan and China [3], a recognition investigation of pictographs was conducted using postgraduate students from Japan and China (eight from each country) as subjects. 112 pictographs were shown to the subjects for the evaluation process (the meaning of the pictographs). As a result, between the Chinese and Japanese subjects, only 4 pictographs differed greatly in recognition (school, house, motor sports, and rice ball). The pictograph for "school", for example, looked like a regular building. Though understandable to Japanese subjects, the Chinese did not see it as a school because

I. Lovrek, R.J. Howlett, and L.C. Jain (Eds.): KES 2008, Part III, LNAI 5179, pp. 167–174, 2008.

Chinese has a grand image like a castle for school. Another example, "Rice ball", although a common snack in Japan, is not a common food in China.

As a result, the recognition of the pictographs was almost the same to the Japanese and Chinese subjects. We then sought to determine whether subjects could understand even if sentences were made using only pictographs, and developed a system in which only pictographs were used for chatting. We had developed a system before [4]. This system is a chat system which can send and receive messages made only from pictographs. 550 pictographs were prepared. The new system (Pictograph Chat Communicator II) has a history tab, which can save and reuse all pictographs used in the previous chat lines. This chat system was experimented by 2 groups of "International students and Japanese students" in the local Japanese University. From the results of the experiment, we considered the possibility of using only pictographs for communication.

2 Pictograph Chat Communicator II

2.1 Composition of System

The hardware of the system is SONY VAIO type-U (OS: Windows XP) computers. Two PCs are linked by LAN. Software was developed by FLASH Professional 8 (Macromedia). It is a program of about 1100 lines. We made pictographs in color. But, a part of pictograph (Monochrome symbols) is made by the PIC-DIC symbols [5]. The pictograph size is 54 pixels * 54 pixels. The pictograph is represented about 5 mm * 5 mm in the screen. All operation is performed by a pen.

2.2 Functions of System

The system consists of the chat log screen, input field where pictographs are selected and sentences are made, and the pictograph selection screen. Figure 1 shows the screen of the system. A pictograph can be added from the pictograph selection screen to the input field with one touch using the pen. There are 8 tabs in the pictograph selection screen. Each tab has about 80 pictographs. The pictograph selection screen also has a history tab (Fig. 2), which can save and reuse all pictographs used in the previous chat lines. In Fig.2, all pictographs in the chat log screen are saved into the history tab. The last line of the chat log screen is saved in the top part of the history tab.

The procedures of the input of the pictographs are as follows.
1)   Selection of tab (including the history tab).
2)   Selection of pictograph in the tab using a pen.
3)   The selected pictograph will be shown in the input field (maximum of 8 symbols).
4)   Pictograph that needs to be deleted is dragged out of the input field using the pen.
5)   By touching the chat input button using the pen, the chosen pictographs are sent to the server and appear in the chat log screen.

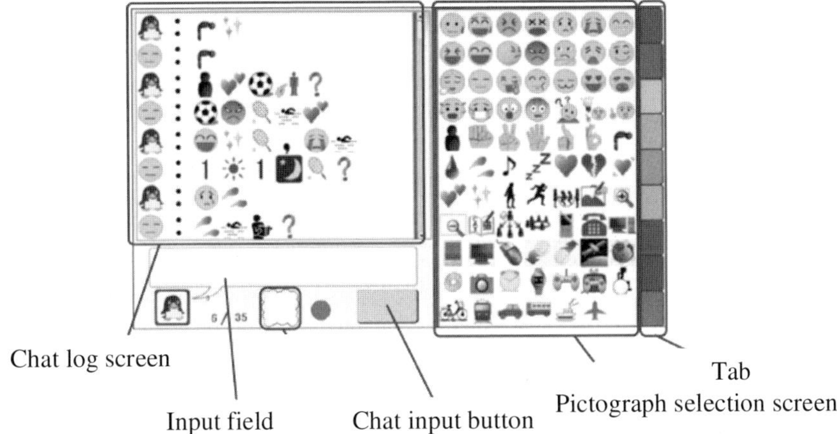

Fig. 1. A screen of the system

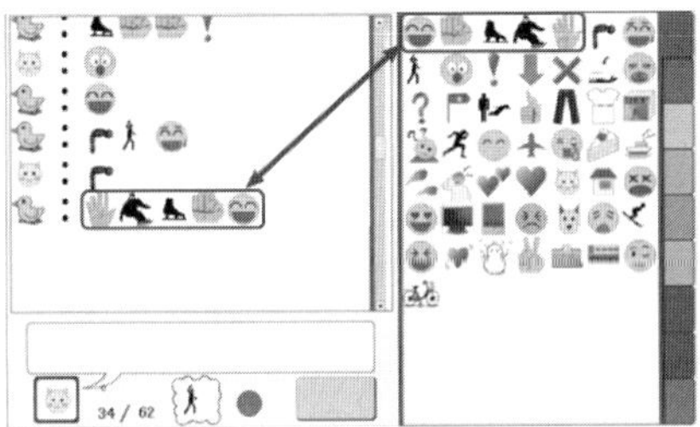

Fig. 2. An example of the history tab

3 Experiment

Two subjects experimented using PCs through LAN in the same room (neighboring). The subjects were all students at Wakayama University including the international students. The international students were two Chinese students, one Malaysian, and one Indonesian. Japanese subjects were third and fourth year students of the university. All subject combinations were a Japanese and an international student. Subjects were 8 in total with four pairs. They talked using the pictograph chat. Each experiment was performed for 30 minutes. After the experiment, subjects wrote the meaning of each line of pictographs they chose and they meaning of each line of pictographs their partner chose, and answered a questionnaire.

4 Experiment Results

Figure 3 shows the pictograph chat communicator II. A result of the chat by Japanese and international student is shown in Fig. 4. They are talking with Fig. 4 concerning

170   J. Munemori et al.

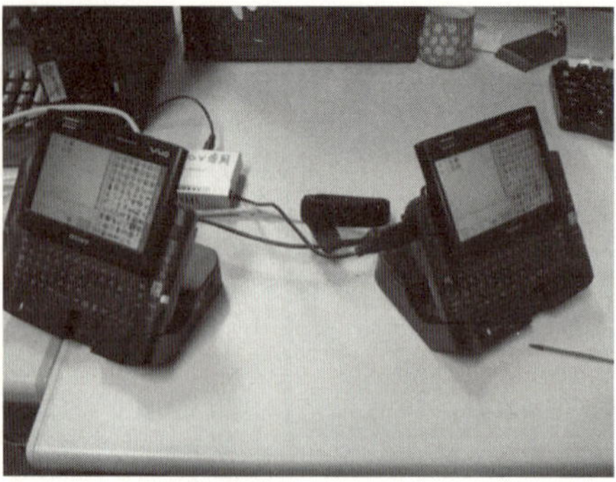

Fig. 3. The screen of the pictograph chat communicator II

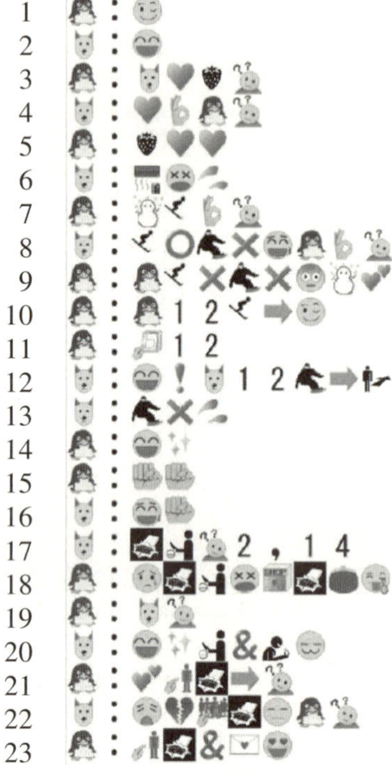

Fig. 4. An example of chat

skiing and Valentine's day. The mark of the first animal of Fig. 4 shows the speaker. Table 1 shows the interpretation of the contents of the pictograph chat in Fig.4. "1-A" means 1st line of chat by subject A (penguin). "2-B" means 2nd line of chat by subject B (dog). We could find that subject B misunderstands "12th February to "December" at 10th line.

Table 1. The interpretation of the contents of the pictograph chat

Subject A (Icon of penguin)	Subject B(Icon of dog)
1-A:Good morning.	Good morning.
2-B:Good morning.	Good morning.
3-A:Do you like strawberry?	Do you like strawberry?
4-B:Yes. Do you like, too?	Yes. Do you like, too?
5-A:I like it very much.	I like it very much.
6-B:It is very cold.	Today is very cold.
7-A:Yes. Can you ski?	Can you ski?
8-B:I can ski. But I can't do snowboarding. Can you?	I can ski. But I can't do snowboarding. Can you?
9-A:I can't ski and do snowboarding. But I like snow..	I can't ski and do snowboarding. But I like snow.
10-A:I will ski at 12th February.	I skied in December.
11-A:---	"12" is December, isn't it?
12-B:Do not go? I will do snowboarding at 12th February.	It is so. I did snowboarding in December.
13-B:In fact, I cannot do snowboarding.	Snowboarding is difficult.
14-A:Is it so? It's great.	It is so.
15-A:Please do its best.	Hang in there.
16-B:I want to do my best.	I do my best somehow.
17-B:By the way, how do you intend to do it on Valentine's Day? Do you make chocolate?	Is the chocolate of Valentine's Day handmade?
18-A:No. Because I cannot make the chocolate, I buy it in a shop.	Because it is hard, I buy in the shop.
19-A:You?	You?
20-B:I make myself and I eat myself.	I make myself and I eat myself.
21-A:Do not you give it to a favorite person?	Do not you give it to a favorite person?
22-B:Because there is not a favorite person, I give it to all. You?	I give it to all. You?
23 A:I hand a love letter to a boyfriend.	I give it to a favorite person and do the confession.

The average number of chat lines was 29 (36, 23, 13, 42 individually) in 30 minutes. The mutual understanding level of the conversation of each experiment was calculated by comparing the meaning of the remark that the subject had written.

Table 2. Ratio of selected pictographs from the history tab

		The number of selected pictographs	The number of selected pictographs from ordinary tab	The number of selected pictographs from history tab	Ratio of selected pictographs from history tab (%)
Exp.1	a	105	62	43	41
	b	45	45	0	0
Exp.2	a	64	41	23	36
	b	57	44	13	23
Exp.3	a	34	33	1	3
	b	41	35	6	15
Exp.4	a	78	49	29	37
	b	96	62	34	35
The mean ratio of using the history tab					24

The average of the understanding level [4] of experiment is 83% (79%, 94%, 70%, 88% individually).

The history tab is used about 24% of all selected pictographs (Table 2).

Part of the results of the questionnaire of five-point scale evaluation (Table 3) and opinions are shown. "5" is the highest score and "1" is the lowest.

Table 3. Results of the questionnaire

Questionnaire items	Evaluation
When the pictograph is tapped once, it is directly added to the input field. Was this operation easy?	4.5
Did you use the history tab?	3.6
Is the history tab convenient?	4.4
Was the sentence making easy?	2.5
Could the meaning of each pictograph be understood?	3.5
Was a target pictograph easily searched?	2.4
Was there a target pictograph?	2.4
Did you understand what the other party said?	3.5
Do you think that you were able to understand the other party?	4.0
Do you think that you can conduct a conversation chatting only with pictographs?	3.9
Was the experiment interesting?	4.6

Moreover, examples of the description-type results of the questionnaire are shown below (Table 4).

Table 4. The description-type results of the questionnaire

In what kind of situation do you think is suitable to use this pictograph chat system?
• Communication with people from different countries that whose spoken language differs from mine. • When I am not able to handle a decent conversation with a foreigner who does not understand my language, I can use the pictograph chat. • Self-introduction conversations. • Very simple conversation. • For fun.
Please write any opinion regarding this system.
• Please increase the number of pictographs. • The categorization of the pictographs was incomprehensible. • It is inconvenient that I was not able to add a pictograph before one that is already in the input field. • A monochrome pictograph is incomprehensible • I need something to display the 5W1H. • It is hard to show place and time. • The history tab was convenient, but I had a hard time when I was looking for a new pictograph.

5   Discussion

5.1   Discussion

(1) The number of lines of the pictograph
The number of output lines of this system is slightly up, comparing to the conventional system (from 26 lines [4] to 29 lines). But the result is not changed essentially.
(2) Understanding degree
We determined the understanding degree of the pictographs before [4]. The understanding degree is slightly up, compared to the conventional system (from 77% [4] to 83%). This system also allows the usage of pictographs chose by the opponent party in the history tab. Therefore, a subject can save time by just tapping on his or partner's used pictographs without having to search for the target pictograph again. In addition, we enlarge the size of the pictographs from 32pixel * 32pixel [4] to 54 pixel * 54 pixel. Therefore, we can recognize the meaning of the pictograph easily. However, it is still hard to show and understand place and time (For example, 10th line in Table 1).
(3) History tab
As a result of the questionnaire, an evaluation of 3.6 was given to the question whether you used the history tab. The history tab displays the pictographs used since the chat begins. During the first half of the chat, pictographs are chose by selecting it from the pictograph selection screen. Therefore, a pair who do not use many pictographs has little opportunities to use the history tab. We thought that would lead to a low evaluation. In addition, the history tab is judged to have been convenient (4.4/5.0). History tab was used frequently in relation to the number of chat lines. The history tab is used 24% on average.
(4) Sentence making
Questionnaire item of "Was the sentence making easy?" is up from 2.0 [4] to 2.5. Some function (for example, the history tab) may give good effects. But it is still difficult to make sentences by pictographs. We should modify the system to make sentences easily.

5.2   Related Works

There is a project by the NHK (Nippon Hoso Kyokai), which is the South Pole kids project [6]. This is a pictograph chat system for children all over the world to communicate using only pictographs. Concerning this system, pictographs which can be lined up to eight are expressible by the chat system of the Web base. Similarly, research involving communication with children in different countries using pictographs was done [7]. However, it is a system using not a real-time chat but a mail base.

Although sign language is a method that may allow communication excluding conversation, there are a lot of dialects of sign languages that differs by country. The sign language used by Japanese is different from one used by Chinese. At present, there is international sign language [8] common to all parts of the world though it is not so general. The comparison of these would be a problem in the future. Moreover, a person in the sphere of Chinese characters can communicate in writing. The comparison with the Chinese characters might also be a problem in the future.

6 Conclusion

We have developed a pictograph chat system, which enables communication using nothing but pictographs. We prepared 550 pictograph symbols. The system has a history tab, which can save and reuse all pictographs used in the previous chat lines. We applied the system for communication to 2 groups of foreign students and Japanese students. We have carried out experiences 4 times. We report the results of the experiments as below. (1) The average chat lines were 29 lines in 30 minutes. (2) The subjects understood 83% of the chat contents. (3) The history tab is used about 24% of all selected pictographs. (4) History tab was used frequently in relation to the number of chat lines. (5) We should re-arrange the categorization of pictographs in the tab and pictographs.

Acknowledgments. This research was partially supported by Japan Society for the Promotion of Science (JSPS), Grant-in-Aid for Scientific Research (B) 18300043, 2006.

References

1. MSN Messenger, http://messenger.msn.com/
2. Kawakami, Y., Kawaura, Y., Ikeda, K., Hurukawa, R.: Passport to social psychology computer communications on electronic network, Seishin Shobo, Tokyo (1993)
3. Munemori, J., Shigenobu, T., Maruno, S., Ozaki, H., Ohno, S., Yoshino, T.: Effects of Applying Multimedia Video Conferencing System to Intercultural Collaboration. Journal of IPSJ (Information Processing Society Japan) 46(1), 26–37 (2005)
4. Munemori, J., Ohno, S., Yoshino, T.: Proposal and Evaluation of Pictograph Chat for Communication. Journal of IPSJ (Information Processing Society Japan) 47(7), 2071–2080 (2006)
5. Godai Embody Co., Ltd. PIC-DIC, http://www.mentek-godai.co.jp
6. NHK the South Pole kids project, http://www.nhk.or.jp/nankyoku-kids/ja/frame.html
7. NPO Pangaean, http://www.pangaean.org/common/
8. The Yomiuri Shimbun,
   http://www.yomiuri.co.jp/komachi/reader/200411/2004111500073.html

Network Forensics on Mobile Ad-Hoc Networks

Akira Otaka, Tsuyoshi Takagi, and Osamu Takahashi

Systems Information Science, Future University-Hakodate
116-2 Kamedanakano-cho, Hakodate Hokkaido, Japan
{m1204155,takagi,osamu}@fun.ac.jp

Abstract. A variety of countermeasures for every eventuality are more important than maintaining a mechanism that does not suffer any kind of assaults in mobile adhoc networks. From the experimental results, the ratio of the number of evidence packets and the total number of nodes was confirmed with the ratios of the transmission range and the objective range. The number of the evidence packets is controlled by the level of reliability. All nodes in mobile ad hoc networks need to be free from all suspicions.

Keywords: mobile ad hoc networks, network forensics, security policy, backward incidence.

1 Introduction

Security in mobile ad hoc networks (MANETs) is more important than security in local area networks (LANs). Defense or avoidance countermeasures are taken in advance. We propose methods for addressing security countermeasures in MANETs.

Defense or avoidance structures are designed to disconnect and change routes when problems are detected. Those methods contribute greatly to problem detection. However, these methods occasionally misidentify information as the number of packets increase.

Figure 1 shows the difference in forensics between Internet and MANET models. In the Internet model, network forensics assumes that attacks from the wide area networks (WAN) to the web server leaks inside information from LAN. All Ethernet frames are saved and encrypted by forensics service vendors. Anyone using in a MANET are vulnerable to attacks. Consequently, the packets that are encrypted by principal are not trustworthy. In network forensics, a company's profit is the top priority. Communication messages are analyzed by collecting Ethernet frames. However, personal privacy needs to be protected in MANETs.

Using transmitted packets as evidence needs to meet following requirements.

- Alteration is impossible
- Guarantor(s)
- Currency of the Data

I. Lovrek, R.J. Howlett, and L.C. Jain (Eds.): KES 2008, Part III, LNAI 5179, pp. 175–182, 2008.

The evidence that can be falsified cannot be trusted. Those packets need to be admissible as evidence and checked by a third party. It is desiable the evidence can referable immediately for a judgment.

Table 1 shows the difference between network and MANET forensics. The aim of network forensics is to pinpoint attacker. The aim of MANET forensics, by contrast, is to prove Relayers who relay data packets correctly.

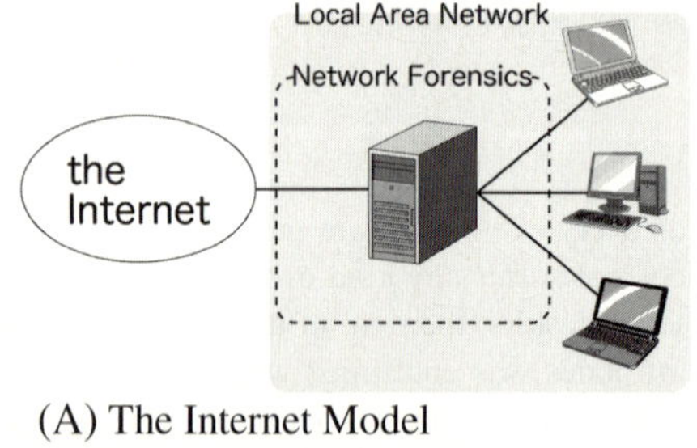

(A) The Internet Model

(B) The MANET Model

Fig. 1. Difference in forensics between Internet Model and MANet Models

Table 1. Difference in forensics between Ethernet and MANET

	Ethernet	MANET
Intended User	Receivers (Server)	Relayers
Aim	Pinpoint Attackers	Proving own Innocence
Evidences	Information of Attackers	Information of guarantors
	Degree of Damage	Evidences of Transmitted Packets
	Routes	
	Methods	
	Timestamp at Attacking	Timestamp at Saving of Evidence

2 System Model

2.1 Overview

Figure 2 shows the target domain of our study. Generally, network forensics monitors when someone breaks bounds (incidents). If a problem is detected, an administrator starts to collect evidence. After that particular communication finishes, the administrator analyzes that communication. The purpose of this paper is on the collection the evidence. We are not concerned here with formulating security policy and analyzing the evidence.

Figure 3 shows the definition of terms. In this paper, we define a MANET as having the following three roles.

Relayer: Relayers are nodes involved in sending or relaying packets. However, destination members are not included.

Eyewitness: Eyewitnesses are nodes near relayers. They can communicate directly with each other.

Guarantor: Guarantors are nodes who return the evidence to the eyewitnesses.

Figure 4 shows the collection flow of evidence. Node A sends data to node C. A route is established by an existing routing protocol. Scheduled removing of packets is skipped in Fig. **4**.

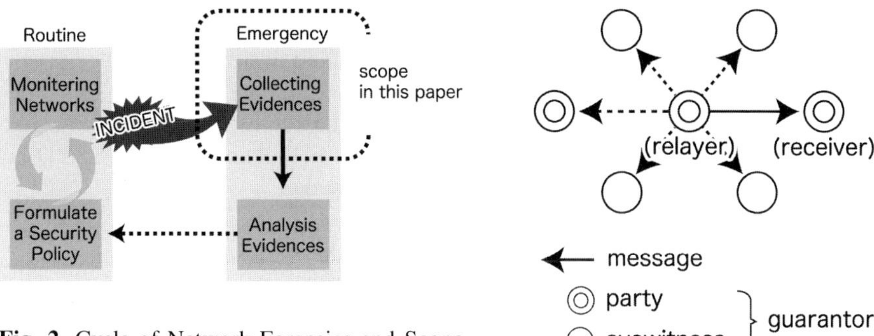

Fig. 2. Cycle of Network Forensics and Scope of this paper

Fig. 3. Definition of terms

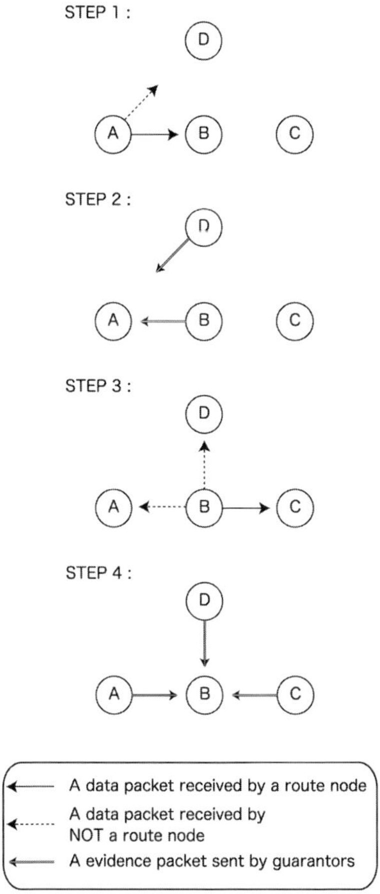

Fig. 4. Collection Flow of Evidence

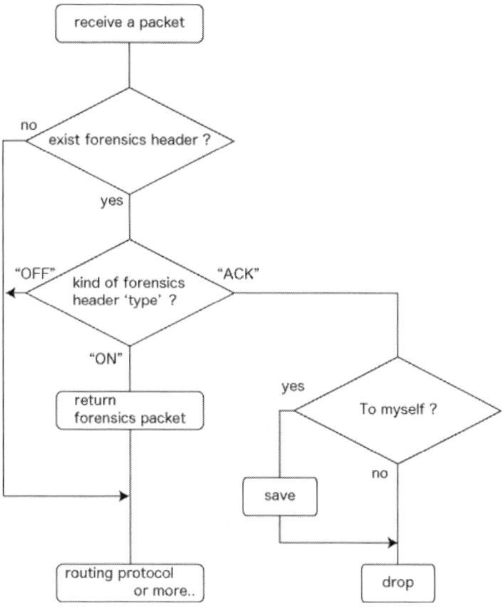

Fig. 5. Proposal Algorithm

STEP 1. Node A sends data to node C. Eyewitnesses of node B and C are received through [/unencrypted?] mode and understand desiring evidence.

STEP 2. Node B and D perform a hash function on the data packet, which is then encrypted by public-key cryptography, and returned to node A. For example, an evidence packet sent by node B is broken by node C and D because the packet was not sent from those nodes.

STEP 3. Node B relays data to node C.

STEP 4. Eyewitnesses of node A, C, and D return the evidence packet.

2.2 Algorithm

Figure 5 shows a proposal algorithm. A received packet checks if there is a header of forensics. If there is no forensics header, the process of packets moves into routing protocols.

The relayers need to save the following contents.

1) Information of guarantors
2) Evidence of transmitted packets
3) Timestamp at saving of evidence

Forensics does not function during communication. After an incident occurs, forensics starts. There are two methods in which forensics starts.

Method 1 : The sender node initiates the forensics as shown in Fig. 6. The evidence can be easily analyzed in this method.

Method 2 : The detector node starts the forensics function as shown in Fig. 7. More evidence can be collected compared to the above method.

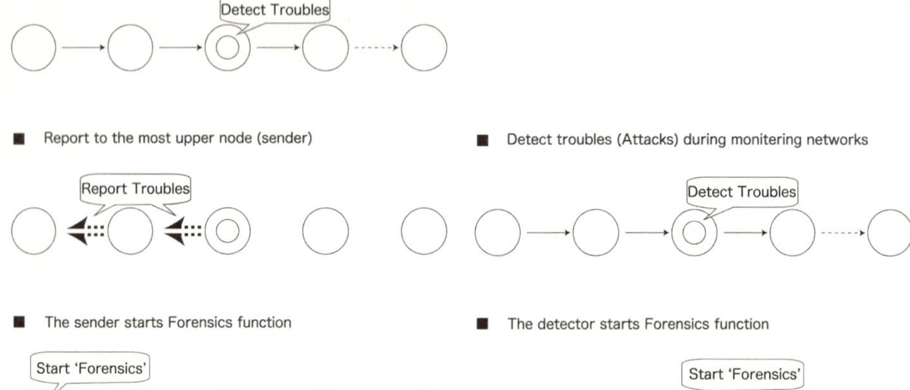

Fig. 6. Case 1 (A sender has a jurisdiction)   **Fig. 7.** Case 2 (A detector has a jurisdiction)

3 Simulation

3.1 Evaluation Metric

(1) Overhead
The overhead is all packets. Controlling packets are removed in the Network. In other words, the overhead is the whole network load. A heavy network load decreases packet delivery ratio, on the other hand, a light network load is provides comfortable communication. The stable communication has to maintain the values of the overhead.

(2) Packet Delivery Ratio
The packet delivery ratio (PDR) is the rate of packets received by the destination node in those packets sent by the source node. The packet delivery ratio is calculated using the following equation.

$PDR =$
(Number of Packets Delivered by Destination Node) /

$$\textit{(Number of Packets Sent by Source Node)}$$

(3) Number of Evidence Packets
The number of evidence packets is the evidence that a relayer obtained per data packet sent by another relayer. The value is the scale of reliability of evidence. However, a large value is required for heavy network loads.

4 Results and Discussion

4.1 Overhead

As shown in Fig. 8, all packets are received by sender and relayers. The overhead is the evidence, removed controlling, and data packets. The data packets were sent by CBR-traffic to the destination node. The number of all the packets increased in that network because of the increase in the total number of nodes in that area.

It should be noted that the graph of the overhead is increases almost monotonically as the total number of nodes increase.

4.2 Packet Delivery Ratio

Figure 10 shows the packet delivery ratio for each number of nodes. In 100 nodes participated, 25% of the packets were lost because different types of packets interfered with each other. These collisions were caused by the increasing overhead. Furthermore, the antenna was busy during receiving the different types of packets.

From 10 to 50 nodes, packet delivery ratios increased whether forensics was enabled or not. Moreover, the highest performance of forensics was obtained when the density of nodes (DN) was 50 or less. The density of nodes is determined from the following expression.

$$DN = \textit{(Total Number of Nodes) / (Planar Dimension } [km^2])$$

However, the fact that packet delivery ratios decreased when 6,000 packets or more were generated (Fig. 8) suggests that packet delivery ratios completely depend on the ability of the nodes.

4.3 Number of Evidence Packets

As shown in Fig. 10, a relayer received the average number of all-packets and evidence-packets per packet sent in the MANET. The results of our experiment clearly show that the number of evidence packets can be determined from the following expressions.

Area Ratio (AR):
AR = (Dimension Space) / (Transmission Range)
Number of Evidence Packets (NEP):
NEP = (Total Number of Nodes) / AR

It is possible to determine the number of evidence packets by controlling the transmission range.

Likewise, the number of evidence packets decreases as the transmission range expands, therefore, we were able to determine the number of evidence packets from the size of the transmission range.

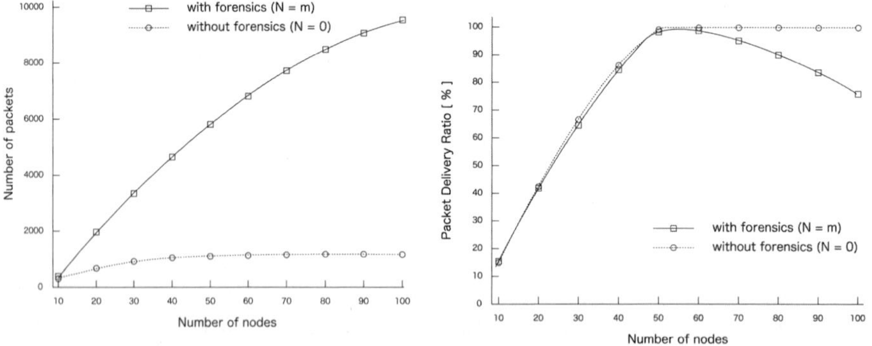

Fig. 8. Overhead   **Fig. 9.** Packet Delivery Ratios

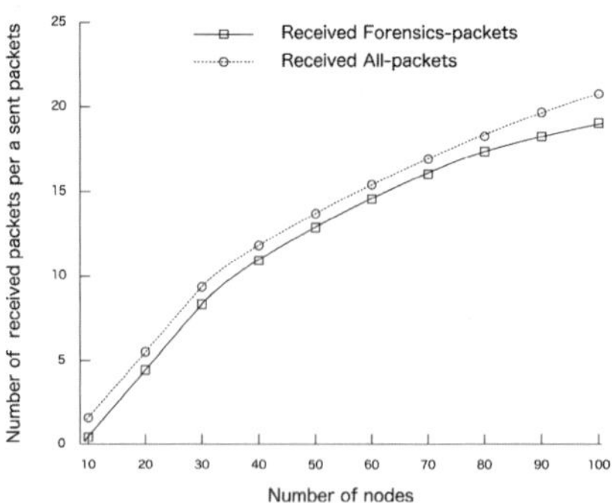

Fig. 10. Number of evidence packets

4.4 Load Reduction of the Networks

4.4.1 Load Reduction of the Networks

The conventional established method of collecting evidence packets from the total number of nodes around the relayer requires a heavy network load. Because evidence packets requires a guarantee of sending contents, the overhead depends on the number of the guarantors. This method is inefficient in the reliability assurance system of packets that the relayer transmitted.

It is possible to improve the reliability assurance system. Evidence packets that are received by the upper and down nodes are needed to ensure that a relayer transmits the correct data packets. Eyewitness nodes are needed to receive evidence packets for improving reliability. A guarantor is chosen by the following methods.

Choosing guarantors:
1. Randomly from routing table
2. A well-known reliable person

The relayer can switch methods depending on the situation. The network load is reduced greatly by this efficient method of collecting evidence packets.

5 Related Work

Routing protocols are proposed for MANETs. Certain types of reactive routing protocols have AODV [1], DSR [2] and so on. Attacking to those routing protocols are existed. Attacks and methods of defenses are explained in [3] [4]. HADOF [5] is one of the defense methods using DSR. Watchdog [6] monitors down nodes and breaks or changes a route when a problem occurs.

6 Conclusion and Future Work

We proposed and evaluated the performance of MANET forensics. It is a method of collecting proven evidence of sent data. Because privacy is protected by hashing the data and encryption by third parties, the reliability of the evidence has greatly improved. The forensics on MANET is the most realizable in the experimental results. In the future, increasing packet delivery ratios will be possible by controlling evidence packets. It is important to analyze the evidence collected by guarantors. Since we thoroughly investigated the methods of analyzing evidence packets, contents of the forensics data were confirmed whether it was complete or not. Establishing methods of collecting and analyzing MANET forensics is an improvement over existing security methods. Additionally, this proposal technique is more required to evaluate reliability, detection rate and delay for the incident report.

References

1. Perkins, C., Belding-Royer, E., Das, S.: Ad Hoc On Demand Distance Vector (AODV) Routing. IETF-Request-for-Comments, rfc3561.txt (2003)
2. Johnson, D., Hu, Y., Maltz, D.: The Dynamic Source Routing Protocol (DSR) for Mobile Ad Hoc Networks for IPv4. IETF-Request-for-Comments, rfc4728.txt (2007)

3. Mori, T., Mori, I., Takahashi, O.: A Classification of Defense Method in Ad-Hoc Networks and Proposal for the Anti Attack Ad-hoc Routing Protocol Architecture, MBL-41. In: MBL, vol. 41, pp. 73–78 (2007)
4. Mori, I., Mori, T., Takahashi, O.: Classification of Attacks and Defense method and Proposal of AODV-based Secure Routing Protocol in Mobile Ad Hoc Networks, MBL-41. In: MBL, vol. 41, pp. 79–84 (2007)
5. Yu, W., Sun, Y., Ray Liu, K.J.: HADOF: Defense Against Routing Disruptions in Mobile Ad Hoc Networks. In: INFOCOM 2005, Miami (2005)
6. Marti, S., Giuli, T., Lai, K., Baker, M.: Mitigating routing misbehavior in mobile ad hoc networks. In: Proceedings of Mobicom 2000, Boston (2000)

Efficient Reliable Data Transmission Using Network Coding in MANET Multipath Routing Environment

Tomonori Kagi and Osamu Takahashi

Future University-Hakodate
116-2, Kameda Nakanocho, Hakodate-shi, Hokkaido, 041-8655, Japan
{g2107006,osamu}@fun.ac.jp

Abstract. We propose an efficient reliability improvement method in a MANET environment by applying network coding encoded by a relay node. Reliability is improved without requiring the source node to send redundant encoding packets. As shown by simulations, the packet delivery ratio of our proposed method is as high as the existing encoding method, packet-level forward error control (FEC), but our method has a longer delay than FEC and split multipath routing (SMR) in an environment with packet errors. In addition, packet overhead in our proposed method is improved by decreasing the number of packets transmitted by the source node. Thus, our proposed method realizes more efficient and reliable data transmission, especially in comparison with FEC.

Keywords: Mobile ad hoc networks, multipath routing, network coding, reliability.

1 Introduction

Recently, wireless modules have been mounted on various devices due to the progress in wireless communication technology. These ad hoc networks are instantly deployable wireless networks which rely on radio waves instead of base stations or communication infrastructure support. Because radio waves have a short propagation range, the route becomes "multihop" when a communication peer is not within range. In general, the reliability is low in ad hoc networks for reasons such as frequent changes of network topology, unstable radio environment, and packet collisions. By "reliability" we mean the probability that a data generated at source node in the network can actually be routed to the intended destination.

Multipath routing in mobile ad hoc networks (MANETs) was originally developed as a means to provide route failure protection. This routing was constructed from plural paths between communication peers in order to provide for unexpected path destruction. The reliability of the data transmission depends on the degree of path independence (disjoint paths). There are two types of disjoint paths: link disjoint paths and node disjoint paths. Regardless of the source and destination, the link disjoint paths have no shared link between paths, and the node disjoint paths have no shared node between paths. Clearly, then, node disjoint paths are also link disjoint paths. If

I. Lovrek, R.J. Howlett, and L.C. Jain (Eds.): KES 2008, Part III, LNAI 5179, pp. 183–192, 2008.

the multiple paths are independent, then the breaking of one path has no influence on another path. Split multipath routing (SMR) [4] is multipath routing that constructs disjoint multiple routes.

Packet-level forward error control (FEC) and automatic repeat request (ARQ) are two methods widely used to recover the lost packets in networks with unreliable links.

ARQ is an error recovery method which uses acknowledgment packets (ACKs) and a timer to achieve reliable data transmission. The acknowledgment packet is a message sent by the receiver to the sender to indicate that it has correctly received a packet. If the sender does not receive an acknowledgment before a specified period of time (timeout), the sender usually retransmits the packet until it receives an acknowledgment or exceeds a predefined number of retransmissions.

However, ARQ is not considered applicable in low reliability and high mobility networks such as the ad hoc network, because the transmission delay increases due to retransmissions by the sender for missing ACKs.

FEC is an error correction method for data transmission, whereby the sender generates an error correction code, adds it to the original packet, and then sends both the error correction code and the original packet. This method allows the receiver to detect and correct errors without the need to ask the sender for retransmission of the packet.

FEC-based methods in ad hoc networks have been studied [6],[7] and found to improve the reliability when used with multipath routing. However, the number of transmissions packets of the source node is increased. For example, in Figure 1, source node S generates a Code from Data1 and Data2 by encoding. Then, the source node sends the Code. In this case, the number of packets transmitted by the source node is three (Data1, Data2, Code). Thus, the transmission frequency at the source node is increased.

In this paper, we propose an efficient and reliable packet transmission method by using multipath routing constructs from multiple node disjoint routes, and by applying network coding, which allows packet encoding at a relay node. Because the encoding packet is generated by a relay node, the source node does not need to encode the packets, and sends only data packets to each route. Thus, the packets transmitted by the source node are not increased. We call this the network control method, or "NC." To evaluate our proposed method, we conducted simulations.

Our proposed method is discussed in Section 2. A prototype implementation of our proposal is described in Section 3. We evaluate our proposal by comparing related protocols in Section 4, and summarize our work in Section 5.

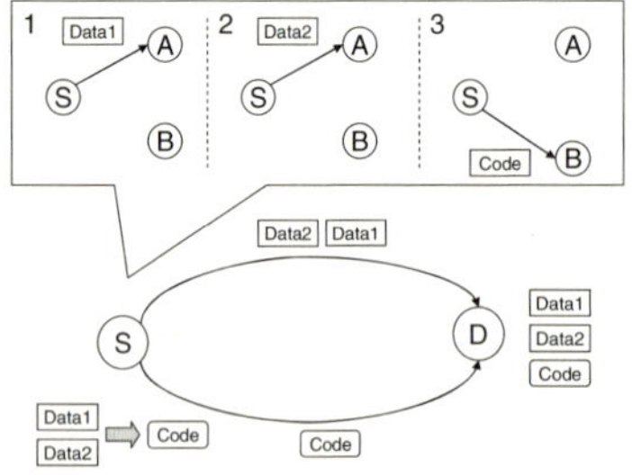

Fig. 1. Number of transmissions by source using FEC

2 Proposed Method

In our proposed method, multiple node disjoint paths are constructed, then the source node sends packets to a neighbor node on each path. The neighbor node generates an encoded packet when it receives the necessary data packets for encoding, then the neighbor node sends the encoded packet.

This method enables us to improve reliability by sending redundant encoded packets to the destination node without increasing the transmission frequency of the source node.

2.1 Basic Operation Model

In our proposed method, NC, multiple node disjoint routes are constructed. Figure 2 shows an example of two node disjoint routes.

We assume the following two routes are constructed.

> Primary path: (S, A, C, D)
> Secondary path: (S, B, E, F, D)

Five kinds of nodes exist in this model. The nodes and their operations are shown as follows.

- Source Node (Node S)

  The source node sends data packets to neighbor nodes on all routes simultaneously.

- The neighbor node on the primary path (Node A)

  This neighbor node on the primary path forwards the data packets from the source node.

- The neighbor node on the secondary path (Node B)

  When this node receives the packets for encoding, it generates an encoded packet and sends it to the next node (Node E).

- Other relay nodes (Node C,E,F)

  These nodes forward the packets toward the destination node.

- The destination node (Node D)

  Even if the data packets are lost, the destination node decodes the coded packets and successfully retrieves the data packets.

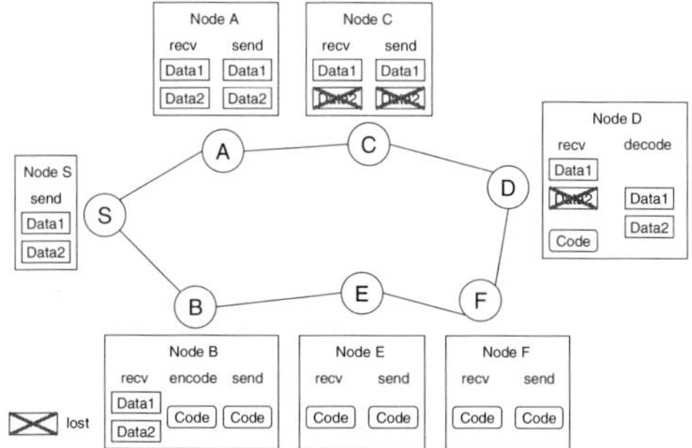

Fig. 2. Sample model using NC

As shown in Figure 2, first, source node S sends a data packet named Data1 to node A and node B. Node A receives Data1 then forwards it to the next node, Node C. Node B receives Data1 then buffers Data1. Next, Source node S sends a data packet named Data2. Again, node A receives Data2 then forwards it to the next node, Node C. Node B receives Data2, and then generates the encoded packet Code by encoding from Data1, which is buffered, and Data2. The other nodes (C, E, and F) forward the packets on their respective routes.

For example, let us assume Data2 is dropped between nodes B and E. However, destination node D receives Data1 and the Code, and so Data2 can be decoded from these packets. Therefore, the number of transmissions does not need to be increased, and reliability can be improved because the encoded packet can be forwarded to destination node D.

2.2 Construction of Multiple Routes

The construction of the multiple route method is shown in this section.

For simplicity, two routes are constructed in our proposal. We use split multipath routing (SMR) to construct the multiple disjoint routes. SMR has the best performance for multipath routing [8]. If multiple node disjoint routes are not constructed, the network coding isn't used and all nodes send only the data packets. Similarly, all nodes send data packets when a single route is constructed.

It is essential that the source node constructs multiple routes in our proposal. Therefore, in order to maintain multiple routes, a source node can again start route discovery when a route is broken.

2.3 Data Packet Transmission Scheduling

A source node transmits the data packets to multiple routes simultaneously. Usually, a broadcast is used for the delivery of data to all nodes simultaneously. However, broadcasting should not be used in our proposal, because it is not suitable in ad hoc networks, since it causes unnecessary congestion and collisions of data packets.

In our proposal, we use unicasting for the transmission of data packets to one node. In this case, all nodes operate in the promiscuous mode. The neighbor nodes that use promiscuous mode can receive data packets sent to other nodes. Thereby, simultaneous transmission of packets to multiple nodes is achieved.

For example, Source node S sends Data1 to Node A by unicast, as shown in Figure 3. Node B can promiscuously receive Data1. Similarly, Node A can receive Data2 which is sent to Node B.

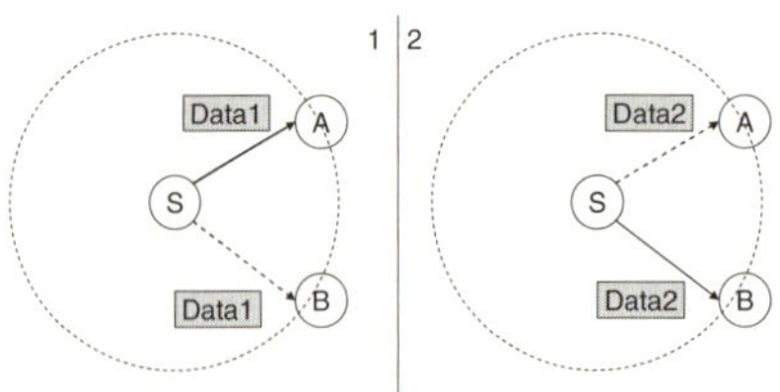

Fig. 3. Packet transmitting scheduling by the promiscuous mode

2.4 Network Coding Method

(1) Network Coding Scheme

The network coding idea was introduced by Ahlswede et al. [2]. Usually, the routers or relay nodes just forward and duplicate the packets in the networks. However, network coding permits routers or relay nodes to encode the packets.

Ref. [5] suggested a two-way traffic integrated transmission technique using a network coding scheme in a multihop wireless network.

(2) Coding Method

In this paper, we use a linear network coding scheme [3]. The linear network coding scheme is an encoding method such that coding vector $g_i = (g_{i1}, g_{i2},..., g_{iN})$ is given, and input packet $M = (M_1, M_2,..., M_N)$ is converted into output packet Pi by the following expression.

$$P_i = \sum_{j=1}^{N} g_{ij} M_j \tag{1}$$

The destination node can decode input packets because the coding vector $G = (g_1, g_2, ... , g_N)$ and output packet data $P = (P_1, P_2, ..., P_N)$ are obtained from the received packets, and an inverse matrix exists in G.

3 Implementation

We implemented our proposal with dynamic source routing (DSR) [1]. A source node adds the source route option header to the data packet in DSR. A source route option header contains route information. In our proposal, the source route option header contains multiple route information because the source node sends one packet to all relay nodes on each route. In addition, the source route option header contains control information for coding. Therefore, we extend the source route option header, shown in Figure 4. (Extension is a shaded region.)

0   16   32 [bit]

Opt Type	Opt Data Len	Reserved	Segs Left
Coding Seqence No[1]		Coding Sequence No[2]	
...		Coding Sequence No[n]	
Coding Vector[1]	Coding Vector[2]	...	Coding Vector[n]
Path1 Address[1]			
Path1 Address[2]			
...			
Path1 Address[n]			
Path2 Address[1]			
Path2 Address[2]			
...			
Path2 Address[n]			
...			
Path N Address[n]			

Fig. 4. Extension of source route option header

The control information contains the following.

- Coding Sequence Number

  This unique number is added to a packet by the source node. It is added to the packet in order to generate redundant packets (encoding packets) by the relay node. Therefore, a destination node can correlate and identify the encoding packets and data packets.

- Coding Vector

  The coding vector is used for packet encoding and decoding. The coding vector is specified by the source node, and it is used for generation of the encoded packet at the neighbor node (the network encoding node) and for decoding at the destination node.

4 Experiment

We evaluate our proposal by a computor simulation. We use ns-2 as the simulator.

4.1 Experiment Description

(1) Simulation Models

We evaluate our proposed method by comparison with the SMR and FEC protocols.

- Protocol Model 1: SMR

  The routing protocol SMR constructs multiple routes. Then, the data packets are alternately sent on a single route (fig.5).

- Protocol Model 2: FEC

  The encoding packets are generated by the source node. The encoding packets are sent on one route, and the data packets are sent on another (fig. 6).

- Protocol Model 3: NC

  Our proposed protocol (NC: SMR with network coding) (fig.7).

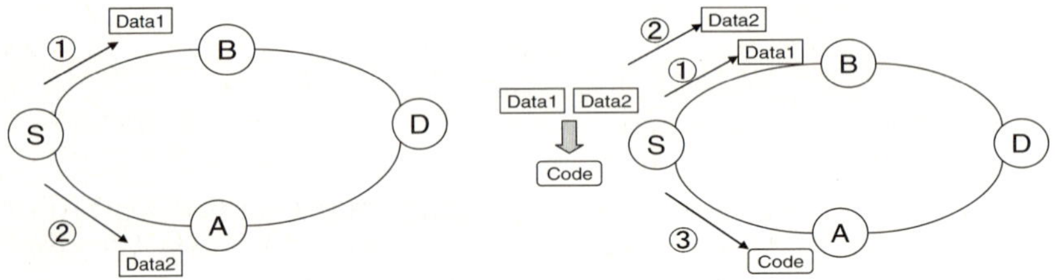

Fig. 5. SMR protocol model   **Fig. 6.** FEC protocol model

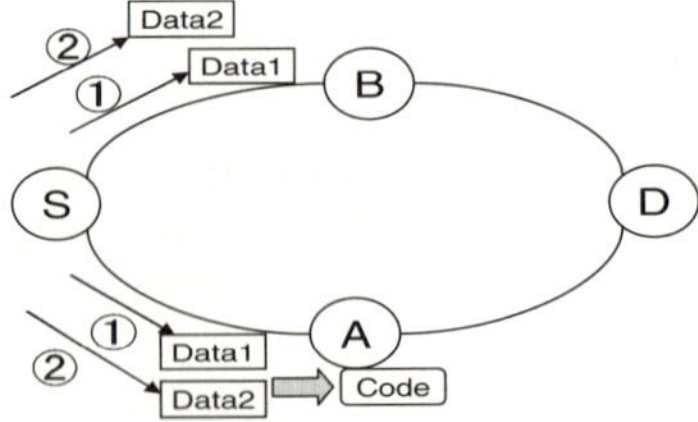

Fig. 7. NC protocol model

(2) Evaluation Items

We evaluate the packet delivery ratio, the packet overhead, and the transmission delay in order to estimate the efficiency of our proposal.

- Packet delivery ratio

  Packet delivery ratio is defined as the number of correctly received data packets at the destination node divided by the number of original data packets sent by the source node.

- Packet overhead

  The packet overhead is defined as the number of all node transmission packets, including data packets and encoded packets.

- Transmission delay

  The transmission delay is defined as the time when the source node generates a packet until the time when the destination node receives it. In the case of the encoding models (Models 2 and 3), the transmission delay is defined as the time when the source node generates a packet until the time the destination node decodes the encoded packets and retrieves the original packets.

4.2 Simulation Environment

The simulation environment is shown in Table 1. In this simulation, one encoding packet is generated by two data packets using the coding method (see Section 2.4).

We use packet loss rate, which is the rate of dropped packets between links to the total number of packets successfully transmitted.

Table 1. Simulation parameters

Field	1000×1000 [m]
Number of nodes	100
Radio range	250 [m]
Connections	1
Speed	5 [km/h]
Move model	Random way point
Simulation time	500 [sec]
Data size	512 bytes
Protocol	UDP
Time between generating packet	0.05 [s]
Packet loss rate	0 ~ 50 [%]

4.3 Simulation Results

The simulation results are shown in Figures 8 through 10.

Figure 8 shows the packet delivery ratio as a function of the packet loss rate for each protocol. The packet delivery ratio decreases as the packet loss rate increases. However, our proposed method (NC) and FEC have a higher ratio of packet delivery in comparison with SMR. In a comparison of the packet delivery ratio by FEC and NC, the simulation results show NC is the same as FEC.

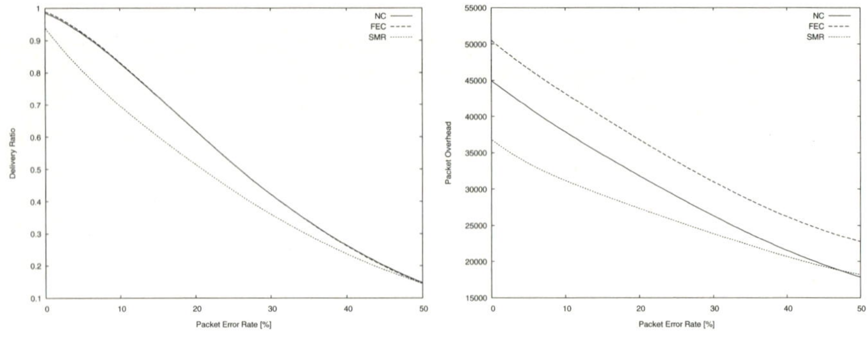

Fig. 8. Packet delivery ratio   **Fig. 9.** Packet overhead

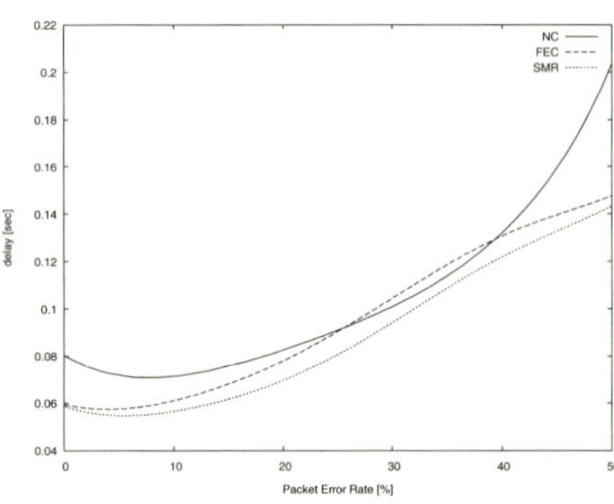

Fig. 10. Transmission delay

Figure 9 shows the packet overhead as a function of the packet error rate for each protocol. The packet overhead decreases as the packet loss rate increases because the number of receivable packets decreases. NC and FEC, which send both the data packets and encoding packets, have higher packet overhead than SMR does, because SMR sends only the data packets

The data packets are encoded by a relay node in the proposed NC protocol, so no source node sends redundant encoding packets. Consequently, NC has a lower packet overhead than does FEC.

Figure 10 shows the evaluation result of the transmission delay. Each protocol has a longer delay that is proportional to the packet loss rate. If the packet error rate is high, the packet loss probability increases, and a longer route setup time is needed (i.e., the exchange time of route request (RREQ) and route reply (RREP) is longer).

NC and FEC have a longer transmission delay than does SMR. This is because the destination node receives the coded packet from the secondary path, then decodes and retrieves the original data packets, as shown in Figure 11 and Figure 12.

NC has a longer delay than FEC, and we explain the reason by using Figure 10. The source node sends Data1 and Data2. The relay node receives and then buffers Data1. Next, the relay node receives Data2 but cannot receive Data2 due to packet

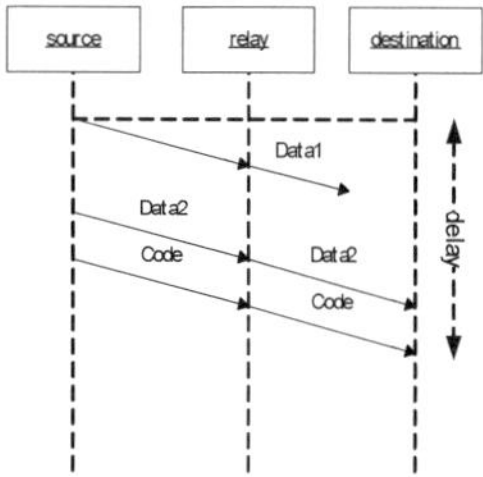

Fig. 11. Delay by encoding (FEC)

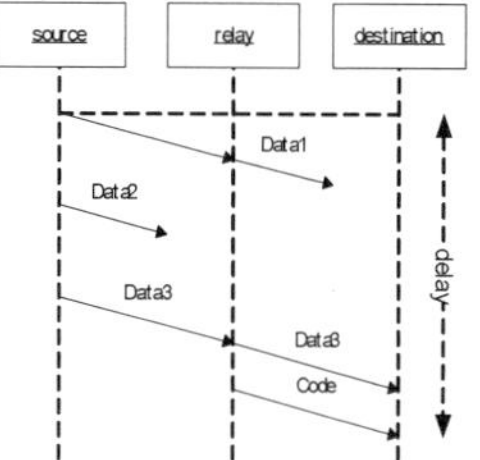

Fig. 12. Delay by encoding (NC)

loss. The relay node receives Data3, sends the Code generated from Data1 and Data3. The destination node receives the Code, and decodes Data1 from Data3 and the Code. Therefore, NC has longer transmission delay than FEC.

5 Conclusion

In this paper, we proposed a reliability improvement method in mobile ad hoc networks by applying network coding encoded by a relay node. Therefore, reliability is improved without requiring the source node to send redundant encoding packets. As shown by simulations, the packet delivery ratio of our proposed method (NC) is as high as the encoding method FEC, but our method has a longer delay than the existing methods (FEC and SMR) in an environment with packet errors. In addition, packet overhead in our NC method is improved by decreasing the number of packets transmitted by the source node. Thus, our proposed method realizes more efficient and reliable data transmission in comparison with FEC.

As future work, (i) we plan to design an encoding and forwarding method for situations in which multiple node disjoint routes are not constructed. (ii) We evaluate which node to encode packet, where path to encode, and the delivery ratio by the number of packets used for encoding.

References

1. Jhonson, D., Hu, Y., Maltz, D.: The Dynamic Source Routing Protocol (DSR) for Mobile Ad Hoc Networks for IPv4. IETF-Request-for-Comments, rfc4728.txt (February 2007)
2. Ahlswede, R., Cai, N., Li, S.R., Yeung, R.W.: Network information flow. IEEE Trans. Information Theory (July 2000)
3. Li, S.R., Yeung, R.W., Cai, N.: Linear network coding. IEEE Trans. Information Theory 49, 371–381 (2003)
4. Lee, S.J., Gerla, M.: Split Multipath Routing with Maximally Disjoint Paths in Ad hoc Networks. In: ICC 2001, pp. 3201–3205 (June 2001)
5. Katti, S., Rahul, H., Hu, W., Katabi, D., Medard, M., Crowcroft, J.: XORs in the air: Practical wireless network coding. In: Proc. of ACM SIGCOMM 2006, Pisa, Italy, September 2006, pp. 243–254 (2006)

6. Lou, W., Liu, W., Zhang, Y.: Performance Optimization using Multipath Routing in Mobile Ad Hoc and Wireless Sensor Networks. In: Cheng, M., Li, Y., Du, D.-Z. (eds.) Combinatorial Optimization in Communication Networks. Kluwer, Dordrecht (2006)
7. Ma, R., Ilow, J.: Regenerating Nodes for Real-Time Transmissions in Multi-Hop Wireless Networks. In: IEEE LCN 2004 (2004)
8. Ueno, Y.: Disjoint Multipath Source Routing Protocol with Route Maintenance. IPSJ 425(12) (December 2004)

Validation at a Small Building with the Mutual Complement Network by Wireless and Wired

Kunihiro Yamada[1], Kakeru Kimura[1], Takashi Furumura[2], Masanori Kojima[3], Kouji Yoshida[4], and Tadanori Mizuno[5]

[1] Tokai University
[2] MegaChips Corpration
[3] Osaka Institute of Technology
[4] Shounan Institute of Technology
[5] Shizuoka University

Abstract. The "mutual complement network by wireless and wired systems" has been studied for possible application to home network. The evaluation of the mutual complement network in a small-scale building has revealed that the home application is reasonably possible. In this network, the communication performance values of both the wireless and wired systems distribute independently from the other over a wide area in a small-scale building compared with those in a conventional home house. The method of power supply, electrical apparatus used and the building structure are responsible for the results.

Keywords: multimedia and communication technology, network, PLC, ZigBee, building.

1 Introduction

The "mutual complement network by wireless and wired systems" produced 100% of its communication performance when it was evaluated in an apartment house, a two-storey prefabricated house and a wooden house, and a three-storey steel reinforced residential building of 200 square meters [1]. The purpose of using the "mutual complement network of wireless and wired systems" in the small-scale building is basically the same as that in a home house. Although the control lines and information communication lines, such as for a power supply, the Internet, a security system, TV, a telephone, a hot-water supply system, lighting, and an elevator, run all around, no network systems interact with others [2][3][4] It comes first to integrate various kinds of information networks in the building. Monitoring various apparatus, sensors and other networks comes next to prevent crimes and disasters.

Secondly, it is the practical operation of private power generation systems such as solar power systems that every building may have in future and the controlled energy-efficient operation of all the apparatus including air conditioning within the building. The third is to make the activity of people more convenient within the building [5].

Possibility to utilize the "mutual complement network by wireless and wired systems" is evaluated in three small-scale buildings for commercial operation and school ranging from 1,500 square meters to 5,600 square meters. In each building, unlike in a

I. Lovrek, R.J. Howlett, and L.C. Jain (Eds.): KES 2008, Part III, LNAI 5179, pp. 193–200, 2008.

conventional home house, the communication performance values of the wireless and wired system distribute independently from the other over a wide area owing to the power supply method, electrical apparatus used and building structure. The results prove that "mutual complement network by wireless and wired systems" is reasonably possible.

2 Mutual Complement Network of Wireless and Wired

The mutual complement network of wireless and wired systems has two or more (two for this study) means of communication. Utilizing the characteristic differences between the two, the network makes them complement with the other to improve the communication performance. Specifically, as the two communication means, a wireless and a wired communication systems are used. The wired communication system uses the existing electric light line that does not require additional expense. This is called PLC (Power Line Carrier Communication) which superimposes communication information on the electric light line as a conveyance signal. By our evaluation, the communication performance of PLC is 70% at a conventional house [1]. The interference due to the noise emitted by the electrical appliances and the signal attenuation attributable to the single-phase, three-wire electrical power supply to the housing are responsible for this characteristic value. Meanwhile, Zigbee (IEEE802.15.4) is used as the wireless communication system[2][3]. Having characteristics of low power consumption and low rate (20Kbps or 250Kbps), its communication performance deteriorates in proportion to the distance. In an area relatively free from obstacles, the communication is possible over the distance of 30 meters or longer. However, the communication is interrupted not only by a concrete object, needless to say, but also by a slightly thick wooden floor or a wall [1]. In a steel-reinforced concrete housing, for example, the communication performance deteriorates to 60% between the first and the second floors. It even lowers to 30% when communicating between the first and the third floors. We evaluate the communication performance of Zig-Bee at 82% in a single, independent house [1]. The characteristic of wired communications PLC and wireless communications Zig-bee are as described above. It is difficult for either one of the systems alone to complete a domestic network system. Now, the communication performance of the mutual complement network by two communication systems, wireless and wired, is evaluated. The data of the communication performance of PLC and Zig-bee which we measured in the home house are grouped into three levels, namely "True", "True?" and "False". The performance of PLC is as follows. True;70%, True?;10% False;20%. The performance of wireless communications Zig-bee is as follows. True;82%, True?;14%, False;4%. (Table 2 maps these values. If either one is "True" or if both are "True?", the communication is considered to be possible. Then, the performance of the mutual complement network is evaluated to be 94.6% for either one to be "True" and 1.4% for both to be "True?", 96% in total.) Mapping these values in Table 2, the evaluation of the performance of the mutual complement network is calculated to be 96.0%.

Next, communication performance of the mutual complement network system of wireless and wired was evaluated in a three-story house [8].

Table 1. Evaluating the communication performance by wireless communications and wired communications

		Wireless		
		Truth 82	Truth? 14	False 4
Wired	Truth 70	57.4	9.8	2.8
	Truth? 10	8.2	1.4	0.4
	False 20	16.4	2.8	0.8

The result of communication performance evaluation is shown in Table.2. The average performance of the wired communication alone is 79.7%. The average performance of the wired communication alone is 80.7%. In comparison with the evaluation of the communication performance shown previously, the wireless is lower by 1.3 points while the wired improves by 10.7 points. The communication performances of the mutual complement network as a whole are 100% at all the evaluation points from the first floor through the third floor. This confirms that the communication is successful through either the wireless or the wired system.

Table 2. Communication performance evaluation between floors in a house

		Overall success rate			Wired single success rate			Wireless single success rate		
		Floor of reception								
		1F	2F	3F	1F	2F	3F	1F	2F	3F
Floor of transmission	1F	100%	100%	100%	98%	80%	74%	100%	83%	55%
	2F	100%	100%	100%	82%	81%	75%	81%	87%	84%
	3F	100%	100%	100%	75%	76%	76%	60%	83%	93%

Table 3. Calculated value of the communication performance by wireless communications and wired communications

		Floor of reception		
		1F	2F	3F
Floor of transmission	1F	100%	97%	88%
	2F	97%	98%	96%
	3F	90%	96%	99%

The rate of communication error of the mutual complement network = The rate of wired communications error x The rate of wireless communications error From the communication performance shown in Table.2, the communication performance of the mutual complement network is obtained as shown in Table.3. From Table 3, the average values of the communication performance between the floors becomes 95.7%, and is mostly in agreement with the evaluated value of 96% acquired from Table.1. Since this evaluation considers that the mutual complement communication is successful if the communication takes place through either the wired system or the wireless system, the performance turns out to be 100% as shown in Table 3.

Thus, the study has found a possibility to establish an effective network in a relatively small space such as a home house or a building of similar size without modifying the communications environment.

A few examinations are in progress to improve the communication performance of the mutual complement network. One of them uses a method that relays each node where two means of communications (wired and wireless systems) are used to study a method that reduces the number of times of communications for each route [9]. In addition, as a means to improve the communication performance in a building with three or more floors, we are studying Two Route Method that sets up two routes and Relay Method that approaches the purpose node one floor at a time [9].

3 Communication Performance Evaluation at a Small Scale Building

At three small-scale buildings, the communication performance evaluation experiment of the mutual complement network system of wireless and wired was conducted to study the effectiveness of the mutual complement network . The communication performance evaluation result in each building is shown in Table.4.

Three buildings are named building A, building B and building C, respectively. The total area of each of those buildings ranges from 1,586 m^2 to 5,365 m^2, about 27 times the size of the home houses evaluated until now. A is a commercial building and B and C are university school buildings. For building A with one transformer, as shown in Table 4, the communication performance of the PLC communication, Zigbee communication and mutual complement network (MCN) are ranked to categories a, b, c, d and x.

The communication performance of PLC and Zig-bee communication systems in each building is shown in Table.5. The building equipment which affects the communication performance of PLC communication and Zigbee communication is shown in Table.6. Use of more pieces of 200V equipment such as kitchen equipment, air conditioners and fluorescent lamps improves the performance of the PLC communication system. It reaches as much as 98% in building A. On the other hand, Zigbee communication performance is very poor at 8.4%. The small windows and the doors made of metal are responsible for the low performance. In addition, the signal waves do not reach the receivers through the walls or floors.

Table 4. The items affecting the communication performance in three buildings for evaluation. a:95% or more b:90% or more c:70% or more d:70% or less Tr:Transform(s) Com;Common Dif:Different -:It does not exist. MCN:Mutual Complement Communication Network.

Building name	Summary	Number of Tr	PLC		Zigbee			MCN
			Tr		Same room	Same floor	Dif floor	
			Com	Dif				
A	1586.4 m^2 B1~7F	1	a	–	a	d	x	b
B	5365 m^2 B1~4F	3	b	x	a	c	c	a
C	1920 m^2 B1~4F	3	c	x	a	c	c	a

The rate of successful communication for the PLC alone is 74.4% in Building B. With the 200V equipment used, the PLC communication between phases are possible between the floors as long as the same transformer is used. The communication performance of Zigbee alone is 60.3%. The building B is relatively in favor of communication because of the spacious corridors and stairways. Particularly, the communication at the stairways is possible even through two floors.

Table 5. The communication performance of PLC and Zigbee (%) in each building

	PLC	Zigbee
A	98	11.1
B	74.4	55.8
C	52	43.8

Table 6. The factor which determines the communication performance in each building

	PLC	Zigbee
A	Communication between phases is possible due to 200V equipment (kitchen equipment and fluorescent lamps) Communication is good due to one transformer.	Small windows, iron door, steel-reinforced concrete walls Communication is possible only within a room.
B	Communication between phases is possible due to 200V equipment (kitchen equipment and fluorescent lamps) Due to 3 transformers, communication is not possible between transformers.	Communication is good at spacious stairways. Corridor is long. Waves do not reach more than 30 m.
C	Communication between phases is possible due to 200V equipment (kitchen equipment and fluorescent lamps). Dependent of switchboard. Due to 3 transformers, communication is not possible between transformers.	Slightly narrower than B, communication quality is lower.

Buildings C and B are similar in their structures. While the communication performance is slightly lower for both the PLC and Zigbee, both systems show fairly similar tendency with the other due to the same nature of school buildings. Building A turns out to be a special case where the PLC alone makes the communication possible in the entire area. For building B and building C, the mutual complement network is confirmed to be useful. At some locations in the two buildings, it is difficult for the PLC or Zigbee alone to establish communications.

3.1 PLC Characteristic in Small Scale Building

Figure 1 shows the power distribution system for Building C that is representative of general small-scale buildings. The 6600v high voltage is changed into single phase 3 lines 100/200V of phase-R, phase-T, and phase-N with the transformer of each building, and is sent to a switchboard through a trunk. From the switchboard, 100V power through RN or NT is supplied to each wall power outlet. The wall power outlets for the 200V apparatus through RT are prepared separately.

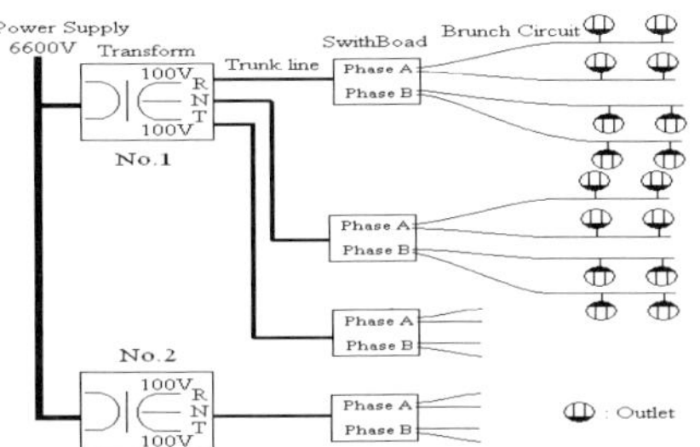

Fig. 1. The power line routing diagram in building C. While there are 3 transformers, one of them is for motor power supply, which is omitted because it is not relevant. The three-phase main power line of R, N and T is distributed through the switchboard to each power outlet.

Phase A: R-N, Phase B: N-T

Table 7. The communication performance of the wired communications in building C
PC : Computer room
_ : it dose not evaluate

Transform	Floor	PC	4F	3F	2F	1F	B1
No.1	PC	100					
No.1	4F	50	100				
No.1	3F	-	18	100			
No.2	2F	-	-	0	100		
No.2	1F	-	-	-	4	100	
No.2	B1	-	-	-	-	0	100
		PC	4F	3F	2F	1F	B1
		No.2			No.1		

Table 7 shows the results of evaluating the PLC in building C. While building C has three transformers, two of them supply electric power to the general electrical apparatus as shown in Fig.1. The experiment shows good a result that the PER is zero percent even between different phases as long as the communication takes place through the same trunk, that is, on the same floor. This is because the 200V fluorescent light which is 200V apparatus forms a bridge of phase-R or phase-T, helping the communication signals to reach a different phase without going to the transformer. Load levels as shown in Table.8 are defined to indicate the communication quality in relation to the routing in a comprehensible manner. Figure 2 shows the relationship between the communication quality and the load level. Four items, namely the transformer, trunk, phase and branch circuit, determine the load levels as shown in Figure 1.

Table 8. The definition of the communication load level in PLC. Tracing the communication route from the wall power outlet to the target wall power outlet provides the load levels. Reference figure Fig.1. Y is in the same wiring and is yes. N is in different wiring and is NO.

Load Level	Transform	Trunk line	Phase	Brunch Circuit
1	Y	Y	Y	Y
2	Y	Y	Y	N
3	Y	Y	N	—
4	Y	N	Y	—
5	Y	N	N	—
6	N	—	—	—

Fig. 2. Relation between the communication load level and a communication performance. "x" mark is an actual measurement

Figure 3 shows the packet error rate distribution in a small-scale building. The experiment took place at 139 locations in the three buildings. Table 9 compares the frequency distribution of PER with that in ordinary homes. The communication performance in a home house is better. Two or more transformers and power distribution to a large area in a small-scale building are considered to be responsible for this result.

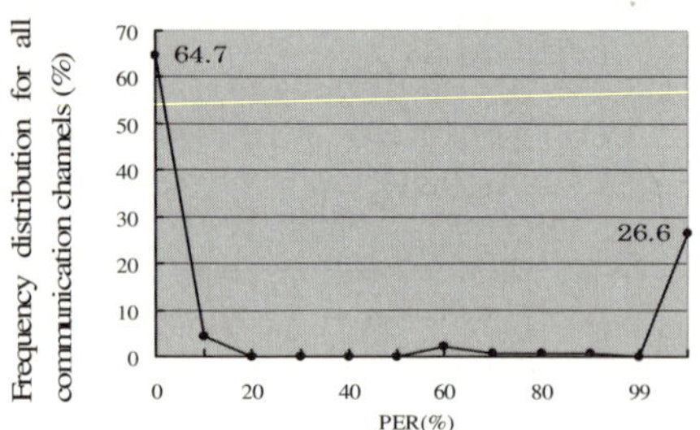

Fig. 3. The evaluation result of the wired communications(PLC) in three buildings

Table 9. The average value comparison of PER # between home and small building. PLC performance in small building is better than that of home network. (%),Diff: Difference.

PER	Home	Building	Diff
100	16.4	26.6	-10.2
0	70.0	64.7	5.3

3.2 Zigbee Characteristic in Small-Scale Building

Commercial building A is located in a densely built-up area, and has narrow stairs made of steel-reinforced concrete and a small window. Since it was divided with the iron door, the communication with the outside hardly came about, staying as low as 8.4%. Building B and building C are school buildings with a large space. B scores 60.3% and C 48.6%. Near the stairs, communication between floors was possible. The survey result of the communication between floors at B building is shown in Fig.4.

The problems associated with the PLC in a small-scale building are attributable to the presence of two or more transformers and with the communications between different phases. As a means to improve the PLC problems, Zigbee system performance is crucial in the communications at the stairways and wells.

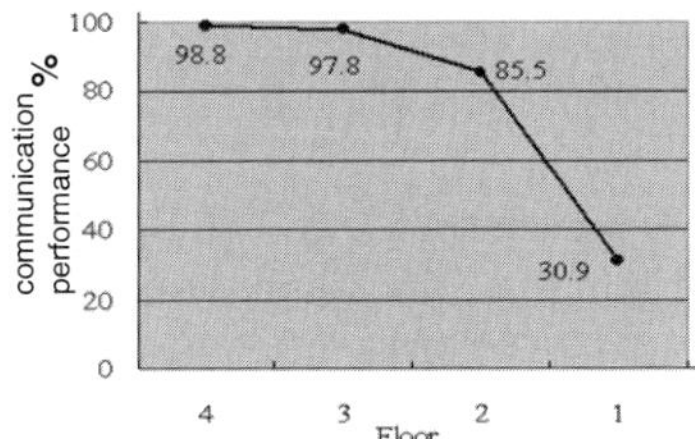

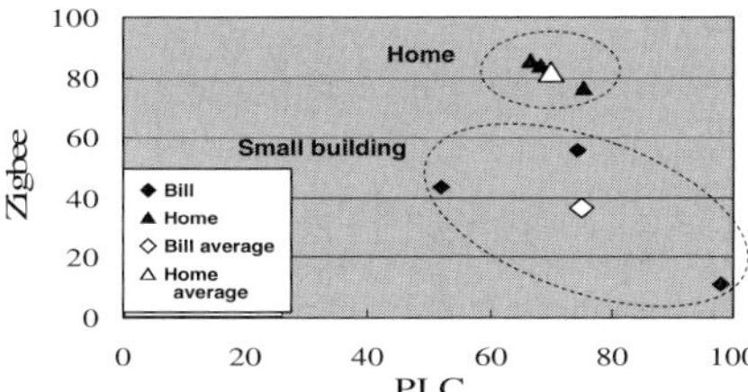

Fig. 4. The evaluation result of the wireless communications(zig-bee) in B building. Evaluation to each floor from four floors.

Fig. 5. A communication performance distribution with a building and a home house

4 Performance Evaluation of Mutual Complement Network

Unlike a home house, a small-scale building has two or more transformers and electricity is supplied in a long trunk from the transformers. These differences make the PLC communication performance nearly zero percent. To cope with the problems, the Zigbee communication at the stairways and between floors through the well is indispensable. If the mutual complement network is employed in a small-scale building, it has to be arranged in a manner that the communications between the floors take place through areas such as the stairways, wells and elevator shaft.

The mutual complement network provides advantages because all it takes is to put the nodes of the wired and wireless systems into the power outlets. At the small-scale building, it is confirmed that the communication performance of wireless and wired systems varied greatly as shown in Fig.5 compared with a home house.

5 Conclusion

As part of our continued study to apply the "mutual complement network by wireless and wired systems" to home use, evaluations took place in three small-scale buildings to check for feasibility. The evaluations revealed that the intended application to small-scale buildings was sufficiently possible. In conventional home houses as well

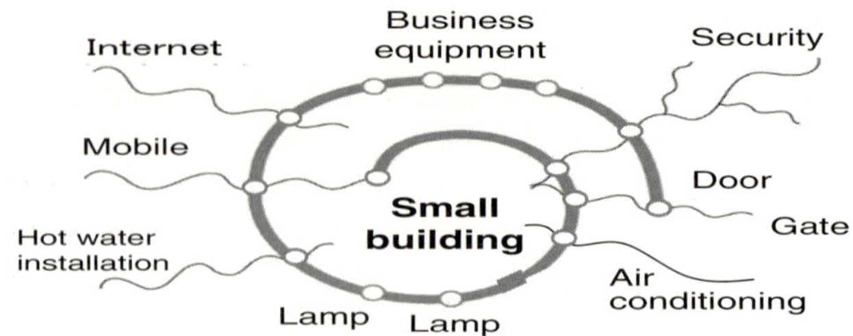

Fig. 6. The conceptual figure of practical use of the mutual complement network by the radio and the cable in a small building

as in small-scale building, it is important in the first place to check the necessity for preventing crimes and disasters, controlling energy and improving people's convenience by monitoring various existing networks and even controlling the apparatus, home electrical appliances and lightings.

Fig.6 shows the interaction of the "mutual complement network by wireless and wired systems" with other networks in a small-scale building. The study looks for a method that is technically less expensive yet carries out a lot of tasks efficiently.

References

1. Yamada, K., Hirata, Y., Naoe, Y., Furumura, T., Inoue, Y., Shimizu, T., Yoshida, K., Kojima, M., Mizuno, T.: Dual Communication System Using Wired and Wireless Correspondence in Small Space. In: Negoita, M.G., Howlett, R.J., Jain, L.C. (eds.) KES 2004. LNCS (LNAI), vol. 3214, pp. 898–904. Springer, Heidelberg (2004)
2. Okuda, T., et al.: Networking GasCentral Heating System. Matushita Technical Jounal 50(3) (June 2004)
3. Mitsubishi Electric Building system incorporated company, http://www.building.melco.co.jp/system/
4. Kubo, M., et al.: Power Line Communication and Development of Power Line Communication Modules. Matushita Technical Journal 49(1), 45–48 (2003)
5. Yamada, K., Furumura, T., Naoe, Y., Kitazawa, K., Shimizu, T., Yoshida, K., Kojima, M., Mizuno, T.: Home-Network of a Mutual Complement Communication System by Wired and Wireless. In: Gabrys, B., Howlett, R.J., Jain, L.C. (eds.) KES 2006. LNCS (LNAI), vol. 4253, pp. 189–196. Springer, Heidelberg (2006)
6. M16C/6S, http://www.renesas.com
7. http://www.zigbee.org
8. Yamada, K., et al.: Dual Communicatio System Using Wired and wireless Correspondence in Home Network. In: Khosla, R., Howlett, R.J., Jain, L.C. (eds.) KES 2005. LNCS (LNAI), vol. 3681, pp. 438–444. Springer, Heidelberg (2005)
9. Yamada, K., Furumura, T., Kimura, K., Kaneyama, T., Yoshida, K., Mineno, H., Mizuno, T.: Dual Communication System Using Wired and Wireless in Home-Network. In: Apolloni, B., Howlett, R.J., Jain, L. (eds.) KES 2007, Part III. LNCS (LNAI), vol. 4694, pp. 469–476. Springer, Heidelberg (2007)

Analysis of Relationships between Smiley and Atmosphere during Chat with Emotional Characters

Junko Itou and Jun Munemori

Faculty of Systems Engineering, Wakayama University,
930, Sakaedani, Wakayama 640-8510, Japan
{itou,munemori}@sys.wakayama-u.ac.jp

Abstract. In this research, we analyze the relationships among smiley, user's mental states represented by smiley, nonverbal expressions of emotional characters, and atmospheres during embodied character chat system. The character behave as agents of users and automatically act on messages of the users and the other character's action. We performed an experiment to investigate the relationships using embodied character.

1 Introduction

Online communication is widely spreading and tools of online communication become diverse, for example e-mail, chat system, remote meeting system, distance learning, and so on. There are also varieties of proposed tools on chat system from conventional text-based one to graphical one that agents in place of users talks in virtual 3D space.

By using embodied character in chat system, the users obtain messages by watching embodied character's actions as well as by reading plain texts. A character in the chat systems plays a role as a agent of a user not only to express the user's emotional states or intentions which cannot be displayed by a chat message, but also to make the chat alive. As the result, it becomes clear what meaning the user implies for the chat messages.

In order to employ embodied characters for chatting users, we need to control their nonverbal expressions which include gestures, eye-gazes, noddings, and facial expressions, and so on.

Previous work on controlling nonverbal expressions of embodied characters mainly discusses the consistency of nonverbal expressions with the speech utterances or the goal of conversation for each agent [1][2]. However, when we consider dialogues between a pair of embodied characters, we need to consider interdependences between nonverbal expressions displayed by those two characters. As explained in more detail in the next section, it is reported in the field of social psychology that nonverbal expressions given by humans during their talks are not independent with each other[3]-[6].

In the remainder of this article, we focus on the relationships between nonverbal expressions and smiley which are used mainly to express the user's emotional

I. Lovrek, R.J. Howlett, and L.C. Jain (Eds.): KES 2008, Part III, LNAI 5179, pp. 201–209, 2008.

state by text. Then we will discuss the chat atmosphere in terms of relationships among nonverbal expressions.

This paper is organized as follows: in section 2, we will describe the knowledge about the interdependences between nonverbal expressions in human communication reported in the previous work on social psychology. In section 3, we will propose a chat system to maintain the interdependences between nonverbal expressions of embodied characters in their chat. We will show the experimental result on the way of smiley and the association between smiley and atmosphere in chat in section 4. Finally, we discuss some conclusions and future work in section 5.

2   Interdependence between Conversation Partners

In order to employ embodied characters for chatting users, we need to control their nonverbal expressions. As far as characters have their own faces and bodies in order to be "embodied", their users read various meanings in the nonverbal expressions displayed by the faces and the bodies of the characters even if the characters are not actually designed to send nonverbal expressions to their users but only to speak. Thus, for any embodied characters, we need to control their nonverbal expressions properly so that they convey appropriate meaning to the users. For example, it would be strange if a character does not smile at all although the user tells a funny story. As another example, when one character talks or smiles to the partner character, if the partner character freeze with no response to the action, it looks also strange.

It has been investigated in social psychology what features are found in nonverbal expressions given by humans during their conversation. Through those investigations, it is known that nonverbal expressions given by humans in real conversation have some interdependences and synchronicity.

For example, it was reported that the test subjects maintained eye contact with their partners in lively animated conversation, whereas they avoided eye contact when they are not interested in talking with their partners [3][4]. In the experiments by Matarazzo, noddings by the listeners in conversation encouraged utterances of the speakers, and as the result, animated conversation between the speakers and the listeners is realized [5]. In another experiments by Dimberg, facial expressions of the test subjects were affected by those of their partners[6]. The subjects smiled when their partners gave smiles to them, whereas they gave expressions of tension when their partners had angry faces.

These results implies positive correlations or synchronicity between the nonverbal expressions given by the conversation partners for eye gazes, noddings and facial expressions. On the other hand, these interdependences are general tendencies in the usual conversation situation and can change by the second according with the conversation. Therefore, we can say that the strength of these interdependences are different by the atmosphere of the scene of talks.

People substitute smiley for nonverbal expressions in the text communication. That means we can estimate the mental state of the user by the usage of the smiley, and analyze relevance of the interdependences mentioned above and the atmosphere of the chat.

3   MEDIAC Messenger Chat System

3.1   Goal

In the most current graphical chat systems, the user should choose and specify the character's action explicitly. By employing the knowledge described in the previous section, we aim to realize automatic actions and reactions of embodied characters in order to make users' chat alive. It is preferable that character's actions and reactions can be produced by the user's message because the user attends to only his/her chat and it is very troublesome for the user to set manually all actions and reactions at each message during his/her chat.

In the remainder of this section, we will propose a system to realize this adjustment process especially for eye gazes, gestures, and facial expressions of the characters for the first step towards our goal.

3.2   Actions of Character

Let us consider a pair of embodied characters A and B to explain our chat system. We denote user A's agent by character$A$ and user B's agent by character$B$. The overview of the character is shown in Fig. 1.

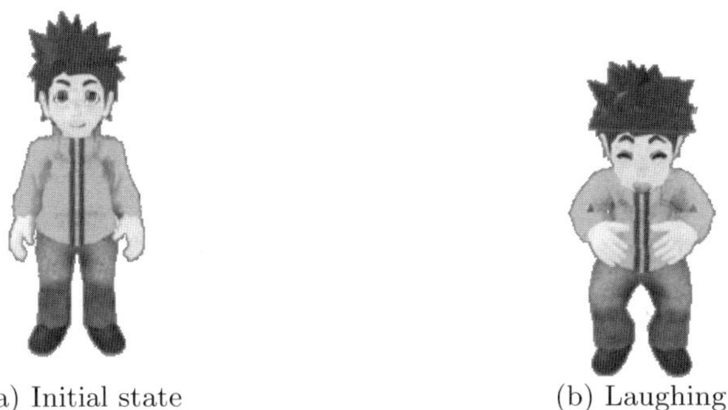

(a) Initial state   (b) Laughing

Fig. 1. 3D Character in MEDIAC Messenger

This character has 34 types of actions, which are related to about 250 words. The example of the mapping is given in Table.1.

As discussed in section 2, the actions of character$A$ are not only determined by the messages of user A but also the actions of character$B$. For example, when character$B$ laughs by the message of user B, character$A$ should smile even if user A's message does not contain the word of happy or smile.

Our system has 21 types of reaction. By Dimberg's experiments in section 2, the action of "smile" which is displayed on one character, responds to the other

Table 1. Character actions and keywords

Action	keyword *
bow	Hi, Hello, Good morning, Thank you, Nice to meet you
laugh	laugh, laughter, boff
smile	smile, happy, :-)
cry	sob, cry, weep, X-(
panic	dismay, uh-oh, nix, oops, darn
surprise	surprise, great, aghast, consternation
think	think, hm, hmm, um

* : In MEDIAC Messenger, all keywords are written in Japanese.

character's reaction "smile". On the other hand, when the action of "angry" is displayed, the reaction of "surprise" is expressed. When a message includes the keywords corresponding to "think", the partner shows the reaction "nodding" to stimulate their dialogue according to the research by Matarazzo. Furthermore, Watanabe[7] pointed we do not only exchange words, but also we share gestures and physical rhythm such as breath, so that we can feel an identification with the conversation partner. From this indication, kinds of synchronicity should be reflected to the relationships between an action and a reaction. Table.2 shows a part of the reaction rules which are defined as noted above.

Table 2. Character reaction rules

Action	Reaction	Action	Reaction	Action	Reaction
bow	bow	point at you	point at me	angry	surprise
laugh	laugh	point at me	point at you	achcha	nod
smile	smile	quake	quake	think	nod
handwave	handwave	no no	no no	cheer	smile
bummage	bummage	frustrate	no no	deny	strand

3.3 System Structure

Our chat system MEDIAC Messenger is constructed of a sever and multiple clients. Users start this client system then some windows come up as shown in Fig. 2

Main Window shows information about login users. In the Chat Window, users chat and User's Agent and Others Agent which is a character of chat partner, variously act according to the users' input chat text. These characters directly render on a user's desktop and the user can move the characters to the various points on the screen.

After user A inputs a message in the field of Chat Window and sends to server, the sever extracts keyword from the sent message by morphologic analysis so that the server determines which action character$A$ should behave. The action and user A's message are sent from the server to each client.

Fig. 2. Overview of MEDIAC Messenger

The client of user B which receives the action and the message, determines the reaction for the character A's action according to the reaction table, and send to the server. Again the server sends the reaction to each clients. In this way, each character's action and reaction are performed.

Each action and reaction takes one or two seconds. A user can input during an action or a reaction but the action or the reaction is not displayed until the last action ends.

4 Experimental Results

We implemented the MEDIAC Messenger described in the previous section. As a next step, we examine the way of using smiley and the relationships between the smiley and the mental state of the users.

4.1 Smileys and Mental States

A smiley can have several means, and the how to use, how to receive are different by users. Therefore we collected generally used smiley, and the emotional words remembered by the smiley.

Experimental subjects were 8 college students. They ordinarily uses a smiley in text communication. The method of the investigation is as follows: at first, we collected in total 126 kinds of smiley from a smiley list registered with IME2007. We indicated the subjects that they wrote the words to associate the smiley, the words associated with the smiley, impression words, Japanese phonetic symbols when the user converted from Kanji to it. Moreover, we asked the subjects to categorize the smiley into 35 emotional word category [8], and into 8 basic feelings[9]. The 8 basic feelings means *delight, grief, affection, aversive, fear,*

Table 3. Emotional word category

dislike	disappointment	pleasure	hatred	humiliation
anger	uneasy	doubt	warmth	pain
intolerance	fear	jealousy	unpleasant	insult
sorrow	pride	loneliness	regret	curiosity
patience	admiration	satisfaction	consideration	boredom
wonder	idly	happiness	respects	love
joy	hesitation	shame	indecent	pity

Table 4. Example of smiley

(^^)	(^o^)	o(^^)o	(LM)	(*^^)v
(E_E;	(*_*)	(ToT)	(>_<)	(LEEM)
m(_ _)m	(EoE)	(^^)v	(EtE|||)	(^^) _U~~

rage, amazement, and, *caution*. Table. 3 shows 35 emotional word category, and Table. 4 shows a part of the smiley.

It was 10 kinds of smiley that more than 5 subjects replied "I often use" and there was 48 kinds of smiley they had not used at all. I show 10 kinds of smiley mentioned above at Table. 5.

Among 8 basic feelings, the subjects had a tendency to choose *delight* and *affection* for (1)'(5), *grief, aversive*, and *fear* for (6), *grief* and *aversive* for (7). As for (8), *grief* was mainly chosen but some chose *affection* and *aversive*. It was (9) that answers was divided most. They selected *delight, affection, grief* and *aversive*. Smiley (10) expresses an aspect that a person waved a hand, but the feelings were chosen not only *delight, affection* but also *aversive*.

Frequently-appearing smiley includes many smiley evoked smile in (1)'(5). Actually, the words expressing friendly such as pleasure, joy and satisfaction are chosen from Emotional word category. In 8 basic feelings, it was classified in *delight* and *affection*, so it became clear that these emotional categories were indispensable in the chat.

About the emotional word category, almost all of subjects chose joy and pleasure in (1)'(3). As for (4) and (5), joy and pleasure were given like (1)'(3), but satisfaction, consideration, and warmth were chosen, too. (4) and (5) show about the same result, but it is revealed that the expression of emotion was stronger in (5). If a smiley expresses same mental state, we should reflect the strength of the emotion in embodied character chat system according to the situation.

Table 5. Frequently-appearing smiley

(1)(^^)	(2)(^_^)	(3)(^-^)	
(4)(^^)/	(5)(^-^)/	(6)(-_-;)	
(7)(T_T)	(8)m(_ _)m	(9)orz	(10)mV

Table 6. Response bias of emotional word category

smiley	Sub.A	Sub.B	Sub.C
(^^)/	joy, pleasure	satisfaction, consideration	warmth
(O[O)m	joy, pleasure, satisfaction	satisfaction, consideration	warmth
o(^^)o	joy, pleasure, satisfaction, warmth,	joy, pleasure, consideration	joy, happiness
(*^^)v	joy, pleasure, satisfaction	joy, pleasure, satisfaction, happiness	joy, satisfaction

The smiley of the (6) had many opinions to express feelings of the *grief*, but there was some subjects chose uneasiness, fear, pity, pain and unpleasant. It is thought that this smiley depends on context greatly when a user really uses it.

They tend to choose sorrow, loneliness and regret for (8), disappointment, uneasy, hesitation, regret and pain for (9), joy, warmth, idly and consideration for (10). The words associated with (8) are "I'm sorry" and "I apologize" such as words to express a feeling of the apology, "Thank you in advance" such as words to depend on, and, "Thank you" such as words to given thanks to. Like (6), information except the smiley is necessary for the estimate of feelings.

In addition, it is clear that every subject generally has a characteristic in choosing the emotional word category. The response bias of emotional word category are shown in Table. 6. Subject A tends to choose joy, subject B tends to choose consideration for joy in this smiley group. Subject C selected warmth that there was few of another person choosing. We may describe the adaptation model to the individual, as the constant pattern from these tendencies.

4.2   Actions and Atmosphere

We will finally reflect these result to 3-D virtual chat system mentioned in section 3. By the current chat system, embodied characters perform various actions based on the keyword extracted from the chat input message of the user. We conducted experimentation using MEDIAC Messenger to examine which word is often appeared and which action is often performed.

Experimental subjects were 6 college students. We instructed them to exchange chat messages about "the most impressive event" for 30 minutes. We

Table 7. Questionnaire specifics

questionnaire item	average
Actions of characters are beneficial to chat? (1:no - 5:yes)	4.3
You feel some atmospheres of chat through characters' actions? (1:no - 5:yes)	4.3

show them the list of smiley which more than 3 subjects answered they frequently use, in previous subsection's experiment. Add to the above direction, we asked them to use smiley on this list. The reaction rules mainly consisted of syntonic actions, for example, when one character showed smile, the other character also showed smile. In the end of the experiment, we asked some questions.

We totally obtained fine rating for the two questions in Table 7. The subjects also answered that they could watch the reaction of dialogue partner so they could convey and receive more easily their mental state or emotional overtone.

The examinees tended to use the words associated with *delight, grief, and amazement*. *Delight* made up 35.0% of the total chat messages, and *grief* was 29.9%, *amazement* was 17.9%. Especially, they felt good if the actions, "smile", "joy", "laugh" were frequently appeared.

We will focus on the 3 types atmosphere, such as *Delight, grief, and amazement* and analyze in depth the relationships between smiley and atmosphere in chat.

5   Conclusion

In this article, we investigated the relationships of smiley and the emotional words to analyze relevance of the atmosphere and nonverbal expressions in embodied character chat system.

In the experiments, we confirm the validity of our approach based on the evaluation by many subjects. There are some comments from test subjects that their mental state or nuance of the chat message can exchange more easily because of characters' actions.

Finally, we should plan to improve the reaction table is proper and to add different types of reaction to the reaction table based on the relationships among smiley, nonverbal expressions and atmosphere.

References

1. Cassell, J., Bickmore, T., Billinghurst, M., Campbell, L., Chang, K., Vilhjalmsson, H., Yan, H.: Embodiment in Conversational Interfaces: Rea. In: CHI 1999, pp. 520–527 (1999)
2. De Carolis, B., Pelachaud, C., Poggi, I., De Rosis, F.: Behavior Planning for a Reflexive Agent. In: Proc. of International Joint Conference on Artificial Intelligence (IJCAI 2001), pp. 1059–1066 (2001)
3. Beattie, G.W.: Sequential patterns of speech and gaze in dialogue. Semiotica 23, 29–52 (1978)
4. Kendon, A.: Some functions of gaze direction in social interaction. Acta Psychologica 26, 22–63 (1967)
5. Matarazzo, J.D., Saslow, G., Wiens, A.N., Weitman, M., Allen, B.V.: Interviewer Head Nodding and Interviewee Speech Durations. Psychotberapy:Theory, Research and Practice 1, 54–63 (1964)
6. Dimberg, U.: Facial Reactions to Facial Expressions. Psychophysiology 19(6), 643–647 (1982)

7. Watanabe, T.: E-COSMIC: Embodied Communication System for Mind Connection. In: Proc. of the 9th International Conference on Human-Computer Interaction (HCI International 2001), vol. 1, pp. 253–257 (2001)
8. Yoshida, M., Kinase, R., et al.: Multidimensional scaling of emotion. Japanese Psychological Research 12(2), 45–72 (1970)
9. Plutchik, R.: The Multifactor-Analytic Theory of Emotion. The Journal of Psychology 50, 154–171 (1960)

Logic of Plausibility for Discovery in Multi-agent Environment Deciding Algorithms

Sergey Babenyshev[1,2] and Vladimir Rybakov[1,2]

[1] Department of Computing and Mathematics,
Manchester Metropolitan University,
John Dalton Building, Chester Street, Manchester M1 5GD, U.K.
[2] Institute of Mathematics, Siberian Federal University,
79 Svobodny Prospect, Krasnoyarsk, 660041, Russia
Sergey.Babenyshev@gmail.com, V.Rybakov@mmu.ac.uk

Abstract. In most popular hybrid logics from the field of Artificial Intelligence, knowledge is usually combined with other logical operations (e.g. awareness). We consider[1] a hybrid logic LPD combining *agent's knowledge operations, linear time, operations for discovering information and plausibility operation*. The *plausibility operation* is seemed to be quite new in this framework. The paper introduces LPD via semantical models based on special Kripke/Hintikka frames. We provide formation rules and rules for computing truth values of formulas in the chosen language inside such models, discuss properties of the described logic comparing to some other already known logics. The main problem we are focused on is finding an algorithm for recognition of satisfiable in LPD formulas, we propose such an algorithm (so we show that LPD is decidable w.r.t. satisfiability). The paper is completed with a short discussion on the essence and features of this algorithm.

Keywords: hybrid logics, plausibility, multi-agent's reasoning, linear temporal logic, decision algorithms, Kripke/Hintikka models.

1   Introduction

The efficient use of techniques from non-classical mathematical logic in Artificial Intelligence (AI) and Computer Science (CS) have yielded a good harvest up to date. Significant part of such applications is connected with multi-agent's logic oriented to reasoning about knowledge, time and computation (cf., for instance, Goldblatt [6], van Benthem [21]). Such logics were often originated by semantic objects based on Kripke/Hinttikka models and temporal algebras (cf. Thomason [20], Goldblatt [7]). In our paper we attempt to develop this technique in order to model and bundle up operations for *chance of discovery, time, knowledge and plausibility of statements*. Chance Discovery (CD) (cf. Ohsawa and McBurney

---

[1] This research is supported by Engineering and Physical Sciences Research Council (EPSRC), U.K., grant EP/F014406/1.

I. Lovrek, R.J. Howlett, and L.C. Jain (Eds.): KES 2008, Part III, LNAI 5179, pp. 210–217, 2008.

[11]) is a contemporary direction in AI and CS which analyzes events with uncertain information, incomplete past data, chance events, which are typically rare or hard to find.

The logic of discovery has a solid prehistory, possibly starting with the monograph *Logic of Discovery and Logic of Discourse* by Jaakko Hintikka and Fernand Vandamme [9]. This logic is of interdisciplinary nature and is influenced by various ideas coming from researchers with diverse background. In particular, the modeling of environmental decision-support systems has been undertaken (cf. Cortés M., Sánchez-Marré and L. Ceccaroni [4], N.M. Avouris [2]), applications of the fuzzy logic were discovered (cf. A.J. Bugarin and S. Barro [3], D.Dubois and H. Prade [5]), tools for sematic web and multi-agent systems has been developed (cf. F.Harmelem and I.Horrocks [8], J. Hendler [10], K.Arisha et al [1]). Our basic idea for this paper is to combine logical operations for *agent's knowledge*, with *time, discovery and plausibility operations*. We introduce a hybrid logic LPD bundling up these operations in a special propositional language. The *plausibility operation* is seemed to be quite new in this framework.

The paper introduces LPD via semantical models based on special Kripke-Hintikka frames $\mathcal{N}_C := \langle \bigcup_{i \in N} C(i), R, R_1, \ldots R_m \rangle$, where N is the set of all integer numbers, all $C(i)$ are some nonempty sets representing time clusters, R and $R_1, \ldots, R_m$ are binary accessibility relations; R is imitating linear time flow, R_i are agent's accessibility relations. We introduce a propositional language to reason about agent's knowledge, time, discovery and plausibility operations. It contains agent's knowledge operations K_i, $1 \leq i \leq m$, standard modal operation $\Diamond$—*is possible*, unary decision operations $\Diamond_{D,l}$, $\Diamond_{D,g}$ for local and global possibility to discover information, and the plausibility operation pl(A, x), where $x \in Q$ (x is a rational number from [0,1]) with meaning: the plausibility of the statement A is x. We discuss the language and rules for computing truth values of formulas in the suggested language in models based on frames $\mathcal{N}_C$. These rules allow to formulate satisfiability problem for our logic LPD—Logic of Plausibility for Discovery. The main question we are dealing with is construction of an algorithm for checking satisfiability of formulas in LPD. We find an algorithm (so we show that LPD is decidable by satisfiability) based on reduction of formulas to rules and converting rules in special reduced normal form, and, then, on checking validity of such rules in models of at most exponential size. The technique which we use is borrowed from our research (cf. [12]–[18]) devoted to study of inference rules in non-classical logics, in particular, it worked well for Logic of Discovery in uncertain situations [19]. At the end we discuss some open problems and briefly comment on the essence and features of the proposed algorithm.

2   Preliminaries, Notation, Basic Facts

In this paper we use standard notation and known facts concerning modal, multi-modal and temporal logics (so, some familiarity with basic facts is assumed, though we give below all necessary definitions to follow the paper). We start by introduction of semantic models which will motivate our choice for language to model multi-agent reasoning about chances to discover necessary facts (knowledge, information, etc). The following Kripke/Hintikka frames form the basis for our models.

$$\mathcal{N}_C := \langle \bigcup_{i \in N} C(i), R, R_1, \ldots R_m \rangle,$$

where N is the set of integers, all $C(i)$ are nonempty sets, R and $R_1, \ldots, R_m$ are binary accessibility relations. For all elements a and b from $\bigcup_{i \in N} C(i)$,

$$aRb \iff [a \in C(i) \text{ and } b \in C(j) \text{ and } i < j] \text{ or } [a, b \in C(i) \text{ for some } i];$$

any R_j is a reflexive, transitive and symmetric relation (S5-relation), and

$$\forall a, b \in \bigcup_{i \in N} C(i), \ aR_j b \implies [a, b \in C(i) \text{ for some } i \ ].$$

Intuition behind this formal definition is as follows. Numbers $i \in N$ model a linear discrete flow of time (or steps, ticks, at discrete computation run, etc), any C(i) is the time cluster of all states at the time i, any R_j is the accessibility relation for the j-th agent between states in any time cluster $C(i)$.

To model multi-agent reasoning about properties of propositions in models based at $\mathcal{N}_C$, their mutual relations, plausibility to discover information, we introduce the following logical language. It contains potentially infinite set of propositional letters P; the logical operations include usual Boolean operations and usual unary agent knowledge operations K_i, $1 \le i \le m$, as well as the modal operation $\diamond$ — *is possible*. We extend the language by taking unary decision operations $\diamond_{D,l}$, $\diamond_{D,g}$ for local and global possibility to discover information, and an additional binary operation $\mathrm{pl}(\varphi, x)$ with arguments to be: first—an arbitrary formula, and second—any rational number from [0,1]. This operation is meant to say that φ from viewpoint of agents is x-plausible. The formation rules for formulas are standard, in particular, for any formula φ and $x \in [0, 1] \subseteq Q$ (i.e. x to be a rational number from [0,1]) $\mathrm{pl}(\varphi, x)$ is a well formed formula. Intuitive meaning of this operations is as follows.

$K_i \varphi$ can be read: agent i knows φ in the current state (informational node) of the current time cluster;

$\diamond_{D,l} \varphi$ has meaning: a state were φ is true is locally discoverable (there is a state in current time cluster where φ is true).

$\diamond_{D,g} \varphi$ is meant to say that φ globally discoverable; in any future time cluster there is a state, where φ is true.

$\diamond \varphi$ says that φ there is a future time cluster and a state within this cluster where φ is true.

For any formula φ and $x \in [0, 1] \subseteq Q$ (x is a rational number from [0,1]) $\mathrm{pl}(\varphi, x)$ says that the plausibility (in accordance with summarized agent's knowledge) of φ to be true is equal to x.

Now we define rules for computing truth values of formulas in models $\mathcal{N}_C$ with valuations V of propositional letters P. Given a frame $\mathcal{N}_C$, and a set of propositional letters P, a valuation V of P in $\mathcal{N}_C$ is a mapping of P into the set of all subsets of the set $\bigcup_{i \in N} C(i)$, in symbols,

$$\forall p \in P, V(p) \subseteq \bigcup_{i \in N} C(i).$$

If, for an element $a \in \bigcup_{i \in N} C(i)$, $a \in V(p)$ we say *the fact p is true in the state a*. In the notation below $(\mathcal{N}_C, a) \Vdash_V \varphi$ is meant to say the formula φ in true at the state a in the model $\mathcal{N}_C$ w.r.t. the valuation V. The rules for computation of truth values of formulas are as follows:

$\forall p \in P, \ \forall a \in \mathcal{N}_C \ \ (\mathcal{N}_C, a) \Vdash_V p \iff a \in V(p);$

$(\mathcal{N}_C, a) \Vdash_V \varphi \wedge \psi \iff [(\mathcal{N}_C, a) \Vdash_V \varphi$ and $[(\mathcal{N}_C, a) \Vdash_V \psi];$

$(\mathcal{N}_C, a) \Vdash_V \varphi \vee \psi \iff [(\mathcal{N}_C, a) \Vdash_V \varphi$ or $[(\mathcal{N}_C, a) \Vdash_V \psi];$

$(\mathcal{N}_C, a) \Vdash_V \varphi \rightarrow \psi \iff [not[(\mathcal{N}_C, a) \Vdash_V \varphi]$ or $[(\mathcal{N}_C, a) \Vdash_V \psi];$

$(\mathcal{N}_C, a) \Vdash_V \neg \varphi \iff$ not $[(\mathcal{N}_C, a) \Vdash_V \varphi];$

$(\mathcal{N}_C, a) \Vdash_V K_i \varphi \iff \forall b \in \mathcal{N}_C[(aR_i b) \implies (\mathcal{N}_C, b) \Vdash_V \varphi];$

$(\mathcal{N}_C, a) \Vdash_V \Diamond \varphi \iff \exists b \in \mathcal{N}_C \, [(aRb)$ and $(\mathcal{N}_C, b) \Vdash_V \varphi];$

$(\mathcal{N}_C, a) \Vdash_V \Diamond_{D,l} \varphi \iff a \in C(i) \,\&\, \exists b \in C(i) \, (\mathcal{N}_C, b) \Vdash_V \varphi;$

$(\mathcal{N}_C, a) \Vdash_V \Diamond_{D,g} \varphi \iff a \in C(i) \,\&\, \forall j \geq i \, \exists b \in C(j)(\mathcal{N}_C, b) \Vdash_V \varphi;$

$$\forall a \in \mathcal{N}, \ \ \mathrm{pr}(\varphi, a, V) := \frac{||\{i \mid (\mathcal{N}_C, a) \Vdash_V K_i \psi\}|| - ||\{i \mid (\mathcal{N}_C, a) \not\Vdash_V K_i \varphi\}||}{m};$$

$\forall a \in \mathcal{N}, \forall x \in Q \ (\mathcal{N}_C, a) \Vdash_V \mathrm{pl}(\varphi, x) \iff \mathrm{pr}(\varphi, a, V) = x.$

The operation $\mathrm{pl}(\varphi, x)$ is seemed to be new in such framework. To illustrate it, assume that $m = 7$ and at a model $\langle \mathcal{N}_C, V \rangle$ with a valuation V, for a state $a \in \mathcal{N}_C$ and a formula φ,

$$(\mathcal{N}_C, a) \Vdash_V K_1 \varphi, \ (\mathcal{N}_C, a) \Vdash_V K_2 \varphi, \ (\mathcal{N}_C, a) \Vdash_V K_3 \varphi,$$

$$(\mathcal{N}_C, a) \not\Vdash_V K_4 \varphi, \ (\mathcal{N}_C, a) \not\Vdash_V K_5 \varphi, \ (\mathcal{N}_C, a) \not\Vdash_V K_6 \varphi, \ (\mathcal{N}_C, a) \not\Vdash_V K_7 \varphi.$$

Then $\mathrm{pl}(\varphi, a, V) = 3/4$, and $(\mathcal{N}_C, a) \Vdash_V \mathrm{pl}(\varphi, 3)$, thus the formula $\mathrm{pl}(\varphi, 3)$ says that, in this model, from summarized viewpoints of all agents the plausibility of the formula φ to be true is $3/4$.

We cannot define a logic based at this semantics in standard way, as set of all formulas which are true in any state of any frame $\mathcal{N}_C$ w.r.t. any valuation, because plausibility operations employ rational numbers, and truth values of plausibility operations vary in the same frame under different valuations. Though, satisfiability problem for our logic LPD has standard formulation.

Definition 1. *A formula φ is satisfiable in the logic LPD if and only if there is a frame $\mathcal{N}_C$ and a valuation V at $\mathcal{N}_C$ such that $\mathcal{N}_C \Vdash_V \varphi$ (i.e., φ is true at any state of $\mathcal{N}_C$ w.r.t. the valuation V)*

For $\square := \neg \Diamond \neg$, the pure $\square$-fragment of LPD (formulas with no other modalities that hold in each frame $\mathcal{N}_C$) is well known modal logic $S4.3$ (this follows directly from the finite model property of $S4.3$). Also the following holds:

Formulas $\Diamond p \to \Diamond_{D,l} p$ and $\Diamond p \to \Diamond_{D,g} p$, are not logical laws of LPD, they may be refuted in some frames $\mathcal{N}_C$ by certain valuations. Formulas $\Diamond_{D,g} p \to \Diamond_{D,l} p$, and $\Diamond_{D,g} p \to \Diamond$, on the contrary, are logical laws, they are true in all frames $\mathcal{N}_C$ w.r.t. all valuations. So, the meaning of the operations well matches the name.

Evidently the language of LPD is much more expressive comparing with the standard hybrid languages combining multi-agent knowledge and time. It uses additional operations for local and global discovery together with evaluation operations for plausibility of statements. For most logics investigated in the literature the prime questions are problems of satisfiability and decidability, and in the next section we will address them.

3   Results, Decidability

We aim to provide an algorithm which would decide by a formula whether it is satisfiable in LPD or not. To construct such an algorithm we will use the technique borrowed from our previous research focused on applying logical consecutions and inference rules to logics originating in AI and CS. (cf. [12] – [19]). In particular, this approach has been successfully implemented for Logic of Discovery in [19] . To apply this technique we need a representation of formulas by rules, and conversion of rules to a special normal reduced form. Recall that a (sequential) *rule* is an expression

$$r := \frac{\varphi_1(x_1,\ldots,x_n),\ldots,\varphi_m(x_1,\ldots,x_n)}{\psi(x_1,\ldots,x_n)},$$

where $\varphi_1(x_1,\ldots,x_n),\ldots,\varphi_m(x_1,\ldots,x_n)$ and $\psi(x_1,\ldots,x_n)$ are some formulas constructed out of letters $x_1,\ldots,x_n$. Letters $x_1,\ldots,x_n$ are called *variables of the rule r*. A formula φ is *valid* in a frame $\mathcal{N}_C$ (symbolically, $\mathcal{N}_C \Vdash \varphi$) if, for any valuation V of $Var(\varphi)$ and for any element a of $\mathcal{N}_C$, $(\mathcal{N}_C, a) \Vdash_V \varphi$. We use notation $\mathcal{N}_C \Vdash \varphi$ for this fact.

Definition 2. *A rule r is said to be* valid *in the Kripke model $\langle \mathcal{N}_C, V \rangle$ with the valuation V (we will use notation $\mathcal{N}_C \Vdash_V r$) if*

$$\forall a \, ((\mathcal{N}_C, a) \Vdash_V \bigwedge_{1 \le i \le m} \varphi_i) \implies \forall a \, ((\mathcal{N}_C, a) \Vdash_V \psi).$$

Otherwise, r is refuted in $\mathcal{N}_C$, *or refuted in $\mathcal{N}_C$ by V, and we write $\mathcal{N}_C \not\Vdash_V r$.*

Definition 3. *A rule r is* valid *in a frame $\mathcal{N}_C$ (notation $\mathcal{N}_C \Vdash r$) if, for any valuation V of $Var(r)$, $\mathcal{N}_C \Vdash_V r$.*

A rule r is said to have the *reduced normal form* if $r = \varepsilon_c / x_1$ where

$$\varepsilon_c := \bigvee_{1 \leq j \leq m} (\bigwedge_{1 \leq i \leq n} [x_i^{t(j,i,0)} \wedge (\Diamond x_i)^{t(j,i,1)} \wedge (\Diamond_{D,l} x_i)^{t(j,i,2)} \wedge$$

$$(\Diamond_{D,g} x_i)^{t(j,i,3)} \wedge \bigwedge_{s \in [1,m]} (\neg K_s \neg x_i)^{t(j,i,0,s)} \wedge \mathrm{pl}(x_i, \alpha_i)^{t(j,i,4)}),$$

and all x_s are certain letters (variables), $\alpha_i \in Q$, $t(j,i,z), t(j,i,k,z) \in \{0,1\}$ and, for any formula α above, $\alpha^0 := \alpha$, $\alpha^1 := \neg\alpha$.

For any formula φ in the language of LPD, we can convert it into the rule $x \to x/\varphi$ and employ technique of reduced normal forms as follows.

Definition 4. *Given a rule r_{nf} in the reduced normal form, r_{nf} is said to be the normal reduced form for a rule r iff, for any frame $\mathcal{N}_C$, $\mathcal{N}_C \Vdash r \iff \mathcal{N}_C \Vdash r_{nf}$.*

Based on proofs of Lemma 3.1.3 and Theorem 3.1.11 from [13], by similar technique, we obtain

Theorem 1. *There exists an algorithm running in (single) exponential time, which, for any given rule r, constructs its normal reduced form r_{nf}.*

It is immediate to see that a formula φ is valid in a frame $\mathcal{N}_C$ iff the rule $x \to x/\varphi$ is valid in $\mathcal{N}_C$, and, consequently, φ is satisfiable in LPD iff the rule $x \to x/\neg\varphi$ is refuted in some $\mathcal{N}_C$ frame. Therefore, from Theorem 1 we obtain

Proposition 1. *A formula φ is satisfiable in LPD iff the rule $(x \to x/\neg\varphi)_{nf}$ is refuted in some frame $\mathcal{N}_C$ by a valuation V.*

Thus, bearing in mind an algorithm recognizing satisfiable in LPD formulas, it is sufficient to find an algorithm recognizing rules in reduced normal form which may be refuted in frames of kind $\mathcal{N}_C$. For this we need the following special finite Kripke models. Take any frame $\mathcal{N}_C$ and some numbers n_1, m_1, where $m_1 > n_1 > 3$. The frame $\mathcal{N}_C(n_1, m_1)$ has the following structure:

$$\mathcal{N}_C(n_1, m_1) := \langle \bigcup_{1 \leq i \leq m} C(i), R, R_1, \ldots R_m \rangle,$$

where R is the accessibility relation from $\mathcal{N}_C$ extended by pairs (x, y), where $x, y \in [n_1, m_1]$, so xRy holds for all such pairs. For any given valuation V of letters from a formula φ in $\mathcal{N}_C(n_1, m_1)$, the truth values of φ can be defined in elements of $\mathcal{N}_C(n_1, m_1)$ by the same rules as for frames $\mathcal{N}_C$ above (actually just similar to ones used previously for the frames $\mathcal{N}_C$). We describe below steps for some logical operations:

$$(\mathcal{N}_C(n_1, m_1), a) \Vdash_V \Diamond\varphi \iff \exists b \in \mathcal{N}_C[(aRb) \text{ and } (\mathcal{N}_C(n_1, m_1), b) \Vdash_V \varphi];$$

$$(\mathcal{N}_C(n_1, m_1), a) \Vdash_V \Diamond_{D,l}\varphi \iff a \in C(i) \,\&\, \exists b \in C(i)\,(\mathcal{N}_C, b) \Vdash_V \varphi;$$

$$(\mathcal{N}_C(n_1, m_1), a) \Vdash_V \Diamond_{D,g}\varphi$$

$$\iff \left[a \in C(i) \,\&\, b \in C(j) \,\&\, aRb \implies \exists b \in C(j)\,(\mathcal{N}_C, b) \Vdash_V \varphi \right];$$

$$\forall a \in \mathcal{N}_C(n_1, m_1), \forall x \in Q : (\mathcal{N}_C(n_1, m_1), a) \Vdash_V \mathrm{pl}(\varphi, x) \iff \mathrm{pl}(\varphi, a, V) = x.$$

Based on this modified Kripke structures $\mathcal{N}_C(n_1, m_1)$ we can derive

Lemma 1. *A rule r_{nf} in the reduced normal form is refuted in a frame $\mathcal{N}_C$ w.r.t. a valuation V if and only if r_{nf} is refuted in a frame $\mathcal{N}_C(n_1, m_1)$ by a valuation V_1, where*

1. *the size of any cluster $C(i)$ in $\mathcal{N}_C(n_1, m_1)$ is linear in r_{nf};*
2. *numbers n and m are exponential in r_{nf};*
3. *the size of the frame $\mathcal{N}_C(n_1, m_1)$ is exponential in r_{nf}.*

Combining Theorem 1, Proposition 1 and Lemma 1 we derive

Theorem 2. *The problem of satisfiability in LPD is decidable.*

The algorithm of verification a formula φ to be satisfiable in LPD consists of the following steps:

(i) reduce the formula φ to the rule $x \to x/\neg\varphi$;
(ii) transform the rule $r := x \to x/\neg\varphi$ to its reduced normal form r_{nf};
(iii) verify if the rule n_{rf} may be refuted at frames of kind $\mathcal{N}_C(n_1, m_2)$ in size exponential from r_{nf}.

A hidden overall complexity also includes the reduction of rules $x \to x/\neg\varphi$ to normal reduced forms, but this complexity is single exponential, the same as for reduction formulas of Boolean logic to disjunctive normal forms.

4   Conclusion

The paper suggests tools for constructing logics combining multi-agent's knowledge, time, discovery operations and plausibility operations. We introduce the logic LPD of plausibility for discovery in multi-agent environment and find an algorithm recognizing satisfiable formulas (thus we prove that the satisfiability problem in LPD is decidable). The suggested approach is proven to be flexible enough to work with a variety of logics originating from AI and CS. An interesting question is whether it is possible to extend the methods of this paper to handle the case of hybrid non-linear temporal logics (e.g. branching time logics, ones based at $\mathrm{S4}_T$, $\mathrm{K4}_T$) with knowledge operations. An interesting problem concerns complexity issues and possible ways of refining complexity bounds in the suggested algorithm. The case of calculating plausibility, being not restricted to agent's knowledge in only the current time cluster, is interesting as well.

References

1. Arisha, K., Ozcan, F., Ross, R., Subrahmanian, V.S., Eiter, T., Kraus, S.: Impact: A platform for collaborating agents. IEEE Intelligent Systems 14(2), 64–72 (1999)
2. Avouris, N.M.: Co-operation knowledge-based systems for environmental decision-support. Knowledge-Based Systems 8(1), 39–53 (1995)
3. Bugarin, A.J., Barro, S.: Fuzzy reasoning supported by Petri nets. IEEE Transactions on Fuzzy Systems 2(2), 135–150 (1994)
4. Cortés, U., Sánchez-Marré, M., Ceccaroni, L.: Artificial intelligence and environmental decision support systems. Applied Intelligence 13(1), 77–91 (2000)
5. Dubois, D., Prade, H.: The three semantics of fuzzy sets. Fuzzy Sets and Systems 90(2), 141–150 (1997)
6. Goldblatt, R.: Logics of Time and Computation. CSLI Lecture Notes 7 (1992)
7. Goldblatt, R.: Mathematical Modal Logic: A View of its Evolution. J. Applied Logic 1(5-6), 309–392 (2003)
8. Harmelem, F., Horrocks, I.: The semantic web and its languages–faqs on oil: The ontology inference layer. IEEE Intelligent Systems 15(6), 69–72 (2000)
9. Hintikka, J., Vandamme, F.: Logic of Discovery and Logic of Discourse. Springer, Heidelberg (1986)
10. Hendler, J.: Agents and the semantic web. IEEE Intelligent Systems 16(2), 30–37 (2001)
11. Ohsawa, Y., McBurney, P. (eds.): Chance Discovery (Advanced Information Processing). Springer, Heidelberg (2003)
12. Rybakov, V.V.: A Criterion for Admissibility of Rules in the Modal System $S4$ and the Intuitionistic Logic. Algebra and Logic 23(5), 369–384 (1984) (Engl. Translation)
13. Rybakov, V.V.: Admissible Logical Inference Rules. Studies in Logic and the Foundations of Mathematics, vol. 136. Elsevier Sci. Publ., North-Holland, Amsterdam (1997)
14. Rybakov, V.V.: Construction of an Explicit Basis for Rules Admissible in Modal System S4. Mathematical Logic Quarterly 47(4), 441–451 (2001)
15. Rybakov, V.V.: Logical Consecutions in Intransitive Temporal Linear Logic of Finite Intervals. Journal of Logic Computation 15(5), 633–657 (2005)
16. Rybakov, V.V.: Logical Consecutions in Discrete Linear Temporal Logic. Journal of Symbolic Logic 70(4), 1137–1149 (2005)
17. Rybakov, V.V.: Linear Temporal Logic with Until and Before on Integer Numbers, Deciding Algorithms. In: Grigoriev, D., Harrison, J., Hirsch, E.A. (eds.) CSR 2006. LNCS, vol. 3967, pp. 322–333. Springer, Heidelberg (2006)
18. Rybakov, V.: Sine-Until Temporal Logic Based on Parallel Time with Common Past. In: Artemov, S.N., Nerode, A. (eds.) LFCS 2007. LNCS, vol. 4514, pp. 486–497. Springer, Heidelberg (2007)
19. Rybakov, V.: Logic of Discovery in Uncertain Situations – Decciding Algorithms. In: Apolloni, B., Howlett, R.J., Jain, L. (eds.) KES 2007, Part III. LNCS (LNAI), vol. 4694, pp. 950–968. Springer, Heidelberg (2007)
20. Thomason, S.K.: Semantic Analysis of Tense Logic. Journal of Symbolic Logic 37(1) (1972)
21. van Benthem, J.: The Logic of Time, vol. 156. Reidel, Dordrecht, Synthese Library (1983)

An Agent-Oriented Dynamic Adaptive Threshold Transmission for XML Data on Networks

Xu Huang and Dharmendra Sharma

Faculty of Information Sciences and Engineering
University of Canberra
Canberra, ACT 2617, Australia
{Xu.Huang,Dhamendra.Shama}@canberra.edu.au

Abstract. Since XML became an official recommendation of the World Wide Web Consortium (W3C) in 1998, it is increasingly being used to transmit data on networks but is a verbose format and needs an efficient encoding to send relatively large amounts of data efficiently. This requirement is particularly important for wireless data communications. It is a common technical challenge for researchers in XML-driven networks to have good performance. One may employ a middleware to enhance performance by minimizing the impact of transmission time [1, 3]. Normally, to reduce the amount of data sent the XML documents are converted to a binary format using a compression routine such as Gzip. However while this would reduce the payload, it results in an increase in the CPU time as the XML document must be compressed before being sent and uncompressed when it is received. In this paper we extended our previous research results [2, 11-13] to an agent-oriented enabling technology, namely Dynamic Adaptive Threshold Transmission (DATT) for XML data on networks. We also show the experimental results obtained from our technique and that from the Network Adaptable Middleware (NAM) established by Ghandeharizadeh et al [1]. Experimental results show that our method is superior to the NAM method [1], supported by the fact that the time taken is 220.6 times better.

Keywords: XML, agent-oriented adaptable middleware, efficient XML, efficient communication on a network.

1 Introduction

XML has become an increasingly important data standard for use in organizations as a way to transmit data [4, 5, 6, 7] and has also attracted the attentions of those people who are working in areas of wireless communications, in particular for so called small wireless devices. Additionally it is being used to enable web services and similar, often custom, RPC functionality to allow greater access to data across multiple systems within an organization and allowing the possibility of future systems to be created from collections of such RPC functionality.

XML is a verbose, text based format with strict requirements on structure and is often criticized for its large space requirements. This large size can be particularly problematic for use in transmission across a network, where network bandwidth

I. Lovrek, R.J. Howlett, and L.C. Jain (Eds.): KES 2008, Part III, LNAI 5179, pp. 218–226, 2008.

restrictions can cause significant delays in receiving the transmission, which has drawn great attention from the wireless communications.

One solution to this problem is to look at reducing the size of these transmissions by rendering them in a binary format, such as by using XMill or Gzip to compress an XML document. However such methods can take longer as compressing and decompressing may take more time than what is saved transmitting the smaller XML document.

One solution to this problem may be the Network Adaptable Middleware (NAM) raised by Ghandeharizadeh et al [8], even though there are some ways to directly compress, such as column-wise compression and row-wise compression for large message sizes [9]. This solution estimates the time it will take to compress, transmit in binary format and decompress a document compared to an estimate of how long it would take to transmit the document as uncompressed text. The estimates are based on a persistent collection of information on how the system has performed in the past and provides an accurate estimate on whether it would be faster too compress the document before transmission or not.

We have introduced another way of determining when to compress an XML document before transmitting it in our *One Pass Technique* (OPT) [2, 11-13]. In this technique we determine a threshold size value for the networks. Any XML document that is smaller than this threshold will be sent uncompressed while any XML document larger than this size will be compressed before it is sent.

It is well known that the performances on networks depend on various parameters, such as traffic situations, bandwidths, transferring rates, etc. We shall use the agent-oriented adapted dynamic threshold to represent the characteristics of the running networks, by which the transferring XML data on networks will be controlled with the optimum condition in terms of transferring decision time defined in the next sections. One agent looks after the traffic situation and another agent will look the system running situation such as CPU, memory size, etc. The agent will compare the predicted results of compressing the data with uncompressed results and then decided if it is efficient to compress or not. The following sections are as follows, in section 2, we shall briefly review the established OPT technique and show that there is possible to improve the OPT technique. In section 3 the Dynamic Adaptive Threshold Transmission (DATT) for XML data on networks will be demonstrated, together with the experimental setup design, which is the natural research project, extended from the previous research results. We shall present the conclusion of this paper in the section 4.

2   Threshold and One Pass Technique (OPT)

Before we establish our Dynamic Adaptive Threshold Transmission (DATT) for XML data on networks, we need briefly to recall our previous method, titled OPT and show OPT needs to be changed if we want it work well on a network. Then, we extend out previous results to current DATT in next section.

In contrast to the five network factors that contribute to the latency time of delivering a query output [1] based on the analysis of the one gigabyte TPC-H benchmark [10], our method presented here is utilizing an established " threshold" for the current working status and then to have "one-pass" transmission. We defined a threshold value for the network such that the transmitted time, for XML documents whose size has been compressed (such as via Gzip) and uncompressed, will be comparable. To determine what this value could be, we first need to determine the networks characteristics. As the networks characteristics will evolve with time the threshold value needs to dynamically change with the network.

Before OPT can be used on a network we need to determine the threshold value by making a number of XML transfers of different sizes across the network. The transmissions need to be made both with the document compressed, using Gzip as an example, (and decompressed where it is received) and by transmitting the document without compression. An estimate of how long it takes to transmit a document of a given size can then be determined by curve fitting to these results. The threshold value is set to be the size when the estimated time to transmit it without compression is equal to the estimated time to transmit it with compression. In some situations this may result in a threshold value that will require compression of all documents or one that will never require compression of a document.

There are a number of factors that can prevent OPT from yielding the best result for all cases. The threshold value will only be valid for the network bandwidth it is calculated for, so if that bandwidth changes a threshold value will give an inaccurate result and a new threshold value will need to be determined.

The compression and decompression times are dependent on the CPU load. If the load on a CPU is heavier (or lighter) than it was when calculating the threshold value it may not make the appropriate decision on whether or not to use compression on the XML document. Similarly the technique works best with a homogenous set of CPUs. Different CPUs will take different time periods to compress and decompress the XML documents. The compression/decompression time of two low end CPUs on a network will be different to the compression/decompression time of two high end CPUs on the same network using the same threshold value. This can also lead to the OPT making a wrong decision on whether or not to compress the document.

OPT can also be affected by changes in the networks traffic density. If the network is under a heavier load than it was when the threshold value was calculated the technique is more likely to transmit an uncompressed XML document when a compressed document would have been faster, and with a lighter network load compressed XML transmissions are more likely to occur when an uncompressed transmission would have been faster. OPT is best used in a homogenous environment where the network bandwidth is well known and network traffic is reasonably stable.

As we discussed that a threshold depends on many factors on the network, if OPT works for a network, it must be changed from time to time depending on the current status of the network, namely it must be dynamically changed to control the transfer date on network. This is the basic idea of our Dynamic Adaptive Threshold Transmission (DATT) for XML data on networks.

In the next section we shall describe the principle of DATT first after that we design an experimental setup for investigating the DATT versus the NAM.

3 DATT for XML Data and Experimental Results

In order to have a Dynamic Adaptive Threshold Transmission (DATT) for XML data on networks, we make a programmed process to check the current network working situation that depends on the current traffic parameters discussed in above sections. Then, the current threshold is worked out based on the same principle described for OPT in section II. The obtained threshold will replace the previous one to work (control) the traffic communications. Since the threshold is monitored dynamically the adaptive threshold will always keep record of the times taken in transferring the data.

Agent-based systems are increasingly being applied in a wide range of areas including telecommunications, business process modeling, computer games, control of mobile robots and military simulations [14-19]. We are going to introduce an agent-oriented enabling technology for a dynamic adaptive threshold transmission for XML data on networks. Agent-based systems, systems based on autonomous software and/or hardware components (agents) that cooperate within an environment to perform some task. An agent can be viewed as a self-contained, concurrently executing thread of control that encapsulates some state and communicate with its environment and possibly other agent via some sort of message passing [14]. A multi-agent system can be considered as a loosely coupled network of problem-solver entities that work together to find answers to problems that are beyond the individual capabilities or knowledge of each entity [15-16]. A multiagent system comprised of multiple autonomous components needs to have certain characteristics [17-18]. In order to make the "transmission" for XML data on networks more effective and efficient, the mutiagent system, operational agent society, is designed as shown in Figure 1.

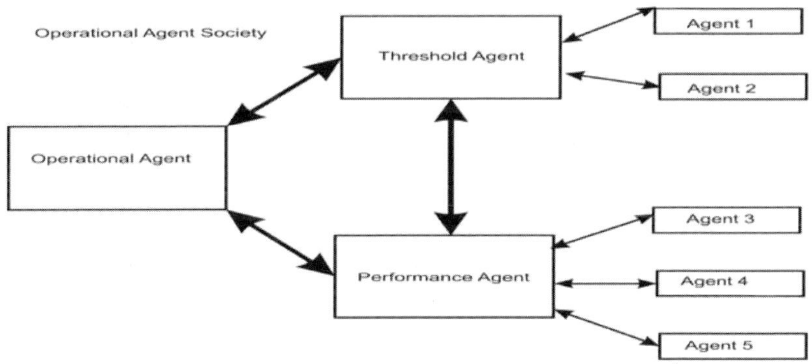

Fig. 1. Block diagram for operational agent society

Agent 1: present the experience threshold; Agent 2: present current data for further decision and the "Threshold Agent" will make the final decision about the current threshold and then sending it to "Performance Agent" for the performance decision and "Operational Agent" to make final transmission decision. Agent 3: dealing with the operation CPU load; Agent 4: collecting the data of the current network traffic; Agent 5: Calculation and comparison and sending the results to performance agent for the final decision.

In order to investigate our Dynamic Adaptive Threshold Transmission (DATT) for XML data on networks, we design our experimental work as shown in Figure 2.

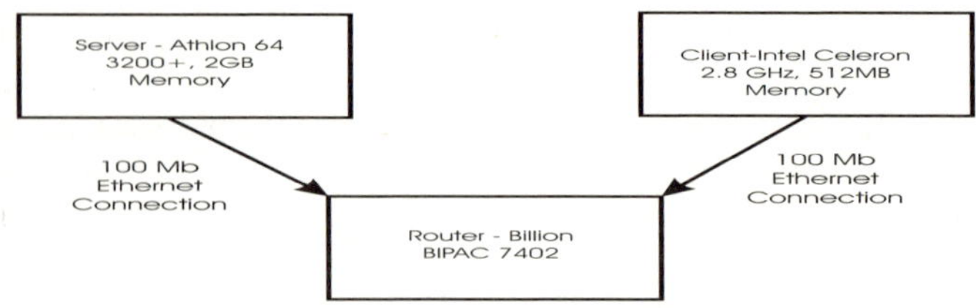

Fig. 2. The experimental setup diagram for DATT

The connection was made across the router using a raw TCP connection created for each transmission. The client used a listener to listen for the incoming files (port 9013). The server was running a Cron style task scheduler to initiate communication and deliver the file.

A number of XML documents (1200 files) were gathered to test using a time based threshold to decide on when to compress a document and when not to. These files were of different sizes. An application program was written to transmit these documents a number of times across a network using a threshold value. Any XML document with a size greater than the threshold value is transmitted compressed while all other XML documents are sent uncompressed. The algorithm used is:

If $\text{Size}_{\text{Document}}$ > $\text{Size}_{\text{Threshold}}$ *Then* transmit_compressed, *Else* transmit_uncompressed

A similar application was set up to transfer the documents using the NAM methodology (Ghandehazrizadeh, 2003). NAM uses measured network and computer characteristics to compare estimates on how long it would take to transmit an uncompressed document against an estimate of how long it would take to transmit a compressed document. The algorithm used is:

If $\text{Time}_{\text{Uncompressed Transmission}}$ > $\text{Time}_{\text{Document Compression}}$ + $\text{Time}_{\text{Compressed Transmission}}$ + $\text{Time}_{\text{Document Decompression}}$ *Then* transmit_compressed, *Else* transmit_uncompressed.

The experiment was conducted using a client PC (754pin Athlon64 3200+@2.05GHz with 1GB RAM), one Server PC (Celeron D 2.8@2.79GHz with 512MB RAM) connected by a Router (Billion BIPAC 7402G) over a 100MBit Ethernet connection.

For the DATT the time taken is calculated by the follows:

calculation time + compression time + transfer time + decompression time + threshold calculation time

In order to obtain good statistics and fair distributed results, a set of twenty-nine runs were carried out for each technique, namely DATT and NAM, sandwiched for one hour. For such a setup the whole running process covers more than 41 hours without breaking.

In order to change the working environments from time to time, the network, while it was processing, has been disturbed by various activities such as "downloading files" in different sizes, browsing the Internet, playing audio on the computer, etc. In order to determine the characteristics of the network before the applications they were run against it, solving the quadratic equations used to get the time and size estimates NAM uses in its decision algorithm and determining the threshold value for the current network traffic load for the OPT. When the threshold value was found at the particular time it will be used for controlling the XML data transferred on the network.

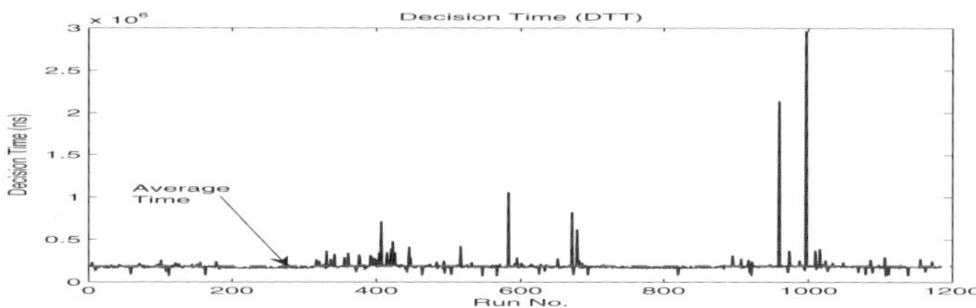

Fig. 3. The experimental results for the average decision time of the DATT described in section 3

The results for the decision time, in terms of average for all the runs, for DATT are shown in Figure 4. As the process of determining the threshold has been passed out to a separate process, the decision times for DATT are very short.

All the running results will be recorded by five types of results, namely the original file size in bytes, decision time (for DATT and NAM, it is the time to decide whether the current file should be compressed and sent or just sent, according to the principle of DATT or NAM respectively), compression time, transferring time, and decompression time.

The results for the decision time, in terms of average for all the runs, for NAM are shown in Figure 4. It is seen that, as NAM accumulates more data from which to make a decision increases, the time it takes to actually make a decision increases to allow all the data to be read.

In the both diagrams, the average decision time were marked as dashed lines.

It is important to highlight two items, one is that the horizontal axis, titled as run number, for the two figures are about 1200, which are the "average" results from 41 hours running as described in above, another one is that the plotted run order should not be meanness due to the fact that we put two results, compressed files and uncompressed files together. Because we do care about the average time taken by the decision rather than when the decision made for the plotting so we just put them together. Also one may find the plotting results seem to be "periodic" results, in particular for Figure 4, which, however, are not real time results.

In these experiments, the average decision time for DATT is 0.187622 milliseconds and for NAM is 41.381926 milliseconds, which means NAM takes 220.56 times longer than DATT to make a decision.

It is also a very interesting to note that the experimental results show the number of compressed files with DATT is 587 files of 1201 running files, which gives the compressing ratio = 0.488. In contrast to DATT, the NAM has 510 files compressed from total 1219 running files, which gives a compressing ratio = 0.418. Therefore, the compressing ratio NAM is about 86.4% as that of DATT. This shows that, in comparison with DATT, NAM is always (or in terms of average) making cautious decisions to keep itself in optimum states but causes heavier network traffic, which means the DATT will make higher quality network transfers for XML data on networks. This improvement for DATT against to NAM is about a quarter percent. In terms of the performances for both the DATT and NAM techniques in this experimental setup they used the same software for compressing and decompressing software, but the

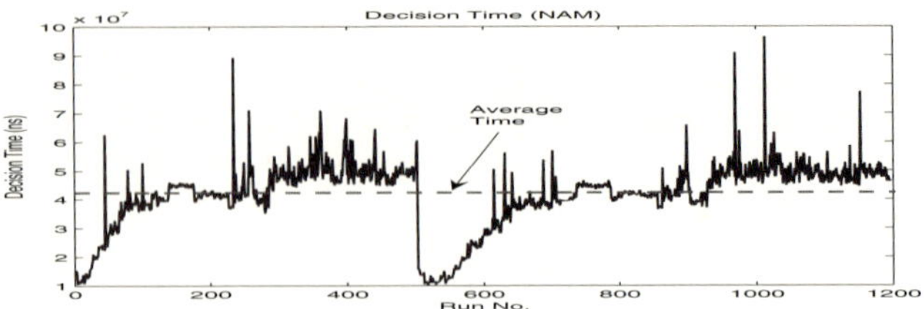

Fig. 4. The experimental results for the average decision time of the NAM described in section 3

processing is different due to the different principles used by each technique. Thus, the data shows, in terms of average, the compressing time and decompressing time are about balanced, shown by the fact that "Time taken by DATT" divided by "Time taken by NAM" is 0.91 for compressing time and 0.895 for the decompressing time.

However, DATT technique is more stable processes for compressing supported by the fact that the standard deviation of compressing time is almost ten times small than that of NAM, even the standard deviation of decompressing is fairly comparable, which is understandable since for DATT when the threshold is obtained the rest of the job is much easier than that of NAM. This is further evidence to show the DATT technique will keep running networks in a better service quality.

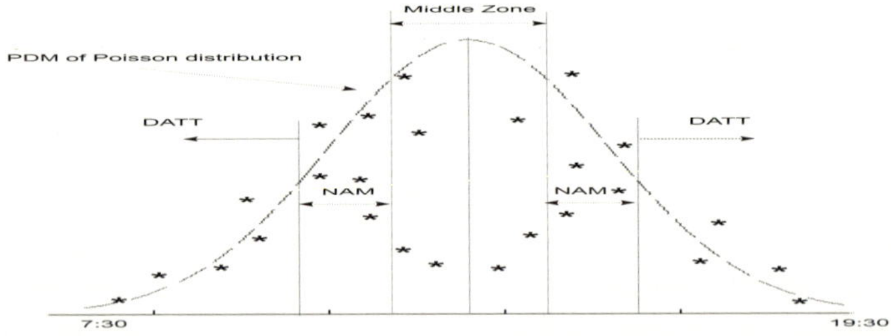

Fig. 5. A traffic distribution averaged from 10 days and suggestions for a hybrid of NAM and DATT

From above description, we know DATT and NAM has its own advantages and backsides, if we combine both technologies what will happen? Figure 5 shows traffic distribution averaged from 10 days from morning 7:30 to 19:30. The distribution suggests that we better to use a hybrid of NAM and DATT as shown by the figure, with which we may have optimum data transmission on a network.

4   Conclusion

We have extended our previous research results to Dynamic Adaptive Threshold Transmission (DATT) for XML data on networks. We compared this technique (DATT) to another control technique, the Network Adaptable Middleware (NAM), and found that the DATT technique is much better than NAM in terms of decision time taken, which was about 220.56 times for the DATT of the decision time less than that of the NAM's. Also the DATT will give running networks better performance by as much as one quarter in comparison with NAM. Finally we introduce a hybrid technology with which we have a optimum data transmission on a network.

References

[1] Ghandeharizadeh, S., Papadopoulos, C., Cai, M., Chintalapudi, K.K.: Performance of Networked XML-Driven Cooperative Applications. In: Proceedings of the Second International Workshop on Cooperative Internet Computing, Hong Kong, China (August 2002)

[2] Ridgewell, A., Huang, X., Sharma, D.: Evaluating the Size of the SOAP for Integration in B2B. In: The Ninth International Conference on Knowledge-Based Intelligent Information & Engineering Systems, Part IV, Melbourne, Australia, p. 29 (September 2005)

[3] Liefke, H., Suciu, D.: XMill: An efficient Compressor for XMLL Data. Technical Report MSCIS-99-26, University of Pennsylvania (1999)

[4] Curbera, F., Duftler, M., Khalaf, R., Nagy, W., Mukhi, N., Weerawarana, S.: Unraveling the web services web: An introduction to SOAP, WSDL, UDDI. IEEE Internet Computing 6(2), 86–93 (2002)

[5] Fan, M., Stallaert, J., Whinston, A.B.: The internet and the future of financial markets. Communications of the ACM 43(11), 83–88 (2000)

[6] Rabhi, F.A., Benatallah, B.: An integrated service architecture for managing capital market systems. IEEE Network 16(1), 15–19 (2002)

[7] Kohloff, C., Steele, R.: Evaluating SOAP for High Performance Business Applications: Real-Time Trading Systems (2003) (accessed March 22, 2005), `http://www2003.org/cdrom/papers/alternate/P872/p872kohlhoff.html`

[8] Ghandeharizadeh, S., Papadopoulos, C., Cai, M., Zhou, R., Pol, P.: NAM: A Network Adaptive Middleware to Enhance Response Time of Web Services. In: MASCOTS 2003, p. 136 (2003)

[9] Iyer, R.R., Wilhite, D.: Data Compression Support in Databases. In: Proceedings of the 20th International Conference on Very Large Dasta Bases (1994)

[10] Poess, M., Floyd, C.: New TPC Benchmarks for Decision Support and Web Commerece. ACM SIGMOD Record 29(4) (December 2000)

[11] Huang, X.: An Imporoved ALOHA Algorithm for RFID Tag Identification. In: Gabrys, B., Howlett, R.J., Jain, L.C. (eds.) KES 2006. LNCS (LNAI), vol. 4253, pp. 1157–1162. Springer, Heidelberg (2006)

[12] Huang, X., Ridgewell, A., Sharma, D.: Efficacious Transmission Technique for XML Data on Networks. International Journal of Computer Science and Network Security 6(3), 14–19 (2006)

[13] Huang, X., Ridgewell, A., Sharma, D.: A Dynamic Threshold Technique for XML Data Transmission on Networks. In: The tenth International Conference on Knowledge-Based Intelligent Information & Engineering Systems, Bournemouth, UK (October 2006)

[14] Wooldridge, M., Jennings, N.R.: Intelligent agent: Theory and practice. Knowledge Engineering Review 10(2), 115–152 (1995)

[15] Durfee, E.H., Lesser, V.R., Corkill, D.D.: Trends in cooperative distributed problem solving. IEEE Transactions on Knowledge and Data Engineering 1(1), 63–83 (1989)

[16] Sajja, P.S.: Multi-layer knowledge-based system: A multi-agent approach. In: Proceedings of the Second Indian International Conference on Artificial Intelligence, Pune, India, pp. 2899–2909 (2005)
[17] Robert, A.F.: Towards standardization of multi-agent systems framework. Crossroads 5(4), 18–24 (1999)
[18] Jennings, N.R., Sycara, K., Wooldridge, M.: A roadmap of agent research and development. Autonomous Agents and Multi-Agent Systems Journal 1(1), 7–38 (1998)
[19] Gianni, D.: Bringing discrete event simulation concepts into multi-agent systems. In: Tenth International Conference on Computer Modeling and Simulation, UKSIM 2008, pp. 186–191 (2008)

An Agent-Oriented Quantum Key Distribution for Wi-Fi Network Security

Xu Huang and Dharmendra Sharma

Faculty of Information Sciences and Engineering
University of Canberra
Canberra, ACT 2617, Australia
{Xu.Huang,Dhamendra.Shama}@canberra.edu.au

Abstract. There are a large variety of kinds of mobile wireless networks, Wi-Fi, based on the IEEE 802.11 standard, is a wireless local area network, mainly used in offices and campus at universities, meeting rooms, halls in hotels or in airports. For such limited coverage area, IEEE 802.11 standard may be observed as building-oriented environment, which potentially offers a chance to let quantum key distribution (QKD) play a role in the security of wireless communications. In fact, secured data transmission is one of the prime aspects of wireless networks as they are much more vulnerable to security attacks. In this paper, we explore the possibility of using an agent-oriented Quantum Key Distribution for authentication and data encryption for IEEE 802.11 standard based on our previous research. It will focus on some basic concept that how QKD merges the wireless communication, in particular the IEEE 802.11 standard. The software implementation of the first two phases of QKD, namely (a) raw key extraction and (b) error estimation, will be carefully investigated in this paper. A Wi-Fi based concept has been extended to the implement the communication between two users in C++ language in this paper.

Keywords: an agent-oriented Quantum Key Distribution, B92 protocol, BB84 protocol, 802.11, Wi-Fi, Socket Programming.

1   Introduction

Wireless security is becoming increasingly important as wireless applications and systems are widely adopted. Numerous organizations have already installed or are busy in installing "wireless local area networks" (WLANs). These networks, based on the IEEE 802.11 standard, are very easy to deploy and inexpensive. Wi-Fi allows LANs to be deployed without cabling for client devices, typically reducing the costs of network deployment and expansion. As of 2007 wireless network adapters are built into most modern laptops. Wi-Fi is a global set of standards, unlike mobile telephones, any standard Wi-Fi device will work anywhere in the world. Other important trends in wireless adoptions are including the introduction of wireless email with devices such as the Blackberry and The Palm VII, rampant digital cell phone use, including the use of short message service (SMWS), and the advent of Bluetooth devices. But the risks associated with the adoption of wireless networking are only

I. Lovrek, R.J. Howlett, and L.C. Jain (Eds.): KES 2008, Part III, LNAI 5179, pp. 227–235, 2008.
© Springer-Verlag Berlin Heidelberg 2008

now coming to light. A number of impressive attacks are possible and have been heavily publicized, especially in the IEEE 802.11b area. As far as base technology is concerned, wireless security appears to be following the usual "penetrate and path" route. Early wireless security focused almost exclusively on cryptography and secure transmission-with unfortunate results thus far. Wi-Fi hacking has been around for some time now, and oddly enough has really received little press. Since 2001, 64 bit WEP has been breakable [1]. That was also around the time that well known tools such as Airsnort gave the ability to break into wireless network to the masses. In all likelihood user's neighbors are probably running a wireless router for their home computer network even though it is not using a wireless card. The wireless communication revolution has been bringing fundamental changes to data networking, telecommunication, and has been making integrated networks a reality. By freeing the user from the cord, personal communications networks, wireless LAN's [2], wireless MAN's, mobile radio networks and cellular systems, harbor the promise of fully distributed mobile computing and communications, any time, anywhere.

There are number of such wireless services widely in use at the moment. Wi-Fi (IEEE 802.11) [3] [4], WiMAX (IEEE 802.16) [5] and Mobile device networks such as GSM, 3G are now cater users across the globe.

Without physical boundaries, a wireless network faces many more security threats than a wired network does. For an example, WEP (Wired Equivalent Privacy) the authentication and data confidentiality definition of IEEE 802.11 standard was found to be vulnerable to security attacks, hence IEEE later came up with its 802.11i [6] to rectify the flaws of WEP. Likewise security flaws of IEEE 802.16 standard too have been exposed [7], [4]. This indicates how important the authentication and data encryption of these wireless networks. One of these is the IEEE 802.1x standard [8]. It is well known that 802.1x is based upon an existing authentication protocol known as the extensible authentication protocol (EAP) which in itself is an extension of PPP (point-to-point protocol). In fact, it is reported that 802.1x is susceptible to session hijacking as well as man-in-the-middle attacks [9], [10].

The uncertainty principle in quantum mechanics created a new paradigm for Quantum Key Distribution (QKD) [11], [12], [13]. The uncertainty principle in quantum mechanics created a new paradigm for cryptography: Quantum cryptography, or more specifically QKD. In fact this is called "No-Cloning" Theorem [14] and implies that a possible eavesdropper cannot intercept, measure and re-emit a photon without introducing a significant and therefore detectable error in the re-emitted signal.

2   Wireless 802.11 and Quantum Cryptography

As we described above that 802.11 security defines WEP [15] for the authentication and data confidentiality of user data over the wireless link. However, WEP was not well designed and presents serious vulnerabilities as a new standard for the 802.11 security. In this context, 802.11i is defined to rectify the flaws of WEP. 80211i received much attention from specialists in cryptography and network security.

Therefore, the former has three elements participating to the authentication and key management are the supplicant (or mobile terminal), authenticator (or access point), and the authentication server. Once having the pairwise master key (PMK), the access point starts the four –way handshake for the mutual authentication and the derivation of the pairwise transient key (PTK) with the mobile terminal.

In contrast to the 802.1x, the preshared key is involved in the authentication and key management using preshared key without "authentication server" and no extensible authentication protocol (EAP)-based authentication.

Following [16], we are using Figure 1 shows the pairwise key hierarchy containing the keys related to the encryption of unicast traffic.

It is noted that 802.11i has many keys at different levels, which becoming a key hierarch as shown Figure 1. At the top level there is the master key titled pairwise master key (PMK) that is used to derive the other keys.

The pairwise transient key (PTK) is created between the access point and the mobile terminal during the 4-way handshake. The PTK is split into three final temporal keys, namely key confirmation key (KCK), key encryption key (KEK), and temporal key (TK).

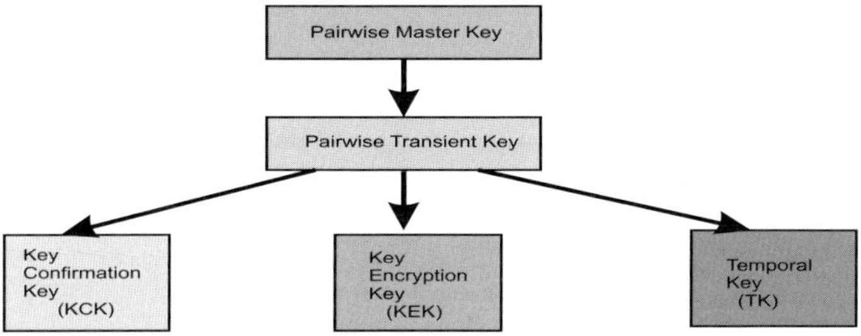

Fig. 1. Pairwise key hierarchy

Quantum Key Distribution systems transmit the secrete key, which are derived from random numbers, one photon (one bit) at a time in a polarized state. If intercepted by an eavesdropper or due to other atmospheric interferences etc, this state will change, and an error will be detected at the receiving side [17].

There are several QKD protocols available. Most widely used is the BB84 [5]. B92 (Charles Bennett), is a slight variation of BB84, is another well known QKD protocol [18]. B92 can be used two non-orthogonal states which represent the bit

values 0 and 1 as shown below: $\qquad |u_0\rangle,$ $\qquad\qquad\qquad\qquad\qquad$ (1)

$$|u_1\rangle,$$

BB84 coding scheme, invented by Charles Bennett and Gilles Brassard, is the first quantum cryptography communication protocol. The corresponding four quantum states can be expressed as equation (2). As an example, this coding system uses four non-orthogonal polarization states identified as *horizontal*, *vertical*, 45° and *135°*.

This protocol operates with transmitting party (say, Alice) sending polarized quantum bits (qubits) to the receiving party (call, Bob) via the quantum channel.

Once the quantum transmission finishes, Bob publicly communicates to Alice which measurements operators he used for each of the received bit. Alice then informs Bob which of his measurement operator choices were correct.

$$| \, 0 \, >,$$

$$| \, 1 \, >,$$

$$| \, \overline{0} \, >= \; \frac{1}{\sqrt{2}} (| \, 0 \, > + \, | \, 1 \, >),$$

$$| \, \overline{1} \, >= \; \frac{1}{\sqrt{2}} (| \, 0 \, > - \, | \, 1 \, >), \qquad (2)$$

The B92 quantum coding scheme is similar to the BB84, but uses only two out of the four BB84 non-orthogonal states, as shown in equation (1). It encodes classical bits in two non-orthogonal BB84 states. In out current paper, we decided to implement B92 protocol as a case study, the whole processing can be easily extended to four states, where BB84 used, and therefore from now on in this paper we are focusing on two quantum states, namely B92 protocol unless otherwise.

The Quantum key transmission happens in two stages that can be shown in Figure 2.

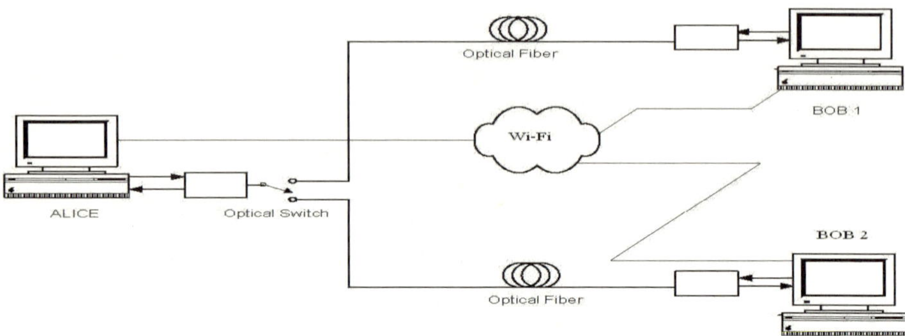

Fig. 2. Simplified block diagram of a point-to-point QKD link in concept, where SS is denoted "subscriber station" and the BS standing for "base station"

It is noted that in Figure 2 the Wi-Fi connections are classical channels and the "optical Fiber" channels are quantum channels.

Those two stages are as follows:

Stage 1: Quantum Channel (One way communication). This transmission could happen in either through free space or optical fiber. At present this implementation is being done at the University of Canberra, Australia.

Stage 2: Classical Channel (Two way communication). This phase deals with recovering identical secrete keys at both ends.

During the stage Alice & Bob communicate over a Classical channel that can be divided further in 4 main phases as shown below:

(a) Raw key extraction (Sifting), (b) Error Estimation, (c) Reconciliation, (d) Privacy Amplification.

For the implementation that we are going to present the wireless Wi-Fi is chosen as the classical channel (Figure 2). The quantum channel is the line of sight (LOS) optical path running by the polarization photon.

We can find, in Figure 2, that the classical channel forms by the standard Wi-Fi wireless and the quantum established by the optical photos. In Figure 2, in order to discuss our implementation more generally, SS is denoted "subscriber station" and the BS standing for "base station".

The Quantum channel is taking the task that using quantum cryptography to establish the key used for the encryption of user data in 802.11i, which is the TK. It is noted that TK is part of the PTK, as shown in Figure 1, which is established during the four-way handshake, we shall modify the four-way handshake to integrate the B92 protocol, as a case study, and make it as quantum handshake.

When the quantum handshake completion the wireless Wi-Fi will either refuse the subscriber station to communicate data via the classical channel or take the subscriber station to access the Wi-Fi and the system becomes "normal" Wi-Fi working states, which will run the communications in the defined classical channels.

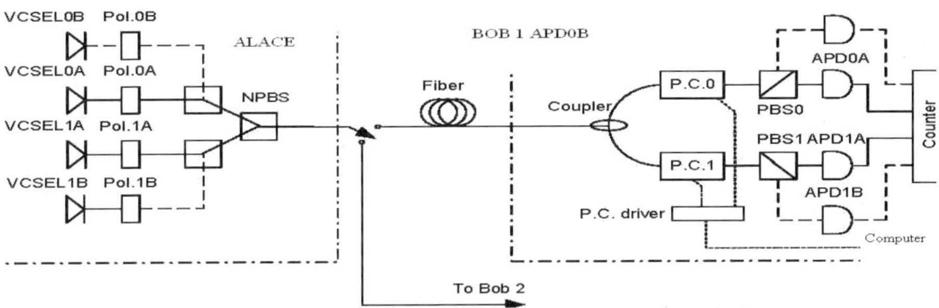

Fig. 3. Schematic diagram of the QKD system with PRAC sub-systems. VCSEL: Vertical Cavity Surface Emitting Laser ; Pol.: Polarizer; NPBS: Non-Polarizing beam splitter; P.C.: Polarization Controller; PBS: Polarizing beam splitter; APD: Silicon avalanche photodiode.

The quantum channel between Alice and Bob1 is shown in Fig. 3, the channel between Alice and Bob2 is similar. At Alice, laser pulses are generated by vertical cavity surface emitting lasers (VCSELs) and attenuated into single photon level. The polarization states of photons are set by polarizers according to corresponding protocol (B92 or BB84). Then photons are combined and sent into a fiber through a non-polarizing beam splitter (NPBS). The polarizers Pol. 0A, 0B, 1A, and 1B are oriented to 0°, 90°, +45°, and -45° respectively. Only two channels, 0A and 1A, are used for B92, while all four channels are used for BB84. At Bob, polarization controllers recover the polarization state of photons to their original state at Alice.

Quantum Network

The quantum network marries a variety of QKD techniques to well established internet technology in order to build a secure key distribution system employed in conjunction with the public internet or, more likely, with private networks that employ the internet protocol suite [2]. At present there are large numbers of such private networks in widespread use around the world with customers' desire secure and private communications.

The merge of QKD technologies to these networks proves feasible and appealing in certain contexts.

Free space QKD uses the air as the medium for the transmission of photons between the quantum sender and receiver. The feasibility of QKD over the air is considered problematic because of a medium with varying properties and a high error rate. In particular for the limited distance and indoor environment the quantum channel would be realized at the reasonable level.

3   System Implementation

The KCK is generated from the PMK to serve the mutual authentication of the supplicant and the authenticator and protect the B92 protocol from the main-in-the-middle attack as described in [16]. Once the mutual authentication finished, the supplicant and the authenticator stats the B92 protocol for the establishment of the Q-PTK. The Q-PTK is splits into the KEK and TK.

It is noted that we can use quantum cryptography to establish the PK, therefore all KEK, KCK, and TK are established using quantum cryptography.

Security provides subscribers with privacy across the broadband wireless network. It achieves security by encrypting connection between BS (Base Station) and SS (Subscriber Station). The protocol for first 2 stages of QKD.

Index Files: The software implementation depends on the key bits recorded at BS and SS. These key bits are to be recorded in set of files, known as "Index Files". Since the original key transmitted by BS in the Quantum Channel could contain many bits (gigabits), there will be multiple index files generated at either ends.

Those index files will act as the input to this software development project.

Index files at BS: All the key bits that BS transmits in the Quantum Channel are to be recorded into index files at her end. These files hold the original key that BS transmitted to SS.

Examples of the bits recorded in those index files:-
1,0,1,1,0,0,0,1,1,0,0,1,0,1,1,1,0,0,0,1,1,0
where "," being the delimiter

Index files at BS: All the key bits that BS transmits in the Quantum Channel are to be recorded into index files at her end. These files hold the original key that BS transmitted to SS.

Examples of the bits recorded in those index files:-
1,0,1,1,0,0,0,1,1,0,0,1,0,1,1,1,0,0,0,1,1,0
where "," being the delimiter

Index files at SS: During the Quantum transmission, SS too records the key bits that he received from BS in Index files. These bits will not be identical to what BS has transmitted due to the random bases used by SS's photon detector, eavesdropper attacks, channel noise, dark counts of the photon detector etc..

Therefore the index files recorded at SS's end will comprise non-receptions. Non-receptions are the bit positions that SS should have received, but not receive a bit. Examples of the bits recorded at SS's index files:-

1,1,,,0,0,,0,1,,1,0,0, ,0, ,1,,1,0,0, ,1,1,0
where "," being the delimiter

With the use of delimiter to separate each key bit, it is easy to identify the of non-reception bits.

Program Structure and Protocol: In order to establish the communication path, BS first *listen*s to a specified port. SS sends the *connect* message to the specified port in BS. Upon receiving the *connect* request, BS sends *accept* call to SS establishing the communication path between them.

Raw Key Extraction (Sifting): During this process, which happens at start up, SS sends the non-erasure bit positions to Alice by running through the index file data loaded into the memory. BS in turn, processes her index files and keeps only those corresponding bits. At the end of this process, both BS and SS will have index files of identical lengths after removing all non-receipt bits. This process is called *Raw Key Extraction* and the keys recovered after this phase is known as *BS Raw Key* and *SS Raw Key*.

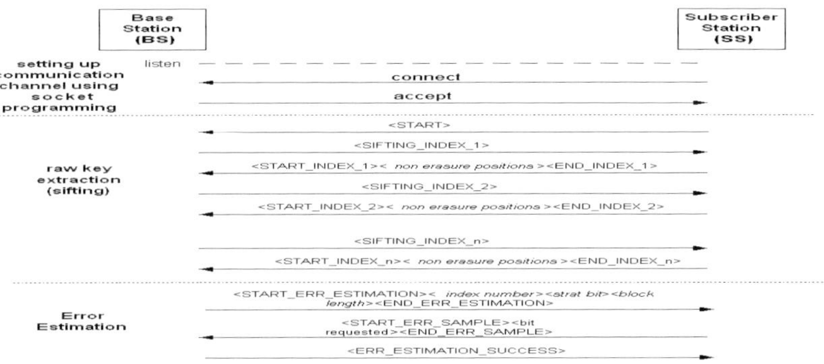

Fig. 4. The protocol for first 2 stages of QKD

This request has the following format: **Error Estimation.**

This process starts with BS requesting SS to send a block of bits of length "L" from a particular index file: *<STRAT_ERR_ESTIMATION> <INDEX FILE NUMBER> <START BIT> <LENGTH> <END_ERR_ESTIMATION>*

BS calculates the Bit Error Rate (**e**) of the Quantum transmission.

$$e = \frac{\text{Number \quad of \quad bits \quad in \quad error}}{\text{Total \quad number \quad of \quad bits \quad in the \quad block}} \times 100 \ \% \qquad (3)$$

BS then compares with the maximum error rate allowed (e_{max}). This value is also known as Quantum Bit Error Rate (QBER).

If $e <= e_{max}$ they accept the quantum transmission and proceeds to the next phase called Reconciliation. Both BS and SS remove those bits which are publicly revealed from their index file(s).

If $e > e_{max}$ BS sends ABORT message to SS indicating the quantum transmission contains errors to a level where they cannot recover the key from the bits received. In this case, they seizes the session by terminating the program.

4 Conclusion

In this paper we present the implementation of first two stages of an agent-oriented KQD for Wi-Fi. At present, the first two stages of B92 (or BB84) protocol has been implemented in C++ language on Linux platform. The KQD can handle multiuser as it benefits from an agent-oriented mechanism.

This caused a heavy overhead as the program consumes considerable amount of time during bit comparisons etc when doing file processing. To avoid this inefficiency, a STL list structure has been implemented to hold the index file data. Due to this modification, most of the computations and bit comparisons are done in-memory. This has resulted in improving the efficiency by about 60%.

With this set up, the program can be operated by setting different values to suit any requirements. One such parameter is the QBER, where this value is used to calculate the error rate of the quantum transmission. QBER of the quantum transmissions could be impacted by various issues (described earlier) causing it to vary per each transmission. Therefore by having the QBER as a configurable parameter, this software can be used to run even for simulation purposes by setting different values.

References

[1] Park, D.: The Lack of WiFi Security (part 1) (December 07, 2006), http://www.windowsecurity.com/articles/WiFi-security-lack-Part1.html?prontversion

[2] IEEE Standard for Local Metropolitan area networks, Part 16: Air Interface for Fixed Broadband Wireless Access Systems

[3] Building the Quantum Network, Chip Elliott, BBN Technologies. New Journal of Physics 4, 46.1–46.12 (2002), http://www.iop.org/EJ/article/1367-2630/4/1/346/nj2146.html

[4] Hasan, J.: Security Issues of IEEE 802.16 (WiMAX)

[5] Bennett, C.H., et al.: Experimental Quantum Cryptography. J. Cryptology 5(1), 3–28 (1992)

[6] IEEE Standrad 802.11i, Part 11: Wireless LAM Medium Access Control (MAC) and Physical Layer (PHY) specifications – Amendment 6: Medium Access Control (MAC) Security Enhancements (July 2004)

[7] Xu, S., Matthews, M., Huang, C.-T.: Security Issues in Privacy Key Management Protocols of IEEE 802.16

[8] Philip Craiger, J.: 802.11, 802.1x, and wireless security, GIAC security essentials certification Practical Asssignment, version 1.4, ©SANS Institute (2002)

[9] Connolly, P.J.: The trouble with 802.1x, InfoWorld (March 8, 2002), http://www.infoworld.com/articles/fe/xml/02/03/11/020311fe8021x.xml

[10] Schwartz, E.: Researchers crack new wireless security spec. InfoWorld (February 14, 2002), http://www.infoworld.com/articies/hn/xml/02/02/14/020214hnwifispec.xml

[11] Bennett, C.H.: Quantum Cryptography: Uncertainty in the Service of Privacy. Science 257, 752–753 (1992)

[12] Lenz, M.: High Bit Rate Quantum Key Distribution Systems 5th Year Project Report 2006/2007, Heriot Watt University (February 16, 2007), http://moritz.faui2k3.org/

[13] Buttler, W.T., Hughes, R.J., Kwiat, P.G., Luther, G.G., Morgan, G.L., Nordholt, J.E., Peterson, C.G., Simmons, C.M.: Free-space quantum key distribution (January 1998) ar Xiv: quant-ph/9801006 v1

[14] Wootters, W., Zurek, W.: A single quantum cannot be cloned. Nature 299, 802–803 (1982)

[15] Edeny, J., Arbaugh, W.A.: Real 802.11 Security-Wi-Fi protected access and 802.11i. Addison Wesley, Reading (2004)

[16] Nguyen, T.M.T., Sfaxi, M.A., Ghernaouti-Hélie, S.: 802.11i Encryption key distribution using quantum cryptography. Journal of Networks 1(5), 9–20 (2006)

[17] Bennett, C.H., Brassard, G.: Quantum Cryptography: Public Key Distribution and Coin Tossing. In: Proceedings of IEEE International Conference on Computers Systems and Signal Processing, Bangalore, India, December 1984, pp. 175–179 (1984)

[18] Bennett, C.H.: Quantum cryptography using any two nonorthogonal states. Phys. Rev. Lett. 68, 3121–3124 (1992)

Developing Intelligent Agent Applications with JADE and JESS

Bala M. Balachandran

Faculty of Information Sciences and Engineering
University of Canberra
ACT 2601, Australia
bala.balachandran@canberra.edu.au

Abstract. Agent systems differ from more traditional software systems because agents are intended to be independent autonomous, reactive, pro-active and sociable software entities. Due to these unique characteristics developing agent systems has been a very challenging task for agent researchers and application developers. JADE (Java Agent DEvelopment Framework) is an agent development tool, implemented in JAVA and FIPA-compliant. Although JADE provides all the mandatory components (FIPA) for the development of autonomous agents, it lacks the ability to include intelligent behaviour to individual agents. JESS (Java Expert System Shell), a rule-based programming environment written in Java, provides a powerful tool for developing systems with intelligent reasoning abilities. This paper examines the use of both JADE and JESS for the development intelligent agent systems and shares the experience of the authors in the development of a Personal Travel Assistant (PTA).

Keywords: Intelligent Agents, AOSE, Multi-Agent Systems, JADE, JESS, FIPA, MaSE, AgentTool.

1   Introduction

An *agent* is a software entity that applies Artificial Intelligence techniques to choose the best set of actions to perform in order to achieve a goal specified by the user. An agent's characteristics include proactive, dynamic, autonomous and goal-oriented. A *multi-agent system* may be defined as a collection of autonomous agents that communicate between them to coordinate their activities in order to be able to solve collectively a problem that could not be tackled by any agent individually [17][18]. In recent times there has been a growing interest in the application of agent-technology to problems in the domain of E-commerce [8]. In this paper our interest is to design and develop a multi-agent system that is capable of producing travel packages for customers based on their personal preferences. We have chosen JADE (**J**ava **A**gent **DE**velopment Framework) which is very popular because of its open source, simplicity and compliant with the FIPA specifications [3][9][14]. JADE conceptualises an agent as an independent and autonomous process that has an identity, possibly persistent, and that requires communication (e.g. collaboration or competition) with other agents in order to fulfill its tasks. This communication is implemented through asynchronous message passing and by

I. Lovrek, R.J. Howlett, and L.C. Jain (Eds.): KES 2008, Part III, LNAI 5179, pp. 236–244, 2008.

using an Agent Communication Language (ACL) with a well-defined and commonly agreed semantics. The development environment incorporates a set of graphic tools, which facilitates the platform management, providing support to agent's execution, debugging and monitoring.

In order to incorporate intelligence with JADE agents, we use JESS (Java Expert System Shell), a rule engine and scripting environment written entirely in Java by Ernest Friedman-Hill at Sandia (Livermore) [10]. JESS was originally inspired by the CLIPS expert system shell, but has grown into a complete, distinct Java-influenced environment. In solving problems with rules, JESS uses the fast and efficient Rete algorithm which involves building a network of pattern-matching nodes. The strength of Rete comes from the fact that it uses a set of memories to retain information about the success or failures of pattern matches during previous attempts.

The rest of this paper is organised as follows. Section 2 describes our application domain and our case study. Section 3 presents the processes involved in the analysis and design of agent systems using MaSE and AgentTool. Section 4 describes the steps involved in developing multi-agents with JADE. Section 5 explains an approach for embedding JESS in JADE agents. Finally, the last section presents our conclusions and some lessons learned form this project.

2   Case Study: A Personal Travel Assistant (PTA)

Multi-agent systems are ideally suited to represent problems that have multiple problem solving entities. Such systems have the traditional advantages of distributed and concurrent problem solving strategies. In this research we aim to design and develop a multi-agent system called **Personal Travel Assistant (PTA).** Recently there have

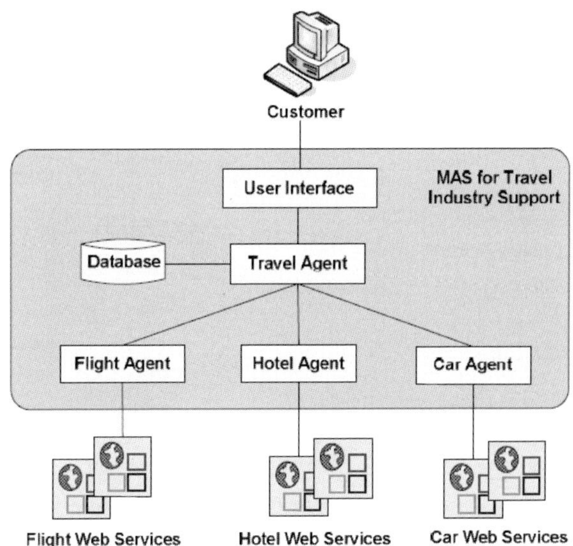

Fig. 1. The software architecture for the Personal Travel Assistant (PTA) System

been some efforts in applying agent technology for travel support systems [2][7][12]. The system will be capable of finding optimum travel packages for the customer depending on their preferences. The travel package is composed of a flight ticket, a hotel accommodation and a car rental. Figure 1 shows the proposed system architecture for our project.

The Travel Agent is the main agent that takes user preferences and requirements as input and produces a set of booking results as output. The Travel Agent communicates with other agents such as Flight Agent, Hotel Agent and Car Agent to perform its tasks. The Travel Agent also handles conflicts between the other agents in regard to fulfilling the user requirements.

3   Analysis and Design of Multi-agent Systems Using MaSE and AgentTool

There have been several proposed methodologies for analyzing, designing and multi-agent systems [13]. The most widely published methods include Gaia methodology, MaSE (Multi-agent Systems Engineering) and Prometheus [19][5][6][16]. Among these methods MaSE is more detailed and provides more guidance to system designers. We have chosen MaSE and its support tool: AgentTool for our system analysis and design. Its goal is to guide a system developer from an initial system specification to a multi-agent system implementation. The MaSE methodology consists of two phases: analysis and design. The first phase, Analysis, focuses on capturing goals, applying use cases and refining roles. The second phase, Design, focuses on creating agent classes, designing interactions, assembling agent classes, and building deployment diagrams. MaSE is supported by the agentTool [1] which is a graphically based

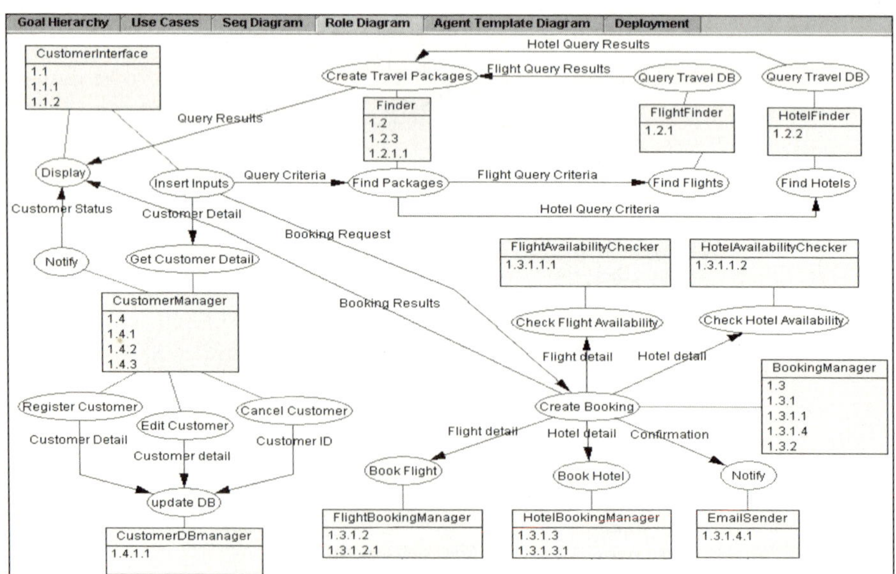

Fig. 2. The role diagram generated using the AgentTool

fully interactive software engineering tool. It also provides automated support for transforming analysis models into design artefacts. Systems designed using MaSE can be implemented using any agent-programming environment such as JADE or JACK. Figure 2 shows the role diagram developed for our case study using the agentTool.

4 Developing Multi-agent Systems with JADE

One goal of JADE is to simplify development while ensuring standard compliance through a comprehensive set of system services and agents. It provides the following mandatory components for agent's management: [5][9]

- AMS (Agent Management System), which besides providing white page services as specified by FIPA, it also plays the role of authority in the platform.
- DF (Directory Facilitator) provides yellow pages services to other agents.
- ACC (Agent Communication Channel) which provides a Message Transport System (MTS) and is responsible for sending and receiving messages on an agent platform.

JADE has been applied to several applications successfully [3][11]. In the following subsections we discuss the various steps involved in developing a multi-agent system using JADE.

4.1 Creating Agents

Creating a JADE agent is as simple as defining a class that extends the jade.core.Agent class and overriding the default implementation of the methods that are automatically invoked by the platform during the agent lifecycle, including SetUp and TakeDown (). Consistent with the FIPA specifications, each agent instance is identified by an 'agent identifier'. In JADE an agent identifier is represented as an instance of the jade.core.AID class. The getAID () method of the Agent class allows retrieval of the local agent identifier.

4.2 Defining Agent Tasks

In JADE, behaviour represents a task that an agent can carry out and is implemented as an object of a class that extends *jade.core.behaviours.Behaviour*. Each such behaviour class must implement two abstract methods. The action () method defines the operations to be performed when the behaviour is in execution. The done () method returns a Boolean value to indicate whether or not a behaviour has completed and is to be removed from the pool of behaviours an agent is executing. To make an agent execute the tasks represented by a behaviour object, the behaviour must be added to the agent by means of the *add Behaviour ()* method of the *Agent* class.

4.3 Agent Discovery Process

The JADE platform provides a yellow pages service which allows any agent to dynamically discover other agents at a given point in time. A specialised agent called the

DF (Directory Facilitator) provides the yellow pages service in JADE. Using this service any agent can both register (publish) services and search for (discover) services.

4.4 Agent Communication

Agent communication is probably the most fundamental feature of JADE and is implemented accordance with the FIPA specifications. The JADE communication paradigm is based on asynchronous message passing. Each agent is equipped with an incoming message box and message polling can be blocking or non-blocking. A message in JADE is implemented as an object of the jade.lang.acl.ACLMessage object and then calling the send () method of the Agent class.

4.5 Defining Ontologies

Agents must share semantics if the communication is to be effective. Therefore, exchanged messages must have written in a particular language and must share the same ontology. An ontology in JADE is an instance of the jade.content.onto.Ontology class to which schemas have been added that define the types of predicates, agent actions and concepts relevant to the addressed domain. These schemas are defined as instances of the PredicateSchema, AgentAction Schema and ConceptSchema classes included in the jade.content.schema package.

5 Integrating JADE Agents with JESS

Rule-based systems have been successfully used for e-commerce applications such as order processing, user preferences, data mining and recommendations [8]. In our case study we have identified two types of knowledge that are necessary: travel-related and customer-related. Rule-based systems have also been used to assist in the negotiation process among agents [4]. Currently JADE alone does not endow agents with specific capabilities beyond those needed for communication and interaction. However,

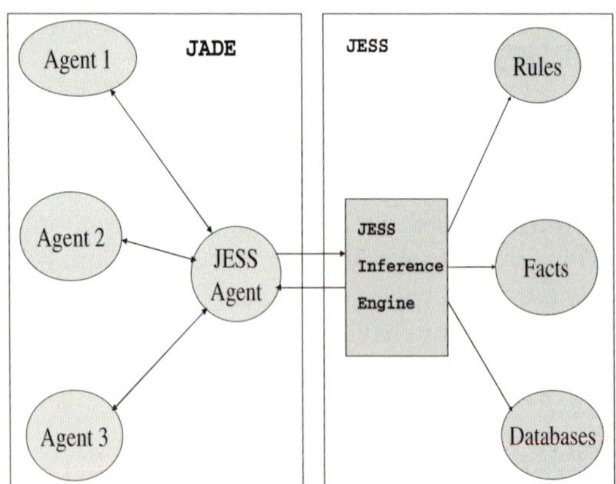

Fig. 3. The integration of JADE and JESS

the behaviour abstraction of the JADE agent model allows simple integration with JESS [15]. Figure 3 illustrates our strategy for the integration of JADE and JESS in order to enhance the JADE's capabilities.

The JESS shell provides the basic elements of an expert system including fact-list, knowledge base that contain all the rules and inference engine which controls overall execution of rules. JESS includes a special class called Rete, which implements the rule-based inference engine. To embed JESS in a JADE agent, we simply need to create a special agent called JESS agent, which incorporates jess.Rete object and manipulates it appropriately. Figure 4 illustrates how *Jess.Rete* class is used inside the JESS agent's behaviour.

```
class JessBehaviour extends CyclicBehaviour {
        private jess.Rete jess;
        JessBehaviour(Agent agent, String jessFile) {
            super(agent);
            jess = new jess.Rete();
            try {
                FileReader fr = new FileReader(jessFile);
                jess.Jesp j = new jess.Jesp(fr, jess);
                try {
                    j.parse(false);
                } catch (jess.JessException je) {
                    je.printStackTrace();
                }
                fr.close();
            } catch (IOException ioe) {
                System.err.println("Error loading, Jess file");
            }
        }
    ...
    ...

    }
```

Fig. 4. A behaviour of the JESS agent includes *jess.Rete* object

The method *Rete.Run()* allows us to run the inference engine. This method will make the engine consecutively fire applicable rules, and will return only when there are no more rules to fire, that is, when the engine stops.

6 An Integrated Intelligent Agent Development Environment

We use the Eclipse platform as our programming environment, which is an open source Integrated Development Environment (IDE) that provides support for range of languages from C++, PHP to Java [20]. The Eclipse environment comes with a plug-in to integrate JADE within Eclipse. Figure 5 shows the JADE **Remote Monitoring Agent** which controls the life cycles of the agent platform and of all registered agents

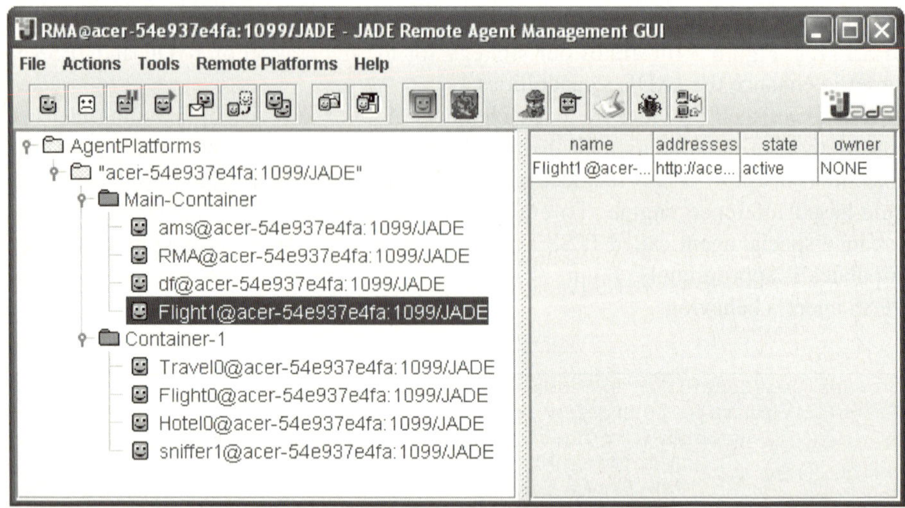

Fig. 5. JADE Remote Monitoring Agent

6.1 Running the PTA System

The PTA system can be started after creating at least one of the each agent (travel, flight, hotel and car). Travel agent is the main component of the system and the customer interacts only with this agent. Other three types of agents can be one or more depending on the web services and they have no direct interaction with the customer. For example, if we have two flight web services to connect, then two Flight agents need to be created and connected to the web services. Figure 6 shows the Travel Agent's user interface screens which is used for entering the user's travel requirements.

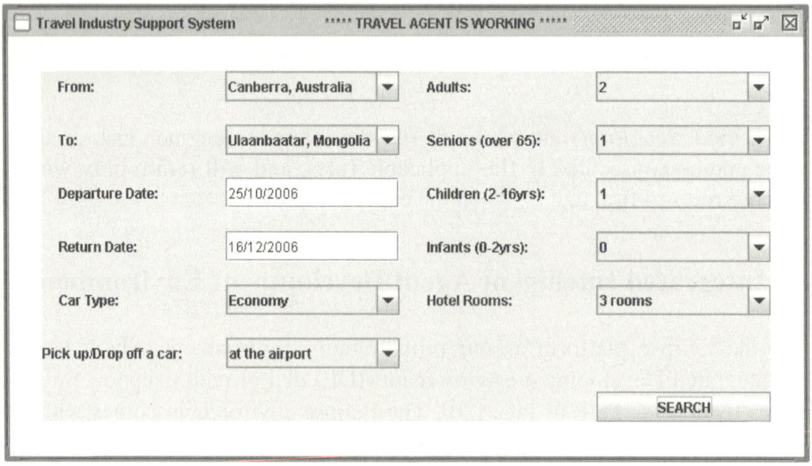

Fig. 6. Travel Agent's user interface

We have carried out some preliminary experiments with the current version of the system. The results so far are promising.

7 Conclusions and Future Work

This paper has shown the use of MaSE and agentTool in the design and development of a multi-agent system which requires communication and collaboration. By analysing the system as a set of roles and tasks, a system designer is naturally led to the definition of autonomous agents that co-ordinate their actions to solve the overall system goals.

The prototype PTA system, which integrates JADE and Jess seamlessly, has been developed and implemented. The JADE development tool provides suitable mechanisms for developing multi-agent systems, making easier the programmers' tasks when implementing agents, tasks, ontologies and communication between them. For intelligence aspects, we have exploited the facilities of JADE relating JESS integration. JADE agents embedded with JESS engines have qualities over and above those found in simple JADE agents. More importantly such hybrid agents have intelligent reasoning abilities. The PTA System has met the aims and goals expected and has been tested for travel booking in terms of hotel, flight and car requirements.

We aim to improve the ability of the agents in regard to conflict resolution using negotiation skills by incorporating more advanced negotiation protocols and negotiation rules. We will report our progress in our future publications.

Acknowledgements

The prototypes described in this paper have been mainly developed by Majigsuren Enkhsaikhan.

References

1. AgentTool, Support tool for MaSE methodology. Multiagent and Cooperative Robotics Laboratory (2006),
   `http://macr.cis.ksu.edu/projects/agentTool/agentool.htm`
2. Balachandran, M.B., Enkhsaikhan, M.: Development of a Multi-agent system for Travel Industry Support (CIMCA 2006 and IAWTIC 2006), Sydney, Australia (2006)
3. Bellifemine, F., Caire, G., Greenwood, D.: Developing Multi-Agent Systems with JADE. John Wiley & Sons, UK (2007)
4. Benyoucef, M., Alj, H., Levy, K., Keller, R.K.: A Rule-Driven Approach for Defining the Behaviour of Negotiating Software Agents. In: Plaice, J., Kropf, P.G., Schulthess, P., Slonim, J. (eds.) DCW 2002. LNCS, vol. 2468, pp. 165–181. Springer, Heidelberg (2002)
5. Cuesta-Morales, P., Gomez-Rodriguez, A., Rodriguez-Martinez, F.J.: Developing a Multi-Agent System using MaSE and JADE. Upgrade 5(4), 27–31 (2004)
6. DeLoach, S.: Multiagent Systems Engineering: a Methodology and Language for Designing Agent Systems. In: Proceedings of Agent Oriented Information Systems 1999, pp. 45–57 (1999)

7. Far BH Sample Project: Travel Agency System (TAS) (2004) (Retrieved on February 2, 2006), http://www.enel.ucalgary.ca/People/far/
8. Fasli, M.: Agent Technology for e-Commerce. John Wiley and Sons, UK (2007)
9. FIPA (2006) The Foundation for Intelligent Physical Agents, http://www.fipa.org/
10. Friedman-Hill, E.: Jess in Action: Java Rule-Based Systems. Manning Publications Co., New York (2003)
11. Ganzha, M., Paprzycki, M., Pirvanescu, A., Badica, C., Abraham, A.: JADE based Multi-agent E-commerce Environment: Initial Implementation (Retrieved on July 30, 2006) (2001), http://www.ganzha.euh-e.edu.pl/agents/index.html
12. Gordon, M., Paprzycki, M.: Designing Agent Based Travel Support System (Retrieved on June 2, 2005), http://agentlab.swps.edu.pl/ISPDC_2005.pdf
13. Iglesias, C., Garijo, M., Gonzalez, J.: A Survey of Agent-Oriented Methodologies. In: Muller, J.P., Singh, M.P., Rao, A.S. (eds.) ATAL 1998. LNCS (LNAI), vol. 1555, pp. 317–330. Springer, Heidelberg (1999)
14. JADE, Java Agent Development Environment (2008), http://jade.tilab.com
15. Lopes, H.: Integrating JADE and Jess (Retrieved on March 13, 2007), http://jade.tilab.com/doc/tutorials/jade-jess/jade-jess.html
16. Padgham, L., Winikoff, M.: Developing intelligent Agent Systems: A Practical Guide. John Wiley & Sons, Chichester (2004)
17. Sycara, K.P.: Multiagent Systems. AI Magazine 19(2), 79–92 (1998)
18. Wooldridge, M.: An Introduction to Multiagent Systems. Wiley Ed., Chichester (2002)
19. Wooldridge M, Jennings NR and Kinny D.: The Gaia Methodology for Agent-Oriented Analysis and Design (Retrieved on 15 February 2006) (2000), http://www.ecs.soton.ac.uk/~nrj/download-files/jaamas2000.pdf
20. The Eclipse Platform, http://www.eclipse.org/

An Agent Based Multifactor Biometric Security System

Girija Chetty, Dharmendra Sharma, and Bala M. Balachandran

{girija.chetty,dharmendra.sharma,
bala.balachandran}@canberra.edu.au

Abstract. A novel agent based protocol for computer and network security based on multifactor fusion of passwords and facial biometrics for user authentication is proposed in this paper. The protocol uses multiple factors for authentication involving a combination of passwords and face biometric template on the smart card and the user workstation. The biometric match is performed by a module running on the user's workstation, authenticated by a mobile agent coming from a reliable server. The proposed protocol uses the current cryptographic tokens without requiring any hardware or software changes.

Keywords: multifactor fusion, biometric, smartcard, mobile agents.

1 Introduction

Biometric technologies allow stronger authentication mechanisms in general by matching a stored biometric template to a live biometric template [1, 2]. Currently, smartcards are being increasingly perceived as secure and tamper-proof devices to store sensitive information such as digital certificates and private keys. The access to smart-cards has historically been regulated by a trivial means of authentication: the Personal Identification Number (PIN). A user can access the card by entering the right PIN. However, as it is well known, the PINs are weak secrets, since more often they are poorly chosen, and they are easy to forget and forge.

One possible alternative is to store the biometric template on the smart card, which will alleviate the need to remember the password. Also, if sufficient computational resources are available on the card, the matching of stored template with live biometric template can be performed. However, this will require high computational resources on the card, which is not possible with currently available smart cards. Further, with both the biometric template and the matching process on the card, it is more risky as it poses a danger of theft of the card, and an attacker sniffing the biometric and later using it to unlock the card in a replay attack. For such scenarios, not much prior work on implementing biometric authentication is available, particularly on commercially available smartcards with limited on-chip resources?

In this paper we propose a new protocol based on mobile agent based multifactor fusion of face biometric and encrypted passwords split between smart cards and the local host/network. The paper is organized as follows: Next section describes the current authentication schemes. The proposed protocol and the implementation details are described in section 3. Section 4 discusses the protocol performance in terms of the extent that the security problem is resolved. Finally, the paper concludes with Section 5 on conclusions and discussion of some future plans.

I. Lovrek, R.J. Howlett, and L.C. Jain (Eds.): KES 2008, Part III, LNAI 5179, pp. 245–251, 2008.

2 Problems with Current Authentication Protocols

Before describing the proposed agent based authentication protocol, problems with using multifactor fusion of biometric and passwords with current smart card authentication technologies is described here. If the biometric template is stored on the card and the matching with live biometric template occurs on either the local host (user's workstation) or the network, there are several points of attack in this authentication protocol, due to unsecure data transmission stream between the face biometric capture device, smart card reader and the local host /network which does the biometric match. Using a biometric with the current authentication protocol on smart cards normally involves the following stages:

1. When the smart card is inserted, a cryptographic application on local host asks the user to authenticate himself to the smart card via a specific cryptographic API call.
2. Verification Module in the local host reads the biometric template from the smart card.
3. A real time live template is acquired from the user using a face biometric scanner or camera.
4. The biometric match between the two templates is performed on the local host.
5. If the biometric template is successful then the local host requests the PIN from the user and submits it to the smart card to unlock the crypto chip

Figure 1 illustrates different stages for this authentication protocol.

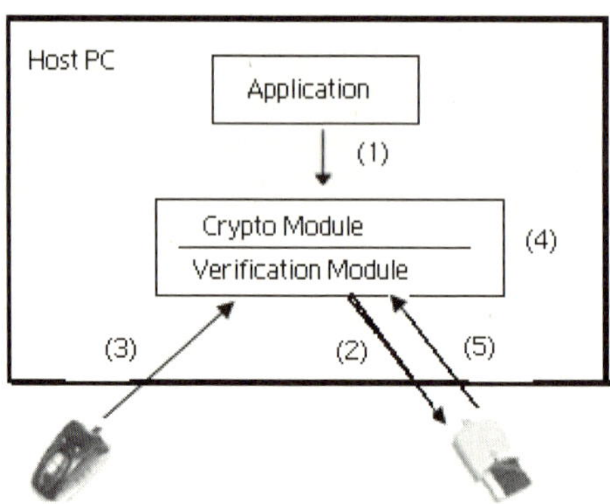

Fig. 1. Technique for using biometric with authentication protocol on smart cards

The major problems with this protocol could be:

-   The enrollment template stored inside the smartcard could be eavesdropped at step 2.
-   The live biometric template could be eavesdropped at step 3.

- The smartcard somehow does not trust the module performing the match at step 4.
- The problem of where to store the secret login data (PIN) used to actually log the user into the cryptographic chip on card. (We don't want to ask the user for it).

The first two issues above are critical if an attacker manages to steal the smart card and makes a replay attack sending again the eaves-dropped data directly to the verification module. In this case, there is no security mechanism to verify that the biometric verification data are derived from an actual live presentation to the biometric sensor. To solve these problems we could encrypt the data exchanged between the devices and the crypto library or find a way to trust the module that acquires the live template.

Smart card is unable to authenticate the verification module. Maybe using a kind of challenge-response protocol: the module requests a random to the smart card and this is returned encrypted with a shared secret key. Now the dilemma is where to store the key on the host.

The only method to unlock the private area of the chip is to supply the exact PIN to the card. If the PIN is correct then the card trusts the module on the local host that has performed the biometric match.

Therefore, the crucial issue is the last one: where to store the secret PIN.

One method is to store the PIN inside the compiled Crypto Module on local host. This may not be a good solution because a malicious user might do reverse engineering on the library and find the secret. Also, every place inside the file system on local host is not secure if a malicious program has manipulated the host, so the safest place is inside a protected remote Server.

Now the problem to solve is how the remote Server can send critical data to an unknown remote host. This is like a black box and the server does not know if a malicious entity is running on that client. A mutual authentication by establishing a SSL connection between the client and the server is a good solution, but like the PIN, there is always the recurrent problem of where to store the certificate/private key of the client (we don't want to use another smartcard [11] and we want to avoid asking the user for another PIN to unblock this private key).

Here we propose a new solution: using a Mobile Agent framework. If the server cannot trust applications running on the client, it will trust the code that it launches to the client: a mobile agent. A remote agent, launched from the secure server, will try to authenticate the module that executes the biometric match on the client; if the result of the authentication is positive then the server sends the secret PIN to the client via a previously opened secure connection. This framework, a Multifactor Biometric Secure Mobile Agent Framework (MBSMAF) based on Java APIs is currently being developed in University of Canberra.

A Mobile Agent is a software entity that is not bound to the host where it begins execution, but has the unique ability to travel across a network and perform tasks on machines that provide agent-hosting capability. Unlike remote procedure calls, where a process invokes procedures of a remote host, process migration allows executable code to travel autonomously and to interact with the hosting machine's resources, including other mobile agents. Therefore, a Mobile Agent framework has to cope with various security threats [8]: malicious agents might try to break into the server in order to harm other agents or to gain unauthorized system access. A malicious host could tamper with agents. Agents might be sniffed while they are transferred over the network. Though many open source agent development frameworks are available on the internet: we decided to develop MBSMAF, as it focuses on agent based secure biometric fusion and tries to solve the above-mentioned problems.

3 Proposed Agent Based Authentication Protocol

3.1 Description of the Entities

Client: the user's workstation

- Application: the user application that requires the access to the smart-card through the Crypto Module (for instance a digital signature application).
- Crypto/Verification Module: the main entity used to access the smart-card and to perform the local biometric match (it implements the client-side protocol).
- MBSMAF Framework: the runtime environment for the Mobile Agent coming from the Server.

Server: the secure host where the secret login data is contained

- MBSMAF Framework: the runtime environment for the Service implementing the protocol.
- Biometric Match on Smart Card (BMSC) Service: the Service which accepts connections and implements the protocol.
- Biometric Match Agent: the Mobile Agent that is launched by BMSC service, gets to the Client and comes back with the result of the authentication.

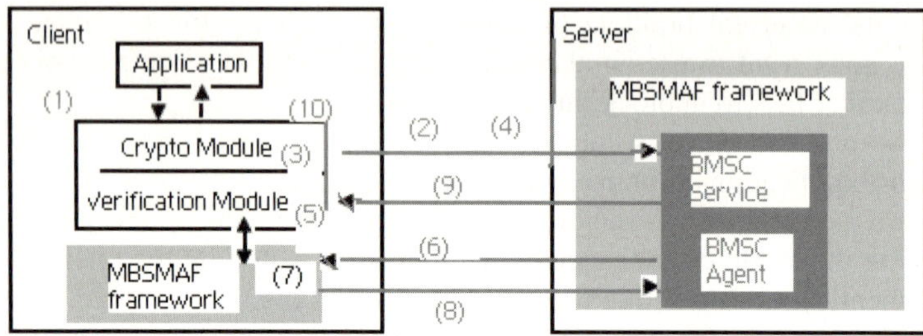

Fig. 2. Framework for Agent based secure multifactor biometric authentication protocol

Figure 2 shows the MBSMAF framework, and the authentication protocol based on this framework is described below:

3.2 Protocol Description

1. User application requires access to the private space of the smartcard through a particular Crypto API call.

2. The Crypto Module (CM) opens an encrypted connection to the BMSC Service running on the Server (BMSC is authenticated via SSL).

3. After the connection has been established, CM generates a large Random value and stores it inside itself (value used by the mobile agent at step 7).

4. Finally, CM sends the Random and the smartcard Serial Number to BMSC: if the subsequent controls succeed, CM will receive the secret login data at step 9, otherwise BMSC will close the connection.

5. In the meantime, the Verification Module executes a biometric match between the template stored inside the smartcard and the live acquired one BMSC stores the Random previously received within a BMSC Mobile Agent and launches it to the Client address over a new encrypted channel; after this, BMSC begins to wait for the return of the Agent.

6. Now the BMSC Agent is migrated on the client:

    a. It tests the validity of the modules residing on the user's workstation checking their digital signature.

    b. It checks that the random inside itself has the same value as the random contained in CM.

    c. It ensures that the biometric match executed at step 5 is successful.

If all the previous controls are positive then we can trust the CM module that has started the protocol.

7. The Agent comes back to the Server and returns the result to BMSC.

8. If the result is positive then BMSC sends, on the same connection opened at step 2, the secret login data (PIN)4 correlated to the Serial Number received at step 4. Otherwise, it closes the connection with the client.

9. In the last step, CM unlocks the private area using the received PIN and confirms to the user application the success of the Crypto API call made at step 1.

3.3 Protocol Implementation

The protocol and all the entities have been developed and deployed on a Windows XP Professional workstation, so some details are particular to this architecture. As regards the hardware, a simple off the shelf webcam is used as the biometric scan device, while the smartcard used is a Cyberflex e-gate produced by Schlumberger [12]. (As a matter of fact, it is possible to employ any kind of biometric device or smartcard without modifying the protocol by only changing the respective library.)

The principal technology employed, besides MBSMAF agent framework , is the PKCS#11 standard [7], which has been used as the Crypto token. We have chosen this solution because this is the most widespread de-facto standard in today's cryptographic tokens. The PKCS#11 standard specifies an API, called "*Cryptoki*" (cryptographic token interface), to interface the devices that hold cryptographic information and that perform cryptographic functions. The Cryptoki is important because it isolates an application from the details of the cryptographic device.

The standard employed to perform all the required biometric operations is BioAPI [4]. This API is intended to provide a high-level generic biometric authentication model, covering the basic functions of Enrollment, Verification, and Identification.

Another technology we have used to implement this protocol has been the Java Native Interface (JNI) [9]. This was used to exchange data between the BMSC service

Mobile Agent (which runs in a Java virtual machine) and the dlls (which are native libraries) at step 7 of the protocol.

4 Protocol Performance

In this section we discuss the performance of the protocol in terms of resolving the security problem.

If the attacker has not stolen the user's smartcard, he could try to get possession of the PINs residing in the safe Server:

- If no smartcard{SC) is inserted, the protocol will not start.
- If a wrong or malicious dll dealing with the communication with the SC is installed, the Mobile Agent will notice it.
- If a different module from CryptoWrapper tries to connect to BMSC service, the Mobile Agent will notice it using the random comparison.
- If she is using a brand new smartcard, with a proper serial number, the server will not return the PIN: the malicious user is not able to pass the local biometric match. (Inside the smartcard there is not an Attribute Certificate containing the face biometric template issued by the Server's CA).

If the attacker has stolen the user's smartcard and the right dlls are installed:
- He could try to change the template stored in the smart card: he cannot do this, because the biometric template is contained in an Attribute Certificate signed by the Server CA.
- Even if a biometric template has been previously sniffed, it is not possible to inject it within the Verification Module: before the verification module acquires a template from the biometric device, it verifies that the Biometric Service Provider dll is the trusted one via digital signature (no more replay attacks).

5 Feasible Attacks

The possible attacks to the implemented protocol concern how the PIN is transmitted in the final step, and the malicious host threats [8] in a mobile agent framework. In the first case, the PIN could be sniffed if the channel between the SC reader and the host is not protected. If the attacker has stolen the user's SC, he could bypass all the protocol and use only the manufacturer PKCS#11 library. We assume that this is not possible because we rely on the smartcard producer's Crypto Module (a trusted path between the SC and the host should be employed).

The other pway to attack this protocol is modifying the MBSMAF framework and/or the Java Virtual Machine on the client's workstation. This is a common problem in the Mobile Agent Systems field [8]. In our case, an attacker succeeds if he is able to alter the Mobile Agent's return value with a positive result even if the checks on the client are negative. This problem can be solved by employing a mutual authentication protocol between the client/server MBSMAF environments, using a secure trusted hardware: it would be used to store the framework private key and to check the validity of the agent runtime environment.

6 Conclusions

In this paper we have proposed a mobile agent based framework for implementing a smart card based multifactor biometric fusion (face biometric with password) for authenticating users. In relies on the actual match being delegated to a module of the card after an authentication performed by a mobile agent coming from a secure server.

The framework we have presented is currently being implemented using open source Java APIs, and through two de-facto standards such as PKCS#11 and BioAPI, respectively used. The use of these standards lead to an implementation where any smartcard and any biometric device can be used. Potential future works will concern addressing the issues described in Section 5.1 (local PIN sniffing and malicious host attack), and adapting the protocol in other areas where the entity performing the biometric match is not trusted.

References

1. Bechelli, L., Bistarelli, S., Martinelli, F., Petrocchi, M., Vaccarelli, A.: Integrating biometric techniques with electronic signature for remote authentication. ERCIM News (49) (2002)
2. Bechelli, L., Bistarelli, S., Vaccarelli, A.: Biometrics authentication with smart- card. Technical Report 08-2002, CNR, IIT, Pisa (2002)
3. Bella, G., Bistarelli, S., Martinelli, F.: Biometrics to Enhance Smartcard Security (Simulating MOC using TOC). In: Proc. 11th International Workshop on Security Protocols, Cambridge, England, April 2-4, 2003 (2003)
4. BioAPI Consortium. BioAPI Specification Version 1.1, `http://www.bioapi.org`
5. Authenticode, `http://msdn.microsoft.com/workshop/security/authcode/signing.asp`
6. Roth, V., Jalali, M.: Concepts and Architecture of a Security-centric Mobile Agent Server. In: IEEE Proceedings of 5th International Symposium on Autonomous Decentralized Systems (ISADS 2001), Dallas, Texas, March 2001, pp. 435–442 (2001)
7. RSA Laboratories. PKCS#11-cryptographic token interface standard
8. Bierman, E., Cloete, E.: Classification of Malicious Host Threats in Mobile Agent Computing. In: Proceedings of SAICSIT 2002, pp. 141–148 (2002)
9. Java Native Interface, `http://java.sun.com/docs/books/jni/index.html`
10. How To Share Data Between Different Mappings of a DLL. Microsoft KB 125677
11. Waldmann, U., Scheuermann, D., Eckert, C.: Protected transmission of biometric user authentication data for oncard-matching. In: Handschuh, H., Hasan, M.A. (eds.) SAC 2004. LNCS, vol. 3357, pp. 425–430. Springer, Heidelberg (2004)
12. Cyberflex E-gate, `http://www.axalto.com`

Classification and Retrieval through Semantic Kernels

Claudia d'Amato, Nicola Fanizzi, and Floriana Esposito

Dipartimento di Informatica, Università degli Studi di Bari
{claudia.damato,fanizzi,esposito}@di.uniba.it

Abstract. This work proposes a family of language-independent semantic kernel functions defined for individuals in an ontology. This allows exploiting well-founded kernel methods for several mining applications related to OWL knowledge bases. Namely, our method integrates the novel kernel functions with a *support vector machine* that can be set up to work with these representations. In particular, we present preliminary experiments where statistical classifiers are induced to perform the tasks of instance classification and retrieval.

1   Ontology Mining

Many application domains require operating on large repositories made up of structured data. In the field of the Semantic Web (SW) [2], knowledge intensive manipulations on complex relational descriptions to be performed by machines are foreseen. In this context, expressive languages borrowed from *Description Logics* (DLs) [1] have been adopted for representing ontological knowledge. Unfortunately, machine learning through logic-based methods is inherently intractable in multi-relational settings [13], unless language bias is imposed to constrain the representation, for the sake of tractability, only very simple DLs have been considered [4, 15, 14]. On the other hand, kernel methods [16] represent a family of statistical learning algorithms, including the *support vector machines* (SVMs), that have been effectively applied to a variety of tasks, recently also in domains that typically require structured representations [10, 11]. They can be very efficient since they map, by means of a kernel function, the original feature space into a high-dimensional space, where the learning task is simplified. Such a mapping is not explicitly performed (*kernel trick*): the usage of a positive definite kernel function (i.e. a *valid* kernel) ensures that the embedding into a new space exists and the kernel function corresponds to the inner product in this space [16]. Two components of kernel methods have to be distinguished: the kernel machine and the kernel function. The kernel machine encapsulates the learning task, the kernel function encapsulates the hypothesis language. The same kernel machine can be applied to several knowledge representations, provided a suitable kernel function for each of them.

As argued in [3], most of the research in the SW context address the problem of learning *for* the SW, less attention has been given to the problem of learning *from* the SW data. Looking in this direction, kernel methods can be adopted for several tasks such as classification, clustering and ranking of individuals. In this work, we exploit a kernel method, specifically a SVM, for performing inductive concept retrieval and query answering. To accomplish this goal, a kernel function for DL representation (henceforth

I. Lovrek, R.J. Howlett, and L.C. Jain (Eds.): KES 2008, Part III, LNAI 5179, pp. 252–259, 2008.

DL-kernel) is proposed. It encodes a notion of similarity of individuals, by exploiting only semantic aspects of the reference representation. Currently, concept retrieval and query answering are performed using merely deductive procedures which easily fail in case of (partially) inconsistent or incomplete knowledge, that can happen when data comes from heterogeneous and distributed resources. We show how the proposed method performs comparably well w.r.t. a standard deductive reasoner, allowing the suggestion of new knowledge that was not previously logically derivable.

In the next section basics of kernel functions for complex representations will be analyzed. In Sect. 3 the *DL-kernel* will be proposed while in Sect. 4 the inductive concept retrieval problem and method will be formally defined. Initial experimental evaluation of the method is presented in in Sect. 5.

2   Kernel Functions

In kernel methods, the learning algorithm (inductive bias) and the choice of the kernel function (language bias) are almost completely independent. The kernel machine encapsulates the learning task, the kernel function encapsulates the hypothesis language. Specifically, the kernel function maps the original feature space of the considered data set into a high-dimensional space, where the learning task is simplified. Such a mapping is not explicitly performed (*kernel trick*): the usage of a positive definite kernel function (i.e. a *valid* kernel) ensures that the embedding into a new space exists, so that the kernel function corresponds to the inner product in this space [16]. In this way, an efficient algorithm for attribute-value instance spaces can be converted into one suitable for structured spaces (e.g. trees, graphs) by merely replacing the kernel function.

Kernels functions are endowed with the closure property w.r.t. many operations, one of them is the convolution [12]: kernels can deal with compounds by decomposing them into their parts, provided that valid kernels have already been defined for them.

$$k_{\mathrm{conv}}(x, y) = \sum_{\substack{\overline{x} \in R^{-1}(x) \\ \overline{y} \in R^{-1}(y)}} \prod_{i=1}^{D} k_i(\overline{x}_i, \overline{y}_i) \tag{1}$$

where R is a composition relationship building a single compound out of D simpler objects, each from a space that is already endowed with a valid kernel. The choice of the function R is a non-trivial task which may depend on the particular application.

On the ground of this property several kernel functions have been defined: for string representations, trees, graphs and other discrete structures [10]. Particularly, in [11] it is shown how to define generic kernels based on type construction, where types are defined in a declarative way. While these kernels are defined as depending on specific structures, a more flexible method is to build kernels parametrized on a uniform representation. Cumby and Roth [5] propose the syntax-driven definition of kernels based on a simple DL representation, the *Feature Description Language*. They show that the feature space blow-up is mitigated by the adoption of efficiently computable kernels. Kernel functions for structured data, parametrized on a description language, allow for the employment of algorithms such as SVM's that can simulate feature generation.

These functions transform the initial representation of the instances into the related active features, thus allowing learning the classifier directly from the structured data.

A notion of kernel for the SW representations has first been proposed in [6]. The kernel function is defined for comparing $\mathcal{ALC}$ concept definitions based on the structural similarity of the AND-OR trees corresponding to the normal form of the input concepts. This kernel is not only structural, since it ultimately relies on the semantic similarity of the primitive concepts on the leaves assessed by comparing their extensions through a set kernel. Moreover, the kernel is actually applied to couples of individuals, after having lifted them to the concept level through realization operators (actually by means of approximations of the most specific concept, see [1]). Since these concepts are constructed on the ground of the same ABox assertions it is very likely that structural and semantic similarity tend to coincide. A more recent definition of kernel functions for individuals in the context of the standard SW representations is reported in [3]. Here the authors define a set of kernels for individuals and for the various types of assertions in the ABox (on concepts, datatype properties, object properties). It is not clear how to integrate such separate building blocks; the preliminary evaluation on specific classification problems regarded single kernels or simple additive combinations.

3   A Family of Kernels for Individuals

In the following we adopt the terminology employed in Description Logics (see [1] for details). We report the basics that are needed for the material in this paper.

Consider a triple $\langle N_C, N_R, N_I \rangle$ made up respectively, of a set of concept names N_C, a set of role names N_R and a set of individual names N_I. An interpretation $\mathcal{I} = (\Delta^{\mathcal{I}}, \cdot^{\mathcal{I}})$ maps (via $\cdot^{\mathcal{I}}$) such names to the corresponding element subsets, binary relations, and objects of the domain $\Delta^{\mathcal{I}}$. A DL language gives the rules for building more complex concept descriptions based on these building blocks by extending the syntax with specific constructs and providing the related interpretation. The *Open World Assumption* (OWA) is made in the underlying semantics. A *knowledge base* $\mathcal{K} = \langle \mathcal{T}, \mathcal{A} \rangle$ contains a *TBox* $\mathcal{T}$ and an *ABox* $\mathcal{A}$. $\mathcal{T}$ is the set of terminological axioms o concept descriptions $C \sqsubseteq D$, meaning $C^{\mathcal{I}} \subseteq D^{\mathcal{I}}$, where C is the concept name and D is its description. $\mathcal{A}$ contains assertions on the world state, e.g. $C(a)$ and $R(a, b)$, meaning that $a^{\mathcal{I}} \in C^{\mathcal{I}}$ and $(a^{\mathcal{I}}, b^{\mathcal{I}}) \in R^{\mathcal{I}}$. Subsumption w.r.t. the models of the KB is the most important inference service. Yet in our case we will exploit *instance checking*, that amounts to deciding whether an individual is an instance of a concept [1].

Grounded on [11], a family of valid kernels for the space X of $\mathcal{ALC}$ descriptions has been proposed [6]. In that case the kernel was defined for pairs of concepts descriptions, then the individuals had to be lifted to the concept level before computation (exploiting approximations of the respective most specific concepts w.r.t. the ABox). The main limitation of the kernel is represented by the dependency on the DL language. In order to overcome this limitation, we propose a set of kernels that can be applied directly to individuals, based on ideas exploited for a family of inductive distance measures [9]. The rationale for this kernels is that similarity between individuals is decomposed along with the similarity w.r.t. each concept in a given committee of features (class definitions). Two individuals are maximally similar w.r.t. a given concept F_i if they exhibit the same behavior, i.e. both are instances of the concept or of its negation. Conversely, the minimal similarity holds when they belong to opposite concepts. Because of the OWA, sometimes a reasoner cannot assess the concept-membership, hence we assign an intermediate value to reflect such uncertainty. The kernel function is formally defined in the following:

Definition 3.1 (family of kernels). *Let* $\mathcal{K} = \langle \mathcal{T}, \mathcal{A} \rangle$ *be a KB. Given a set of concept descriptions* $\mathsf{F} = \{F_1, F_2, \ldots, F_m\}$, *a family of kernel functions* $k_p^{\mathsf{F}} : \mathsf{Ind}(\mathcal{A}) \times \mathsf{Ind}(\mathcal{A}) \mapsto [0, 1]$ *is defined as follows:*

$$\forall a, b \in \mathsf{Ind}(\mathcal{A}) \quad k_p^{\mathsf{F}}(a, b) := \frac{1}{|\mathsf{F}|} \left[\sum_{i=1}^{|\mathsf{F}|} \mid \sigma_i(a, b) \mid^p \right]^{1/p}$$

where $p > 0$ *and* $\forall i \in \{1, \ldots, m\}$ *the* similarity *function* σ_i *is defined:* $\forall a, b \in \mathsf{Ind}(\mathcal{A})$

$$\sigma_i(a, b) = \begin{cases} 1 & (F_i(a) \in \mathcal{A} \wedge F_i(b) \in \mathcal{A}) \vee (\neg F_i(a) \in \mathcal{A} \wedge \neg F_i(b) \in \mathcal{A}) \\ 0 & (F_i(a) \in \mathcal{A} \wedge \neg F_i(b) \in \mathcal{A}) \vee (\neg F_i(a) \in \mathcal{A} \wedge F_i(b) \in \mathcal{A}) \\ \frac{1}{2} & otherwise \end{cases}$$

Note that the kernel functions can be model-theoretically defined by substituting, in the formulation above, the following expression $\mathcal{K} \models F_i(a)$ to $F_i(a) \in \mathcal{A}$ (resp. $\mathcal{K} \models \neg F_i(a)$) to each belongingness expression $F_i(a) \in \mathcal{A}$ (resp. $\neg F_i(a) \in \mathcal{A}$), both for a and b.

Instance-checking is to be employed for assessing the value of the simple similarity functions. This is known to be computationally expensive (also depending on the specific DL language of choice). Alternatively, especially for largely populated ontologies, a simple look-up may be sufficient, as suggested by the formal definition of the σ_i functions. If it is required that $k(a, b) = 0$ even though the selected features are not able to distinguish the individuals, one might make the *unique names assumption* on all individuals occurring in the ABox $\mathcal{A}$, and employ a special feature based on equality: $\sigma_0(a, b) = 1$ iff $a = b$ (and 0 otherwise). Alternatively, equivalence classes might be considered instead of individuals.

The most important property of a kernel function is its validity (it must correspond to a dot product in a certain embedding space).

Proposition 3.1 (validity). *Given an integer* $p > 0$ *and a committee of features* F, *the function* k_p^{F} *is a valid kernel.*

This result can be assessed by proving the function k_p^{F} definite-positive. Alternatively it is easier to prove the property by showing that the function can be obtained by composing simpler valid kernels through operations that guarantee the closure w.r.t. this property [12]. Specifically, since the various σ_i functions ($i = 1, \ldots, n$) correspond to a *matching kernel*, the property follows from the closure w.r.t. sum, multiplication by a constant and kernel multiplication.

It is also worthwhile to note that this is indeed a family of kernels parametrized on the choice of features. As for the semantic pseudo-metric that inspired the kernel definition [9], a preliminary phase may concern finding an optimal choice of features. This may be carried out by means of randomized optimization procedures, similar to the developed for the pseudo-distance [8].

4   Inductive Classification and Retrieval through Kernel Methods

SVMs are classifiers based on kernel functions that, exploiting a kernel function, map the training data into a higher dimensional feature space where they can be classified by means of a linear classifier. This is done by constructing a separating hyperplane with the maximum margin in the new feature space, which yields a nonlinear decision boundary in the input space. By the use of a kernel function, the separating hyperplane is computed without explicitly carrying out the mapping into the feature space. A SVM, as any other kernel method, can be applied to whatever knowledge representation, provided a valid kernel function suitable for the chosen representation. Given the *DL-kernel* defined in Sect. 3, we use the SVM to solve the following problem:

Definition 4.1 (classification problem). *Let* $\mathcal{K} = \langle \mathcal{T}, \mathcal{A} \rangle$ *be a KB, let* $\mathsf{Ind}(\mathcal{A})$ *be the set of all individuals in* $\mathcal{A}$ *and let* $C = \{C_1, \ldots, C_s\}$ *be the set of all concepts (both primitive and defined) in* $\mathcal{T}$. *The* classification problem *to solve is:* **given** *an individual* $a \in \mathsf{Ind}(\mathcal{A})$, **determine** *the set of concepts* $\{C_1, \ldots, C_t\} \subseteq C$ *to which* a *belongs to.*

In the general setting of SVMs, the classes, w.r.t. which the classification is performed, are disjoint. This is not generally verified in the SW context, where an individual can be instance of more than one concept. To solve this problem, a different answering procedure is proposed. The multi-class classification problem is decomposed into smaller binary classification problems (one per class). Specifically, given an individual in the KB, instead of returns the set of concepts to which it belongs to, it is classified w.r.t. to each concept, namely for each concept, the classifier assesses if the individual is instance or not. Therefore, a simple binary value set ($V = \{-1, +1\}$) can be employed, where $(+1)$ indicates that an individual x_i is instance of the concept C_j; (-1) indicates that x_i is not instance of C_j. Anyway, this is not enough. Indeed, in the general setting, an implicit *Closed World assumption* (CWA) is made, while in the SW context, the *Open World Assumption* (OWA) is usually adopted. To deal with the OWA, the absence of information on whether a certain individual x_i belongs to the extension of the concept C_j should not be interpreted negatively, as seen before, rather, it should count as neutral information. Thus, we consider another value set: $V = \{+1, -1, 0\}$, where the three values denote, respectively, assertion occurrence ($C_j(x_i) \in \mathcal{A}$), occurrence of the opposite assertion ($\neg C_j(x) \in \mathcal{A}$) and assertion absence in $\mathcal{A}$. Hence, given a query instance x_q, for every concept $C_j \in C$, the classifier will return $+1$ if x_q is an instance of C_j, -1 if x_q is an instance of $\neg C_j$, and 0 otherwise. The classification is performed on the ground of a set of training examples from which such information can be derived.

Considered a knowledge base $\mathcal{K} = \langle \mathcal{T}, \mathcal{A} \rangle$, all the individuals in $\mathcal{A}$ can be classified w.r.t. one or more concepts in $\mathcal{T}$, thus performing the concept retrieval inductively. In the same way, the classifier can be employed for solving a query answering task, by determining the extension of a new concept built from concepts and roles in $\mathcal{T}$. As it will be shown in the next section, the classifier behavior, in performing concept retrieval, is comparable with the one of a standard reasoner. Moreover, the classifier may be able to induce new knowledge that is not logically derivable.

Table 1. Ontologies employed in the experiments

ontology	DL	#concepts	#obj. prop	#data prop	#individuals
S.-W.-M.	$\mathcal{ALCOF}(D)$	19	9	1	115
SCIENCE	$\mathcal{ALCIF}(D)$	74	70	40	331
NTN	$\mathcal{SHIF}(D)$	47	27	8	676
WINES	$\mathcal{ALCIO}(D)$	112	9	10	149

5   Experimental Evaluation

For performing the classification problem defined in Sect. 4 and for assessing the validity of the *DL-kernel* (see Def. 3.1), a SVM from the LIBSVM library[1] has been considered. The instance classification has been performed on four different OWL ontologies: SURFACE-WATER-MODEL, NEWTESTAMENTNAMES, SCIENCE, and WINES from the Protégé library[2]. Details about such ontologies are reported in Tab. 1. The classification method was applied to all the individuals in each ontology; namely, for each ontology, the individuals were checked to assess if they were instances of the concepts in the ontology through the SVM and the *DL-kernel*[3] embedded in it. A similar experimental setting has been considered in [3] with an exemplified version of the GALEN Upper Ontology[4]. The ontology has been randomly populated and only seven concepts have been considered while no roles have been taken into account[5]. Differently from this case, we did not apply any changes on the considered ontologies.

The SVM-based classifier performance was evaluated by comparing its responses to those returned by a standard reasoner[6] used as baseline. The experiment has been performed by adopting the ten-fold cross validation procedure. For each concept in the ontology, the following parameters have been measured for the evaluation: *match rate*: number of cases of individuals that got exactly the same classification by both classifiers with respect to the overall number of individuals; *omission error rate*: amount of unlabeled individuals (namely the method returns 0 as classification results) while they were to be classified as instances of the considered concept; *commission error rate*: amount of individuals labeled as instances of a concept, while they (logically) belong to the negation of that concept or vice-versa; *induction rate*: amount of individuals that were found to belong to a concept or its negation, while this information is not logically derivable from the knowledge base. The average rates obtained over all the concepts in each ontology are reported in Tab. 2, jointly with their range.

Looking at the table, it is important to note that, for every ontology, the commission error is null. This means that the classifier did not make critical mistakes, i.e. cases when an individual is deemed as an instance of a concept while it really is an instance

---

[1] http://www.csie.ntu.edu.tw/~cjlin/libsvm

[2] See the webpage: http://protege.stanford.edu/plugins/owl/owl-library

[3] The feature set for computing the *DL-kernel* was made by all concepts in the considered ontology (see Def. 3.1).

[4] http://www.cs.man.ac.uk/ rector/ontologies/simple-top-bio/

[5] Due to the lack of information for replicating the ontology used in [3], a comparative experiment with the proposed kernel framework cannot be performed.

[6] PELLET: http://pellet.owldl.com

Table 2. Results (average and range) of the experiments using *DL-kernel*

ONTOLOGY		match rate	induction rate	omis. err. rate	comm. err. rate
WINES	avg.	**0.952**	**0.006**	**0.042**	**0.000**
	range	0.585 - 0.993	0.000 - 0.415	0.000 - 0.343	0.000 - 0.00
SCIENCE	avg.	**0.971**	**0.018**	**0.011**	**0.000**
	range	0.947 - 1.000	0.000 - 0.053	0.000 - 0.021	0.000 - 0.000
S.-W.-M.	avg.	**0.959**	**0.000**	**0.041**	**0.000**
	range	0.836 - 1.000	0.000 - 0.000	0.000 - 0.164	0.000 - 0.000
N.T.N.	avg.	**0.982**	**0.002**	**0.016**	**0.000**
	range	1.000 - 0.932	0.000 - 0.055	0.000 - 0.068	0.000 - 0.000

Table 3. Results (average and range) of the experiments with $\mathcal{ALC}$ kernel ($\lambda = 1$)

ONTOLOGY		match rate	induction rate	omis. err. rate	comm. err. rate
WINES	avg.	**0.956**	**0.004**	**0.040**	**0.000**
	range	0.650 - 1.000	0.000 - 0.270	0.010 - 0.340	0.000 - 0.000
SCIENCE	avg.	**0.942**	**0.007**	**0.051**	**0.000**
	range	0.800 - 1.00	0.000 - 0.040	0.000 - 0.200	0.000 - 0.000
S.-W.-M.	avg.	**0.871**	**0.067**	**0.062**	**0.000**
	range	0.570 - 0.980	0.000 - 0.420	0.000 - 0.400	0.000 - 0.000
N.T.N.	avg.	**0.925**	**0.026**	**0.048**	**0.001**
	range	0.660 - 0.990	0.000 - 0.320	0.000 - 0.220	0.000 - 0.030

of the negation of that concept. Furthermore, a very high match rate is registered for every ontology. Particularly, looking at Tab. 2, it can be observed that the match rate increases with the increase of the number of individuals in the considered ontology. This is because, being the SVM a statistical method, its performance improves with the augmentation of the set of the available examples. Almost always the SVM-based classifier is able to induce new knowledge. However, a conservative behavior has been also registered, indeed the omission error rate is not null (even if it is very close to 0). To decrease the tendency to a conservative behavior of the method, a threshold could be introduced for the consideration of the "unknown" (labeled with 0) training examples.

We have compared the results obtained by performing the inductive concept retrieval exploiting the *DL-kernel*, with the one obtained applying the SVM-based classifier jointly with the $\mathcal{ALC}$ kernel (see [6, 7]). The outcomes of the second set of experiments are reported in Tab. 3. By comparing Tab. 2 and Tab. 3 it is possible to note that *DL-kernel* improves both match rate and omission rate with respect to $\mathcal{ALC}$ kernel. Consequently a decrease of the induction rate is observed. The commission rate for the $\mathcal{ALC}$ kernel is almost null as for the *DL-kernel*.

6   Conclusions

We have defined a novel family of semantic kernel functions for individuals in the context of populated ontologies based on their behavior w.r.t. a number of features (concept definitions). The kernels are language-independent (they simply require instance-checking or ABox look-up) and can be easily integrated with a kernel machine (a SVM in our case) for performing a broad spectrum of activities. In this paper we focused on the application to statistical classification of individuals in an ontology. The resulting classifier has been used to perform instance classification in an inductive way which may be more efficient and noise-tolerant w.r.t. the standard deductive procedures. It has been experimentally shown that its performance is not only comparable to the one of a standard reasoner, but the classifier is also able to induce new knowledge, which is not logically derivable. Particularly, an increase in prediction accuracy was observed when the instances are homogeneously spread, as expected from statistical methods. The realized classifier can be exploited for predicting/suggesting missing information about individuals, thus completing large ontologies.

References

[1] Baader, F., Calvanese, D., McGuinness, D., Nardi, D., Patel-Schneider, P. (eds.): The Description Logic Handbook. Cambridge University Press, Cambridge (2003)

[2] Berners-Lee, T., Hendler, J., Lassila, O.: The semantic web. Scientific American 284(5), 34–43 (2001)

[3] Bloehdorn, S., Sure, Y.: Kernel methods for mining instance data in ontologies. In: Aberer, K., Choi, K.-S., Noy, N., Allemang, D., Lee, K.-I., Nixon, L., Golbeck, J., Mika, P., Maynard, D., Mizoguchi, R., Schreiber, G., Cudré-Mauroux, P. (eds.) ISWC 2007. LNCS, vol. 4825, pp. 58–71. Springer, Heidelberg (2007)

[4] Cohen, W.W., Hirsh, H.: Learning the CLASSIC description logic. In: Torasso, P., et al. (eds.) Proc. of the 4th Int. Conf. on the Principles of Knowledge Representation and Reasoning, pp. 121–133. Morgan Kaufmann, San Francisco (1994)

[5] Cumby, C.M., Roth, D.: On kernel methods for relational learning. In: Fawcett, T., Mishra, N. (eds.) Proceedings of the 20th International Conference on Machine Learning, ICML 2003, pp. 107–114. AAAI Press, Menlo Park (2003)

[6] Fanizzi, N., d'Amato, C.: A declarative kernel for $\mathcal{ALC}$ concept descriptions. In: Esposito, F., et al. (eds.) ISMIS 2006. LNCS (LNAI), vol. 4203, pp. 322–331. Springer, Heidelberg (2006)

[7] Fanizzi, N., d'Amato, C.: Inductive concept retrieval and query answering with semantic knowledge bases through kernel methods. In: Esposito, F., et al. (eds.) KES 2007, Part I. LNCS (LNAI), vol. 4692, pp. 148–155. Springer, Heidelberg (2007)

[8] Fanizzi, N., d'Amato, C., Esposito, F.: Evolutionary conceptual clustering of semantically annotated resources. In: Proc. of the Int. Conf. on Semantic Computing. IEEE, Los Alamitos (2007)

[9] Fanizzi, N., d'Amato, C., Esposito, F.: Induction of optimal semi-distances for individuals based on feature sets. In: Calvanese, D., et al. (eds.) Working Notes of the 20th Int. Description Logics Workshop. CEUR Workshop Proceedings, vol. 250 (2007)

[10] Gärtner, T.: A survey of kernels for structured data. SIGKDD Explorations 5(1), 49–58 (2003)

[11] Gärtner, T., Lloyd, J.W., Flach, P.A.: Kernels and distances for structured data. Machine Learning 57(3), 205–232 (2004)

[12] Haussler, D.: Convolution kernels on discrete structures. Technical Report UCSC-CRL-99-10, Department of Computer Science, University of California – Santa Cruz (1999)

[13] Haussler, D., Kearns, M.J., Schapire, R.E.: Bounds on the sample complexity of bayesian learning using information theory and the vc dimension. Machine Learning 14(1), 83–113 (1994)

[14] Iannone, L., Palmisano, I., Fanizzi, N.: An algorithm based on counterfactuals for concept learning in the semantic web. Applied Intelligence 26(2), 139–159 (2007)

[15] Lehmann, J.: Concept learning in description logics. Master's thesis, Dresden University of Technology (2006)

[16] Schölkopf, B., Smola, A.J.: Learning with Kernels. The MIT Press, Cambridge (2002)

Semantic Bookmarking and Search in the Earth Observation Domain

Francesca Fallucchi[1], Maria Teresa Pazienza[1],
Noemi Scarpato[1], Armando Stellato[1],
Luigi Fusco[2], and Veronica Guidetti[2]

[1] DISP, University of Tor Vergata,
Via del Politecnico 1, Rome, Italy
{fallucchi,pazienza,scarpato,stellato}@info.uniroma2.it
http://ai-nlp.info.uniroma2.it
[2] ESA/ESRIN, Frascati, Italy
{luigi.fusco,veronica.guidetti}@esa.int

Abstract. This paper describes the experience of introducing a service for semantic bookmarking and search in the Earth Observation (EO) domain. To perform the work reported such a service has been integrated and customized in the framework of the DILIGENT project which concluded in 2007. The service proposed is a web browser ontology based extension: at the application layer the interaction between DILIGENT end-users and this service takes place from within the DILIGENT web portal. The service described allows end users to capture and annotate on-demand the information stored in the DILIGENT infrastructure and to organize it according to the user's representation of the specific ontological domain. The effort spent meets today end users' requirements in the EO application domain, applying an innovative solution to make specific actors keep track of the information of interest in a personalized and easy way. End users are able to store, access and play only with the information of interest and under their own (ontological) perspective.

1 Introduction

Nowadays, images of our planet from orbit are acquired continuously; they have become powerful scientific tools to enable better understanding and improved management of the earth and its environment. Space taken Earth Observation (EO) images show the world through a wide-enough frame so that complete large-scale phenomena can be observed with great accuracy. EO provides objective coverage across both space and time. The same space-based sensor gathers data from sites across the world, including places too remote or otherwise inaccessible for ground-based data acquisition. In the long term, this monitoring of the earth's environment will enable a reliable assessment of the global impact of human activity and the likely future extent of climate change. Due to the flexibility (and complexity) of the instruments, it has to be noted that one single EO instrument supports many different domains (e.g. ocean and land application) and, at the same time, one application domain needs to access many different sources of information from space (EO) and from ground measurements.

I. Lovrek, R.J. Howlett, and L.C. Jain (Eds.): KES 2008, Part III, LNAI 5179, pp. 260–268, 2008.

Another important aspect to note is that different space and ground instruments are operated under the responsibility of different entities (e.g. space agencies, research institutions…), so that the interoperability of the different available data systems to generate data products and information to users is not at all a trivial task. Definitely, the semantic web technology can play a very important role to easy the various intrinsic interoperability problems for data discovery, access and use. As immediate example of the help that semantic technology can help, we refer to the handling of time and geographic location associated with each image dataset, which can be expressed in many different ways, units, relative to other events etc.

Furthermore, the user services which ensure data and information access, like those offered and shown through the ESA's EO Portal[1], have to deal with continuous streams of data coming from satellites, which need to be collected, organized and presented to their actual and potential consumers. Managing this data is not a trivial task, not only for the huge quantity of available material, but also considering the several, diverse scenarios in which its content can be used as well as the different categorization schemas which can be thought over it. So far, providers of geographical information have archived their knowledge in huge databases and found complex projections of their raw data for populating web sites and portals with rich and navigable descriptions of their content. Specific consumers' exigencies are instead been dealt, case by case, with the knowledge and expertise possessed by educated and trained personnel.

Modern web technologies, pushed by Web 2.0 and its leading objective of a read/write Web, and strong knowledge representation standards offered by the Semantic Web [2], offer now viable alternatives which could be used to fill the gap between data archiving and data publication, by allowing for the latter to being driven through proper, multi-modal representations of the raw data as offered by ontology based approaches. RDF/OWL marked-up versions of the acquired data could not only provide a layer which can be easily ported to the web, but may offer diverse perspectives on the same raw data which can thus be easily aggregated, selected and composed according to different needs and exigencies.

2 Context and Scenario

ESA/ESRIN reached today several years of world-wide experience in cooperating on ICT topics concerning the exploitation and integration of digital libraries and Grid technologies, putting the basement to exploit such technologies in earth science and EO domains. DILIGENT[2] is one of the European projects in which ESA/ESRIN participated during the past three years with the role of end-user and services/data provider, exposing requirements from the EO community to gain from innovative ICT based solutions.

The test-bed scenario lead in the context of DILIGENT based on concepts like virtual organization, digital library, user workspace, collection, compound service, report with the aim of creating on-demand ad hoc digital libraries and then aggregating

---

[1] http://www.eoportal.org/
[2] http://www.diligentproject.org/

pertained information into complex report documents (e.g. reports on the environment status), browsing, searching and managing private workspaces from within *virtual research environments*. During the project lifetime, together with the furthering of the activities and the development of the DILIGENT platform through the definition of its middleware *gCube*[3] and a pan-European Grid infrastructure, the need to exploit knowledge representation techniques, annotation and ontology management raised up in the context of the EO scenario.

To this end ESA/ESRIN required the support of University of Rome, Tor Vergata, which shown a certain experience in the development of ontology based service for web pages annotation. In the scope of DILIGENT such a joint effort moved towards two directions: a) to define a suitable ontology for demonstration purposes in the project and b) to integrate the ontology based service in gCube. The former point involved the definition of an ontology based on the ESA's EO Portal structure together with the evaluation of ontology publicly available in the EO domain, as the SWEET ontologies developed by NASA [12], while the latter saw the proposal and implementation of a technical solution as described in the next sections.

The main motivation which lead the project, and the ESA participation in particular, in this direction resides in the lack of a user-centric approach in the majority of web portals and access points for EO information and services: currently, people enter the EO portal and browse its content like in any standard web site; the portal offers a few search functionalities and its standard html pages are evidently produced automatically from templates + content regularly fed from a database. The EO user portal, which is being developed in DILIGENT, is based on slightly more recent technologies, essentially dynamic html, and features interaction modalities based on "content objects": they are moveable interactive knowledge units, generated upon searches submitted by the user, which can be inspected, aggregated and moved into sort of bookmark folders which are saved and accessed on a per-user basis. In both cases however, users could beneficiate of the underlying knowledge organization which completely drives the structure of both portals.

Unfortunately, in our present context, we were limited to research on a lightweight solution providing interesting features for potential users of the EO portal, without requiring heavy modification or introduction of new components inside the business logic of the systems behind the portal: we thus had to concentrate on some sort of customizable desktop solution, able to provide fast content collect&retrieve functionalities for browsed data, as well as maintaining a strong connection with the source where it has been extracted. This rich client would have dedicated functionalities thought (or customized) for the eoPortal, though not relying on any kind of preferential access to its content, but standard http interaction.

3 Approach

Given the above constraint, which required no substantial change in the existing system, we decided to adopt a completely client-side navigation mechanism: a browser-on-semantic-steroids, which, beneficiating of the expressive power of Semantic Web

---

[3] http://www.gcube-system.org/

W3C formalisms for providing both ontology editing and annotation/bookmarking functionalities, could at least support users in the process of collecting and organizing information observed through traditional navigation in the portal, implicitly maintaining references to its resources. Ontologies assume thus a two-fold role in this approach, on the one side, the domain ontology established by the user, on the other hand, a mere ontological translation of the original metadata defined inside the eoPortal. With this approach, and by developing the platform described above, users could depict their personal perspective on the interested data, implicitly defining filters, projections and data transformations, through the act of developing an own representation of their specific domain of interest. This representation would remain deeply coupled with the underlying standard eoPortal data, for search purposes.

Following these premises, we defined the following two working steps:

- the first requiring no intervention on the portal, with a client application just being "aware" of portal content, structure and navigation modalities, and allowing for collecting and organizing its content according to user needs.
- The second step involves minor operations on the portal side, by allowing the explicit exposure of metadata inside the same pages which are browsed through our enhanced client, which can then be easily harvested and reorganized into the client platform

Fulfilling the above steps required, first of all, to design the rich client platform, or to delineate its characteristics and then identify a proper candidate for providing the new eoPortal service.

3.1 The Client Platform

A recent work from the University of Tor Vergata, the Semantic Bookmarking platform Semantic Turkey [8], was elected by ESA/ESRIN as a potential application for achieving their objectives. Its main advantages with respect to other possible candidates were: the fact that it is based on a real Web Browser (Mozilla Firefox), thus guaranteeing robustness with respect to any sort of document which can be found on web pages, its native datamodel, based on the OWL language, and its flexible extension mechanism (based on the standard OSGi), which left the door open for customization to ESA needs.

Semantic Turkey (ST), in its original version, is a "Semantic Bookmarking" platform: an hybrid between a Web Browser, an Annotation tool and and Ontology Editor (see fig. 1). The expression Semantic Bookmarking was coined to indicate the process of annotating information from (web) documents, to acquire new knowledge and represent it through representation models. Its basic functionalities allow for:

1. capturing information from web pages – both by considering the page as a whole, as well as by selecting portions of their text – and annotating it with respect to a personal ontology.
2. editing the above ontology for classifying the annotated information and for better characterizing its interests according to its descriptive properties (attributes and relations).
3. navigating the structured information as an underlying semantic net which, populated with the many relationships which bind the annotated objects between them, eases the process of retrieving the knowledge which was buried by the past of time.

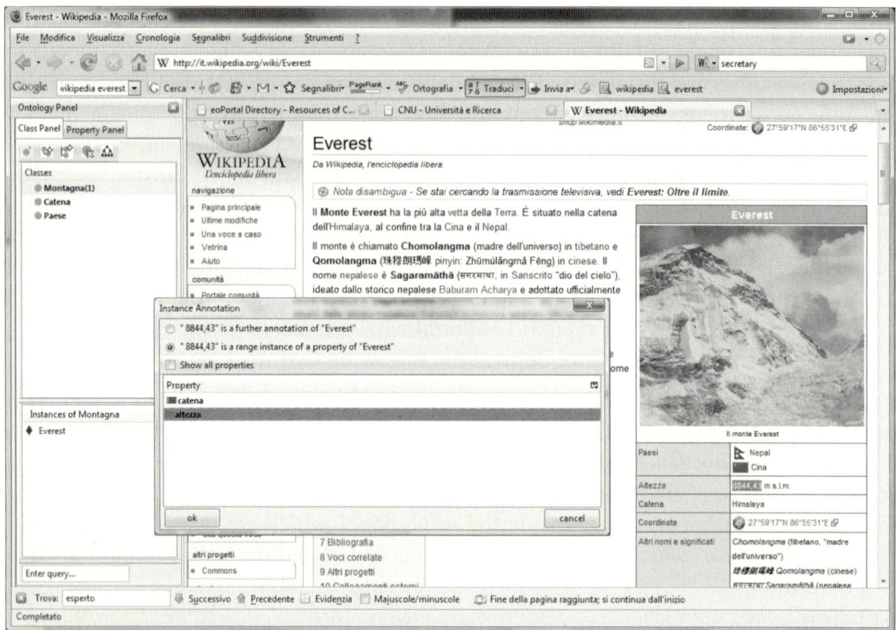

Fig. 1. Semantically bookmarking a reference ontology with objects from web pages

Its architectural and functional design make Semantic Turkey differentiate from similar, existing annotation tools, as it offers a lightweight structure, which completely exploits the infrastructure of the hosting web browser (with respect to, for example, the complex completely-web based interface of Piggy-Bank [9]) and which grants the user a good control over its personal domain representation (while traditional semantic annotation tools like Magpie [6] and Melita [4], are only able to import and adopt ontologies which have been defined elsewhere).

3.2 Preserving the Source Metadata

The data used to produce pages of the eoPortal is organized after a general-purpose model for resources management and documentation, reported and documented in [13]. This model has been implemented inside a Database Service (details in [11]) which hosts both the pure metadata entries as well their processed information which is directly accessed for populating the presentation pages of the eoPortal.

In Semantic Turkey there is a neat separation between the explicit knowledge managed by the user (*user layer*) and the one which guides system's behavior (*application layer*). The role of this latter layer is to keep track of all the information which is required by the application to perform its functionalities and is typically hidden form the user; in the original version of ST it includes the *Annotation Ontology*, a small ontology containing a set of concepts used to keep track of user annotations from the web (annotated web pages, selected textual entries, timestamps etc...). In order to provide an homogeneous data layer for our framework which is based on technologies and languages of the RDF family, we realized an almost straight 1-1 porting of the

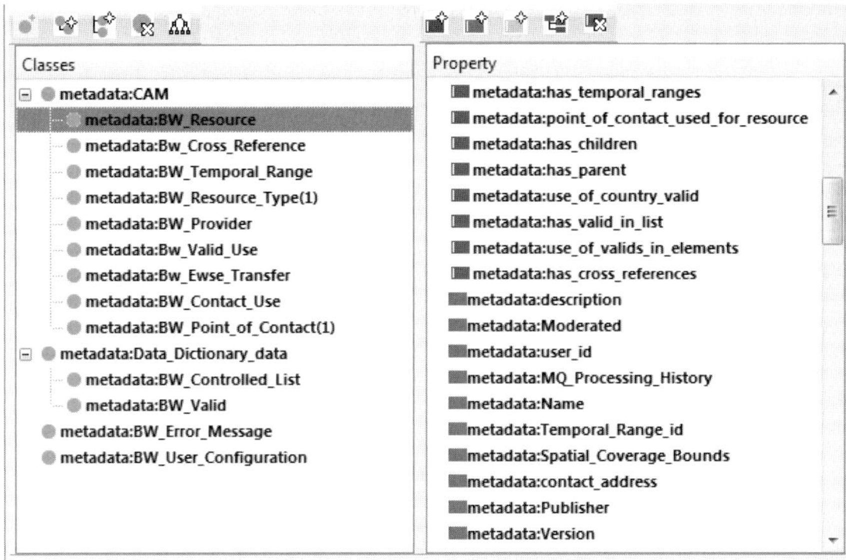

Fig. 2. Partial view of the ESA metadata in their straight ontological translation

eoPortal metadata model to an ontological representation (see fig. 2) and added it to the application layer of Semantic Turkey. This preserves consistent reference to the eoPortal native data model while guaranteeing interoperability with the specific ontologies that different users may adopt for their personal semantic bookmarks.

3.3   Automatic Data Harvesting: Exploiting RDFa

The second step of our effort on introducing Semantic Web technologies in the EO domain went in the direction of making data explicit (whenever possible and if available) inside the same traditional web pages used for presentation: we have thus considered the new possibilities offered by RDFa and Microformats .

Microformats (http://microformats.org/) emerged from the work of blog aggregator Technorati's developer community (http://www.technorati.com/), following the principles of the Global Multimedia Protocols Group (http://gmpg.org/): they are a proposed format for making recognizable data items (such as events, contact details or geographical locations) capable of automated processing by software, as well as directly readable by end-users. Microformats consist in specific HTML extensions intended for carrying commonly published semantics, such has contact information, events, reviews, episodic content, which go beyond the range of their hosting language. RDFa [1, 10] is the natural answer from W3C, associated to Semantic Web RDF-based models and technologies. The key aspect in both technologies is the same: embedding data inside web pages, which is exactly what we needed to achieve. The main difference (syntactic sugar apart) between the two resides in the vocabulary generation. While there exists one specific microformat for each of the addressed domains (contacts, events etc…), the only existing RDFa vocabulary is based on the

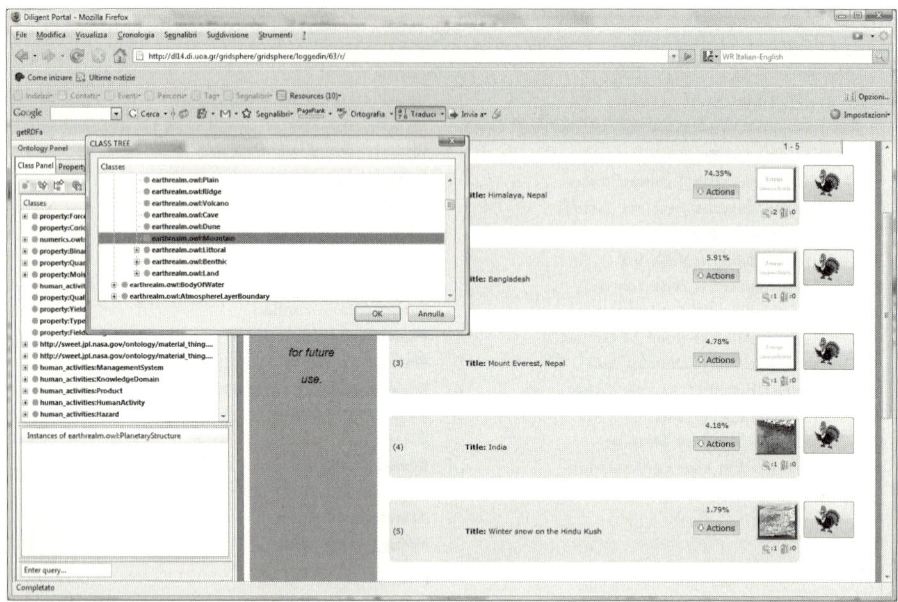

Fig. 3. recognizing, extracting and bookmarking RDFa data in the DILIGENT portal

RDF family of web languages (RDF, RDFS, OWL), thus extending existing HTML tags with attributes like: instanceof, property, etc.. and relying on specific content (the values of these attributes) driven by referenced ontologies.

We have thus developed an extension for Semantic Turkey specifically designed for recognizing information coded inside web pages according to these languages, and use their content both for populating the user ontology or, again, for submitting it to the semantic repository. Due to its W3C standardization and its flexibility with respect to microformats, we chose RDFa as the main language whose content can be recognized and extracted from web pages, and mapped to the user ontologies. For allowing a wider compatibility with available data on the web, we identified those microformats which could fall in the domain of Earth Observation, like: hCalendar (http://microformats.org/wiki/hcalendar), for describing time events, and geo (http://microformats.org/wiki/geo), for marking up WGS84 geographic coordinates (latitude and longitude) and enabled the conversion of their content through dedicated GRDDL (Gleaning Resource Descriptions from Dialects of Languages [5]) gleaners, to match the specification of standard W3C ontologies (hCalendar is by default translated to iCal, http://www.w3.org/2002/12/cal/ical, while geo is converted to WGS84 ontology [3]).

Fig. 3 shows a typical user session with RDFa data: the page in the browser comes from the DILIGENT portal, and contains data coded in RDFa after an OWL translation of the DILIGENT Metadata model. Semantic Turkey has recognized these data and added buttons (they are on the right, with the turkey symbol on top) in correspondence of their position in the web page. The user can thus press the buttons and get all the related RDFa (it generally involves multiple RDF triples describing a single object) collected in one click: he can then decide if the data will be imported as they are (i.e. referring the ontology of the RDFa data source), or if they will be projected upon the domain ontology (like the SWEET ontologies showed in figure) adopted by him. In the latter case, each object retrieved from the RDFa description can be projected towards a concept of the user domain ontology. 1-to-1 or directed 1-to-many mappings between concepts and properties of the ESA Metadata ontology and those of the domain ontology adopted by the user can be defined a-priori or they can be declared "inline" for each new harvested object. Property reification may also be used to map properties of one ontology towards classes and roles of the other one. This mapping capability has proved to be sufficient in most of the case, partially facilitated by the lack of specification on data format present in many currently available ontologies (such as the already cited SWEET ontologies). Actually, in our case study, predefined mappings proved to be helpful mostly for submitting info to the portal (though this aspect has not been investigated extensively), while their usefulness were limited to very few cases (for example, the already cited calendar or geospatial ontologies and microformats) when projecting harvested data from the web to the user ontology. The main reason for that resides in the ESA Metadata ontology, which contains few top-level concepts providing a perspective focused over mere data organization, with no evident relation to domain information, so that it was pretty easy to define in advance data projections from the rich user domain ontologies towards the general concepts of the ESA model, while relying on on-the-fly mappings for doing the contrary.

4 Conclusions and Future Work

This paper reports on experimenting with technologies coming from both worlds of Web 2.0 and the Semantic Web in the domain of earth science and EO specifically

Our research in this field will continue inside the new D4Science project (http://www.d4science.org/) - which started at the beginning of 2008 as a natural follow-up of DILIGENT - and will focus on improving ontology based content organization, and search and retrieval functionalities even on the service side.

We will probably extend current mapping capabilities of the framework, mainly a) to extend the range of achievable mappings, also including raw datatype transformations and more complex mapping patterns and b) to facilitate the user in harvesting several objects and in relating them to the adopted domain ontology, by exploiting results from pattern based ontology design [7] and applying them to our data projection facility. These additions (especially for the second purpose), will require methodological analysis of user activity on a range of realistic use-cases, considering different domain ontologies as a test basis, and keeping track of desired mappings which cannot easily be projected according to available models. User-friendliness, on the other side, has to be taken into account (though hardly measurable if not through user questionnaires), to avoid approaches proving to be potentially "more complete", but of unrealistic applicability in the hands of the average user.

Moreover a deeper attention will be paid to the exploitation of the ESA's EO Portal that aims to open the door to the world of EO resources, in its role of single access point to EO information and services provider.

Acknowledgements

The joint effort described in this paper has been partially funded by DILIGENT (Call Identifier: FP6-2003-IST-2 Project number: 004260).

References

1. Adida, B., Birbeck, M.: RDFa Primer. Retrieved from W3C (October 26, 2007), `http://www.w3.org/TR/xhtml-rdfa-primer/`
2. Berners-Lee, T., Hendler, J.A., Lassila, O.: The Semantic Web: A new form of Web content that is meaningful to computers will unleash a revolution of new possibilities. Scientific American 279(5), 34–43 (2001)
3. Brickley, D. (n.d.).: Basic RDF Geo Vocabulary (Retrieved December 10, 2007), `http://www.w3.org/2003/01/geo/`
4. Ciravegna, F., Dingli, A., Petrelli, D., Wilks, Y.: User-system cooperation in document annotation based on information extraction. In: Gómez-Pérez, A., Benjamins, V.R. (eds.) EKAW 2002. LNCS (LNAI), vol. 2473, Springer, Heidelberg (2002)
5. Connolly, D.: Gleaning Resource Descriptions from Dialects of Languages (GRDDL). Tratto da World Wide Web Consortium (September 11, 2007), `http://www.w3.org/TR/grddl/`
6. Dzbor, M., Motta, E., Domingue, J.B.: Opening Up Magpie via Semantic Services. In: McIlraith, S.A., Plexousakis, D., van Harmelen, F. (eds.) ISWC 2004. LNCS, vol. 3298, pp. 635–649. Springer, Heidelberg (2004)
7. Gangemi, A.: Ontology Design Patterns for Semantic Web Content. In: Gil, Y., Motta, E., Benjamins, V.R., Musen, M. (eds.) Proceedings of the Fourth International Semantic Web Conference, Galway, Ireland. Springer, Heidelberg (2005)
8. Griesi, D., Pazienza, M.T., Stellato, A.: Semantic Turkey - a Semantic Bookmarking tool (System Description). In: Franconi, E., Kifer, M., May, W. (eds.) ESWC 2007. LNCS, vol. 4519, Springer, Heidelberg (2007)
9. Huynh, D., Mazzocchi, S., Karger, D.: Piggy Bank: Experience the Semantic Web Inside Your Web Browser. In: Gil, Y., Motta, E., Benjamins, V.R., Musen, M.A. (eds.) ISWC 2005. LNCS, vol. 3729, pp. 413–430. Springer, Heidelberg (2005)
10. Lassila, O., Hendler, J.: Embracing Web 3.0. IEEE Internet Computing (May-June 2007)
11. Pedersen, G.S.: Database Design Document. Centre for Earth Observation (CEO). CEO programme Office (TBD copies) (1999)
12. Raskin, R.: Semantic web for earth and environmental terminology (2005), `http://sweet.jpl.nasa.gov/index.html`
13. SSSA, Recommendations on Metadata: Describing the data, services and information you have available! Ispra (VA), Italy: Strategy and Systems for Space Applications Unit (SSSA), Space Applications Institute (SAI) / Joint Research Centre, European Commission (2000)

VSB: The Visual Semantic Browser

Ismael Navas-Delgado, Amine Kerzazi, Othmane Chniber,
and José F. Aldana-Montes

Computer Languages and Computing Science Department, University of Málaga,
Málaga, Spain
`{ismael,kerzazi,chniber,jfam}@lcc.uma.es`

Abstract. The Semantic Web not only covers ontology definitions, but also
their relationships and instances. This paper describes an adaptable tool for the
visualization of all these Semantic Web elements. The tool includes a set of interfaces to enable the inclusion of different visualization tools as plug-ins. Thus,
it is divided into four views: ontology groups, ontology mappings, ontologies
and instances. Some algorithms are included but we are planning to develop
new ones to improve the tool capabilities (the current version is available at
http://khaos.uma.es/VSB where new plug-ins will be made public). The tool has
also been successfully applied to develop a graphical query interface that takes
advantage of the ontology and instance levels.

Keywords: Semantic Web, Semantic Browsing.

1 Introduction

The main aim of the Semantic Web is to try to solve some of the problems of the
Contents Web. Thus, the Semantic Web can be seen as an evolution of the current
Web, in which users will be able to deal with information based on its semantics.
Software applications will be able to interoperate more efficiently and produce better
results by means of reasoning.

The core of the Semantic Web is the use of ontologies to describe the concepts that
applications must tackle to perform daily tasks. In this way, ontologies are formal
descriptions of specific domains. These ontologies are intended to be shared knowledge in the target domain, and contribute to the production of new knowledge. There
have been several languages proposed to describe ontologies, but the most widely
used today is the W3C recommendation, OWL.

As the main goal of the Semantic Web is to enable application interoperability,
there are only a few proposals oriented to users. Protège (http://protege.stanford.edu/)
is the most used ontology editor, and it includes some basic visualization algorithms
[1][2][3][4][5]. Other ontology editors include more limited visualization characteristics (http://sisinflab.poliba.it/owled/ and http://oiled.man.ac.uk/) and there are no
proposals for visualizing groups of interrelated ontologies or ontologies with large
amounts of instances. Furthermore, few works provide the visualization of relationships (mappings) between pairs of ontologies. Such mappings are usually included in
matching tools [6] as an additional help for researchers searching for mappings.

I. Lovrek, R.J. Howlett, and L.C. Jain (Eds.): KES 2008, Part III, LNAI 5179, pp. 269–276, 2008.

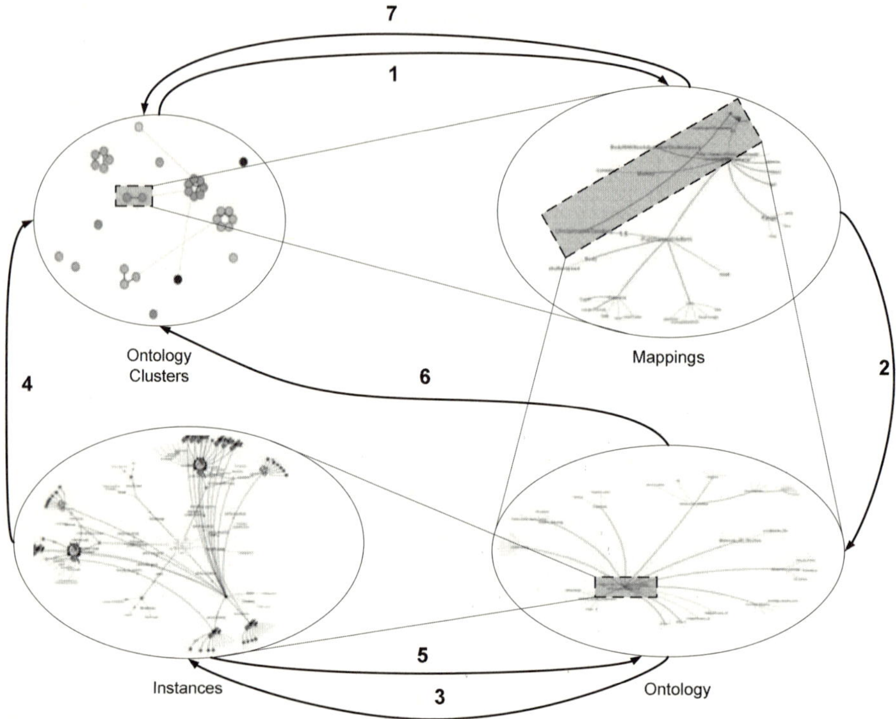

Fig. 1. Level Navigation. Each level provides a way of changing to the lower level. Besides, some additional navigation paths are provided, as shown in the "Work Flow".

This paper presents a tool (the Visual Semantic Browser, VSB) for visualizing different views of semantics as described previously. The main aim of the tool is to enable user interaction, to locate and use semantics usually only available to computers. It provides the necessary elements to enable the inclusion of new algorithms with little effort. Also, some algorithms have been adapted in this prototype for the visualization of ontology groups, mappings, ontologies and instances. Furthermore, we will describe the use of this tool to build another more specific application: a graphical query interface.

2   VSB Design

The proposal is based on the development of a tool which will enable the inclusion of new visualization algorithms that adapt better to specific application needs. This tool is divided into four levels that represent four types of applications (Figure 1). The requirements of these levels are described in following sections.

2.1   Level 4: Ontology Clusters

This level aims to visualize global relationships between ontologies. These relationships are the global similarity between ontologies and the main requirement is that the

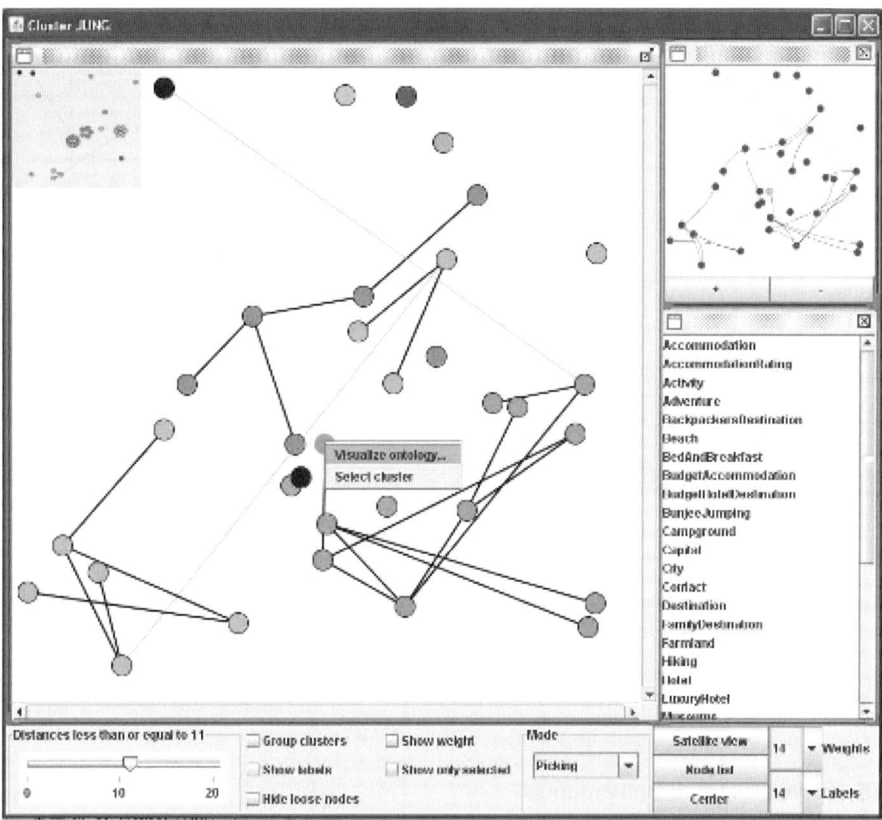

Fig. 2. Jung visualization for a cluster of ontologies, in which distances greater than 3 have been filtered. The cluster view on the right allows the user to know how the ontologies are grouped.

distance between the representations of the ontologies should be representative of these relationships. In this case the input is a list of ontologies and the distance between them. These distances could be provided by external tools such as the Semantic Field Tool (SemFiT) [7].

Level 4 has been implemented using JUNG (http://jung.sourceforge.net/) which is a Java API commonly used for visualization in Java applications. The input data for this algorithm is described in Pajek [8]. This format includes a list of nodes and the description of the arcs between nodes. The distribution of nodes has been modified from the original distribution technique of Fruchterman-Reingold [9] for another based on the attraction and repulse force [10] (based on the laws of physics). Thus, the arc length is proportional to the distance. This visualization also includes a set of options that allow the user to filter nodes depending on their distances and also to group those that are too near (see Figure 2).

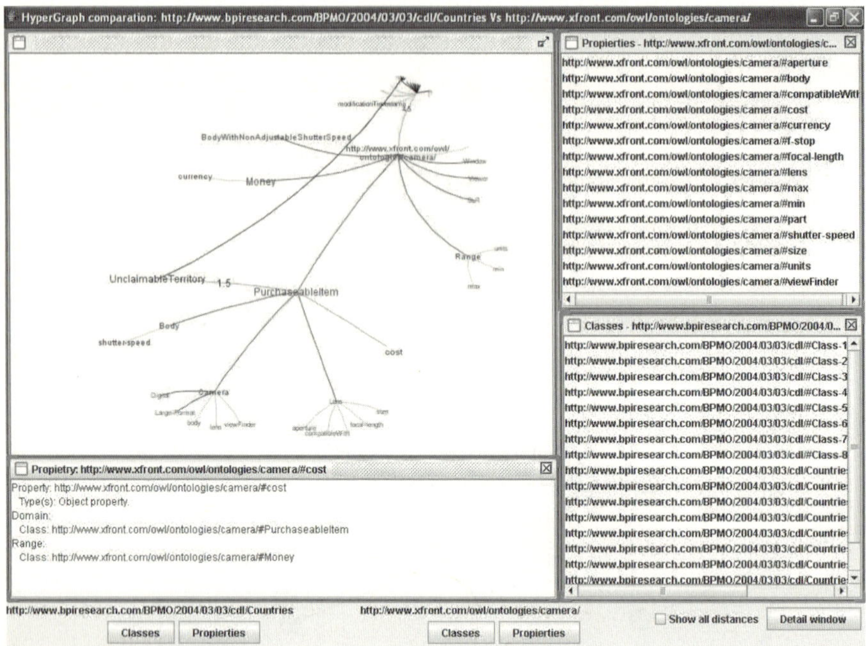

Fig. 3. Ontology mappings view. This view provides users with a way of knowing the similarities between pairs of ontologies.

2.2 Level 3: Ontology Mappings

This level shows the relationships between the internal concepts of two ontologies. Mappings between concepts show how similar two concepts are. The input of this level could be provided by external tools such as ontology matching tools [11][12][13][14][15][16][17][18][19]. The management of both ontologies requires parsing them from their specifications.

Levels 3, 2 and 1 have been implemented using the HyperGraph algorithm (http://hypergraph.sourceforge.net/), which provides hyperbolic views and is designed to represent hierarchies. The view of Level 3 is divided into two sub-graphs representing both ontologies (Figure 3). The inputs of this algorithm are two ontologies and a list of mappings. Arcs between concepts of both ontologies are shown, defining their similarity. Each ontology is represented in a different colour, and those nodes farthest from the focus are seen to be smaller. The lists of concepts shown on the right allow users to select nodes that are not in the centre of the view.

The representation used to deal with ontologies, their relationships and instances, aims to provide an efficient way of accessing them, because the visualization process should be as fluid as possible. As a starting point we have used Jena (http://jena. sourceforge.net/index.html) as the ontology parser for Levels 2 and 3. However, Jena has a slow response time for information retrieval and for this reason we have designed an internal representation that provides better response time values. This representation is based on the use of *HashMap* or *TreeMap*, and has been included as two implementations of a common interface (*Storage*). Thus, it is possible to change from one to the other if so required. A valid model will contain a set of structures for the different elements that we can find in an ontology: classes, properties, restrictions and instances.

From these levels (level 3, 2 and 1) the use of Jena to parse the ontologies also provides a way of using reasoning. The reasoning provided by Jena is limited but at least it allows complex concepts to be classified in the ontology hierarchy.

2.3 Level 2: Ontologies

This level should provide the visualization of the ontology concepts. One main requirement is that the visualization algorithm should provide a way of visualizing all the elements of an ontology: classes, properties and their relationships. The management of the ontology requires parsing it from its specifications. In the previous level Jena was also used to parse the ontologies. In addition, the user interface provides users with the choice of applying simple reasoning to classify the concepts.

The visualization with Hypergraph uses the fisheye view. Thus, concepts that are in the centre are seen clearly, while the others are shown smaller as they are nearer the edge. This kind of visualization is the most appropriate for large ontologies, because the user can decide which part of the ontology they wish to see.

2.4 Level 1: Ontology Instances

This level should provide efficient mechanisms to visualize large numbers of instances. Furthermore, personalization is possible, so non-expert users can define representative icons for each instance type in order to better understand the view provided.

The personalization has been done for each concept or property. Thus, all the instances of a concept could be represented with the same icon, helping users to understand the resulting graph.

2.5 Inter-level Navigation

The tool provides a way of communicating the different levels, so it is possible to select two ontologies from a cluster view to show their mappings (Step 1 in the work flow shown in Figure 1). Users can select to view one of these two ontologies, by clicking on the root node of the target ontology (Step 2 in the work flow shown in Figure 1).

If the ontology contains instances, a way of showing them is provided by clicking again on the root node (Step 3 in the work flow shown in Figure 1).

Some paths to return to previous visualizations are also provided. Thus, from the instance view we can directly return to the cluster view (Step 4 in the work flow shown in Figure 1) and to the ontology view (Step 5 in the work flow shown in Figure 1). Finally, users can return to the cluster view also from the ontology view (Step 6 in the work flow shown in Figure 1) and the mapping view (Step 7 in the work flow shown in Figure 1).

2.6 Application Example: Graphical Query Interface

This section describes a tool that allows editing conjunctive queries (it is useful to have a sound knowledge of the ontology). In addition, it can help us through the ontology visualization to construct the query automatically, by clicking on class and property nodes. Queries are expressed as conjunctive predicates, with three main categories: classes that are represented as *Class_name(Variable_name)*, datatype properties that link individuals to data values represented as *DatatypeProperty_name(Variable_name, value_or_variable)* and object properties that link individuals to individuals represented as *ObjectProperty_name(Variable_name1, Variable_name2)*. This tool automatically assigns variable names to both classes (generating a new variable name) and properties (automatically assigning the variable name that corresponds to its domain and/or range class). In the case where a property is added and the domain or range class does not exist, the missing class is added automatically and a new variable name assigned to it (see Figure 4).

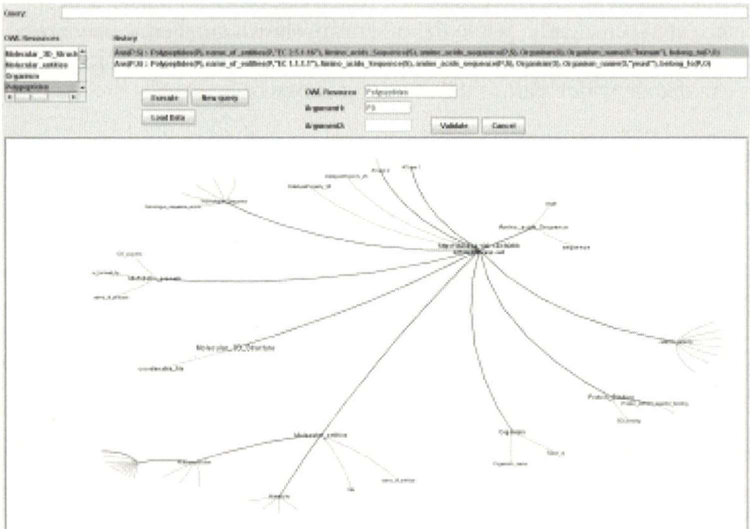

Fig. 4. Graphical Query Tool. This tool uses Hypergraph to show the ontology elements, and provides an algorithm to automatically build the query.

Once the query is executed, in a data integration system or a reasoner able to resolve conjunctive queries (searching for instances in the ontology), the result can be viewed as a graph (see Figure 5). The graphical view of instances can help users to find various kinds of semantic associations between two entities. Providing a graph with attributes would help users to recognize the type of node and therefore assist them in their analysis. The tool allows icons or names to represent the type of instance nodes.

In order to view a graph with a large number of nodes and edges, it is better to view smaller portions of it at a time. Our purpose is to provide a fisheye view with basic

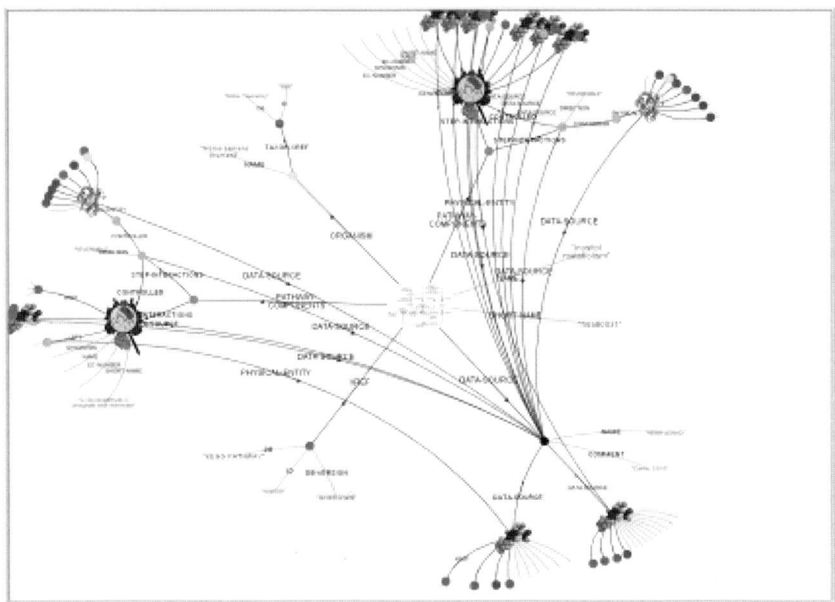

Fig. 5. Instances view. This view provides links between instances (through object properties) and the possibility of showing different icons for different instance classes.

navigational techniques like dragging the graph and centring nodes and edges. Hence, the user would be able to view small sections of the large graphs in the centre window. The size of the graph being viewed at any time depends on the screen resolution.

4   Discussion

This paper describes a tool for browsing semantic information. The tool has been divided into four levels which represent the different types of semantic elements detected. The main advantage of this tool is that it enables navigation among the different levels, allowing users to locate information based on semantics. In addition, the tool provides an easy way of adding new algorithms.

Finally, we have used the tool to develop a graphical query tool. This tool provides users with a representation of an ontology loaded from a file or an URL. This representation is an active graph that recognizes different actions, and reacts creating different parts of a query. Finally, if this interface is connected to a system able to evaluate these queries, a graph representing the instances is shown. The instance graph is configurable to enable the use of representative icons for the target users.

As future work we are developing new plug-ins to allow users to determine ontology distances (connecting the tool to the Semantic Field Tool [6]) and to identify the mappings between two ontologies (using different ontology matching tools).

Acknowledgments. Supported by the ICARO Project Grant, TIN2005-09098-C05-01 (Spanish Ministry of Education and Science), and Applied Systems Biology Project, P07-TIC-02978 (Innovation, Science and Enterprise Ministry of the regional government of the Junta de Andalucía).

References

1. Jiang, G., et al.: FCAViewTab: A concept-oriented view generation tool for clinical data using formal concept analysis. In: Proceedings of the 8th International Protégé Conference, Madrid, Spain (July 2005)
2. Growl, `http://www.uvm.edu/~skrivov/growl/`
3. OntoViz, `http://protege.cim3.net/cgi-bin/wiki.pl?OntoViz`
4. OWLViz, `http://www.co-ode.org/downloads/owlviz/`
5. Alani, H.: TGVizTab: An Ontology Visualisation Extension for Protégé. In: Proceedings of Knowledge Capture (K-Cap 2003), Workshop on Visualization Information in Knowledge Engineering, Sanibel Island, Florida, USA (2003)
6. Euzenat, J., Shvaiko, P.: Ontology matching. Springer, Heidelberg (DE) (2007)
7. Navas, I., Sanz, I., Aldana, J.F., Berlanga, R.: Automatic Generation of Semantic Fields for Resource Discovery in the Semantic Web. In: Andersen, K.V., Debenham, J., Wagner, R. (eds.) DEXA 2005. LNCS, vol. 3588, pp. 706–715. Springer, Heidelberg (2005)
8. Program for Large Networks Analysis, `http://www.ucm.es/info/pecar/pajek.pdf`
9. `http://www.cs.ubc.ca/local/reading/proceedings/spe91-95/spe/vol21/issue11/spe060tf.pdf`
10. Force-based algorithms, `http://en.wikipedia.org/wiki/Force-based_algorithms`
11. Doan, A., Domingos, P., Halevy, A.: Learning to match the schemas of data sources: A multistrategy approach. Machine Learning 50(3), 279–301 (2003)
12. Do, H.-H., Rahm, E.: Coma - a system for flexible combination of schema matching approaches. In: Bressan, S., Chaudhri, A.B., Li Lee, M., Yu, J.X., Lacroix, Z. (eds.) VLDB 2002. LNCS, vol. 2590, pp. 610–621. Springer, Heidelberg (2003)
13. Madhavan, J., Bernstein, P.A., Rahm, E.: Generic schema matching with cupid. In: Proceeding of the 27th International Conference on Very Large Data Bases, pp. 49–58. Morgan Kaufmann Publishers Inc., San Francisco (2001)
14. Ehrig, M., Staab, E.: Qom - quick ontology mapping. In: Proceeding of the 3rd International Semantic Web Conference, Hiroshima, Japan (November 2004)
15. Giunchiglia, F., Shvaiko, P., Yatskevich, M.: S-match: An algorithm and an implementation of semantic matching. In: Proceeding of the European Semantic Web Conference, Heraklion, Greace, pp. 61–75 (2004)
16. Noy, N., Musen, M.: Anchor-prompt: Using non-local context for semantic matching. In: Proceedings of the IJCAI. Workshop on Ontology and Information Sharing, Seatle, USA, pp. 63–71 (2001)
17. McGuinness, D.L., Fikes, R., Rice, J., Wilder, S.: An environment for merging and testing large ontologies. In: Proceedings of KR, pp. 483–493 (2000)
18. Doan, A.-H.: Learnign to Map Between Structured Representations of Data. PhD thesis, University of Washington (2002)
19. Doan, A.-H., Madhavan, J., Domingos, P., Halevy, A.: Handbook of Ontologies. In: Ontology Matching: A Machine Learning Approach, pp. 385–404. Springer, Heidelberg (2004)

Lexical and Semantic Resources for NLP:
From Words to Meanings

Anna Lisa Gentile, Pierpaolo Basile, Leo Iaquinta, and Giovanni Semeraro

Dipartimento di Informatica, Università di Bari
Via E. Orabona, 4 - 70125 Bari - Italia
{al.gentile,basilepp,l.iaquinta,semeraro}@di.uniba.it

Abstract. A user expresses her information need through words with a precise meaning, but from the machine point of view this meaning does not come with the word. A further step is needful to automatically associate it to the words. Techniques that process human language are required and also linguistic and semantic knowledge, stored within distinct and heterogeneous resources, which play an important role during all Natural Language Processing (NLP) steps. Resources management is a challenging problem, together with the correct association between URIs coming from the resources and meanings of the words.

This work presents a service that, given a lexeme (an abstract unit of morphological analysis in linguistics, which roughly corresponds to a set of words that are different forms of the same word), returns all syntactic and semantic information collected from a list of lexical and semantic resources. The proposed strategy consists in merging data with origin from stable resources, such as WordNet, with data collected dynamically from evolving sources, such as the Web or Wikipedia. That strategy is implemented in a wrapper to a set of popular linguistic resources that provides a single point of access to them, in a transparent way to the user, to accomplish the computational linguistic problem of getting a rich set of linguistic and semantic annotations in a compact way.

1   Introduction

Natural Language Processing (NLP) steps include text normalization, tokenization, stop words elimination, stemming, Part Of Speech tagging, lemmatization. Further steps, as Word Sense Disambiguation (WSD) or Named Entity Recognition (NER), are aimed to enrich texts with semantic information. The majority of these steps require sources of linguistic knowledge. To support this processing steps, expecially the semantic-intensive ones, we suggest a wrapper, namely Lexical Collector.

The work is structured as follows: in Section 2 we give a brief description of all used resources, both static resources, as WordNet [11] to cite the most popular one, and dynamic resources, such as a Web Search Service and a Wiki Search Service. In Section 3 we give a sketch of the System Architecture. Finally we depict some conclusions and some future work about the usage of Proton Ontology as an additional exploited resource.

I. Lovrek, R.J. Howlett, and L.C. Jain (Eds.): KES 2008, Part III, LNAI 5179, pp. 277–284, 2008.

2 Resources

In the followings subsections we give a description of all used resources. At this step our system encompasses WordNet, WordNet Domains, WordNet Affect, WordNet Clustering, WordNet Mapping and Topinc Signitures as static resources, whether it queries the web and Wikipedia for dynamic information.

2.1 Wiki Search

An important role in the work is played by Wikipedia [1] which is a free, multilingual, open content encyclopedia project. Its name is a portmanteau of the words wiki (a technology for creating collaborative websites) and encyclopedia: articles have been written collaboratively by volunteers around the world. Unlike traditional encyclopedias, such as Encyclopedia Britannica, no article in Wikipedia undergoes formal peer-review process and changes to articles are made available immediately. Launched in 2001 by Jimmy Wales and Larry Sanger, it is the largest, fastest-growing and most popular general reference work currently available on the Internet. As of December 2007, Wikipedia had approximately 9.25 million articles in 253 languages.

Wikipedia open nature gives rise to criticisms concerning reliability and accuracy, such as the addition of spurious or unverified information; uneven quality, systemic bias and inconsistencies. To ensure the quality, the project relies on its community members to remove vandalism or identify problems such as violation of neutrality and factual errors in its articles. Articles in Wikipedia are subject to several policies and guidelines; they need to be on "notable" topics, contain "no original research" and only "verifiable" facts and must be written from a "neutral point of view". Contents that fail to meet those guidelines and policies are modified or get deleted.

Almost every article in Wikipedia may be edited anonymously through a user account, while only registered users may create a new article. Since Wikipedia aims to be an encyclopedia, each article is structured as the explaination of a proper concept, the real world object the article is describing and for this reason most articles represent a Named Entity. Although it cannot be used as gazetteers directly since it is not intended as a machine readable resource, extracting knowledge such as gazetteers from Wikipedia will be much easier than from raw texts or from usual Web texts because of its structure. It is also important that Wikipedia is updated every day and therefore new named entities are added constantly.

Wikipedia has steadily gained status as a general reference website and its content has also been used in academic studies, books and conferences [2]. Many studies that try to exploit Wikipedia as a knowledge source have recently emerged [17] [19] [13] [23]. In particular Bunescu and Pasca [4] presented a method for disambiguating ambiguous entities by exploiting internal links in Wikipedia as training examples and Toral and Munoz [21] tried to extract gazetteers from Wikipedia by focusing on the first sentences.

Our system includes a Wiki Search Service that allows to return a set of information, collected from Wikipedia, that are connected to an input query string. The service builds

---

[1] http://en.wikipedia.org/wiki/Wikipedia
[2] http://en.wikipedia.org/wiki/Wikipedia:Wikipedia_in_academic_studies

an object that contains the most appropriate wikipedia pages (and their URLs) and the list of links and related Categories of the Wikipedia pages.

2.2 Web Search

The World Wide Web is a mine of language data of unprecedented richness and ease of access [7] and it has been used and proved to be useful in many linguistic tasks, spreading from collecting frequency data [22], to the building of corpus both with specific features [5] or even with a general large scale scope [20]. Question Answering systems can also take advantage from the Web as a source of answers [15] [8] or validation [10]. The list can also continue with Word Sense Disambiguation, where the Web was proved to be more effective when integrated with other resources rather than stand-alone [16]. For this reasons we decided to integrate in our system a service that simply find out the first n links provided by a web search engine for using the lexeme as query. As web search engine we used Yahoo! Search APIs [3].

2.3 WordNet

WordNet [11] is a semantic lexicon for English language. It groups English words into sets of synonyms called synsets. It provides short, general definitions and it records the various semantic relations between these synonym sets. The purpose is twofold: producing a combination of dictionary and thesaurus that is more intuitively usable, and supporting automatic text analysis and artificial intelligence applications. WordNet distinguishes between nouns, verbs, adjectives and adverbs because they follow different grammatical rules. Every synset contains a group of synonymous words or collocations (a collocation is a sequence of words that go together to form a specific meaning, such as "car pool"); different senses of a word are in different synsets. The meaning of the synsets is further clarified with short defining glosses. A typical example synset with gloss is:

good, right, ripe – (most suitable or right for a particular purpose; "a good time to plant tomatoes"; "the right time to act"; "the time is ripe for great sociological changes")

2.4 WordNet Related Resources

The system includes a set of WordNet related resources, such as WordNet Domains, WordNet Affect, WordNet Clustering, Topic Signitures and WordNet Mapping.

WordNet Domains [9] provides semantic domains, as a natural way to establish semantic relations among word senses, which can be profitably used for Computational Linguistics. Semantic domains are areas of human discussion, such as POLITICS, ECONOMY, SPORT, which exhibit their own terminology and lexical coherence. WordNet Domains was created by augmenting the Princeton English WordNet with domain labels. Information brought by domains is complementary to what is already in WordNet. A domain may include synsets of different syntactic categories and from different WordNet sub-hierarchies. Domains may group senses of the same word into homogeneous

---

[3] http://developer.yahoo.com/faq/

clusters, with the side effect of reducing word polysemy in WordNet. We use WordNet Domains 2.0 developed by TCC Division of ITC-irst (Trento-ITALY).

WordNet Affect [18] is an affective lexicon and was developed as an extension of WordNet in order to collect affective concepts and correlate them with words. We use the version 1.0 of WordNet Affect which refers to synsets in WordNet 1.6. Each synset is labeled with a particular 'affective-state' such as emotion, cognitive state, hedonic signal, emotional response and so on. The resource contains *core synsets* that are manually labeled from expert and other synset obtained using WordNet relation like *similar-to* and *antonym*.

WordNet clustering is a method for reducing the granularity of the WordNet sense inventory based on the mapping to a manually crafted dictionary encoding sense hierarchies, namely the Oxford Dictionary of English. We use WordNet clustering based on WordNet 2.1. A detailed description of WordNet Clustering can be found in [12].

Another included resource is Topic Signature [3], that is a list of words that co-occur with the words in a particular synset in WordNet automatically built using the Web. The topic signatures are built for the word senses of english nouns (monosemous and polysemous) in WordNet version 1.6.

Our system also provides a mapping between synsets offsets in various WordNet versions. The current supported mappings are:

1. from WordNet 1.5 to WordNet 1.6
2. from WordNet 1.6 to WordNet 1.7.1
3. from WordNet 1.7.1 to WordNet 2.0
4. from WordNet 2.0 to WordNet 2.1

3   System Architecture

The system architecture is sketched in figure 1. The main component is the Collector that processes the user query and collects all informations over different resources. The system has two kinds of interfaces, one for what we called *static resource* (in figure 1, at the right side of the main component, Collector) and one for the *dynamic ones* (in figure 1, at the left side). In particular, at the current state of the work, the dynamic side consists of two interfaces that allow to query the Web and Wikipedia (respectively Web Search Interface and Wiki Interface). On the other hand, the static side consists of interfaces that supply the access to local resources, in detail:

- Mapping and WordNet interfaces provide the access respectively to WordNet mapping and different versions of WordNet;
- Domain Interface supplies the WordNet domain informations;
- Affective Interface provides the access to WordNet Affective;
- Topic Signature Interface retrieves a topic signatures for each synset.

Finally the output is a XML document that contains all the collected information.

The flexibility of the Collector Architecture permits to plug new resources in a simple way: the only effort consists of providing the actual resource and to implement an Interface that performes the effective access to that. To plug a new resource, for instance

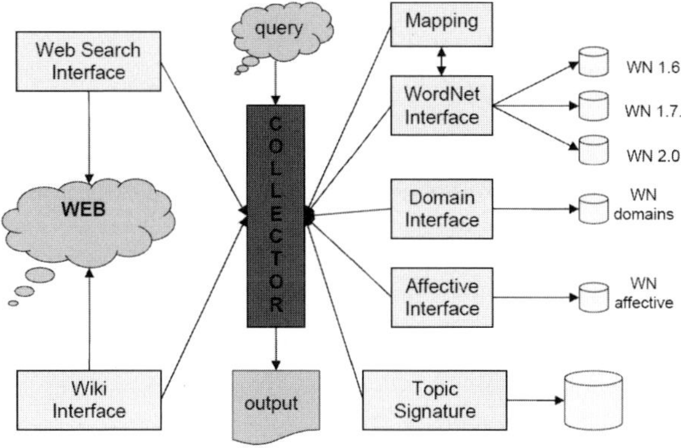

Fig. 1. System architecture

a domain ontology, we need to develop xyzOntology Interface for defining relevant information of the ontology that we need to extract and for designing the structure of the XML node we want to append to the final document.

3.1   Lexical Browser

Lexical Browser is a web application, developed using the Java technology, that provides a web access to Lexical Collector. Given a lexeme, Lexical Browser invokes the Collector to obtain an XML document containing all information about the lexeme; furthermore the web application permits to obtain the XML document or part of it, through the graphic interface or through web services.

It also gives the possibility to store information about users and their searches to collect statistics.

A demo version of the Lexical Browser (Figure 2) is available at the URL http://193.204.187.223:8080/LexicalBrowser/.

3.2   System Output

Elaboration results are stored in an XML document. The root node (tagged as Entry) has the attribute Id valued with the lexeme. Each child node contains information obtained quering employed resources using the lexeme.

The information derived from Wordnet concern offset of the synset in different versions of Wordnet, single words, domain and gloss.

```
<wnentry posTag="2">
 <offsets>
   <offset id="n07739854" version="1"/>
   <offset id="n09157420" version="2"/>
   <offset id="n10436898" version="3"/>
   <offset id="n10998387" version="4"/>
 </offsets>
 <words>
   <word>Mary</word>
   <word>Virgin_Mary</word>
```

Fig. 2. Search results of the Lexical Browser

```
    <word>the_Virgin</word>
    <word>Madonna</word>
  </words>
  <domains>
    <domain>person</domain>
    <domain>religion</domain>
  </domains>
  <gloss>
    The mother of Jesus; Christians refer to her as the
    Virgin Mary; she is especially honored by Roman Catholics
  </gloss>
</wnentry>
```

For each Wikipedia entry, Lexical Collector crawls data about URL, title, a short definition for the content of the page.

```
<wikientry url="..."
    title="Blessed_Virgin_Mary">
  <definition text="This article is about the Catholic,
  Orthodox and Anglican understanding of Mary; for other
  views, see Mary (mother of Jesus) and Islamic view of Virgin Mary.../>
  ...
</wikientry>
```

Additionally there are data about categories related to the page, all captions of images, all related pages within Wikipedia and external links cited within the page.

```
<categories>
  <category label="Titles of Mary"/>
</categories>

<imageCaptions>
  <imageCaption text="Immaculate Heart"/>
  <imageCaption text="Our Lady of Guadalupe.
        Highly venerated image in Mexico."/>
  <imageCaption text="Fedorovskaya"/>
  ...
```

```
</imageCaptions>

<relatedPages>
  <relatedPage reference=
    "Mary (mother of Jesus)"/>
  <relatedPage reference=
    "Islamic view of Virgin Mary"/>
  <relatedPage reference=
    "Immaculate Heart of Mary"/>
  ...
</relatedPages>

<webLinks/>
```

4   Conclusions and Future Work

The task of collecting dictionary-like data from various sources and integrating them into a unique resource is useful but not new. The innovation of this work concerns mainly the flexibility of the Collector Architecture, that gives us the possibility to plug new resources in a simple way. A future attempt is to plug into the service different ontological resources, as suggested in [14] and a legacy ontological repository, based on Proton upper ontology [4], automatically populated from the web. For this aim we will consider to use the standard LMF [2]. The challenge we want to achieve is putting together lexical and semantic information, without relying on a user with semantic web technology skills.

As a future work we intend to use information obtained from Lexical Collector within a Named Entity Recognition system, exploiting not only Wikipedia Information, as suggested in [6], but also all other resources.

References

1. Proceedings of the 2001 ACM CIKM International Conference on Information and Knowledge Management. ACM Press, New York (November 5-10, 2001)
2. Language resource management –Lexical markup framework (LMF) (March 2008), `http://www.lexicalmarkupframework.org/`
3. Agirre, E., de Lacalle Lekuona, O.L.: Publicly available topic signatures for all wordnet nominal senses. In: The 4rd International Conference on Languages Resources and Evaluations (LREC), Lisbon, Portugal (2004)
4. Bunescu, R., Pasca, M.: Using encyclopedic knowledge for named entity disambiguation. In: Proceedings of the 11th Conference of the European Chapter of the Association for Computational Linguistics (EACL 2006), Trento, Italy, pp. 9–16 (April 2006)
5. Ghani, R., Jones, R., Mladenic, D.: Mining the web to create minority language corpora. In: CIKM [1], pp. 279–286
6. Kazama, J., Torisawa, K.: Exploiting wikipedia as external knowledge for named entity recognition. In: Joint Conference on Empirical Methods in Natural Language Processing and Computational Natural Language Learning, pp. 698–707 (2007)
7. Kilgarriff, A., Grefenstette, G.: Introduction to the special issue on the web as corpus. Computational Linguistics 29(3), 333–348 (2003)

---

[4] http://proton.semanticweb.org/

8. Kwok, C.C.T., Etzioni, O., Weld, D.S.: Scaling question answering to the web. ACM Trans. Inf. Syst. 19(3), 242–262 (2001)
9. Magnini, B., Cavaglià, G.: Integrating subject field codes into wordnet. In: 2nd International Conference on Language Resources and Evaluation (LREC 2000), pp. 1413–1418 (2000)
10. Magnini, B., Negri, M., Prevete, R., Tanev, H.: Is it the right answer? exploiting web redundancy for answer validation. In: ACL, pp. 425–432 (2002)
11. Miller, G.: Wordnet: An on-line lexical database. International Journal of Lexicography 3(4) (1990) (Special Issue)
12. Navigli, R.: Meaningful clustering of senses helps boost word sense disambiguation performance. In: Annual Meeting of the Association for Computational Linguistics joint with the 21st International Conference on Computational Linguistics (COLING-ACL 2006), pp. 105–112 (2006)
13. Ponzetto, S.P., Strube, M.: Exploiting semantic role labeling, wordnet and wikipedia for coreference resolution. In: Moore, R.C., Bilmes, J.A., Chu-Carroll, J., Sanderson, M. (eds.) HLT-NAACL. The Association for Computational Linguistics (2006)
14. Prévot, L., Borgo, S., Oltramari, A.: Interfacing ontologies and lexical resources. In: Proceedings of OntoLex 2005 - Ontologies and Lexical Resources, Jeju Island, Republic of Korea (October 15, 2005)
15. Radev, D.R., Qi, H., Zheng, Z., Blair-Goldensohn, S., Zhang, Z., Fan, W., Prager, J.M.: Mining the web for answers to natural language questions. In: CIKM [1], pp. 143–150
16. Rosso, P., y Gómez, M.M., Buscaldi, D., Pancardo-Rodríguez, A., Pineda, L.V.: Two Web-Based Approaches for Noun Sense Disambiguation. In: Gelbukh, A. (ed.) CICLing 2005. LNCS, vol. 3406, pp. 267–279. Springer, Heidelberg (2005)
17. Ruiz-Casado, M., Alfonseca, E., Castells, P.: From wikipedia to semantic relationships: a semi-automated annotation approach. In: Völkel, M., Schaffert, S. (eds.) SemWiki. CEUR Workshop Proceedings, CEUR-WS.org, vol. 206 (2006)
18. Strapparava, C., Valitutti, A.: Wordnet-affect: an affective extension of wordnet. In: 4th International Conference on Language Resources and Evaluation (LREC 2004), pp. 1083–1086 (2004)
19. Strube, M., Ponzetto, S.P.: Wikirelate! computing semantic relatedness using wikipedia. In: AAAI. AAAI Press, Menlo Park (2006)
20. Terra, E.L., Clarke, C.L.A.: Frequency estimates for statistical word similarity measures. In: HLT-NAACL 2003 (2003)
21. Toral, A., Munoz, R.: A proposal to automatically build and maintain gazetteers for Named Entity Recognition by using Wikipedia. In: EACL 2006 (2006)
22. Turney, P.D.: Mining the web for synonyms: Pmi-ir versus lsa on toefl. In: Raedt, L.D., Flach, P. (eds.) ECML 2001. LNCS (LNAI), vol. 2167, pp. 491–502. Springer, Heidelberg (2001)
23. Zesch, T., Gurevych, I., Mühlhäuser, M.: Analyzing and accessing wikipedia as a lexical semantic resource. In: Biannual Conference of the Society for Computational Linguistics and Language Technology (2007)

A Semantic Similarity Measure for the SIMS Framework

Roberto Pirrone[1], Giuseppe Russo[1], Pierluca Sangiorgi[1], Nunzio Ingraffia[2],
and Claudia Vicari[2]

[1] DINFO - Universitá degli Studi Palermo
`pirrone@unipa.it, russo@csai.unipa.it`
[2] Engineering - Ingegneria Informatica S.p.A., R&D Lab
`{nunzio.ingraffia,claudia.vicari}@eng.it`

Abstract. The amount of currently available digital information grows rapidly. Relevant information is often spread over different information sources. An efficient and flexible framework to allow users to satisfy effectively their information needs is required. The work presented in this paper describes SIMS (Semantic Information Management System), a reference architecture for a framework performing semantic annotation, search and retrieval of information from multiple sources. The work presented in this paper focuses on a specific SIMS module, the SIMS Semantic Content Navigator, proposing an algorithm and the related implementation to calculate a semantic similarity measure inside an OWL ontology. This measure is based on ontology structure and on the information provided by attributes and relations that are defined inside the ontology. This work is the result of a collaborative effort between the DINFO (Department of Computer Science and Engineering) and the research team of Engineering - Ingegneria Informatica.

1 Introduction

The amount of currently available digital information grows rapidly. Searching for relevant information through traditional keyword based search engines is a very time-consuming activity and can lead to inaccurate search results [1]. These issues are important especially inside organizations and small or medium enterprise clusters, where relevant information is often spread over a number of different applications. As knowledge is widely recognized as an important asset for sustaining competitive advantage of an organization, the use of innovative technologies for improving the management of organizational knowledge represents a key factor in order to improve business performance. In response to this, Semantic Web technologies enable machines to understand the meaning and the context of the information and process it in an automated way, through the use of mark-up languages, ontologies [2] and rules, and can aid to improve traditional knowledge management systems. The work presented in this paper describe a semantic similarity measure inside the SIMS framwork. The SIMS (Semantic Information Management System) [11], a reference architecture for a

I. Lovrek, R.J. Howlett, and L.C. Jain (Eds.): KES 2008, Part III, LNAI 5179, pp. 285–292, 2008.

framework performing semantic annotation, search and retrieval of information from multiple sources. SIMS provides users with a uniform view of information items that belong to heterogeneous and already existing information systems by use of Semantic Web technologies and standards. This paper focuses on a specific SIMS module, the SIMS Semantic Content Navigator, proposing for it an algorithm and an implementation to calculate a semantic similarity measure inside an OWL ontology [3]. The rest of the paper is structured as follows. Section 2 presents Semantic Knowledge Management Systems. Section 3 presents an overview of the SIMS framework. Section 4 illustrates the SIMS architecture, describing briefly its layers and modules. Section 5 discusses the Semantic Similarity in the Semantic Content Navigator Module. Some consideration about the implementation and the evaluation are also reported. Finally, section 6 presents conclusions and future work.

2   Semantic-Based Knowledge Management Systems

The purpose of Knowledge Management is to provide strategies, technologies and processes to leverage information and expertise. Traditional knowledge management systems are not so flexible. To overcome this problem it is necessary an agile and modular knowledge management architecture, which is able to add semantic value to information systems. Ontologies can improve three key aspects in Knowledge Management: communication, integration and reasoning [4]. The SIMS framework focus is on communication and integration aspects. However, data model on SIMS could be a suitable starting point to enable reasoning to gain new knowledge from integrated sources. Different approaches have been proposed in literature to define architectural solutions for knowledge management systems. Most approaches focus on agent-based technologies [5] while others are more focalized on SOA approach [6] and grid computing [7]. Many architectures are multi-level ones. The OBSA [8] architecture is divided into five layers and tries to define a top-down method to collect structured or unstructured information from different data sources. KoMIS [9] is a multi-tier system that is focused on retrieval of information through an indexing process.

3   SIMS Framework Functionalities

The SIMS framework is designed to provide the following main functionalities:

Integration of heterogeneous informative sources at a conceptual level. SIMS enables to link different and autonomous information management systems such as Wikis, Blogs, Document Management Systems and Content Management Systems which are connected to the framework by means of specific modules called proxy-wrappers. This approach allows an uniform and semantic-based access to retrieve the corresponding information items. An *information item* is an atomic element of a generic information management system. SIMS proposes to integrate the information sources only on a conceptual level.

Semantic Enrichment of information items. SIMS proposes two different but complementary ways to enrich information items. The first one is domain independent, and associates a specific subset of Dublin Core Metadata [10] to each information item. The second one is domain dependent and it is based on the association between the information items and the domain ontologies expressed in a formal language, such as OWL (Ontology Web Language). This association is obtained as a result of a semantic annotation process.

Search and Presentation of information. Starting from user requests, SIMS is able to retrieve and present personalized information. This process is based on queries over the ontologies. Ontologies and the process of semantic annotation allow the SIMS framework to provide the users with more accurate results. The SIMS provides functionalities based on semantic similarity between concepts to improve this interaction.

4   SIMS Architecture

The SIMS architecture is structured in three logical layers, relying on two lower level layers (see Fig. 1). They are briefly described in this section.

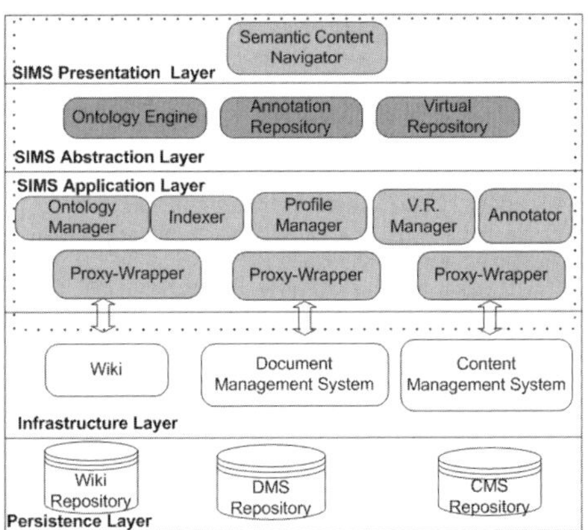

Fig. 1. The SIMS Framework Architecture

The **Persistence and Infrastructure Layer** represent respectively the physical repositories and the business logic of the information sources, whose information items are made available to the SIMS final users by means of specific software components called Proxy-Wrapper. The SIMS **Application Layer** is composed by several modules to permit the unified view of the information sources and manage modules of the upper layers. It allows integration with external information sources (Proxy-Wrapper), metadata storage for information items retrieval (Virtual repository), indexing functions (Indexer), ontology management (Ontology Manager) and semantic annotation (Annotator). The

Abstraction Layer can be considered as the whole knowledge base for the framework. The information is homogeneously structured and ready to be accessed through **Presentation Layer**. The highest architecture layer, the SIMS **Presentation Layer** manages the end user interaction, by means of Semantic Content Navigator, a module that enables end users to interact with SIMS Framework and retrieve information. This component is described, in detail, in the next section.

5   Semantic Similarity in the Semantic Content Navigator Module

Within the SIMS, the Semantic Content Navigator module has a web-based front end, which enables the end users to navigate and access the knowledge base, giving to them a transparent access and a unified view of the underlying information sources. Users can make their request or directly selecting ontological concepts or entering a natural language query (in this case the system will associate to it one or more ontological concepts). As a result users receive back the list of information items related to the directly or indirectly selected concepts. Therefore, the information presented to end users is related not only to selected concepts, but it is enriched with new information connected to similar concepts.

5.1   Semantic Similarity Measure

The implementation of the Semantic Content Navigator module needs, in addition to an user friendly interface, a core algorithm for semantic similarity between concepts. Estimation of semantic similarity measure in the same ontology or between distinct ontologies is a central need for the processes involving the information exchange over the Web. There are two fundamental categories of measures. The first one is based on the topological distance between concepts in the ontology graph [12] [13], while the second one makes use of their information content and it is adopted in information theory based methods [14] [15]. Some other approaches try to combine these two aspects. A similarity measure algorithm is defined and implemented within the SIMS. The approach is based on a fuzzy combined semantic similarity measure applied on OWL-DL ontologies. This measure is based on ontology structure and on the information provided by attributes and relations that are defined inside the ontology.

5.2   Fuzzy Combined Similarity Measure

The algorithm is based on the assumption that "two concepts are much more similar how less distant and more correlated among them they are". This simple definition makes intuitive the use of a fuzzy system [16] to produce a similarity measure of two concepts as a function of distance and correlation. The measure will range in [0, 1]. The distance component is defined by the following:

$$dist(c_1, c_2) = ShortestWeightedPath(c_1, c_2). \tag{1}$$

This term is based on a structural approach (Graph Distance Model). It is expressed as the minimum weighted sum of relations crossed to reach a concept c_2 from a concept c_1. Instead, the correlation component is given by:

$$corr(c_1, c_2) = \left(1 + \frac{N_i}{T_a}\right) * \left(1 + \frac{N_r}{T_r}\right) \qquad (2)$$

where N_i is the number of not-inherited common attributes (NICAs) of c_1 and c_2, T_a is the total number of attributes of c_1 and c_2, N_r is the number of the common relations (CRs) of the two first children of the most recent common ancestor (FCAs) of c_1 and c_2, T_r is the total number of these relations. This term is based on a behavioral approach and denotes the correlation between two ontology concepts in terms of their specialization considering the functional aspects given by attributes and relations. In this scenario, the basic assumption is that two concepts are correlated "if they are able to express something similar". As a consequence, the measure is the product of two gain terms. Correlation increases if c_1 and c_2 share many attributes not inherited from their most recent common ancestor and they share many relations with their ancestors or children. In this way it is possible to measure Distance and Correlation. These two values, are used as crisp inputs in a fuzzy system. The membership functions of the system are depicted in Fig. 2. The distance component is weighted according to the depth level of the concepts. The formulation recursively weights this depth starting from the weight of ancestor of the concepts by adding a value representing the depth of each concept in the ontology structure. Correlation component doesn't need to be normalized because the three related membership functions for this component are always distributed in [1,4] range for every ontology in use. The output membership functions for similarity have not been manipulated. Standard triangular functions equally distributed over the whole range have been used. This approach is useful to determine the semantic aspects, through a measure related to the context and considering the domain of the ontology. In fact, while traditional methods give low values of similarity between generally dissimilar concepts, this method is able to evaluate if they are semantically similar (related) in the specific utilization domain where the results could be more different. This is possible if the relations and the attributes inside the ontology that allow to highlight possible correlations between concepts are considered.

5.3 Algorithm Implementation and Experimental Results

The algorithm implementation has been realized using some predefined Java packages. To manage OWL ontologies the Jena framework has been used [17]. The shortest weighted path has been obtained through the SimPack libraries [18] while the correlation component has been implemented from scratch. The used fuzzy library is the FuzzyJToolkit [19]. As always, the implementation is the result of a trade-off between the accuracy of measure and the execution time. The system is able to calculate the semantic similarity between sets of concepts belonging to the same ontology. The time consuming steps of the algorithm are

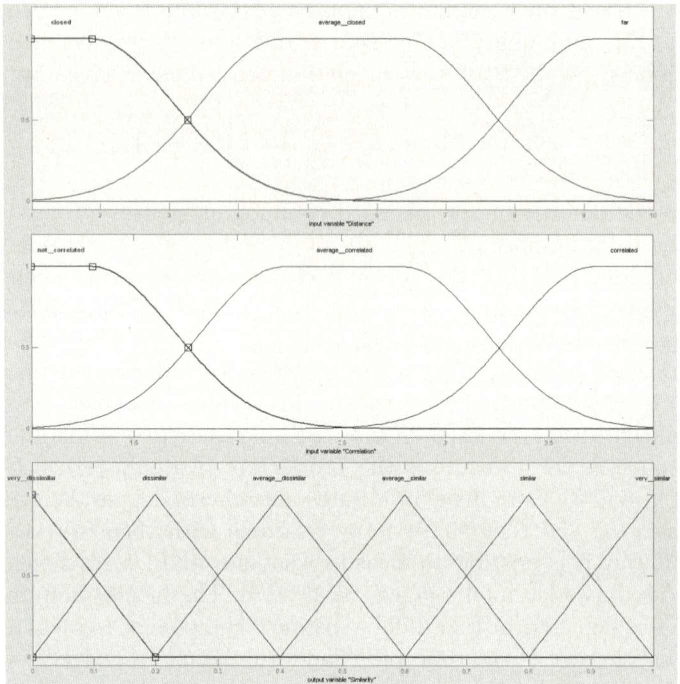

Fig. 2. The membership functions of the Fuzzy System

the calculation of the NICAs and the calculation of FCAs. To better integrate the algorithm it is possible to define different modalities: it is possible to use only Common Relations instead of NICAs or to use Common Relations instead of FCAs. Hybrids combinations can be used too. Time execution is depending on two main factors that are the extension of the ontology and the density of attributes and relations. For average ontologies the execution time of the algorithm is in the order of hundred of milliseconds or few seconds in the worst cases. To better show the presented approach a test on an ontology about the tourism domain contributed by Holger Knublauch is presented. This ontology has been specifically developed for the Semantic Web and provided by the Protege' team of the Stanford Medical Informatics at the Stanford University School of Medicine. A comparison of the results with Resnik's algorithm [12] and Lin's algorithm [14] is presented. The obtained values need to be explained in detail because of the different interpretation with respect to the other methods. The proposed approach is able to carry out the semantic aspect considering the domain of the ontology and making a measure related to the context. The result of semantic similarity between Accommodation and Destination concepts is a good example. Generally, this two terms are dissimilar and traditional methods confirm that, but if the terms are evaluated in an appropriate context, they could be semantically similar (related). In the domain about tourism they share

Table 1. Experimental Results

Concept 1 - Concept 2	Resnick Similarity	Lin Similarity	Fuzzy Combined Sim.
Capital - National Park	0.640	0.640	0.404
Bunjee Jumping - Surf	0.502	0.502	0.454
Destination - Accomodation	0.012	0.015	0.490
Museum - Town	0.012	0.012	0.195
Beach - City	0.640	0.503	0.511

the relation hasAccommodation - hasDestination and are more similar respect to traditional methods that neglect the relations.

6 Conclusions and Future Work

In this paper a semantic similarity measure used to enrich search results inside the SIMS framework has been presented. The different functionalities and an overview of the SIMS architecture have been discussed. At the top of the architecture, in the SIMS Presentation layer, a core algorithm for the Semantic Content Navigator has been defined. The algorithm implementation has been presented. This algorithm measures the semantic similarity between two set of concepts inside an OWL ontology and list in a decreasing order the sets for the users. Semantic similarity allows to provide the users with information, which is not only related to selected ontology concepts connected directly or indirectly to a user request, but also coming from similar concepts that are found automatically. The focus of future works for the Semantic Content Navigator is to test the system in order to refine the algorithm. An extension of the presented algorithm able to find semantic similarities in different ontologies is under investigation.

Acknowledgments

The SIMS is a framework designed by Engineering Ingegneria Informatica S.p.A and eBMS ISUFI University of Salento within the common research projects, cofounded by the Italian Department "Ministero dell'Universitá e della Ricerca". Thanks are expressed to research team of eBMS ISUFI University of Salento for their work.

References

1. Fensel, D., Ding, Y., Stork, H.-G.: The Semantic Web:from Concept to Percept. OGAI-Journal 22(1) (2003)
2. Gruber, T.: A translation approach to portable ontology specifications. Knowledge Acquisition 5(2), 199–220 (1993)
3. McGuinness, D.L., van Harmelen, F.: Owl Web Ontology Language: Overview. W3C Recommendation (February 2004), http://www.w3.org/TR/owl-features/

4. Mika, P., Akkermans, H.: The Knowledge Engineering Review. Knowledge Acquisition 19(4), 317–345 (2004)
5. Abar, S., Abe, T., Kinoshita, T.: A Next Generation Knowledge Management System Architecture. In: Proceedings of the 18th International Conference on Advanced Information Networking and Application (AINA 2004), vol. 2, pp. 191–195 (2004)
6. Pappas, N., Kazasis, F.G., Anestis, G., Gioldasis, N., Christodoulakis, S.: A Knowledge Management Platform for Supporting Digital Business Ecosystems based on P2P and SOA Technologies. In: IEEE International Conference on Digital Ecosystems and Technologies (IEEE DEST 2007), pp. 196–202 (2007)
7. Chen, L., Shadbolt, N.R., Goble, C.A.: A Semantic Web-Based Approach to Knowledge Management for Grid Application. IEEE Transactions on Knowledge and data Engineering 19(2), 283–295 (2007)
8. Gu, J., Chen, H., Yang, L., Zhang, L.: OBSA: Ontology based Information Processing Architecture. In: Proceedings of the IEEE/WIC/ACM International Conference on Web Intelligence, pp. 607–610 (2004)
9. Assali, A.A., Lenne, D., Debray, D.: KoMIS: An Ontology-based Knowledge Management System For Industrial Safety. In: Proceedings of the 18th Workshop on Database and Expert System Application, pp. 475–479 (2005)
10. Dublin Core Metadata Home Page, http://dublincore.org/
11. Corallo, A., Ingraffia, N., Vicari, C., Zilli, A.: SIMS: An Ontology-based Multi-source Knowledge Management System. In: 11th World Multiconference on Systemics, Cybernetics and Informatics (WMSCI 2007) (2007)
12. Resnik, P.: Using Information Content to evaluate Semantic Similarity in a Taxonomy. In: Proceedings of the International Joint Conference on Artificial Intelligence (IJCAI), vol. 19, pp. 448–453 (1995)
13. Wu, Z., Palmer, M.: Verb Semantics and Lexical Selection. In: Proceedings of 32nd Annual Meeting of the Associations for Computational Linguistics, pp. 133–138 (1994)
14. Lin, D.: An Information-theoretic Definition of Similarity. In: Proceedings of the 15th International Conference on Machine Learning, pp. 296–304 (1998)
15. Jiang, J.J., Conrath, D.: Semantic Similarity based on Corpus Statistics and Lexical Taxonomy. In: Proceedings of International Conference on Research in Computational Linguistics (1997)
16. Zadeh, L.A.: Knowledge Representation in Fuzzy Logic. An Introduction to Fuzzy Logic Applications in Intelligent Systems, 1–25 (1992)
17. Carroll, J.J., Dickinson, I., Dollin, C., Reynolds, D., Seaborne, A., Wilkinson, K.: Jena: Implementing the Semantic Web Recommendations. In: WWW Alt. 2004: Proceedings of the 13th international World Wide Web conference, pp. 74–83 (2004)
18. Ziegler, P., Kiefer, C., Sturm, C., Dittrich, K.R., Bernstein, A.: Generic Similarity Detection in Ontologies with the SOQA-SimPack Toolkit (Demo Paper). In: ACM SIGMOD International Conference on Management of Data (SIGMOD 2006) (2006)
19. FuzzyJToolkit Home Page,
http://ai.iit.nrc.ca/IRpublic/fuzzy/fuzzyJToolkit2.html

Fuzzy Bilateral Matchmaking in e-Marketplaces

Azzurra Ragone[1,4], Umberto Straccia[2], Fernando Bobillo[3], Tommaso Di Noia[4],
and Eugenio Di Sciascio[4]

[1] University of Michigan, Ann Arbor, MI, USA
`aragone@umich.edu`
[2] ISTI-CNR, Pisa, Italy
`straccia@isti.cnr.it`
[3] Department of Computer Science and Artificial Intelligence, University of Granada
`fbobillo@decsai.ugr.es`
[4] SisInfLab, Politecnico di Bari, Bari, Italy
`{t.dinoia,disciascio}@poliba.it`

Abstract. We present a novel Fuzzy Description Logic (DL) based approach to automate matchmaking in e-marketplaces. We model traders' preferences with the aid of Fuzzy DLs and, given a request, use utility values computed w.r.t. Pareto agreements to rank a set of offers. In particular, we introduce an expressive Fuzzy DL, extended with concrete domains in order to handle numerical, as well as non numerical features, and to deal with vagueness in buyer/seller preferences. Hence, agents can express preferences as *e.g.*, *I am searching for a passenger car costing about 22000€ yet if the car has a GPS system and more than two-year warranty I can spend up to 25000€*. Noteworthy our matchmaking approach, among all the possible matches, chooses the mutually beneficial ones.

1 Introduction

In an e-marketplace, a transaction can be organized in three different stages [21]: *discovery*, *negotiation* and *execution*. During the discovery phase, the marketplace helps the buyer to look for promising offers best matching her request. The result of this **matchmaking** phase is a ranked list of offers (usually ranked with respect to buyer's preferences). In the eventual *negotiation* phase, the marketplace guides the buyer and the seller to reach an agreement. With the *execution* of the transaction, the buyer and the seller exchange the good. Usually negotiation and matchmaking are two distinct processes executed sequentially. First, the marketplace ranks offers for the buyer taking into account her request, *i.e.*, her preferences expressed w.r.t. some utility function, then, usually, a negotiation starts with the seller having the best ranked supply, in order to reach an agreement that satisfies both traders. That is, the marketplace tries to find an agreement which is Pareto efficient [1], as well as mutually beneficial for both traders. In other word, the marketplace, among all the actual Pareto solutions, looks for the ones maximizing the traders' utility value w.r.t. some criteria, *e.g.*, the Nash bargaining

---

[1] An agreement is Pareto efficient when it is not possible to improve the utility of one trader, without lowering the utility of the opponent's one.

I. Lovrek, R.J. Howlett, and L.C. Jain (Eds.): KES 2008, Part III, LNAI 5179, pp. 293–301, 2008.

solution, [2] [14]. A typical marketplace uses only buyer's preferences for discovery and both traders' preferences for negotiation. In a few words we can say that discovery is "unilateral", while negotiation is "bilateral". Due to this difference, it might occur often that an offer resulting promising for the buyer *i.e.*, with a good satisfaction degree for her preferences, does not lead to an agreement because, on the other side, seller's preferences are not adequately satisfied. The idea behind the approach we propose in this paper is to merge the discovery and negotiation phase in a **bilateral matchmaking**. In our bilateral matchmaking scenario given a buyer's request and a set of supplies, the matchmaker computes for each supply a Pareto-efficient agreement maximazing the degree of satisfaction of the traders (see Section 3), and then ranks all these agreements w.r.t. the utility of the buyer.

We propose here a *fuzzy Description Logic* (see, [3] for an overview) endowed with *concrete domains* to model relations among issues and as a communication language between traders. In our proposal, concrete domains allow to deal with numerical features, which are mixed, in preferences, with non numerical ones.

In our framework it is possible to model *positive* and *negative* preferences (I would like a car black or gray, but not red), as well as *conditional preferences* (I would like leather seats if the car is black) involving both numerical features and non numerical ones (If you want a car with GPS system you have to wait at least one month) or only numerical ones (I accept to pay more than 25000€ only if there is more than a two-year warranty).

The rest of the paper is structured as follows: next section discusses the fuzzy language we adopt in order to express traders' preferences. In Section 3 we set the stage of the the bilateral matchmaking problem in fuzzy DL and then we illustrate how to compute Pareto agreements. In Section 3.1 the whole process is highlighted with the aid of a simple example. Related Work and discussion close the paper.

2 A Fuzzy DL to Express Preferences

In a bilateral matchmaking scenario traders express preferences involving numerical as well as non numerical issues, in some way interrelated. The variables representing numerical features are either involved in *hard constraints* or *soft constraints*. In hard constraints, the variables are always constrained by comparing them to some constant, like $(\leqslant price\ 20.000)$, or $(\geqslant month\_warranty\ 60)$, and such constraints can be combined into complex requirements, *e.g.*, $Sedan \sqcap (\leqslant price\ 25.000) \sqcap (\leqslant deliverytime\ 30)$, or $AlarmSystem \sqcap (\geqslant price\ 26.000)$. Vice-versa when numerical features are involved in *soft constraints*, also called *fuzzy constraints*, the variables representing numerical features are constrained by so-called fuzzy membership functions, as shown below.

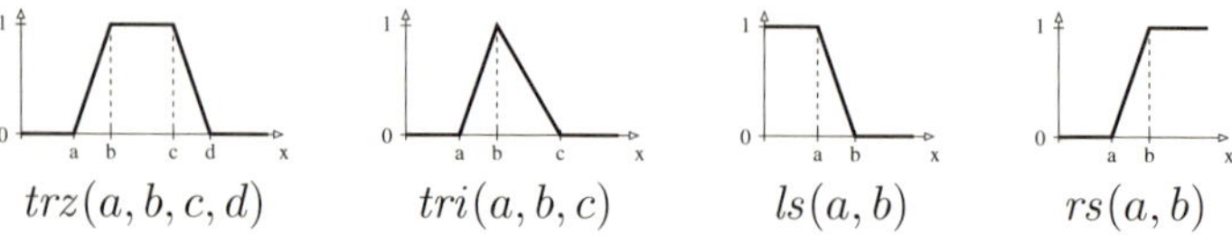

$$trz(a,b,c,d) \qquad tri(a,b,c) \qquad ls(a,b) \qquad rs(a,b)$$

For instance, $(\exists price.ls(18000, 22000))$ dictates that given a price it returns the degree of truth to which the constraint is satisfied. Essentially, $(\exists price.ls(18000, 22000))$ states that if the price is no higher than 18000 then the constraint is definitely satisfied, while if the price is higher than 22000 then the constraint is definitely not satisfied. In between 18000 and 22000, we use linear interpolation, given a price, to evaluate the satisfaction degree of the constraint.

Fuzzy DL Now, we specify the syntax of our fuzzy DL for matchmaking. The fuzzy DL considers the salient features of the fuzzyDL reasoner *fuzzyDL* [3] (see [3]). The basic fuzzy DL we consider is the fuzzy DL $\mathcal{SHIF}(D)$ [3], *i.e.*, $\mathcal{SHIF}$ with concrete data types. But, for our purpose, we do not need individuals and assertions. So, let us consider an alphabet for *concepts names* (denoted A), *abstract roles names* (denoted R), *i.e.*, binary predicates *concrete roles names* (denoted T), and *modifiers* (denoted m). $\mathbf{R}_a$ also contains a non-empty subset $\mathbf{F}_a$ of *abstract feature names* (denoted r), while $\mathbf{R}_c$ contains a non-empty subset $\mathbf{F}_c$ of *concrete feature names* (denoted t). Features are functional roles. Concepts in fuzzy $\mathcal{SHIF}$ (denoted C, D) are build as usual from atomic concepts A and roles R: $\top, \bot, A, C \sqcap D, C \sqcup D, \neg C, \forall R.C$ and $\exists R.C$. Now, Fuzzy $\mathcal{SHIF}(D)$ extends $\mathcal{SHIF}$ with concrete data types [1], *i.e.*, it has the additional concept constructs $\forall T.d, \exists T.d$ and DR, where

$$d \rightarrow ls(a,b) \mid rs(a,b) \mid tri(a,b,c) \mid trz(a,b,c,d)$$
$$DR \rightarrow (\geqslant t\ val) \mid (\leqslant t\ val) \mid (= t\ val)$$

and val is an integer or a real depending on the range of the concrete feature t. Finally, we further extend $\mathcal{SHIF}(D)$ as follows:

$$C, D \rightarrow (w_1 C_1 + w_2 C_2 + \ldots + w_k C_k) \mid C[\geqslant n] \mid C[\leqslant n]$$

where $n \in [0, 1], w_i \in [0, 1], \sum_{i=1}^{k} w_i = 1$. The expression $(w_1 C_1 + w_2 C_2 + \ldots + w_k C_k)$ denotes a weighted sum, while $C[\geqslant n]$ and $C[\leqslant n]$ are threshold concepts.

A *fuzzy DL ontology* (also Knowledge Base, KB) $\mathcal{K} = \langle \mathcal{T}, \mathcal{R} \rangle$ consists of a fuzzy TBox $\mathcal{T}$ and a fuzzy RBox $\mathcal{R}$. A *fuzzy TBox* $\mathcal{T}$ is a finite set of *fuzzy General Concept Inclusion axioms* (GCIs) $\langle C \sqsubseteq D, n \rangle$, where $n \in (0, 1]$ and C, D are concepts. If the truth value n is omitted then the value 1 is assumed. Informally, $\langle C \sqsubseteq D, n \rangle$ states that all instances of concept C are instances of concept D to degree n, that is, the subsumption degree between C and D is at least n. For instance, $\langle Sedan \sqsubseteq PassengerCar, 1 \rangle$ states that a sedan is a passenger car. We write $C = D$ as a shorthand of the two axioms $\langle C \sqsubseteq D, 1 \rangle$ and $\langle D \sqsubseteq C, 1 \rangle$. Axioms of the form $A = D$ are called *concept definitions* (e.g., , $InsurancePlus = DriverInsurance \sqcap TheftInsurance$). A *fuzzy RBox* $\mathcal{R}$ is a finite set of role axioms of the form: (i) $(fun\ R)$, stating that a role R is functional, *i.e.*, R is a feature; (ii) $(trans\ R)$, stating that a role R is transitive; (iii) $R_1 \sqsubseteq R_2$, meaning that role R_2 subsumes role R_1; and (iii) $(inv\ R_1\ R_2)$, stating that role R_2 is the inverse of R_1 (and vice versa). A simple role is a role which is neither transitive nor has a transitive subroles. An important restriction is that functional needs to be simple.

---

[3] http://gaia.isti.cnr.it/~straccia/software/fuzzyDL/fuzzyDL.html

The main idea is that concepts and roles are interpreted as fuzzy subsets of an interpretation's domain. Therefore, axioms, rather than being "classical" evaluated (being either true or false), they are "many-valued" evaluated, *i.e.*, their evaluation takes a degree of truth in $[0, 1]$.

A *fuzzy interpretation* $\mathcal{I} = (\Delta^{\mathcal{I}}, \cdot^{\mathcal{I}})$ relative to a concrete domain $D = \langle \Delta_D, C(D) \rangle$ consists of a nonempty set $\Delta^{\mathcal{I}}$ (the *domain*), disjoint from Δ_D, and of a *fuzzy interpretation function* $\cdot^{\mathcal{I}}$ that assigns: (i) to each abstract concept C a function $C^{\mathcal{I}} : \Delta^{\mathcal{I}} \to [0, 1]$; (ii) to each abstract role R a function $R^{\mathcal{I}} : \Delta^{\mathcal{I}} \times \Delta^{\mathcal{I}} \to [0, 1]$; (iii) to each abstract feature r a partial function $r^{\mathcal{I}} : \Delta^{\mathcal{I}} \times \Delta^{\mathcal{I}} \to [0, 1]$ such that for all $x \in \Delta^{\mathcal{I}}$ there is an unique $y \in \Delta^{\mathcal{I}}$ on which $r^{\mathcal{I}}(x, y)$ is defined; (iv) to each concrete role T a function $R^{\mathcal{I}} : \Delta^{\mathcal{I}} \times \Delta_D \to [0, 1]$; (v) to each concrete feature t a partial function $t^{\mathcal{I}} : \Delta^{\mathcal{I}} \times \Delta_D \to [0, 1]$ such that for all $x \in \Delta^{\mathcal{I}}$ there is an unique $v \in \Delta_D$ on which $t^{\mathcal{I}}(x, v)$ is defined. In order to extend the mapping, the interpretation function $\cdot^{\mathcal{I}}$ is extended to roles and complex concepts, we need functions to define the negation, conjunction, disjunction (called norms), etc of values in $[0, 1]$. The choice of them is not arbitrary. Some well-known specific choices are described in the table below.

	Łukasiewicz Logic	Gödel Logic	Product Logic	Zadeh
$\ominus x$	$1 - x$	if $x = 0$ then 1 else 0	if $x = 0$ then 1 else 0	$1 - x$
$x \otimes y$	$\max(x + y - 1, 0)$	$\min(x, y)$	$x \cdot y$	$\min(x, y)$
$x \oplus y$	$\min(x + y, 1)$	$\max(x, y)$	$x + y - x \cdot y$	$\max(x, y)$
$x \Rightarrow y$	if $x \leqslant y$ then 1 else $1 - x + y$	if $x \leqslant y$ then 1 else y	if $x \leqslant y$ then 1 else y/x	$\max(1 - x, y)$

For the sake of space we do not report the formal semantics of the logic we adopt. The interested reader may refer to [3].

Proposition 1. *If the maxima of $x \otimes_L y$, with $\langle x, y \rangle \in S \subseteq [0, 1] \times [0, 1]$, where $\otimes_L$ is Łukasiewicz t-norm, is positive then the maxima is also Pareto optimal.*

As we will see later on, relying on Łukasiewicz logic will guarantee that the solutions of the bilateral matchmaking process are then Pareto optimal ones. Note also that the maxima of $x \otimes_G y$, with $\langle x, y \rangle \in S$, is not Pareto optimal.

3   Multi Issue Bilateral Matchmaking in Fuzzy DLs

Marketplaces are typical scenarios where the notion of fuzziness appears frequently. The concept of Cheap or Expensive are quite usual. In a similar way it is common to have a fuzzy interpretation of numerical constraints. If a buyer looks for a car with a price lesser than 15,000 €and a supplier selling his car for 15,500 €, we can not say they do not match at all. Actually, they match with a certain degree. Hence, a fuzzy language, as the one we presented in the previous sections, would be very useful to model demands and supplies in matchmaking scenarios.

In bilateral matchmaking scenarios, both buyer's request and seller's offer can be split into *hard constraints* and *soft constraints*. *Hard constraints* represent what has to be (necessarily) satisfied in the final agreement; *soft constraints* represent traders' preferences.

Example 1. Consider the example where buyer's request is: "I am searching for a Passenger Car equipped with Diesel engine. I need the car as soon as possible, and I can not wait more than one month. Preferably I would like to pay less than 22,000 € furthermore I am willing to pay up to 24,000 € if warranty is greater than 160000 km. I won't pay more than 27,000 €".

hard constraints: I want a Passenger Car provided with a Diesel engine. I can not wait more than one month. I won't pay more than 27,000 €.
soft constraints: I would like to pay less than 22,000 € furthermore I am willing to pay up to 24,000 € if warranty is greater than 160000 km.

Definition 1 (Demand, Supply, Agreement). *Given an ontology* $\mathcal{K} = \langle \mathcal{T}, \mathcal{R} \rangle$ *representing the knowledge on a marketplace domain*

- *a* demand *is a concept definition* β *of the form* $B = C[\geqslant 1.0]$ *(for Buyer) such that* $\langle \mathcal{T} \cup \{\beta\}, \mathcal{R} \rangle$ *is satisfiable.*
- *a seller's* supply *is a concept definition* σ $S = D[\geqslant 1.0]$ *(for Seller) such that* $\langle \mathcal{T} \cup \{\sigma\}, \mathcal{R} \rangle$ *is satisfiable.*
- $\mathcal{I}$ *is a* possible deal *between* β *and* σ *iff* $\mathcal{I} \models \langle \mathcal{T} \cup \{\sigma, \beta\}, \mathcal{R} \rangle$. *We also call* $\mathcal{I}$ *an* agreement.

σ and β represent the minimal requirements needed in the final agreement. As they are mandatory the threshold value is set to 1.0, meaning that they have to be in the agreement. In a bilateral matchmaking process, besides hard constraints, both traders can express preferences on some (bundle of) issues. In our fuzzy DL framework preferences can be represented as weighted formulae (see Section 2). More formally:

Definition 2 (Preferences). *The buyer's preference* $\mathcal{B}$ *is a weighted concept of the form* $n_1 \cdot \beta_1 + \ldots + n_k \cdot \beta_k$, *where each* β_i *represents the subject of a buyer's preference, and* n_i *is the weight associated to it. Analogously, the seller's preference* $\mathcal{S}$ *is a weighted concept of the form* $m_1 \cdot \sigma_1 + \ldots + m_h \cdot \sigma_h$, *where each* σ_i *represents the subject of a seller's preference, and* m_i *is the weight associated to it.*

For instance, the Buyer's request in Example 1 is formalized as:

$$\beta \text{ is } B = (PassengerCar \sqcap Diesel \sqcap (price \leqslant 27,000) \sqcap (deliverytime \leqslant 30))[\geqslant 1.0]$$
$$\beta_1 = (\exists price.ls(22000, 25000))$$
$$\beta_2 = (\exists km\_warranty.rs(140000, 160000)) \rightarrow (\exists price, ls(24000, 27000))$$

where *price* and *km_warranty* are concrete features. We normalize the sum of the weights of both agents' to 1 to eliminate outliers, and make the set of preferences comparable. The utility function, that we call **preference utility**, is then a weighted sum of the preferences satisfied in the agreement. Dealing with concrete features, we always have to set a **reservation value** [18] represented as a *hard constraint*. Reservation value is the maximum (or minimum) value in the range of possible feature values to reach an agreement, *e.g.*, the maximum price the buyer wants to pay for a car or the minimum warranty required, as well as, from the seller's perspective the minimum price he will accept to sell the car or the minimum delivery time. Usually, each participant knows its own reservation value and ignores the opponent's one. In the following, given a concrete feature f we refer to reservation values of buyer and seller on f with $r_{\beta,f}$ and $r_{\sigma,f}$ respectively. Since *reservation values* represent *hard constraints* then buyer's ones are added to β and seller's ones to σ (see Example 1). The last elements we have to introduced in order to formally define an agreement in a bilateral matchmaking process are *disagreement thresholds*, also called disagreement payoffs, t_β, t_σ. They represent the minimum utility that the agent need to reach to accept the agreement. Minimum utilities may incorporate an agent's attitude toward concluding the transaction, but also overhead costs involved in the transaction itself, *e.g.*, fixed taxes.

Definition 3. *Given an ontology* $\mathcal{K} = \langle \mathcal{T}, \mathcal{R} \rangle$, *a demand* β, *a set of buyer's preferences* $\mathcal{B}$ *and a disagreement threshold* t_β, *a supply* σ *and a set of seller's preferences* $\mathcal{S}$ *and a disagreement threshold* t_σ, *an agreement in a bilateral matchmaking process is a model* $\mathcal{I}$ *of*

$$\bar{\mathcal{K}} = \langle \mathcal{T} \cup \{\sigma, \beta\} \cup \{Buy = (\mathcal{B}[\geqslant t_\beta]), Sell = (\mathcal{S}[\geqslant t_\sigma])\}, \mathcal{R} \rangle .$$

Clearly, not every agreement $\mathcal{I}$ is beneficial both for the buyer and for the seller. We need a criterion to find the optimal mutual agreement. Given a demand and a set of supplies, for each of them we will compute the optimal agreement with the demand and we will rank them with respect to the buyer's utility value in the optimal agreement itself.

To compute an optimal agreement we rely on the notion of Pareto agreement. Given an ontology $\mathcal{K}$ representing a set of constraints, we are interested in agreements that are Pareto-efficient, in order to make traders as much as possible satisfied. In our fuzzy DL based framework, in order to compute a *Pareto agreement* we procede as follows. Let $\mathcal{K}$ be a fuzzy DL ontology, let β be the buyer's demand, let σ be the seller's supply, let $\mathcal{B}$ and $\mathcal{S}$ be respectively the buyer's and seller's preferences. We define $\bar{\mathcal{K}}$ as the ontology

$$\bar{\mathcal{K}} = \langle \mathcal{T} \cup \{\sigma, \beta\} \cup \{Buy = (\mathcal{B}[\geqslant t_\beta]), Sell = (\mathcal{S}[\geqslant t_\sigma])\}, \mathcal{R} \rangle .$$

In $\bar{\mathcal{K}}$, the concept Buy collects all the buyer's preferences. Hence, the higher is the maximal degree of satisfiability of Buy (*i.e.*, $bsb(\bar{\mathcal{K}}, Buy)$), the more the buyer is satisfied. Similarly, the concept $Sell$ collects all the seller's preferences in such a way that the higher is the maximal degree of satisfiability of $Sell$ (*i.e.*, $bsb(\bar{\mathcal{K}}, Sell)$), the more the seller is satisfied. Now, it is clear that the best agreement among the buyer and the seller is the one assigning the maximal degree of satisfiability to the conjunction $Buy \sqcap Sell$ (remember we use Łukasiewicz semantics). In formulae, once we determine

$$v_P = bsb(\bar{\mathcal{K}}, Buy \sqcap Sell) ,$$

we can say that a *Pareto agreement* is a model $\bar{\mathcal{I}}$ of $\bar{\mathcal{K}}$ such that

$$v_P = \sup_{x \in \Delta^{\mathcal{I}}} (Buy \sqcap Sell)^{\mathcal{I}}(x) > 0 ,$$

that is the *Pareto agreement value* is attained at $\bar{\mathcal{I}}$ and has to be positive.

3.1   The Matchmaking Process

Summing up, given a demand and a set of supplies, the bilateral matchmaking process is executed covering the following steps:

Initial Setting. The buyer defines *hard constraints* β and preferences (*soft constraints*) $\mathcal{B}$ with corresponding weights for each preference $n_1, n_2, ..., n_k$, as well as the threshold t_β. The same the sellers did when they posted the description of their supply within the marketplace[4]. Notice that for numerical features involved in the negotiation process, both in β and σ their respective reservation values are set either in the form $(\leqslant f\ r_f)$ or in the form $(\geqslant f\ r_f)$.

Find and Rank Agreements. For each supply in the marketplace, the matchmaker computes the corresponding Pareto agreement (see Section 3). Given a supply σ_i and the corresponding optimal agreement $\bar{\mathcal{I}}_i$, we rank σ_i w.r.t. the value of $Buy^{\bar{\mathcal{I}}_i}$, *i.e.*, w.r.t. the buyer's degree of satisfiability.

Let us present a tiny example in order to better clarify the approach. For the sake of simplicity, we will consider only one seller, clearly, in a real case scenario, the whole process will be repeated for each seller's supply posted in the e-marketplace. Given the toy ontology $\mathcal{K} = \langle \mathcal{T}, \emptyset \rangle$, with

$$\mathcal{T} = \begin{cases} Sedan \sqsubseteq PassengerCar \\ ExternalColorBlack \sqsubseteq \neg ExternalColorGray \\ SatelliteAlarm \sqsubseteq AlarmSystem \\ InsurancePlus = DriverInsurance \sqcap TheftInsurance \\ NavigatorPack = SatelliteAlarm \sqcap GPS\_system \end{cases}$$

The buyer and the seller specify their *hard* and *soft constraints*. For each numerical feature involved in *soft constraints* we associate a fuzzy function. If the bargainer has stated a reservation value on that feature, it will be used in the definition of the fuzzy function, otherwise a default value will be used.

β is $B = PassengerCar \sqcap (\leqslant price\ 26000)[\geqslant 1.0]$

$\beta_1 = ((\exists HasAlarmSystem.AlarmSystem) \rightarrow (\exists Has\ Price.L(22300, 22750)))$

$\beta_2 = ((\exists HasInsurance.DriverInsurance) \sqcap$
$\qquad \sqcap((\exists HasInsurance.TheftInsurance) \sqcup (\exists HasInsurance.FireInsurance)))$

$\beta_3 = ((\exists HasAirConditioning.Airconditioning) \sqcap (\exists HasExColor.(ExColorBlack \sqcup$
$\qquad ExColorGray)))$

$\beta_4 = (\exists price.ls(22000, 24000))$

$\beta_5 = (\exists km\_warranty.rs(150000, 175000))$

$\mathcal{B} = (0.1 \cdot \beta_1 + 0.2 \cdot \beta_2 + 0.1 \cdot \beta_3 + 0.2 \cdot \beta_4 + 0.4 \cdot \beta_5)[\geqslant 0.7]$

σ is $S = Sedan \sqcap (\geqslant price\ 22000)[\geqslant 1.0]$

$\sigma_1 = ((\exists HasNavigator.NavigatorPack) \rightarrow (\exists Has\ Price.R(22500, 22750))))$

$\sigma_2 = (\exists HasInsurance.InsurancePlus)$

$\sigma_3 = (\exists km\_warranty.ls(100000, 125000))$

$\sigma_4 = (\exists HasMWarranty.L(60, 72))$

$\sigma_5 = ((\exists HasExColor.ExColorBlack) \rightarrow (\exists Has\ AirConditioning.AirConditioning))$

$\mathcal{S} = (0.3 \cdot \sigma_1 + 0.1 \cdot \sigma_2 + 0.3 \cdot \sigma_3 + 0.1 \cdot \sigma_4 + 0.2 \cdot \sigma_5)[\geqslant 0.6]$

---

[4] An investigation on how to compute t_β, t_σ, n_i and m_i is out of the scope of this paper. We can assume they are determined in advance by means of either direct assignment methods (Ordering, Simple Assessing or Ratio Comparison) or pairwise comparison methods (like AHP and Geometric Mean).

Let $\bar{\mathcal{K}} = \langle \mathcal{T} \cup \{\sigma, \beta\} \cup \{Buy = (\mathcal{B}[\geqslant t_\beta]), Sell = (\mathcal{S}[\geqslant t_\sigma])\}, \mathcal{R}\rangle$, it can be verified that the Pareto optimal agreement value is

$$v_P = bsb(\mathcal{K}, Buy \sqcap Sell) = 0.7625 \,,$$

with a Pareto agreement $\bar{\mathcal{I}}$ that maximally satisfies

$$(= HasPrice\, 22500.0) \sqcap (= HasKMWarranty\, 100000.0) \sqcap (= HasMWarranty\, 60.0)\,.$$

4   Related Work and Discussion

Automated bilateral negotiation has been widely investigated, both in artificial intelligence and in microeconomics research communities. AI-oriented research has usually focused on automated negotiation among agents, and on designing high-level protocols for agent interaction [6,11]. As stated in [13], negotiation mechanisms often involve the presence of a mediator, which collects information from bargainers and exploits them in order to propose an efficient negotiation outcome. Various recent proposals adopt a mediator, including [10,7]. However in these approaches no semantic relations among issues are investigated. Several recent logic-based approaches to negotiation are based on propositional logic. In [4], Weighted Propositional Formulas (WPF) are used to express agents preferences in the allocation of indivisible goods, but no common knowledge (as our ontology) is present. The work presented here builds on [17], where a basic propositional logic framework endowed of a logical theory was proposed. In [16] the approach was extended and generalized and complexity issues were discussed. We are currently investigating other negotiation protocols, without the presence of a mediator, allowing to reach an agreement in a reasonable amount of communication rounds. The use of aggregate operators is also under investigation.

Acknowledgement.   We wish to thank Francesco M. Donini for fruitful suggestions and discussions.

References

1. Baader, F., Hanschke, P.: A schema for integrating concrete domains into concept languages. In: Proc. of IJCAI 1991, pp. 452–457 (1991)
2. Binmore, K.: Fun and Games. A Text on Game Theory. D.C. Heath and Company (1992)
3. Straccia, U., Bobillo, F.: fuzzyDL: An Expressive Fuzzy Description Logic Reasoner. In: Proc. of the 2008 International Conference on Fuzzy Systems (FUZZ 2008) (2008)
4. Bouveret, S., Lemaitre, M., Fargier, H., Lang, J.: Allocation of indivisible goods: a general model and some complexity results. In: Proc. of AAMAS 2005, pp. 1309–1310 (2005)
5. Chevaleyre, Y., Endriss, U., Lang, J.: Expressive power of weighted propositional formulas for cardinal preference modeling. In: Proc. of KR 2006, pp. 145–152 (2006)
6. Chevaleyre, Y., Endriss, U., Lang, J., Maudet, N.: Negotiating over small bundles of resources. In: Proc. of AAMAS 2005, pp. 296–302 (2005)
7. Gatti, N., Amigoni, F.: A decentralized bargaining protocol on dependent continuous multi-issue for approximate pareto optimal outcomes. In: Proc. of AAMAS 2005, pp. 1213–1214 (2005)

8. Hájek, P.: Metamathematics of Fuzzy Logic. Kluwer, Dordrecht (1998)
9. Jennings, N.R., Faratin, P., Lomuscio, A.R., Parsons, S., Wooldridge, M.J., Sierra, C.: Automated negotiation: prospects, methods and challenges. Int. J. of Group Decision and Negotiation 10(2), 199–215 (2001)
10. Klein, M., Faratin, P., Sayama, H., Bar-Yam, Y.: Negotiating complex contracts. In: Proc. of AAMAS 2002, pp. 753–757 (2002)
11. Kraus, S.: Strategic Negotiation in Multiagent Environments. The MIT Press, Cambridge (2001)
12. Lukasiewicz, T., Straccia, U.: Tutorial: Managing uncertainty and vagueness in semantic web languages. In: Twenty-Second Conference on Artificial Intelligence (AAAI 2007) (2007)
13. MacKie-Mason, J.K., Wellman, M.P.: Automated markets and trading agents. In: Handbook of Computational Economics. North-Holland, Amsterdam (2006)
14. Nash, J.F.: The Bargaining Problem. Econometrica 18(2), 155–162 (1950)
15. Parsons, S., Sierra, C., Jennings, N.: Agents that reason and negotiate by arguing. Journal of Logic and Computation 8(3), 261–292 (1998)
16. Ragone, A., Di Noia, T., Di Sciascio, E., Donini, F.M.: A logic-based framework to compute pareto agreements in one-shot bilateral negotiation. In: Proc. of ECAI 2006, pp. 230–234 (2006)
17. Ragone, A., Di Noia, T., Di Sciascio, E., Donini, F.M.: Propositional- logic approach to one-shot multi issue bilateral negotiation. ACM SIGecom Exchanges 5(5), 11–21 (2006)
18. Raiffa, H.: The Art and Science of Negotiation. Harvard University Press (1982)
19. Rosenschein, J.S., Zlotkin, G.: Rules of Encounter. MIT Press, Cambridge (1994)
20. Salkin, H., Kamlesh, M.: Foundations of Integer Programming. North-Holland, Amsterdam (1988)
21. Wellman, M.P.: Online marketplaces. In: Practical Handbook of Internet Computing. CRC Press, Boca Raton (2004)
22. Wooldridge, M., Parsons, S.: Languages for negotiation. In: Proc. of ECAI 2004, pp. 393–400 (2000)
23. Zhang, D., Zhang, Y.: A computational model of logic-based negotiation. In: Proc. of the AAAI 2006, pp. 728–733 (2006)

Machine Vision Application to the Detection of Micro-organism in Drinking Water

Hernando Fernandez-Canque[1], Sorin Hintea[2], Gabor Csipkes[2], Allan Pellow[1], and Huw Smith[3]

[1] Glasgow Caledonian University, School of Engineering & Computing
Cowcaddens Road, Glasgow. G4 0BA. United Kingdom
`hfe@gcal.ac.uk`
[2] Faculty of Electronics and Telecommunication, Technical University of Cluj Napoca, Str. Baritiu, Nr. 26-28, 3400 Cluj Napoca, Romania
Fax: +4 064 191 340; Tel.: +4 064 196 285
[3] Stobhill General Hospital, Scottish Parasite Diagnostic Laboratory
Springburn, Glasgow. G21 3UW. United Kingdom

Abstract. The work presented in this paper uses a novel Machine Vision application to detect and identify micro-organism oocysts on drinking water. This new concept of water borne micro-organism detection uses image processing to allow detailed inspection of parasite morphology to nanometre dimensions. The detection results are more reliable than existing manual methods. Combining Normarski Differential Interface Contrast (DIC) and fluorescence microscopy using Fluorescein Isothiocyanate (FITC) and UV filters, the system provides a reliable detection of micro-organisms with a considerable reduction in time, cost and subjectivity over the current labour intensive time consuming manual method.

Keywords: Machine vision, image processing, water quality.

1 Introduction

The problem of having a water supply free of pathogens micro-organism has been of great concern to public health and water authorities. Many countries in the world have had fatal contamination due to infection produced by their base drinking water. The problem is not confined to economically poor countries but to developed countries as well. Europe and the US have had dozens of water borne outbreaks affecting more than a million individuals in the last 15 years [1].

To reduce the risk of more outbreaks, some national regulatory bodies and public health authorities have created sets of regulations and protocols. The European Union has produced a Drinking Water Directive [2] to tackle this problem. This directive states that 'drinking water should be free from micro –organisms to a certain level that does not constitute a potential danger to human health'. Some countries have put a figure on this water directive. The Secretary of State for the Environment in the UK has issued an amendment to the water supply regulations which states, in the case of cryptosporidium, that the water intended for human consumption should contain no

I. Lovrek, R.J. Howlett, and L.C. Jain (Eds.): KES 2008, Part III, LNAI 5179, pp. 302–309, 2008.

more than one cryptosporidium oocyst in 10 litres of water. [3]. In order to ensure the implementation of water regulations, a Standard Operating Protocol (SOP) for the monitoring of drinking water has been issued. [4]

The Machine Vision System presented here has been designed to follow this SOP and mimic the current manual analysis procedure in order to simplify and improve detection and reliability of micro-organism detection and enumeration.

2   Current Manual Detection of Micro-organisms

Large micro-organisms can be filtered with relatively ease, others can be neutralised by chemical treatment. The main problem of drinking water is infection with smaller micro-organisms that are difficult to eliminate due to their size. In the UK, the monitoring procedure consists of taking 40 litres of water per hour from each point at which water leaves the treatment works.

2.1   Current Detection

The simplest technique is Nomarski Differential Interference Contrast (DIC) brightfield microscopy; it does not require any special laboratory sample preparation. Under DIC microscopy, the oocyst shape can be defined, and some sporozoites may be distinguished, although in some cases with poor reliability [5].

For smaller micro-organisms of complex morphology a more detailed analysis is required. Samples are stained and dried onto microscope slides and micro-organisms are detected using fluorescence microscopy. The drying process causes oocysts to collapse which in some cases leads to shape distortion and the physical release of sporozoites. Fluorescein isothiocyanate (FITC) labelled monoclonal antibodies assist in the detection and measurement of oocysts in environmental samples by defining the wall of the oocyst, hence defining shape. For a more detailed analysis the use of the fluorogenic dye 4'6 diamidino-2-phenylindole (DAPI) staining, highlight the sporozoite nuclei in blue, provides supportive evidence.

2.2   Problem Associated with SOP

The detection of micro-organisms is done following the SOP. DIC detection can be used to eliminate certain micro organism and algae based on their dimension and shape. The current manual method uses well trained operators to observer these small micro-organisms under a microscopy to detect their morphology using FITC and DAPI. The process is expensive, labour intensive, time consuming and unreliable. Also duplicate counting may occur as each slide well is scanned.

3   Machine Vision Detection

The Machine Vision System presented here has been designed and implemented to simplify and improve the detection and analysis procedures required by the SOP in term of laboratory and analytical procedures. Our system follows the manual procedure exactly, making the system compatible with other manual laboratory findings.

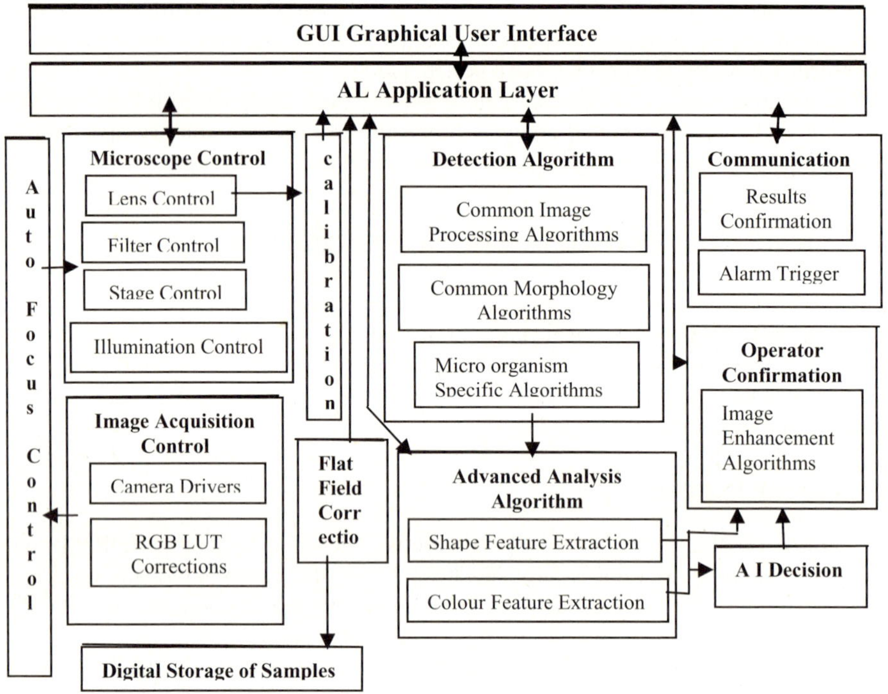

Fig. 1. Machine Vision Software Architecture

Detection can be confirmed at different stages. Some research groups have reported limited results of larger micro organism [6,7] using different methods. The SOP is implemented according to the software architecture presented in Figure 1.

3.1 Microscope Control

The Microscope Control Block provides the interface to the motorised focus and stage controller (such as Prior Scientific Proscan Advanced Controller), prompts the user for the correct microscope set-up, generates and stores the calibration data for every lens (needed to convert the size information from pixels in real world units) and controls the illumination to avoid camera overload.

3.2 Image Acquisition Control

The image Acquisition Control Block provides the software interface to the frame grabber and corrects the acquired image to compensate camera aberrations and non-linear colour encoding.

3.3 Auto Focus Control

The auto-focus algorithm automatically controls the microscope.

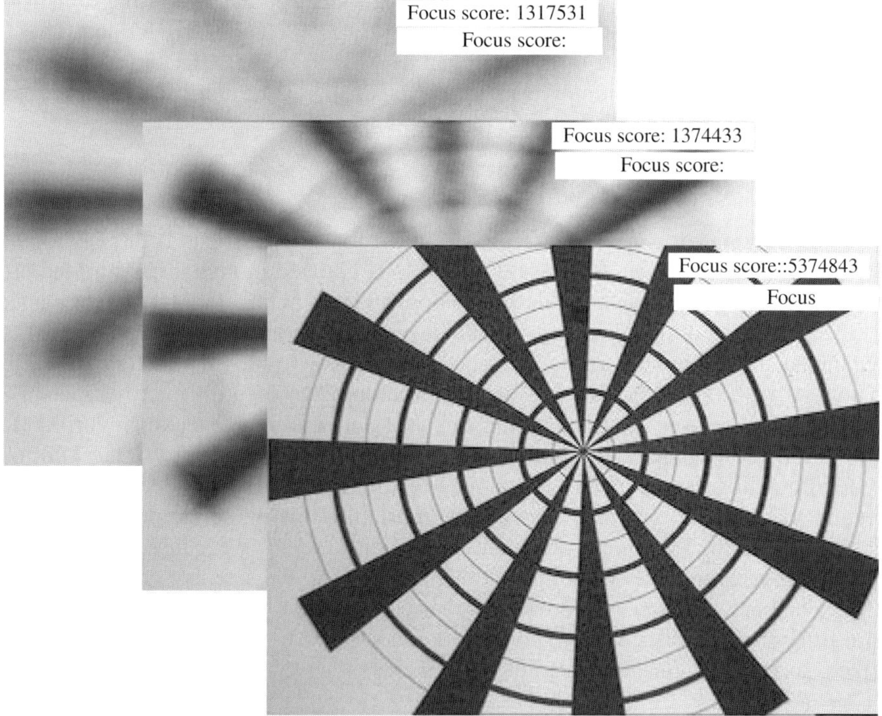

Fig. 2. Three test frames with a focus score indication of the sum of all pixel values

Auto-focussing is achieved by computing a focus score, and then trying to maxi-mise this score by adjusting the Z stage of the microscope. The focus score provides a measure of the overall focus quality; therefore in the case of an uneven focus plane this algorithm would find the best compromise. Figure 2 shows an example of focus control. The algorithm for computing the focus score is fast enough for a reliable "on the fly" auto-focus.

4 Detection of Micro-organisms

The machine vision system can be used to obtain a digital picture using their frame grabber allowing the manipulation of the picture at the pixel level.

4.1 DIC Detection

Detection and identification of micro organisms under DIC microscopy is the most desirable procedure as it does not involve any laboratory preparation of samples. This method is not used under the current SOP because it is very unreliable as little con-trast exists between object of interest and background. With a special highlight details filter using a kernel size of 7 the contrast has been dramatically increased in our system [8] and it can be used to eliminate objects that are not within the dimension range of micro-organisms of interest. Our improved method for DIC analysis, although not accepted as part of the SOP can be used to create a data base containing the position of each presumptive micro-organism. It also eliminates Binary Large Objects (BLOB) outside the searched range.

4.2 FITC Prepared Sample Detection

Samples are stained using FITC to prepare them for analysis. This is the method accepted by the SOP. Once a frame for a sample is grabbed in digital form by the proposed Machine Vision the analysis can be implemented mimicking the current manual method. One aim of this project is to include as much as possible the human expertise accumulated at Scottish Parasite Diagnostic Laboratory to provide a reliable machine vision solution. Our Machine Vision can be tailored to identify micro-organism in general. The proposed Machine Vision can use select different detection algorithm for different micro-organisms as considered in the software architecture (figure 1) of our system. An illustration of the advantages of this method is considered in the next section using as an example one of the most difficult micro-organism to detect: Cryptosporidium. This micro-organism is difficult to identify because of its size and morphology as there are some algae with similar shape and size, but cryptosporidium has a distinct feature: the oocyst is spherical or slightly ovoid, being very small (4-6 µm), smooth, thick walled, colourless and refractive; containing four elongated, naked sporozoites. Each sporozoite contains one nucleus.

4.2.1 Detection Algorithm

Detection of FITC stained Cryptosporidium using Epi-illumination fluorescence microscopy is considered the best method available, being the one employed in the current Standard Operating Protocol. The separation between the background and the objects of interest is defined to the degree that threshold algorithm can be performed. Figure 3 shows a sample containing cryptosporidium oocysts. Threshold is performed, in two phases, one for the lighter component, and one for the darker component, then the results are combined using an OR function.

All objects truncated by the acquisition process, namely objects which are partially in the image and partially outside the image boundaries, are eliminated.

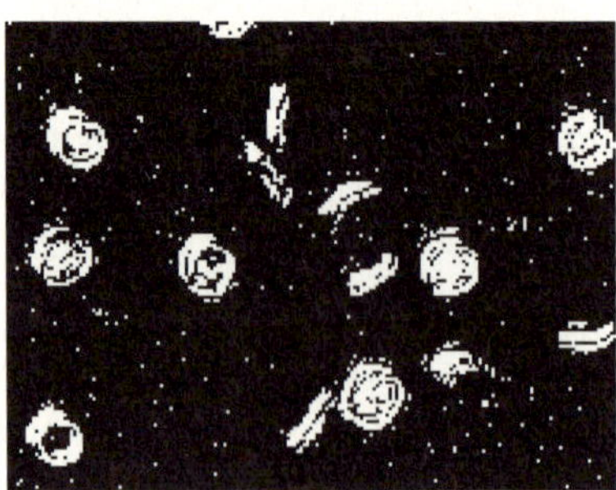

Fig. 3. Sample of water containing cryptosporidium with threshold images combined

The binary noise present is efficiently eliminated using the following algorithm:

(i) A buffer copy of the image to be cleaned up is generated. (ii) Two successive erode functions are applied on the original image. (iii) All pixels from the buffer copy 8-connected to the non-zero pixels from the image are added to the image. (iv) Step (iii) is repeated until no pixel is added. The result is presented in Figure 4. It can be noted that, unlike other binary noise reduction algorithms, the objects of interest are unaffected.

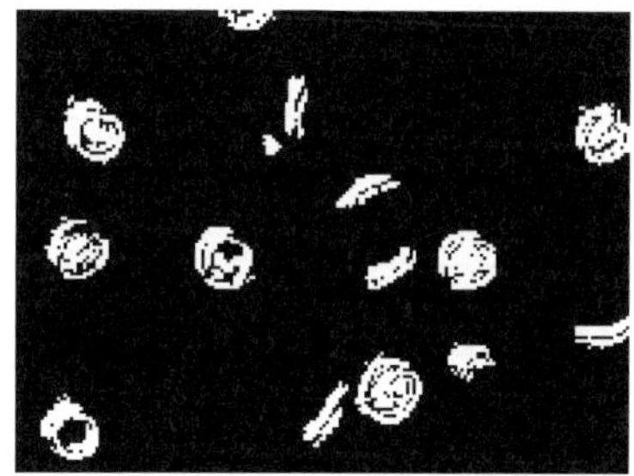

Fig. 4. Sample with binary noise clean-up

The next step is to reinsert the missing pixels within the objects boundaries. A closing algorithm is performed, using a kernel size of 5. A NOT function is performed, followed by a labelling function. The result is an image which has a value of 0 associated with all objects, a value of 1 associated with the background and a value greater than 1 for every hole in the objects. By replacing the values greater than 1 with 0 and negating the image again we achieve holes filling. All objects too small to be a Cryptosporidium are eliminated. This is achieved using the same algorithm as for binary noise removal, but with 7 erosion functions applied. Then a distance function is applied, and Danielsson [9] circle detection algorithm is used for Cryptosporidium detection. The result is presented in Figure 5.

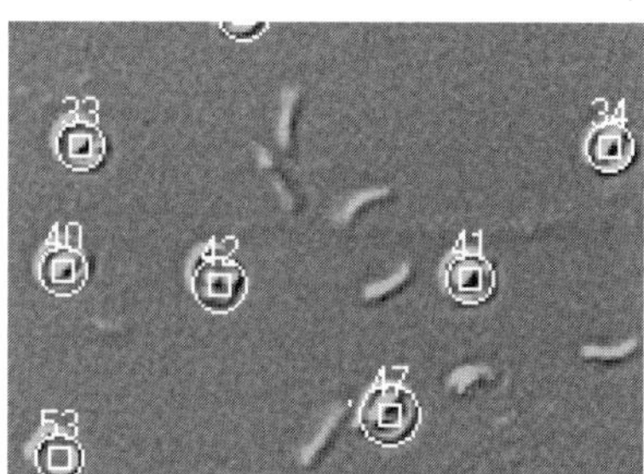

Fig. 5. Sample final result oocysts identified

4.2.2 Advanced Analysis Algorithm

Each presumptive oocyst is centred and focussed, using in the first stage the FITC filter block and a total magnification of 1000x. The advance analysis algorithm is an application mentioned above but this time we can use different colour planes. This allow to determining the existence of the 4 nucleus in an oocyst hence identification of a cryptosporidium. On the green plane, both the wall and the nucleons of the Cryptosporidium are detected. On the red plane, only the nucleons were acquired. This provides a method of separation. Firstly the green plane was taken and the highlight details filter was applied. A threshold within the value 95 to 255 was used, the binary noise is eliminated using an erode function followed by a dilation. The BLOB was enclosed using dilation followed by erosion. On the red plane a look-up table function is applied. Finally, threshold and binarisation is used, followed by noise removal and BLOB closing. At this stage the number of nucleons present in this micro-organism can be determined. This is achieved using the same algorithm, which inscribed circles. By subtracting the nucleons from the green plane in binary format the wall was obtained.

5   Classifications and Enumeration

In our laboratory test results, detection using this algorithm was 100% successful.

The Advanced Analysis algorithm is a feature extraction algorithm, extracting relevant information under FITC and DAPI filters at 1000x total magnification, for each potential Cryptosporidium oocyst tagged by the Detection Algorithm. These parameters are stored together with the images acquired for each potential Cryptosporidium oocyst for each mode. Table 1 show morphological analysis results for a series of test cryptosporidium oocysts.

Table 1. Test Results for Cryptosporidium oocysts

Object #	1	2	3	4	5	6
Area (unit)	22.45	25.38	46.44	26.79	24.66	21.72
Perimeter (unit)	17.63	21.34	30.85	19.57	19.92	17.92
Width (unit)	5.49	6.11	8.7	5.89	5.45	5.43
Height (unit)	5.16	5.89	7.63	5.68	6.32	5.21
Waddel disk diameter (unit)	5.35	5.68	7.69	5.84	5.6	5.26
Compactness factor	0.79	0.7	0.7	0.8	0.72	0.77
Heywood circularity factor	1.05	1.19	1.28	1.07	1.13	1.08

5.1   AI Application

After the completion of the Advanced Analysis the system goes into the next stage – Artificial Intelligence Decision Making. Expert System have problem with flexibility for this application and Artificial Neural Network encounter problem with training. We found that the use a Fuzzy Logic suit better this applications to mimic human knowledge based decision making. The classification is based on the features extracted by the Advanced Analysis Algorithm and a customisable rule base.

At the completion of the Artificial Intelligence Decision Making, all the acquired images can be presented to an expert, if required, sorted accordingly to their classification, together with the relevant information about each object such as precise size, FITC staining analysis and DAPI confirmation, and their position on the sample. An experienced operator can quickly review the classification process, or can mark any object for an expert to confirm the analysis if required.

6 Conclusions

The Machine Vision approach presented in this paper can perform automated analysis to determine whether or not micro-organism oocysts are present in treated water. The system can reliably determine the presence of micro-organisms and enable samples to be accurately and efficiently reviewed by an operator.

The proposed approach allows a reliable detection of waterborne micro-organisms in large quantities of water. This provides an industrial system to monitor micro-organisms in the water industry. The implemented algorithm complies with the standard operating protocol provided by the water authorities in UK and satisfies the EU directive on drinking water quality. The Machine Vision proposed provides a 100% detection of cryptosporidium micro-organism as test case and outperforms the current manual detection of micro-organisms in term of cost, time of results, accuracy and reliability.

References

1. Smith, H.V., Rose, J.B.: Waterborne Cryptosporidiosis: Current Status. Parasitology Today 14(1) (1998)
2. Council Directive 98/83/EC on the quality of water intended for human consumption. Official Journal L 330, 32–50 (1998)
3. The Water Supply (Water Quality) (Amendment) Regulations 1999 No. 1524, Water Industry, England and Wales (1999)
4. Standard Operating Protocol for the Monitoring of Cryptosporidium Oocysts in Treated Water Supplies to satisfy The Water Supply Regulations 1999 No. 1524, UK (1999)
5. Fricker, C.R.: Protozoam Parasites and Water, pp 91–96. The Royal Society of Chemistry, Cambridge
6. Alvarez, T., Martin, Y., Perez, S., Santos, F.: Classification of Microorganism using Image Processing Techniques. In: Proceedings of the International Conference on Image Processing, 2001, vol. 1, pp. 329–332 (2001)
7. Rodenacker, K., Gais, P., Jutting, U., Hense, B.A.: (Semi-)automatic Recognition of Micro-organisms in Water. In: Proceedings of the International Conference on Image Processing, 2001, October 7-10, 2001, vol. 3, pp. 30–33 (2001)
8. Fernandez- Canque, H.L., Bota, S., Beggs, B., Smith, E., Hintea, S., Smith, H.V.: A Novel Approach for Industrial Cryptosporidium Monitoring. In: The 20th International Manufacturing Conference IMC20 Cork, Ireland, pp. 465–470 (2003) ISBN 0954573609
9. Nilson, F., Danielsson, P.-E.: Finding the Minimal Set of Maximum Disks for Binary Objects. Graphical Models and Image Processing 1, 55–60 (1997)

Adaptive and Evolvable Hardware and Systems: The State of the Art and the Prospectus for Future Development

Mircea Gh. Negoita[1], Lukas Sekanina[2], and Adrian Stoica[2]

[1] KES International, 2nd Floor, 145-157 St John Street, London, EC1V 4PY,
United Kingdom
mnegoita@hotmail.com
[2] Faculty of Information Technology, Brno University of Technology, Bozetechova 2,
612 66Brno, Czech Republic
sekanina@fit.vutbr.cz
[3] NASA Jet Propulsion Laboratory (JPL), USA, 4800 Oak Grove Drive MS 303-300,
Pasadena, CA 91109
adrian.stoica@jpl.nasa.gov

Abstract. This paper is an overview on the Evolvable Hardware (EHW) – th exciting and rapidly expanding industrial application area of the Evolutionary Computing (EC), of the Genetic Algorithms especially. The content of the work has the following structure: the first part includes generalities on industrial applications of EC and the importance of EHW in this frame; the second part presents the outstanding technological support making possible the implementation of system adaptation in hardware. Different kind of programmable circuits arrays are introduced. The third part tackles the most known EC based methods for EHW implementation; the fourth part deals with some concrete elements of the EHW design, including the current limits in evolutionary design of digital circuits. The last part is focused on some concluding remarks with regard to future perspectives of the area. A list of references used in this work was inserted at the end.

1 Introduction

Nevertheless, it is hardware implementation of the most benefit for the society and indeed most revolutionizing application of **EC** by leading to the so-called **EHW**. These new **EC** based methodologies make possible the hardware implementation of both genetic encoding and artificial evolution, having a new brand of machines as a result.

This type of machines is evolved to attain a desired behaviour that means they have a *behavioural computational intelligence*. There is no more difference between adaptation and design concerning these machines, these two concepts representing no longer opposite concepts. A dream of technology far years ago currently became reality: adaptation transfer from software to hardware is possible by the end. Much more, the electronics engineering as a profession was radically changed: the most based on soldering assembling manufacturing technologies are largely replaced now by programming circuitry-based technologies, including **EHW** technologies.

I. Lovrek, R.J. Howlett, and L.C. Jain (Eds.): KES 2008, Part III, LNAI 5179, pp. 310–318, 2008.
© Springer-Verlag Berlin Heidelberg 2008

1.1 Definitions, General Consideration and Classification of EHW

A *definition of* the **EHW** may be as follows: a sub-domain of artificial evolution represented by *a design methodology* (consortium of methods) involving the application of **EC** to the synthesis of digital and analogue electronic circuits and systems. A more agreed definition among the practitioners might be: **EHW** means *programmable hardware* that can be evolved [2]. But some members of the scientific community acting in the area consider the term *evolutionary circuit design* more descriptive for **EHW** features. Much more, another term is used nowadays for the same work - *evolware* concerning to this evolvable ware with hardware implementation. This leads to a future perspective of using the term *bioware* concerning to a possible evolving ware with biologic environments implementation. Even some other environments are seen as possible evolvable media: *wetware* – real chemical compounds are to be used as building blocks or *nanotechnology* – relied on molecular scale engineering.

This new design methodology for the electronic circuits and systems is not a fashion. It is suitable to the special uncertain, imprecise or incomplete defined real-world problems, claiming a continuous adaptation and evolution too. An increased efficiency of the methodology may be obtained by its application in the soft-computing framework that means in aggregation with other intelligent technologies [3; 4] such as fuzzy logic (**FS**) and neural networks (**NN**)., Artificial Immune Systems (**AIS**) , evolutionary algorithms (**EA**) . The reason of **EHW** using the above mentioned type of application is relied on the main advantage over the traditional engineering techniques for the electronic circuit design, namely the fact that the designer's job is very much simplified following an algorithm [5] with a step sequence as below:

STEP 1 – *problem specification* - requirements specification of the circuit to be designed – specification of basic (structural) elements of the circuit
STEP 2 – *genome codification* -an adequate (genotypic) encoding of basic elements to properly achieve the circuit description
STEP 3 – fitness *calculation* - specification of testing scheme used to calculate the genome fitness
STEP 4 – *evolution (automatically generation of the required circuit)* - generation of the desired circuit

Despite of the above-mentioned advantage of **EHW** methodology, its disadvantage is not to be neglected: (sometimes) suboptimal results are got over those ones of the classical methods that remain preferable in case of well-defined problems. The designer himself is involved by acting directly during the first three steps, while the fourth step is an *automatically generation* of the circuit. The flow modality of both step *3* and step *4* leads to same categorizing classes criteria for **EHW** (see [5]).

The practical **EHW** classification [6], can be adopted by: hardware type; controller type; Objective Function.

Based on the type of hardware and type of changes, **EHW** is classified as follows:

> -**R***econfigurable* **d***igital* **h***ardware* (**RDH**) - means digital hardware that changes quasi-instantly (short transition time compared with to processing time) and directly (single transition, no meaningful intermediary transition states) from the initial to the final configuration.

-**M**orphable **d**igital **h**ardware (**MDH**) – means digital hardware that changes from one complete configuration to another through a set of intermediate states , each having a functional role.
-**R**econfigurable **a**nalog **h**ardware (**RAH**) – is analog hardware with a dynamics of changes that is similar to **RDH** from the initial to final configuration
-**M**orphable **a**nalog **h**ardware (**MAH**) – is analog hardware that changes gradually without switches.
-**A**djustable/**T**unable/**P**arametric **HW** (**ATPHW**) - is hardware in which the changes influence the parameters of a function
Classification by controller type leads to :
-**E**volutionary **h**ardware – **EHW** in which the controller employs evolutionary algorithms
-**E**mbryonic **h**ardware – **EHW** in which the controller employs a embryonic-initiated growth mechanism
The classification by Objective Function is a more general one , covering all kind of Adaptive Systems, not the **EHW** only:

-*External adaptation* - featured by adaptive behaviour in the presence of stimuli originating in the surrounding environment
-*Internal adaptation* – featured by adaptive behaviour in the presence of a disturbance located in the system itself
-*Darwinian adaptation* – featured by adaptive behaviour when the response is directed toward modifying the object
-*Singerian adaptation* – adaptive behaviour when the response is directed towards modifying the environment
-**S**mart **S**ystem (**SS**) – a system aware of its state , operation and changing environment. It can predict what will happen at a certain future . This knowledge can lead to adaptation.
-**S**mart **a**daptive **s**ystem (**SAS**) – can adapt to a changing environment, a similar setting without being "ported" to it, adapt to a new/unknown application

Dramatic changes happen in the relation between hardware and the application environment, and this in case of *malicious faults* or need for *emergent new functions* that claim for in-situ synthesis of a totally new hardware configuration. **EHW** is suitable for flexibility and survivability of autonomous systems as that ones developed by NASA JPL. **EHW** *survivability* means to maintain functionality coping with changes in hardware characteristics under the circumstances of adverse environmental conditions as for example: temperature variations, radiation impacts, aging and malfunctions. **EHW** *flexibility* means the availability to create new functionality required by changes in requirements or environment.

The application developer may meet different design tasks to be evolved. As the case, the design to be evolved could be: a program, a model of hardware or the hardware itself. Algorithms that run outside the reconfigurable hardware, mainly feature the actual **EHW** state of the art. but also some chip level attempts were done. The path from chromosome to behavior data-file is different in case of intrinsic and *extrinsic* (see more in [1]). **EHW** Evolution in Simulation has some typical features and parameters value as follows [5]: computationally intensive (640,000 individuals for about 1000 generations); runs over tens of hours, expected about 3 min in 2010 on desktop PC for experiments with netlists of about 50 nodes; SPICE scales badly , namely time increases nonlinearly with as a function of nodes in netlist , in about a sub-quadratic to quadratic way; no existing hardware resources allow porting the technique to evolution directly in hardware (and not sure will work in hardware). See in [1] more details on the huge advantage of the on-chip **EHW** versus CAD/synthesis tools.

1.2   The Technological Support of EHW

The appearance of programmable integrated circuits, especially their new generation – *field programmable gate arrays* (**FPGA**s) and most recently reconfigurable *analogue arrays* (**FPAA**s) and *field-programmable interconnection circuits* (**FPIC**s) or configurable digital chips at the functional block level, (*open-architecture* **FPGA**s) make possible for most companies to evolve circuits as the case would be. The appearance of reconfigurable *analogue arrays* (**FPAA**s) was crucial for **EHW** technological support. The analog reconfigurable hardware allows prevention or removal of essential fabrication mismatches and other refined technological problems by evolving circuits. The programmability of **FPAA**s is limited to just only allow configuration around op-amp level. But the applications require also for many interesting circuit topologies to be evolved below the op-amp level. This application requirement led to another kind of programmable (reconfigurable) hardware   relied on *evolution-oriented devices* that have some advantages over **FPAA**, namely: can reprogram many times, can understand what's inside and are featured by a flexible programmability. This is the so called *custom made* **EHW**-*oriented reconfigurable hardware..* See in [1] more details regarding the types of *custom made* **EHW**-*oriented reconfigurable hardware.*

2   The Structure of an EHW System. EC Based of EHW Implementation

The language for programming reconfigurable hardware must define: an *alphabet,* expressing choices of cells and, a *vocabulary/grammar* , expressing the rules of interconnect.An **EHW** system, any is its destination, either for demonstrations, prototype experiments or real time implementation, must be structured in from of two main components: the *reconfigurable hardware* (**RH**) and the *reconfigurable mechanism* (**RM**) [7]. Regarding the practical ways of implementing an evolvable system, real-world applications requirements are toward a reliable solution featured by compactness, low-power consumption and autonomy. The evolution on **JPL SABLES** (*Stand-Alone Board-Level Evolvable System*) [8] is a solution that proved to be effective in various applications, see Fig. 1.

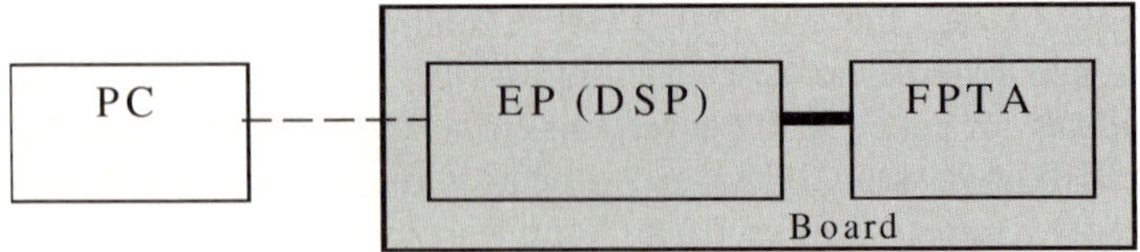

Fig. 1. The simplified bloc diagram of JPL **SABLES** [14]

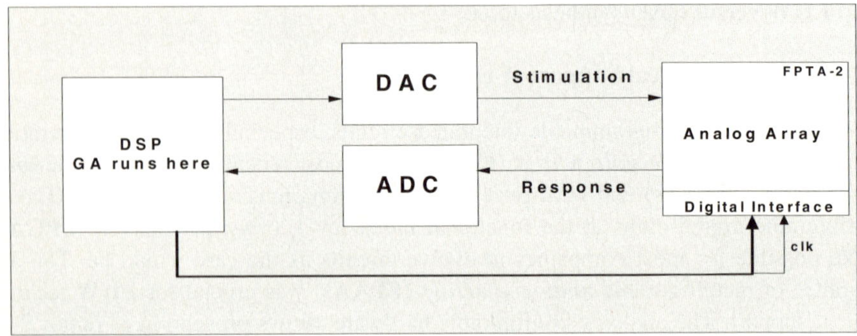

Fig. 2. The information flow between **DSP** and **FPTA** in **JPL SABLE**

The main integrated components by the JPL **SABLES** are: the transistor–level **RH** - a JPL FPTA, and the **RM** - the evolutionary algorithm, implemented by a TI DSP acting as a controller for reconfiguration. The information flow and implementation of **JPL SABLES** is featured –see Fig 2 - both by autonomy and speed in providing on-chip circuit reconfiguration: about 1000 circuit evaluations are performed per second. Another parameters of **JPL SABLES** performance are the followings: 1 – 2 orders of magnitude reduction in memory and about 4 orders of magnitude improvement in speed compared to systems evolving in simulations.

The final aim of **EC** techniques development and of their silicon implementation is to create architectures towards an artificial brain, a computer having its own ability of reasoning or decision making, but also being able of emergent functionality creation, or having the possibility of self-creation and evaluation of its own structure. The main types of **EHW** architectures with intrinsic **EC** logic elements are: *embryological* architectures; *emergent functionality* architectures; *evolvable fault tolerant* systems; *parallel evolvable architectures* (of Higuchi type) [1].

Other practical considerations regarding the concept of **EHW** architectures and hardware **GA** implementation, including the engineering compromise between the performances and the **GA** architectural complexity is to be found in [9]. *Gate Level Evolution* in **FPGA**s at the intrinsic level was applied in [10]. In the *Functional Level Evolution* the circuits are composed of high-level functional blocks such as adders and multiplier [11]. *Incremental Evolution* was applied in order to evolve circuits of high complexity. A *divide-and conquers* approach for the evolution of systems (also known under the name of "increased complexity evolution") is introduced in [12]. **EHW** *Architectures at Functional Level* This aim is feasible on a specialized dedicated **FPGA** architecture proposed by Higuchi [13] and called **F²PGA**. This architecture is relied on a network of *switch settings* that allow the basic cells inside the device to be connected as the case. A basic cell in this kind of **FPGA** is called **P**rogrammable **F**loating **U**nit *(PFU)* because of its availability of performing a large function variety: adding, subtracting, multiplication, cosine and sine using floating point numbers. A **F²PGA** programming word is a variable length chromosome containing both the **PFU**'s function programming and the crossbar switch settings.

3 Evolutionary Design of Digital Circuits: Where Are Current Limits?

It is important to understand that *evolutionary circuit design* and *evolvable hardware* (**EHW**) are two different and distinct approaches [14]. *Evolutionary circuit design* performs the evolution (the design) of a single circuit. The aim is typically to design novel implementations that are better (in terms of area, speed, power consumption) than conventional deigns and/or to design circuits with additional features such as fault-tolerance, testability, polymorphic behaviour, that are difficult to design by conventional methods. *Evolvable hardware* (**EHW**) involves an **EC** responsible for continual adaptation. **EHW** is applied to high-performance and adaptive systems in which the problem specification is unknown beforehand and can vary in time [15], [16].

The main issues in the evolutionary circuit design are : the bias in the design method; the chromosome size versus complexity of circuits; the fitness calculation as a bottleneck; the level of innovation [14].

It is important to explore the relation between the size of chromosome and the complexity of evolved circuits. Long chromosomes imply large search spaces that are difficult to search. Computing resources that are currently available determine the size of the search space that can be explored. This is known as the problem of *scalability of representation*. In case of evolving the combinational circuits, the evaluation time of a candidate circuit grows exponentially with the increasing number of inputs. Hence, the evaluation time becomes a main bottleneck of the evolutionary approach even if the circuit consists of a few components. This is the problem of *scalability of evaluation*. In order to reduce this time of evaluation, only a subset of all possible input vectors can be utilized. It is impossible to measure the level of innovation in an evolved circuit; the level of innovation does not depend on the approach utilized to evolve the circuit and on the complexity of the evolved circuits.

The *Gate-Level Evolution* can produce circuits of complexity up to about 100gates.The *Application-specific Encoding* (computer & swapping medians or sorting networks in **FPGA**) can be used to obtain more complex circuits .The *Functional Level Approaches* produce circuits of an unlimited complexity and depends on the complexity of components used as building blocks. *Development Approaches* produce arbitrarily complex solutions using chromosomes of a short length. The advantage is that the area on a chip is minimized. The limitation is that the obtained solutions are featured by regularity; this causes redundancy. *Transistor Level* evolution produces only small digital circuits. These elementary circuits are implemented using a large range of unconventional approaches. The complexity of evolved circuits

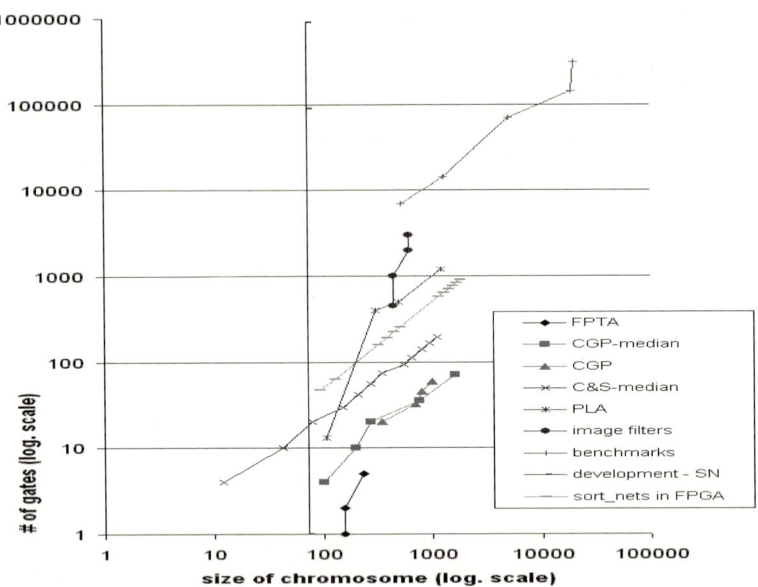

Fig. 3. The complexity of the evolved circuits versus the size of chromosomes

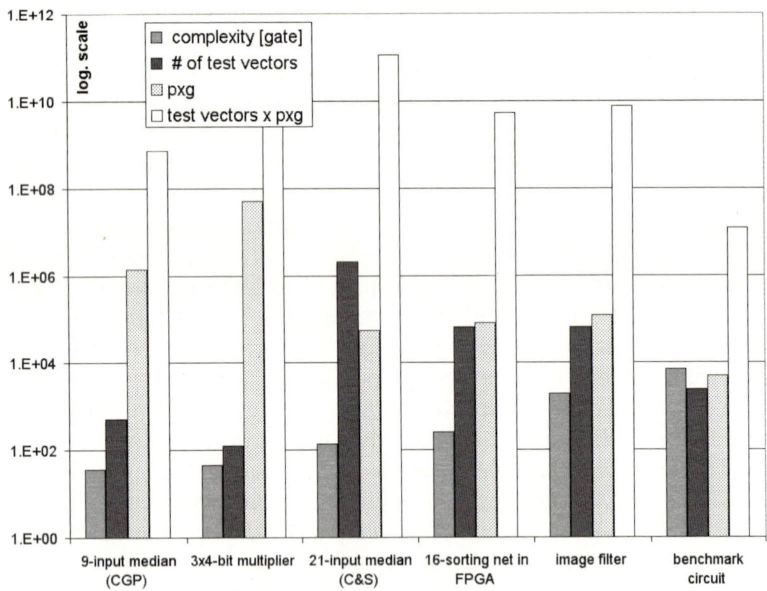

Fig. 4. Evolved circuits complexity versus the configuration bitstream

only partially depends on the size of the chromosomes. It mainly depends on the size of objects encoded in the chromosome or handled by instructions encoded in the chromosome (see the developmental approaches), Fig.3, after [14].

Another relevant view is made by the comparison of computational effort considering circuits that have similar size of the configuration bitstream [14], see Fig 4, where the chromosomes are of a size in the range of 512 – 868. The legend of this table is as follows: *in* – the number of inputs; *out* – the number of outputs; *chrom-size* – the size of chromosome; *gates* – the number of gates the circuit is composed; *test-vect* – the number of test vectors used in the fitness function to evaluate the circuit; *popsize* – the population size; *gnrs* – the average number of generations to evolve the circuit. Some interpretations of different evolutionary design methods are as follows [14]:

- the evaluation of a 21-input median circuit is the most time consuming
- *pxg = popsize* x *gnrs* , is a maximal parameter for CGP although the complexity of these gate-level circuits is not so high ; surprisingly, pxg decreases with the complexity of evolved circuits.
- the *"time of evolution"* = *pxg* x *test-vect* ; this parameter is very similar for almost all circuits. This is important, because the time of evolution indicates how much time/resources the designers are able to invest to the evolutionary design process.

5 Concluding Remarks

The most spectacular **EC** application in the **CI** framework is the (**EHW**). It opened a revolutionary eve in technology and in social life development by its radical implications on engineering design and automation. A dream of humanity became reality - the *systems adaptivity* was implemented (transferred) by **EHW** *from software to hardware*. A drastic time saving way from design to real world application of intelligent hardware is used, *no more difference* exists *between design* and *adaptation* concerning **EHW** based machines having a *behavioural computational intelligence*. *Electronics engineering was fundamentally changed* as a profession by using **EHW** custom design technologies instead of soldering based manufacturing.

A survey of the state of the art of evolutionary design of digital circuits was made. The investigation was mainly focused on level of complexity of the evolved circuits with respect to the size of the search space. The practical conclusion is that innovative results can be obtained by all mentioned approaches. There is no reason to prefer some approaches.

Acknowledgements

The support of the **EHW Group at NASA JPL** was crucial for tackling research objectives surveyed in this work that was suggested by the **KES International Advisory Board.**

References

[1] Negoita, Gh. M. (ed.): What is Evolvable Hardware, a Lecture, KES 2008 Post-Doc School on EHW and AHS, Zagreb, Croatia (September 2008)
[2] Torresen, J.: Evolvable Hardware – The Coming Hardware Design Method? In: Kasabov, N. (ed.) Neuro-fuzzy Tools and Techniques. Springer, Heidelberg (1997)

[3] Negoita, M.Gh., Neagu, C.D., Palade, V.: Computational Intelligence. In: Engineering of Hybrid Systems. Springer, Heidelberg (2004)

[4] Negoita, M.Gh.: A Modern Solution for the Technology of Preventing and Alarm Systems: EC in EHW Implementation. In: Proceedings of The Second Dunav Preving International Conference on Preventing and Alarm Systems, Belgrade, pp. 201–209 (1997)

[5] Negoita,M.Gh., Stoica, A.: Evolvable Hardware (EHW) in the Framework of Computational Intelligence-Implications on Engineering Design and Intelligence of Autonomous Systems. In: Tutorial Course, KBCS 2004, Hyderabad, India, pp. 4–52 (December 2004)

[6] Stoica, A., Radu, A.: Adaptive and Evolvable Hardware – A Multifaceted Analysis. In: Arslan, T., Stoica, A., et al. (eds.) Proceedings of 2007 NASA/ESA Conference on Adaptive Hardware and Systems, Edinburgh, UK, August 5-8, pp. 486–496 (2007)

[7] Stoica, A., Zebulum, R.S., et al.: Evolving Circuits in Seconds: Experiments with a Stand-Alone Board-Level Evolvable System. In: Proceedings of The 2002 NASA/DoD Conference on EHW, Alexandria, Virginia, July 15-18, pp. 67–74 (2002)

[8] Zebulum, R.S., Keymeulen, D., Duong, V., Guo, X., Ferguson, M.I., Stoica, A.: Experimental Results in Evolutionary Fault-Recovery for Field Programmable Analog Devices. In: Proceedings of The 2003 NASA/DoD Conference on Evolvable Hardware, Chicago, Illinois, July 9-11, pp. 182–188 (2003)

[9] Mihaila, D., Fagarasan, F., Negoita, M.Gh.: Architectural Implications of Genetic Algorithms Complexity in Evolvable Hardware Implementation. In: Proceedings of the European Congress EUFIT 1996, September, Aachen, Germany, vol. 1, pp. 400–404 (1996)

[10] Thompson, A., Layzell, P., Zebulum, S.: Exploration in Design Space: Unconventional Electronics Design Through Artificial Evolution. IEEE Transactions on Evolutionary Computation 3(3), 167–196 (1999)

[11] Murakawa, M., et al.: Evolvable Hardware at Functional Level. In: Ebeling, W., Rechenberg, I., Voigt, H.-M., Schwefel, H.-P. (eds.) PPSN 1996. LNCS, vol. 1141, pp. 62–71. Springer, Heidelberg (1996)

[12] Torresen, J.: A Scalable Approach to Evolvable Hardware. Genetic Programming and Evolvasble Machines 3(3), 259–282 (2002)

[13] Higuchi, T.: Evolvable Hardware with Genetic Learning. In: Proceedings of IEEE International Symposium on Circuits and Systems, ISCAS 1996, Atlanta, USA (May 13) (1996)

[14] Sekanina, L.: Evolutionary Design of Digital Circuits: Where are Current Limits? In: Stoica, A., Arslan, T. (eds.) Proceedings of First NASA/ESA Conference on Adaptive Hardware and Systems, Istanbul, Turkey, June 15-18, pp. 171–177 (2006)

[15] Sekanina, L.: Evolvable Components: From Theory to Hardware Implementation. Springer, Heidelberg (2004)

[16] Torresen, J., Bake, W.J., Sekanina, L.: Recognizing Speed Limit Sign Numbers by EHW. In: Raidl, G.R., Cagnoni, S., Branke, J., Corne, D.W., Drechsler, R., Jin, Y., Johnson, C.G., Machado, P., Marchiori, E., Rothlauf, F., Smith, G.D., Squillero, G. (eds.) EvoWorkshops 2004. LNCS, vol. 3005, pp. 682–691. Springer, Heidelberg (2004)

Component Adaptation Architectures
A Formal Approach

Andreea Vescan

Department of Computer Science
Faculty of Mathematics and Computer Science
Babeş-Bolyai University, Cluj-Napoca, Romania
avescan@cs.ubbcluj.ro

Abstract. Component selection and component adaptation are crucial problems in Component Based Software Engineering. In any component selection method, it is unrealistic to expect a perfect match between components needed and components available. Adapting the selected components to meet the specific need of the user is not an easy task. The aim of this research is to provide a formal mathematical model for component function adaptation. We introduce four component adaptation architectures for component functional adaptation. The behavior adaptation constraints for each architecture are discussed.

1   Introduction

Component Based Software Engineering (CBSE) [1] is concerned with the assembly of pre-existing software components that leads to a software system that responds to client-specific requirements. Component selection [2], [3], [4] and component adaptation [5], [6], [7] are widely recognized to be a crucial problem in component-based software engineering. Component selection problem is define as chosen a set of components from a repository such that the final system fulfills the specification of the needed system. When composing a system out of pre-existing components it is often the case of needing adaptation to complete the functionality of the composed system in order to satisfy the user needed requirements. There are three different type of adaptation [8]: signature adaptation, function adaptation and behavior adaptation.

In this paper, we address the problem of component (function) adaptation. We introduce four architectures for component function adaptation. The novelty of the approach consist in the mathematical formalism. An extension of program correctness of Floyd [9] and Hoare [10] is proposed. The formal description of components and behavior constraints adaptation (for different architectures) are the basis for verifying the obtained system(s).

We discuss the proposed approach as follows. A formal definition of component is given in Section 2. A short current state of Component Selection and Component Adaptation is given in Section 3. Section 4 presents and details our approach for component function adaptation. The conditions for each adaptation architecture are discussed. After presenting an example of our proposal in Section 4.2, we conclude our paper and discuss future work.

I. Lovrek, R.J. Howlett, and L.C. Jain (Eds.): KES 2008, Part III, LNAI 5179, pp. 319–326, 2008.

2   Correctness and Component Formal Description

The concept of program correctness was introduced by Floyd [9]. In the same paper he presented a method for proving program correctness. The correctness of individual algorithms was proved by Hoare [10].

In a program P we distinguish [11] three types of variables, grouped as three vectors X, Y and Z. The input vector $X = (x_1, x_2, ..., x_n)$ consists of the input variables. The input variables denote the known data of the given problem solved by the program P. We may suppose they do not change during computation. The output vector $Z = (z_1, z_2, ..., z_m)$ consists of those variables which denote the results of the problem. Two predicates [11] are associated to the program P: an input predicate and an output predicate. The input predicate $\varphi(X)$ is TRUE for those values a of X for which the problem may be solved. The input predicate $\varphi(X)$ is FALSE for those values a of X for which the problem doesn't have sense. The output predicate $\psi(X, Z)$ shows the relation between the results Z and the input values X. It is TRUE for those values a and b of the vectors X and Z for which the results of the problem are b when the initial/input data is a (when the input predicate is TRUE). The output predicate is FALSE if the input data is not valid or does not conforms to the input predicate.

The specification of the program P is the pair formed from the input predicate $\varphi(X)$ and the output predicate $\psi(X, Z)$: $Specification : [\varphi(X), \psi(X, Z)]$.

The program P terminates [11] in respect to the input predicate $\varphi(X)$ if for each value $a = (a_1, a_2, ..., a_n)$ of the vector X for which the predicate φ is TRUE, the execution of P terminates. In this case, the computation done by P is the sequence of states passed during the execution, and the value b of the vector Z in the final state is the result of the execution. We may write $b = P(a)$, i.e. P implements a function. The program P is partially correct [11] with respect to the specification if for the value a for which $\varphi(a)$ is TRUE and the execution terminates with the results $b = P(a)$ then $\psi(a, b)$ is TRUE. The program P is totally correct with respect to $\varphi(X)$ and $\psi(X, Y)$ if the program P terminates in respect to $\varphi(X)$ and it is partially correct with respect to $\varphi(X)$ and $\psi(X, Y)$.

In the following an example is provided to illustrate TRUE and FALSE cases for the output predicate. Consider a subalgorithm that computes the square root of n. The precondition and postcondition are: $\varphi(n) ::= n \geq 0$ and $\psi(n, r) ::= r^2 \leq n < (r + 1)^2$. For the value $n = 4$ the result of the subalgorithm is 2, results that conforms to the output predicate: $2^2 \leq 4 < (2 + 1)^2$. In this case the output predicate is TRUE. If the subalgorithm output the result of 14 for the same input data $n = 4$ (and the subalgorithm eventually terminates) this is not a valid solution of the problem because the output predicate is FALSE: $14^2 \leq 4 < (14 + 1)^2$.

We have extended the above definitions to formally specify components. A component specification is given by: the set of input variables - X; the set of output variables - Z; input predicate - $\varphi(X)$ and output predicate - $\psi(X, Z)$. In this proposal we discuss situations when a component implements a single method (with one provided interface). We name such a component a *simple* component. Future work will research cases with multiple component interfaces.

Definition 1 *A Component specification is given by: $ComponentName = [X = \{inputVar\}, Z = \{outputVar\}, \varphi(X), \psi(X, Z)]$.*

3   Component Selection and Component Adaptation

Component-Based Software Engineering (CBSE) is concerned with composing, selecting and designing components. As the popularity of this approach and hence number of commercially available software components grows, selecting a set of components to satisfy a set of requirements is becoming more difficult.

Component Selection Problem: Given a repository of components and a specification of the component system that we want to construct, we need to choose components and to connect them such that the target component system fulfills the specification. For this, we need a specification, if and how services (methods and interfaces) can be mapped to each other.

Component selection methods [12], [13] are traditionally done in an architecture - centric manner. Another type of component selection [14], [15] approaches is built around the relationship between requirements and components available for use.

In any component selection method, it is unrealistic to expect a perfect match between components needed and components available. A group of components that compose a system may have overlaps and gaps in required functionality. A gap represents a lack of functionality; an overlap can cause a confusion of responsibility and degrade nonfunctional properties.

Component adaptation can be divided into three families [8]: to adapt component signature properties, function adaptation (how to integrate different components both meeting partly the users requirement to satisfy the final requirement) and behavior adaptation that is how to mediate the component behavior when behavior mismatches occur.

Our current work concerns only component function adaptation. Behavior adaptation is not considered here.

4   Component Function Adaptation Architectures

This section presents four adaptation architectures to be used when selecting components to obtain a larger system and there is a gap between the required and the provided functionalities in the selected components. There are cases when an adaptation of the involved components can provided the required functionality. For example, we have a component that has an operation available, but we would like to call that operation several times. We need to adapt the way we use the operation by calling it several times in order to accomplish our goal.

4.1   Adaptation Architectures

The following adaptation architectures offer a new mechanism to compose different selected components in order to obtain a new component that fulfills the user requirements. Each new component is composed by using two selected components (except the repetitive adaptation architecture) and by taking into consideration the specification of the contained components.

Let us denote the specification of the new component (after applying adaptation architecture - Component Adaptation Architecture CAA) by: $CAA = [X, Z, \varphi(X), \psi(X, Z)]$ and by $CS_1 = [X_1, Z_1, \varphi_1(X_1), \psi_1(X_1, Z_1)]$ and $CS_2 = = [X_2, Z_2, \varphi_2(X_2), \psi_2(X_2, Z_2)]$ the specification of the selected components to be composed.

In the following the behavior adaptation constraints are given for each architecture. The specification of the composed component for each architecture is presented.

Serial (or Sequential) Adaptation Architecture - SAA. The serial behavior adaptation constraints are:

1. $\forall a \in X = X_1 | \varphi(a) \Rightarrow \varphi_1(a)$;
2. $\forall a \in X, \exists b \in Z_1 = X_2 | \varphi(a) \wedge \psi_1(a, b) \Rightarrow \varphi_2(b)$;
3. $\forall a \in X, \exists b \in Z_1 = X_2, \exists c \in Z | \varphi(a) \wedge \psi_1(a, b) \wedge \psi_2(b, c) \Rightarrow \psi(a, c)$.

The serial adaptation architecture is based on the assumption that the output of the C_1 component is the input of the C_2 component. The behavior conditions state that for those values a of X for which the problem may be solved: the precondition of C_1 holds, the postcondition of C_1 makes the the precondition of C_2 to hold and from the previous two conditions, the postcondition of C_2 will make the postcondition of the architecture to be true.

The new component composed using SAA from the two selected components is described as following:
$CAA = [\{i\}, \{o\}, \varphi_S(X), \psi_S(X, Z)]$, where
$\varphi_S : \varphi_1(i), i \in X_1$ and $\psi_S : \exists z \in Z_1 | \psi_1(i, z) \wedge \psi_2(z, o)$.

Parallel Adaptation Architecture - PAA. The parallel behavior adaptation constraints are:

1. $\forall a \in X, \exists a_1 \in X_1, \exists a_2 \in X_2 | a = (\{a_1\} \cup \{a_2\}) \wedge (\varphi(a) \Rightarrow \varphi_1(a) \wedge \varphi_2(a_2))$;
2. $\forall a \in X, \exists z_1 \in Z_1, \exists z_2 \in Z_2, \exists a_1 \in X_1, \exists a_2 \in X_2 | a = (\{a_1\} \cup \{a_2\}) \wedge (\varphi(a) \wedge \psi_1(a_1, z_1) \wedge \psi_2(a_2, z_2) \Rightarrow \psi(a, \{z_1\} \cup \{z_2\}))$, where z_1 and z_2 can be the same value.

The assumption of the parallel adaptation architecture is that the selected components involved in the composition satisfy only partially the user requirements. The union of the outputs of the two selected components can satisfy the user requirements. The behavior conditions state that the user requirements (precondition) can be decomposed into two requirements (preconditions), and that the union of the two outputs of the selected components makes the postcondition of the adapted component to hold (for those values for which the problem may be solved).

The new component composed using PAA from the two selected components is described as following:

$PAA = [\{i_1, i_2\}, \{o_1, o_2\}, \varphi_P(X), \psi_P(X, Z)]$, *where*

$\varphi_P : \varphi_1(i_1) \wedge \varphi_2(i_2)$ *and* $\psi_P : \psi_1(i_1, o_1) \wedge \psi(i_2, o_2)$.

Alternative Adaptation Architecture - AAA. The alternative behavior adaptation constraints are:

1. $\forall a \in X | (\varphi(a) \Rightarrow \varphi_1(a)) \vee (\varphi(a) \Rightarrow \varphi_2(a))$;
2. $\forall a \in X, \exists z \in Z | (\varphi_1(a) \wedge \psi_1(a, z) \Rightarrow \psi(a, z)) \vee (\varphi_2(a) \wedge \psi_2(a, z) \Rightarrow \psi(a, z))$;
3. $\forall a \in X | (\varphi_1(a) \Rightarrow \neg\varphi_2(a)) \vee (\varphi_2(a) \Rightarrow \neg\varphi_1(a))$.

The alternative adaptation architecture is based on the assumption that the precondition of the adapted components is only a part of the precondition of the user requirements. The requirements of the user are implemented using two separate cases. The behavior adaptation constraints state that: if the precondition of the adapted component is true then one precondition of the C_1 or C_2 components could be satisfied, if the precondition and postcondition of one of the components are satisfied then the postcondition of the adapted components will be true, and the last constraint, only one of the two preconditions of the two selected components may be true at the same time.

The new component composed using AAA from the two selected components is described as following:

$AAA = [\{i\}, \{o\}, \varphi_A(X), \psi_A(X, Z)]$, *where*

$\varphi_A : \varphi_1(i) \vee \varphi_2(i)$ *and* $\psi_A : \psi_1(i, o) \vee \psi_2(i, o)$, *where* $X = X_1 \cup X_2$ *and* $Z = Z_1 \cup Z_2$.

Repetitive Adaptation Architecture - RAA. Let rc be the repetitive condition and $bFalse$ and $bTrue$ values that will keep rc (*RepetitiveCondition*) unmodified (in the current state).

The repetitive behavior adaptation constraints are:

1. $\forall a \in X | \varphi(a) \Rightarrow \varphi_1(a)$;
2. $\forall a \in X, \exists bFalse \in Z | \varphi(a) \wedge \psi(a, bFalse) \wedge rc \Rightarrow \varphi(a)$;
3. $\forall a \in X, \exists bTrue \in Z | \varphi(a) \wedge \psi(a, bTrue) \wedge \neg rc \Rightarrow \psi(a, bTrue)$.

In the repetitive adaptation architecture an extra condition for the repetitive structure is provided. The behavior conditions state that if the precondition of the adapted component holds then the precondition of the selected component will also hold; if the precondition of the adapted component holds and the postcondition of the selected component is true and the extra condition rc is *true* then the precondition of the adapted component will remain *true*, and (the last behavior condition) if the precondition of the adapted component holds and the postcondition of the selected component is *true* and the extra condition rc does not hold then the postcondition of the adaptive component will be *true*.

The new component composed using RAA from the two selected components is described as following:

$RAA = [\{i\}, \{o\}, \varphi_R(X), \psi_R(X, Z)]$, *where*

$\varphi_R : \varphi_1(i), i \in X$ *and* $\psi_R : \exists o \in Z : \psi_1(i, o) \wedge \neg rc.$

4.2   Example

Problem: Develop a system that computes the numbers less than n given that have the sum of the digits a prime number.

In Fig. 1 the structure of the problem is given: five simple components to solve the problem with different adaptation architectures were applied: three repetitive adaptation architectures, two sequential (serial) adaptation architectures and one alternative.

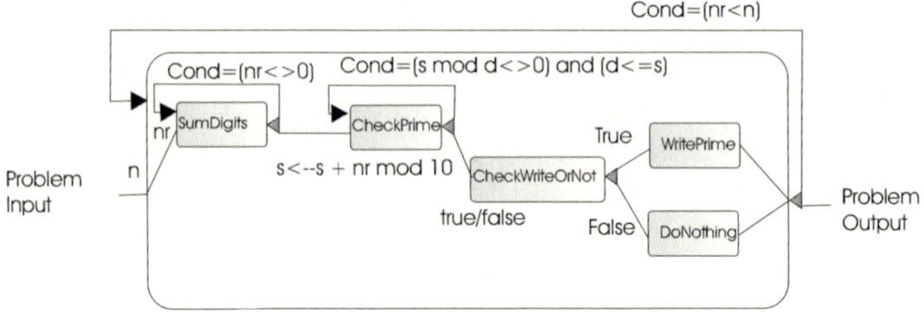

Fig. 1. Component adaptation architectures. Example

For each type of adaptation architecture a short discussion is following using Tab. 1 for serial adaptation architecture, Tab. 2 for alternative and repetitive adaptation architectures.

Table 1. Serial Adaptation Architecture example

SAA	Input	Output	$\varphi(X)$	$\psi(X, Z)$
SumDigits	nr	s	$nr \geq 0$	s=sum of digits of nr $\wedge\ s \geq 0.$
CheckPrime	s	True ∨ False	$s \geq 0$	if s is prime then True, else False.
CSerial	nr is s	True ∨ False	$s \geq 0$ ∧ s is sum of digits of the number nr	s=sum of digits of nr $\wedge\ s \geq 0$ ∧ if s is prime then True, else False.

Between *SumDigits* and *CheckPrime* we have applied serial adaptation architecture: the result of the execution of the *SumDigits* component is send as input for the *CheckPrime* component. The behavior adaptation constraints are satisfied.

Table 2. Alternative and Repetitive Adaptation Architecture example

AAA	Input	Output	$\varphi(X)$	$\psi(X,Z)$
WritePrime	val	write	$val = true$	write.
DoNothing	val	don't write	$val = false$	do nothing.
CAlternative	val	write $\vee$ don't write	val=true $\vee$ val=false	if $val = true$ then write, else do nothing.
RAA	Input	Output	$\varphi(X)$	$\psi(X,Z)$
CheckPrime	s	True $\vee$ False	$s \geq 0 \wedge rc$	if s is prime then true, else false.
RepeatCond=rc		(s mod 2 $\neq$0) $\wedge$ (d $\leq$ s)		
CRepetitive	s	True or False	$s \geq 0 \wedge rc$	$(\neg rc) \wedge$ (if s is prime then true, else false).

Alternative adaptation architecture is used for *WritePrime* and *doNothing* components: only one of the two components preconditions is *true* at a time. The others behavior adaptation constraints are easily to be checked (see Tab. 2 for details). Repetitive adaptation architecture is used for the *CheckPrime* component.

5   Conclusion and Further Work

Component function adaptation has been investigated in this paper. When developing system using pre-existing components the provided functionalities of the available set of components may not cover all the user requirements and adaptation is used. Serial (Sequential) Adaptation, Parallel Adaptation, Alternative Adaptation and Repetitive Adaptation are the adaptation architectures introduced in this paper to provide a formal mathematical model for component function adaptation. Behavior adaptation constrains are discussed for each architecture.

Further work directions will be focused on: extending component specification with multiple interfaces (provided methods) so that we could be able to deal with more complex problems. Another future work will discuss the current proposal using a real case study.

References

1. Crnkovic, I., Larsson, M.: Building Reliable Component-Based Software Systems. Artech House publisher (2002)
2. Fox, M.R., Brogan, D.C., Paul, F., Reynolds, J.: Approximating component selection. In: 36th conference on Winter simulation, Winter Simulation Conference, pp. 429–434 (2004)
3. Haghpanah, N., Moaven, S., Habibi, J., Kargar, M., Yeganeh, S.H.: Approximation algorithms for software component selection problem. In: 14th Asia-Pacific Software Engineering Conference, Washington, DC, USA, pp. 159–166. IEEE Computer Society, Los Alamitos (2007)
4. Baker, P., Harman, M., Steinhofel, K., Skaliotis, A.: Search based approaches to component selection and prioritization for the next release problem. In: 22nd IEEE International Conference on Software Maintenance, Washington, DC, USA, pp. 176–185. IEEE Computer Society, Los Alamitos (2006)
5. Ines Mouakher, A.L., Souquieres, J.: Component adaptation: Specification and verification. In: 11th Intl. Workshop on Component Oriented Programming, HAL - CCSD, pp. 23–30 (2006)
6. Penix, J., Alexander, P.: Toward automated component adaptation. In: 9th International Conference on Software Engineering and Knowledge Engineering, Knowledge Systems Institute, pp. 535–542 (1997)
7. Hemer, D.: A formal approach to component adaptation and composition. In: 28th Australasian conference on Computer Science, Darlinghurst, Australia, pp. 259–266. Australian Computer Society, Inc. (2005)
8. Xiong, X., Weishi, Z.: The current state of software component adaptation. In: SKG 2005: Proceedings of the First International Conference on Semantics, Knowledge and Grid, Washington, DC, USA, p. 103. IEEE Computer Society, Los Alamitos (2005)
9. Floyd, R.: Assigning meaning to programs. In: Schwartz, J.T. (ed.) Symposium on Applied Mathematics, pp. 19–32. A.M.S (1967)
10. Hoare, C.A.R.: An axiomatic basis for computer programming. Commun. ACM 12(10), 576–580 (1969)
11. Frentiu, M.: Correctness: a very important quality factor in programming. Studia Universitas Babes-Bolyai, Seria Informatica L(1), 11–20 (2005)
12. Fox, M.R., Brogan, D.C., Paul, F., Reynolds, J.: Approximating component selection. In: 36th conference on Winter simulation, Winter Simulation Conference, pp. 429–434 (2004)
13. Baker, P., Harman, M., Steinhofel, K., Skaliotis, A.: Search based approaches to component selection and prioritization for the next release problem. In: 22nd IEEE International Conference on Software Maintenance, Washington, DC, USA, pp. 176–185. IEEE Computer Society, Los Alamitos (2006)
14. Haghpanah, N., Moaven, S., Habibi, J., Kargar, M., Yeganeh, S.H.: Approximation algorithms for software component selection problem. In: APSEC, pp. 159–166. IEEE Computer Society, Los Alamitos (2007)
15. Gesellensetter, L., Glesner, S.: Only the best can make it: Optimal component selection. Electron. Notes Theor. Comput. Sci. 176(2), 105–124 (2007)

Adaptive Microarray Image Acquisition System and Microarray Image Processing Using FPGA Technology

Bogdan Belean, Monica Borda, and Albert Fazakas

Technical University of Cluj Napoca, Faculty of Electronics, Communication and
Information Technology, G. Baritiu. 28, 400027 Cluj Napoca, Romania
{Bogdan.Belean,Monica.Borda}@com.utcluj.ro,
Albhert.Fazakas@bel.utcluj.ro

Abstract. The present paper proposes an adaptive hardware implementation for
a microarray image acquisition system, which is mandatory for implementing
hardware algorithms for processing microarray images. Processing techniques
for microarray image are also described, together with a hardware implementation of a spot border detection algorithm. The hardware implementation takes
advantage of parallel computation capabilities offered by FPGA technology.
Results which prove time and cost efficiency are presented for both hardware
implementations.

Keywords: cDNA Microarray, Image Processing, Parallel Processing, FPGA
Technology.

1 Introduction. Microarray Image Processing and FPGA Technology

Gene expression represents the transformation of gene's information into proteins.
For identifying a gene in a biological sequence and predicting the function of the
identified gene, quantifying gene expression is mandatory. Some of these methods for
detecting and quantifying gene expression levels are: Diferential Display – CR (DD –
Polymeric Chain Reaction), Serial Analysis of Gene Expresion (SAGE) and Complementary DNA (cDNA) microarray.

The last mentioned technology is based on creating DNA microarrays, which represent gene specific probes arrayed on a matrix such as a glass slide or microchip.
Usually, samples from two sources are labeled with two different fluorescent markers
and hybridized on the same array (glass slide). After the hybridization, the array is
scanned using two light sources with different lengths (red and green) to determine
the amount of labeled sample bound to each spot through hybridization process. The
light sources induce fluorescence in the spots, which is captured by a scanner and a
composite image is produced [1]. Further on, image processing techniques are used to
quantify gene expression levels present in the captured microarray image, in order to
identify a gene in a biological sequence and to predict the function of the identified
gene. The flow of processing a microarray image [2] is generally separated in the
following tasks: addressing, segmentation, intensity extraction and preprocessing to

I. Lovrek, R.J. Howlett, and L.C. Jain (Eds.): KES 2008, Part III, LNAI 5179, pp. 327–334, 2008.

improve image quality and enhance weakly expressed spots. The first step associates an address to each spot of the image. In the second one, pixels are classified either as foreground, representing the DNA spots, or as background. The last step calculates the intensities of each spot and also estimates background intensity values.

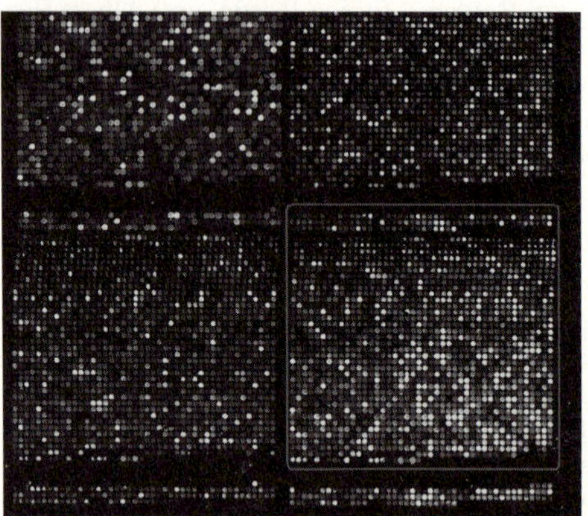

Fig. 1. Microarray image with underlined independent image element; Stanford University Medical Center, Department of Genetics

The major tasks of microarray image processing, which contributes in fulfilling the last mentioned steps, are to identify the array format including the array layout, spot size and shape, spot intensities and distances between spots. The main parameters taken into consideration in image processing are accuracy and time. The accuracy is given by the quality of image processing techniques. The second parameter, time, is critical due to the large amount of data contained in a microarray image (microarray image size being as high as hundreds of MB). Tacking into consideration the parameters and the main tasks of microarray image processing, this paper proposes a hardware implementation of an automatic spot border estimation algorithm to decrease processing time. Also a strategy for estimating distances between spots is discussed due to the fact that spot distance estimation is a mandatory step in microarray image processing. Also a cost efficient hardware implementation for microarray image acquisition trough a general purpose interface is described.

The microarray image obtained as a result of a cDNA microarray experiment is analyzed and processed using FPGA technology. The proposed hardware implementation of the automatic spot border estimation algorithm makes use of the features of FPGA, which allow accessing at the same time hundreds of memory addresses. In this way, calculations specific to microarray image processing are made in parallel, increasing the processing speed. Being reprogrammable and also cost efficient, FPGA technology was chosen also for fulfilling tasks such as realizing hardware implementation for microarray image acquisition systems. A solution for microarray image acquisition on an FPGA device is presented in the next chapter. The acquisition methods are adapted to the structure, size and shape of microarray image in order to increase the processing speed. A part of a microarray image is shown in Figure 1.

Both the hardware implementation of the spot distance estimation algorithm and the hardware implementation of microarray image acquisition are part of the system on a chip designed for parallel microarray image processing described in the previous paper [3].

2 Microarray Image Acquisition on Hardware Platform

The employment of hardware algorithms represents a new approach in microarray image processing, which uses parallel capabilities of hardware technologies like FPGA to decrease computational time of the specific algorithms of microarray image processing. For this to be possible, first we have to transport the microarray image on the hardware platform. This chapter proposes a user interface, its task being microarray image acquisition on the hardware platform used for further processing. There will also be presented a strategy for data transfer from RAM to FPGA, adapted to the structure of the microarray images.

2.1 Hardware Platform Description

The hardware platform used for implementation described in this chapter is a Nexys board produced by Digilent. This board offers us the following components:

- USB 2.0 port together with a USB controller, produced by Cypress, for high speed data transfer;
- a Micron asynchronous cellular RAM with capacity of 16 MB used for image acquisition;
- a Spartan3 XC3S200 FPGA from Xilinx [4] with 200K gates for implementing the spot distance estimation algorithm;
- a quartz oscillator with a frequency of 50 MHz used to generate the clock signal.

2.2 Microarray Image Acquisition

Using the components described above, a microarray image acquisition system was implemented, together with a Visual C++ interface for the user to have access and control to the system. In this paragraph, the experimental results of the implementation are presented. Before describing the user interface, it is to be mentioned that the file format used for microarray images is TIFF (Tagged Image File Format). Because it is an uncompressed file format, image decompressing procedures are avoided.

As you can see in the Figure 2, the user has access to the hardware platform trough Visual C++ interface. He can select from the hard drive the 16 bits TIFF image to be written or read in or from the memory, depending on the "Rd/Wr" value. The "adr" value represents the address to which or from which 16 bits data values are written or read. This user interface has access to the data from the memory trough the "Read Buffer" and "Write Buffer".

The "Read Buffer" is used when the user wants to access an address from the Micron memory and to read the data from the address specified by "adr". In this way, when the memory is read, the user interface reads the read result of the operation from

330   B. Belean, M. Borda, and A. Fazakas

"Read Buffer". In the "Write Buffer", the user writes using the Visual C++ interface the 16 bits data which will be written in the Micron memory. Reading and writing operations are controlled by a finite state machine incorporated in the "Control Unit". Another important role of the "Control Unit" is to generate the "clk" signal for the Micron memory so that the write and read operations be accomplished. In other words, the "Control Unit" incorporates a clock divider circuit so that the resulted clock period T_{CLK} will be:

$$T_{CLK} \geq max\,(\,t_{WC},\,t_{RC}\,) \tag{1}$$

where t_{WC}, t_{RC} represent Micron memory parameters: write cycle time and read cycle time respectively.

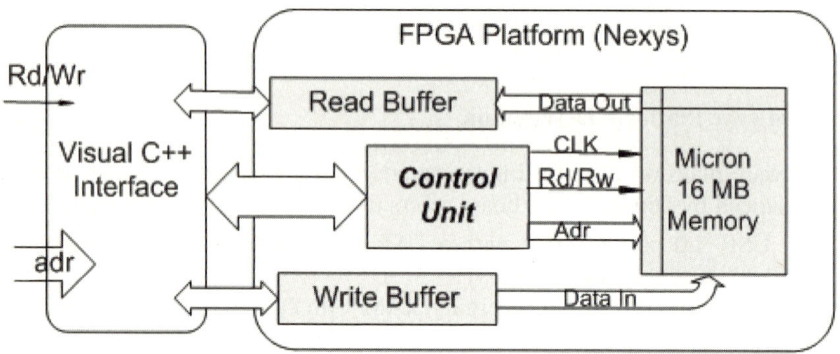

Fig. 2. Block diagram for microarray image acquisition

2.3 Parallel Approach for Microarray Image Processing

Once a copy of a microarray image is written in RAM, the next step is to find hardware architectures and hardware algorithms which use parallel capabilities of FPGA technology to decrease the computational time for the large amount of data contained in a microarray image. The ideal case would be to have the possibility to copy the entire microarray image in FPGA, so parallel processing strategies would be applied to the whole image. Due to the large amount of data, pieces of the image will be copied in the FPGA for parallel processing. After processing, the result will be transferred back into RAM. If we choose this approach for microarray image processing, we have to take in consideration that the pieces of the microarray image copied into FPGA have to be independent.

Now, if we look at Figure 1, we see that a piece of the microarray image is delimited. This piece constitutes an independent microarray sub-image. Based on the possibility of cropping a microarray image into independent sub-images, another strategy for microarray image processing emerges. The idea is to design an architecture which processes in parallel two or more independent sub-images. The number of microarray sub-images that can be processed at the same time depends on the FPGA dimension.

Also, the transfer speed between memory and FPGA depends on the memory parameters (read and write cycle time) and on the memory type. For instance, we can have dual port memories, appropriate for a multiprocessor implementation of a microarray image processing system.

A cropping step is introduced at the beginning of microarray image processing. This processing step determines the positions of each independent block and its dimension. In this way results a cropped microarray image, which is represented in the next figure (Figure 3) and which consists in a matrix of independent sub-images called A(i,j). The numbers from 1 to 3 and from 1' to 3' represent the operations that are applied to the microarray sub-images. This way, operation 1 stands for copying the sub-image from RAM to the FPGA. Operation 2 represents the parallel processing of the sub-image using FPGA technology. Operation 3 copies the microarray processed sub-image back into the RAM, on the exact place from which it was transferred into FPGA. The operations 1, 2 and 3 are done for each microarray sub-image. In the end we will have in the RAM memory a processed microarray image.

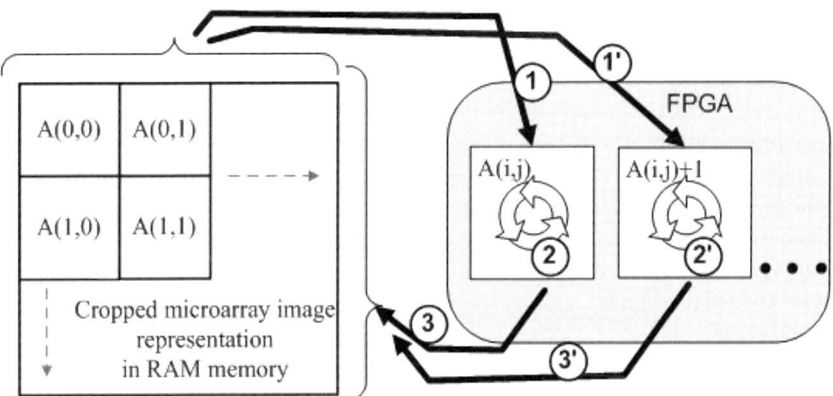

Fig. 3. Block diagram for microarray image acquisition

Since time is a critical issue in microarray image processing, we will focus on decreasing the computational time. A solution is to use a dual port memory and an FPGA that allows processing at least 2 independent sub-images at the same time using pipelining techniques. For example, if it is possible to write 2 sub-images in the FPGA, the processing time can be decreased to half of its regular value.

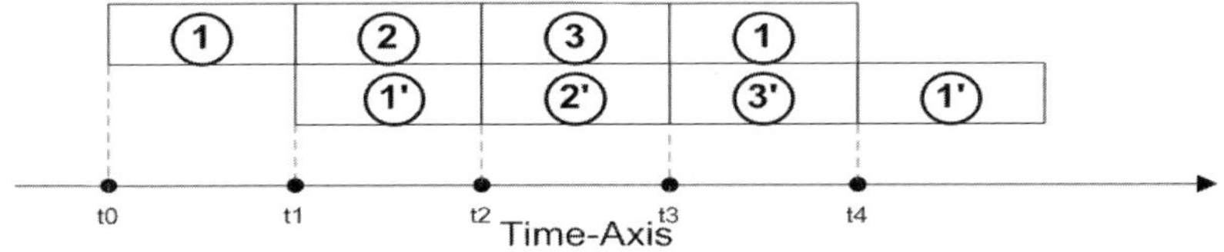

Fig. 4. Time representation of the operations 1, 2 and 3 in pipelining

The previous figure (Figure 4) presents how operations 1, 2 and 3 are arranged on a time axis to decrease the processing time, by realizing 2 operations on different sub-images in the same time.

3 Microarray Image Processing. Hardware Implementation of Automatic Spot Border Detection

In the next paragraphs cDNA microarray image processing strategies are described in order to eliminate the shortcoming of the previous algorithms, which needed user intervention to manually adjust a grid to determine the spot addresses in the image. This strategy is based on generating image vertical and horizontal profiles by summing up intensities in each row and column. By calculating the minimum and the maximum values of these values we can determine a grid positioned around each spot, and also we can estimate spot dimensions. This method works well except in case that weakly expressed spots exist. The problem of weakly expressed spots is eliminated using image transformations to enhance spot intensities. Another strategy used in microarray image processing is to take advantage of the fact that a microarray images can be divided by a global addressing step into different independent images. This strategy was discussed in Chapter 2.3. It provides the benefit of parallelizing further processing of microarray image as the due to the fact that the whole microarray image consists of many independent sub-images. After addressing and image enhancement, image segmentation can be done. In our case image segmentation is realized using spot border detection algorithm.

Taking into consideration the strategies mentioned above, a microarray image $I(x,y)$ is treated as follow:

Step 1: Median filters are applied on the image to eliminate shot noise and the following transformation is used to enhance weekly expressed spots:

$$I_N = \frac{I(x,y)^{\lambda_x+\lambda_y}}{\lambda_x+\lambda_y} \text{ with } \lambda_x = \frac{1}{\frac{1}{y} \cdot \sum_{i=0}^{Y-1} I(x,n)} \, , \, \lambda_y = \frac{1}{\frac{1}{x} \cdot \sum_{i=0}^{X-1} I(n,y)} \tag{2}$$

where $x = 1,2,...X-1$ and $y = 1,2,...Y-1$ are the dimensions of the image $I(x,y)$ and $I_N(x,y)$ is the enhanced image.

Step 2: A mandatory step in microarray image processing is called addressing and its task is to assign coordinates to each spot of the image, or, in other words, it estimates a target mask around each spot. After a prior preprocessing to enhance image quality, the first task in order to determine the target mask around each spot is to estimate the spot distances.

In recent papers ([6] [7]), microarray image addressing is realized using image projections obtained by averaging pixel intensities on rows and columns. The horizontal (HP) and the vertical projections (VP) are calculated as follows:

$$HP(y) = \frac{1}{X}\sum_{x=0}^{X-1} I(x, y) \qquad (3)$$

$$VP(x) = \frac{1}{Y}\sum_{y=0}^{Y-1} I(x, y) \qquad (4)$$

where x = 1,2,...X-1 and y = 1,2,..Y-1 are the dimensions of the image I(x,y).

Step 3: To determine spot distances, the HP and VP are considered to be periodical signals. To find out the periodicity, the signal is cross-correlated with itself. Since the HP and VP signals are discrete ones, the cross-correlation is defined as follows:

$$pv(i) = \sum_{j=0}^{X-1} HP(j) \cdot HP((j+i)\bmod X) \qquad (5)$$

$$ph(l) = \sum_{j=k}^{Y-1} VP(l) \cdot VP((l+k)\bmod Y) \qquad (6)$$

where j = i = 0,1,...X-1 is the first size of the image I(x,y) and k = 1 = 0,1,...Y-1 is the second size of the image I(x,y). Now the first derivative of the signals pv(i) and ph(l) is determined. The distance between the zeros of the first derivative represents the spot distance.

Step 4: Following image addressing, segmentation is the next step which occurs in microarray image processing. This paper puts forward a hardware implementation of an image segmentation method based on border detection of a spot. To detect the border of a spot, the Mexican Hat algorithm was used. The idea is that for each pixel, based on the values of the surrounding pixels, we determine if the center pixel belongs to the contour of the spot or not.

The results of the parallel hardware implementation of the previous image segmentation method are described as follows: the processing of a 12x12 pixels image requires 60 ns (around 4 clock periods). Table 1 presents the FPGA resource used for such an implementation.

Table 1. Statistics of FPGA resource (XC3S1000) used for the implementation of microarray spot border detection

Logic utilization	Used	Utilization
Number of slice Flip flops	1.288	8%
Number of 4 input LUTs	5.932	38%
Number of occupied slices	3.933	51%
Number of slices containing only related logic	3.933	100%
Number of bonded IOBs	12	6%
Number of GCLKs	1	12%

4 Conclusions

The acquisition of microarray images on a hardware platform trough a general purpose interface (USB 2.0) represents an essential step for developing microarray image processing algorithms based on hardware implementation. The results of the hardware implementation of a microarray image segmentatii method presented above emphasize the importance of using hardware methods in cDNA microarray image processing. The technology chosen to implement a digital system for microarray image processing was FPGA, due to its parallel computation capabilities and to the possibility of reconfiguration.

An idea for future implementations presented in this paper is to take into consideration the possibility of cropping a microarray image into independent sub-images. In this way, a method for parallel processing of two or more microarray sub-images is described.

To sum up, the experimental results made this hardware technology a solution for realizing a fast, cost efficient and accurate automated system for cDNA microarray image processing.

Acknowledgement

This work was supported by PNII IDEI Nr. 909/2007 grant and also by CNCSIS (National University Research Council) BD scholarship.

References

1. Bajcsy, P.: An Overview of DNA Microarray Image Requirements for Automated Processing. IEEE Transactions on Image Processing 13(1), 15–25 (2004)
2. Yang, Y.H., Buckley, M.J., Dudoit, S., Speed, T.P.: Comparison of methods for image analysis on cdna microarray data. Technical Report 584, Department of statistics, University of California, Berkeley (2001)
3. Belean, B., Malutan, R., Gomez, P., Borda, M.: FPGA based digital device and microarray image processing. Buletinul stiintific al Universitatii Politehnica din Timisoara (January 2008)
4. Digilent inc. support website, http://www.digilentinc.com/
5. Xilinx Spartan 3-E support website,
   http://www.xilinx.com/support/documentation/spartan-3e.htm
6. Blekas, K., Galatsanos, N.P., Likas, A., Lagaris, I.E.: Mixture model analysis of dna microarray images. IEEE Transactions on Medical Imaging, 901–909 (July 2005)
7. Li, Q., Fraley, C., Baumgarner, R.E., Yeung, K.Y., Raftery, A.E.: Donuts, scratches and blanks: Robust model-based segmentation of microarray images. Technical Report 473, Department of statistics, University of Washington (2005)
8. Berman, K.A., Paul, J.L.: Algorithms: Sequential, Parallel, and Distributed, Thomson Learning (2005)
9. Parhami, B.: Introduction to Parallel Processing. Plenum Series in Computer Science (1999)

A Digitally Reconfigurable Low Pass Filter for Multi-mode Direct Conversion Receivers

Gabor Csipkes[1], Sorin Hintea[1], Doris Csipkes[1], Cristian Rus[1], Lelia Festila[1], and Hernando Fernandez-Canque[2]

[1] Technical University Cluj Napoca, Romania
gabor.csipkes@bel.utcluj.ro
[2] Caledonian University, Glasgow, United Kingdom

Abstract. This paper presents a novel approach to the design of integrated reconfigurable and programmable analog filters for commercial mobile applications. The method is described in the specific context of an analog filter, dedicated to multi-mode receiver front ends employing direct frequency conversion. The filter has been designed using state variables and leap-frog OTA-C techniques, features programmable order, digitally variable frequency parameters and wide linear range, while supporting several standard approximations. The fundamental OTA cell has been implemented with fully balanced second generation current conveyors, suitable for low voltage operation. The novel method is demonstrated by presenting the simulation results of the filter, designed for a 0.18 μm CMOS technology.

1 Introduction

The widespread of mobile communication systems in the last two decades has led to an increase in the diversity of the radio access intefaces the portable terminals must integrate. The advent of the GSM standard with its many derivations and of the CDMA-based UMTS family of standards, have created an unprecedentedly diverse distribution of particular specifications in terms of carrier frequency, modulation, analog-digital signal processing and hardware requirements.

The ultimate goal of the industry is the development of the software defined radio technology that gives the context for the implementation of different analog and digital reconfigurable building blocks. In the first acceptance the software defined radio has been envisioned as a signal processing chain in which the data converters are placed directly in the vicinity of the antenna, allowing all the specific radio functions to be realized in the digital domain. In spite of the advantages offered by the ideal architecture, the technology induced limitations concerning the consumption of a data converter capable of sampling the signal at the carrier frequency with sufficiently high dynamic range, may render a real implementation impossible. Therefore, the ideal model must be extended in order to include some analog interface circuitry [1]. One of the most widely used analog interfaces, considered mainly due to its simplicity and well understood behavior, is the so called zero-IF or direct conversion architecture shown in Fig.1.

I. Lovrek, R.J. Howlett, and L.C. Jain (Eds.): KES 2008, Part III, LNAI 5179, pp. 335–342, 2008.

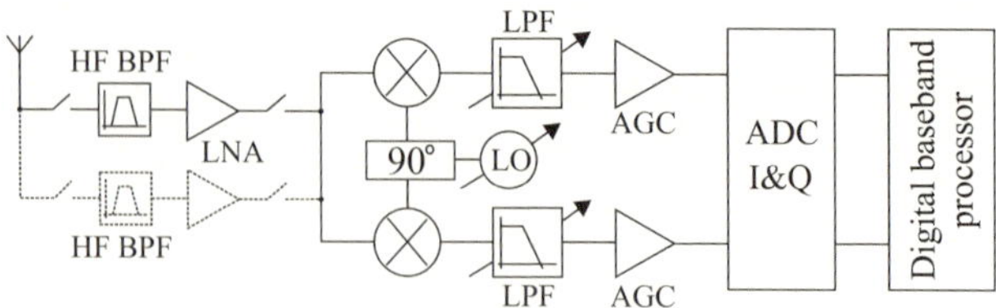

Fig. 1. A typical reconfigurable direct conversion receiver architecture

In a zero-IF receiver the radio frequency signal is converted directly to the baseband by multiplication with a complex local oscillator signal having the same frequency with the RF carrier. The LO signal is generated in quadrature to circumvent the image problem and to allow the correct demodulation of the wanted signal. The advantages of the direct conversion approach over the classical heterodyne architecture are straightforward: there is no need for multiple high-frequency mixing, amplifying and filtering stages, and the channel selection is simply done by means of low pass filters. Its simplicity makes the direct conversion architecture a good candidate for the implementation of practical analog software radio interfaces [2].

A survey of the existing literature shows that the reconfigurable circuits of the analog interface are clearly the main bottleneck in the development of a practically feasible SDR transceiver. In spite of their importance, only few theoretical studies have been published on reconfigurable analog circuits [3][4] and no generalized methods in developing such circuits exist. In the remainder of this paper, one modular and easily scalable method is offered to support the design of reconfigurable and programmable analog low pass filters compatible with the specifications of a SDR built upon zero-IF analog interfaces.

2   The Reconfigurable State Variable OTA-C Filters

State variable filters stand out among several well known filter structures due to their low sensitivity performance, inherited from doubly terminated passive LC prototypes. The implementation based on operational transconductance amplifiers (OTA-s) allows the operation at high frequencies, making these filters suitable for signal conditioning in analog transceiver front-ends. Furthermore, the modular structure creates the premise for easy topological reconfiguration, a prerequisite for the design of fully reconfigurable filters. The versatility in operation is achieved by creating a circuit template that implements the signal flow graph associated with the passive ladder prototype [5].

The graph corresponding to a generalized passive ladder is shown in Fig.2. The block diagram has been obtained by writing Kirchoff's theorems and Ohm's law for the nodes and branches of the LC ladder. The dummy resistance R transforms branch currents into voltages allowing voltage only state variables.

Filters designed for communications systems exhibit mainly a low pass or band pass frequency response. In particular, when the imposed operating frequencies

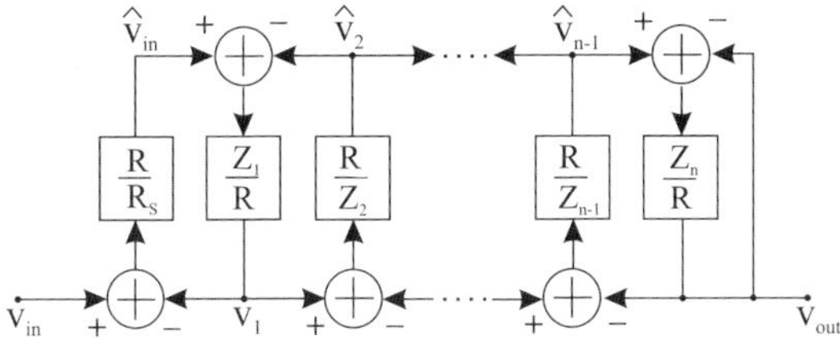

Fig. 2. The signal flow graph corresponding to the ladder filter in Fig.2

are not prohibitive and quadrature signal paths are available, band pass filters may also suppress the image signal in the intermediate frequency stage. In these cases the band pass response is obtained from the low pass transfer function by performing a linear frequency transformation. Therefore, the impedances in the generalized ladder should be particularized in order to accommodate first of all with a low pass response. The resulting implementation of the low pass OTA-C filter for an arbitrary order n is shown in Fig.3 [5].

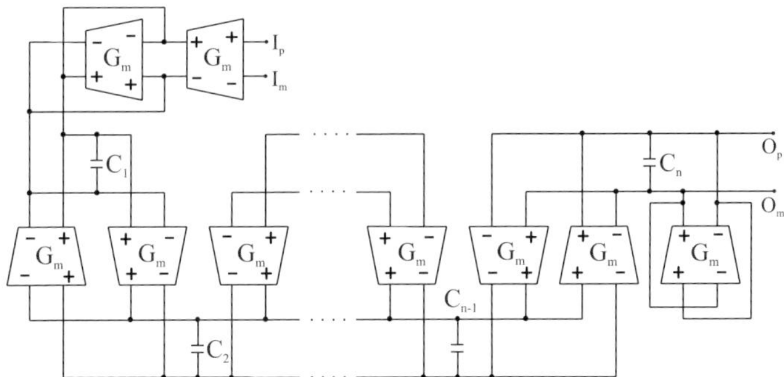

Fig. 3. Schematic of a fully balanced odd order OTA-C filter

The values of the inductances and capacitances can be calculated after denormalisation and impedance scaling for the components of the low pass LC ladder prototype. If the characteristic resistance of the ladder is $R = R_S = R_L = 1/G_m$ and f_t is the desired corner frequency, the capacitances in the active implementation may be calculated according to the equations (1) [5].

$$C_{odd} = \frac{C_{2k+1(n)} \cdot G_m}{2\pi \cdot f_t}, \quad C_{even} = \frac{L_{2k(n)} \cdot G_m}{2\pi \cdot f_t} \tag{1}$$

The complete reconfiguration of the low pass filter implies freely changing the frequency parameters, the filter order and the approximation while dynamically adjusting the current consumption. A careful examination of the filter topology for two consecutive orders may be done when considering the corresponding termination networks for odd and even orders. The comparison between the two circuits suggests that the fundamental module should be implemented with switches that connect or disconnect the target transconductance cells depending on some digital control. Additionally, a negative feedback path must be created around the second OTA that effectively implements the load resistance of the ladder [6]. The resulting reconfigurable module is illustrated in Fig.4.

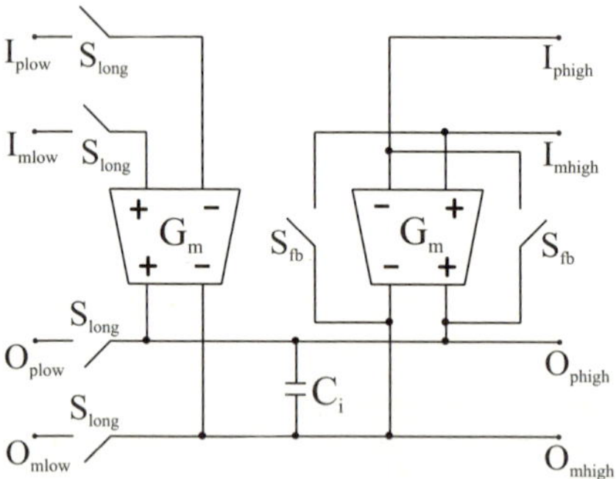

Fig. 4. The proposed elementary module of the reconfigurable low pass filter

A simple cascade connection of identical modules, as many as required by the highest desired filter order, controlled by a decode logic, leads to a fully reconfigurable low pass filter implementation. When the filter order is decreased by one, the longitudinal switches S_{long} in the last module of the cascade are turned OFF separating the unit from the rest of the ladder. Meanwhile, the switches S_{fb} are also turned OFF and the OTA cells enter in a power-down state. The new outputs of the filter will be the O_{phigh} and O_{mhigh} of the previous module. Furthermore, the switches S_{fb} of the previous module must be turned ON in order to shift the load resistance to the output of the lower order filter. The detailed implementation with the state of every switch has been described in [6]. The termination network of the filter implemented as a cascade of three identical reconfigurable modules is shown in Fig.5.

3   The Programmable Fully Balanced Transconductance Amplifier

The most important high frequency analog filter implementation techniques enconuntered in the literature are based on OTA-C structures. However, transconductance amplifiers are reknown to exhibit a limited range for the transconductance parameter, low-to-average linear range and relatively high current consumption. Classical linearization methods make use of the non-linear transistor equations and various circuit topologies in order to effectively cancel odd order non-linearities, while even order harmonics may be reduced by fully balanced designs. This paper proposes an alternative to a wide linear range OTA.

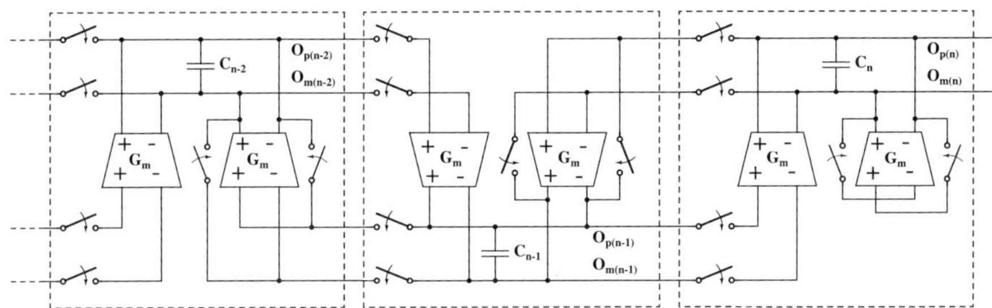

Fig. 5. Cascaded reconfigurable modules in the filter termination network

The fundamental building block of the OTA is a fully differential second generation current conveyor.

Current conveyors play the same role in current mode signal processing as opamps in voltage mode circuits, namely they hold virtual ground for the wanted signal. Their versatility in applications is mainly defined by the configuration of the terminals. A classical method to build a transconductance amplifier with a CCII cell is to use the Y terminal as a voltage input. The input voltage is then copied to the X terminal. The voltage V_X determines a current through a resistor connected between the X terminal and the ground. This current is then copied to the Z terminal, used as current output [7]. The overall transconductance of the circuit will be equal to the reciprocal of the passive resistance. The implementation of the fully balanced OTA is somewhat more complicated since it requires two conveyors and a common mode control circuit that sets the DC voltages at the high impedance output terminals. In the proposed design the output stage is programmable in order to effectively change the transconductance of the circuit and implicitly the frequency parameters of the entire filter. The programmability is achieved through binary weighted current mirrors. The block diagram of the OTA is shown in Fig.6.

In the ideal case the equivalent transconductance G_m is $1/R$. Real current conveyors exhibit a non-zero parasitic resistances of the X terminal which must be added twice to the passive resistance R when calculating the transconductance

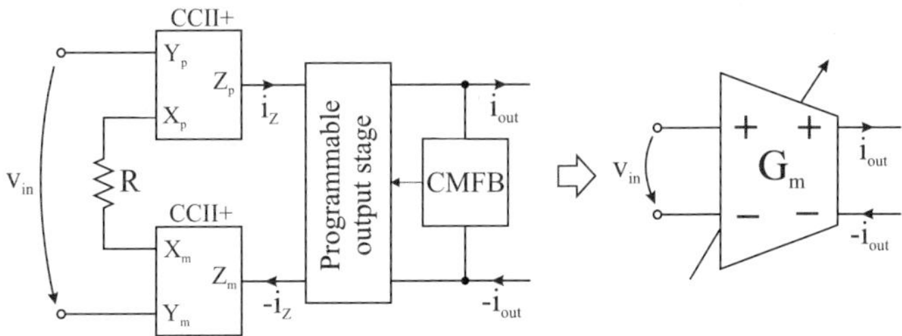

Fig. 6. Block diagram of the OTA cell built around second generation current conveyors

340   G. Csipkes et al.

[7]. Furthermore, R_X is often frequency dependent, and introduces unwanted ripples in the overall filter transfer function.

The solution that reduces the relatively high parasitic resistance is to use negative feedback. There are various low input resistance conveyor implementations proposed in the literature. These are mainly using a differential amplifier on the negative feedback path. The conveyor used to design the filter presented in this paper has been introduced and described by Liu in [8]. Liu's CCII is essentially an unbuffered opamp with Miller compensation, whose second gain stage is replicated for copying the X terminal current to the Z output. Its main advantages are the simple structure, low current consumption and the potential to operate at supply voltages as low as 1.2V. A special care must be taken of the opamp stability through an adequate compensation. The lead-lag type of compensation, implemented with the resistor R_C in series with the capacitor C_C, yields good results in insuring the stability and in extending the range of the conveyor operating frequencies [7]. The schematic of the resulting fully balanced OTA is shown in Fig.7.

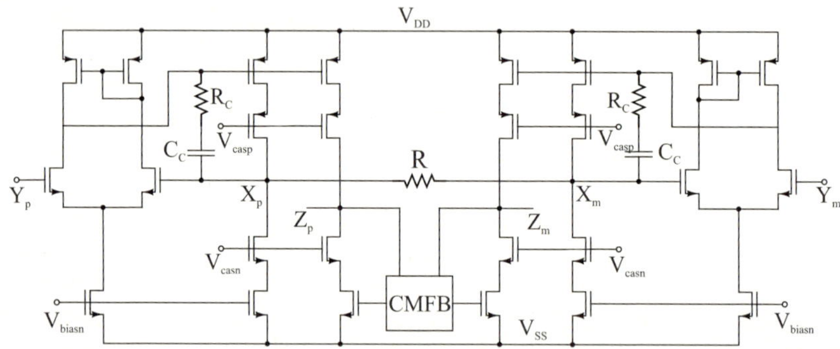

Fig. 7. Schematic of a fully balanced OTA built around Liu's CCII

A detailed analysis of the large signal behavior shows that the current flowing through the X terminal exhibits a parabolic dependence on the X terminal voltage, according to (2), where A is the gain of the differential amplifier, β_p is the intrinsic transconductance of the p-channel transistors and I_B is the bias current of the X terminal input stage.

$$I_X = 2A\sqrt{\beta_p I_B} \cdot V_X - \beta_p A^2 V_X^2 \tag{2}$$

This second order equation may be considered the main source of non-linearity in CCII based circuits, but it can be effectively canceled by the fully differential architecture. Therefore, the proposed OTA has the potential for very linear operation along with relatively low current consumption.

4   Simulation Results

The proposed reconfigurable filter can be considered as the natural follow-up of the architecture described in [6]. The simulations performed on the filter at transistor level have the main goal to demonstrate the functionality of the circuit in the presence of the inherent non-idealities compared to the concept level implementation in [6]. The main emphasis here lays on the performance indicators concerning frequency and order variability, non-linearity and a dynamically adjustable current consumption.

The corner frequency programming strategy implies changing the transconductances through a binary weighted mirror array. The four programming bits are able to set a double value transconductance and implicitly double corner frequency. Fig.8 shows the magnitude response of the filter, designed for the lowest possible corner frequency equal to 4MHz, standard Butterworth approximation, orders 4-5-6-7, various programming codes and corner frequencies.

The linearity of the filter has been extensively simulated for all the corner frequency-order combinations. The frequency of the input signal has been chosen

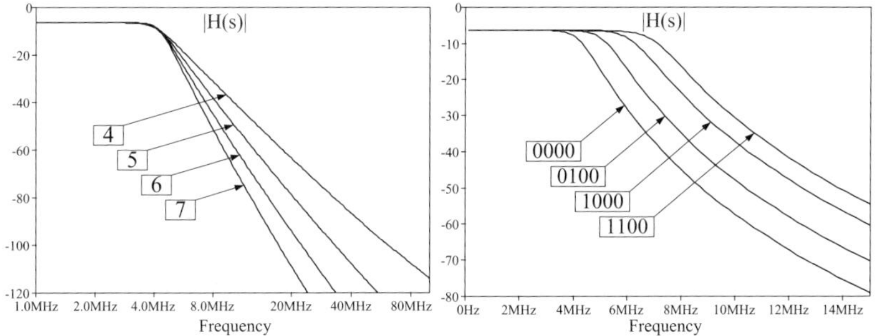

Fig. 8. The filter magnitude response for different orders and corner frequencies

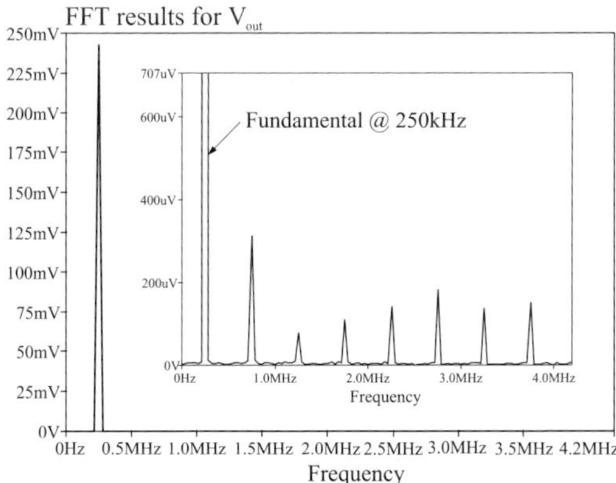

Fig. 9. The output spectrum of the 6th order filter

250kHz, such that the lowest ten harmonics lay in the filter pass band. The worst case total harmonic distortion was approximately -53dB for a 1V peak to peak differential input sine wave. Fig.9 shows the typical output spectrum of the filter and the higher order harmonics for the 6th order configuration.

The built in power down algorithm allows the dynamical reduction of the filter consumption when the order is lowered. The 4th order filter draws 8.4mA from a single 3V supply, while the 7th order configuration draws 13.6mA. The consumption has been measured for the highest possible corner frequency, when all the branches of the programmable mirror are functional.

5 Conclusions

The reconfigurable low pass filter described in this paper is suitable for integration into multi-mode receiver front-ends that employ a form of direct conversion architecture. The modular design is easily scalable, allowing the extension to higher orders and wider programming range of the corner frequency by simple replication of the fundamental reconfigurable module. Furthermore, the circuit can be extended to implement a polyphase band pass response, adapting the transfer function to the needs of other receiver configurations, such as the low-IF architecture. The simulations performed at transistor level have proven the functionality of the reconfiguration concept. The filter achieves good performances in terms of linearity and features dynamically adjustable current consumption. In all the cases the consumption is lower compared to classical OTA-C implementations with similar linear range.

References

1. Mittola, J.: The Software Radio Architecture. IEEE Communications Magazine, 26–38 (1995)
2. Maurer, L., et al.: On the Architectural Design of Frequency Agile Multi-standard Wireless Receivers. In: 14th IST Mobile and Wireless Communications Summit, Dresden (June 2005)
3. Selex Communications: Wideband Structural Antenna Operating in the HF Range, Particularly for Naval Installations, patent (2006)
4. Reconfigurable SDR Equipment and supporting Networks. Reference Models and Architectures, Wireless World Research Forum white paper (2004)
5. Csipkes, D., et al.: Synthesis method for state variable Gm-C filters with a reduced number of active components. In: MIXDES 2003, pp. 292–297 (2003)
6. Hintea, S., et al.: On the Design of a Reconfigurable OTA-C Filter for Software Radio. In: NASA/ESA Conference on Adaptive Hardware and Systems (AHS 2007) (2007)
7. Koli, K., Halonen, K.: Cmos Current Amplifiers: Speed Versus Nonlinearity. Kluwer, Dordrecht (2002)
8. Liu, S., Tsao, H., Wu, J.: CCII Based Continuous-Time Filters with Reduced Gain-Bandwidth Sensitivity. IEE Proceedings-G 138, 210–216 (1991)

Novel Image Rejection Filter
Based on Neural Networks

Botond Sandor Kirei, Marina Topa, Irina Dornean, and Albert Fazakas

Technical University from Cluj-Napoca, Department of Basis of Electronics,
C. Daicoviciu. 15, 400020 Cluj Napoca, Romania
{botond.kirei,marina.topa,irina.dornean}@bel.utcluj.ro

Abstract. In this paper a novel image rejection algorithm based on neural networks is proposed. The low-IF receiver architecture and the phenomena of I/Q imbalance (also referred as image interference) are described. The proposed filter is an enhancement of a complex LMS adaptive filter, which separates the desired and image signals, but the recovered signal still suffers from the effects of imbalance parameters. This is corrected by the proposed filter. Simulink simulations were performed in order to prove the functionality of the novel filter. The simulations prove the convergence and stability of the filter. The necessary sample number is given to achieve -60 dB image rejection.

Keywords: low-IF receiver, I/Q imbalance, image rejection algorithm, neural networks.

1 Introduction

The increased interest of the consumers for wireless communication devices resulted in attractive technical developments in the last two decades in this field. Nowadays rigorous efforts are invested in development of multistandard receivers with low cost, single chip implementation, low power consumption etc. In becoming the 20th century multistandard receiver architecture, a great dispute is done between the zero-IF (or heterodyne) and low-IF (or superheterodyne) receiver architectures. Both architectures can be integrated on a single chip allowing the fabrication of analog front-ends, but each has serious drawbacks. In the low-IF receiver, due to the local oscillator non-ideality, amplitude and phase errors are introduced to the IQ paths. Thus the desired signal can be interfered by the adjacent band's signal, making mandatory the usage of image rejection filters. The desired signal in the zero-IF receiver is perturbed by the mixer stage leakage resulting in a time variant DC offset, which is hard to be eliminated.

1.1 Prior Image Rejection Techniques

In the literature various image rejection algorithms/filters for low-IF receivers are described. A recent solid state circuit solution for image rejection is presented in [1]. This adaptive filter based on sign detection LMS algorithm allows a simple hardware implementation at the cost of lower estimation accuracy. Despite this, depending on the quantization noise it can achieve an image rejection ratio (IRR) of approximately 60dB.

I. Lovrek, R.J. Howlett, and L.C. Jain (Eds.): KES 2008, Part III, LNAI 5179, pp. 343–350, 2008.

Paper [2] presents a non-data-aided image rejection algorithm. Exploring the mutual independence property of the desired and image signals the measure of the interference can be determined and the initial signals restored. From the simulation model presented in the paper, in the desired signal's expression remains a phase mismatch dependent scaling factor, making the filter vulnerable to phase errors. In [3] a modified complex LMS filter is used in order to reject the image frequencies. The amplitude errors are corrected by the filter, but jut like in [2] the phase errors are not cancelled because of a remaining scaling value. If small phase mismatch is present, this LMS filter can achieve an IRR of 30dB [4].

Other solutions are based on blind source (or signal) separation. The measure of image interference can be estimated by evaluating some statistical properties of the received signals [5].

The earliest implementations make use of a tone signal applied on the front-end of the low-IF receiver. By measuring the distortion introduced by the mixing stage image rejection can be achieved just like in [6]. These kinds of solutions are obsolete, because of the additional hardware used for the tone generation.

1.2 Proposed Image Rejection Filter

The proposed image rejection filter is based on the one described in [3]. As already mentioned the major drawback of this implementation is a distortion in the recovered signal, due to a scaling value dependent on the imbalance parameters. This paper proposes to correct this handicap by combining neural networks, accomplishing the LMS adapting function, in order to eliminate the remaining scaling value. Although the low-IF architecture can contain several noise sources (noise introduced by the analog mixer, quantization noise due to the analog-digital conversion, image interference, etc.) this paper analyzes only the effect of image interference

In Section 2 and 3 the low-IF receiver architecture and the image interference also referred as I/Q imbalance are presented. Section 4 describes the enhanced image rejection filter built on neural networks. For better understanding, first the combination of two neurons resulting in the filter studied in [3] and [4] is presented. Then the combination of 4 neurons achieving the enhanced image rejection is pointed out. Section 5 contains the simulation results carried out in Simulink and last but not least conclusions shall be drawn in Section 6.

2 The Low-IF Receiver Architecture

In Fig. 1 the Low-IF receiver architecture is depicted. The radio frequency (RF) signal collected from the antenna is processed by the band-pass filter BPF. Image cancellation can be achieved at this point, but it requires narrow band-filtering, and thus significantly increases the cost of the device. The RF signal is down converted to an intermediary frequency (IF) using a voltage-controlled oscillator VCO (x_{LO}) generating a complex signal. Thus the first mixing stage will result in a complex signal physically represented by the in-phase and quadrature signals, also referred as *I/Q* signal (*I(t) and Q(t)*). The IF complex signal is sampled by the ADCs (*I(n) and Q(n)*).

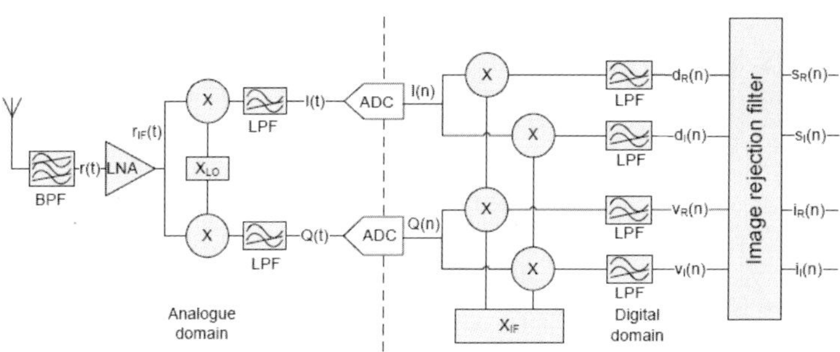

Fig. 1. Low-IF receiver architecture

The state of the art solutions prefer delta-sigma converters [7]. The baseband demodulation and image rejection are done in the digital domain.

3 I/Q Imbalance

The spectrum of the filtered RF signal $r(t)$ from Fig. 1 is depicted in Fig 2. a) and its expression is given by:

$$r(t) = z(t) \cdot e^{+j2\pi f_{LO}t} + z^*(t) \cdot e^{-j2\pi f_{LO}t} , \tag{1}$$

where $z(t)$ is the complex signal and $z^*(t)$ is its conjugate. The signal $z(t)$ is formed by the desired signal $s(t)$ and the image signal $i(t)$ coming from an adjacent band:

$$z(t) = s(t) \cdot e^{+j2\pi f_{IF}t} + i(t) \cdot e^{-j2\pi f_{IF}t} , \tag{2}$$

Due to the imperfections of the electronic components the VCO generates the local oscillation:

$$x_{LO}(t) = \cos(2\pi f_{LO}t) - j \cdot g \cdot \sin(2\pi f_{LO}t + \varphi) , \tag{3}$$

where g is the amplitude and φ the phase error. At down-conversion these mismatches will result in the interference between the I/Q paths. This effect is also called I/Q imbalance. In order to write in a friendlier manner the resulting signal, the I/Q imbalance parameters are defined:

$$k_1 = \frac{1 + g \cdot e^{-j\varphi}}{2} ; k_2 = \frac{1 - g \cdot e^{+j\varphi}}{2} . \tag{4}$$

The IF signal can be expressed using the I/Q imbalance parameters as follows:

$$I(t) + jQ(t) = LPF\{r(t) \cdot x_{LO}\} = k_1 z(t) + k_2 \cdot z^*(t) , \tag{5}$$

where LPF stands for low-pass filtering. The spectrum of the IF signal is depicted in Fig 2 b).

After the sampling produced by the ADC, the IF signal is downconverted into the baseband, thus providing the following complex signals:

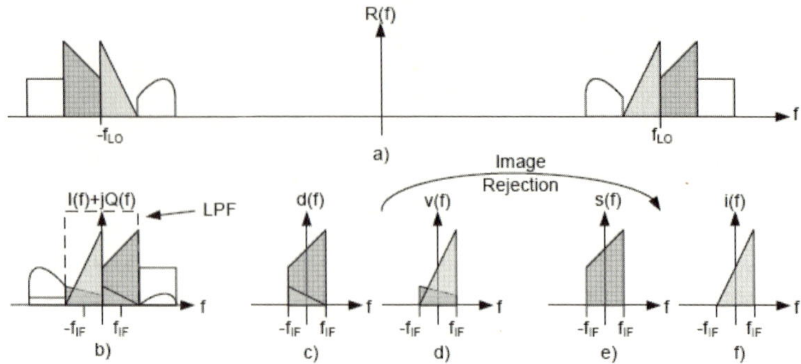

Fig. 2. a). Spectrum of the RF signal b). Spectrum of IF signal c). Spectrum of the mixture signal d(n) d). Spectrum of the mixture signal v(n) e). Spectrum of desired signal s(n) f) Spectrum of interferer signal i(n).

$$d(n) = LPF\{I(n) \cdot e^{-j2\pi f_{IF} nT}\} = k_1 \cdot s(n) + k_2 \cdot i^*(n)$$
$$v(n) = LPF\{Q(n) \cdot e^{+j2\pi f_{IF} nT}\} = k_1 \cdot i(n) + k_2 \cdot s^*(n)$$

$$(6)$$

The signal $d(n)$, depicted in Fig 2. c)., contains the desired signal $s(n)$ (Fig 2. e).) and the conjugate of the interferer $i^*(n)$. Likewise $v(n)$, depicted in Fig 2. d)., contains the interferer signal $i(n)$ (Fig 2. f).) and the conjugate of the desired signal $s^*(n)$. By conjugating the second equation, the equation system (6) can be written in matrix form as follows:

$$\begin{bmatrix} d(n) \\ v^*(n) \end{bmatrix} = \begin{bmatrix} k_1 & k_2 \\ k_2^* & k_1^* \end{bmatrix} \cdot \begin{bmatrix} s(n) \\ i^*(n) \end{bmatrix}$$

$$(7)$$

4 Novel Image Rejection Filter

Let us consider the signal flow chart in Fig. 3. The image separation has two outputs, one for the desired and one for the interfering signal. In this case the neurons are not performing a LMS adaptation but an adaptive prediction operation. Thus we write two cost functions (8) to force the outputs to represent the desired and interfered signals.

$$\xi_1 = E\left\{\left|r_1(n) - w_1^* \cdot x_1(n)\right|^2\right\}$$
$$\xi_2 = E\left\{\left|r_2(n) - w_2^* \cdot x_2(n)\right|^2\right\}$$

$$(8)$$

Replacing in equation (8) the desired signals $r_1(n)$ and $r_2(n)$ with the corresponding values of $d(n)$ and $v^*(n)$ and the pair of input signal $x_1(n)$ and $x_2(n)$ with $v^*(n)$ and $d(n)$ it follows that:

$$\xi_1 = E\left\{\left|d(n) - w_1^* \cdot v^*(n)\right|^2\right\} \rightarrow i^*(n)$$
$$\xi_2 = E\left\{\left|v^*(n) - w_1^* \cdot d(n)\right|^2\right\} \rightarrow s(n)$$

$$(9)$$

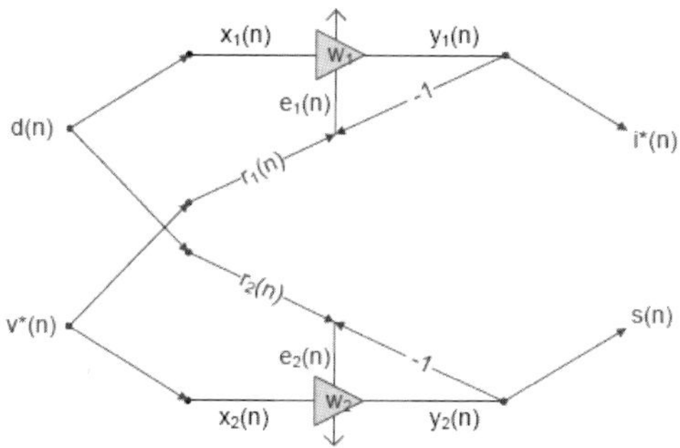

Fig. 3. Signal flow graph of image rejection

For writing the recursive process similar to LMS weight adaptation several considerations have to be done: 1.One pair of adapting neurons implies the usage of two weight values. 2. Because the neuron performs a prediction operation, instead of the input signal $x_1(n)$ and $x_2(n)$ should be written the pair error $e_2(n)$, respectively $e_1(n)$, representing a value that is rejected. The resulting recursive process for the neurons coefficients is (13):

$$\begin{cases} w_1(n+1) = w_1(n) + 2 \cdot \mu_1 \cdot e_1^*(n) \cdot e_2(n) \\ w_2(n+1) = w_2(n) + 2 \cdot \mu_2 \cdot e_2^*(n) \cdot e_1(n) \end{cases} \tag{10}$$

When w_1 and w_2 are adapted the next relation is fulfilled:

$$\begin{cases} k_2 = w_1^*(n) \cdot k_1^* \\ k_1^* = w_2^*(n) \cdot k_2 \end{cases} \tag{11}$$

For the predicted signals the next relations were found:

$$y_1(n) = \left(k_2 - w_2^*(n) \cdot k_1^* \right) \cdot i^*(n)$$
$$y_2(n) = \left(k_2^* - w_1^*(n) \cdot k_1 \right) \cdot s(n) \tag{12}$$

The imbalance parameters k_1 and k_2 still contain the phase error introduced in the mixing stage, but the signals are successfully separated.

The enhanced image rejection filters flow chart is depicted in Fig. 4. As visible, the enhanced image rejecter contains four neurons, thus four cost functions will be processed. A pair of the neurons is following the low given equation in (12.) and the complementary neurons shall adapt for the following objective:

$$\xi_{11} = E\left\{ \left| d(n) - w_{11}^* \cdot v^*(n) \right|^2 \right\} \rightarrow i^*(n)$$

$$\xi_{12} = E\left\{ \left| v^*(n) - w_{12}^* \cdot d(n) \right|^2 \right\} \rightarrow s(n)$$

$$\xi_{21} = E\left\{ \left| d(n) - w_{21} \cdot v^*(n) \right|^2 \right\} \rightarrow s(n) \tag{13}$$

$$\xi_{22} = E\left\{ \left| v^*(n) - w_{22} \cdot d(n) \right|^2 \right\} \rightarrow i^*(n)$$

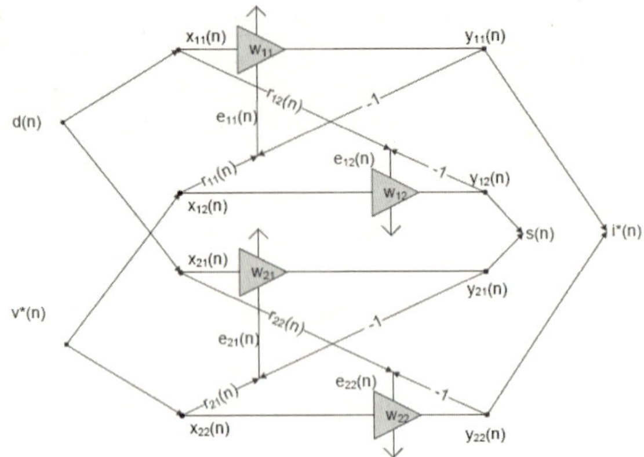

Fig. 4. Signal flow graph of proposed filter

Using the same considerations, the resulted recursive processes for the neurons co-efficients are:

$$\begin{cases} w_{11}(n+1) = w_{11}(n) + 2 \cdot \mu_{11} \cdot e_{12}(n) \cdot e_{11}^*(n) \\ w_{12}(n+1) = w_{12}(n) + 2 \cdot \mu_{12} \cdot e_{11}(n) \cdot e_{12}^*(n) \\ w_{21}(n+1) = w_{21}(n) - 2 \cdot \mu_{21} \cdot e_{21}(n) \cdot e_{22}^*(n) \\ w_{22}(n+1) = w_{22}(n) - 2 \cdot \mu_{22} \cdot e_{22}(n) \cdot e_{21}^*(n) \end{cases}$$

(14)

For the predicted signals the next relations were found:

$$y_{11}(n) = \left(k_2 - w_{12}^*(n) \cdot k_1^*\right) \cdot i^*(n); \; y_{12}(n) = \left(k_2^* - w_{11}^*(n) \cdot k_1\right) \cdot s(n)$$

$$y_{21}(n) = \left(k_1 - w_{22}^*(n) \cdot k_2^*\right) \cdot s(n); \; y_{22}(n) = \left(k_1^* - w_{21}^*(n) \cdot k_2\right) \cdot i^*(n)$$

(15)

From this point the recovery of desired signal consist in a gain correction given by the weight values and since $k_1 + k_2^* = k_2 + k_1^* = 1$ the next relation is found:

$$\frac{y_{11}(n)}{1 - w_{12}^*(n) \cdot w_{11}^*(n)} + \frac{y_{22}(n)}{1 - w_{21}^*(n)^2} = \left(k_2 + k_1^*\right) \cdot i^*(n) = i^*(n)$$

$$\frac{y_{12}(n)}{1 - w_{11}^*(n) \cdot w_{12}^*(n)} + \frac{y_{21}(n)}{\left(1 - w_{22}^*(n)^2\right)} = \left(k_1 + k_2^*\right) \cdot s(n) = s(n)$$

(16)

5 Simulations

Since the presented image rejection filter is an adaptive one, stability, convergence and performance issues are plotted. In the Simulink simulation the input signals of the filter are mixtures of 8-QAM and 6-QAM modulated signals, where the amplitude error is $g=1.2$ and the phase error is $\varphi=10°$. In practice these values are highly exaggerated[1].

---

[1] Usually the amplitude mismatch between the I/Q branch is 1-2% and the phase mismatch is 1-2 degree.

The characteristics are depicted in Fig. 5a, where the number of necessary samples are given to reach a target IRR (-20, -40 and -60dB) for different learning rates. From this picture an optimum value for the learning rate can be chosen, making a trade off between adaptation time and performance. Analyzing the characteristic, the conclusion is that by decreasing the learning rate, the time to reach a given IRR increases exponentially. On the other hand, by decreasing the learning rate higher precision filtering is achieved.

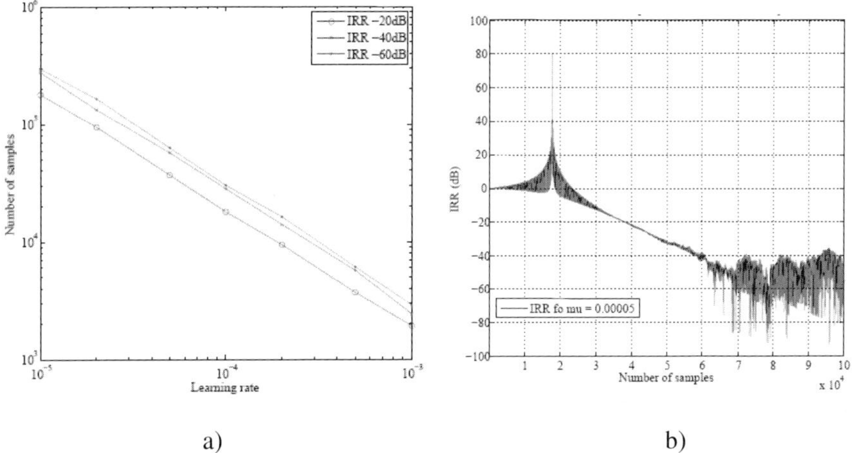

Fig. 5. a) Necessary number of samples function of the learning rate to reach a given IRR b). IRR evolution in time for μ=0.0005.

In Fig. 5b. is depicted the evolution of IRR from the first sample until the filter is adapted. From previous simulation results, a learning rate of $\mu=0.0005$ is considered. The simulation result is interpreted as follows: in the first third of the characteristics the filter is not stabilized, the image is not rejected, thus the values of IRR are senseless; when adaptation begins the IRR is decreasing monotonously; the filter finds the resting point, the IRR tends to vary between two values. This is because the weight values of the neurons may vary around the optimal solution with the freedom given by the learning rate. Although no mathematical proof of convergence was found, the multiple simulations performed suggest that the filter can find a stability point.

6 Conclusions

In this paper the low-IF receiver and the I/Q imbalance are presented. From previous implementations of image rejection filters a new one is derived using neural networks. The drawback of simple LMS filtering is that the recovered signal still suffers from the effect of imbalance parameters. The proposed filter in this paper is able to compensate this.

Simulink model was developed in order to prove the availability of the filter. Simulation results show that there are no convergence problems and depending on the selected learning rate IRRs above -60dB are reachable. A trade off has to be made between adaptation time and IRR.

Further developments consist of implementing the filter on FPGA. The design process is eased by the Simulink model created to serve like an RTL description of the filter. Another direction is to speed up the learning process by the usage of variable learning rate or other methods.

Acknowledgments. This work was supported partly the Romanian National University Research Council under Grant TD 428/2007 entitled "Contributions to Designing and Implementing Integrated Polyphase Filters for Wireless Applications".

References

1. Lerstaveesin, S., Song, B.-S.: A Complex Image Rejection Circuit with Sign Detection Only. IEEE Journal on Solid-State Circuits 41(12), 2693–2702 (2006)
2. Gil, G.-T., Kim, Y.-D., Lee, Y.H.: Non-Data-Aided Approach to I/Q Mismatch Compansation in Low-IF Receivers. Transaction on Signal Processing 55(7), 3360–3365 (2007)
3. Yu, L., Snelgrove, W.M.: A Novel Adaptive Mismatch Cancellation System for Quadrature IF Radio Receivers. IEEE Trans. On. Circuits and Systems II: Analog and Digital Signal Processing 46(6), 789–801 (1999)
4. Kirei, B.S., Dornean, I., Topa, M.: Image Rejection Filter Based on Complex LMS Filter for Low-IF Receiver. In: Proceedings of IEEE - ELMAR 2008 (submitted for publication, 2008)
5. Windisch, M., Fettweis, G.: Blind I/Q Imbalance Parameter Estimation And Compensation In Low-If Receivers. In: IEEE Proceedings of 1st International Symposium on Control, Communications and Signal Processing (ISCCSP 2004), Hammamet, Tunisia (March 2004)
6. Glas, J.P.F.: Digital I/Q Imbalance Compensations in a Low-IF Receiver, Global Telecommunications Conference. In: GLOBECOM 1998. The Bridge to Global Integration, vol. 3, pp. 1461–1466. IEEE, Los Alamitos (1998)
7. Kiss, P., Arias, J., Li, D., Boccuzzi, V.: Stable high-order delta-sigma DACs. In: IEEE Proceeding of the 2003 International Symposium on Circuits and Systems, May 2003, vol. 1, pp. 985–988 (2003)
8. Farhang-Boroujeny, B.: Adaptive Filters – Theory and Applications, pp. 178–182. John Wiley & Sons, Chichester (1998)
9. Haykin, S.: Neural Networks – A Comprehensive Foundation, pp. 118–137. Prentice Hall, Englewood Cliffs (1999)

Towards Reconfigurable Circuits Based on Ternary Controlled Analog Multiplexers/Demultiplexers

Emilia Sipos, Lelia Festila, and Gabriel Oltean

Technical University from Cluj-Napoca, Bases of Electronics Department,
C-tin Daicoviciu 15, 400020 Cluj-Napoca, Romania
{Emilia.Sipos,Lelia.Festila,Gabriel.Oltean}@bel.utcluj.ro

Abstract. The multiplexers are used in a wide range of applications. At present, they are controlled by binary signals. To reduce the number of interconnections, ternary signals are proposed to be used to control analog multiplexers/demultiplexers. The analog multiplexers/demultiplexers are realized with CMOS transmission gates, their control circuits being designed using CMOS ternary inverters. The technology used to design ternary inverters is SUS-LOC. A 3-to-1 analog multiplexer is proposed in the design of a reconfigurable ternary inverter.

Keywords: Analog multiplexer, reconfigurable circuit, transmission gate, CMOS transistors, SUS-LOC.

1 Introduction

Almost any company manufacturing integrated circuits has some binary controlled analog multiplexers. For example, Fairchild Semiconductor produce analog multiplexers with two, four and eight input addresses, called CD4053, CD4052, respectively CD4051 [1]. Also, On Semiconductor produce the MC1406x family [2], and ST Microelectronics the HCF 405x family [3].

Multi-valued logic compared to binary one, works with more than two logical levels. The minimum cost or complexity C, in a numerical system, is obtained for $R=e$ (2.718), where R represents the radix. Since R must be an integer the minimum cost C is obtained for R=3, rather than for R=2 [4]. This paper deals with ternary logic, with logic levels "0" (corresponding to "0" logic in binary), "1" (an intermediary level) and "2" (corresponding to "1" logic in binary). The number of input terminals for ternary controlled multiplexers Ni, is given by formula $Ni=3^{Ci}$, where Ci represent the number of control signals.

The multiplexers are used in many applications. In communications - the telephony network is an example of a very large virtual multiplexer built from many smaller discrete ones. In building digital semiconductors – CPUs and graphics controllers are built with multiplexers [5]. In the design of large-scale systems - the multiplexers are used to reduce the number of integrated circuits required in some designs [6].

In literature, many papers deal with ternary circuits, but they are not dedicated to multiplexers. Generally, the multiplexers do not represent the subjects of papers; they are rather used to prove some theoretical aspects discussed in papers [7], [8], [9] and [10]. Recently, the minimization of ternary multiplexers becomes an interesting subject for many researchers. In [11] and [12] some new minimized ternary multiplexers

I. Lovrek, R.J. Howlett, and L.C. Jain (Eds.): KES 2008, Part III, LNAI 5179, pp. 351–359, 2008.

are proposed using new transistors types such multiple-junction surface tunnel transistors and carbon nanotube transistors. In [13] a design minimization method of ternary multiplexer networks based on the RDSOP forms is presented.

The aims of the paper are to develop a ternary controlled 3-to-1multiplexer/1-to-3 demultiplexer based on CMOS transmission gates in SUS-LOC technology, using a minimized design method,, and to design a reconfigurable ternary inverter as an application of our proposed multiplexer. These analog multiplexers/demultiplexers are realized with transmission gates (TG). The circuits are simulated in Orcad, in $0.25\mu m$ CMOS technology. The obtained ternary controlled circuits are used to complete the author's ternary libraries, in order to design more complex ternary systems (arithmetic circuits, flip-flops). The advantages of ternary controlled multiplexers/demultiplexers are: reduced number of interconnections, less chip area and less complexity comparing with the binary controlled ones.

All circuits are designed in SUS-LOC (Supplementary Symmetrical Logic Circuit Structure) technology. The SUS-LOC was created by D. Olson and patented in 2000, Unites State Patent No.6133754 [14]. It is a self-sustaining architecture, meaning that it has the capability to realize redundant functionally complete sets of logic by any radix [15]. Moreover, SUS-LOC has won an international award from World Economic Forum [16], named Technology Pioneer for 2007, recognizing it as a fundamental advancement in circuit technology.

The paper is organized as follows. In Section 2 and Section 3, the 3-to-1 ternary controlled analog multiplexer /1-to-3 analog demultiplexer are designed. In section 4, the 3-to-1 analog multiplexer is used to design the reconfigurable ternary inverter circuits. The paper ended with some concluding remarks, presented in Section 5.

2   Designing the 3-to-1 Ternary Controlled Multiplexer

A multi-valued (ternary, in our case) multiplexer is to multi-valued (ternary) logic design as the NAND gate is for binary logic design, namely, a universal gate [17]. The operating table of 3-to-1 ternary controlled multiplexer (see Fig.3) is presented in Table 1. The S_0 represent the control (address) signal.

Table 1. Operating table of 3 –to-1 ternary controlled multiplexer

S_0	Out
0	In_0
1	In_1
2	In_2

Our analog multiplexer is realized with three transmission gates (TG). Classical TG (like 4066) is designed to work between $+V_{PS}$ and $-V_{PS}$; it is in on state for control signal equal with $+V_{PS}$ and in off state for control signal equal with $-V_{PS}$.

The proposed ternary controlled TGs are supplied between 5V and ground, and they will be in on state for $V_{S0}=5V$ (corresponding 2 logic) and in off state for both $V_{S0}=2.5V$ (corresponding 1 logic) and $V_{S0}=0V$ (corresponding 0 logic), where V_{S0} represent the electrical values of control signal. In order to do this, the transistors with threshold voltages V_{THM3}= 2.4V (n type transistor in Fig.1) and V_{THM4}= -2.4V (p type transistor in Fig.1) are used.

To design the ternary controlled transmission gate, we replaced the binary inverter from the classical transmission gate circuit with a simple ternary inverter designed using SUS-LOC. The terminals of transmission gate are: In (input terminal), Vcmd (control terminal), V_{ALH} (supply terminal), GND (ground terminal) and Out (output terminal). The In/Out ports of designed transmission gate are interchangeable; the signal can be transmitted from input to output, or from output to input. The input signal must be a positive one, in the range 0V to 5V, having any shape.

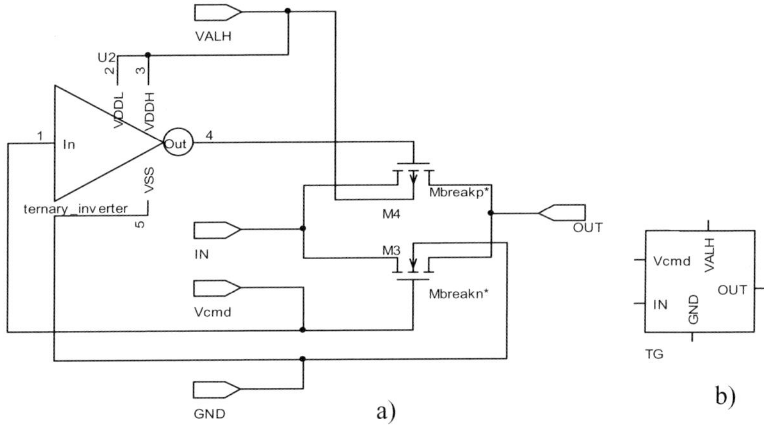

Fig. 1. Transmission gate. a) Circuit; b) Symbol

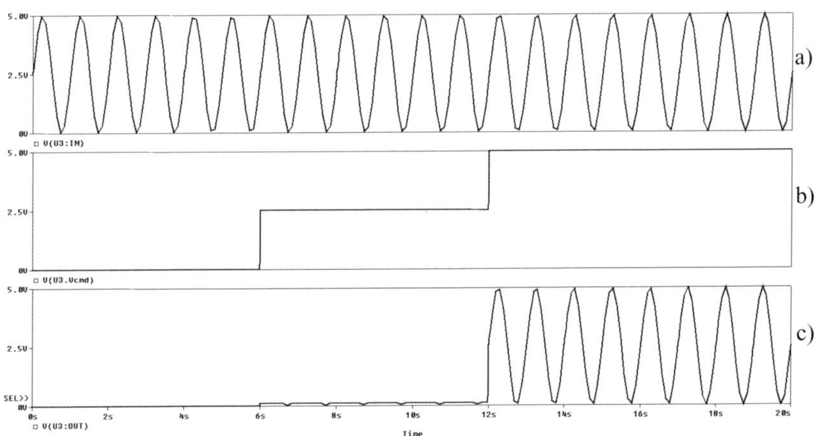

Fig. 2. Simulation results for ternary controlled TG a) Input signal; b) Control signal; c) Output signal

Fig.1 presentes the ternary controlled transmission gate, and Fig.2 presents the Orcad simulation results obtained for ternary controlled TG.

The three-input analog multiplexer can be designed using only one control signal and three ternary controlled TGs. Every TG must be controlled to transmit to the output only the input corresponding to logical value of control signal (see Fig.3). For example, when the control signal is 0 logic, the TG used to transmit the In_0 input must be in on state, and the other two TGs must be in off state. To connect the In_0 input to the output we control the corresponding TG (U13 from Fig.3) with C[200] circuit and to connect In_1 input to the output we control the corresponding TG (U15 from Fig.3) with C[020] circuit. The output value of the circuit C[200] is 2 logic only when the control signal is 0 logic and the output value of the circuit C[020] is 2 logic only when the control signal is 1 logic; in the rest, the output of these circuits is 0. Theoretically, another circuit, denoted C[002], is need to control the TG corresponding to In_2 (U14 from Fig.3) (the output value of C[002] circuit is 2 logic only when the control signal is 2 logic, in the rest the output is 0). But, using the proposed TG, the C[002] circuit has no relevance in functionality of MUX (the TG is already in off state when the control signal is 0 and 1 logic). So, the control signal of multiplexer is directly applied to the control terminal of U14. The initial circuit at transistor level designed in SUS-LOC technology for both C[200] and C[020] presented in [18] was modified in order to properly operate for our 3-to-1 multiplexer.

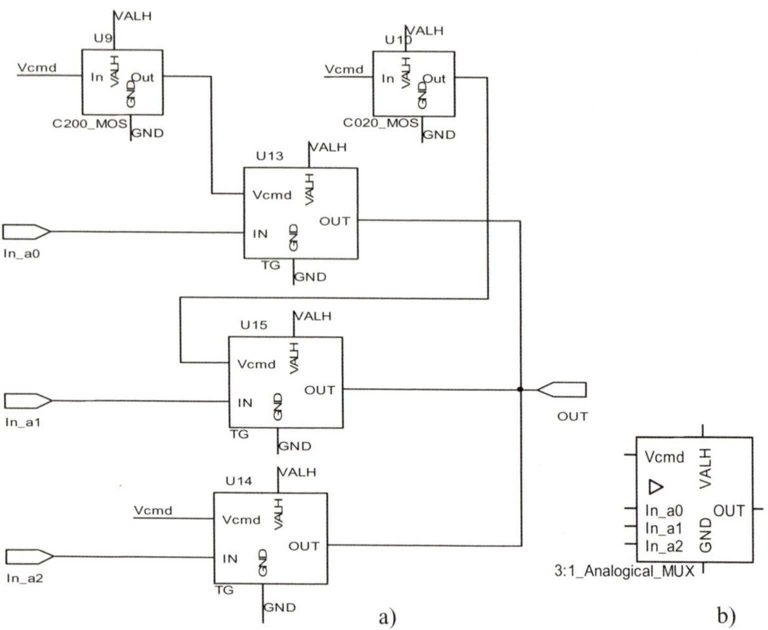

Fig. 3. The 3-to1 ternary controlled multiplexer. a) Block diagram; b) Symbol

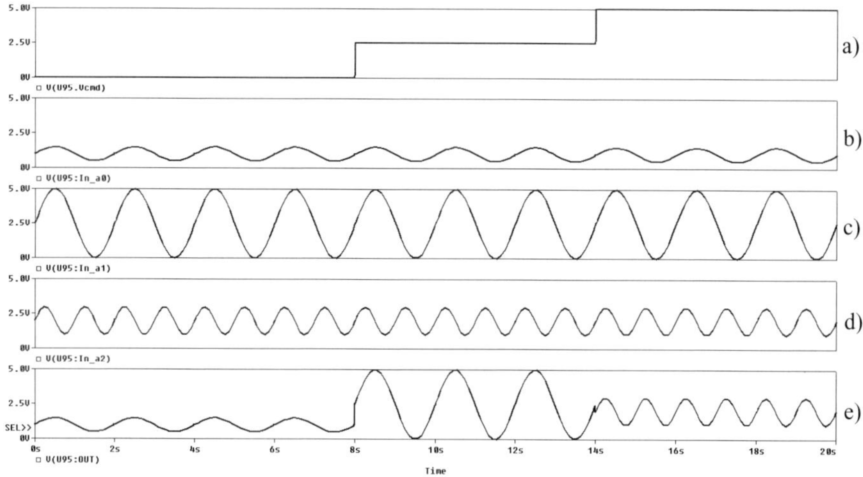

Fig. 4. Waveforms for 3-to1 ternary controlled analog multiplexer. a) Vcmd control signal; b) In_a0 input; c) In_a1 input; d) In_a2 input; e) OUT output.

The control circuit of the proposed multiplexer has only 8 transistors (2 transistors for C[200] and 6 transistors for C[020]). The number of transistors decrease, compared to the case when C[002] circuit is used (with 4 transistors). Moreover, comparing to [19], the number on transistors drops to the half (the control circuit is realised with some arithmetic circuit, having 16 transistors).

In Fig.3 the block diagram (a) and the symbol (b) of 3-to-1 ternary controlled analog multiplexer are presented and Fig.4 shows the Orcad simulation results.

3   Designing the 1-to-3 Ternary Controlled Demultiplexer

The operating table of 1-to-3 ternary controlled demultiplexer is presented in Table 2.

Table 2. Operating table of 1-to-3 ternary controlled demultiplexer

S_0	Out		
	Out_0	Out_1	Out_2
0	In	0	0
1	0	In	0
2	0	0	In

Connecting together the input terminals of all three TGs of ternary controlled multiplexer, the 1-to-3 ternary controlled demultiplexer is obtained. The common terminal represents the input of the demultiplexer. The control circuit of the demultiplexer is identical with the control circuit of the multiplexer.

In Fig.5 the block diagram and the symbol of the designed 1-to-3 ternary controlled demultiplexer are presented. The simulation results are shown in Fig.6.

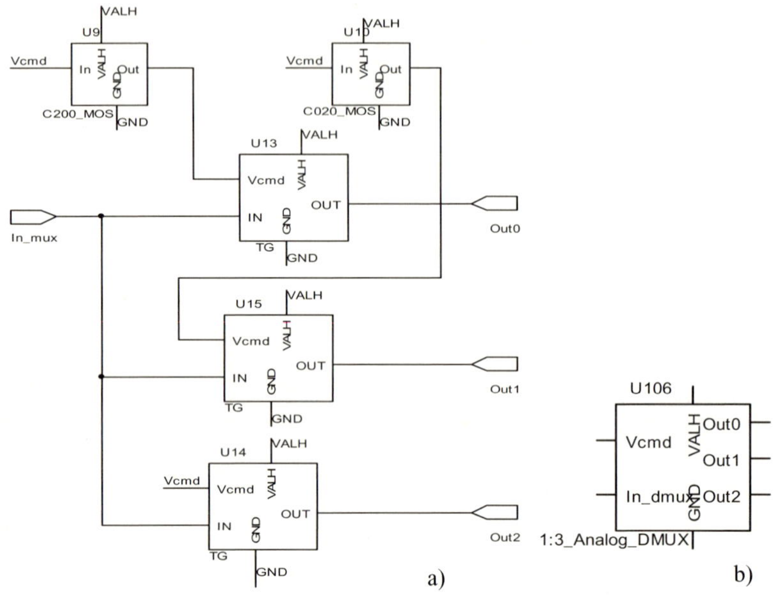

Fig. 5. The 1-to-3 ternary controlled analog DMUX. a) Block diagram; b) Symbol.

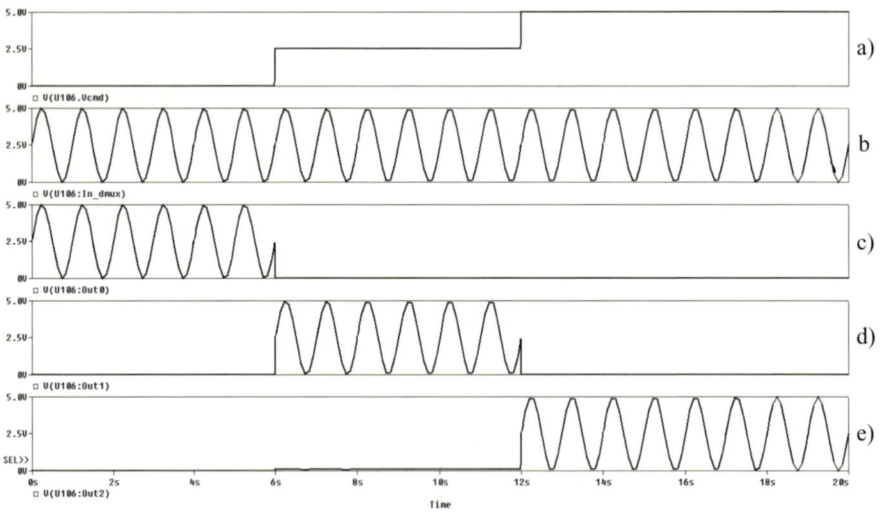

Fig. 6. Waveforms for 1-to-3 analog demultiplexer ternary controlled. a) Vcmd control signal; b) In_dmux input; c) Out_0 output; d) Out_1 ; e) Out_2.

4 The Reconfigurable Ternary Inverter Using Ternary Controlled Analog Multiplexers

In this section, a reconfigurable ternary circuit designed with 3-to-1 ternary controlled multiplexer is presented. The simple application is a reconfigurable ternary inverter (ternary operations of one variable). Depending on the logical values of control signal, our reconfigurable inverter can be the negative ternary inverter (NTI), the simple ternary inverter (STI) or the positive ternary inverter (PTI). These elements form a complete set of algebraic operators. Table 3 represents the truth table for all three types of inverters.

Also, the operating table of the reconfigurable ternary inverter circuit is given in Table 4.

<table>
<tr><td colspan="4">

Table 3. Truth table of basic ternary inverters

</td><td colspan="3">

Table 4. Operating table of ternary recon-figurable circuit

</td></tr>
</table>

IN	NTI	STI	PTI		Vcmd	In_ai	Out
0	2	2	2		0	In_a0=NTI	NTI
1	0	1	2		1	In_a1=STI	STI
2	0	0	0		2	In_a2=PTI	PTI

The reconfigurable ternary inverter circuit realised using 3-to-1 ternary controlled multiplexer is presented in Fig.7 and the waveforms obtained by simulation for a given input are shown in Fig.8.

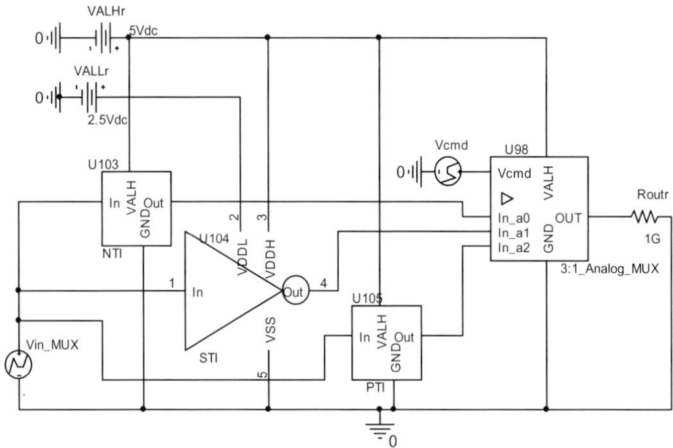

Fig. 7. The reconfigurable ternary inverter circuit

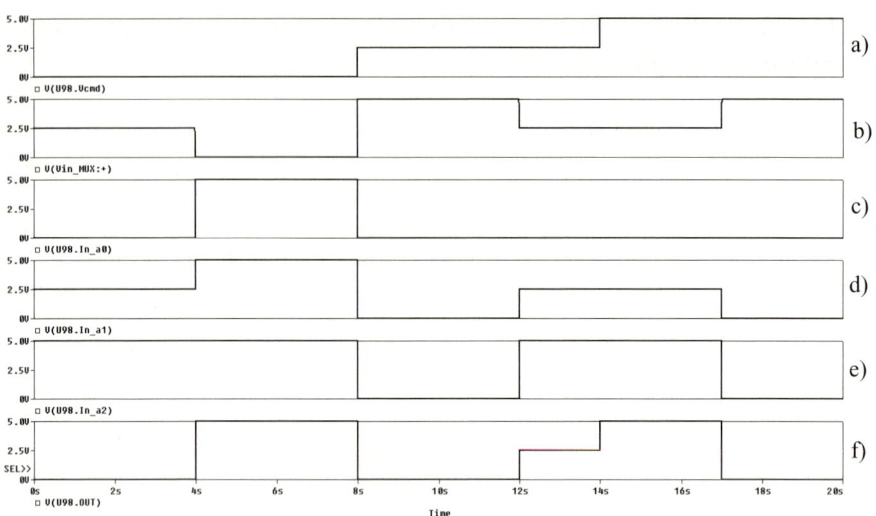

Fig. 8. Waveforms of reconfigurable ternary inverter circuit. a) Control signal; b) Input signal c) In_a0; d) In_a1; e) In_a2; f) Output signal.

5 Conclusions

A 3-to-1 ternary controlled multiplexer and a 1-to-3 ternary controlled demultiplexer was proposed. Both circuits are realized with TGs designed to be in on state only when control signal is 2 logic. Using our proposed TGs, the control circuit of MUX/DEMUX has fewer transistors. We used the designed multiplexer to implement a reconfigurable inverter circuit for ternary signals. All circuits are implemented in SUS-LOC technology and simulated in Orcad, using models for real transistors (PWRMOS). The simulation results validate the correct operation of our circuits.

Using ternary controlled multiplexer with at least two inputs, other reconfigurable circuits can be further designed (a combination of MIN, MAX, NegMIN, NegMAX).

As future research, some practical realizations of ternary multiplexer/demultiplexer and new reconfigurable ternary circuits represent our goals.

References

1. Fairchild Semiconductor, `http://www.fairchildsemi.com/`
2. On Semiconductor, `http://www.onsemi.com/`
3. STicroelectronics, `http://www.st.com/stonline/`
4. Hurst, S.L.: Multiple-valued logic - its status and its future. IEEE Transactions on Computers C-33(12), 1160–1179 (1984)
5. WiseGeek, `http://www.wisegeek.com/what-is-a-multiplexer.htm`
6. University of Surrey, Department of Electronic Engineering, `http://www.ee.surrey.ac.uk/Projects/Labview/multiplexer/index.html`

7. Khan, A.I., Nusrat, N., Khan, S.M., Hasan, M., Khan, M.H.A.: Quantum Realization of Some Ternary Circuits Using Muthukrishnan-Stroud Gates. In: Proc. of the 37th International Symposium on Multiple-Valued Logic ISMVL, p. 20 (2007)
8. Khan, M.H.A.: Design of reversible/quantum ternary multiplexer and demultiplexer. Engineering Letters 13(2), 65–69 (2006)
9. Chew, B.P., Mouftah, H.T.: On the design of CMOS ternary logic circuits using T-gates. International Journal of Electronics 63(2), 229–239 (1987)
10. Dhande, A.P., Ingole, V.T.: Design Of 3-Valued R-S & D Flip-Flops Based on Simple Ternary Gates. Proc. of World Academy of Sc., Eng.&Tech. 4, 1307–6884 (2005)
11. Uemura, T., Baba, T.: Demonstration of a novel multiple-valued T-gate usingmultiple-junction surface tunnel transistors and its application tothree-valued data flip-flop. In: Proc. of the 30th IEEE International Symposium on Multiple-Valued Logic, pp. 305–310 (2000)
12. Al-Rabadi, A.N.: Carbon Nano Tube (CNT) Multiplexers for Multiple-Valued Computing, Facta Universitatis. Ser.: Elec. Energ. 20(2), 175–186 (2007)
13. Jiang, E.H., Jiang, W.B.: Algebra Theory of RDSOP Forms of Ternary Logic Functions and Its Implementation with T-Gates. Chinese Journal of Comp. (7), 1132–1137 (2007)
14. Patent Storm http://www.patentstorm.us/patents/6133754-description. html
15. Olson, E.: Supplementary symmetrical logic circuit structure. In: Proceedings of the 29th IEEE International Symposium on Multiple-Valued Logic, pp. 42–47 (1999)
16. Omnibase Logic, http://www.omnibaselogic.com/
17. Wos, L., Pieper, G.W.: A Fascinating Country in the World of Computing: Your Guide to Automated Reasoning, p. 137. World Scientific, Singapore (1999)
18. Irisa, http://www.irisa.fr/cosi/SEMINAIRE/transparents/MVL-juin% 202001.pdf
19. Srivastava, A., Venkatapathy, K.: Design and Implementation of a Low Power Ternary Full Adder. VLSI Design 4(1), 75–81 (1996)

Analog Multiplying/Weighting VLSI Cells for SVM Classifiers

Lelia Festila, Lorant Andras Szolga, Mihaela Cirlugea, and Robert Groza

Basis of Electronics Department, Technical University of Cluj-Napoca
G.Baritiu 26-28 Street, Romania,
lelia.festila@bel.utcluj.ro

Abstract. VLSI support vector machine classifiers require a large amount of calculations, therefore their implementation needs high density, high speed and low power circuits. In a SVM architecture based on a multiplying law the main building blocks are multipliers. We propose in this paper multiplying and weighting cells, developed by using a model consisting of a compound of two inverse non-linear functions. This procedure is suitable for VLSI implementation because it permits the use of simple nonlinearized standard DA cells that compensate each other nonlinearities to obtain an extended domain of operation. The resulted weighting/multiplying cells were analyzed and tested by simulations.

Keywords: weighting circuits, analog multipliers, th domain, square-root domain, current controlled amplifiers.

1 Introduction

Support vector machines (SVM) can be used in classifying problems to classify objects into one or more classes. As an example a binary classifier classifies positively an object if it belongs to the class C and negatively if it does not.[4]

A binary SVM classifier is trained with positive and negative labeled vectors of features (objects) belonging or not to the given class C. The training results are M support vectors (SVs) $\mathbf{X_m}$, m=1,2…M, which include relevant features of the training vector set and Lagrange coefficients α_m and labels $y_m \in \{-1,1\}$ assigned to each SV.

The already trained classifier characterized by SVs $\mathbf{X_m}$, m=1,…,M and their corresponding Lagrange coefficients and labels is given a vector $\mathbf{X}$ to be classified.

$$X = \left[X_1, X_2 ... X_N \right]^T \quad ; \quad X_m = \left[X_{m1}, X_{m2} ... X_{mN} \right]^T . \tag{1}$$

The classifier will calculate a decision function, as an example of the form[6]:

$$y = sign\left(\sum_{m=1}^{M} y_m \alpha_m X^T X_m + b \right) = sign\left(\sum_{m=1}^{M} y_m \alpha_m \left(\sum_{j=1}^{N} X_{mj} X_j \right) + b \right). \tag{2}$$

I. Lovrek, R.J. Howlett, and L.C. Jain (Eds.): KES 2008, Part III, LNAI 5179, pp. 360–367, 2008.

It is in fact the label **y** to be assigned to the tested vector of features X, which is positive classified if $y \geq 0$.

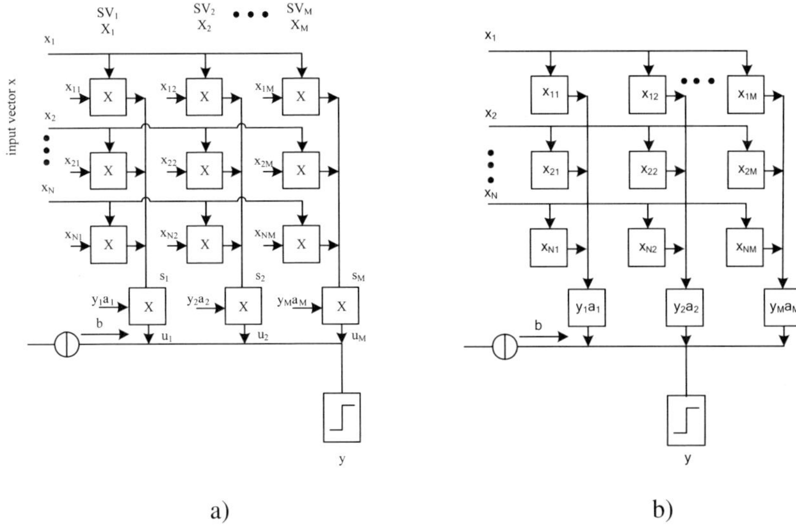

a)   b)

Fig. 1. Block diagram of a cellular parallel SVM classifier with multiplying law: $(x_k \, x_{kj})$ k = 1,...,N; j = 1,...,M; N – vector length; M – SV number a) with multiplying cells; b) with weighting cells

Examining relation (2) one can see that to implement such a SVM classifier we need multipliers as basic functional unities. The $\mathbf{X^T X_m}$ matrix of multiplying cells performs the multiplication of the input vector of features $\mathbf{X}$ to be tested with each of the given SVs, $\mathbf{X_m}$ (m= 1,2...M) as Fig.1,a shows. Each product $\mathbf{X^T X_m}$ is then multiplied by its corresponding coefficient $\alpha_m y_m$. If the resulted u_m, m=1,...,M signals are currents, they can be summed, by simple connections. A constant current b is added then and finally a current comparator delivers the decision function y.[2]

If a SVM classifier is dedicated to a specific application, SVs, X_m and coefficients could be fixed parameters, set by design and the basic functional unities become simple weighting elements, as Fig.1,b shows.

As SVM classifiers usually require a large amount of calculations, their VLSI implementation needs high density, high speed and low power circuits. Because the main building blocks in the SVM architecture in Fig.1 are multipliers (a)) or weighting elements (b)) we have studied different possibilities to implement such blocks[5]. We propose in this paper multiplying and weighting cells, developed by using a model consisting of a compound of two inverse non-linear functions. This procedure is suitable for VLSI implementation because it permits the use of simple nonlinearized standard DA cells that compensate each other nonlinearities to obtain an extended domain of operation – procedure characteristic for ELIN circuits[3].

2 Multiplier Models Based on F-F^{-1} Functions

Let define an invertible function F: x $\rightarrow$ y, where variables x and y are nondimensional, expressed by normalized voltages and currents respectively:

$$\frac{i}{I} = F\left(\frac{v}{V}\right) \qquad and \qquad \frac{v}{V} = F^{-1}\left(\frac{i}{I}\right). \tag{3}$$

Consider that the above functions can be implemented by two basic building blocks F and F^{-1} represented in Fig.1

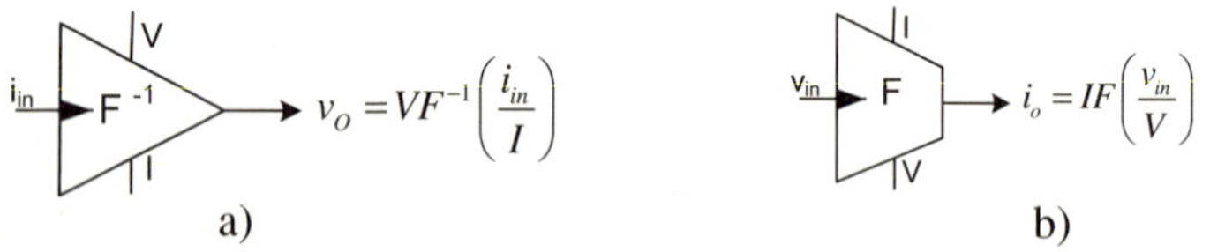

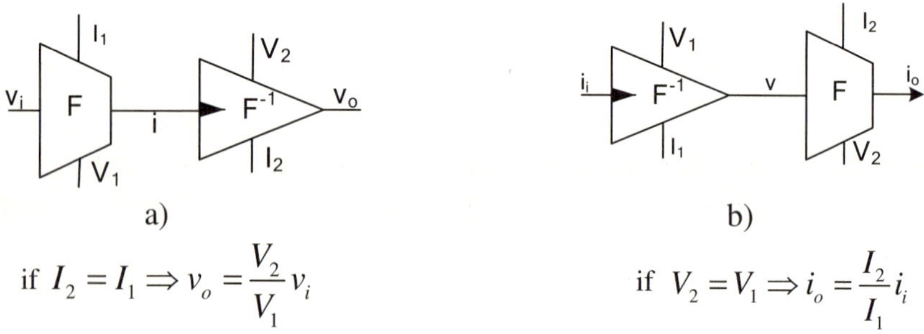

Fig. 2. F^{-1} (Fig1,a) and F (Fig1,b) building block symbols

One connects two F and F^{-1} blocks as in Fig.3 a, and b. If one of the requirements $I_1=I_2=I$(a) or $V_1=V_2=V$(b) may be fulfilled, different types of circuits, voltage or current followers, weighting cells or multipliers can be realized as figures 3 show:

Fig. 3. Weighting cell or multiplier models a) voltage-mode b) current-mode

If currents I can be set independently of voltages V or viceversa, multipliers result. The type of the multiplier depends on the input port type. There are F-F^{-1} circuits with single or differential input ports, so the input signal may admit one or two directions respectively. Scaling currents I or V are usually onedirectional in the basic building blocks so by using the block diagrams from figure 3, 1-Q or 2-Q multipliers result. For a four quadrant multiplier an extra connection is needed as Fig.4 shows as an example.

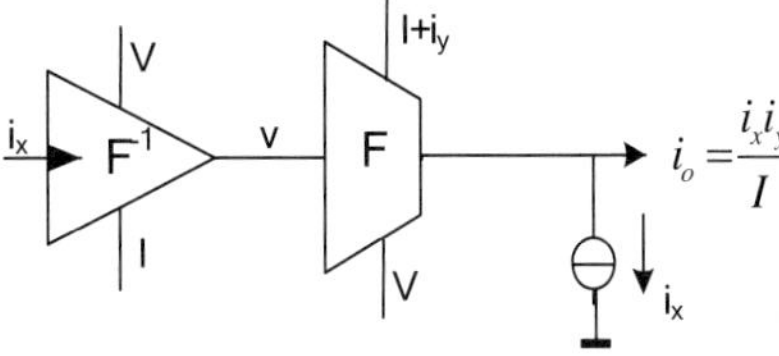

Fig. 4. Basic 4-Q current mode multiplier model: i_x and i_y are bidirectional input signals

An observation will be made about the possibility to implement block F^{-1} with an inverting block F, as Fig.5 a) and b) show.

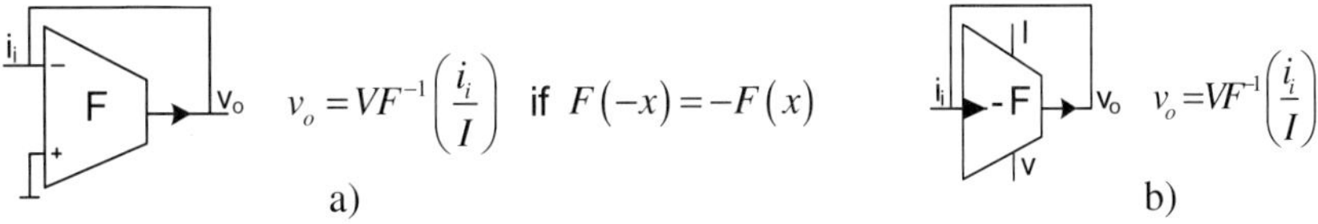

$$v_o = VF^{-1}\left(\frac{i_i}{I}\right) \quad \text{if } F(-x) = -F(x) \qquad v_o = VF^{-1}\left(\frac{i_i}{I}\right)$$

a)   b)

Fig. 5. Implementation of function F^{-1} with a) a differential F block b) an inverting F block

In the following we analyze the opportunity to use the simple basic bipolar or CMOS DA cell as an **F** building block of the proposed m/w cells.

3 Multiplying Cells Realized with Bipolar DAs

The large signal model of a DA in bipolar technology is described by a hyperbolic tangent function as Fig.6,a shows. In Fig.6,b the connection implementing F^{-1} function is given. The input signal in the F cell is $v_i = v_i^+ - v_i^-$.

One can see that the requirement $V_1 = V_2 = V_T$ are fulfilled and a 2Q current mode multiplier may be realized using the model from Fig.3b. Fig.7 shows the schematic of such a

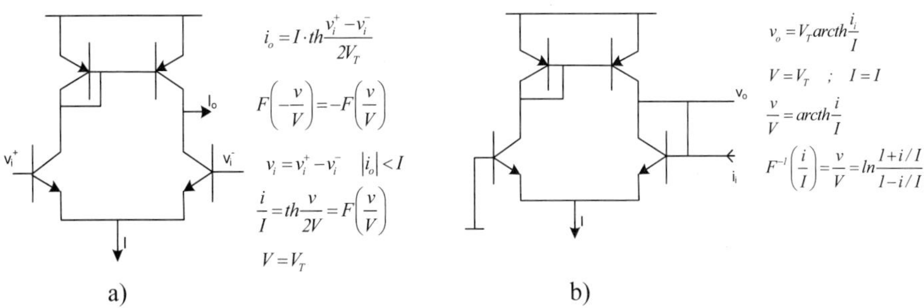

a)   b)

Fig. 6. Bipolar differential amplifier used in the large signal domain a) F block: th cell; b) F^{-1} block: arcth cell

364   L. Festila et al.

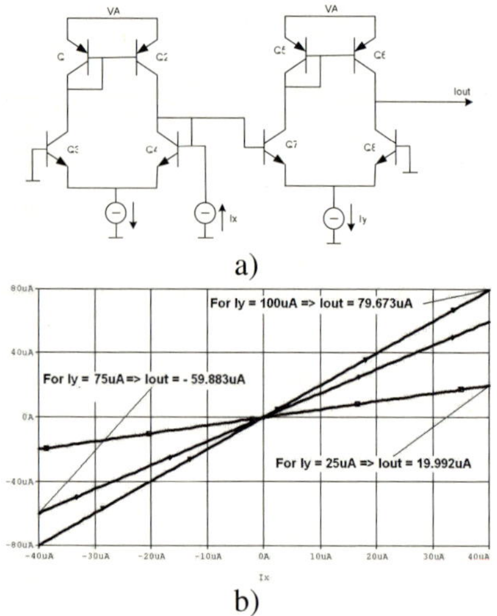

a)

b)

Fig. 7. 2Q multiplier a) schematics b) in-out characteristics for Iy= ct and I=50µA

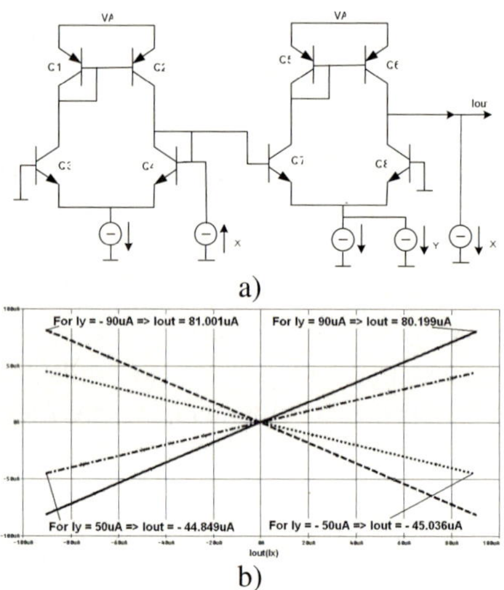

a)

b)

Fig. 8. 4Q multiplier for Iy= ct and Ix=100µA a) schematics b) in-out characteristics

2Q multiplier and $i_{out}=f(i_x)$ characteristics for $i_y=ct$ as a parameter. In Fig.8 a 4Q multiplier is shown and also its characteristics. Simulations have proved the validity of the models shown in Fig.3,b and 4 and also the linearity on the whole domain $|i_x|,|i_y| < I$.

The same conclusions result if schematics from Figs 6-8 are realized with MOS transistors in weak inversion, because functions $F-F^{-1}$ remain of the same type.

4   DA Cells with MOSFETs Saturated in Strong Inversion

Fig. 9 a) and b) present two F-F^{-1} functions implemented by simple differential amplifiers (DAs) with saturated MOS transistors in strong inversion.

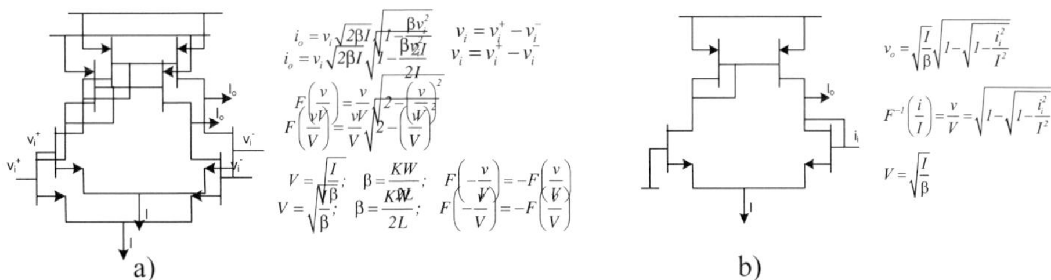

Fig. 9. Square-root F-F^{-1} cells a) F cell ; b) F^{-1} cell

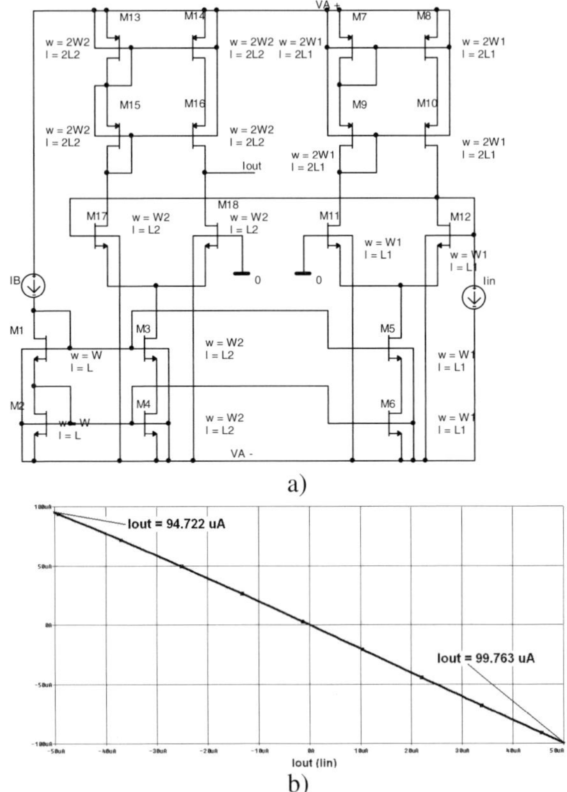

b)

Fig. 10. Current amplifier for $W_2/L_2 = 2*W_1/L_1$ a) schematics b) in-out characteristic

In the same figures the functions F and F^{-1} are deduced on the base of the input output relations $i_O = f_1\left(v_i^+ - v_i^-\right)$ Fig.9,a and $v_O = f_2\left(i_i\right)$ Fig.9,b. The circuit having the model in Fig.3,b is shown in Fig.10

In this case considering Fig3 b), relation $V_1 = V_2$ has to be fulfilled, that is:

$$\frac{I_1}{\beta_1} = \frac{I_2}{\beta_2} \;\Rightarrow\; i_o = \frac{I_2}{I_1} i_i = A_i i_i \;;\; A_i = \frac{W_2/L_2}{W_1/L_1}. \tag{4}$$

One can see that the current gain A_i can not be controlled. Therefore the current amplifier from fig.10 could only be used to realize weighting signals with fixed weights.

Using the model in Fig3,a if one sets $I_1 = I_2 = I$ a voltage mode amplifier results and could be also used for weighting signals with a fixed valued weight. In this case the voltage gain is smaller than in the previous case of the current-mode amplifier:

$$v_o = \frac{V_2}{V_1} v_i = \sqrt{\frac{W_1/L_1}{W_2/L_2}}\, v_i. \tag{5}$$

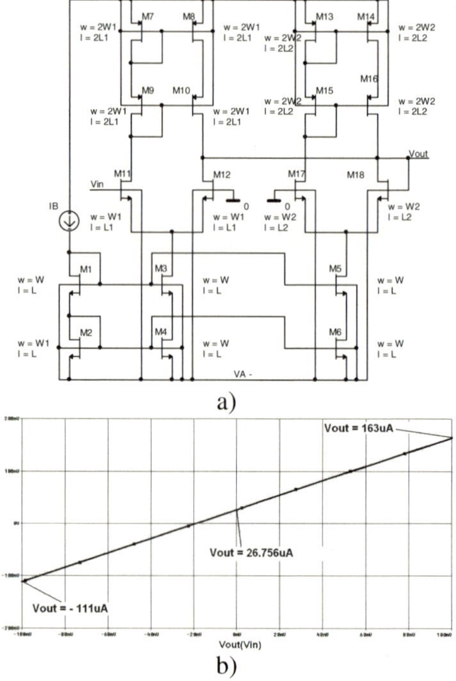

Fig. 11. Voltage amplifier schematic b) in/out characteristic

Fig 11 a) shows the simulated schematic of this VM amplifier and b) an input output characteristic.

This last resulted V-M variant of an amplifier prove the validity of the proposed general model. It is in fact a known connection used in linear G_m-C designs. The association to the F-F^{-1} models explains the extended domain of linearity of this schematic without the need of linearity of each transconductor cell.

5 Conclusion

The general modular models proposed in this paper use simple nonlinear F-F^{-1} cells that compensate each other nonlinearities. In this manner the large signal domain can be used extending the linear behaviour of input output characteristics. Replacing the F blocks by simple nonlinearized bipolar or CMOS DA building blocks very simple m/w cells result. They are suitable to be used in VLSI neuron-like networks in particular SVM classifiers because of their simplicity, efficiency and good performance for a large domain of signal variations.

References

1. Genov, R., Chakrabartty, S., Cauwenberghs, G.: Silicon support vector machine with on-line learning. Int. J. Pattern Recognition Artificial Intell. 17(3), 385–404 (2003)
2. Cirlugea, M., Festila, L., Albert, F.: Considerations About Hardware Architectures of SVM Classifiers. In: Inter-Ing 2005, Targu-Mures, Romania, pp. 691–696 (2005)
3. Tzividis, Y.: Externally Linear, Time Invariant Systems and Their Application to Companding Signal Processing. IEEE Trans. On CS II 44(2), 65–84 (1997)
4. Gordan, M.: Doctoral Thesis: Applications of the Fuzy technics and the SVMs in the image processing, Technical University of Cluj-Napoca (2004) (in Romanian)
5. Festila, L., Groza, R., Cirlugea, M., Topa, M.: Modular Log-Domain Multipliers for SVM Classifiers. In: International Conferrence on AQTR, Cluj-Napoca, 1-4244-0360-x (2006)
6. Joachims, T.: SVM Light Version: 6.01, University of Dortmund, Informatik, AI-Unit (02.09.2004)

Log-Domain Binary SVM Image Classifier

Robert Groza, Lelia Festila, Sorin Hintea, and Mihaela Cirlugea

Technical University of Cluj-Napoca, Basis of Electronics Department,
G. Baritiu street, no. 26-28, 40027 Cluj-Napoca, Romania
`{lelia.festila,robert.groza,sorin.hintea,`
`mihaela.cirlugea}@bel.utcluj.ro`

Abstract. We propose in this paper a binary log-domain Support Vector Machine classifier based on a polynomial decision function. To implement such a classifier log-domain multipliers proposed by the authors are used. For the parallel-serial implementation a log-domain summing amplifier and a current mode comparator are also needed. Current mode log-domain design is used for its low voltage, low power and high frequency characteristics. The resulted classifier is simulated taking into account real parameters of transistors in BiCMOS technology.

Keywords: Log-Domain, SVM classifier, current multiplier.

1 Introduction

SVM (Support Vector Machines) are self learning computational systems used for classification and regression tasks. They offer an attractive alternative to ANN's (Artificial Neural Networks). From the hardware point of view, SVM's are very advantageous because of their simple functional model that leads naturally to parallel real-time computation.

The SVM classification goal is to assign an object to a class containing similar objects. The SVM is trained with a set of positive and negative examples given as vectors of features representing points in the n-dimensional space. After learning, a small number of vectors is retained from the whole training data, vectors that include the relevant characteristics of the whole learning set. These are the support vectors [1], [2], [3]. The classifier response is computed using a decision function, expressed by a linear combination of functions called kernels. The coefficients of those functions are derived in the learning process. Most of the kernels are used in polynomial classification. They are based on internal products and therefore they can be implemented as multilayer perceptrons [1],[4],[5].

A linear classifier classifies an unknown pattern $\mathbf{X}$ to one of the two classes $y \in \{-1,1\}$, based on a given set of M predefined patterns $\mathbf{X}_i$ (i=1…M), called support vectors (SVs) and their corresponding labels y_i, by calculating the decision function. We consider in this paper a decision function of the form:

$$y = \text{sign}\left(\sum_{i=1}^{M} a_i y_i \left(\sum_{j=1}^{N} x_{ij} x_j \right) + b \right) \tag{1}$$

I. Lovrek, R.J. Howlett, and L.C. Jain (Eds.): KES 2008, Part III, LNAI 5179, pp. 368–375, 2008.

where $\mathbf{X}=[x_1,x_2,\ldots x_N]$; $\mathbf{X_i}=[x_{i1},x_{i2},\ldots,x_{iN}]$, $i=1,\ldots,M$. Coefficients $\mathbf{a_i}$, $i=1\ldots M$, the bias term b and SV's $\mathbf{X_i}$ are known from the learning process. The resulted decision function is a label y corresponding to the vector of features under test, which will be positive classified if $y \geq 0$. A parallel architecture of a classifier based on relation (1) is given in Fig. 1 [2].

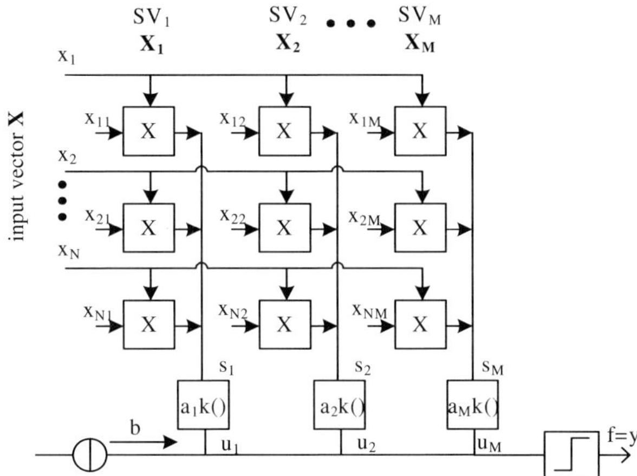

Fig. 1. Block diagram of a cellular fully parallel current mode SVM classifier with a multiplying law: $(x_k\,x_{kj})$; $k = 1,2,\ldots,N$; $j = 1,2,\ldots,M$; N – vector length; M – number of SV's

One can see that for implementing such a current mode circuit we need multipliers, and comparator circuits. Summation is performed by routing currents to a common low impedance node. If necessary, current followers (mirrors) or amplifiers are used. As SVM classifiers usually require a large amount of calculations, their implementation needs high density, high speed and low power circuits. Very simple and efficient modular structures offer the log-domain design.

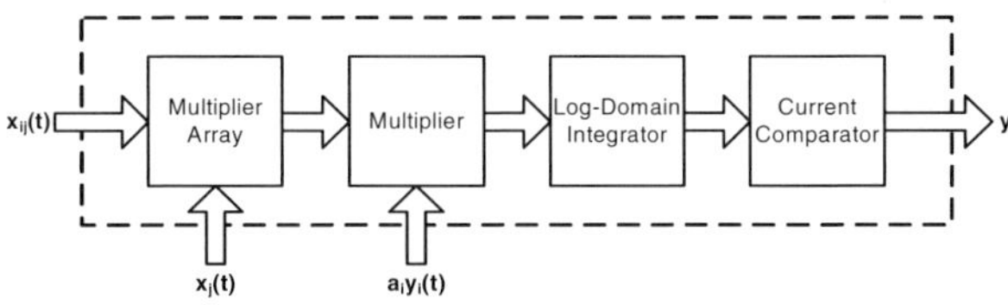

Fig. 2. Proposed Log-Domain SVM classifier block diagram

Fig. 2 shows the block diagram of the serial-parallel classifier presented in this paper. The multiplier array has the structure of a column in the NxM array in Fig. 1. Support vectors are serially introduced and also coefficients a_iy_i. To add each vector product $\mathbf{XX_i^T}$ to the partial sum a log-domain integrator is used. The circuit also contains at the output a signal comparator to classify the input signal. In the following we will present the component circuits used for the SVM classifier and then the results of amplification for image recognition.

2   Multiplier Cell

The multiplier cell used for the multiplying operations is a four quadrant log-domain current multiplier proposed by the authors [6]. The block schematic is presented in Fig. 3. By simulations we proved the validity of this circuit.

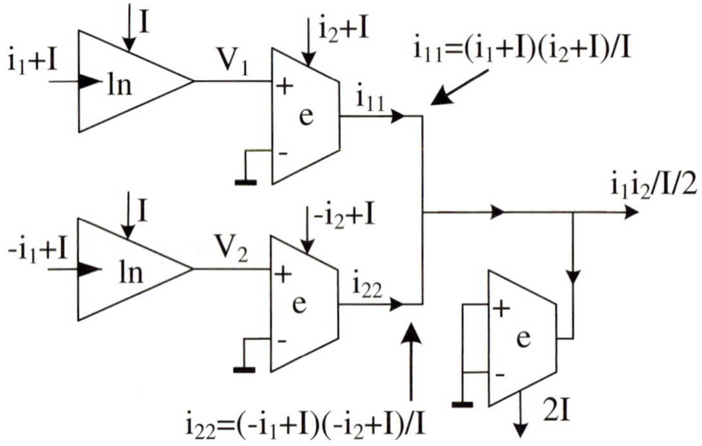

Fig. 3. Log-domain four quadrant current multiplier

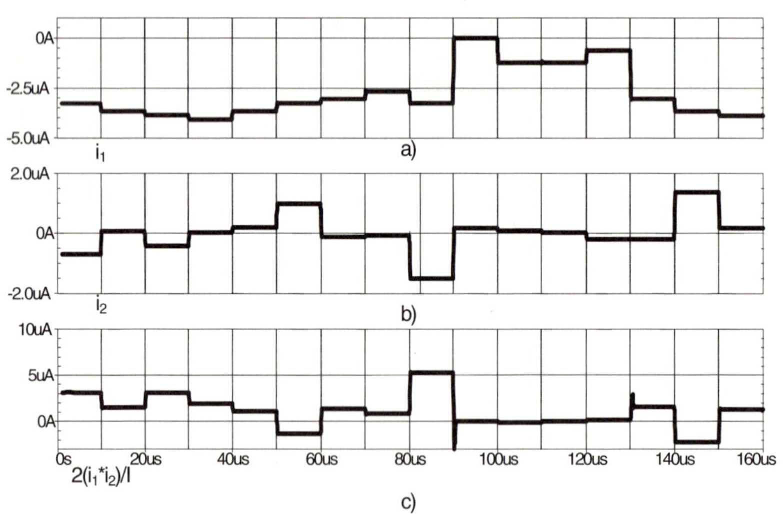

Fig. 4. Input signals i_1 and i_2 a); b) and output signal c)

Fig. 4 shows a PSpice simulation of the proposed circuit from Fig. 3 where **ln** and **e** BiCMOS circuits from our cell library contain real transistors models. Input i_1 is the output sequence of the multiplier array for a given image (Fig. 2), i_2 represents the vector containing the SV a y coefficients obtained from the learning process. They are represented in Fig. 4 a) and b). The output signal of the Log-Domain multiplier is shown in Fig. 4, c).

3 Log-Domain Current Summator (Log-Domain Integrator)

The circuit needed to realize the sum of signals serially collected in time is a log-domain integrator [7]. It is presented in Fig. 5.

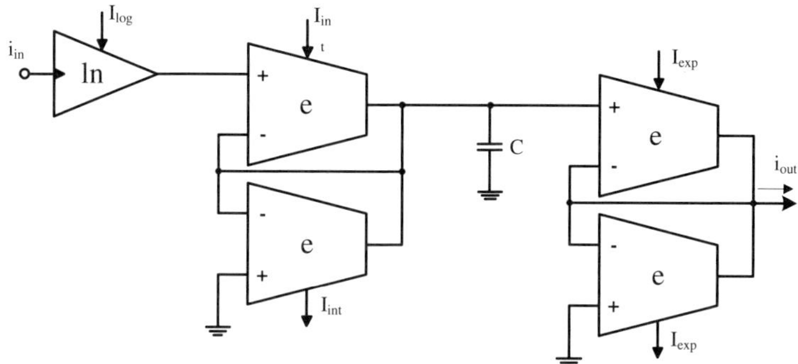

Fig. 5. Log-Domain Integrator

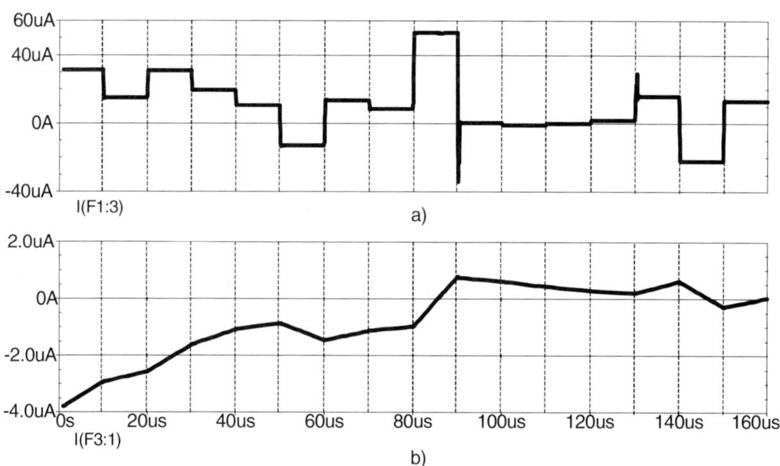

Fig. 6. Time domain simulation: input signal a); output signal b)

Fig. 6 shows a PSpice simulation of the proposed circuit where the **ln** and **e** cells are the same, used to implement the log-domain multiplier. To prove its functionality simulations were performed. We applied at the input a time variant signal presented in Fig. 6 a), this signal is the output of the multiplier block presented in Fig.2. The resulted signal at the output of the Log-Domain integrator is shown in Fig. 6 b).

4 Current Comparator

The schematic of the current comparator used in our design is presented in Fig. 7 [8]. We tested this comparator by simulations. In Fig.8 we present the input (Fig. 8 a)) and the output signal from the comparator (Fig. 8 b)). The input signal is the one obtained at the output of the Log-Domain integrator from Fig. 2.

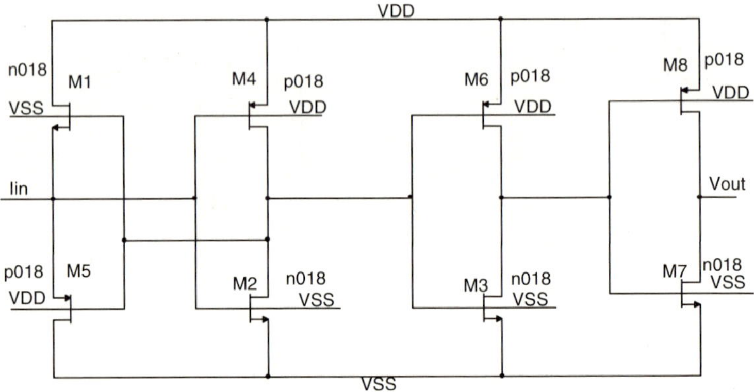

Fig. 7. Current comparator circuit

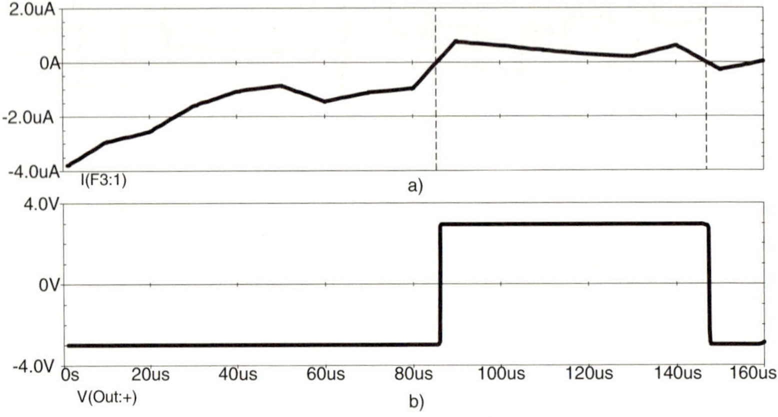

Fig. 8. Simulations in time domain: input signal a); current comparator output signal b)

5 Proposed Binary SVM Classifier

We realized in SPICE the circuit based on the block diagram from Fig. 2 using the circuits from Fig. 3, Fig. 5 and Fig. 7.

Simulations have been performed taking into account applications in image classifications. As an example a SVM was trained by software to identify the letter **F** in a 6x7 image. A number of 16 positive and 16 negative examples were considered in the training. The results are 16 SV's as Fig. 9 shows and also coefficients $a_i y_i$, i=1,2…6 given in Table1.

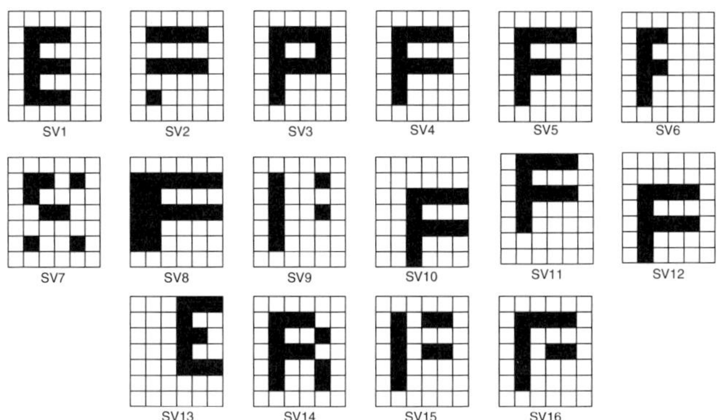

Fig. 9. SV's resulted after the training process

Table 1. Support Vector Coefficients

Support Vector	$a_i y_i$	Support Vector	$a_i y_i$
SV1	-0.700	SV9	-1.493
SV2	0.067	SV10	0.187
SV3	-0.410	SV11	0.098
SV4	0.038	SV12	0.025
SV5	0.213	SV13	-0.199
SV6	0.989	SV14	-0.200
SV7	-0.113	SV15	1.382
SV8	-0.070	SV16	0.187

The images to be tested are given in Fig. 10. The resulting output currents corresponding to each tested image are shown in Table 2.

The input current for a black pixel was considered $-1\,\mu A$ and the value for the white one $+1\,\mu A$, the bias current b is $4\,\mu A$. At the output of the inverting current comparator, the positive values of the decision function indicate the images close to the given reference image, in our case letter **F**.

In Fig. 11, we present the signals from the outputs of the main building blocks, for a given test image. In this case the image is Tst1 from Fig. 10. The following output

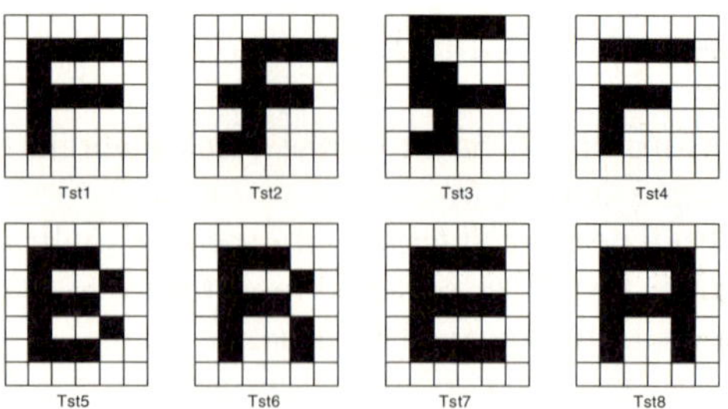

Fig. 10. Test images

Table 2.

Test Image	Iout_integ	Classified	Test Image	Iout_integ	Classified
Tst1	1.34u	+	Tst5	-2.47u	-
Tst2	0.45u	+	Tst6	-1.95u	-
Tst3	-3.51u	-	Tst7	-0.4u	-
Tst4	-4.85u	-	Tst8	-0.42u	-

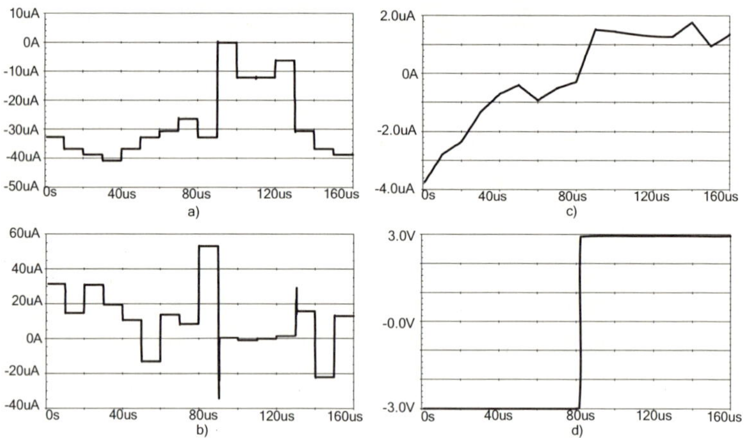

Fig. 11. Simulation results: a) output signal from the multiplier array, b) output signal of the second multiplier, c) output signal from the log-domain current summator, d) output signal from the current comparator

signals are considered: the output current of the multiplier array, a), the output current from the second multiplier block which realize the product between the first signal and the SV's coefficients presented in Table 1, b), the resulting current at the log-domain current summator block, c) and finally the output voltage of the current comparator d). The entire process of classification takes 160us. This is the time necessary to realize the classification operation with the given number of SV's.

6 Conclusions

We extend the externally linear – internally nonlinear design to modular nonlinear circuits like multipliers. They proved to be the very appropriate for implementing SVM classifiers. Modularity, current programmability of the parallel-serial SVM classifiers proposed in this paper recommend it for standard cell design and real time operation in image classification tasks.

References

1. Genov, R., Chakrabartty, S., Cauwenberghs, G.: Silicon SVM with on Line Learning. J.J. of Patern Recognition and Artificial Intelligent Systems, 43–48 (2002)
2. Cîrlugea, M., Feştilă, L., Albert, F.: Considerations About Hardware Architectures of SVM Classifiers. In: Inter-Ing 2005, Tîrgu Mureş, Romania, pp. 691–696 (2005)
3. Feştilă, L., Groza, R., Cîrlugea, M., Ţopa, M.: Modular Log-domain Multipliers for SVM Classifiers. In: AQTR 2006, Cluj-Napoca, Romania, pp. 271–274 (2006)
4. Gordan, M.: Applications the fuzzy technics and the machines with support vectors in the image processing. Doctoral Thesis (2004)
5. Joachims, T.: Support Vector and Kernel Methods. Cornell University, Computer Science Department, SIGIR (2003)
6. Festila, L., Groza, R., Szolga, L., Hintea, S.: Log-Domain multipliers for VLSI architectures. Scientific Bulletin of the "Politechnica" University of Timisoara. Timisoara, pp. 121–125 (September 2006)
7. Groza, R., Festila, L., Fazakas, A.: A Log-Domain Summing-Amplifier for Serial Signal Flows. In: Inter-Ing 2007, Tîrgu-Mures, November 15-16, 2007, pp. IV 4-1—IV4-4 (2007)
8. Dominguez-Castro, T., Rodriguez-Vazquez, A., Huertas, J.L.: High resolution CMOS current comparators. In: Proc. 1992 European Solid State Circuits Conf., pp. 242–245 (1992)

An Adaptive GP Strategy for Evolving Digital Circuits

Mihai Oltean[1] and Laura Dioşan[1,2]

[1] Department of Computer Science, Faculty of Mathematics and Computer Science,
Babeş-Bolyai University, Cluj-Napoca, Romania
[2] Laboratoire d'Informatique, de Traitement de l'Information et des Systèmes, EA 4108
Institut National des Sciences Appliquées, Rouen, France
{moltean,lauras}@cs.ubbcluj.ro

Abstract. The aim of this research is to develop an adaptive system for designing digital circuits. The investigated system, called Adaptive Genetic Programming (AdGP) contains most of the features required by an adaptive GP algorithm: it can decide the chromosome depth, the population size and the nodes of the GP tree which are the best suitable to provide the desired outputs. We have tested AdGP algorithm by solving some well-known problems in the field of digital circuits. Numerical experiments show that AdGP is able to perform very well on the considered test problems being able to successfully compete with standard GP having manually set parameters.

1   Introduction

The problem of designing digital circuits by evolutionary means has been intensely analyzed in the recent past [11]. It is hopped that evolution can derive unconventional designs, which are more efficient than the human designed ones. A considerable effort has been spent on evolving very efficient circuits (regarding the number of gates). Genetic Programming (GP) [3,8] and its variants have been the main techniques utilized for this purpose due to the chromosome structure which can be easily translated into a circuit layout.

When building the system, our first concerns were related to the human independence and the speed of problem solving. A system is human-independent if it requires little or no intervention from human beings when a new situation arises in the system. Our system should also obey this rule because it has to be capable of evolving a wide range of digital circuits (with different number of inputs, outputs or gates) without any human intervention. In order to achieve this purpose, our system uses an adaptive mechanism, which could discover by itself the chromosome depth, the population size and the number of generations

The Adaptive GP (AdGP) system is used to evolve different digital circuits. The results show that our system can perform very well on the considered test problems.

The paper is organized as follows: related work is reviewed in Section 2. The proposed AdGP system is presented in Section 3. Several numerical experiments about evolving digital circuits are performed in Section 4, where we also discuss their results. Conclusions are outlined in Section 5.

I. Lovrek, R.J. Howlett, and L.C. Jain (Eds.): KES 2008, Part III, LNAI 5179, pp. 376–383, 2008.

2   Related Work

Developing automated problem solvers is one of the central themes of mathematics and computer science. In the early stages, Evolutionary Algorithms have been seen as problems solvers that exhibit the same performance over a wide range of problem without too much user interference [5,7]. The modern view say that there is no guarantee that an algorithm (having the same parameter settings) will perform similarly well on multiple problems [5,19]. This is why many techniques for adapting the parameters have been proposed. Mainly, there are two taxonomy schemes [1,5] which group adaptive computations into distinct classes - distinguishing by the type of adaptation (i.e., how the parameters are changed [2,17]), and by the level of adaptation (i.e., where the changes occur [1,6,9,10,16,17,18]). Another effective way to boost the performance is to switch between multiple algorithms depending on the current stage of the optimization process [4].

3   Adaptive Genetic Programming

A variant of Genetic Programming (GP) [8,10] is utilised as underlying mechanism for our approach. Usually, the root node gives the result of a GP tree. However, in our case we need multiple nodes to provide the results because a digital circuit can have multiple outputs.

3.1   Representation

Standard GP chromosomes are represented as trees. The maximal number of levels of a tree (the depth of a tree) is restricted to a given value. Each internal node of a tree contains a function and each leaf contains a terminal (or the argument(s) of a function).

The problem inputs are used as terminals.

The function set utilised to evolve different problem solutions depends on the type of data involved (inputs and outputs) in the problem. In our case of evolving digital circuits, the Boolean function set is used as it was suggested in [15]. The decision on which function set to use is usually made based on the inputs of the problem. The function set is kept fixed during the search process. This is somehow inefficient because a large function set (such that are used for Boolean problems) means a large search space. In the near future, we plan to use an adaptive strategy for searching the best decision. In this case, we will encode (into each GP tree) a set of functions, which are allowed to be used by that chromosome.

Because we want to obtain a system, which is able to solve problems with any number of outputs, we will describe the way in which Genetic Programming may be efficiently adapted to deal with problems having multiple outputs (such as designing digital circuits [11]). In this paper, NI denotes the number of inputs and NO denotes the number of outputs.

In standard GP [8], the root of a chromosome is chosen to provide the output of the program. Thus, when a single value is expected as output, we simply choose the root of the chromosome. When the problem has multiple outputs, we have to choose the number of outputs (*NO*) nodes that will provide the desired outputs. It is obvious that the genes must be distinct unless the outputs are redundant. A similar situation is encountered in the case of Cartesian Genetic Programming [11], where the genes providing the output of the program are evolved in the same way as all the other genes.

In our case, the best sub-trees are chosen to provide the outputs. In fact, the quality of each sub-tree is computed and the best *NO* sub-trees (according to each output) are selected to provide the solution of the problem. If the problem has only one output, the best sub-tree of the chromosome will always provide it.

3.2   Fitness Assignment

The quality of a GP individual is usually computed by using a set of fitness cases [8,10]. For instance, we can consider a problem with NI Boolean inputs: x_1, x_2, ... x_{NI} and NO Boolean outputs f_1, f_2, ..., f_{NO}. Each fitness case is given as a one-dimensional array of $(NI + NO)$ values:

$$v_1^k, v_2^k, \ldots, v_{NI}^k, f_1^k, f_2^k, \ldots, f_{NO}^k$$

where v_j^k is the value of the jth attribute, x_j, with $j = 1, \ldots, NI$, in the kth fitness case and f_i^k is the ith output (or the target) with $i = 1, \ldots, NO$ for the kth fitness case ($k = 1, \ldots, m$, where m represents the number of examples used by GP for learning).

The set of fitness cases for a digital circuit problem consists of several combinations of NI Boolean arguments. The quality of an expression (encoded into a tree or a sub-tree) is the sum, over all these fitness cases, of the Hamming distance (error) between the returned value by the current expression and the correct value of the Boolean output. The sum over all outputs provides the fitness of a GP individual.

Finding the best sub-tree(s) of a tree (which corresponds to the ith output) is an easy and inexpensive task. We know that, when the fitness of an expression (encoded into a GP tree) is computed, the values of all expressions (encoded by the sub-trees of the original tree) are also computed. This operation is done in $O(n)$ steps, where n is the number of nodes of the GP tree. Thus, computing the value of an expression and of all expressions (encoded by sub-trees) requires the same number of operations as the evaluation of the entire expression.

Once we have the values of each sub-expression, it is very easy to compute their quality. We only have to take the difference (in absolute value) between the actual and expected i^{th} output and then we sum these results for all fitness cases and for each output i, $i = 1, \ldots, NO$.

In order to find the best sub-tree for a particular output we employ a bottom up approach: first, we compute the fitness of the smallest expressions (sub-trees) and then we compute the fitness of increasingly bigger sub-expressions (trees). The last expression whose quality is computed is the one encoded by the entire tree. Note that this bottom up strategy is identical with that of computing the value of an expression encoded into a tree.

The lowest fitness (corresponding to each output) indicates which the best sub-tree of the chromosome is. The fitness of the chromosome will be the sum of all individual fitness (corresponding to all the outputs) [13].

This representation has the advantage that it can deal with both cases when the outputs are interconnected and when the outputs are truly independent one of the others. In the first case, the outputs may share the same sub-tree. In the second case, another sub-tree may provide each output (with no common parts).

Example. Let us consider a problem with 2 inputs and 2 outputs (in fact a *Half Adder* circuit – it has two inputs, generally labelled A and B, and two outputs, the sum S and carry C). S is the 2-bit XOR of A and B, and C is the AND of A and B. Essentially the output of a *Half Adder* is the sum of two one-bit numbers, with C being the most significant of these two outputs. Therefore, we have 4 fitness cases (see Table 1)) and a GP chromosome encoding the solution, which is depicted in Figure 1.

Table 1. Four fitness cases for *Half Adder* problem

Fitness case	A	B	C	S
i	0	0	0	0
ii	0	1	0	1
iii	1	0	0	1
iv	1	1	1	0

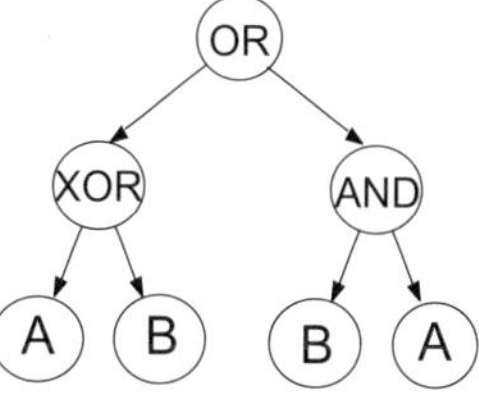

Fig. 1. GP with multiple outputs. Nodes are numbered from top to bottom, left to right starting with the root (which has index 1). The first output $C = o_1$ is provided by the 2nd node (XOR(A,B)) and the second output $S = o_2$ by the 3rd node (AND(B, A)).

3.3   Search Operators

Search operators mainly used within the GP algorithm are crossover and mutation. These operators preserve the chromosome structure. All the tree offspring are syntactically correct expressions.

By crossover, two parents are selected and recombined. For instance, within the cutting-point recombination two sub-trees (one from each parent) are exchanged.

In order to perform a mutation operation a sub-tree of the chromosome is deleted and a new sub-tree is grown by using the same procedure like that one used during the initialization stage. The maximal depth of a tree limits the growth process.

3.4 Algorithm

There is a basic requirement for standard GP that regards the size of a chromosome and the searching time: both must be minimal.

In order to meet these criteria we have chosen to implement an adaptive GP algorithm (AdGP). This algorithm will try to find a good population size and a good chromosome depth for the problem that is being solved.

AdGP algorithm will start with a small population size and a small value for the maximal chromosome depth. These values will be increased as the search process advances [12]. Note that this is not a true adaptation (as in [5]) because the population size never decreases. We have chosen not to reduce the population size because Genetic Programming techniques usually require large populations [8] for solving problems.

From experimental trials, we have deduced some good values for these parameters (the number of the individuals added to the current population and the number of genes added to the current chromosome). However, the parameter settings are really problem dependent and a general strategy cannot be devised.

Another problem is that different strategies to change the population size and the number of genes in a chromosome will affect the number of generations required to find a good solution to the problem that is being solved.

After many trials we have devised the following strategy:

1. The algorithm starts with a population of 1 GP individual which has the depth limited to 5 levels. The choice of these parameters (the initial population size and the initial chromosome depth) is problem independent.
2. The size of the population should doubled and the chromosome maximal depth should increase with one level if no improvements of the best individual occur for $A = 10$ generations. The newly inserted individuals are randomly generated. The size of the trees may also increase due to the effect of genetic operators.
3. The algorithm will stop when there are no improvements of the best individual in the population for $B = 50$ consecutive generations.

The values A and B should be carefully chosen because the behaviour of the algorithm depends on them largely. An inappropriate choice of these values can lead to very poor results.

4 Numerical Experiments

We perform several numerical experiments in order to evolve digital circuits. Several test problems are chosen for these experiments:

1. Even 3 - The even-n parity problem has n Boolean arguments and it returns T (True) if an even number of its arguments are T. Otherwise the even-n-parity function returns F (False). Therefore, the even-3-parity problem has three Boolean inputs and one Boolean output. The number of fitness cases is $2^3 = 8$;

2. Even 4 - The even-4-parity problem has four Boolean inputs and one Boolean output. The number of fitness cases is $2^4 = 16$;
3. Even 5 - The even-5-parity problem has five Boolean inputs and one Boolean output. The number of fitness cases is $2^5 = 32$;
4. 2-bit Adder - The circuit implementing this problem adds 4 bits (two numbers represented using 2 bits each) and gives a three-bit number representing the output. The training set consists of 16 fitness cases with 4 inputs and 3 outputs.
5. 2-bit Multiplier - The two-bit multiplier implements the binary multiplication of two 2-bit numbers to produce a possible 4-bit number. The training set for this problem consist of 16 fitness cases, each of them having 4 inputs and 4 outputs.

In order to compute some performance metrics we perform more runs that are independent. Several statistics were computed for all these experiments: Success rate, Average Evaluations and Average tree depth of the best chromosome. Note that this comparison is not fair because in our AdGP algorithm the population size and the chromosome depth are adapted during the search process. These comparison drawbacks can be reduced by performing the same number of fitness function evaluations.

The values from Tables 2 and 3 indicate the performance of our AdGP. For instance, we can observe that for the even-3-parity problem we obtain 89 successful runs (out of 100). This means that in 89 runs (out of 100) AdGP is able to evolve a perfect digital circuit.

Table 2. Comparative results (averaged over 100 runs) obtained for even-parity problems. We give the average number of fitness evaluations because standard GP performs less fitness evaluations, since it starts with larger chromosomes then AdGP. All 16 Boolean gates with two inputs and one output are used.

	AdGP			GP		
	Success Rate (%)	Average Evaluations	Average Depth	Success Rate (%)	Average Evaluations	Average Depth
Even-3	89	22580.60	4.57	100	2066.14	1.97
Even-4	50	33487.82	4.89	73	15913.81	2.84
Even-5	3	33682.34	3.29	27	25362.30	3.27

Table 3. Results (averaged over 100 runs) obtained by AdGP algorithm for 2-bit Adder and 2-bit Multiplier problems. Only AND, OR and XOR Boolean gates with two inputs and one output are actually used for these two problems.

AdGP	Success Rate (%)	Average Evaluations	Average Depth
2-bit Adder	33	67483.28	13.48
2-bit Multiplier	69	52269.37	11.75

As a comparison, we performs some experiments with a standard GP [8] algorithm containing 100 individuals for the first three problems. The initial chromosome depth is set to 3, 4 and 5 for even-3-parity, even-4-parity and even-5-parity problem, respectively. The success rates obtained with this GP algorithm are presented in Table 2 also. Note that AdGP has performed more functions evaluations than the standard GP.

5 Conclusions

An adaptive version of GP method, called AdGP has been investigated in this paper. The evolutionary algorithm is able to adapt the population size and the depth of the chromosome to the problem that has to be solved. Furthermore, a special way of selecting the sub-tree providing the solution of GP chromosomes was investigated. Instead of using the root of the tree as solution provider, any other sub-tree was seen as a potential solution of the problem (either with a single or with more outputs).

Numerical experiments have shown that AdGP was able to handle a wide range of problems (with one or more outputs) yielding good results. A comparison with the standard GP was performed. The results have shown that standard GP performs better than AdGP. However, the experimental conditions were very different: GP parameters were set by hand after many experimental trials, whereas AdGP had to discover the parameters by itself.

References

1. Angeline, P.J.: Adaptive and Self-adaptive Evolutionary Computations. In: Palaniswami, M., Attikiouzel, Y. (eds.) Computational Intelligence: A Dynamic Systems Perspective, pp. 152–163. IEEE Press, Los Alamitos (1995)
2. Back, T.: Self-adaptation in Genetic Algorithms. In: Varela, F.J., Bourgine, P. (eds.) Toward a Practice of Autonomous Systems: Proceedings of the First European conference on Artificial Life, pp. 263–271. MIT Press, Cambridge (1992)
3. Banzhaf, W., Nordin, P., Keller, R.E., Francone, F.D.: Genetic Programming - An Introduction; On the Automatic Evolution of Computer Programs and its Applications, 3rd edn. Morgan Kaufmann, San Francisco (2001)
4. Colaco, M.J., Dulikravich, G.S., Martin, T.J.: Control of unsteady solidification via optimized magnetic fields. Materials and manufacturing processes 20(3), 435–458 (2005)
5. Eiben, A.E., Hinterding, R., Michalewicz, Z.: Parameter Control in Evolutionary Algorithms. IEEE Transactions on Evolutionary Computation 3(2), 124–141 (1999)
6. Fogel, L.J., Fogel, D.B., Angeline, P.J.: A Preliminary Investigation on Extending Evolutionary Programming to Include Self-adaptation on Finite State Machines. Informatica 18, 387–398 (1994)
7. Goldberg, D.E.: Genetic Algorithms in Search, Optimization, and Machine Learning. Addison-Wesley, Reading (1989)
8. Koza, J.R.: Genetic Programming: On the Programming of Computers by Means of Natural Selection. MIT Press, Cambridge (1992)
9. Grosan, C., Oltean, M.: Adaptive Representation for Single Objective Optimization. Soft Computing 9(8), 594–605 (2005)
10. Koza, J.R.: Genetic Programming II: Automatic Discovery of Reusable Subprograms. MIT Press, Cambridge (1994)

11. Miller, J.F., Job, D., Vassilev, V.K.: Principles in the Evolutionary Design of Digital Circuits - Part I. Genetic Programming and Evolvable Machines 1(1), 7–35 (2000)
12. Oltean, M., Diosan, L.: An autonomous GP-based system for regression and classification problems. Applied Soft Computing (in press, 2008)
13. Muntean, O., Dioşan, L., Oltean, M.: Solving the even-n-parity problems using Best Sub Tree Genetic Programming. In: AHS 2007, pp. 511–518 (2007)
14. Oltean, M., Groşan, C.: Evolving Digital Circuits using Multi Expression Programming. In: Zebulum, R., et al. (eds.) NASA/DoD Conference on Evolvable Hardware, Seatle, pp. 87–90. IEEE Press, NJ (2004)
15. Poli, R., Page, J.: Solving high-order Boolean parity problems with smooth uniform crossover, sub-machine-code GP and demes. Genetic programming and evolvable machines 1, 37–56 (2000)
16. Rosca, J.P., Ballard, D.H.: Genetic Programming with Adaptive Representations, Technical Report 489, University of Rochester, Computer Science Department (1994)
17. Shaefer, C.G.: The ARGOT System: Adaptive Representation Genetic Optimizing Technique. In: Grefenstette, J.J. (ed.) Proc. of the Second International Conference on Genetic Algorithms. Lawrence Erlbaum, Hillsdale (1987)
18. Teller, E.: Evolving programmers: The co-evolution of intelligent recombination operators. In: Angeline, P., Kinnear, K. (eds.) Advances in Genetic Programming, vol. 2 (1996)
19. Wolpert, D.H., McReady, W.G.: No Free Lunch Theorems for Optimisation. IEEE Transaction on Evolutionary Computation 1, 67–82 (1997)

Computational Intelligence in Analog Circuits Design

Gabriel Oltean, Sorin Hintea, and Emilia Sipos

Technical University of Cluj-Napoca, C. Daicoviciu Street, 15, Cluj-Napoca, Romania
`{Gabriel.Oltean,Sorin.Hintea,Emilia.Sipos}@bel.utcluj.ro`

Abstract. This paper presents an optimization algorithm based on computational intelligence techniques, applied to the design of analog circuits. Our approach combines the best qualities of these techniques. Fuzzy sets assure the desired flexibility for the objective functions and a known range of objective function values. High accuracy and low computational complexity are simultaneously present in the neuro-fuzzy models of performances. A powerful global optimization engine is implemented by means of a genetic algorithm. Finally, the proposed algorithm is validated by the design optimization of a CMOS amplifier.

Keywords: analog circuit design, optimization, genetic algorithm, fuzzy sets, neuro-fuzzy systems.

1 Introduction

The goal of analog circuit synthesis is to create a sized circuit schematic from given circuit requirements. This is a very challenging task for several reasons: most analog circuits require a custom optimized design; the design problem is typically underconstrained with many degrees of freedom and it is common that many (often conflicting) performance requirements must be taken into account; and tradeoffs must be made that satisfy the designer [1].

Optimizations tools appear, naturally, as the key factor for the tremendous effort of finding design parameters, which satisfy a complex, high-dimensional, multi-objective and multi-constrained problem [2]. An optimization algorithm for analog circuit design has three key components: formulation of the optimization problem, performance evaluation engine, and optimization engine.

Research efforts on circuit synthesis involving a broad spectrum of computational intelligence (CI) techniques have begun to appear in the literature over the past few years. Fuzzy sets are used to formulate the objective functions, in this way obtaining the possibility to consider different degrees for requirement achievements and acceptability degrees for a particular solution. One approach [3]-[5], is to consider that the membership degree μ represents the degree of fulfillment of the objective.

One way to reduce the computation time in the iterative optimization loop is to use simple models of circuit performances. In order to satisfy both main requirements (accuracy and speed), many researches proposed several CI-based methods to evaluate circuit performances. In [6] and [7] least-square support vector machines are used to model circuit performances. Fuzzy systems are very useful to model the circuit performances because they imply simple mathematical operations and can model any

I. Lovrek, R.J. Howlett, and L.C. Jain (Eds.): KES 2008, Part III, LNAI 5179, pp. 384–391, 2008.

complex, multivariable and nonlinear function at any level of accuracy. Such fuzzy models are used in [8]-[10].

To search the whole solution space, a powerful global optimization technique should be considered. Genetic algorithms (GA) have many desirable characteristics and offer significant advantages over traditional methods. They are inherently robust and have been shown to efficiently search large solution spaces containing discrete or discontinuous parameters and non-linear constraints, without being trapped in local minima. GA have already been employed in many CAD applications [2], [8], [11], and [12].

The objective of this work is to develop an efficient algorithm for design optimization of analog circuits based on computational intelligence techniques. Our approach tries to combine the best qualities of these techniques: flexibility in formulation the objective functions and a known range of their values using fuzzy sets, accuracy and low computational complexity of performance models based on neuro-fuzzy systems and a powerful global optimization engine based on GA.

2 CI-Based Optimization Algorithm

The novelty introduced in this paper is the utilization of different CI techniques in all phases of the optimization algorithm, as it is shown in Fig. 1.

Fuzzy sets are used to define the objective function in formulation of the optimization problem. Neuro–fuzzy systems address the performance evaluation issue (evaluation engine). Finally, in the optimization engine, a genetic algorithm is responsible for the evolution of the population to finally produce the (near) optimum solution.

2.1 Formulation of the Optimization Problem

To solve a multiobjective optimization problem, as is the design optimization of analog circuits, the optimization problem can be formulated as a single-objective optimization, where different performance objectives are combined to form a single scalar objective, and which produces one solution.

The authors proposed the utilization of fuzzy sets to define the objective functions in previous papers [13] and [14]. By contrast with the existing approaches where membership degree represents the degree of fulfillment, in our approach the membership degree μ represents the error degree in the fulfillment of the objective. A value $\mu=1$ means the objective is not satisfied at all, while a value $\mu=0$ means that the objective is fully satisfied. With this approach we know the range of possible values for objective functions as being [0, 1]. When the value of a certain objective function (unfulfillment degree - UD) is 0, we know that the corresponding requirement is fulfilled, no further effort being necessary to improve the associated performance.

For our single-objective optimization approach, we combine the individual objective functions into a cost function by means of a weighted sum. The formulation of the multiobjective optimization problem became:

$$\text{Find } x \text{ that minimizes } F(x) = \sum_{k=1}^{n} w_k \mu_k \left(f_k(x) \right) \ . \tag{1}$$

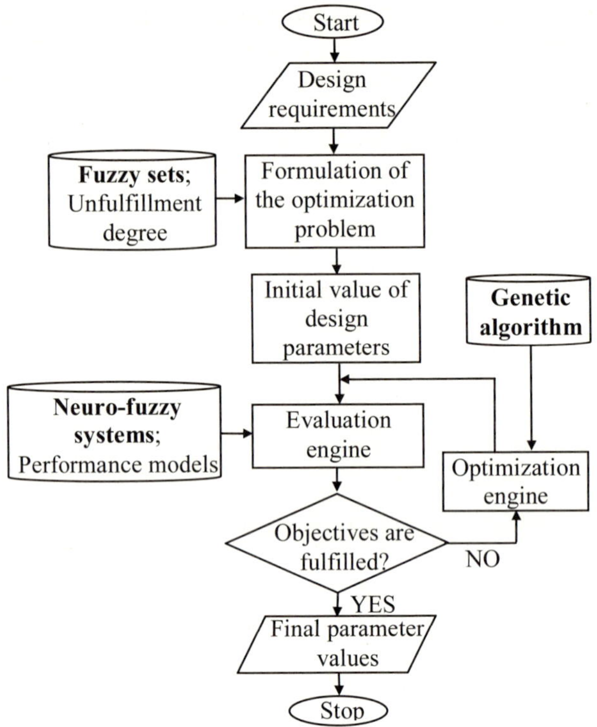

Fig. 1. CI-based optimization algorithm

where n is the number of requirements, f_k is the kth objective function, and w_k is the weight or relative preference associated with the kth objective function.

2.2 Evaluation Engine

Fuzzy systems are very useful to model circuit performances because they are considered universal approximators. The authors have been synthesized a method to build neuro-fuzzy models of circuit performances [13] and [15]. These models are automatically built up based on input-output data sets, using ANFIS (Adaptive Neuro-Fuzzy Inference System) framework [16] to develop first order Takagi-Sugeno fuzzy systems. ANFIS framework contains a six layer architecture for its artificial neural network trained by a mixture of backpropagation and least mean square estimation learning algorithm [16].

The parameter set (the combination of the parameter values) should be chosen to be representative for the performance to be modeled (covers thoroughly the parameter space and embeds all the specific characteristics of the performance). For each input vector (one combination of the parameter values), the performance value has to be found, in our case by SPICE simulation. Two data sets, a training set and a checking set are generated. A subtractive clustering procedure generates an initial first order

Takagi-Sugeno fuzzy system. Next the fuzzy set is trained and validated using previously generated data sets.

2.3 Optimization Engine

The heart of the whole algorithm is the optimization engine. In our approach, a GA is responsible for the exploration of solution space in quest of the optimal one.

Our design parameters are real variables so we chose real numbers as genes to compose a chromosome. This way our population individuals are real valued vectors.

The underlying procedure of our GA is shown in Fig. 2.

> *1.   Population initialization*
> *2.   Fitness assignment*
> *3.   Selection*
> *4.   Recombination*
> *5.   Mutation*
> *6.   Reinsertion*
> *7.   Loop to step 2 until optimal solution is found*

Fig. 2. GA procedure

Each individual in the selection pool receives a reproduction probability depending on the own objective value and the objective value of all other individuals in the selection pool. This fitness is used for the actual selection step afterwards. Our implementation uses rank-based fitness assignment with linear ranking [17].

Roulette-wheel method is used for individuals selection. For each individual a selection probability is computed. The individuals are mapped to contiguous segments of a line, such that each individual's segment is equal in size to its selection probability. A uniformly distributed random number is generated and the individual whose segment spans the random number is selected. The process is repeated until the desired number of individuals is obtained.

Recombination produces new individuals in combining the information contained in two or more parents. For our real valued variables the intermediate recombination method was chosen. Offspring are produced according to the rule [17]:

$$Var_j^O = a_j Var_j^{P1} + \left(1 - a_j\right) Var_j^{P2}, \quad j = 1, 2, ..., Nvar \; . \tag{2}$$

where Var_j^O represents the jth variable of the offspring, Var_j^{P1} represents the jth variable of the first parent, while Var_j^{P1} represents the jth variable of the second parent. The scaling factor a_j is chosen uniformly at random over an interval $[-d, 1+d]$ for each variable anew.

By mutation, individuals are randomly altered. Mutation of real variables means that randomly created values are added to the variables with a low probability. Thus, the probability of mutating a variable (mutation rate) and the size of the changes for each mutated variable (mutation step) must be defined. The probability of mutating a variable is inversely proportional to the number of variables (dimensions). In our implementation only one variable per individual is changed/mutated per mutation.

The size of the mutation step is difficult to choose. A good mutation operator should often produce small step-sizes with a high probability and large   step-sizes with a low probability. Such an operator [17] was considered for our algorithm:

$$Var_j^{Mut} = Var_j + s_j r_j a_j, \quad j = 1, 2, ..., Nvar$$

$$s_j \in \{-1, +1\} \text{ uniform at random}$$

$$r_j = r \cdot domain_i, \quad r \text{ - the mutation range (standard 10\%)}$$

$$a_j = 2^{-uk}, u \in [0, 1] \text{ uniform at random, } k \text{ - the mutation precision}$$

(3)

Our GA uses the pure reinsertion scheme: produce as many offspring as parents and replace all parents by the offspring. Every individual lives one generation only.

3 Design Optimization of a CMOS Amplifier

The proposed CI-based design optimization algorithm is implemented in the Matlab environment.

We used our algorithm for the design optimization of a simple operational transconductance amplifier, shown in Fig.3. The design parameters of the circuit are the dimensions of the transistors $(W/L)_1=(W/L)_2$, $(W/L)_3=(W/L)_4$, $(W/L)_5=(W/L)_6$ and bias current I_b. As performances, the important ones were considered: voltage gain A_{vo}, unity gain bandwidth GBW, and common mode rejection ratio $CMRR$.

The design optimization is illustrated here for the set of requirements presented in Table 1., for equal weighted objective functions. The optimization was run several

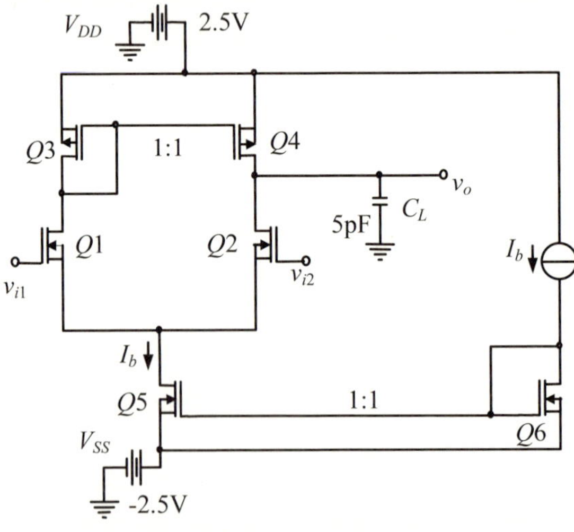

Fig. 3. Simple operational transconductance amplifier

Table 1. Requirements and performances in four optimization runs

Circuit functions	Requirements	Performances			
		run1	*run2*	*run3*	*run4*
A_{v0}	≥ 50	52.17	52.01	51.83	51.97
GBW [kHz]	$\geq 4\ 600$	4 601	4 620	4612	4608
$CMRR$	$\geq 1\ 000\ 000$	1 011 413	1 008 935	1 004 844	1 003 761
Number of iterations		94	90	87	63

Table 2. Solutions in four optimization runs

Design parameter	run1	run2	run3	run4
$(W/L)_{12}$	39.28	39.73	39.33	39.07
$(W/L)_{34}$	4.00	3.99	3.98	3.99
$(W/L)_{56}$	6.90	6.86	7.21	7.23
I_b [µA]	96.33	97.22	94.74	94.00

times, for a population of 100 individuals. The algorithm proved to be a robust one, always a solution being found that fulfills all the requirements. Different numbers of iterations are necessary to search for the optimum solution depending on the initial population and on the evolution process. Table 1 gives also the final performances of the circuit after four different optimization runs, while Table 2 shows the solutions (values of the design parameters). The solutions appear to be slightly different from each other. At a closer look we can see that the values of the design parameters arc calculated with two decimals. In practical implementations these values should be rounded toward some discrete values, so our resultant solutions are in fact small variations around one solution - our solutions are near optimum solution.

For the first optimization run (*run1*), the dynamic behavior of the algorithm is presented in Fig. 4.

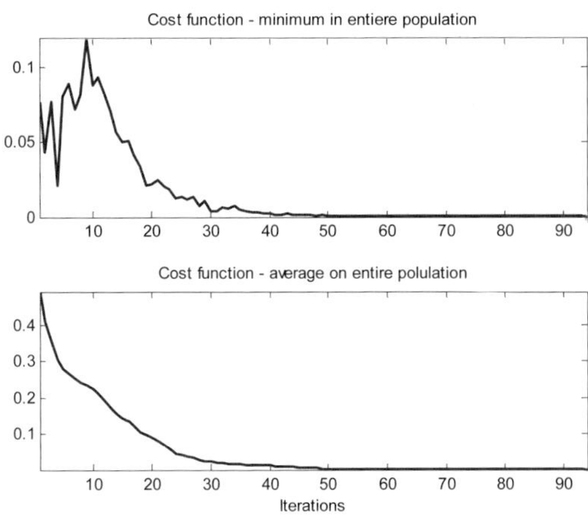

Fig. 4. Minimum and average cost function evolution for *run1*

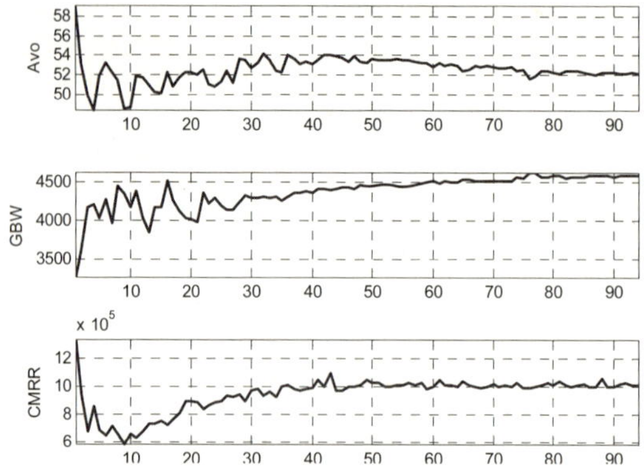

Fig. 5. Dynamic behavior of performances in *run1*

In the first iterations (up to 10), due to a high diversity of individuals, a new population does not contain always a fittest individual than in the previous population (minimum value of the cost function increases). On the other hand, the population as a whole is improved continuously, the average value of the cost function on the entire population decreasing in time. As the population improves during evolution, all individuals move toward the optimal solution, decreasing both minimum and average values of cost function.

The evolution of all performances during optimization is presented in Fig. 7. The optimization effort is mainly spend to continuously increase *GBW*, while keeping the values for A_{vo} and *CMRR* large enough to satisfy the requirements (see Table 1).

4 Conclusions

In this paper a new hybrid optimization algorithm based on computational intelligence techniques was presented. The design of a CMOS simple transconductance amplifier showed encouraging results. The algorithm is a robust one being able to converge to the optimal solution regardless the initial population. The multiobjective optimization problem, specific to the analog circuit design, was reformulated as a single-objective optimization. For each optimization run the proposed algorithm produces one optimum solution. A further research direction is to use a real multiobjective optimization method to produce solutions on the Pareto front. To address this issue, an elitist non-dominated sorting GA might be considered.

References

1. Rutenbar, R.A., Gielen, G.G.E., Roychowdhury, J.: Hierarchical Modeling, Optimization, and Synthesis for System-level Analog and RF Designs. Proc. of the IEEE 95(3), 640–669 (2007)
2. Barros, M., Guilherme, J., Horta, N.: GA-SVM Optimization Kernel Applied to Analog IC Design Automation. In: Proc. of 13th IEEE International Conference on Electronics, Circuits and Systems, ICECS 2006, pp. 486–489 (2006)

3. Sahu, B., Dutta, A.K.: Automatic Synthesis of CMOS Operational Amplifiers: A Fuzzy Optimization Approach. In: Proc. of 15th Int. Conference on VLSI Design, pp. 366–371 (2002)
4. Fares, M., Kaminska, B.: FPAD: A Fuzzy Nonlinear Programming Approach to Analog Circuit Design. IEEE Trans. on Computer-Aided Design of Integrated Circuits and Systems 14(7), 785–793 (1995)
5. Rodriguez, B.R.S., Styblinski, M.A.: Adaptive Hierarchical Multi-Objective Fuzzy Optimization for Circuit Design. In: IEEE International Symposium on Circuits and Systems, ISCAS 1993, pp. 1813–1816 (1993)
6. Ren, X., Kazmierski, T.: Performance Modeling and Optimization of RF Circuits Using Support Vector Machines. In: Proc. of 14th International Conference on Mixed Design of Integrated Circuits and Systems, MIXDES 2007, pp. 317–321 (2007)
7. Kiely, T., Gielen, G.: Performance Modeling of Analog Integrated Circuits Using Least-Squares Support Vector Machines. In: Proc. of Design, Automation and Test in Europe Conference and Exhibition, vol. 1, pp. 448–453 (2004)
8. Balkir, S., Dundar, G., Alpaydin, G.: Evolution Based Synthesis of Analog Integrated Circuits and Systems. In: Proc. of the 2004 IEEE NASA/DoD Conference on Evolution Hardware, EH 2004, 0-7695-2145-2/04 (2004)
9. Torralba, A., Chavez, J., Franquelo, L.G.: A Collection of Fuzzy Logic-Based Tools for the Automated Design, Modeling and Test of Analog Circuits. IEEE Micro 16(4), 61–68 (1996)
10. Venkata, N.D., Evans, B.L.: An Automated Framework for Multicriteria Optimization of Analog Filter Design. IEEE Trans. on Circuits and System-II: Analog and Digital Signal Processing 46(8), 981–990 (1999)
11. Somani, A., Chakrabarti, P.P., Patra, A.: An Evolutionary Algorithm-Based Approach to Automated Design of Analog and RF Circuits Using Adaptive Normalized Cost Functions. IEEE Transactions on Evolutionary Computation 11(3), 336–353 (2007)
12. Eeckelaert, T., Schoofs, R., Gielen, G., Steyaert, M., Sansen, W.: Hierarchical Bottom-up Analog Optimization Methodology Validated by a Delta-Sigma A/D Converter Design for the 802.11a/b/g standard. In: Design Automation Conf., 43rd ACM/IEEE, pp. 25–30 (2006)
13. Oltean, G.: FADO - A CAD Tool for Analog Modules. In: Proc. of the International Conference on Computer as a Tool EUROCON 2005, Belgrade, pp. 515–518 (2005) ISBN 1-4244-0050-3, IEEE catalog number: 05EX1255C
14. Oltean, G.: Multiobjective Fuzzy Optimization Method. Scientific Bulletin of the Politechnica University of Timisoara, Trans.on Electronics and Communications 49(63)(1), 220–225 (2004)
15. Oltean, G., Miron, C., Crasi, M., Carlugea, M.: Evaluation of the Analog Circuits Performances Using Fuzzy Models. Scientific Bulletin of the Politechnica University of Timisoara, Trans.on Electronics and Communications 47(61)(1), 12–17 (2002)
16. Jang, R.J.-S.: ANFIS, Adaptive-Network-Based Fuzzy Inference System. IEEE Transaction on System, Man, and Cybernetics 23(3), 665–685 (1993)
17. Pohlheim, H.: GEATbx Introduction to Evolutionary Algorithms: Overview, Methods and Operators, version 3.7 (November 2005), http://www.geatbx.com/

Dimensionality Reduction and Linear Discriminant Analysis for Hyperspectral Image Classification

Qian Du and Nicolas H. Younan

Department of Electrical and Computer Engineering
Mississippi State University, USA
{du,younan}@ece.msstate.edu

Abstract. In this paper, we investigate the application of Fisher's linear discriminant analysis (FLDA) to hyperspectral remote sensing image classification. The basic idea of FLDA is to design an optimal transform so that the classes can be well separated in the low-dimensional space. The practical difficulty of applying FLDA to hyperspectral images includes the unavailability of enough training samples and unknown information for all the classes present. So the original FLDA is modified to avoid the requirements of complete class knowledge, such as the number of actual classes present. We also investigate the performance of the class of principal component analysis (PCA) techniques prior to FLDA and find that the interference and noise adjusted PCA (INAPCA) can provide the improvement in the final classification.

Keywords: Fisher's Linear Discriminant Analysis, Dimensionality Reduction, Classification, Hyperspectral Imagery.

1 Introduction

Pattern recognition techniques have been widely used in remote sensing image analysis [1]. As a standard technique for dimensionality reduction in pattern recognition, Fisher's linear discriminant analysis (FLDA) projects the original high-dimensional data onto a low-dimensional space, where all the classes can be well separated [2]. Assume there are n training sample vectors given by $\{\mathbf{r}_i\}_{i=1}^n$ for p-classes $\{C_1, C_2, ..., C_p\}$, and there are n_j samples for the jth class, i.e., $\sum_{j=1}^p n_j = n$. Let $\boldsymbol{\mu}$ be the mean of the entire training samples, i.e., $\frac{1}{n}\sum_{i=1}^n \mathbf{r}_i = \boldsymbol{\mu}$, and $\boldsymbol{\mu}_j$ be the mean of the jth class, i.e., $\frac{1}{n_j}\sum_{\mathbf{r}_i \in C_j} \mathbf{r}_i = \boldsymbol{\mu}_j$. Then, the within-class scatter matrix $\mathbf{S}_W$ and between-class scatter matrix $\mathbf{S}_B$ are defined as

$$\mathbf{S}_W = \sum_{\mathbf{r}_i \in C_j} \left(\mathbf{r}_i - \boldsymbol{\mu}_j\right)\left(\mathbf{r}_i - \boldsymbol{\mu}_j\right)^T \tag{1}$$

I. Lovrek, R.J. Howlett, and L.C. Jain (Eds.): KES 2008, Part III, LNAI 5179, pp. 392–399, 2008.

$$\mathbf{S}_B = \sum_{j=1}^{p} n_j \left(\boldsymbol{\mu}_j - \boldsymbol{\mu}\right)\left(\boldsymbol{\mu}_j - \boldsymbol{\mu}\right)^T \tag{2}$$

respectively. The goal is to find a transform vector $\mathbf{w}$ such that the Raleigh quotient is maximized, which is defined as

$$q = \frac{\mathbf{w}^T \mathbf{S}_B \mathbf{w}}{\mathbf{w}^T \mathbf{S}_W \mathbf{w}}. \tag{3}$$

The $\mathbf{w}$ can be determined by solving a generalized eigen-problem specified by

$$\mathbf{S}_B \mathbf{w} = \lambda \mathbf{S}_W \mathbf{w}, \tag{4}$$

where λ is a generalized eigenvalue. Since the rank of $\mathbf{S}_B$ is $p-1$, there are $p-1$ eigenvectors associated with $p-1$ nonzero eigenvalues. Therefore, an $L \times (p-1)$ matrix $\mathbf{W}$ can be found to transform the original L-dimensional data into a $(p-1)$-dimensional space. In this low-dimensional space, it is expected that the p classes can be well separated.

The major problem when applying FLDA to remote sensing imagery is the difficulty in finding enough training samples for all the classes. In particular, for an L-band hyperspectral image, L linearly independent samples are required to make $\mathbf{S}_W$ full-rank. This problem may be resolved by a dimensionality reduction approach such as principal component analysis (PCA) before FLDA [3-5]. In this paper, we will also investigate other dimensionality reduction approaches, such as noise-adjusted principal component analysis (NAPCA) [6-7], interference and noise adjusted principal component analysis (INAPCA) [8], and their performance in conjunction with FLDA for hyperspectral image classification.

When applying FLDA, it is assumed that the number of classes, p, present in the image scene is known. For remote sensing imagery, this information may be difficult to know due to scene complexity. In many practical situations, we may know the number of foreground classes but not the number of background classes. We will investigate the performance of FLDA in this situation and propose an approach to overcome this difficulty.

2 PCA-Class Techniques for Dimensionality Reduction

2.1 PCA

Consider the observation model

$$\mathbf{z} = \mathbf{s} + \mathbf{n}, \tag{5}$$

where $\mathbf{z}$ is a pixel vector in a hyperspectral image with data dimensionality L (i.e, the number of spectral bands is L), $\mathbf{s}$ is a signal vector, and $\mathbf{n}$ represents the uncorrelated additive noise. Let $\mathbf{V} = [\mathbf{v}_1, \mathbf{v}_2, \cdots, \mathbf{v}_L]$ and $\boldsymbol{\Lambda} = diag\{\lambda_1, \lambda_2, \cdots, \lambda_L\}$ be the eigenvector and eigenvalue matrices of the data covariance matrix $\boldsymbol{\Sigma}$, where $\mathbf{v}_1, \mathbf{v}_2, \cdots, \mathbf{v}_L$ are

394   Q. Du and N.H. Younan

L eigenvectors of size $L \times 1$ and $\lambda_1, \lambda_2, \cdots, \lambda_L$ are the corresponding L eigenvalues, i.e.,

$$\mathbf{V}^T \mathbf{\Sigma} \mathbf{V} = \mathbf{\Lambda}. \tag{6}$$

Then, the PC images can be calculated from

$$\mathbf{z}_{PCA} = \mathbf{\Lambda}^{-1/2} \mathbf{V}^T (\mathbf{z} - \mathbf{m}), \tag{7}$$

where $\mathbf{m}$ is the mean vector. Assume $\lambda_1 \geq \lambda_2 \geq, \cdots, \geq \lambda_L$, the variances of the L PC images of the transformed data using $\mathbf{z}_{PCA}$ are $\lambda_1, \lambda_2, \cdots, \lambda_L$, respectively.

2.2 NAPCA

NAPCA ranks principal components (PCs) in terms of the signal-to-noise ratio (SNR). It can be performed with two steps. The first step conducts the noise-whitening to the original data, and the second step performs the ordinary PCA to the noise-whitened data. Since the noise variance is unity in the noise-whitened data, the resultant PCs are in the order of SNR. Let $\mathbf{\Sigma}_n$ be the noise covariance matrix and $\mathbf{F}$ be the noise-whitening matrix such that

$$\mathbf{F}^T \mathbf{\Sigma}_n \mathbf{F} = \mathbf{I}, \tag{8}$$

where $\mathbf{I}$ is the identity matrix. Transforming $\mathbf{\Sigma}$ by $\mathbf{F}$, i.e.,

$$\mathbf{F}^T \mathbf{\Sigma} \mathbf{F} = \mathbf{\Sigma}_{n\_adj}, \tag{9}$$

where $\mathbf{\Sigma}_{n\_adj}$ is the covariance matrix with the noise being whitened. Finding a matrix $\mathbf{G}$ such that

$$\mathbf{G}^T \mathbf{\Sigma}_{n\_adj} \mathbf{G} = \mathbf{I}. \tag{10}$$

Then, the operator for NAPCA can be constructed by

$$\mathbf{z}_{NAPCA} = \mathbf{G}^T \mathbf{F}^T (\mathbf{z} - \mathbf{m}). \tag{11}$$

The major difficulty in performing NAPCA is having an accurate noise covariance matrix $\mathbf{\Sigma}_n$, which is difficult in general. The following method is adopted in our research for its simplicity and effectiveness [9]. Let $\mathbf{\Sigma}$ be decomposed as

$$\mathbf{\Sigma} = \mathbf{DED}, \tag{12}$$

where $\mathbf{D} = diag\{\sigma_1, \sigma_2, \cdots, \sigma_L\}$ is a diagonal matrix with σ_l^2 being the diagonal elements of $\mathbf{\Sigma}$, which is the variance of the l-th original band, and $\mathbf{E}$ is the correlation coefficient matrix whose ij-th element represents the correlation coefficient between the i-th and j-th bands. Similarly, in analogy with the decomposition of $\mathbf{\Sigma}$, its inverse $\mathbf{\Sigma}^{-1}$ can be also decomposed as

$$\boldsymbol{\Sigma}^{-1} = \mathbf{D}_{\Sigma^{-1}} \mathbf{E}_{\Sigma^{-1}} \mathbf{D}_{\Sigma^{-1}} \,, \tag{13}$$

where $\mathbf{D}_{\Sigma^{-1}} = diag\{\zeta_1, \zeta_2, \cdots, \zeta_L\}$ is a diagonal matrix with ζ_l^2 being the diagonal elements of $\boldsymbol{\Sigma}^{-1}$ and $\mathbf{E}_{\Sigma^{-1}}$ is a matrix similar to $\mathbf{E}$ with the diagonal elements being one and all the off-diagonal elements being within $(-1, 1)$. It turns out that ζ_l^2 is the reciprocal of a good noise variance estimate of the l-th band [9]. Therefore, the noise covariance matrix $\boldsymbol{\Sigma}_n$ can be estimated by $\boldsymbol{\Sigma}_n = diag\{\zeta_1^{-2}, \zeta_2^{-2}, \cdots, \zeta_L^{-2}\}$, which is a diagonal matrix.

2.3 INAPCA

INAPCA ranks PCs in terms of the signal-to-interference-plus-noise ratio (SINR), where the interference can be considered as any unwanted signals. Inspired by the two-step procedure of NAPCA, here we introduce a similar two-step procedure for INAPCA, which can be easily implemented without requiring prior interference information. Let $\boldsymbol{\Sigma}_{i+n}$ be the interference-noise covariance matrix. Then, a matrix $\mathbf{B}$ is determined via the eigen-decomposition of $\boldsymbol{\Sigma}_{i+n}$ such that

$$\mathbf{B}^T \boldsymbol{\Sigma}_{i+n} \mathbf{B} = \mathbf{I} \,, \tag{14}$$

which is used to transform the data as

$$\mathbf{B}^T \boldsymbol{\Sigma} \mathbf{B} = \boldsymbol{\Sigma}_{i+n\_adj} \,. \tag{15}$$

Here, $\boldsymbol{\Sigma}_{i+n\_adj}$ is the covariance matrix with interference and noise being whitened. If the resultant data is transformed by an ordinary PCA, then, the net effect is to order the PCs in terms of SINR. So, a matrix $\mathbf{C}$ can be found via the eigen-decomposition of $\boldsymbol{\Sigma}_{i+n\_adj}$ such that

$$\mathbf{C}^T \boldsymbol{\Sigma}_{i+n\_adj} \mathbf{C} = \mathbf{I} \,. \tag{16}$$

Then, the PC images using INAPCA can be computed by

$$\mathbf{z}_{INAPCA} = \mathbf{C}^T \mathbf{B}^T (\mathbf{z} - \mathbf{m}) \,. \tag{17}$$

$\boldsymbol{\Sigma}_{i+n}$ can be determined by estimating individual interference signatures using some unsupervised techniques. But $\boldsymbol{\Sigma}_{i+n}$ can be more easily computed via orthogonal subspace projection without the need of interference and noise estimation, which is explained as follows. Let the desired signature matrix be $\mathbf{S} = [\mathbf{s}_1 \, \mathbf{s}_2 \cdots \mathbf{s}_p]$ with p desired object signatures. In order to get a data set with interference and noise only, all these p signatures can be annihilated by projecting the original data onto the subspace that is orthogonal to these desired signals using projection operator $\mathbf{P}^{\perp}$, where

$\mathbf{P}^{\perp} = \mathbf{I} - \mathbf{S}(\mathbf{S}^T \mathbf{S})^{-1} \mathbf{S}^T$. Let $\hat{\mathbf{z}} = \mathbf{P}^{\perp} \mathbf{z}$. Note that, only interference and noise are present in $\hat{\mathbf{z}}$. Then, $\boldsymbol{\Sigma}_{i+n}$ can be easily estimated by using the sample covariance matrix of $\hat{\mathbf{z}}$.

3   Variant of Fisher's Linear Discriminant Analysis

The total scatter matrix $\mathbf{S}_T$ is defined as

$$\mathbf{S}_T = \sum_{i=1}^{n} (\mathbf{r}_i - \boldsymbol{\mu})(\mathbf{r}_i - \boldsymbol{\mu})^T \tag{18}$$

and it can be related to $\mathbf{S}_W$ and $\mathbf{S}_B$ by [2]

$$\mathbf{S}_T = \mathbf{S}_W + \mathbf{S}_B. \tag{19}$$

So the maximization of Eq. (3) is equivalent to maximizing

$$q' = \frac{\mathbf{w}^T \mathbf{S}_B \mathbf{w}}{\mathbf{w}^T \mathbf{S}_T \mathbf{w}}. \tag{20}$$

Following the same idea of FLDA, the solution will be the eigenvectors of the generalized eigenproblem: $\mathbf{S}_B \mathbf{w} = \lambda \mathbf{S}_T \mathbf{w}$.

$\mathbf{S}_T$ in Eq. (20) can be replaced by the data covariance matrix $\boldsymbol{\Sigma}$, i.e.,

$$\hat{\mathbf{S}}_T = \boldsymbol{\Sigma} = \sum_{i=1}^{N} (\mathbf{r}_i - \mathbf{m})(\mathbf{r}_i - \mathbf{m})^T , \tag{21}$$

where $\mathbf{m}$ is the sample mean of the entire data set with N pixels, i.e., $\frac{1}{N}\sum_{i=1}^{N} \mathbf{r}_i = \mathbf{m}$.

Then, the solution is the eigenvectors of the generalized eigenproblem: $\mathbf{S}_B \mathbf{w} = \lambda \boldsymbol{\Sigma} \mathbf{w}$ or $\boldsymbol{\Sigma}^{-1} \mathbf{S}_B$.

Regardless of the actual classes present in the data, replacing $\mathbf{S}_T$ with $\boldsymbol{\Sigma}$ represents an extreme case, which means all the pixels are separated into the classes they belong to and selected as samples. Here, the implicit assumption is that pixels are put into all the existing classes including unknown background classes (i.e., even when the actual number of classes, p_T, is unknown and greater than the p used in FLDA). In this paper, it will be shown that the term $\boldsymbol{\Sigma}^{-1}$ in this FLDA variant is very effective in background suppression so as to support the desired foreground classification.

4   Experiment

The AVIRIS Cuprite subimage scene of size 350×350, shown in Fig. 1, was collected in Nevada in 1997. The spatial resolution is 20m. Originally, it has 224 bands with

0.4-2.5 μm spectral range. After water absorption and low SNR bands were removed, 189 bands were used. This image scene is well understood mineralogically. At least five minerals were present: alunite (A), buddingtonite (B), calcite (C), kaolinite (K), and muscovite (M). Their approximate spatial locations of these minerals are marked in Fig. 1(b), but no pixel level ground truth is available. Due to the scene complexity, the actual number of classes, p_T, is much greater than five. The number of training samples for the five classes was 63, 59, 69, 72, 63, respectively.

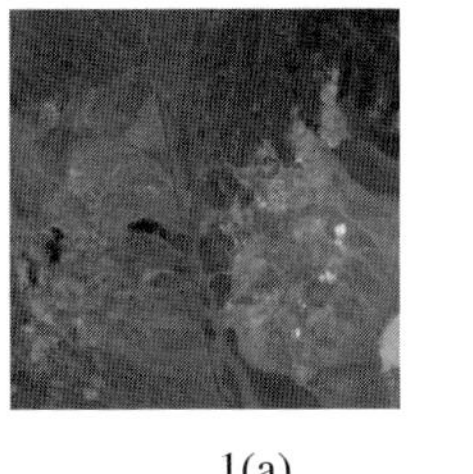
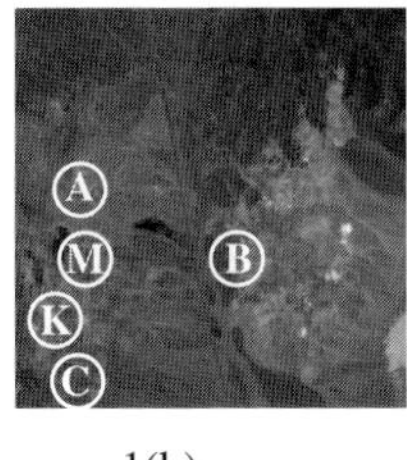

1(a)   1(b)

Fig. 1. AVIRIS Cuprite image scene (a) a spectral band image; (b) spatial locations of five pure pixels corresponding to minerals: alunite (A), buddingtonite (B), calcite (C), kaolinite (K), and muscovite (M)

Fig. 2(a) is the result using Target-Constrained Interference-Minimized Filter (TCIMF) on the original image [10], where the five classes were well separated. Fig. 2(b) is the result when TCIMF was applied to the FLDA-transformed images, where we can see there were a significant number of misclassified pixels, particularly when classified B, C, K, and M. Fig. 2(c), 2(d), and 2(e) are the results using PCA, NAPCA, and INAPCA before FLDA, where no obvious improvement can be perceived. This demonstrates that FLDA provides very poor results if the information on the number of classes is inaccurate. Under this circumstance, another dimensionality reduction process before FLDA cannot provide much help.

Fig. 2(f) is the result using the variant of FLDA, denoted as modified FLDA (MFLDA), where the classification is much better than that in Fig. 2(b) using the original FLDA. Fig. 2(g), 2(h), 2(i) are the results when using the three PCA-class techniques for dimensionality reduction before MFLDA and TCIMF. It seems that using INAPCA for dimensionality reduction, as illustrated in Fig. 2(i), may provide comparable results to that without dimensionality reduction, as shown in Fig. 2(f), while using PCA and NAPCA may not bring improvement.

To perform a quantitative comparison, Table 1 lists the spatial correlation coefficients between the results obtained from the use of LDA-transformed data and that from the TCIMF on the original data, where a value closer to one is associated with better classification. It is obvious that MFLDA outperforms the corresponding FLDA counterparts. This means Σ is a better term than $\mathbf{S}_W$ when the actual number of classes and their information are difficult or even impossible to obtain. It also confirms that using PCA and NAPCA for dimensionality reduction does not necessarily improve classification, but using INAPCA can improve classification particularly when MFLDA is applied.

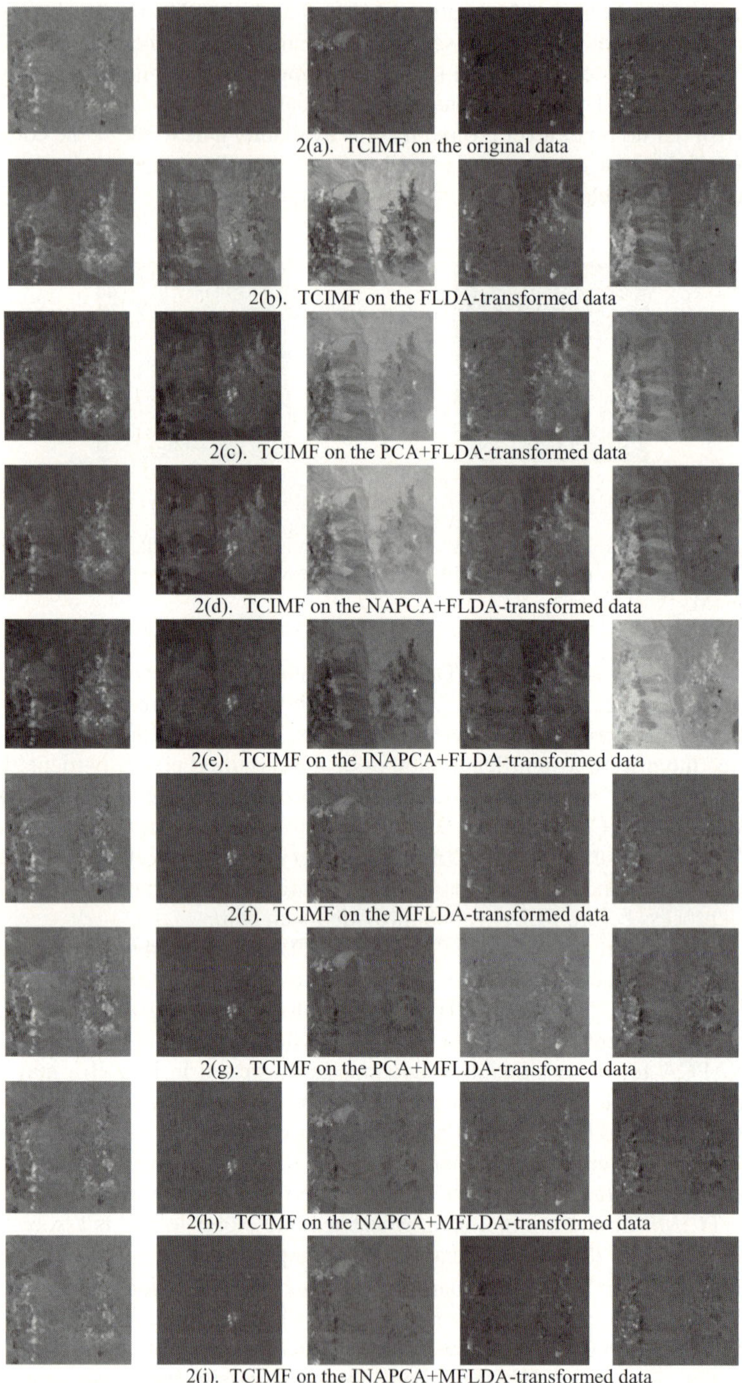

2(a). TCIMF on the original data

2(b). TCIMF on the FLDA-transformed data

2(c). TCIMF on the PCA+FLDA-transformed data

2(d). TCIMF on the NAPCA+FLDA-transformed data

2(e). TCIMF on the INAPCA+FLDA-transformed data

2(f). TCIMF on the MFLDA-transformed data

2(g). TCIMF on the PCA+MFLDA-transformed data

2(h). TCIMF on the NAPCA+MFLDA-transformed data

2(i). TCIMF on the INAPCA+MFLDA-transformed data

Fig. 2. Classification Results (left to right A, B, C, K, and M)

Table 1. Classification results compared with the TCIMF result using original data (correlation coefficient) in AVIRIS experiment

	A	**B**	**C**	**K**	**M**	**Average**
FLDA+TCIMF	0.49	0.16	0.08	0.49	0.13	0.27
PCA+FLDA+TCIMF	0.54	0.33	0.11	0.56	0.16	0.34
NAPCA+FLDA+TCIMF	0.54	0.30	0.11	0.55	0.14	0.33
INAPCA+FLDA+TCIMF	0.52	0.52	0.07	0.61	0.16	0.38
MFLDA+TCIMF	0.75	0.93	0.76	0.87	0.72	0.80
PCA+MFLDA+TCIMF	0.66	0.81	0.74	0.76	0.50	0.70
NAPCA+MFLDA+TCIMF	0.74	0.90	0.73	0.87	0.71	0.79
INAPCA+MFLDA+TCIMF	0.76	0.95	0.78	0.92	0.84	0.85

5 Conclusion

In this paper, we discuss how to effectively implement Fisher's linear discriminant analysis for remotely sensed hyperspectral imagery in the situation when the total number of classes is unknown. We find the variant, which uses the data covariance matrix Σ to replace the within-class scatter matrix $\mathbf{S}_W$, can significantly improve the classification performance. In this case, using INAPCA as another dimensionality reduction approach prior to FLDA can further improve the classification performance, while the typical PCA or NAPCA cannot.

References

1. Landgrebe, D.A.: Signal Theory Methods in Multispectral Remote Sensing. John Wiley & Sons, Chichester (2003)
2. Duda, R.O., Hart, P.E.: Pattern Classification and Scene Analysis. John Wiley & Sons, New York (1973)
3. Belhumeur, P.N., Hespanha, J.P., Kriegman, D.J.: Eigenfaces vs. Fisherfaces: Recognition Using Class Specific Linear Projection. IEEE Transactions on Pattern Analysis and Machine Intelligence 19, 711–720 (1997)
4. Etemad, K., Chellappa, R.: Discriminant Analysis for Recognition of Human Face Images. Journal of Optical Society of America A 14, 1724–1733 (1997)
5. Casasent, D., Chen, X.-W.: Feature Reduction and Morphological Processing for Hyperspectral Image Data. Applied Optics 43, 227–236 (2004)
6. Green, A.A., Berman, M., Switzer, P., Craig, M.D.: A Transformation for Ordering Multispectral Data in Terms of Image Quality with Implications for Noise Removal. IEEE Transactions on Geoscience and Remote Sensing 26, 65–74 (1988)
7. Lee, J.B., Woodyatt, A.S., Berman, M.: Enhancement of High Spectral Resolution Remote Sensing Data by a Noise-Adjusted Principal Components Transform. IEEE Transactions on Geoscience and Remote Sensing 28, 295–304 (1990)
8. Chang, C.-I., Du, Q.: Interference and Noise-Adjusted Principal Components Analysis. IEEE Transactions on Geoscience and Remote Sensing 37, 2387–2396 (1999)
9. Roger, R.E., Arnold, J.F.: Reliably Estimating the Noise in AVIRIS Hyperspectral Imagers. International Journal of Remote Sensing 17, 1951–1962 (1996)
10. Ren, H., Chang, C.-I.: Target-Constrained Interference-Minimized Approach to Subpixel Target Detection for Hyperspectral Images. Optical Engineering 39, 3138–3145 (2000)

Endmember Extraction Methods: A Short Review

Miguel A. Veganzones and Manuel Graña

Computational Intelligence Group, Basque Country University
http://www.ehu.es/computationalintelligence

Abstract. The analysis of hyperspectral images on the basis of the spectral decomposition of their pixels through the so called spectral unmixing process, has applications in thematic map generation, target detection and unsupervised image segmentation. The critical step is the determination of the endmembers used as the references for the unmixing process. We give a comprehensive enumeration of the methods used in practice, because of its implementation in widely used software packages, and those published in the literature. We have structured the review according to the basic computational approach followed by the algorithms: those based on the computational geometry formulation, the ones following lattice computing ideas and heuristic approaches with a weak formal foundation.

1 Introduction

In the field of hyperspectral image processing, Spectral Unmixing [15] is the computation of the fractional contribution of elementary spectra, called endmembers because they constitute the vertices of a convex polytope covering (most of) the image data points in high dimensional space. Spectral Unmixing can be used for target detection, thematic map building and unsupervised segmentation. The underlying image model is a linear mixture of the endmembers, with positive coefficients that sum up to one. Given the endmembers, enforcing these conditions when performing Spectral Unmixing involves constrained non-negative least squares estimation, which can be a very computationally expensive process by itself. When the convex polytope defined by the provided endmembers does not cover all the data points, it is not possible to force these conditions. Solving the problem of providing the appropriate set of endmembers is a precondition to the realization of the Spectral Unmixing. Early approaches to endmember determination were based on human expertise. The prior knowledge about the contents of the imaged terrain was used by the expert to select some candidate endmember spectra from a provided library. The spectra in the library must have some correspondence with the sensor characteristics, in order to perform the matching and unmixing. Besides the methodological questions, this approach is not feasible when trying to process large quantities of image data.

Current approaches try to induce the endmembers from the image data. They either try to select some image pixel spectra as the best approximation to the endmembers in the image [20,10], or to compute estimations of the endmembers

I. Lovrek, R.J. Howlett, and L.C. Jain (Eds.): KES 2008, Part III, LNAI 5179, pp. 400–407, 2008.

on the basis of the transformations of the image data (i.e. [6,14]). The latter is the predominant class of tecniques in the literature. Previous reviews found in the literature [21] make some emphasis on the degree of automation to classify the algorithms. In this review the emphasis will be on the computational foundations, assuming that user interaction must be minimal or null. We distinguish three fundamental approaches:

- Geometric approaches, that try to find a simplex that covers the image data.
- Lattice computing approaches, that use some kind of lattice theoretic formalism or mathematical morphology approach.
- Heuristic approaches, that are not very rigorously formalized under a theoretical framework.

There are some problems that we will not touch in deep in this review. The problem of the endmember induction algorithms initialization is discussed in [19], where an Endmember Initialization Algorithm (EIA) is proposed. The number of spectral signatures that form an hyperspectral image is usually unknown. Recently, a new concept denoted Virtual Dimensionality (VD) [5,4,19] has been used for automated search of the optimal number of endmembers in an image. Most endmember induction algorithms are quite computationally expensive, so a mention is due to the efforts to obtain distributed implementations [22] that may help them to be a feasible approach for real-life applications. The use of off-the-shelf Graphical Processing Units [24] are a low cost way to obtain substantial speed-ups. The outline of the paper is as follows: section 2 describe some geometrically oriented methods, section 3 describe methods based on lattice computing or mathematical morphology, section 4 describe some heuristic methods. Finally, we give some concluding remarks in section 5.

2　Geometric Endmember Induction Methods

Geometric methods follow the formal definition of the endmembers, they search for the vertices of a convex set that covers the image data. Because the distribution of the data in the hyperspace is usually tear-shaped they look for the minimum simplex that covers all the data. Unless said otherwise, the algorithms search for a prefixed number of endmembers, defined by the user. The first such methods is the Minimum Volume Transform, proposed by [8] that introduces two non-orthonormal transforms, the dark-point-fixed (DPF) transform and the fixed-point-free (FPF) transform that map the data onto the minimal simplex that contain all the data points.

One of the earliest approaches is the N-FINDR algorithm proposed in [26]. The N-FINDR algorithm is a selection algorithm. Its works are described as follows: it starts with a random collection of image pixel spectra, corresponding to the initial set of endmembers. Then, each of the remaining image pixels is considered as a candidate to replace each endmember, if doing so the volume of the simplex increases, then it is accepted as the new endmember. The process ends when no more replacements are possible. The N-FINDR algorithm requires a dimension reduction step, originally an orthogonal subspace projection (OSP) to an space of dimension N-1, where N is the number of endmembers. This set of endmembers found by the N-FINDR would not allow the nonnegative unmixing of the pixel spectra in general.

The Convex Cone Analysis (CCA) [14] is based on the fact that the vectors formed by discrete radiance spectra are linear combinations of nonnegative components, and they lie inside a nonnegative convex region. The object of CCA is to find the boundaries of this convex region, which can be used as endmember spectras. The algorithm performs a Principal Component Analysis (PCA) dimension reduction based on the sample spectral correlation matrix of the image. In this reduced space, the endmembers must define a convex cone on the positive hyperquadrant of the space, whose apex is in the space origin. Endmembers are points with exactly $c-1$ zero coefficients in the PCA decomposition, being c the number of eigenvectors selected.

The approach followed in [1] searchs for the optimal simplex using a simulated annealing algorithm (SA) whose state configuration is given by the partition of the faces of the convex hull of the image pixel spectra, after a reduction to N-1 dimensions by the Minimum Noise Fraction (MNF) algorithm. The partition in the configuration space defines a simplex convering the image data whose vertices are the candidate endmembers. The objective function minimized is the simplex volume. This approach is followed by the generation of endmember bundles that allow the computation of bounds on the abundance images.

The Iterated Constrained Endmembers (ICE) [2] algorithm performs the minimization of a regularized residual sum of squares (RSS). The regularization term is the volume of the simplex. The name of the algorithm comes from the minimization schema applied. Given that the free parameters are the endmembers and the proportions (abundances) for each pixel the algorithm iterates the solution of the tow interleaved and interdependent minimization problems (much like in an Expectation Maximization process): first the proportions are computed by quadratic programming problem solving assuming that the endmembers are known, then the endmembers are computed as the direct minimization of the RSS functional. The addition of an sparsity promoting term in the RSS functional gives way to SPICE [27]. This sparsity promoting term is derived as the substitution of a Gaussian prior by a Laplacian prior in a bayesian formulation of the RSS functional. The SPICE algorithm allows the selection of the appropriate number of endmembers based on the sparsity measure. The ICE algorithm does need a dimension refuction step, performed by the MNF algorithm.

The Vertex Component Analysis algorithm (VCA) is presented in [18]. The algorithm is unsupervised and exploits that the affine transformation of a simplex is also a simplex. It works with projected and unprojected data. The algorithm iteratively projects data onto a direction orthogonal to the subspace spanned by the endmembers already determined. The new endmember signature corresponds to the extreme of the projection. The algorithm iterates until all endmembers are exhausted.

In [7] a simplex-based endmember extraction algorithm, called Simplex Growing Algorithm (SGA), is presented. It is a sequential algorithm to find a simplex with the maximum volume every time a new vertex is added. Virtual Dimensionality (VD) is applied as stopping rule to determine the number of vertices required. SGA improves N-FINDR by including a process of growing simplexes one vertex at a time until the desired number of vertices is reached, which results in a high computational complexity reduction; and by selecting an appropiate initial vector to avoid the use of random vectors as initial condition, which produces different sets of final endmembers if different sets of randomly generated initial endmembers are used.

In [16] a method for endmember extraction for highly mixed data, when there are not pure pixels in the hyperspectral image, is presented. The proposed method, called Minimum Volume Constrained Nonnegative Matrix Factorization (MVC-NMF) takes advantage of the fast convergence of NMF schemes and at the same time eliminates the pure-pixel assumption. It consists in the reformulation of an NMF cost function introducing a volume regularization term, much like the ICE, substituting the RSS by the NMF criteria.

3   Lattice Computing Endmember Induction Methods

Lattice computing can be defined as the collection of computational methods that either are defined on the algebra of lattice operators inf and sup, with the addition, or employ lattice theory to generalize previous approaches. Mathematical Morphology is a very successfull case of this paradigm, but it also encompasses some fuzzy systems approaches and neural networks. The Automated Morphological Endmember Extraction (AMEE) method [20] is a mathematical morphology inspired algorithm for the extraction of the endmembers from the data. It is based on the definition of multispectral erosion and dilation operators, which are then used to compute the Morphological Eccentricity Index (MEI) over kernels of increasing size that are computed over all the pixels in the image. The result is a MEI image whose maxima correspond to the endmember pixels. The method does not need a dimension reduction step.

The concept of morphological independence, later reformulated as lattice independence, was the basic tool in the approach proposed in [10,13,12,11]. The set of endmembers was formulated as a set of morphologically independent vectors, either in a dilative or erosive sense, or both. There the Associative Morphological Memories, later renamed Lattice Associative Memories, are proposed as detectors of morphologically independent vectors. The algorithm works in a single pass over the sample data.

This approach has been followed by the one proposed in [9]. The relationship between strong lattice independence and affine independence was proven. Then it was found that most vectors in the erosive and dilative lattice memories are strong lattice independent. Therefore, the mere construction of the lattice memories provide a way to obtain the convex hull of the data. Provided an endmember selection mechanism, the algorithm can obtain in a single pass over the image a set of endmembers.

4   Heuristic Endmember Extraction Methods

The heuristic methods collects a set of heterogeneus endmember extraction methods that use different approaches not grouped under a strict theorical background for endmember induction. The most famous and widely used method, due to its inclusion in the ENVI software package, it is the Pixel Purity Index (PPI) [3]. The algorithm reduces the data dimensionality and makes a noise whitened process by MNF method, and then it determines the pixel purity by repeatedly projecting data onto random unit vectors. The extreme pixel in each projection is counted, identifying the purest pixels in scene. PPI requires the human intervention to select those extreme pixels that best satisfy the target spectrum.

Although PPI has been intesively used, its implementation aspects are kept unknown due to the limited published results. In [6] PPI is investigated and a fast iterative algorithm to implement PPI is proposed. The Fast Iterative PPI algorithm (FIPPI) improves PPI in several aspects. FIPPI produces an appropiate initial set of endmembers to speed up the process. Additionally, it estimates the number of endmembers to be generated by Virtual Dimensionality (VD). FIPPI is also an unsupervised and iterative algorithm, where an iterative rule is developed to improve each of the iterations until it reaches a final set of endmembers.

In [25] the well known Independent Component Analysis (ICA) method is the base of the proposed approach for endmember extraction and abundance quantification. The algorithm, called ICA-based Abundance Quantification Algorithm (ICA-AQA), is a high-order statistics-based technique, that can accomplish endmember extraction and abundance quantification simultaneously in one-shot operation. [17] analyzes the use of ICA and Independent Factor Analysis (IFA) for unmixing tasks, showing that the statistically independent of the sources, assumed by ICA and IFA, is violated in the hyperspectral unmixing, compromising the performance of ICA/IFA algorithms for this purpose. It concludes that the accuracy of this ICA/IFA-based methods tends to improve with the increase of the signature variability and the signal-to-noise ratio.

The Spatial-Spectral Endmember Extraction algorithm (SSEE) proposed in [23] is another projection based method that works by analyzing a scene in parts (subsets), such that it increases the spectral contrast of low contrast endmembers, thus improving the potential for these endmembers to be selected. The SSEE method uses a singular value decomposition (SVD) to determine a set of basis vectors that describe most of the spectral variance for subsets of the image. Then the full image dataset is projected onto the locally defined basis vectors to determine a set of candidate endmember pixels from where the final endmembers are selected. For that, it searchs for spectrally similar but spatially independent endmembers. This is realized by imposing spatial constraints for averaging spectrally similar endmembers.

5   Conclusions

The field of hyperspectral image processing has been an applicaction domain for many pattern recognition techniques. Among them, spectral unmixing offers the appealing of a physical image formation model with an easy interpretation. It also allows subpixel resolution results. Therefore, increasingly Spectral Unmixing will be a tool of hyperspectral image analysis. The requisite for this analysis is the determination of the endmembers. The current approaches reviewed in this paper favor the endmember induction from the image data. It is desirable that the endmembers have some physical meaning, which is more likely in the case of approaches that perform a selection from the image pixel spectra. However, these approaches usually do not produce convex polytopes that cover all the image data points, so that the candidate set of endmembers do not fit into the formal definition of endmembers. Geometrically oriented methods are the best theoretically grounded ones, however they ask for great computational resources and the endmembers that they obtain do not have a clear physical meaning.

Acknowledgements

This work is partially supported by grant MEC DPI2006-15346-C03-03. Miguel A. Veganzones has a predoctoral grant from the Basque Goverment. The Computational Intelligence Group has earned a six year grant as University Research Group from the Basque Goverment.

References

1. Bateson, C.A., Asner, G.P., Wessman, C.A.: Endmember bundles: a new approach to incorporating endmember variability into spectral mixture analysis. IEEE Transactions on Geoscience and Remote Sensing 38, 1083–1094 (2000)
2. Berman, M., Kiiveri, H., Lagerstrom, R., Ernst, A., Dunne, R., Huntington, J.F.: Ice: a statistical approach to identifying endmembers in hyperspectral images. IEEE Transactions on Geoscience and Remote Sensing 42, 2085–2095 (2004)
3. Boardman, J., Kruse, F., Green, R.: Mapping target signatures via partial unmixing of aviris data. In: Summaries of the Fifth Annual JPL Airborne Geoscience Workshop, vol. 1 (1995)
4. Chang, C.-I.: Hyperspectral Imaging: Techniques for Spectral Detection and Classification. Springer, Heidelberg (2003)
5. Chang, C.-I., Du, Q.: Estimation of number of spectrally distinct signal sources in hyperspectral imagery. IEEE Transactions on Geoscience and Remote Sensing 42, 608–619 (2004)
6. Chang, C.-I., Plaza, A.: A fast iterative algorithm for implementation of pixel purity index. Geoscience and Remote Sensing Letters 3, 63–67 (2006)

7. Chang, C.-I., Wu, C.-C., Liu, W., Ouyang, Y.-C.: A new growing method for simplex-based endmember extraction algorithm. IEEE Transactions on Geoscience and Remote Sensing 44, 2804–2819 (2006)

8. Craig, M.D.: Minimum-volume transforms for remotely sensed data. IEEE Transactions on Geoscience and Remote Sensing 32, 542–552 (1994)

9. Schmalz, M.S., Ritter, G.X., Urcid, G.: Autonomous single-pass endmember approximation using lattice auto-associative memories. In: 10th Joint Conference on Information Sciences. Elsevier, Amsterdam (preprint, 2008) (Special Issue)

10. Grana, M., Gallego, J.: Associative morphological memories for endmember induction. In: Proceedings of IEEE International Geoscience and Remote Sensing Symposium. IGARSS 2003, vol. 6, pp. 3757–3759 (2003)

11. Grana, M., Gallego, J.: Hyperspectral image analysis with associative morphological memories. In: Proceedings of International Conference on Image Processing. ICIP 2003, volume 3, vol. 2, pp. III–549–552 (2003)

12. Grana, M., Gallego, J., Hernandez, C.: Further results on amm for endmember induction. In: 2003 IEEE Workshop on Advances in Techniques for Analysis of Remotely Sensed Data, pp. 237–243 (2003)

13. Grana, M., Sussner, P., Ritter, G.: Associative morphological memories for endmember determination in spectral unmixing. In: The 12th IEEE International Conference on Fuzzy Systems. FUZZ 2003, vol. 2, pp. 1285–1290 (2003)

14. Ifarraguerri, A., Chang, C.-I.: Multispectral and hyperspectral image analysis with convex cones. IEEE Transactions on Geoscience and Remote Sensing 37, 756–770 (1999)

15. Keshava, N., Mustard, J.F.: Spectral unmixing. Signal Processing Magazine 19, 44–57 (2002)

16. Miao, L., Qi, H.: Endmember extraction from highly mixed data using minimum volume constrained nonnegative matrix factorization. IEEE Transactions on Geoscience and Remote Sensing 45, 765–777 (2007)

17. Nascimento, J.M.P., Dias, J.M.B.: Does independent component analysis play a role in unmixing hyperspectral data? IEEE Transactions on Geoscience and Remote Sensing 43, 175–187 (2005)

18. Nascimento, J.M.P., Dias, J.M.B.: Vertex component analysis: a fast algorithm to unmix hyperspectral data. IEEE Transactions on Geoscience and Remote Sensing 43, 898–910 (2005)

19. Plaza, A., Chang, C.-I.: Impact of initialization on design of endmember extraction algorithms. IEEE Transactions on Geoscience and Remote Sensing 44, 3397–3407 (2006)

20. Plaza, A., Martinez, P., Perez, R., Plaza, J.: Spatial/spectral endmember extraction by multidimensional morphological operations. IEEE Transactions on Geoscience and Remote Sensing 40, 2025–2041 (2002)

21. Plaza, A., Martinez, P., Perez, R., Plaza, J.: A quantitative and comparative analysis of endmember extraction algorithms from hyperspectral data. IEEE Transactions on Geoscience and Remote Sensing 42, 650–663 (2004)

22. Plaza, A., Valencia, D., Plaza, J., Chang, C.-I.: Parallel implementation of endmember extraction algorithms from hyperspectral data. Geoscience and Remote Sensing Letters 3, 334–338 (2006)

23. Rogge, D.M., Rivard, B., Zhang, J., Sanchez, A., Harris, J., Feng, J.: Integration of spatial-spectral information for the improved extraction of endmembers. Remote Sensing of Environment 110, 287–303 (2007)

24. Setoain, J., Prieto, M., Tenllado, C., Plaza, A., Tirado, F.: Parallel morphological endmember extraction using commodity graphics hardware. Geoscience and Remote Sensing Letters 4, 441–445 (2007)
25. Wang, J., Chang, C.-I.: Applications of independent component analysis in endmember extraction and abundance quantification for hyperspectral imagery. IEEE Transactions on Geoscience and Remote Sensing 44, 2601–2616 (2006)
26. Winter, M.E.: N-findr: an algorithm for fast autonomous spectral endmember determination in hyperspectral data. In: Proceedings of SPIE: Imaging Spectrometry, vol. 3753 (1999)
27. Zare, A., Gader, P.: Sparsity promoting iterated constrained endmember detection in hyperspectral imagery. Geoscience and Remote Sensing Letters 4, 446–450 (2007)

Super Resolution of Multispectral Images Using TV Image Models

Miguel Vega[1,*], Javier Mateos[2], Rafael Molina[2], and Aggelos K. Katsaggelos[3]

[1] Dpto. de Lenguajes y Sistemas Informáticos,
Universidad de Granada, Granada 18071, Spain
[2] Dpto. de Ciencias de la Computación e Inteligencia Artificial,
Universidad de Granada, Granada 18071, Spain
[3] Department of Electrical Engineering and Computer Science,
Northwestern University, Evanston, IL 60208, USA

Abstract. In this paper we propose a novel algorithm for the pansharpening of multispectral images based on the use of a Total Variation (TV) image prior. Within the Bayesian formulation, the proposed methodology incorporates prior knowledge on the expected characteristics of multispectral images, and uses the sensor characteristics to model the observation process of both panchromatic and multispectral images. Using real and synthetic data, the pansharpened multispectral images are compared with the images obtained by other pansharpening methods and their quality is assessed both qualitatively and quantitatively.

1 Introduction

Remote sensing systems include sensors able to capture, simultaneously, several low resolution images of the same area on different wavelengths, forming a multispectral image, along with a high resolution panchromatic image. Throughout this paper the term *multispectral image reconstruction* will refer to the joint processing of the multispectral and panchromatic images in order to obtain a new multispectral image that, ideally, exhibits the spectral characteristics of the observed multispectral image and the resolution of the panchromatic image.

A few approaches to the super resolution of multispectral images problem, also called pansharpening, have been proposed in the literature. See, for instance, [1,2] and the comparison of algorithms in [3]. Recently a new Bayesian variational framework for TV-based image restoration problems has been presented [4]. Here we explore and adapt this framework to the pansharpening of multispectral images.

The paper is organized as follows. In section 2 the Bayesian modeling and inference for super resolution reconstruction of multispectral images is presented. Section 3 describes the variational approximation of the posterior distribution of the high resolution multispectral image and how inference is performed. Section 4 presents experimental results and section 5 concludes the paper.

* This work has been supported by the "Comisión Nacional de Ciencia y Tecnología" under contract TIN2007-65533 and the Consejería de Innovación, Ciencia y Empresa of the Junta de Andalucía under contract P07-TIC-02698.

I. Lovrek, R.J. Howlett, and L.C. Jain (Eds.): KES 2008, Part III, LNAI 5179, pp. 408–415, 2008.
© Springer-Verlag Berlin Heidelberg 2008

2   Bayesian Modeling and Inference

Let us assume that $\mathbf{y}$, the unknown high resolution multispectral image we would have observed under ideal conditions, has B bands $\mathbf{y}_b$, $b = 1, \ldots, B$, each of size $p = m \times n$, that is, $\mathbf{y} = [\mathbf{y}_1^t, \mathbf{y}_2^t, \ldots, \mathbf{y}_B^t]^t$, where each band of this image is expressed as a column vector by lexicographically ordering the pixels in the band, and t denotes the transpose of a vector or matrix. The observed low resolution multispectral image $\mathbf{Y}$ has B bands $\mathbf{Y}_b$, $b = 1, \ldots, B$, each of size $P = M \times N$ pixels, with $M < m$ and $N < n$. These images are also stacked into the vector $\mathbf{Y} = [\mathbf{Y}_1^t, \mathbf{Y}_2^t, \ldots, \mathbf{Y}_B^t]^t$, where each band of this image is also expressed as a column vector by lexicographically ordering the pixels in the band. The sensor also provides us with a panchromatic image $\mathbf{x}$ of size $p = m \times n$, obtained by spectrally averaging the high resolution images $\mathbf{y}_b$.

The objective of the high resolution multispectral image reconstruction problem is to obtain an estimate of the unknown high resolution multispectral image $\mathbf{y}$ given the panchromatic high resolution observation $\mathbf{x}$ and the low resolution multispectral observation $\mathbf{Y}$. The Bayesian formulation of this problem requires the definition of the joint distribution $\mathrm{p}(\mathbf{y}, \mathbf{Y}, \mathbf{x})$. We define this joint distribution as $\mathrm{p}(\mathbf{y}, \mathbf{Y}, \mathbf{x}) = \mathrm{p}(\mathbf{y})\mathrm{p}(\mathbf{Y}, \mathbf{x}|\mathbf{y})$ and inference is based on $\mathrm{p}(\mathbf{y}|\mathbf{Y}, \mathbf{x})$. Let us now describe those probability distributions.

In this paper we use TV priors [5] for each band (the correlation among high resolution bands is not taken into account), thus defining the multispectral image prior

$$\mathrm{p}(\mathbf{y}) = \prod_{b=1}^{B} \mathrm{p}(\mathbf{y}_b) \propto \prod_{b=1}^{B} \alpha_b^{p/2} \exp\left[-\alpha_b \mathrm{TV}(\mathbf{y}_b)\right], \tag{1}$$

with $\mathrm{TV}(\mathbf{y}_b) = \sum_{i=1}^{p} \sqrt{(\Delta_i^h(\mathbf{y}_b))^2 + (\Delta_i^v(\mathbf{y}_b))^2}$, where $\Delta_i^h(\mathbf{y}_b)$ and $\Delta_i^v(\mathbf{y}_b)$ represent the horizontal and vertical first order differences at pixel i respectively, α_b is the model parameter of the band b, and the partition function has been approximated using the approach in [4].

We assume that $\mathbf{Y}$ and $\mathbf{x}$, for a given $\mathbf{y}$, are independent and write

$$\mathrm{p}(\mathbf{Y}, \mathbf{x}|\mathbf{y}) = \mathrm{p}(\mathbf{Y}|\mathbf{y})\mathrm{p}(\mathbf{x}|\mathbf{y}). \tag{2}$$

For each multispectral image band, we consider the model $\mathbf{Y}_b = \mathbf{H}\mathbf{y}_b + \mathbf{n}_b$, $b = 1, \ldots, B$, where the degradation matrix $\mathbf{H}$ can be written as $\mathbf{H} = \mathbf{D}\mathbf{B}$, with $\mathbf{B}$ a $p \times p$ blurring matrix and $\mathbf{D}$ a $P \times p$ decimation operator, and $\mathbf{n}_b$ is the noise term assumed to be independent white Gaussian of known variance β_b^{-1}. The distribution of the observed image $\mathbf{Y}$ given $\mathbf{y}$ is

$$\mathrm{p}(\mathbf{Y}|\mathbf{y}) = \prod_{b=1}^{B} \mathrm{p}(\mathbf{Y}_b|\mathbf{y}_b) \propto \prod_{b=1}^{B} \beta_b^{P/2} \exp\left\{-\frac{1}{2}\beta_b \parallel \mathbf{Y}_b - \mathbf{H}\mathbf{y}_b \parallel^2\right\}. \tag{3}$$

The panchromatic image $\mathbf{x}$ is modeled as

$$\mathbf{x} = \sum_{b=1}^{B} \lambda_b \mathbf{y}_b + \mathbf{v},$$

where $\lambda_b \geq 0$, $b = 1, 2, \ldots, B$, are known quantities that can be obtained from the sensor spectral characteristics, and $\mathbf{v}$ is the capture noise that is assumed to be Gaussian with zero mean and known variance γ^{-1}. Based on this model, the distribution of the panchromatic image $\mathbf{x}$ given $\mathbf{y}$, is given by

$$p(\mathbf{x}|\mathbf{y}) \propto \gamma^{p/2} \exp \left\{ -\frac{1}{2}\gamma \parallel \mathbf{x} - \sum_{b=1}^{B} \lambda_b \mathbf{y}_b \parallel^2 \right\}. \tag{4}$$

3   Bayesian Inference and Variational Approximation of the Posterior Distribution

As already known, the Bayesian paradigm dictates that inference on $\mathbf{y}$ should be based on

$$p(\mathbf{y}|\mathbf{Y}, \mathbf{x}) = p(\mathbf{y}, \mathbf{Y}, \mathbf{x})/p(\mathbf{Y}, \mathbf{x}), \tag{5}$$

with $p(\mathbf{y}, \mathbf{Y}, \mathbf{x}) = p(\mathbf{y})p(\mathbf{Y}|\mathbf{y})p(\mathbf{x}|\mathbf{y})$, where $p(\mathbf{y})$, $p(\mathbf{Y}|\mathbf{y})$ and $p(\mathbf{x}|\mathbf{y})$ have been defined in Eqs. (1), (3) and (4), respectively.

Since $p(\mathbf{y}|\mathbf{Y}, \mathbf{x})$ can not be found in closed form, we apply variational methods to approximate this distribution by a distribution $q(\mathbf{y})$. The variational criterion used to find $q(\mathbf{y})$ is the minimization of the Kullback-Leibler (KL) divergence, given by [6]

$$\begin{aligned} C_{KL}(q(\mathbf{y})\|p(\mathbf{y}|\mathbf{Y}, \mathbf{x})) &= \int q(\mathbf{y}) \log \left(\frac{q(\mathbf{y})}{p(\mathbf{y}|\mathbf{Y}, \mathbf{x})} \right) d\mathbf{y} \\ &= \int q(\mathbf{y}) \log \left(\frac{q(\mathbf{y})}{p(\mathbf{y}, \mathbf{Y}, \mathbf{x})} \right) d\mathbf{y} + \mathrm{const} \\ &= \mathcal{M}(q(\mathbf{y})) + \mathrm{const}, \end{aligned} \tag{6}$$

which is always non negative and equal to zero only when $q(\mathbf{y}) = p(\mathbf{y}|\mathbf{Y}, \mathbf{x})$.

Due to the form of the TV prior, the above integral can not be evaluated. We can however majorize the TV prior by a function which renders the integral easier to calculate. Let us consider the following inequality, also used in [7], which states that, for any $w \geq 0$ and $z > 0$

$$\sqrt{w} \leq \frac{w + z}{2\sqrt{z}}. \tag{7}$$

Let us define, for $\mathbf{y}_b$ and $\mathbf{u}_b$, where $\mathbf{u}_b$ is any p-dimensional vector $\mathbf{u}_b \in (R^+)^p$, with components $\mathbf{u}_b(i)$, $i = 1, \ldots, p$, the following functional

$$M(\mathbf{y}_b, \mathbf{u}_b) = \alpha_b^{p/2} \exp \left[-\frac{\alpha_b}{2} \sum_i \frac{(\Delta_i^h(\mathbf{y}_b))^2 + (\Delta_i^v(\mathbf{y}_b))^2 + \mathbf{u}_b(i)}{\sqrt{\mathbf{u}_b(i)}} \right]. \tag{8}$$

Now, using the inequality in Eq. (7) with $w = (\Delta_i^h(\mathbf{y}_b))^2 + (\Delta_i^v(\mathbf{y}_b))^2$ and $z = \mathbf{u}_b(i)$, and comparing Eq. (8) with Eq. (1), we obtain $p(\mathbf{y}) \geq c \cdot \prod_{b=1}^{B} M(\mathbf{y}_b, \mathbf{u}_b)$.

As will be shown later, vectors $\mathbf{u}_b$ are quantities that need to be computed and have an intuitive interpretation related to the unknown images $\mathbf{y}_b$. This leads to the following lower bound for the joint probability distribution

$$p(\mathbf{y}, \mathbf{Y}, \mathbf{x}) \geq c \cdot \left[\prod_{b=1}^{B} M(\mathbf{y}_b, \mathbf{u}_b) \right] p(\mathbf{Y}|\mathbf{y}) p(\mathbf{x}|\mathbf{y}) = F(\mathbf{y}, \mathbf{Y}, \mathbf{x}, \mathbf{u}), \qquad (9)$$

where $\mathbf{u} = [\mathbf{u}_1^t, \mathbf{u}_2^t, \ldots, \mathbf{u}_B^t]^t$.

Hence, by defining

$$\tilde{\mathcal{M}}(q(\mathbf{y}), \mathbf{u}) = \int q(\mathbf{y}) \log\left(\frac{q(\mathbf{y})}{F(\mathbf{y}, \mathbf{Y}, \mathbf{x}, \mathbf{u})} \right) d\mathbf{y}, \qquad (10)$$

and using Eq. (9), we obtain

$$\mathcal{M}(q(\mathbf{y})) \leq \min_{\mathbf{u}} \tilde{\mathcal{M}}(q(\mathbf{y}), \mathbf{u}). \qquad (11)$$

Therefore, by finding a sequence of distributions $\{q^k(\mathbf{y})\}$ that monotonically decreases $\tilde{\mathcal{M}}(q(\mathbf{y}), \mathbf{u})$ for a fixed $\mathbf{u}$, a sequence of an ever decreasing upper bound of $C_{KL}(q(\mathbf{y})\|p(\mathbf{y}|\mathbf{Y}, \mathbf{x}))$ is also obtained due to Eq. (6). However, also minimizing $\mathcal{M}(q(\mathbf{y}))$ with respect to $\mathbf{u}$, generates a sequence of vectors $\{\mathbf{u}^k\}$ that tightens the upper-bound for each distribution $q^k(\mathbf{y})$. Therefore, the two sequences $\{q^k(\mathbf{y})\}$ and $\{\mathbf{u}^k\}$ are coupled. We develop the following iterative algorithm to find such sequences. We note that the process to find the best posterior distribution approximation of the image in combination with $\mathbf{u}$ is a very natural extension of the Majorization-Minimization approach to function optimization [8].

Algorithm 1. *Posterior image distribution estimation.*

Given $\mathbf{u}^1 \in (R^+)^{Bp}$, *for* $k = 1, 2, \ldots$ *until a stopping criterion is met:*

1. *Find*
$$q^k(\mathbf{y}) = \arg\min_{q(\mathbf{y})} \int q(\mathbf{y}) \times \log\left(\frac{q(\mathbf{y})}{F(\mathbf{y}, \mathbf{Y}, \mathbf{x}, \mathbf{u}^k)} \right) d\mathbf{y}. \qquad (12)$$

2. *Find*
$$\mathbf{u}^{k+1} = \arg\min_{\mathbf{u}} \int q^k(\mathbf{y}) \times \log\left(\frac{q^k(\mathbf{y})}{F(\mathbf{y}, \mathbf{Y}, \mathbf{x}, \mathbf{u})} \right) d\mathbf{y}. \qquad (13)$$

Set $q(\mathbf{y}) = \lim_{k \to \infty} q^k(\mathbf{y})$.

To calculate $\mathbf{u}_b^{k+1}$, for $b = 1, \ldots, B$, we have from Eq. (13) that

$$\mathbf{u}_b^{k+1} = \arg\min_{\mathbf{u}_b} \sum_i \frac{\mathbf{E}_{q^k(\mathbf{y})}\left[\Delta_i^h(\mathbf{y}_b))^2 + (\Delta_i^v(\mathbf{y}_b))^2 \right] + \mathbf{u}_b(i)}{\sqrt{\mathbf{u}_b(i)}}, \qquad (14)$$

and consequently

$$\mathbf{u}_b^{k+1}(i) = \mathbf{E}_{q^k(\mathbf{y})}\left[\Delta_i^h(\mathbf{y}_b))^2 + (\Delta_i^v(\mathbf{y}_b))^2 \right], \quad \text{for} \quad i = 1, \ldots, p. \qquad (15)$$

It is clear from Eq. (15) that the vector $\mathbf{u}_b^{k+1}$ is a function of the spatial first order differences of the unknown image $\mathbf{y}$ under the distribution $q^k(\mathbf{y})$ and represents the local spatial activity of $\mathbf{y}_b$.

To calculate $q^k(\mathbf{y})$, we observe that differentiating the integral on the right-hand side of Eq. (12) with respect to $q(\mathbf{y})$ and setting it equal to zero, we obtain

$$q^k(\mathbf{y}) = \mathcal{N}\left(\mathbf{y} \mid \mathbf{E}_{q^k(\mathbf{y})}[\mathbf{y}], \mathbf{cov}_{q^k(\mathbf{y})}[\mathbf{y}]\right), \tag{16}$$

with

$$\mathbf{cov}_{q^k(\mathbf{y})}[\mathbf{y}] = \mathcal{A}^{-1}(\mathbf{u}^k), \quad \text{and} \quad \mathbf{E}_{q^k(\mathbf{y})}[\mathbf{y}] = \mathbf{cov}_{q^k(\mathbf{y})}[\mathbf{y}]\phi^k. \tag{17}$$

In Eq. (17) ϕ^k is the $(B \times p) \times 1$ vector

$$\phi^k = \left(\text{diag}(\beta) \otimes \mathbf{H}^t\right)\mathbf{Y} + \gamma\left(\text{diag}(\lambda) \otimes \mathbf{I}_p\right)\left(\mathbf{x}^t, \mathbf{x}^t, \ldots, \mathbf{x}^t\right)^t,$$

and

$$\mathcal{A}(\mathbf{u}^k) = \begin{pmatrix} \alpha_1 \mathcal{G}(\mathbf{u}_1^k) & \mathbf{0}_p & \cdots & \mathbf{0}_p \\ \mathbf{0}_p & \alpha_2 \mathcal{G}(\mathbf{u}_2^k) & \cdots & \mathbf{0}_p \\ \vdots & \vdots & \ddots & \vdots \\ \mathbf{0}_p & \mathbf{0}_p & \cdots & \alpha_B \mathcal{G}(\mathbf{u}_B^k) \end{pmatrix} + \text{diag}(\beta) \otimes \mathbf{H}^t\mathbf{H} + \gamma(\lambda\lambda^t) \otimes \mathbf{I}_p,$$

where $\otimes$ is the Kronecker product, $\beta = (\beta_1, \beta_2, \ldots, \beta_B)^t$, $\lambda = (\lambda_1, \lambda_2, \ldots, \lambda_B)^t$, and

$$\mathcal{G}(\mathbf{u}_b^k) = (\Delta^h)^t W(\mathbf{u}_b^k)(\Delta^h) + (\Delta^v)^t W(\mathbf{u}_b^k)(\Delta^v), \quad \text{for} \quad b = 1, \ldots, B, \tag{18}$$

with Δ^h and Δ^v representing $p \times p$ convolution matrices associated with the first order horizontal and vertical differences, respectively, and $W(\mathbf{u}_b^k)$ a $p \times p$ diagonal matrix of the form $W(\mathbf{u}_b^k) = \text{diag}\left(\mathbf{u}_b^k(i)^{-\frac{1}{2}}\right)$, for $i = 1, \ldots, p$. This matrix $W(\mathbf{u}_b^k)$ can be interpreted as a spatial adaptivity matrix since it controls the amount of smoothing at each pixel location depending on the strength of the intensity variation at that pixel, as expressed by the horizontal and vertical intensity gradients. That is, for pixels with high spatial activity the corresponding entries of $W(\mathbf{u}_b^k)$ are very small, which means that no smoothness is enforced, while for pixels in a flat region the corresponding entries of $W(\mathbf{u}_k^b)$ are very large, which means that smoothness is enforced.

4   Experimental Results

Although we performed a wide set of experiments to assess the quality of the proposed approach, here we report results only on a synthetic color image and a real Landsat multispectral image. In our first experiment, the synthetic multispectral observations are obtained from the color image displayed in Fig. 1(a) by first convolving it with mask $0.25 \times \mathbf{1}_{2\times2}$ to simulate sensor integration, and then downsampling it by a factor of two by discarding every other pixel in each

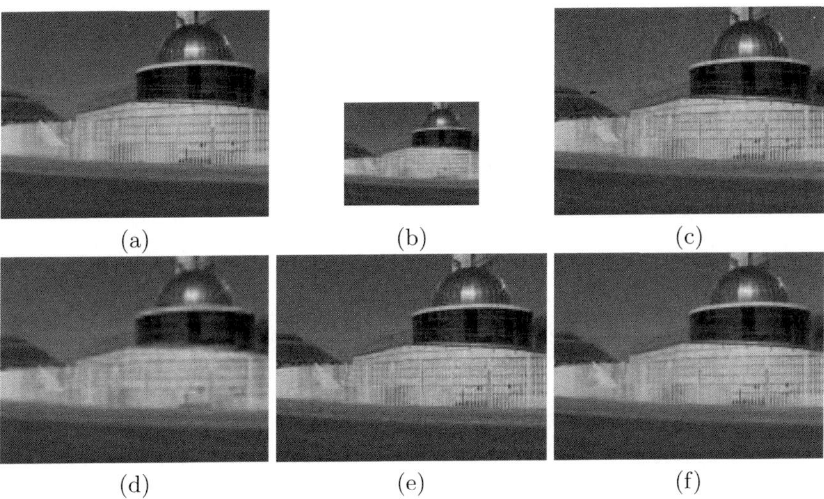

Fig. 1. (a) Original image; (b) Observed multispectral image; (c) Observed panchromatic image; (d) Bicubic interpolation of (a); (e) Reconstruction using the method in [2]; (f) Reconstruction using the proposed model

direction and adding zero mean Gaussian noise with variance 16. The panchromatic image was obtained from the original high resolution color image using the model in Eq. (4), with $\lambda_b = 1/3$, for $b = 1, 2, 3$, and additive zero mean Gaussian noise of variance 25. The observed multispectral and panchromatic images are depicted in Figs. 1(b) and 1(c), respectively.

Spatial improvement was assessed by means of the correlation of the high frequency components (COR) [9] which takes values between 0 and 1 (the higher the value the better the quality of the reconstruction), while spectral fidelity was assessed by means of the peak signal-to-noise ratio (PSNR) between each band of the reconstructed and original multispectral images, and the standard ERGAS index [10]. The lower the value of this index, especially a value under 3.0, the higher the quality of the multispectral reconstructed image. We have also included *Structural Similarity Index Measure* (SSIM) values for each band (see [11]). Its maximum value is one when the original and reconstructed bands coincide.

The proposed algorithm was ran until the criterion $\|\mathbf{y}^k - \mathbf{y}^{k-1}\|^2/\|\mathbf{y}^{k-1}\|^2 < 10^{-4}$ was satisfied, where $\mathbf{y}^k$ denotes the mean of $q^k(\mathbf{y})$. It typically required 4 iterations. The value of the parameters were experimentally chosen to be $\alpha_b = 0.001$ and $\beta_b = 1/16$, for $b = 1, 2, 3$, and $\gamma = 1/25$. Table 1 shows the resulting PSNR, COR, SSIM and ERGAS values for the reconstructions of the image using the proposed method, bicubic interpolation and the pansharpening method in [2]. The reconstructed images corresponding to those methods are displayed in Fig. 1(d)–1(f). For all the studied cases, the bicubic interpolation produces over-smoothed reconstructions. The method in [2] produces noisy reconstructions and some color bleeding at the edges, as can be observed at the center of the fence in Fig. 1(e). However the proposed model produces both

Table 1. Values of PSNR, COR and ERGAS for the color image in Fig. 1.

Method / Band	PSNR 1	2	3	COR 1	2	3	SSIM 1	2	3	ERGAS
Bicubic interpolation	29.09	29.01	29.03	0.49	0.50	0.49	0.80	0.80	0.80	4.62
Method in [2]	32.36	32.55	32.24	0.89	0.90	0.89	0.83	0.83	0.82	3.14
Proposed model	35.31	35.31	35.05	0.92	0.93	0.92	0.92	0.93	0.92	2.26

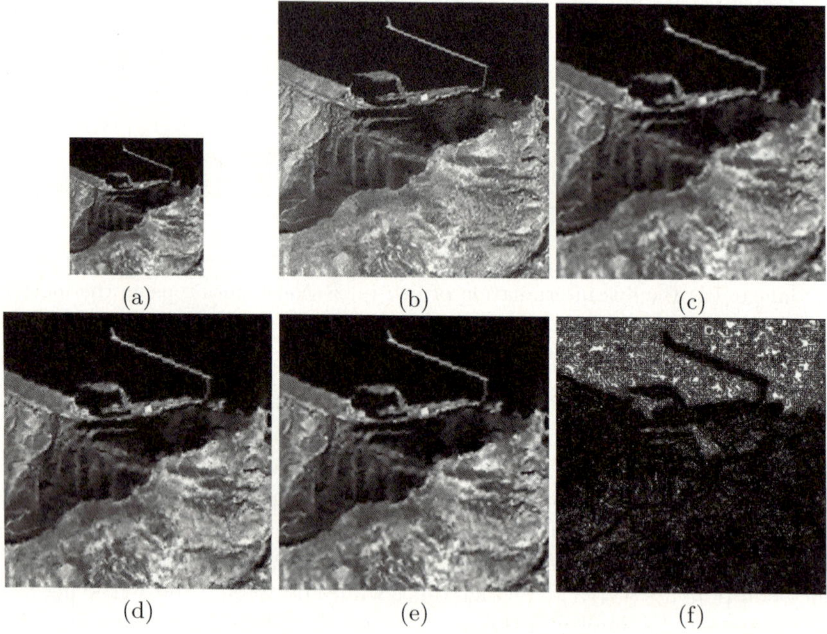

(a)   (b)   (c)

(d)   (e)   (f)

Fig. 2. (a) Observed multispectral image; (b) Observed panchromatic image; (c) Bicubic interpolation of (a); (d) Reconstruction using the method in [2]; (e) Reconstruction using the proposed model; (f) Spatial adaptivity matrix $W(\mathbf{u}_4)$ in log scale.

numerically (see table 1) and visually better results recovering the high resolution structures of the original image while reducing at the same time the amount of noise.

In a second experiment, the method was tested on a real Landsat ETM+ image. Figures 2(a) and 2(b) depict a 128×128 pixels false RGB color region of interest composed of bands 3, 4, and 2 of a Landsat ETM+ multispectral image, together with their corresponding 256×256 panchromatic image. According to the ETM+ sensor spectral response, the panchromatic image covers only the spectrum of a part of the first four bands of the multispectral image. Hence, we apply the proposed method with $B = 4$. The values of λ_b, $b = 1, 2, 3, 4$, calculated from the spectral response of the ETM+ sensor, are equal to 0.0078, 0.2420, 0.2239, and 0.5263, for bands one to four, respectively [12]. Visual inspection of the results displayed in Figs. 2(c)–2(e) reveals similar conclusions to the ones for the synthetic image. The proposed method provides the best result, preserving the spectral properties of the multispectral image while incorporating the high frequencies from the panchromatic image and controlling the noise in the image. Figure 2(f) depicts the value of $W(\mathbf{u}_4)$ (in log scale), that is, the spatial adaptivity matrix for band 4 in Eq. (18), in which high spatial activity areas are clearly depicted corresponding to lower image values.

5   Conclusions

We have presented a new method for TV-based pansharpening of multispectral images using a super resolution approach. The proposed method takes into account the sensor characteristics in the image formation model. We have used the variational approach to approximate the posterior distribution of the pan-sharpened multispectral image. Based on the presented experimental results, the proposed method outperforms bicubic interpolation and the method in [2].

References

1. Nuñez, J., Otazu, X., Fors, O., Prades, A., Pala, V., Arbiol, R.: Multiresolution-based image fusion with additive wavelet decomposition. IEEE Trans on Geosc. & Rem. Sens. 37(3), 1204–1211 (1999)
2. Price, J.: Combining multispectral data of different spatial resolution. IEEE Trans. on Geosc. & Rem. Sens. 37(3), 1199–1203 (1999)
3. Alparone, L., Wald, L., Chanussot, J., Thomas, C., Gamba, P., Bruce, L.: Comparison of Pansharpening Algorithms: Outcome of the 2006 GRS-S Data-Fusion Contest. IEEE Trans. on Geosc. & Rem. Sens. 45(10), 3012–3020 (2007)
4. Babacan, S., Molina, R., Katsaggelos, A.: Parameter estimation in TV image restoration using variational distribution approximation. IEEE Trans. on Image Proc. 17(3), 326–339 (2008)
5. Rudin, L.I., Osher, S., Fatemii, E.: Nonlinear total variation based noise removal algorithms. Physica D 63(6), 259–268 (1992)
6. Kullback, S., Leibler, R.A.: On information and sufficiency. Annals of Math. Stat. 22, 79–86 (1951)
7. Bioucas-Dias, J., Figueiredo, M., Oliveira, J.: Total-variation image deconvolution: A majorization-minimization approach. In: ICASSP 2006 (2006)
8. Lange, K.: Optimization. In: Springer texts in Statistic. Springer, New York (2004)
9. Vijayaraj, V.: A quantitative analysis of pansharpened images. Master's thesis, Mississippi St. University (2004)
10. Wald, L., Ranchin, T., Mangolini, M.: Fusion of satellite images of different spatial resolutions: assessing the quality of resulting images. Phot. Eng. Rem. Sens. 63(6), 691–699 (1997)
11. Wang, Z., Bovik, A.C., Sheikh, H.R., Simoncelli, E.P.: Image quality assessment: From error measurement to structural similarity. IEEE Trans. on Img Proc. 13(4), 600–612 (2004)
12. Molina, R., Vega, M., Mateos, J., Katsaggelos, A.: Variational posterior distribution approximation in Bayesian super resolution reconstruction of multispectral images. Applied and Computational Harmonic Analysis 24(2), 251–267 (2008)

Classification of Hyperspectral Images Compressed through 3D-JPEG2000

Ian Blanes[1], Alaitz Zabala[2], Gerard Moré[3],
Xavier Pons[2,3], and Joan Serra-Sagristà[1]

[1] Department of Information and Communications Engineering
[2] Department of Geography
[3] Centre for Ecological Research and Forestry Applications (CREAF)
Universitat Autònoma de Barcelona
Cerdanyola del Vallès 08290, Spain
Ian.Blanes@uab.cat, Alaitz.Zabala@uab.cat, G.More@creaf.uab.cat,
Xavier.Pons@uab.cat, Joan.Serra@uab.cat

Abstract. Classification of hyperspectral images is paramount to an increasing number of user applications. With the advent of more powerful technology, sensed images demand for larger requirements in computational and memory capabilities, which has led to devise compression techniques to alleviate the transmission and storage necessities.

Classification of compressed images is addressed in this paper. Compression takes into account the spectral correlation of hyperspectral images together with more simple approaches. Experiments have been performed on a large hyperspectral CASI image with 72 bands. Both coding and classification results indicate that the performance of 3d-DWT is superior to the other two lossy coding approaches, providing consistent improvements of more than 10 dB for the coding process, and maintaining both the global accuracy and the percentage of classified area for the classification process.

Keywords: JPEG2000 standard, 3-dimensional coding, hyperspectral images, classification.

1   Introduction

With an ever increasing amount of data being gathered from the real world, the trend is that just a piece of them will only be used by the final application. This is the case in Remote Sensing, where it is a common practise to present data directly to a computer for it to perform an automatic process [1].

Still, since these data are valuable and expensive to produce, they need to be stored and distributed efficiently. To deal with the vast amounts of data gathered, compression is used. Lossy compression, *i.e.*, where compression performance is traded for quality, may be used if the decoder is still able to recover most of the relevant information. This relevant information is usually a small part of the whole image and can be coded efficiently. However, how much information loss can be allowed is application dependent.

I. Lovrek, R.J. Howlett, and L.C. Jain (Eds.): KES 2008, Part III, LNAI 5179, pp. 416–423, 2008.
© Springer-Verlag Berlin Heidelberg 2008

In this paper we evaluate the influence of a lossy compression process on the computer-based classification of hyperspectral images. In the compression stage, basic JPEG2000 and 3-dimensional JPEG2000 [2] encodings have been considered.

This issue has been recently studied in the literature. In [3], the effects of the classic JPEG are examined on single component images. Later, effects of the new JPEG2000 on automated digital surface model extraction using aerial photographs are studied on [4]. And in [5], a JPEG and JPEG2000 comparison on single component is presented. In [6], classic JPEG, Vector Quantization, and a Wavelet-based compression technique are combined with a Karhunen-Loève Transform (KLT) to compare their effects on classification of 6-component imagery. In [7], the impact of using a Discrete Wavelet Transform (DWT) or a KLT with JPEG2000 is carefully examined, however they only present results for relatively small images.

Our work examines the performance penalty introduced due to the use of lossy compressed images (instead of the original images) on the classification process. Three distinct compression strategies are tested on a very large hyperspectral image obtained by a flight performed by the Cartographic Institute of Catalonia (ICC) using a Compact Airborne Spectrographic Imager (CASI) sensor. Due to the size of the data to compress, over 29 Gigabytes and 72 components, special care on constraining resources has had to be taken and some techniques could not be applied successfully (namely KLT). However, the three successfully tested techniques have been proved to be applicable in a production environment.

2   Material and Study Area

CASI and Laser Imaging Detection and Ranging (LIDAR) images over the Montseny area (a densely forested area on Catalonia) are used.

CASI is an optic sensor of hyperspectral scanning based on a small CCD bar that can be programmed to obtain a maximum of 288 bands among the 411 and 958 nm (red to near infrared). In this study a single image is used, recorded on 19-06-2007 (starting at 9.00 am) with 72 radiometric bands and an average bandwidth of 7.6 nm. The sensor flew on the airplane Cessna-Citation I from the ICC. The original image (without geometric correction) has 512 columns and 21133 rows. Lines captured by the sensor are affected by the movement of the airplane; therefore, it is necessary to geometrically correct the image line by line using a Global Positioning System (GPS) receiver and an Inertial Naviga-tion System (INS). This geometric correction was made using the SISA system (developed by the ICC) [8]. Flight orientation is SW-NE, so the geometrically corrected image is very large (18366 columns and 11918 rows) and has a great proportion of NODATA pixels (only 3.98% of the image contains data pixels, see Fig. 1). Radiometric resolution of CASI images is unsigned 16 bits/pixel.

The airborne LIDAR is a sensor that consists of a laser and a mirror that diverts a beam in various directions perpendicularly to the trajectory of the air-plane. These lateral displacements combined with the trajectory of the airplane allow carrying out a ground sweeping. The system measures the distance from the sensor to the ground based on the time that the ray of light takes to reflect on the ground and reach back to the sensor. The result of a LIDAR flight is a collection of points with known coordinates. For every emitted pulse it can detect up to two echoes and for each of them register the reflected intensity. Over forested areas it is possible to obtain a Digital Terrain Model (DTM) and a Digital Vegetation Height Model (DVHM, *i.e.*, elevation of the vegetation covering the study area, obtained from the Digital Surface Model and the DTM). A Digital Slope Model (DSM) can be derived from the DTM. The original spatial resolution of the LIDAR images was 2 m, so they were resampled to 3 m to match the CASI resolution. The DTM had a range of 400-1500 m, the DVHM a range of 0-50 m while the DSM had a range of 0-100%. LIDAR-derived images were rescaled to a range similar to that of the CASI images in order to avoid under-representation during the unsupervised stage of classification.

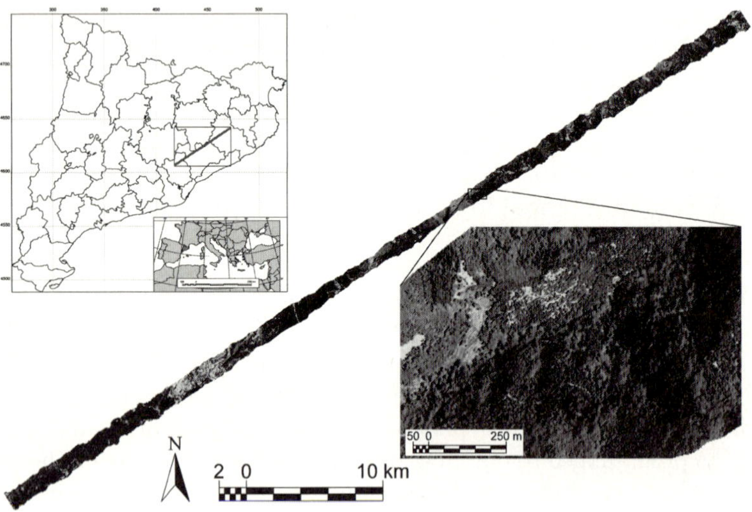

Fig. 1. RGB composition of bands 40 (707.3 nm), 66 (908.2 nm) and 12 (495.8 nm) of CASI image over Montseny (19-06-2007), area located in the western Mediterranean

3   Hyperspectral Coding

Among the many compression techniques now available to perform hyperspectral image compression [9,10,11], JPEG2000 is the most prominent one. JPEG2000 is the latest ISO image compression standard, that achieves very good compression results and provides a lot of other useful features, such as embeddedness and selective decoding. In addition, JPEG2000 has been devised to work in memory constrained environments.

Various techniques exist that perform a hyperspectral-aware compression, all generating a JPEG2000-compliant file. The following three have been tested in our experiments:

Band-Independent Fixed-Rate (BIFR)
In this method each hyperspectral band is independently compressed (in the sense that no inter-band redundancy is removed) and the bit-rate is equally split among all the hyperspectral bands. This approach reduces the encoder complexity and is suitable for memory constrained environments.

Volumetric Post-Compression Rate-Distortion (VolPCRD)
This technique may be thought of as the natural evolutionary step after BIFR. As in BIFR, bands still preserve inter-band redundancy, because no decorrelation transform has been applied. In this case though, rate is not equally shared, but "invested" where it helps to reduce further the distortion.

3 Dimensional Discrete Wavelet Transform (3d-DWT)
The DWT can be used as an inter-band decorrelation technique, allowing the removal of duplicate information across multiple bands (the 3rd dimension). This transform is applied on the input image, leading to an spectrally decorrelated image, and then a typical JPEG2000 compression follows in each band. Due to the nature of this method, an equally allocated rate must not be used, so the rate in the 3d-DWT must also be optimized.

All the presented techniques have in common that they allow selected zones of the images to be decompressed almost independently. This independence on the decoder is a very desirable characteristic, as a full decoding is a very memory and time consuming task. This allows human inspection of particular zones for human validation and improves the perceived decompression time.

Another well known technique exists, the KLT, where the optimal decorrelating linear transform is found for a Gaussian source. This search is extremely computationally expensive and not easily applicable on large images.

4   Classification

The classification is carried out with 72 bands of the CASI and with the DTM, the DVHM and the DSM obtained from the LIDAR data. Classification is performed on the CASI and the LIDAR images of forest land covers in order to obtain a land cover map with a six-category legend: *Fagus sylvatica, Castanea sativa, Quercus suber, Quercus ilex, Pinus halepensis*, and coniferous plantations. Combination between radiances (from CASI images) and topographic variables (from LIDAR images) improves the accuracy of the classification [12].

A mask obtained from the Land Cover Map of Catalonia (LCMC) was applied over the original and compressed images to only classify the forest areas. LCMC is a map obtained by means of photo-interpretation of 1:5000 color orthoimages captured in 2000, distinguishing up to 21 land covers.

A hybrid classification methodology is carried out using the MiraMon IsoMM and ClsMix modules [13]. The first part, IsoMM, is an IsoData implementation [14]. The second part of the hybrid classification is based on the ClsMix

MiraMon module. This module reclassifies each spectral class of an unsupervised classified image into thematic classes through spatial correspondence with training areas. The quality of this spectral to thematic class assignation is acquired with the use of two statistical thresholds called fidelity and representativeness. If the spatial correspondence between a spectral (obtained in the unsupervised stage) and a thematic (given by the supervised training areas) class does not fulfill both statistical thresholds, the spectral class will remain unclassified.

Training and test areas are obtained from field work (ground truth). From September to December 2007 the study area was visited several times and plots of forest inventory were generated (10 m radius parcels). This field work was complemented with digital photo-interpretation of 1:5000 color orthoimages of the ICC and ancillary data (Habitat Map of Catalonia, national forests inventories) to obtain areas of ground-truth. Thirty percent of those areas where reserved as independent test areas (to validate the final classification) and the remaining ones were used on the second part of the hybrid classification.

The obtained classification is validated using a confusion matrix between the classified image and the test areas reserved for this purpose.

5   Experimental Results and Discussion

5.1   Compression Results

The three reviewed coding approaches, BIFR, VolPCRD, and 3d-DWT, have been applied to encode the CASI image. Two distortion metrics have been computed, PSNR and SNR_Variance. Because of brevity, the coding performance is only reported for SNR_Variance, but PSNR results provide similar performance.

As expected, see Fig. 2a, 3d-DWT provides the best overall performance, followed by VolPCRD and then BIFR. When the distortion in individual bands is considered, see Fig. 2, all three approaches yield consistent results, improving the quality of the recovered image as the compression ratio (CR) increases (CR as compressed file size over uncompressed file size, NODATA pixels excluded in the uncompressed file size).

5.2   Classification Results

Global accuracy of the classification and percentage of classified area related to CR are shown in Fig. 3 and Table 1. As mentioned before, during the classification, the fidelity and representativeness parameters control the spectral to thematic class assignation. Spectral confusion can cause that some statistical classes remain unclassified, obtaining a classified area smaller than the total study area. Therefore, the classified area can be used as a quality index of the classification: the less the spectral confusion, the larger the classified area.

As could be expected, as CR decreases, the global accuracy also decreases. For high values of CR there are not important differences between compressed images and the images without compression. The decrease of global accuracy

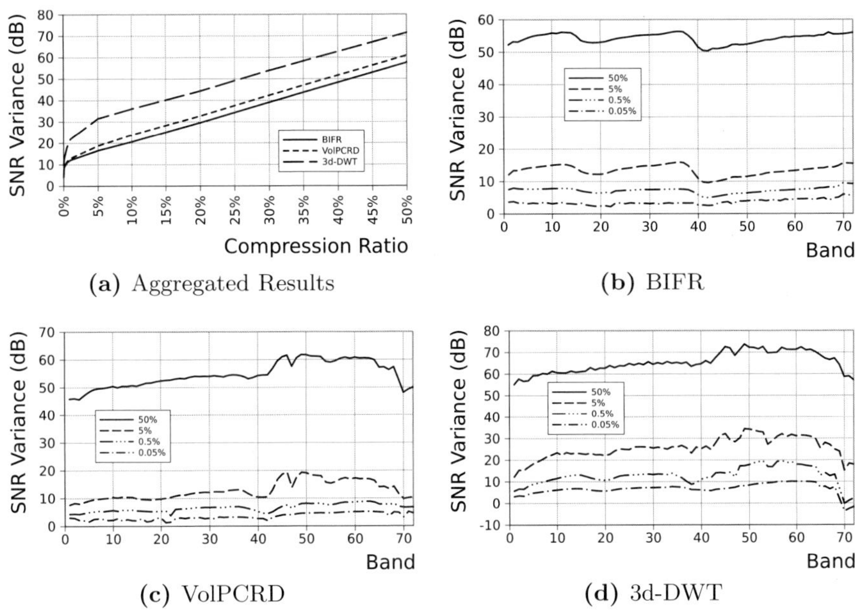

(a) Aggregated Results

(b) BIFR

(c) VolPCRD

(d) 3d-DWT

Fig. 2. Aggregated results of SNR_Variance over all components are shown in 2a. Per band results of the three studied methods are shown in 2b, 2c and 2d.

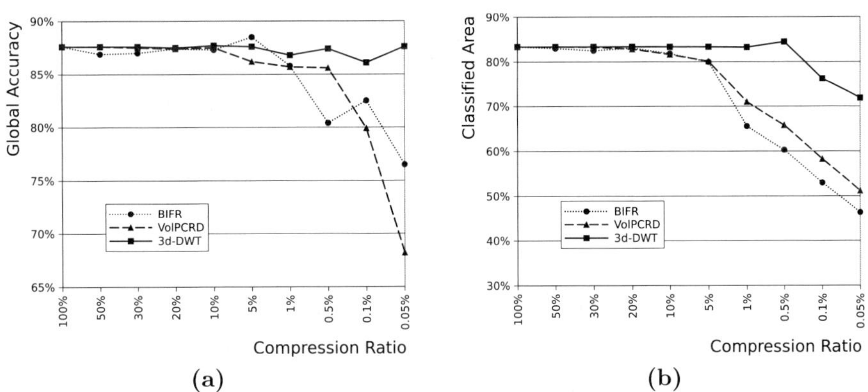

(a)

(b)

Fig. 3. Global accuracy (3a) and classified area (3b) of the obtained classifications using three compression techniques (BIFR, VolPCRD and 3d-DWT) related to CR

depends on the compression technique used. 3d-DWT brings the best results of global accuracy for the same CR values. According to classified area, the same tendencies are observed: there is no important influence when CR values are high, and 3d-DWT brings more classified area when CR values are low.

Table 1. Global accuracy of the obtained classifications and percentage of classified area using three compression techniques (BIFR, VolPCRD and 3d-DWT) related to CR

CR (%)	Global accuracy (%)			Classified area (%)		
	BIFR	VolPCRD	3d-DWT	BIFR	VolPCRD	3d-DWT
100.00	87.60	87.60	87.60	83.36	83.36	83.36
50.00	86.90	87.60	87.60	82.99	83.36	83.36
30.00	87.00	87.50	87.60	82.49	83.28	83.36
20.00	87.40	87.40	87.50	83.06	82.82	83.36
10.00	87.30	87.50	87.70	81.84	81.63	83.34
5.00	88.50	86.20	87.60	79.96	80.11	83.34
1.00	85.80	85.70	86.80	65.57	71.01	83.25
0.50	80.40	85.60	87.40	60.30	65.80	84.46
0.10	82.50	79.90	86.10	52.97	58.27	76.23
0.05	76.50	68.20	87.60	46.36	51.17	71.87

6   Conclusions

Modern sensors generate hyperspectral data at an unprecedented increasing pace. Coding techniques have to be devised to help disseminating and storing the sensored images; in order to improve the CR, lossy coding techniques are preferred over the lossless techniques. This approach poses the problem of assessing the influence of the compression process on an *a posteriori* classification process.

In this paper we investigate three JPEG2000-based lossy coding approaches: BIFR, VolPCRD, and 3d-DWT. 3d-DWT performs a DWT decorrelation in the spectral direction, and then a 2d-DWT in the spatial domain. After compression, a hybrid classification of the recovered images is performed. The hybrid classification combines an unsupervised module (IsoData), and a supervised module (ClsMix) that takes into account the spatial relationship among the thematic and spectral classes.

Experiments have been performed on a large hyperspectral CASI image with 72 bands. Both coding and classification results indicate that the performance of 3d-DWT is superior to the other two lossy coding approaches, providing consistent improvements of more than 10 dB for the coding process, and maintaining both the global accuracy and the classified area for the classification process.

Acknowledgments. This work was supported in part by the Ministry of Education and Science and the FEDER funds through the research projects Wavelet image compression for Remote Sensing and GIS applications (TIC2003-08604-C04) and Interactive coding and transmission of high-resolution images: applications in Remote Sensing, Geographic Information Systems and Telemedicine (TSI2006-14005-C02-01, TSI2006-14005-C02-02), and by the Catalan Government, under Grants 2008FI 473 and SGR2005-00319.

We would like to express our gratitude to the ICC for the availability of Remote Sensing data.

References

1. Jensen, J.: Introductory Digital Image Processing. A Remote Sensing Perspective. Pearson Prentice Hall, London (2005)
2. Taubman, D.S., Marcellin, M.W.: JPEG 2000: Image Compression Fundamentals, Standards, and Practice. Kluwer Academic Publishers, Dordrecht (2002)
3. Li, Z., Yuan, X., Lam, K.W.: Effects of JPEG compression on the accuracy of photogrammetric point determination. Photogrammetric Engineering and Remote Sensing 68(8), 847–853 (2002)
4. Shih, T.Y., Liu, J.K.: Effects of JPEG 2000 compression on automated dsm extraction: evidence from aerial photographs. The Photogrammetric Record 20, 351–365 (2005)
5. Zabala, A., Pons, X., Diaz-Delgado, R., Garcia, F., Auli-Llinas, F., Serra-Sagrista, J.: Effects of JPEG and JPEG2000 lossy compression on remote sensing image classification for mapping crops and forest areas. In: IGARSS 2006, pp. 790–793. IEEE, Los Alamitos (2006)
6. Tintrup, F., De Natale, F., Giusto, D.: Automatic land classification vs. data compression: a comparative evaluation. In: Proceedings of IGARSS 1998, vol. 4, pp. 1751–1753. IEEE, Los Alamitos (1998)
7. Penna, B., Tillo, T., Magli, E., Olmo, G.: Transform coding techniques for lossy hyperspectral data compression. IEEE Trans. Geoscience Remote Sensing 45(5), 1408–1421 (2007)
8. Palà, V., Alamús, R., Pérez, F., Arbiol, R., Talaya, J.: El sistema CASI-ICC: un sensor multiespectral aerotransportado con capacidades cartográficas. In: Revista de Teledetección, Asociación Española de Teledetección, vol. 12, pp. 89–92 (1999)
9. Tang, X., Pearlman, W.A.: Three-Dimensional Wavelet-Based Compression of hyperspectral Images. In: Hyperspectral Data Compression, pp. 273–308. Springer, USA (2006)
10. Yeh, P.S., Armbruster, P., Kiely, A., Masschelein, B., Moury, G., Schaefer, C., Thiebaut, C.: The New CCSDS Image Compression Recommendation. In: Aerospace Conference, vol. 5-12, pp. 4138–4145. IEEE, Los Alamitos (2005)
11. Ramakrishna, B., Plaza, A., Chang, C.I., Ren, H., Du, Q., Chang, C.C.: Spectral/Spatial Hyperspectral Image Compression. In: Hyperspectral Data Compression, pp. 309–346. Springer, Heidelberg (2006)
12. Serra, P., Pons, X., Saurí, D.: Post-classification change detection with data from different sensors. Some accuracy considerations. International Journal of Remote Sensing 24(16), 3311–3340 (2003)
13. Pons, X., Moré, G., Serra, P.: Improvements on Classification by Tolerating No-Data Values. Application to a Hybrid Classifier to Discriminate Mediterranean Vegetation with a Detailed Legend Using Multitemporal Series of Images. In: IEEE IGARSS and 27th CSRS, Denver, pp. 192–195 (2006)
14. Duda, R.D., Hart, P.E.: Pattern Classification and Scene Analysis. John Wiley & Sons, New York (1973)

A New Band Selection Strategy for Target Detection in Hyperspectral Images

Marco Diani, Nicola Acito, Mario Greco, and Giovanni Corsini

Dipartimento di Ingegneria dell'Informazione,
Via G. Caruso, 14, 56122 Pisa, Italy
{m.diani,mario.greco,n.acito,g.corsini}@iet.unipi.it

Abstract. This manuscript presents a novel methodology for band selection (BS) for hyperspectral sensors (HSs) tailored to target detection applications. The new selection strategy chooses a subset of bands that maximizes an *objective function* suitable for target detection. In particular, it extracts the subset of bands that optimizes the probability of detecting a target (P_D) in a given background, for a fixed probability of false alarm (P_{FA}). An experimental example of the methodology effectiveness is given. In the example the well-known Adaptive Matched Filter (AMF) detector and synthetic data derived from an AVIRIS hyperspectral image are considered. The results obtained show that the the new strategy outperforms two existing BS algorithms.

Keywords: target detection, band selection, hyperspectral images.

1 Introduction

The high spectral resolution provided by HSs enables the user to identify spectral features of a certain material and to discriminate different surfaces better than multispectral sensors. Thus, HSs are exploited in many remote sensing applications, such as automatic target detection and recognition (ATD&R), search-and-rescue and wide area surveillance [1].

The high dimensionality of hyperspectral data and the related increased computational complexity are not in compliance with hardware (HW) constraints for real time processing, which is typically required by ATD&R systems and search-and-rescue missions. Such a issue motivates the great interest in dimensionality reduction of hyperspectral data.

Dimensionality reduction is often performed by binning adjacent spectral channels (via HW) [2] or by using general purpose fast algorithms such as Principal Component Analysis (PCA) [3], [4]. Both these approaches have serious limitations when they are applied to target detection problems. In fact, binning reduces spectral resolution and often masks absorption lines that possibly characterize the spectrum of the material of interest. PCA extracts a subset of features that explains a given percentage of data variance estimated from the image pixels themselves. Target pixels in the image are generally rare in comparison with the background ones. Thus, PCA-based dimensionality reduction addresses mostly the background variability rather than the target one [5].

I. Lovrek, R.J. Howlett, and L.C. Jain (Eds.): KES 2008, Part III, LNAI 5179, pp. 424–431, 2008.
© Springer-Verlag Berlin Heidelberg 2008

The paper focuses on data dimensionality reduction for target detection. The term "target detection" refers to spectral signature-based target detection or spectral matching, where the target signature must be known a priori [5].

Data dimensionality reduction has been thoroughly investigated in the field of classification of hyperspectral images ([2, 3, 6-8]). Instead, only few works in the field of target detection have been presented in the literature. They are based on the selection of a set of bands/features that improves or preserves (according to a given metric) detectability of a target against a background. Some examples of metrics are the Spectral Angle Mapper (SAM) in [10], the Mahalanobis Distance (MD) in [11] and, more recently, the band correlation in [12].

In this paper we present a novel approach to data dimensionality reduction that is tailored to target detection applications. It allows us to choose the subset of sensor bands that maximizes the probability of detection (P_D) for a fixed probability of false alarm (P_{FA}), when a target with a known spectral signature must be detected in a given scenario. Detection performance relates to the statistical adaptive algorithm ([4, 5]) that will be exploited for the target detection task.

Unlike the other existing band/feature selection methods for target detection applications, the proposed strategy directly accounts for detection performance.

The manuscript is organized as follows. In Sections 2 the BS technique is mathematically formalized and its implementation is illustrated. In Section 3 a numerical example is proposed in which the results obtained by the new methodology are discussed and compared with the ones of other BS strategies.

2 Band Selection for Target Detection

Let X be a set of bands with cardinality l. BS consists in searching for the subset Ψ_d of X ($\Psi_d \subseteq X$), with dimensionality d ($d \leq l$), that maximizes a given *objective* or *criterion function* $J(\cdot)$ [10]:

$$\Psi_d = \arg\left\{ \max_{Y \subseteq X, |Y|=d} J(Y) \right\}. \tag{1}$$

In eq. (1) Y is a subset of X and $|\cdot|$ is the cardinality operator.

We assume that a target with a known spectrum must be detected in a given scenario and the detection task is accomplished by an adaptive statistical detection algorithm (reference detector - RD, hereinafter) whose performance in terms of P_D and P_{FA} is known. The BS problem is stated according to eq. (1) where the objective function is the theoretical P_D of the RD. It generally depends on d, P_{FA} and on a set of parameters Ω_Y, related to Y, the target spectrum and the background.

$$\Psi_d = \arg\left\{ \max_{Y \subseteq X, |Y|=d} P_D(d, P_{FA}, \Omega_Y) \right\}. \tag{2}$$

In order to define the detection strategy we start by considering the case of a known target embedded in a single background class. Furthermore, we assume that the RD is derived by resorting to the Local Gaussian Model (LGM), which is adopted to statistically characterize the background. Such a model is used in most the adaptive statistical detection algorithms proposed in the literature [5, 14]. LGM-based detectors exploit decision rules that involve the estimates of the background mean vector $\boldsymbol{\mu}$ and covariance matrix $\boldsymbol{\Gamma}$. The estimates are obtained by using a set of K secondary data, which are realizations of K independent random vectors with the same distribution as the background. Secondary data are actually taken from the image itself by using a moving window centered on the pixel under test (PUT) [14].

For a given subset Y of the sensor bands, the theoretical performance of an LGM-based detector depends on 1) K which determines the accuracy of the background parameters estimates, 2) the MD between the target and the background classes defined as

$$MD(\mathbf{s},\boldsymbol{\mu},\boldsymbol{\Gamma})_Y = \sqrt{(\mathbf{s}_Y - \boldsymbol{\mu}_Y)^T \boldsymbol{\Gamma}_Y^{-1}(\mathbf{s}_Y - \boldsymbol{\mu}_Y)}. \tag{3}$$

where $\mathbf{s}_Y$, $\boldsymbol{\mu}_Y$ and $\boldsymbol{\Gamma}_Y$ are the target spectrum, the background mean vector and the background covariance matrix evaluated on the subset of sensor bands Y, respectively.

Therefore, Ω_Y is uniquely determined by K and $MD(\mathbf{s},\boldsymbol{\mu},\boldsymbol{\Gamma})_Y$, and the BS problem can be rewritten as:

$$\Psi_d = \arg\left\{\max_{Y \subseteq X, |Y|=d} P_D\left(d, P_{FA}, K, MD(\mathbf{s},\boldsymbol{\mu},\boldsymbol{\Gamma})_Y\right)\right\}. \tag{4}$$

The BS strategy (4) will be referred to as P_D - *based BS*, hereinafter. The strategy can be generalized to the more realistic case of a target embedded in a complex scenario that contains different background classes (*multi-class background*) by modifying the objective function (P_D) according to the total probability theorem:

$$P_D\left(d, P_{FA}, \{\Omega_Y^{(n)}\}_{n=1}^N, N\right) = \sum_{n=1}^N P_D\left(d, P_{FA}, \Omega_Y^{(n)} \mid n\right) P(n). \tag{5}$$

In eq. (5) N denotes the number of background classes, $P_D(\cdot|n)$ the probability of detection conditioned on the n-th class, $P(n)$ the *a priori* probability of the n-th class and $\Omega_Y^{(n)}$ the set of parameters of the n-th class. By adopting the objective function in eq. (5) the selection strategy for LGM-based detectors is rewritten as:

$$\Psi_d(N) = \arg\left\{\max_{Y \subseteq X, |Y|=d} \sum_{n=1}^N P_D\left(d, P_{FA}, K, MD(\mathbf{s},\boldsymbol{\mu}_n,\boldsymbol{\Gamma}_n)_Y \mid n\right) P(n)\right\} \tag{6}$$

where $\boldsymbol{\mu}_n$ and $\boldsymbol{\Gamma}_n$ are the mean spectrum and the covariance matrix of the n-th background class. Often the *a priori* probability $P(n)$ is unknown or no useful information is available to reliably estimate it. In such cases, a possible solution is to assume an equal prior probability $P(n) = 1/N$, as suggested in [3].

Note that the P_D - *based BS* can be used to solve two distinct problems:

- to find the best subset Ψ_d of sensor bands for a fixed cardinality d. This problem arises in those applications where target detection must be performed on board and in real time. In fact, in this case the number of bands that can be processed is limited by the HW constraints;

- to find the number d_H and the correspondent subset Ψ_H of sensor bands to achieve the best detection performance (the highest P_D for a given P_{FA}). In formulas, with reference to the multi-class background case, the problem becomes:

$$\Psi_H(N) = \arg\left\{\max_{\{\Psi_d\}_{d=1}^N} \sum_{n=1}^N P_D\left(d, P_{FA}, K, MD(\mathbf{s}, \boldsymbol{\mu}_n, \boldsymbol{\Gamma}_n)_{\Psi_d} \mid n\right) P(n)\right\}$$

$$d_H = |\Psi_H|$$

(7)

This latter problem concerns off-line applications, where target detection is accomplished on the hyperspectral image after its acquisition. In this case, the existence of an optimum value of d (and the correspondent subset of sensor bands) is justified by the "curse of dimensionality" or "Hughes phenomenon" [6]. It arises in hyperspectral image processing when a statistical detection algorithm, which requires the estimates of the unknown background parameters, is adopted. In fact, in such a case the estimates are obtained by a limited number K of secondary data, which determines the accuracy of the estimation method. For a given value of K, if we increase data dimensionality, the estimation accuracy will degrade and the detection performance will not necessarily raise.

3 Application of the Band Selection Strategy: A Numerical Example

In this section we discuss an example where the P_D - *based BS* is applied by considering as RD the well known AMF algorithm. The numerical results have been obtained with reference to the multi-class background case.

The AMF algorithm is an LGM-based adaptive detector and it is widely used in hyperspectral image processing ([4, 5]). It was originally derived in [13] by considering a typical radar detection problem. Recently, in [5], it has been adapted to target detection in hyperspectral images. It is based on the following decision rule:

$$\Lambda_{AMF}(\mathbf{x}) = \frac{\left[(\mathbf{s} - \hat{\boldsymbol{\mu}})^T \hat{\boldsymbol{\Gamma}}^{-1} (\mathbf{x} - \hat{\boldsymbol{\mu}})\right]^2}{(\mathbf{s} - \hat{\boldsymbol{\mu}})^T \hat{\boldsymbol{\Gamma}}^{-1} (\mathbf{s} - \hat{\boldsymbol{\mu}})} \underset{H_0}{\overset{H_1}{\gtrless}} \lambda$$

(8)

where $\mathbf{x} = [x_1,...,x_l]^T$ is the PUT; $\hat{\boldsymbol{\mu}}$ and $\hat{\boldsymbol{\Gamma}}$ are the estimates of the mean vector and the covariance matrix of the background, which are adaptively obtained from K secondary data; λ is the test threshold; H_1 and H_0 represent the "target present" and the "target absent" hypotheses, respectively. In practice, secondary data are obtained from a neighborhood of the pixel under test and selected by a suitable spatial window.

Exploiting the results on the AMF theoretical performance in [13], the P_D can be expressed as a function of $MD(n)_Y = MD(\mathbf{s}, \boldsymbol{\mu}_n, \boldsymbol{\Gamma}_n)_Y$, K, d and λ. The test threshold λ can be chosen so as to obtain a desired value of P_{FA}.

Note that to apply the P_D - *based BS* the parameters P_{FA}, K, $\{\boldsymbol{\mu}_n, \boldsymbol{\Gamma}_n\}_{n=1}^N$ and $\mathbf{s}$ must be set. Furthermore, when the procedure is adopted to select the best bands for a fixed cardinality (exploiting eq. (7)), the parameter d must be also defined.

As concerns the target spectrum it is usually available in terms of reflectance and must be properly transformed into radiance units keeping into account for the specific acquisition geometry and atmospheric conditions. In target detection applications, this is usually accomplished by empirical methods, physical models or by in-scene compensation techniques (see [15] and references therein). The value of K is chosen as a compromise: small to span a homogenous region and large enough to accurately estimate the local background of the image that will be processed. P_{FA} is chosen on the basis of the reliability required to the detection process; typical values fall in the interval $\left[10^{-6},10^{-4}\right]$. Finally, the background class parameters $\left\{\mu_n,\Gamma_n\right\}_{n=1}^{N}$ can be obtained following different procedures that depend on the particular application of interest. For instance, if we are interested in real-time target detection the background parameters could be set during the mission planning in an off-line manner. In particular, they could be retrieved by resorting to hyperspectral images acquired during a previous survey of the site of interest. If we are interested to apply the detection process to the image after its acquisition, the background parameters could be obtained from the hyperspectral data themselves. In this case, first a classification method should be adopted to determine the background classes in the scene, then, a specific estimation technique should be used to derive the parameters of each class provided by the previous procedure.

It is worth noting that an *exhaustive search* of the subset $\Psi_d(N)$ enables the optimal solution to be found, but it is not feasible when the number of sensor bands (l) is in the order of magnitude of 10^2 (typical for HSs). Hence, several sub-optimal search algorithms have been proposed in the literature. Such algorithms represent a trade-off between computational load and accuracy of the solution Ψ_d or $\Psi_d(N)$. Examples of search strategies proposed in the literature are the sequential forward selection (SFS), the sequential forward floating selection (SFFS), the branch-and-bound (BB) method and the steepest ascent (SA) algorithm. An interested reader can find a detailed review in [7], [8] and [9]. In this paper the SFS has been adopted because it is fast from a computational viewpoint and it has been widely adopted in the literature [7]. The SFS can be summarized as follows:

1. starting form the l-dimensional set of bands X, the SFS selects the band that maximizes $J(\cdot)$, i.e. ψ_1;

2. starting from the remaining (l-1) bands, the SFS appends to ψ_1 the band that maximizes $J(\cdot)$, i.e. $\left\{\psi_1,\psi_2\right\}$;

3. the SFS adds one band at a time until the subset with the desired cardinality is found: $\Psi_d = \left\{\psi_1,\psi_2,...,\psi_d\right\}$.

To give a numerical example we consider a case study where the AVIRIS sensor is assumed as acquisition device. In our example we refer to a scenario containing $N=3$ background classes and we set the *a priori* class probability $P(n)$ equal to $1/3$ (with $n=1,2,3$). In order to use realistic values of the background parameters we used the AVIRIS data set called *Indian Pine* [6]. Such data set has been selected because it includes a detailed ground truth that has been exploited to estimate the background parameters $\left\{\mu_n,\Gamma_n\right\}_{n=1}^{3}$, and to simulate the target spectrum $\mathbf{s}$. The Indian Pine data

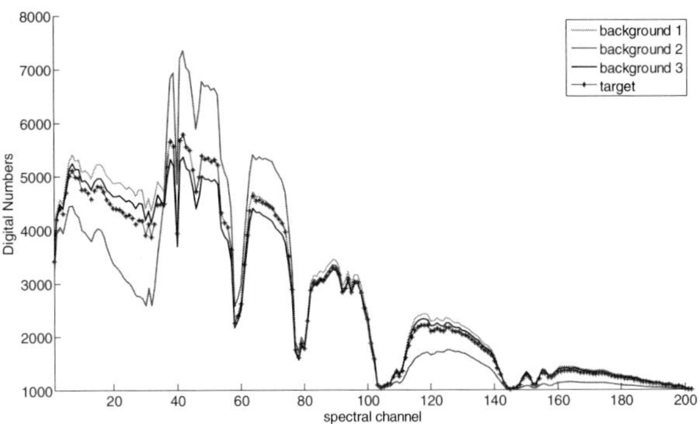

Fig. 1. Target and mean background spectra

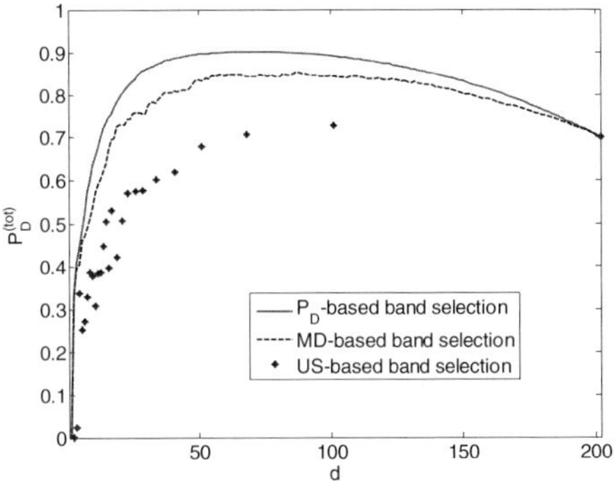

Fig. 2. AMF theoretical performances: P_D vs. number of selected features/bands d

have l=202 spectral channels (22 noisy channels out of the 224 available from in the original data have been discarded). In our experiment, the parameters $\{\mu_n, \Gamma_n\}_{n=1}^3$ were set equal to the mean value and the covariance matrix of the following Indian Pine classes (from the ground truth): *Corn-notill*, *Grass/Pasture*, *Soybeans-notill*, respectively. The target spectrum $\mathbf{s}$ is simulated as a linear combination of the mean spectra of the above cited classes ($\mathbf{s} = 0.1\mu_1 + 0.2\mu_2 + 0.7\mu_3$). The coefficients of the linear model were chosen so as to simulate a challenging situation in which the target is spectrally similar to one of the background classes (Class 3, *Soybeans-notill*). In Fig. 1 $\{\mu_n\}_{n=1}^3$ and $\mathbf{s}$ are shown.

In the proposed example we compare the P_D - *based BS* procedure with two of the existing BS methodologies: the MD-based strategy proposed in [12] and the one obtained by uniform sampling of the spectral channels (US-based BS). The comparison is carried out by evaluating the $P_D^{(tot)}(d)$ for each technique and the results are shown in Fig. 2. In particular, we plot $P_D^{(tot)}$ as a function of number of sensor bands d selected by each *BS* strategy. The novel method performs clearly better than the MD-based BS and the US-based BS. The results show that $P_D^{(tot)}(d)$ is not a monotonically increasing function of d. This is an experimental evidence of the Hughes phenomenon which arises because of the poor accuracy of the background parameter estimates, which are computed with a small number of secondary data (compared to d). A direct consequence of the Hughes phenomenon is the existence of an optimum number of bands for the HS and the correspondent subset of bands. For the PD-based BS the optimum value of d is d_H=69, which yields $P_D^{(tot)}(d_H)$=0.9 and a dimensionality reduction in the order of 66%. The MD-based strategy reaches the best value of $P_D^{(tot)}$ =0.85 with d =87 bands, while the US-based method gets $P_D^{(tot)}$ =0.73 with d =101 bands. In both cases, the best detection performance is lower than the one obtained by the P_D - *based BS* ($P_D^{(tot)}$ =0.9), while the resulting data dimensionality is higher. Thus, in this experiment, the P_D - *based BS* strategy selects the smallest band subset that achieve the best detection performance.

For low values of d, for example d =21, the proposed methodology continues to provide $P_D^{(tot)}(d)$ in the order of 0.8, while it decreases to 0.73 and 0.5 with the MD-based and the US-based technique, respectively.

4 Conclusion

In this paper, we have presented a novel BS methodology for target detection in hyperspectral images. It is based on an objective function that takes into account the theoretical probability of detecting a known target in a complex scene, for a fixed P_{FA}. The methodology can be applied with reference to all the statistical algorithms whose theoretical detection performance is known. The BS strategy has been formalized and illustrated by considering the P_D of the AMF detector. The technique for BS and its impact on the AMF detection performance have been investigated with reference to a target embedded in a complex scene. Target and background parameters have been obtained from an AVIRIS data set. The effectiveness of the method has been shown for both real time and off-line target detection applications. In the former case, where d is fixed by the HW limitations, the P_D - *based BS* chooses the subset of d bands that provides detection performance higher than the one given by the MD-based strategy and the uniform sampling. In the latter, our experiments have shown the existence of a subset of bands yielding the best detection performance. Also in this case the P_D - *based BS* outperformed the others BS techniques.

It is worth noting that the methodology does not claim to guess the true performance of a detector; thus, real performance may differ from the theoretical one. Here, the theoretical P_D was used as an objective function for BS. Opposite to existing objective functions such as the MD, the band correlation and the SAM, the one proposed in this work allows the user to account for the joint relationship between the number of secondary data (K) and the estimation accuracy of the background parameters.

References

1. Stevenson, B.: The Civil Air Patrol ARCHER Hyperspectral Sensor System. In: Proc. Int. SPIE Conf. Airborne ISR Systems, Bellingham, WA, May 2005, pp. 17–28 (2005)
2. Miaohong, S., Healey, G.: Hyperspectral texture recognition using a multiscale opponent representation. IEEE Trans. Geosci. and Remote Sensing 41(5), 1090–1095 (2003)
3. Richards, J.A., Jia, X.: Remote sensing digital image analysis: an introduction, 3rd edn. Springer, Heidelberg (1999)
4. Farrell, M.D., Mersereau, R.M.: On the impact of PCA dimension reduction for hyperspectral detection of difficult targets. IEEE Trans. Geosci. and Remote Sensing Letters 2(2), 192–195 (2005)
5. Manolakis, D.: Taxonomy of detection algorithms for hyperspectral imaging applications. Opt. Eng. 44(6), 1–11 (2005)
6. Landgrebe, D.: Signal Theory Methods in Multispectral Remote Sensing. John Wiley & Sons Press, Chichester (2003)
7. Fukunaga, K.: Introduction to Statistical Pattern Recognition, 2nd edn. Academic Press, London (1990)
8. Serpico, S.B., Bruzzone, L.: A New Search Algorithm for Feature Selection in Hyperspectral remote Sensing Images. IEEE Trans. Geosci. and Remote Sensing 39(7), 1360–1367 (2001)
9. Jain, A., Zongker, D.: Feature selection: evaluation, application, and small sample performance. IEEE Trans. Pattern Analysis and Machine Intelligence 19(2), 153–158 (1997)
10. Keshava, N.: Distance metrics and Band Selection in hyperspectral processing With Application to Material Identification and Spectral Libraries. IEEE Trans. Geosci. and Remote Sensing 42(7), 1552–1565 (2004)
11. Kanodia, A., Hardie, R.C., Johnson, R.O.: Band Selection and Performance Analysis for Multispectral target Detectors Using Truthed Bomem Spectrometer Data. In: Proc. Int. IEEE Conf. National Aerospace and Electronics, Dayton, OH, May 1996, pp. 33–40 (1996)
12. Chang, C.-I., Wang, S.: Constrained band selection for hyperspectral imagery. IEEE Trans. Geosci. and Remote Sensing 44(6), 1575–1585 (2006)
13. Robey, F.C., Fuhrmann, D.R., Kelly, E.J., Nitzberg, R.: A CFAR adaptive matched filter detector. IEEE Trans. on Aerospace and Electronic Systems 28(1), 208–216 (1992)
14. Acito, N., Corsini, G., Diani, M.: Adaptive Detection Algorithm for Full Pixel Targets in Hyperspectral Images. IEEE Proc. Vision, Image and Signal Processing 152(6), 731–740 (2005)
15. Bernstein, L.S., Adler-Golden, S.M., Sundberg, R.L., Levine, R.Y., Perkins, T.C., Berk, A., Ratkovski, A.J., Felde, G., Hoke, M.L.: Validatio of the QUick Atmospheric Correction (QUAC) algorithm for VNIR-SWIR multi- and hyperspectral imagery. In: Proceedings SPIE, vol. 5806, pp. 668–679 (2005)

Fatal Mutations in HIV Population as an Influential Factor for an Onset of AIDS

Kouji Harada and Yoshiteru Ishida

Toyohasi University of Technology,
1-1, Tenpaku, Toyohashi-shi, Aichi 441-8585, Japan
{harada,ishida}@tutkie.tut.ac.jp

Abstract. It is well known that mutations occur at high rate at the replication of HIV genome. Mutations on the gene encoding requisite proteins for infection to host cells deprive ability to infect to them, thus such mutations are fatal for HIV population. We have extended Nowak and May's HIV dynamical model into more realistic one by considering the fatal mutations of HIV and an evolution of a mutation rate. Through analyses of the model, we have derived a critical condition controlling an onset of AIDS. The most significant consequence from the derived condition is to have clarified the antigenic diversity threshold theory proposed by Nowak and May succeeds in a characterless and homogeneous HIV population; the theory does not in an inhomogeneous HIV population.

Keywords: HIV, Fatal Muation, Anitignic Diversity Threshold.

1   Introduction

HIV (Human Immunodeficiency Virus) is the causal agent of AIDS (Acquired Immune Deficiency Syndrome). Its particle possesses a RNA genome and a reverse transriptase for translation of its RNA into DNA. HIV infects its host cell (which is called CD4 positive T cell) by GP-120 proteins on its membrane binding to CD4-cell receptor. Once inside the cell, the reverse transcriptase starts to copy the viral RNA into DNA then the double stranded DNA is inserted into the genome of the host cell. This state is called "pro-virus". Subsequently, the genetic machinery of the host cell is manipulated to produce mRNA copies of the viral genome, some enzymes (such as reverse transcriptase, protease and integrase) and structural proteins (such as GP-120, GP-41 etc) necessary for the synthesis of new HIV particles. The new HIV particles are assembled on the inside of the host cell and finally they bud from the host cell; the host cell usually dies.

A distinct feature of HIV is a high transcription error rate of the reverse transcriptase. The error rate is about 10^{-4} per base [1]. GP-120 protein is a requisite molecule for the infection in the host cell as well as a target molecule (i.e.antigen) for the immune system. As GP-120 gene is encoded on a reading frame called "env", at a reverse transcription from the viral RNA to DNA, it is expected some errors would occur on the gene. This mutation induces GP-120's structural changes which enable HIV to escape from immune reactions, at the

I. Lovrek, R.J. Howlett, and L.C. Jain (Eds.): KES 2008, Part III, LNAI 5179, pp. 432–439, 2008.

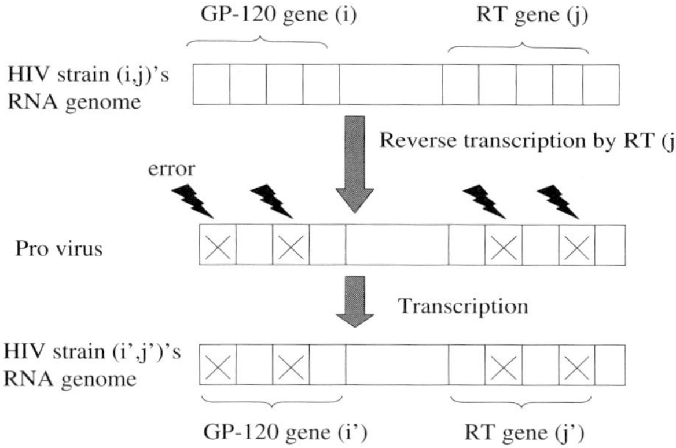

Fig. 1. A mutation mechanism due to an inaccurate transcription of a reverse transcriptase (RT)

same time, which make hard for HIV to infect its host cell. That is, mutations on GP-120 gene put HIV in so-called "double-edged sword" situation.

The reverse transcriptase gene is encoded on the reading frame called "pol" in RNA genome of HIV. At a reverse transcription process, some mutations on a particular base sequence encoding active sites of the reverse transcriptase result in producing reverse transcriptases different from the original one on the transcription error rate (Fig.1). If we consider a variation of GP-120 protein (HIV antigen) and transcription error rate (mutation rate) of reverse transcriptase, a HIV strain (i, j) with an antigen i and a mutation rate j evolves into HIV strain (i', j') through a reverse transcription by a reverse transcriptase with the mutation rate j (RT_j) in an infected cell:

$$HIV_{i,j} + RT_j \rightarrow HIV_{i',j'} + RT_{j'}. \tag{1}$$

Nowak and May proposed a basic model describing a replicating HIV which is opposed by strain-specific immune responses. They showed a critical transition: onset of AIDS defined as divergence of the total HIV population occurs when an antigenic diversity exceeds a threshold determined by some parameters of the model. They called it "antigenic diversity theory" [2,3,4]. We have extended their basic model by considering antigenic variation on GP-120, a fatal mutation (a lack of viral binding ability with CD40) and an evolution dynamics of a mutation rate. Detailed analyses of the extended model have clarified a specific condition determining an onset of AIDS. The derived conditional formula clarified an antigen diversity can determine an onset of AIDS only when a probability of the fatal mutation not occurring $(\bar{Q})$ is independent of a type of HIV strain. Nowak and May's result corresponds to a case of when the probability $\bar{Q}$ is equal to one. Furthermore, we have analyzed more general case of when the probability $\bar{Q}$ is dependent of the type of HIV strain. The analysis has clarified a critical position of a hyperplane designated by not only the antigen diversity but also some parameters of the model determines an onset of AIDS, which is not controlled by sole antigenic diversity as Nowak and May suggested.

2   Model

Our proposed model is the following system of ordinary differential equations:

$$\dot{v}_{i,j} = -px_i v_{i,j} + \sum_{i'=1}^{M} \sum_{j'=1}^{L} Q_{(i,j)(i',j')} bv_{i',j'} \qquad i = 1 \ldots M, j = 1 \ldots L, \qquad (2)$$

$$\dot{x}_i = kv_i - uvx_i \qquad i = 1 \ldots M, \qquad (3)$$

$$v_i = \sum_{j=1}^{L} v_{i,j}, \qquad (4)$$

$$v = \sum_{i=1}^{M} \sum_{j=1}^{L} v_{i,j}. \qquad (5)$$

Let $v_{i,j}$ denote the population size of HIV strain (or mutant) with the antigen i and the mutation rate j and let x_i denote the magnitude of the specific immune response against HIV strains with the antigen i. The differential equation (2) describes a population dynamics of HIV strain (i, j). The first term means an immune response against the antigen i eliminates HIV strain (i, j) at the rate $px_i v_{i,j}$. The second term means HIV strain (i', j') mutates into HIV strain (i, j) with the mutation rate j', and let $Q(i, j)(i', j')$ be the probability of strain (i', j') mutating to strain (i, j) with the mutation rate j':

$$v_{i',j'} + RT_{j'} \xrightarrow{Q(i,j)(i',j')} v_{i,j} + RT_j. \qquad (6)$$

Also an error free replication of HIV strain (i, j) is given by $Q_{(i,j)(i,j)} bv_{i,j}$.

In this model, there are N HIV strains concerning to the type of antigen (i) and M strains $(1 \leq i \leq M)$ out of them are survival ones (holding an ability to infect a host cell); the left strains $(M + 1 \leq i \leq N)$ are fatal ones (Fig.2). On the other hand, there are L levels on the mutation rate j of HIV and the index j takes an integer value of 1 to L.

Also the probability $Q_{(i,j)(i',j')}$ satisfies the following condition:

$$\sum_{i=1}^{N} \sum_{j=1}^{L} Q_{(i,j)(i',j')} = 1 \qquad (7)$$

The differential equation (3) describes the time evolution of the magnitude of immune responses specific for HIV strain with the antigen i. The first term means an immune response is stimulated at the rate kv_i, which is proportional

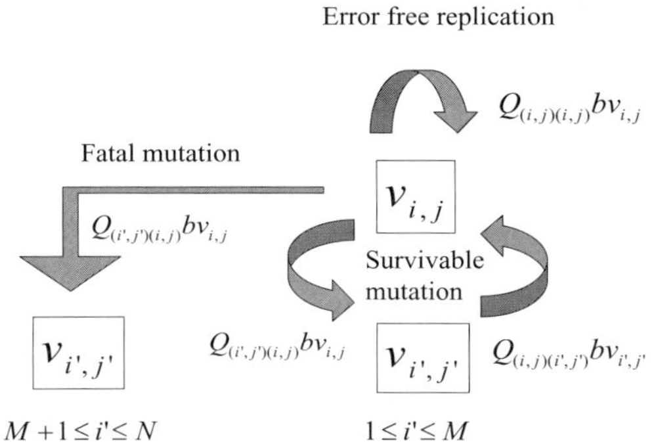

Fig. 2. A schematic representation of fatal mutations and survivable ones

to the abundance of HIV strain with the antigen i. The second term means the immune response weakens through the decrease of CD4 positive T cells infected by any HIV strains.

3   Specific Conditions for an Onset of AIDS

We derive conditional equations controlling an onset of AIDS defined as divergence of the total HIV population. It assumes the individual x_i converges to steady-state levels. Let $\dot{x}_i$ be zero then we obtain $x_i = \frac{kv_i}{uv}$. Substituting x_i into the formula (2), we obtain this equation:

$$\dot{v}_{i,j} = -\frac{pk}{uv}v_i v_{i,j} + \sum_{i'=1}^{M}\sum_{j'=1}^{L} Q_{(i,j)(i',j')}bv_{i',j'}. \tag{8}$$

Figuring out a sum of the obtained formula (8) on the suffixes, i, j, we obtain an equation for the rate at which the total HIV population v changes:

$$\dot{v} = -\frac{pkv}{u}\sum_{i=1}^{M}(\frac{v_i}{v})^2 + \sum_{i'=1}^{M}\sum_{j'=1}^{L} bv_{i',j'}\sum_{i=1}^{M}\sum_{j=1}^{L} Q_{(i,j)(i',j')}. \tag{9}$$

Let the new index D represent the formula $\sum_{i=1}^{M}(\frac{v_i}{v})^2$. D is called "Simpson index" which is an inverse measure for "antigenic diversity": D takes a value between $1/M$ and 1. Furthermore let the new probability $\overline{Q}_{i',j'}$ represent the formula $\sum_{i=1}^{M}\sum_{j=1}^{L} Q_{(i,j)(i',j')}$. $\overline{Q}_{i,j}$ means a probability of HIV strain (i,j) not occurring fatal mutations. If M equals to N (i.e. no fatal mutants), $\overline{Q}_{i,j}$ is one. Updating the equation (9) using these two indexes, we obtain the following:

$$\dot{v} = -\frac{pkv}{u}D + \sum_{i'=1}^{M}\sum_{j'=1}^{L} bv_{i',j'}\overline{Q}_{i',j'}. \tag{10}$$

Here when $\dot{v}$ is greater than zero, the total HIV population v diverges to the positive infinity. "$\dot{v} > 0$" leads to the following inequality expression:

$$\sum_{i'=1}^{M}\sum_{j'=1}^{L}\overline{Q}_{i',j'}\frac{v_{i',j'}}{v} > \frac{pk}{bu}D. \tag{11}$$

Here let $\bar{v}_{i,j}$ represent $\frac{v_{i,j}}{v}$ in the inequality (11), then we obtain

$$\sum_{i=1}^{M}\sum_{j=1}^{L}\overline{Q}_{i,j}\bar{v}_{i,j} > \frac{pk}{bu}D. \tag{12}$$

It notes that $\bar{v}_{i,j}$ must satisfy the condition: $0 \le \bar{v}_{i,j} \le 1$. By further deformation of the inequality expression (12), we obtain

$$\sum_{i=1}^{M}\sum_{j=1}^{L}\frac{\bar{v}_{i,j}}{\frac{pkD}{Q_{i,j}bu}} > 1. \tag{13}$$

The inequality (13) means upside of a hyperplane in LM-dimensional space. The hyperplane denoted by Σ is given by:

$$\Sigma: \quad \sum_{i=1}^{M}\sum_{j=1}^{L}\frac{\bar{v}_{i,j}}{\frac{pkD}{Q_{i,j}bu}} = 1. \tag{14}$$

The feature of the hyperplane Σ is that its normal vector $\boldsymbol{n}$ is determined by the probability $\bar{Q}_{i,j}$ as follows:

$$\boldsymbol{n} = (\bar{Q}_{1,1}, \bar{Q}_{1,2}, \ldots, \bar{Q}_{M,L}). \tag{15}$$

Also, $\bar{v}_{i,j}$-intercept of the hyperplane Σ is given by:

$$\frac{pkD}{\bar{Q}_{i,j}bu}. \tag{16}$$

Meantime, from the equation (5), another hyperplane we call Σ_0

$$\Sigma_0: \quad \sum_{i=1}^{M}\sum_{j=1}^{L}\bar{v}_{i,j} = 1, \tag{17}$$

is obtained. This hyperplane is also in LM-dimensional space and its normal vector $\boldsymbol{n}_0$ is

$$\boldsymbol{n}_0 = (1, 1, \ldots, 1). \tag{18}$$

It notes that a value of $\bar{v}_{i,j}$ is limited in the hyperplane Σ_0.

3.1 A Case of Antigenic Diversity Being Sole Factor for an Onset of AIDS

Here we consider a case of the normal vector $\boldsymbol{n}$ of the hyperplane Σ being proportional to the normal vector $\boldsymbol{n}_0$ of the hyperplane Σ_0. This means $\bar{Q}_{i,j}$ is independent of the type of HIV strain, in other words, $\bar{Q}_{i,j}$ is equal to a constant value: $\bar{Q}$. Then, from the inequality (12), we can obtain

$$\bar{Q} > \frac{pk}{bu}D. \tag{19}$$

That is,

$$D < \frac{pk\bar{Q}}{bu}. \tag{20}$$

This inequality mentions the Simpson index D, in other words, antigenic diversity is sole factor to determine divergence of the total HIV population.

3.2 Some Cases of Antigenic Diversity Being Not Sole Factor for an Onset of AIDS

Here we consider some cases of the normal vector $\boldsymbol{n}$ of Σ is not proportional to the normal vector $\boldsymbol{n}_0$ of Σ_0. In this case, the divergence of the total HIV population v is determined by not only the Simpson index, D but also a positional relationship between the two hyperplanes: Σ_0 and Σ.

1. In the case of all $\bar{v}_{i,j}$-intercepts of the hyperplane Σ is larger than one even when the Simpson index D takes the minimum: $\frac{1}{M}$ (Fig.3 (1)).
   Since the hyperplane Σ_0 is located on the downside of the hyperplane Σ, $\dot{v}$ is less than zero from the inequality (13). Therefore AIDS does not develop.
2. In the case of all $\bar{v}_{i,j}$-intercepts of the hyperplane Σ is less than one even when the Simpson index D takes the maximum: 1 (Fig.3 (2)).
   Since the hyperplane Σ_0 is located on the upside of the hyperplane Σ, $\dot{v}$ is larger than zero from the inequality (13). Therefore AIDS develops.
3. In the case of some of $\bar{v}_{i,j}$-intercepts of the hyperplane Σ are less than one, the others are larger than one (Fig.3 (3)).

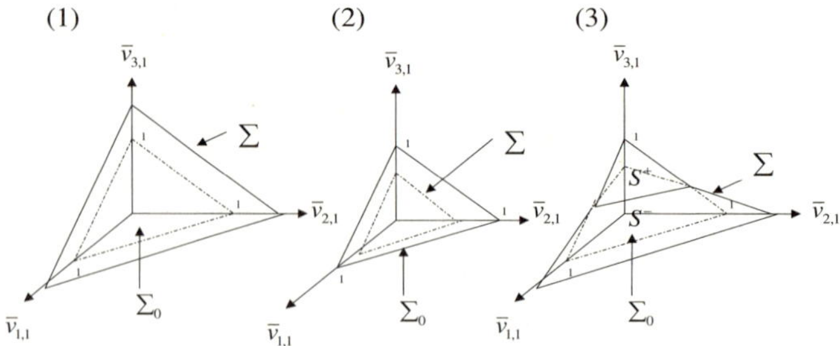

Fig. 3. Positional relations between two planes, Σ_0 and Σ ($M = 3, L = 1$): (1) The case of all $\bar{v}_{i,j}$-intercepts of the hyperplane Σ ($D = \frac{1}{M}$) is larger than one. (2) The case of all $\bar{v}_{i,j}$-intercepts of the hyperplane Σ ($D = 1$) is less than one. (3) The case of $\bar{v}_{3,1}$-intercept of the hyperplane Σ are less than one, the others are larger than one.

An intersection of the two hyperplanes: Σ_0 and Σ is ML-2 dimensional hyperplane, which separates the hyperplane Σ_0 into two subspaces: S^+ and S^- corresponding to $\dot{v}$ is positive and negative, respectively. Then there is a potential of an orbit of the variable v iterates between the two subspaces S^+ and S^-, thus in this case we can expect a time evolution of v would not be monotonous behavior such as divergence and distinction.

4   Conclusions

This study has extended Nowak and May's basic model into more realistic one by considering the fatal mutations of HIV and the evolution of the mutation rate due to an inaccurate transcription of the reverse transcriptase. The model analysis has clarified the antigenic diversity is sole factor determining the onset of AIDS when the probability $\bar{Q}_{i,j}$ of HIV strain (i,j) not occurring fatal mutations is independent of the type of HIV strain. It is significant point that our result includes Nowak and May's results and it is more general than their result. Meanwhile, when the probability $\bar{Q}_{i,j}$ depends on the type of HIV, we have clarified an onset of AIDS is determined by not only the Simpson index D but also a positional relationship between the two hyperplane: Σ_0 and Σ. In this case, we have indicated the time evolution of the total population of HIV is not monotonous behavior such as divergence and extinction but more complex behaviors such as oscillating ones. We can expect that this case might represent the incubation period observed in some HIV infected patients

Acknowledgment

This study has been supported by the Grant-in-Aid for Young Scientists (B) No.18700293 of the Ministry of Education, Culture, Sports, Science and Technology (MEXT) from 2006 to 2008. We are grateful for their support.

References

1. Preston, B.D., Poiesz, B.J., Loeb, L.A.: Fidelity of HIV-1 reverse transcriptase. Science 242, 1168–1171 (1988)
2. Nowak, M.A., May, R.M.: Virus dynamics. Oxford University Press, Oxford (2000)
3. Nowak, M.A., May, R.M.: Mathematical biology of HIV infections: antigenic variation and diversity threshold. Mathematical BioSciences 106, 1–21 (1991)
4. Nowak, M.A., Anderson, R.M., McLean, A.R., Wolfs, T.F., Goudsmit, J., May, R.M.: Antigenic diversity threshold and the development of AIDS. Science 254, 963–969 (1991)

Asymmetric Interactions between Cooperators and Defectors for Controlling Self-repairing

Yoshiteru Ishida and Masahiro Tokumitsu

Department of Knowledge-Based Information Engineering,
Toyohashi University of Technology
Tempaku, Toyohashi 441-8580, Japan
http://www.sys.tutkie.tut.ac.jp

Abstract. In an information network composed of selfish agents pursuing their own profits, undesirable phenomena such as spam mail occur if the profit sharing and other game structures permit such equilibriums. This note focuses on applying the spatial Prisoner's Dilemma to control a network of selfish agents by allowing each agent to cooperate or to defect. Cooperation and defection respectively correspond to repair (using the self resource) and not repair (thus saving the resource) in a self-repair network. Without modifying the payoff, the network will be absorbed into the state where all the agents become defectors and abnormal. Similarly to *kin selection*, agents favor survival of neighbors in organizing these two actions to prevent the network from being absorbed if payoffs are measured by summing all the neighboring agents. Even with this modification, the action organization exhibits spatial and temporal adaptability to the environment.

Keywords: asymmetric interaction, spatial prisoner's dilemma, maintenance of cooperating clusters, kin selection, autonomous distributed systems.

1 Introduction

The Prisoner's Dilemma (PD) has motivated studies on why cooperation emerges in many autonomous systems with selfish agents. The Spatial Prisoner's Dilemma (SPD) [1], in turn, has led to studies on why clusters of cooperators emerge when selfish agents interact with a spatially restricted world. We noted that the spatial Prisoner's Dilemma may be used for organizing an autonomous and distributed network if cooperation and defection are properly mapped to mutual operations among agents.

Spam mail is a phenomenon that can be understood as agents implementing a defective strategy. As shown by the case of spam mail, SPD or the game theoretic approach in general requires some modification to be applied to information networks such as the Internet. A game theoretic approach assumes that each player will sense, decide, act, and receive the reward/penalty as a distinct identity. The first modification would be to relax these assumptions. This note focuses on relaxing the last one: each player receives the reward (positive or negative) not exclusively but can share it with neighbors. Indeed, virus infection and intrusion to neighboring nodes is a crucial concern for most nodes. Sharing the reward/penalty with neighboring nodes amounts

I. Lovrek, R.J. Howlett, and L.C. Jain (Eds.): KES 2008, Part III, LNAI 5179, pp. 440–447, 2008.

to generalizing the unit of reward-receiving. This wider scope of reward-receiving would have a similar effect on the evaluation of profit over a longer time span, and hence space-time interplay is an interesting issue; however, this interplay will be discussed elsewhere.

As for mapping cooperation/defection to control operations, this note focuses on only repair/not-repair in a self-repairing network [2, 3]. However, there are many ways of mapping to deal with other issues of autonomous distributed networks, such as mapping to sharing resources/not sharing; and communicating messages/not communicating.

2 Basic Model and Assumptions

We consider a problem of cleaning up a network composed of autonomous agents with binary state: normal and abnormal. Network cleaning is carried out by mutually repairing agents. Repairing is done by each agent transferring own content to other agents. We assume that the state of agents (normal/abnormal) cannot be known to the agent itself or to the other agents. Thus, repairing by abnormal agents could harm other agents rather than repairing, thus mutual repairing involves a "double-edged sword" aspect. An appropriate strategy would lead to further contamination of the network. Furthermore, repairing uses resources of the repairing agent, and so it should be decided according to the resources used and remaining in the system and the network environment.

The selfish agents will determine their actions by considering their payoffs. To implement this selfish framework, a game theoretic approach deserves investigation. Since repair/not-repair corresponds to cooperate/defect in the SPD preserving payoff matrix, we will use SPD to organize the actions of agents. In this note, we will restrict ourselves to two actions, repair and not repair, in the self-repairing network. Repairing actions are carried out asynchronously by neighboring agents. Every agent has its own strategy that will determine its actions.

The SPD has been studied to investigate when, how, and why cooperation emerges among selfish agents when they are spatially arranged, hence interactions are limited only to their neighbors [1]. Each player is placed at each lattice of the two-dimensional lattice. Each player has an action and a strategy, and receives a score. Each player plays the Prisoner's Dilemma (PD) with the neighbors, and changes its strategy to the strategy that earns the highest total score among the neighbors. We will use this deterministic SPD.

In sum, our model has the following assumptions (Fig. 1):

- Each agent has a binary state: normal and abnormal; and a binary strategy: always cooperate (All-C) and always defect (All-D).
- Agents with the All-C strategy (cooperators or C) will repair the agents in the neighbor, whereas agents with the All-D (defectors or D) will not repair.
- Each agent becomes abnormal with a failure rate.
- Each agent is given a fixed amount of resources updated in a unit of time.
- Repairing requires some fixed amount of resources of the repairing agents.
- Repairing (and hence state updating) will be done asynchronously.

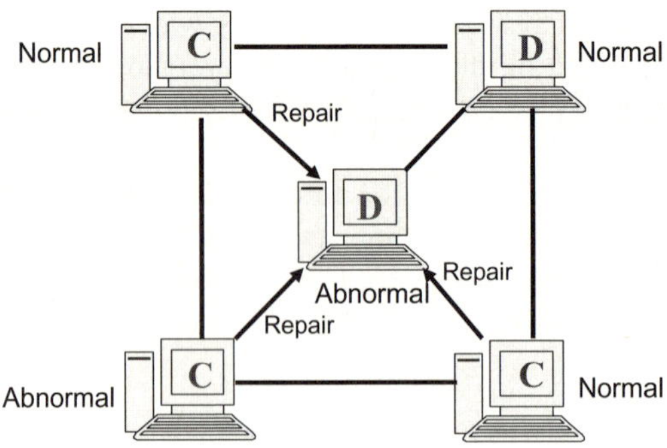

Fig. 1. A neighbourhood of the self-repairing network whose actions are organized by SPD

3 Simulations in Heterogeneous and Dynamic Environments

Every agent is placed at each cell of a two-dimensional lattice with a periodic boundary condition. (Thus the lattice size coincides with the number of agents.) They will interact with the agents in the neighboring cells. In this study, the *Moore* neighborhood (eight neighboring agents) is used.

Simulations are conducted using the following parameters. Each agent has a failure rate (λ). This failure rate will be varied to implement spatially and temporally varied environments. The repair will be done by a repair rate (α), and the successfully repaired agents are made normal. The adverse impact in the repair caused by abnormal

Table 1. List of parameters for simulations

	Description	Value
$L \times L$	Size of the space	50×50
N	Number of agents	2500
$N_f(0)$	Initial number of abnormal agents	100
λ	Failure rate	0.01
α	Repair rate	0.1
δ	Damage rate	0.1
r	Strategy update cycle	100
$R_{\max}$	Maximum resource	25
R_λ	Resource used for repairing	1

agents is implemented by raising the failure rate (by the amount of damage rate δ) of the target agents. Further, the agents are assumed to consume some resources (R_λ) in repairing. This amounts to a cost for cooperation, and hence encourages selfish agents to free-ride. The agents have to do the tasks assigned to them; but without repairing, the number of abnormal agents increases leading to a performance degradation in the system, and hence a dilemma. The agent is able to repair more than one agent in the neighbor, provided that the quantity of maximum resource R_{max} is not exceeded. Abnormal agents have no resource. We consider the available resource (the resource that is not used for repairing) as the score of an agent.

The agents update their own strategy after r (called the strategy update cycle) steps from the previous change. The next strategy will be chosen from the strategy that earned the highest score among neighboring agents. The total available resource is updated in each step by adding a fixed resource and subtracting the consumed resource from the current resource value. Table 1 lists the simulation parameters used in this study.

3.1 Repair Rate Control with Systemic Payoff

Here we compare the case when the payoff is evaluated as usual and that when the payoff is evaluated collectively in the neighborhood. Without the collective evaluation, defectors (All-D) will eradicate cooperators (All-C) (Fig. 2 (b)); hence all the agents will remain silent without repairing any agents. Thus, eventually all the agents will be abnormal with a positive failure rate (Fig. 2 (a)).

We devised a payoff to prevent all the agents from taking D actions and from being abnormal by incorporating not only their own remaining resources but also all the resources in the neighborhood. This modified payoff has an impact on making agents more attentive by caring for neighboring agents that might be able to repair the agent in the future. This modification reminds us of *kin selection* theory [4, 5] that emphasizes the survival of close relatives. As already noted, the neighboring nodes are important in

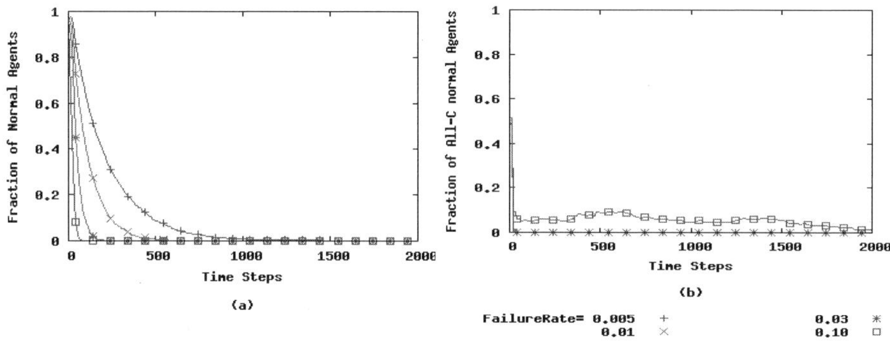

Fig. 2. SPD with simple payoff measured by available resources of the agent [2]. Parameters are: failure rate 0.005 - 0.10, repair success rate 0.1, damage rate 0.1, strategy update cycle 20, max resources 9, cost for repair 1. Randomly chosen 100 agents are made abnormal and randomly chosen a half of agents take all-D initially.

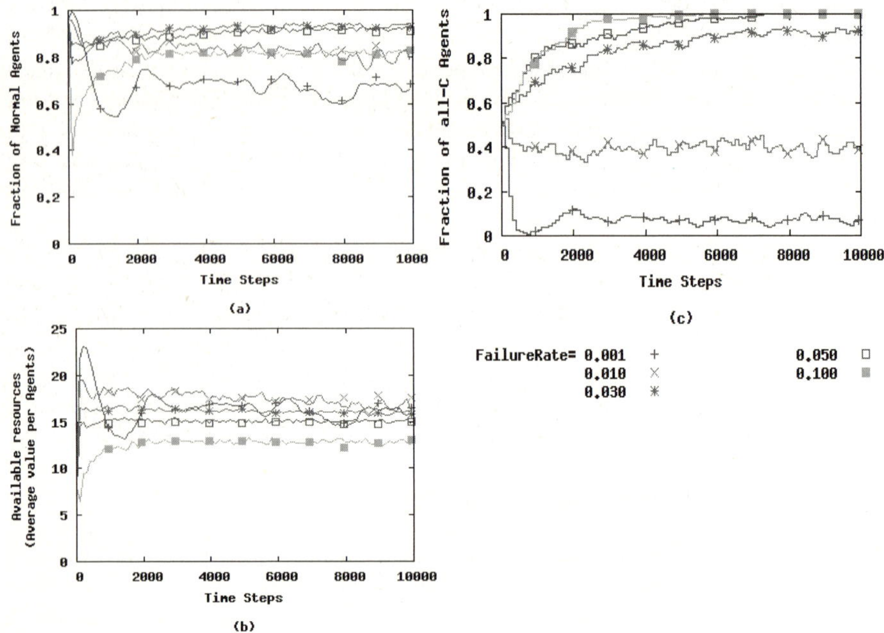

Fig. 3. SPD with strategic control with the modified payoff (available resources of the neighbor agents are added to payoff) [2]. (a) fraction of normal agents, (b) available resources, (c) fraction of All-C agents.Parameters are as in Table 1 and initial configuration with half of All-D agents and 100 failure agents randomly chosen.

information networks, since security issues (infection and intrusion) and the performance of the neighboring nodes are critical to every node.

Simulations are conducted for the strategic repair with the modified payoff. Figure 3 plots the time evolution of the fraction of normal agents (a), available resources left in the system (b), and the fraction of agents with All-C (c).

It can be seen that the strategic repair with the modified payoff is capable of selecting an appropriate strategy, hence the system can adapt to different failure rates: when the failure rate is low the fraction of All-C agents is kept small (Fig. 3 (c)) and thus unnecessary repairs are limited, while the fraction of All-C agents is made high when the failure rate is high. As a result of the flexible change of repair strategy, the fraction of normal agents (Fig. 3 (a)) as well as available resources (Fig. 3 (b)) are made stable and the difference in the failure rate is absorbed.

3.2 Adaptation to a Heterogeneous Environment

In the simulations of this heterogeneous environment and the next dynamic environment, the agents flip their action C or D by an action error rate (μ=0.01). The action error is introduced to prevent the network from becoming stuck at a local minima.

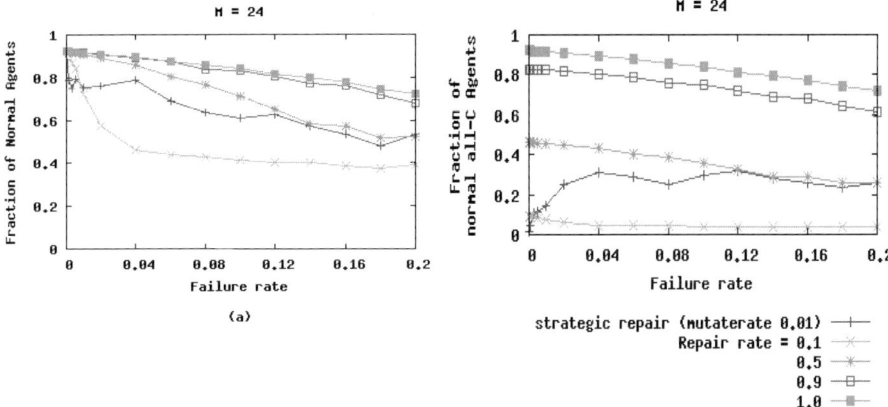

Fig. 4. SPD with strategic control with the modified payoff (available resources of the neighbor agents are added to payoff); fraction of normal agents (left) and fraction of All-C agents (right). Parameters are as in Table 1 and initial configuration with half of All-D agents and 100 a failure gents randomly chosen.

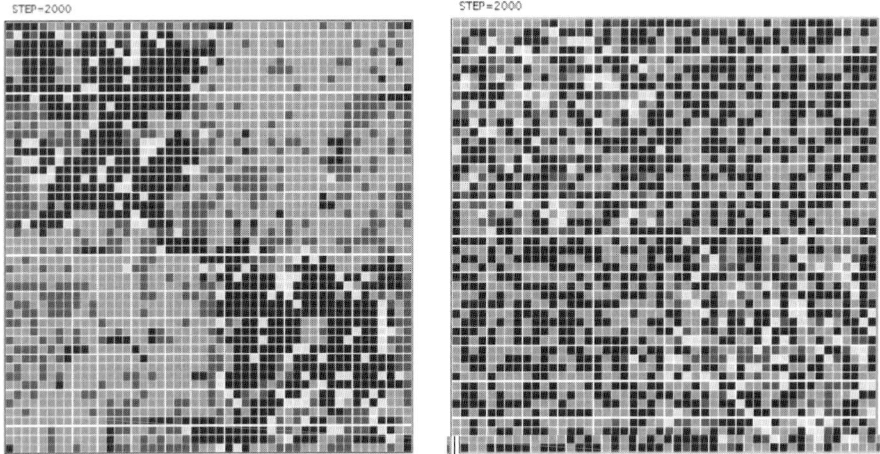

Fig. 5. A snapshot of agent configurations at 2000 time step by strategic repair (left) and uniform probabilistic repair (right), when simulation is carried out with the same condition as that in Fig. 4. Light gray (green) is Normal Defector, dark gray (red) is Abnormal Defector, white (yellow) is Abnormal Cooperator and black (blue) is Normal Cooperator.

Further, the heterogeneous environment is implemented with a checkerboard arrangement where the upper left and lower right 24×24 agents have a failure rate varying with relatively high values, while the upper right and lower left 24×24 agents have a failure rate fixed at a relatively low value of 0.001. Figure 4 plots the performance with varying failure rate on the horizontal axis.

Figure 5 (left) shows a snapshot of agent configurations at the 2000th time step when simulation is carried out with the same conditions as those in Fig. 4 with the strategic repair. It can be seen that areas with higher failure rate of 0.1 (above left and below right) are cooperators because they require repair. On the other hand, areas with lower failure rate of 0.001 are defectors. Figure 5 (right), on the other hand, uses a uniform repair rate of 0.5.

3.3 Adaptation to a Dynamic Environment

Since the strategic repair allows each agent to take its own action, it generally provides better performance than the case when the repair rate is fixed in time and is uniform among agents, particularly in a heterogeneous environment where agents fail with different failure rate. The strategic repair is also expected to perform well in a dynamic environment with a time-varying failure rate. In this simulation also, the action error is involved (action error rate $\mu = 0.1$).

A time-varying function is used to specify the failure rate. Agents are assumed to fail with a dynamic failure rate $\lambda(t)$. The failure rate oscillates with a triangular waveform as shown in Fig. 6 (d). In this simulation, the following parameters are also needed: cycle of the failure rate function (T: 10–10000) and amplitude of the failure rate function (A: 0.001–0.9).

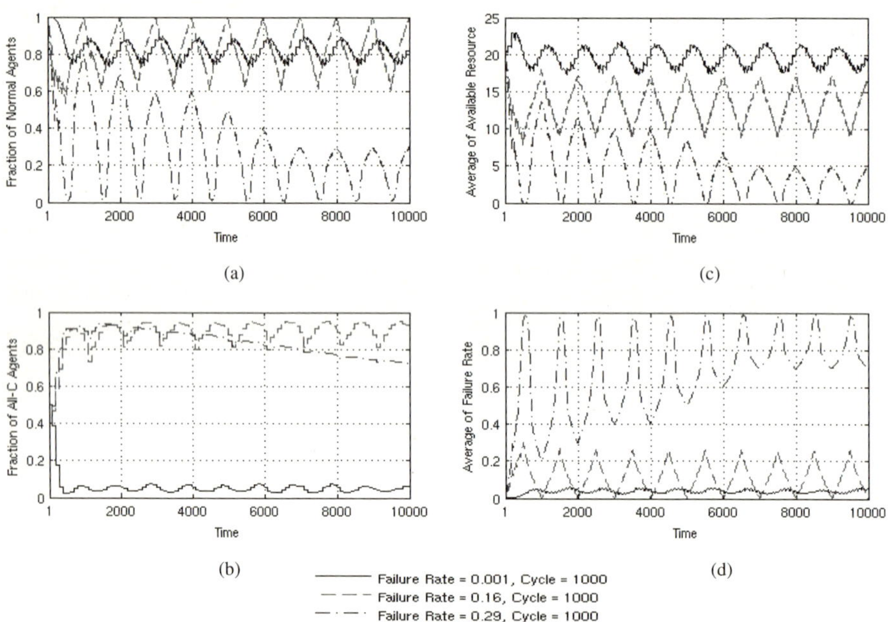

Fig. 6. The performance development when the failure rate changes as a triangle wave [3]. (a) fraction of normal agents (b) fraction of All-C agents (c) the average of available resource per agent (c) the average of failure rate in the network Parameters are as in Table 1 and initial configuration with the strategy each agents and 100 failure agents randomly chosen. The failure rate is between 0.001 to 0.90.

4 Discussion

In an autonomous network that leaves the decision of whether to repair neighboring agents to each selfish agent, the strategic repair exhibits adaptation to the environment. The game theoretic framework suits the autonomous and distributed decision-making context for regulation and maintenance of large-scale information systems. A major problem of using the spatial Prisoner's Dilemma in regulating the repair rate of agents is that agents tend to remain silent and stuck at the Nash equilibrium of mutual defection.

This paper presents a new solution to this problem: that is, involving more systemic payoff incorporating not only its own resources but also all the resources in the neighborhood. With this modified payoff, agents not only have an adaptive decision-making dependent on the environmental parameters such as failure rate, but also have more favorable resource allocation when compared with a uniform regulation of repair rate.

The simulations indicated that an appropriate uniform rate could be set when parameters were identified correctly. However, it is often the case that parameters are difficult to identify, or they may change dynamically. In such cases, strategic rate control can be used.

It is worth noting, however, that defection (remaining silent without repairing the neighboring agents) may not always be unfavorable actions that should be avoided. Although the global state of all agents defecting should be avoided, remaining silent can be a wise strategy when many agents are contaminated. Whether this is true even for biological systems or whether it is specific to the network cleaning problem with an assignment of cooperate and defect to repair and not repair, remain as issues for further studies.

Acknowledgments. This work was supported by The Global COE Program "Frontiers of Intelligent Sensing", from the Ministry of Education, Culture, Sports, Science and Technology. This work was also supported in part by Grants-in-Aid from Toyohashi University of Technology and CASIO Science Promotion Foundation.

References

1. Nowak, M.A., May, R.M.: Evolutionary games and spatial chaos. Nature 359, 826–829 (1992)
2. Oohashi, M., Ishida, Y.: A Game Theoretic Approach to Regulating Mutual Repairing in a Self-Repairing Network. In: Innovative Algorithms and Techniques in Automation, Industrial Electronics and Telecommunications, pp. 281–286. Springer, Netherlands (2007)
3. Tokumitsu, M., Ishida, Y.: Self-Repairing Network in a Dynamic Environment with a Changing Failure Rate. In: Innovative Algorithms and Techniques in Automation, Industrial Electronics and Telecommunications. Springer, Netherlands (to appear, 2008)
4. Hamilton, W.D.: The genetical evolution of social behaviour. Journal of Theoretical Biology 7(1–16), 17–52 (1964)
5. Maynard Smith, J.: Group Selection and Kin Selection. Nature 201, 1145–1147 (1964)

A Note on Space-Time Interplay through Generosity in a Membrane Formation with Spatial Prisoner's Dilemma

Yoshiteru Ishida and Yuji Katsumata

Department of Knowledge-Based Information Engineering,
Toyohashi University of Technology
Tempaku, Toyohashi 441-8580, Japan
http://www.sys.tutkie.tut.ac.jp

Abstract. The spatial Prisoner's Dilemma is divided into two stages: strategy selection and action (cooperation/defection) selection. This renewal allows a spatiotemporal strategy that determines the player's next action based not only on the adversary's history of actions (temporal strategy) but also on neighbors' configuration of actions. Several space-time parallelisms and dualisms would hold in this spatiotemporal generalization of strategy. Among them, this note focuses on the generosity (how many defections are tolerated). A temporal strategy involving temporal generosity, such as Tit for Tat (TFT), exhibited good performance such as noise tolerance. We report that a spatial strategy with spatial generosity can maintain a cluster of cooperators by forming a membrane that protects against defectors. The condition of membrane formation can be formulated with the spatial generosity exceeding a certain threshold determined by the number of neighborhoods.

Keywords: spatial prisoner's dilemma, spatiotemporal strategy, generosity, second-order cellular automata, membrane formation, maintenance of cooperation.

1 Introduction

The Prisoner's Dilemma has motivated studies in many domains such as international politics and evolutionary biology since the seminal work by Axelrod [1]. The spatial Prisoner's Dilemma invented by Nowak and May [2] also provides another dimension that these originally game theoretic studies can be related to the field of cellular automata [3].

Many possible mechanisms for the maintenance and protection of the cooperators' cluster have been proposed [4-6]. In a spatiotemporal generalization of the Prisoner's Dilemma, we observed that a membrane at the perimeter of the cooperators' cluster protects the cluster from invasion by defectors where the cooperators' cluster would be invaded otherwise.

In an asymmetric interaction between cooperators and defectors, defectors can exploit cooperators if the cooperators cannot escape or recognize the exploitation. Defectors can gain more as the number of exploitations increases. Usually in social interactions, cooperators recognize that they are losing each time, and escape from exploitation or become defectors. From the defectors' viewpoint, they can gain much

I. Lovrek, R.J. Howlett, and L.C. Jain (Eds.): KES 2008, Part III, LNAI 5179, pp. 448–455, 2008.

benefit only when the adversary cooperators do not escape. In a spatial extension defectors can exploit N cooperators at once, since they need not play the game N times with a player, but the game once with N players. This is similar to spam mail; this exploitation is made possible by the Internet which allows a player to interact with so many players in one action with almost no cost.

Thus, to deal with interactions through information networks such as the Internet where one-to-many or many-to-one interactions can be done instantaneously with almost no cost, the game framework of the Prisoner's Dilemma must be extended from the conventional constraint such that a player is a unit of action, recognition, decision making, and receiving reward/penalty. (The extension seems necessary to deal with biological systems as well regarding several levels of biological units such as DNA, RNA, cells, individuals, and species as profit-pursuing players.) The extension may be theoretically possible in the following ways:

1.   Scope of players on which the strategy is based;
2.   Scope of players which the action will affect;
3.   Scope of players whose (weighted) sum will be the payoff.

Among the three options above, this note will focus on the first and second ones. The scope of agents on which the strategy is based and which the action will affect is usually set to be one, however, we will extend this to many (called a "neighborhood"). This extension can be recognized as generalizing the conventional temporal strategy to spatiotemporal strategies. The third extension amounts to extension of players' identity to a collective identity composed of players in the neighborhood. This third extension will be reported elsewhere [7], but it is worth mentioning that extending players' identity will enhance and protect clusters of cooperators and bias the benefit of a set of players (neighbors) rather than a single player. Hence, with the perspective that space and time are related, the extension creates a bias toward a longer time evaluation of accumulated benefits than the conventional single player's payoff.

Section 2 presents our spatiotemporally generalized framework used in this note. Section 3 defines the spatial version of strategy (in contrast to the conventional temporal version) and spatial version of generosity. Section 4 presents the main results of conditions for the membrane formation with simulation results.

2 Spatiotemporal Generalization of Prisoner's Dilemma

In pursuit of a mechanism that allows cooperators' clusters to be preserved, we are studying a spatial version of generosity: how many defectors in the neighborhood are tolerated rather than how many previous defections were made by the adversary, in a spatiotemporally generalized context.

The action of each player is determined in two stages: the strategy is determined based on the scores in the neighborhood, then the action is determined based on the strategy and action configurations (spatial patterns) in the neighborhood. This two-stage determination of actions is similar to a second-order cellular automaton (CA) [3], however, our model involves also two layered states (strategies k-D and actions C/D).

Our SPD is done in the following way with N players simultaneously interacting with $(2r+1)^2$ neighbors.

(0) **Initial arrangement:** The action and strategy of each player are determined (see the next section for two types of simulations).

(1) **Renewal of action:** The next action will be determined by its strategy based on neighbors' (excluding the player itself) actions.

(2) **Score:** The score for each player is calculated by summing up all the scores received from PD with 8 neighboring players and the player itself with the payoff matrix in Table 1, and then adding the sum to the current score of the player. (The score will be added until the strategy is renewed.)

(3) **Renewal of strategy:** After (1) - (2) is repeated q (= 1 in our simulations) times, the next strategy will be chosen from the strategy with the highest score among the neighbors and the player itself. (1) - (3) are counted as one time step.

In our model, each player determines the next action by the current action of the neighbors. The strategy may be regarded as a "spatial strategy" in contrast to a "temporal strategy" in IPD. In Fig. 1 (the periodic boundary in the space direction, while the discrete time is infinite to the future and past), with the temporal strategy, the current player at space 0 and time 0 determines the action based on the action history of the

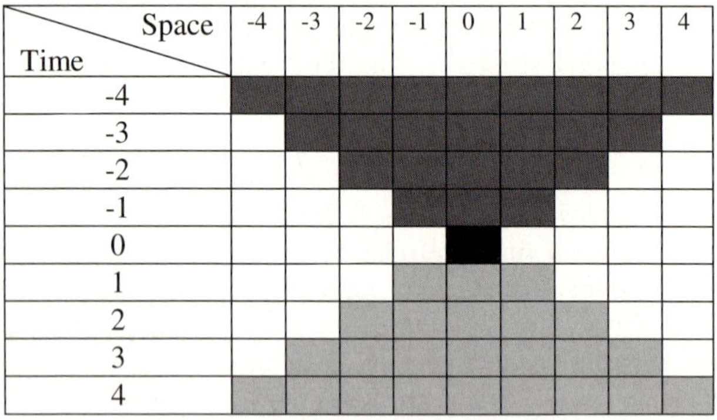

Time \ Space	-4	-3	-2	-1	0	1	2	3	4
-4									
-3									
-2									
-1									
0									
1									
2									
3									
4									

Fig. 1. A space-time diagram of 1-dim SPD when the propagation speed is 1 (neighborhood radius r = 1). The current center cell (black) is causally related to the cells within the cone (triangle). The cell can affect the cells in the cone of the future and is affected by the cells in the cone of the past.

Table 1. The payoff matrix of the Prisoner's Dilemma Game. Payoffs are to the player. A single parameter b ($1 < b < 2$) is used following Nowak-May's simulations [2].

Player \ Adversary	C	D
C	1	0
D	b	0

adversary (say, space coordinate 1) column 1 within the past cone (say, time coordinate from -1 to -3). With the spatial strategy, the current center player determines the action based on the action configuration of players -1 and 1 in row -1 within the past cone. The spatiotemporally generalized strategy is based on the action patterns in some volume of the past cone.

3 Spatial Strategies and Spatial Generosity

To specify a spatial strategy, actions of the eight neighbors (i.e., *Moore* neighbors or the neighborhood radius $r = 1$) and the player itself must be specified, hence 2^9 rules are required.

For simplicity, we restrict ourselves to a "totalistic spatial strategy" that depends on the number of D (defect) actions of the neighbors, not on their positions. To represent a strategy, let l be the number of Ds of the neighboring players excluding the self.

As a typical (totalistic) spatial strategy, we define k-D that takes D if $l \geq k$ and C otherwise (Fig. 2). For example, All-C is 9-D and All-D is 0-D. Note that these context independent strategies can be considered both spatial strategies and temporal strategies as extremes and bases.

This k-D can be regarded as a spatial version of TFT (1-D) where k seems to indicate the generosity (how many D actions in the neighborhood are tolerated).

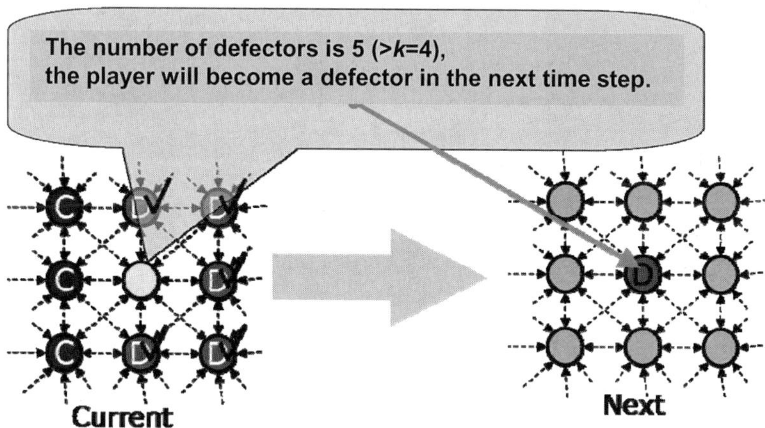

Fig. 2. The k-D strategy as a spatial strategy. When $k = 4$ for example, the player tolerates up to three defectors in the neighborhood. However, when the number exceeds the spatial generosity k, the player will become a defector as well.

4 Conditions for Membrane Formation

With the generalized SPD, we studied interactions between All-D and k-D instead of those between All-D and All-C (as in Nowak-May's SPD). Simulations revealed that clusters of k-D form a membrane of action D that protects the inner cluster of action C (note that k-D can take both C and D depending on the number of Ds in the neighborhood). We observed that this membrane formation occurs (Fig. 3) for a certain scope of parameters k (spatial generosity), r (neighborhood radius) and b (bias for defectors in the payoff matrix in Table 1).

We conducted two kinds of simulations which differ in initial configuration: a random configuration and a single (symmetric centered) seed. The random configuration simulation (Fig. 3) is to observe membrane formation which is robust against the initial seed; and the single seed simulation (Fig. 4) is to investigate the conditions for membrane formation.

In both simulations, the parameter b in the payoff matrix (Table 1) is set to be the smallest value that allows the All-D strategy (hence defectors) to expand. That is, b is the smallest value satisfying $5b > 9$ ($r = 1$), $16b > 25$ ($r = 2$), or $33b > 49$ ($r = 3$). In the case of $r = 1$, for example, $5b$ is the maximum score that the corner of cluster D will earn and 9 is the maximum score of C which is in cluster C and located in the neighborhood of player C adjacent to corner D. Ohtsuki *et al.* [8] derived an elegant formula to balance C and D when the lattice is a network (hence more general), however, we are unable to use it due to the form of the payoff matrix (Table 1).

Fig. 3. Random configuration simulation for membrane formation in a generalized Prisoner's Dilemma. In this snapshot, All-D and k-D ($k = 6$) strategies as well as C/D actions are allocated in random positions initially. Lattice size is 150×150. Time steps are 5 (left), 30 (middle), and 200 (right). Black cells are All-D; white and gray cells are respectively C and D states of k-D.

We will focus on the condition for the membrane formation with respect to these parameters. Throughout this note, simulations are conducted in a square lattice with periodic boundary condition with the following parameters listed in Table 2.

Table 2. List of parameters for simulations

Name	Description	Value		
$L{\times}L$	Size of the space	150×150 (Fig. 3, Random configuration with maps, Fig. 4, Single seed with maps);		
T	Number of steps	200 (Fig. 3, Random configuration with maps);		
r	Neighborhood radius	1	2	3
b	Bias for defectors	1.81	1.57	1.485

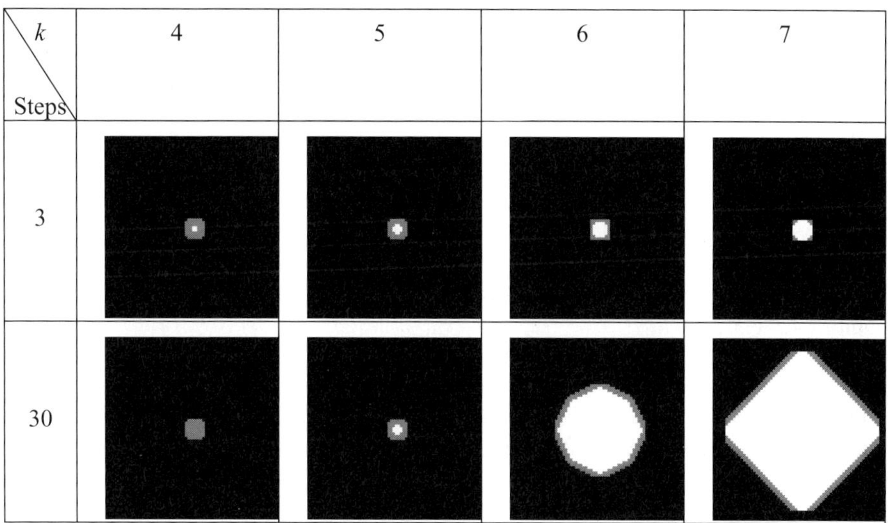

Fig. 4. Single seed simulations to investigate membrane formation. The k-D strategy with C actions is initially set in the center with the size 4×4. The membrane is indeed formed when k exceeds 4.5 (9/2). Otherwise (as in $k = 4$), the membrane grows inside, and eventually replaces the cooperators. If k is too big (as in $k = 7$), however, the membrane will be broken and eventually cooperators inside will no longer be protected.

After the membranes are formed, the following three phenomena are observed depending on the k value.

1. k is too small: The membrane will grow toward the center of cluster k-D and will corrode the C state of k-D.
2. k is small: Cluster k-D covered by the membrane will stay stable.
3. k is large: Cluster k-D covered by the connected membrane will expand.
4. k is too large: Cluster k-D covered by the broken membrane will expand and the cluster will eventually collapse.

Since we are interested in conditions for cooperators to be preserved, we focus on the conditions on cases 2 and 3 above. The spatial generosity k increases as the case proceeds downward from 1 to 4.

For a membrane formation, the spatial generosity k must exceed half of the number of players in the neighborhood:

$$k > (2r + 1)^2/2.$$

Otherwise, the membrane will grow inside toward the center of cluster k-D as in case 1.

For the cluster protected by the membrane to expand, the spatial generosity k must further exceed a larger threshold:

$$k > (2r + 1)^2/2 + r.$$

Otherwise, the cluster does not expand although it is indeed protected by the membrane as long as it exceeds $(2r + 1)^2/2$.

In case 4, the membrane is broken if k exceeds a threshold (for example, $k = 7$ when $r = 1$ as shown in Fig. 4). This threshold has not been formulated yet.

This membrane formation as a phenomenon can be considered as a "double-edged sword", since the membrane (k-D strategy with action D) can not only protect k-D players from being invaded by All-D players outside but also could replace all the players inside (k-D strategy with action C). Although this double-edged sword character may be shared by the immune system, the immune system is a multi-faceted double-edged sword, since it not only protects a material identity (protein type) but also functional identities.

The double-edged sword character of the k-D strategy will provide two threshold values: one for the membrane not growing toward the inside and another for the players inside the membrane not growing too fast so that it would not break the membrane.

Membrane formation can be observed in an original SPD where the membrane is composed of cooperators following former defectors and the breached part is composed of defectors following former cooperators, when the bias for defectors is relatively high ($1.8 < b < 2$) [2] and variants (e.g. [9]). These phenomena indicate that the cooperators' expansion or defectors' invasion occurs in one direction for a period of time like wave propagation. In variants of SPD with an additional third strategy such as loners in the public goods game [6] and the TFT [10], it is reported that these third strategies prevent cooperators from being exploited by defectors. It is, however, not clear that these protective functions come from geometrical origins such as the membrane. It is noted that TFT also has generosity (temporal one), and a parallelism to the membrane formation involving spatial generosity can be recognized.

It is conjectured that the membrane works as a "buffer area" which can be deployed either in space (as in k-D) or in time (as in TFT or loners).

5 Discussion

This note focuses on a space-time interplay through the eyes of generosity in a spatio-temporal Prisoner's Dilemma. However, many other results presented independently by researchers on the spatial Prisoner's Dilemma and by those on the temporal (conventional) Prisoner's Dilemma could be related in the spatiotemporal framework. As another example, we modified the payoff by adding not only the players' resources but also counting the neighbors' resources [7]. This modified payoff has an effect of favoring cooperators. Considering more collective players as a reward receiving unit is somewhat similar to considering longer term profit in evaluating payoff.

The membrane formation and its protective function can be related to many physical phenomena such as crystal growth, and biological phenomena such as the origin of life or formation of biological units in an early developmental stage.

The mechanisms and factors that determine the shape (e.g. whether the polygon is tetragon, hexagon or octagon when the boundary is convex; fractal dimension when the boundary is a concave-convex complex) in symmetric seed simulations are left untouched.

The membrane formation in the interaction between defectors and cooperators by spatial strategies with *spatial generosity* seems robust against changes in parameters in the payoff matrix. The membrane formation can be observed even if we use payoff in the Snow Drift (chicken game) [4].

Acknowledgments. This work was supported by the Global COE Program "Frontiers of Intelligent Sensing" of the Ministry of Education, Culture, Sports, Science and Technology. This work was also supported in part by Grants-in-Aid from Toyohashi University of Technology and CASIO Science Promotion Foundation.

References

1. Axelrod, R.: The Evolution of Cooperation. Basic Books, New York (1984)
2. Nowak, M.A., May, R.M.: Evolutionary games and spatial chaos. Nature 359, 826–829 (1992)
3. Wolfram, S.: A New Kind of Science. Wolfram Media, 437–440 (2002)
4. Hauert, C., Doebeli, M.: Spatial structure often inhibits the evolution of cooperation in the spatial Snowdrift game. Nature 428, 643–646 (2004)
5. McNamara, J.M., Barta, Z., Houston, A.I.: Variation in behaviour promotes cooperation in the Prisoner's Dilemma game. Nature 428, 745–748 (2004)
6. Szabó, G., Hauert, C.: Phase Transitions and Volunteering in Spatial Public Goods Games. Phys. Rev. Lett. 89, 118101 (2002)
7. Ishida, Y., Tokumitsu, M.: Asymmetric Interactions between Cooperators and Defectors for Controlling Self-Repairing. LNCS (LNAI), vol. 5179 (2008)
8. Ohtsuki, H., Hauert, C., Lieberman, E., Nowak, M.A.: A simple rule for evolution of cooperation on graphs and social networks. Nature 441, 502–505 (2006)
9. dos Santos Soares, R.O., Martinez, A.S.: The geometrical patterns of cooperation evolution in the spatial prisoner's dilemma: An intra-group model Physica A: Statistical Mechanics and its Applications, vol. 369, pp. 823–829 (2006)
10. Szabó, G., Antal, T., Szabó, P., Droz, M.: On the Role of External Constraints in a Spatially Extended Evolutionary Prisoner's Dilemma Game. In: Garrido, P.L., Marro, J. (eds.) Proceedings of Modeling Complex Systems, AIP Conference, vol. 574, pp. 38–55 (2001)

Dynamic Updating of Profiles for an Immunity-Based Anomaly Detection System

Takeshi Okamoto[1] and Yoshiteru Ishida[2]

[1] Dept. of Network Engineering, Kanagawa Institute of Technology,
1030 Shimo-ogino, Atsugi, Kanagawa, 243-0292, Japan
take4@nw.kanagawa-it.ac.jp
[2] Dept. of Knowledge-Based Information Engineering, Toyohashi University of Technology,
Tempaku, Toyohashi , Aichi, 441-8580, Japan
ishida@tutkie.tut.ac.jp

Abstract. Our immunity-based anomaly detection system aims to detect anomalous behavior of users on a computer. To improve the detection accuracy, we introduced the framework of dynamically updating profiles into our system. Our system enables agents to update not only self profiles, but also nonself profiles. Briefly, our system enables agents to adapt to new behavior of the original users and of others. The receiver operating characteristic (ROC) analysis of our system indicated that the updating of both profiles markedly decreased both the false alarm rate and the missed alarm rate.

Keywords: anomaly detection, adaptation, immunity-based system, intrusion detection, receiver operating characteristic, ROC.

1 Introduction

Intrusion detection in user behavior can be dived into two approaches: misuse detection and anomaly detection. A misuse detection system monitors and scans for known misuses, while an anomaly detection system monitors and detects deviations from normal behavior. If a masquerader does not commit a misuse, it will fail to be detected by the misuse detection system. In contrast, the anomaly detection system can detect deviations from normal behavior. However, anomaly detection sometimes gives rise to false alarms. Many false alarms cause "the boy who cried wolf" syndrome. Hence, the objectives of anomaly detection are "no false alarms" and "no missed alarms." There is a tradeoff between false alarm rate and detection rate. Our goal is to increase detection rate with no false alarms.

Many anomaly detection methods developed to date have been restricted to the reference of a single user profile, which accounts for normal user behavior [1–3]. One drawback of these methods is many false alarms that arise when valid users carry out new operations that they have not carried out previously in the training data set. In contrast, our immunity-based anomaly detection system refers to multiple profiles specifying each user, and our system outperforms some single profile methods [4–7]. However, the detection accuracy of our system is still insufficient for actual use. Our goal is to improve the detection accuracy further.

I. Lovrek, R.J. Howlett, and L.C. Jain (Eds.): KES 2008, Part III, LNAI 5179, pp. 456–464, 2008.

In this study, we investigated the time series of likelihood for operation sequences, and concluded that improvement of detection accuracy requires reduction of untrained operations. To reduce the untrained operations, we introduced a framework of dynamically updating profiles with test sequences (*i.e.*, not training sequences) into our immunity-based system. The updating of profiles has been applied to anomaly detection methods [2]. These methods update only the profile of normal behavior, *i.e.*, only the self profile, as they treat only the self profile. On the other hand, methods that update information on the nonself side have also been reported [8–10]. These methods are based on the mechanism of positive and negative selection in the thymus [11]. In contrast to these methods, our system updates not only the self profiles, but also the nonself profiles. We expect that the updating of self profiles will decrease false alarms, and updating of nonself profiles will decreases missed alarms. In our experiments, we evaluated the extent to which profile updating improves the detection rate in the actual web histories of 12 users.

2 Immunity-Based Anomaly Detection System

Our immunity-based system was developed to detect anomalous behavior of users on a computer, inspired by discrimination between self and nonself in the immune system. Our immunity-based system has three functions: generation of agents, diversity generation of agents, and discrimination of self and nonself [7]. The function of diversity generation improved the detection accuracy of our system. However, we excluded the function of diversity generation, to allow us to focus on the extent to which profile updating can improve the detection accuracy of our system. In this section, we describe our immunity-based system, except for mechanisms for diversifying agents and establishing self-tolerance.

2.1 Definitions of "Self" and "Nonself"

The heart of the immune system is the ability to distinguish between "self" (*i.e.*, the body's own molecules, cells, and tissues) and "nonself" (*i.e.*, foreign substances, such as viruses or bacteria). Similarly, we defined the behavior of an original user (a legitimate user) on his/her computer as "self," and all other behaviors as "nonself." User behavior is expressed by operation sequences, such as web page browsing, document writing, compiling code, *etc.*

2.2 Generation of Agents

Our immunity-based anomaly detection system generates a user-specific agent for every user on every computer. Each agent has a unique profile that is expressed by a parameter of the detection method, *i.e.*, hidden Markov model (HMM), neural network, *etc.* Our previous study indicated that the HMM performs well [3, 4]. The parameters of the HMM are given by $\lambda = [\pi, A, B]$, where π is the initial state distribution, A is the state-transition probability distribution, and B is the observation symbol probability distribution. These parameters are estimated from training data composed of operation sequences obtained previously from each original user. In our performance evaluations, we used the HMM for which the number of states is 1 due to the lowest computational cost and the best accuracy of other states [6]. The parameter estimation is conducted using the Baum-Welch algorithm. The maximum iteration count is set to 1000, and the minimum probability is set to e^{-200}.

The agent computes the likelihood $P(O|\lambda)$ of an operation sequence O with profile λ. The likelihood represents the possibility that the operation sequence O was performed by the original user corresponding to the agent (*i.e.*, profile λ). The agent would compute a high score (*i.e.*, a high likelihood) for only the sequences of the original user corresponding to the agent. In this way, the agent is specialized to recognize the user.

In this paper, the experimental data for user operation sequences are IP address sequences. The IP address space is too large to allow construction of a profile, and therefore pre-processing of the sequence is done to scale down the IP address space. The number of different IP addresses to which a user transmits a packet more than once is very small. Hence, each agent assigns a unique number v to each IP address to which the user transmits a packet more than once, where v begins with 0. The assignment table is created at profile construction, and all the sequences are replaced according to this table. If the IP address does not exist in this table, the IP address is assigned a unique number $v_{max}+1$, where $v_{max}+1$ is equal to the number of all IP addresses included in the table.

2.3 Discrimination of "Self" and "Nonself"

In every operation, all the agents on the computer compute their own scores for the most recently executed operation sequence. Only the original agent on the computer is activated and it compares its own score with those of all other agents. If the user that executes the operation on the computer is the same as the original user, the score computed by the original agent will be relatively high, but not necessarily the highest score compared with those of other agents. Thus, we define a threshold, T, which is the percentage difference between the minimum score *Min* and the maximum score

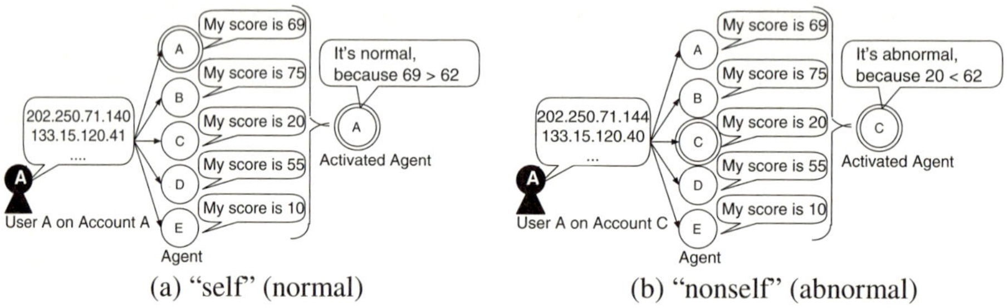

(a) "self" (normal)   (b) "nonself" (abnormal)

Fig. 1. Discrimination between "self" (a) and "nonself" (b). If we set the threshold value to 80% and the different agents compute 10, 20, 55, 69, and 75, the effective threshold value is calculated to be 62 (= 10+(75–10)×0.80). In the case of (a), if user A browses a website on his/her own account, agent A that is specialized to recognize user A is activated, and decides that the operation sequence is normal. In the case of (b), if user A browses a website on the account of user C, the agent C that is specialized to recognize user C is activated, and it decides that the operation sequence is abnormal.

Max. If the activated agent computes a score higher than the effective threshold obtained by the equation *Min + (Max – Min) × T*, the activated agent classifies the operation sequence as normal (*i.e.*, self). Otherwise, the agent classifies the operation sequence as abnormal (*i.e.*, nonself) and raises an alarm. Furthermore, provided that all scores are equal to the computable minimum value of $P(O|\lambda)$, the sequence is regarded as abnormal. Conversely, the operation sequence is regarded as normal if all scores are equal to the computable maximum value of $P(O|\lambda)$. Examples of user discrimination are shown in Fig. 1.

3   Experimental Data and Evaluation Model

To evaluate detection performance, we captured network traffic from 12 users for about one month. These experimental data were the same as those used in our previous work [6]. We focused on web traffic, as this accounts for the majority of network traffic. Hence, we extracted only outgoing TCP SYN packets with destination port 80. The web traffic of each user contained more than 3000 requests. The first 500 requests for each user were used as training data to allow construction of a profile. The next 1000 requests were test data to evaluate the detection performance. The test for the sequence was performed every 100 requests.

For evaluation of user traffic, we simulated anomalous behavior by testing one user's request sequence against another user's profile. This simulation corresponded to the evaluation of masquerader detection.

4   Analysis of User Behavior

In our immunity-based system, the rate of false alarms could be controlled by varying the threshold *T*. At the 100% threshold, only if the activated agent computes the maximum score of all agents, the sequence of the original user is regarded as normal (*i.e.*, "self"). On the other hand, at the 0% threshold, if the activated agent computes more than the minimum score of all agents, the sequence of the original user is regarded as normal. Briefly, the smaller the threshold, the fewer false alarms.

However, in actual web histories of users, false alarms arose even at the 0% threshold. At the time, the score of the activated agent is the minimum score of all agents. Figure 2 shows the scores of all agents for all test sequences of user No. 5. The vertical axis shows the scores of all agents, and the horizontal axis is the sequence number. The activated agent (*i.e.*, Agent 5) computes the minimum score in the 3rd sequence and the 9th sequence despite the sequences of user No. 5. The analysis showed that these sequences consisted of only untrained websites.

Figure 3 shows the average score of all agents and the average frequency of untrained websites for all test sequences of all users. The left axis is the average score of all agents, and the right axis is the average frequency of untrained websites in all the test sequences of all users. The horizontal axis is the sequence number. Figure 3 shows that the average score tended to decrease with time. On the other hand, the average number of untrained websites tended to increase with time. In addition, Fig. 3

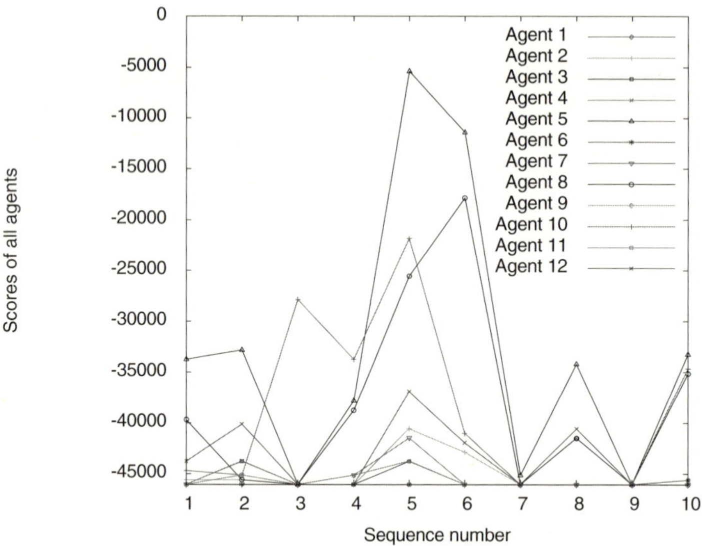

Fig. 2. Scores of all agents for all the test sequences of user No. 5. Each score is the log likelihood for each test sequence.

shows that the average score was connected with the average number of the untrained websites. Therefore, the untrained websites were responsible for the decrease in the agent scores.

5 Dynamic Updating of Profiles

The reduction of false alarms requires a decrease in untrained websites. It is possible to speculate that the way to decrease the untrained websites in test sequences is to extend the training sequences. However, we predict that extension of the training sequence will not be sufficient to reduce false alarms, because the number of untrained websites increases with time as stated above. Instead of extension of the training sequence, we introduce the framework of dynamically updating profiles with test sequences into our immunity-based system. Each agent dynamically updates its own profile at every test sequence examined, based on its decision regarding whether the test sequence is normal or abnormal. The profile is newly estimated from the test sequence just examined and all the sequences trained previously, and the new profile is replaced with the current profile. At the same time, using the assignment table described in Section 2.2, the assignment table is updated.

Our system has not only self agents specific to the original user (*i.e.*, self), but also nonself agents specific to strangers (*i.e.*, nonself). Thus, our system can update both self and nonself profiles. If the activated agent classifies the sequence as normal, the activated agent updates its own profile. If not, the agent that computed the maximum score of all agents excluding the activated agent updates its own profile. Updating self profiles is expected to decrease false alarms, while updating nonself profiles would decrease missed alarms.

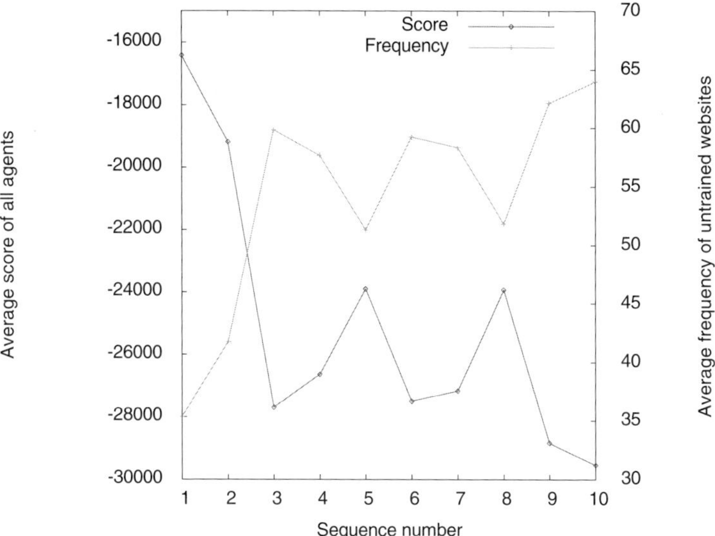

Fig. 3. Average score of all agents and average frequency of untrained websites for all test sequences of all users. The average score is the log likelihood for all test sequences of all users.

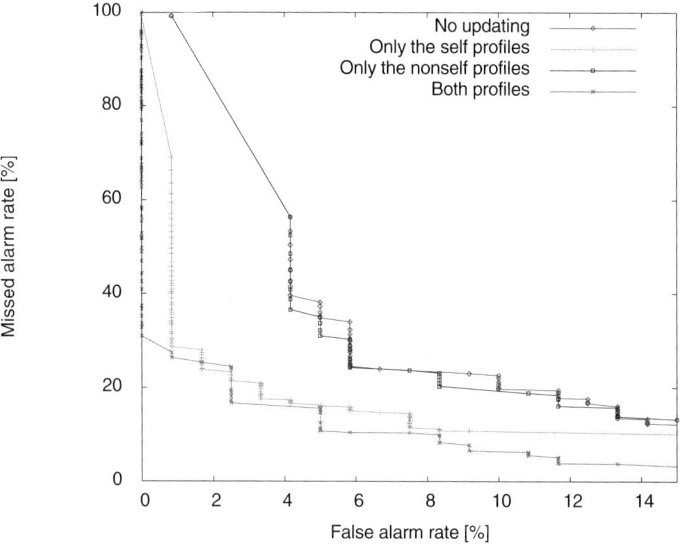

Fig. 4. ROC curves for our system without updating any profiles, with updating only self profiles, with updating only nonself profiles, and with updating both profiles

However, it is possible that profile updating may cause deterioration of the detection accuracy. If the activated agent falsely classified a stranger's sequence as normal, the activated agent would learn the stranger's sequence. In addition, if the activated agent falsely classified the original user's sequence as abnormal, another agent that computes the maximum score would learn the sequence of the original user. Once some agent falsely learns one sequence, the agent would falsely discriminate between self and nonself.

We conducted evaluation of the detection accuracy of the profile updating. The metrics of detection accuracy are based on the false alarm rate (*i.e.*, false positive rate) and missed alarm rate (*i.e.*, false negative rate). In general, there is a trade-off between the false alarm rate and the missed alarm rate. The relation of these rates can be described visually by a receiver operating characteristic (ROC) curve, which is parametric curve generated by varying the threshold T from 0% to 100%, and computing these rates at each threshold T. The lower and further left the curve, the better.

Figure 4 shows ROC curves for our system without updating any profiles, our system with updating only the self profiles, with updating only the nonself profiles, and with updating both profiles. Although we had concerns about deterioration of the detection accuracy, the results indicated improvement in the detection accuracy. Our system with updating both profiles indicated the best detection accuracy of all curves. This accuracy appeared to be dependent mainly on the updating of the self profiles, as updating of only the self profiles markedly improved the detection accuracy, whereas updating of only the nonself profiles slightly decreased only the false alarm rate. It should be noted that updating of both profiles achieved a missed alarm rate of 29.5% with a false alarm rate of 0% at the threshold 56.99%.

Figure 5 shows the average scores and the frequencies of untrained websites for the cases with and without updating. The results without updating were the same as those shown in Fig. 3. The results with updating were computed by all agents for all test sequences of all users at the threshold 56.99%. Figure 5 shows that the average score for updating tended to slightly increase with time, while the frequency of updating tended to slightly decrease. That is, our system could adapt to untrained websites with time.

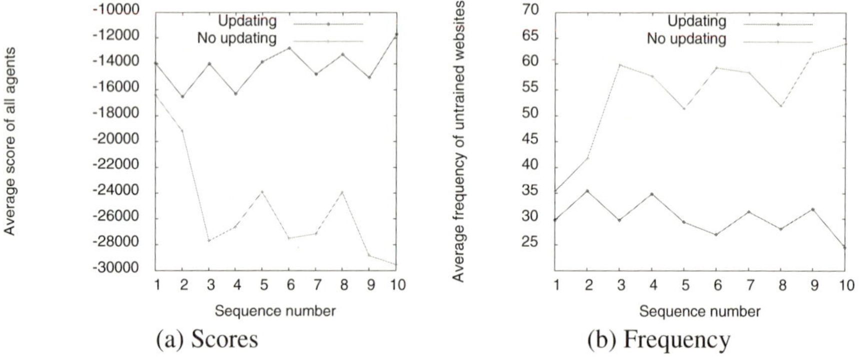

Fig. 5. Comparison of average scores (a) and untrained website frequencies (b) with and without updating. The results without updating were the same as those in Fig. 3. The results with updating were computed by all agents for all test sequences of all users at the threshold 56.99%. Both average scores are log likelihoods for all test sequences of all users.

6 Conclusions

We investigated the time series of likelihood for histories of websites, and concluded that improvement of detection accuracy requires reduction of the untrained websites. For this purpose, we introduced a framework of dynamically updating profiles into our immunity-based system. The ROC analysis indicated that updating of both profiles markedly decreased both the false alarm rate and the missed alarm rate. The success is probably dependent on the updating of self profiles. In addition, it is noteworthy that updating of both profiles achieved a missed alarm rate of 29.5% with a false alarm rate of 0%.

We are currently evaluating the missed alarm rate for sequences of external users and computer worms. Further studies are required to reduce the computational costs for updating profiles and for integration with the function of diversity generation.

Acknowledgements

This work was supported by Grant-in-Aid for Young Scientists (B) No. 19700072 from the Ministry of Education, Culture, Sports, Science and Technology of Japan.

References

1. Lane, T.: Hidden Markov models for human/computer interface modeling. In: IJCAI 1999 Workshop on Learning about Users, pp. 35–44 (1999)
2. Schonlau, M., DuMouchel, W., Ju, W., Karr, A., Theus, M., Vardi, Y.: Computer intrusion: Detecting masquerades. Statistical Science 16(1), 58–74 (2001)
3. Okamoto, T., Watanabe, Y., Ishida, Y.: Test statistics for a masquerader detection system – a comparison between hidden Markov model and other probabilistic models. Transactions of the ISCIE 16(2), 61–69 (2003)
4. Okamoto, T., Watanabe, T., Ishida, Y.: Towards an immunity-based system for detecting masqueraders. In: Palade, V., Howlett, R.J., Jain, L. (eds.) KES 2003. LNCS, vol. 2774, pp. 488–495. Springer, Heidelberg (2003)
5. Okamoto, T., Watanabe, T., Ishida, Y.: Mechanism for generating immunity-based agents that detect masqueraders. In: Negoita, M.G., Howlett, R.J., Jain, L.C. (eds.) KES 2004. LNCS (LNAI), vol. 3214, pp. 534–540. Springer, Heidelberg (2004)
6. Okamoto, T., Ishida, Y.: Towards an immunity-based anomaly detection system for network traffic. In: Gabrys, B., Howlett, R.J., Jain, L.C. (eds.) KES 2006. LNCS (LNAI), vol. 4252, pp. 123–130. Springer, Heidelberg (2006)
7. Okamoto, T., Ishida, Y.: Framework of an immunity-based anomaly detection system for user behavior. In: Apolloni, B., Howlett, R.J., Jain, L.C. (eds.) KES 2007, Part III. LNCS (LNAI), vol. 4694, pp. 821–829. Springer, Heidelberg (2007)
8. Williams, P.D., Anchor, K.P., Bebo, J.L., Gunsch, G.H., Lamont, G.D.: CDIS: towards a computer immune system for detecting network intrusions. In: Lee, W., Me, L., Wespi, A. (eds.) Fourth International Symposium, Recent Advances in Intrusion Detection, pp. 117–133 (2001)

 9. Ayara, M., Timmis, J., de Lemos, R., de Castro, L.N., Duncan, R.: Negative selection: how to generate detectors. In: Timmis, J., Bentley, P.J. (eds.) ICARIS, pp. 89–98. University of Kent at Canterbury (2002)
10. Esponda, F., Ackley, E., Forrest, S., Helman, P.: On-line negative databases. Journal of Unconventional Computing 1(3), 201–220 (2005)
11. Forrest, S., Hofmeyr, S., Somayaji, A., Longstaff, T.: A sense of self for Unix processes. In: IEEE Symposium on Security and Privacy, pp. 120–128. IEEE Computer Society Press, Los Alamitos (1996)

OGRE-Multimedia: An API for the Design of Multimedia and Virtual Reality Applications

Paulo N.M. Sampaio, Roberto Ivo C. de Freitas, and Gonçalo Nuno P. Cardoso

Laboratory for Usage-centered Software Engineering (LabUSE)
Centro de Ciências Matemáticas (CCM)
University of Madeira (UMA)
Campus da Penteada 9000-390
Funchal, Madeira, Portugal
Tel.: +351 291 705291
`psampaio@uma.pt`, `robertoicf@hotmail.com`,
`balizeiro@netmadeira.com`

Abstract. Most of the tools and languages for modeling Virtual Reality environments, such as VRML, X3D, Java3D, etc. do not provide means of describing the synchronized presentation of multimedia content inside these environments. Multimedia has demonstrated its capabilities of motivating users and capturing their attention, which is an interesting characteristic when we want to provide a higher degree of immersion inside Virtual Reality applications. This paper presents a robust and generic solution for the integrated presentation of different kinds of media objects inside virtual environments based on the Graphical Engine OGRE.

1 Introduction

Multimedia has been applied in different applications as a helpful tool for providing insight about a subject being presented. At the same time, it has been proved that human is more receptive to new information and construct easier cognitive models if this information is presented in different modalities [1]. The integration of multimedia content inside Virtual Environments (VEs) is a promising and interesting trend in the development of Virtual Reality (VR) applications. Indeed, Multimedia captivates users´ attention inside the VE enhancing interaction, promoting user´s interest, facilitating learning and improving user´s immersion.

Some important issues must be considered for the integration of multimedia content inside a VE such as the specification of the temporal and logical synchronization of different media objects (with at least one audio or video) to be rendered inside the 3D environment, and determining which events (e.g., user interactions) will be applied for the communication between the 2D/3D worlds. Unfortunately, most of the existing languages for describing 3D environments (such as VRML [2], X3D [3] or Java3D[4]) are monomedia and non-interactive since they support only the presentation of isolated media objects without any synchronization relations among them. One exception to this is MPEG-4 which by means of BIFS allows the creation of rich

I. Lovrek, R.J. Howlett, and L.C. Jain (Eds.): KES 2008, Part III, LNAI 5179, pp. 465–472, 2008.

2D/3D graphical scenarios with synchronized multimedia [5]. However, the authoring of the MPEG-4 BIFS is still too complex and intuitive tools and approaches are still lacking.

Many VR systems have been proposed in the literature addressing different application domains: e-learning [6], [7], [8], collaboration among workgroups [9], augmented collaborative spaces [10], multimodal VR applications [11], among others. The rapid prototyping, modeling and authoring of VEs has been a major concern to many authors, as presented in [12], [13], and [14]. Although, most of the systems propose the development of VEs, few of them explore the presentation of integrated multimedia content inside VEs [15].

This paper presents a solution to provide the integration of multimedia content inside a VE based on the Graphical Engine OGRE [16]. The API implemented is called OGRE-Multimedia, and can be applied to any VR application to allow their customization with multimedia content. OGRE (Object-Oriented Graphics Rendering Engine) is a scene-oriented, flexible 3D engine written in C++ designed to make it easier and more intuitive for developers to produce applications using hardware-accelerated 3D graphics. When comparing OGRE with other existing languages and approaches for describing virtual worlds, we decided to adopt this platform based on its design quality, flexibility and clear documentation.

This paper is organized as follows: Section 2 presents a solution to customize a VE with multimedia presentations; Section 3 presents the main architecture of the API developed; Section 4 illustrates a Multimedia and Virtual Reality application, and; Finally, Section 5 presents some conclusions.

2   Customizing Multimedia Presentation Inside a VE

The main goal of our work is to propose and develop a solution for providing the presentation of multimedia content within an OGRE´s virtual environment (VE). This multimedia content is related to: the output of any embedded multimedia player (such as RealPlayer [17], GRiNs [18], etc.); a Flash executable content [19], or; a web-browser content. The main idea is to present the multimedia content as textures over any 3D object inside a virtual environment.

When proposing the integration of multimedia content inside a VE, we had to come up with a customized solution to cope with the need for specifying synchronization relations among the media objects, supporting user interactions with these objects, ensuring interoperability of multimedia players, and mapping 2D objects into the 3D world. Unfortunately, the existing languages and models for describing multimedia presentation such as SMIL [20] do not support the description of three-dimensional channels, that is, the specification of the x, y and z coordinates for the presentation of the multimedia content inside the 3D environment. For this reason, the solution relied on the proposal of a simpler XML-based meta-language for describing multimedia documents to be presented inside virtual environments, or as we called the *meta-multimedia document*.

The *meta-multimedia document* was strongly inspired on the syntax of SMIL and can be applied as a multimedia authoring language where users can customize the virtual environment and describe what is going to be presented, where and when they will be presented. Briefly, it describes all the components of the multimedia presentation and their temporal and logical synchronization. The particularity about this presentation is that all the media objects (multimedia documents, flash, web-browsers, and primitive media objects such as video, image, text, audio, etc.) are synchronized and rendered anywhere inside the virtual environment. The interpretation and coordination of this document presentation inside the VE is done by the API developed for OGRE, the *OGRE-Multimedia*.

As presented in Figure 1, the structure of the *meta-multimedia document* is composed of four main elements: *panel*, *trigger*, *eventHandler* and *event*.

Each *panel* element describes a presentation panel for media objects inside the virtual environment. The container called *panels* is a set of the panel objects that must be rendered inside a VE.

```
<multimediaControl>
  <panels>
    <panel name='MainUMa' width='1024' height='768' scale='0.2' position='-745, -150, 0' verRotation='90°' />
    <panel name='LeftUMa' width='640' height='480' scale='0.3' position='-745, -150, 225' verRotation='90°' />
    <panel name='TopLeftUMa' width='640' height='480' scale='0.3' position='-700, 25, 225' verRotation='90°' horRotation='45°' />
    <panel name='TopUMa' width='640' height='480' scale='0.3' position='-700, 25, 0' verRotation='90°' horRotation='45°' />
    <panel name='TopRightUMa' width='640' height='480' scale='0.3' position='-700, 25, -225' verRotation='90°' horRotation='45°' />
    <panel name='RightUMa' width='640' height='480' scale='0.3' position='-745, -150, -225' verRotation='90°' />
  </panels>
  <triggers>
    <trigger name='TriggerUMa' position='-535, -174, 0' scale='2.5' verRotation='45°' />
  </triggers>
  <eventHandlers>
    <eventHandler triggerName='TriggerUMa' action='click' loopEvents='true'>
      <event source='Moby.ogg' volume='25' />
      <event panelName='MainUMa' source='http://www.uma.pt/' />
      <event panelName='LeftUMa' source='Cantina2.swf' start='3s' stop='14s' fadeOut='3s' />
      <event panelName='TopLeftUMa' source='FachadaInf.swf' start='3s' stop='14s' fadeOut='3s' />
      <event panelName='TopUMa' source='Biblioteca1.jpg' start='3s' fadeIn='4s' stop='14s' fadeOut='3s' />
      <event panelName='TopRightUMa' source='ESC.swf' start='3s' stop='14s' fadeOut='3s' />
      <event panelName='RightUMa' source='Biblioteca2.jpg' start='3s' fadeIn='4s' stop='14s' fadeOut='3s' />
      <event panelName='RightUMa' source='Anfiteatro1.swf' start='18s' fadeIn='4s' stop='30s' />
    </eventHandler>
  </eventHandlers>
</multimediaControl>
```

Fig. 1. Example of a meta-multimedia document

Each element *trigger* characterizes an object inside the VE which controls the activation and deactivation of a multimedia presentation. The container called *triggers* is a set of all the trigger objects that will be used to control the multimedia presentations inside a VE.

Each element *EventHandler* characterizes how the presentation of the media objects associated with a given *trigger* will be controlled (e.g., start their presentation when the user clicks on the trigger or when he approximates it). The container called *eventHandlers* is a set of all multimedia presentation described by all the elements *eventHandler*.

Each element *event* characterizes how and when the presentation of each media object of a given *eventHandler* will be carried out.

The structure of the meta-multimedia document was defined to make the process of authoring the multimedia document easier and intuitive. We consider the *meta-multimedia document* as the key-solution for the integration of multimedia content inside VEs. Indeed, with this document, the author of the application is able to customize his virtual environment with new multimedia content without changing a single line of his code.

3 OGRE-Multimedia: Integrating Multimedia within Virtual Environments

Most of the libraries available for creating Virtual Reality applications do not have appropriate APIs for the integration of multimedia content inside a VE. Some languages and platforms such as VRML, X3D, Java3D and OGRE, provide only APIs for the presentation of single media objects (such as video or audio) without integrating or synchronizing these objects. The solution proposed with OGRE-Multimedia is to provide an API to integrate the presentation of different multimedia objects (rendered by different plug-ins or APIs) around the definition of the *meta-multimedia document*. In this sense, the meta-multimedia document describes all the synchronization relations among all the components (media objects, Flash presentation, web-browsers, etc.) of the document. Taking advantage of the OGRE´s component-based architecture, this API can be easily instantiated and integrated with the remaining available library. This section presents the main architecture of OGRE-Multimedia

The architecture of OGRE-Multimedia describes the integration of the *meta-multimedia document* with the Virtual Environment, which are supported by the implemented software modules and some existing APIs. This architecture is depicted in Figure 2.

The architecture of *OGRE-Multimedia* (Figure 2) is composed of four main components:

(1) *External modules*, which are represented by those APIs developed by other projects, or which were already provided by the OGRE´s library, such as (i) TinyXML [21] (XML syntactic analyzer parser), (ii) OIS [22] (Interactions management), (3) OgreAL [23] (Presentation of audio objects), (4) DevIL [24] (Presentation of images inside VEs), (5) Navi [25] (Presentation and interaction of a web browser inside the VE), and (6) OGRE graphic engine which is the main module of the system being responsible for creating, managing and updating the tri-dimensional model.
(2) *Elementary module*, which describes the non-functional components of the architecture used as a support for the application (*Meta-multimedia document* and *Virtual Environment*).
(3) *System startup*, which is in charge to set up the presentation of multimedia content inside VEs.
(4) *System Update*, which is in charge to control and update the multimedia presentation inside the VE according to possible user interactions.

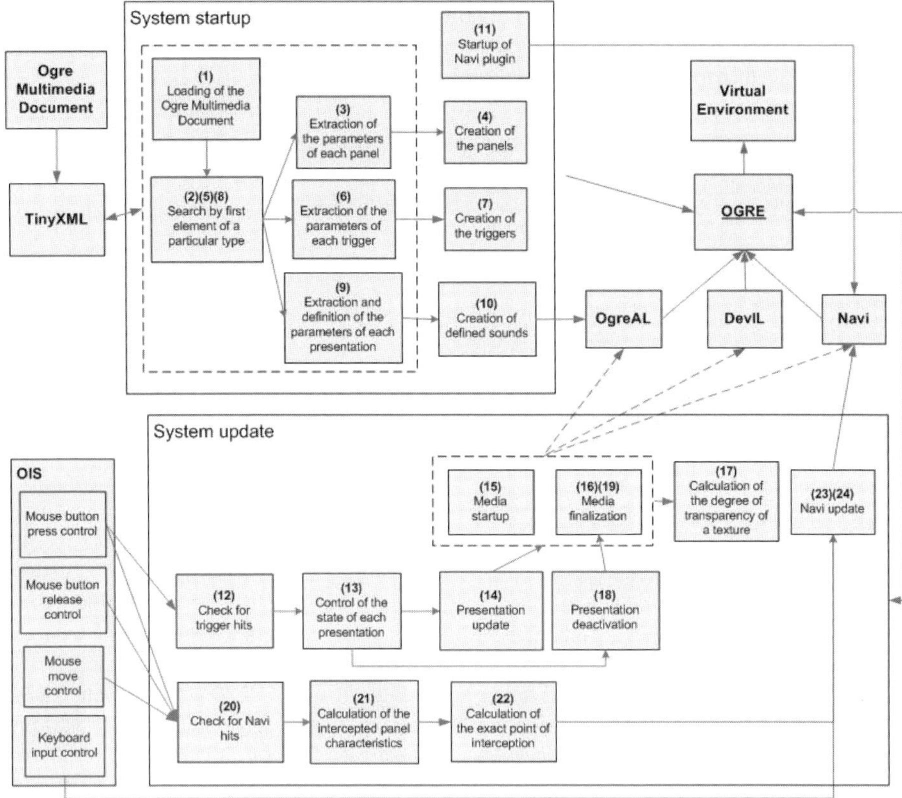

Fig. 2. Architecture of OGRE-Multimedia

The implemented prototype is further discussed on the next section.

4 The Prototype Implemented

This section illustrates the prototype implemented by the presentation of an OGRE-Multimedia application. This application applied a virtual world (also called map) for the presentation of multimedia content in an exhibition style. This environment can be applied for educational or art-exhibit purposes, for instance. OGRE-Multimedia was easily integrated to the OGRE application, where the methods of OGRE-Multimedia were invoked to enable the presentation of multimedia content previously defined on the *meta-multimedia document*, called "MMDocument.mmc". In the case of this VR application, the virtual world describes an open-wide area, where in front of each wall there is a sensitive column which can be activated by the user's click or by his proximity (depending on its previous configuration on the *meta-multimedia document*) in order to trigger the multimedia presentation on the wall (Figure 3a).

Figure 3(b) illustrates the interactive column in the virtual environment. In particular, this column (which is represented by a *trigger* element on the *meta-multimedia document*) is able to launch a presentation by a user´s click.

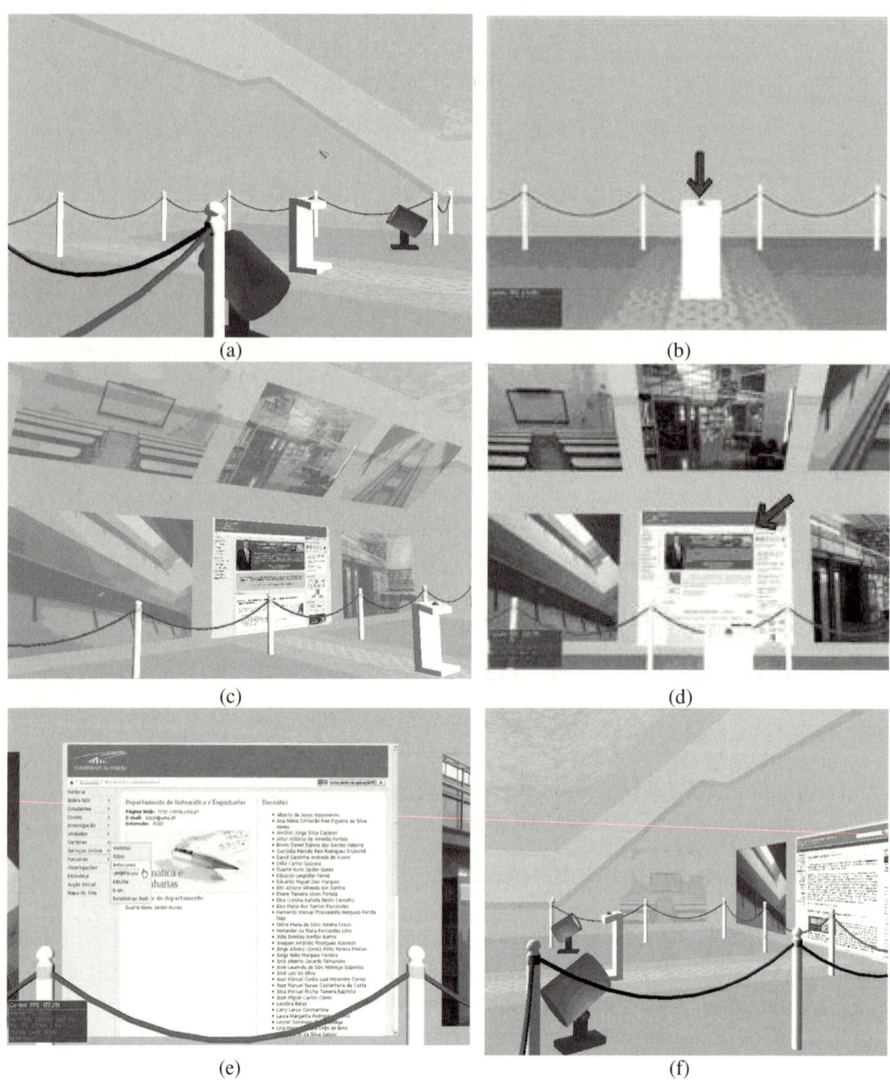

Fig. 3. Presentation of the multimedia content inside the virtual environment

The multimedia presentation is launched after the user interacts (by clicking) with the sensitive column (Figures 3c). When the media objects start to be presented, their level of transparency is gradually changed producing the effect of fading-in. As we can see in Figure 3(d), a web-browser is also presented as one of the multimedia textures. This browser is rendered by the API Navi which allows the user to navigate on the Web. Figures 3(d), 3(e) and 3(f) present an example of this navigation.

All the media objects presented inside the VE are synchronized and managed by OGRE-Multimedia which keeps pace of each presentation starting and interrupting all the media objects according to their previous configuration on the *meta-multimedia document*. OGRE-Multimedia enables the multimedia presentation inside the VE making of it a more realistic environment and, above all, keeping the user´s focus.

5 Conclusions

This paper presented the development of an API for the presentation of integrated multimedia content inside Virtual Environments based on the Graphical Engine OGRE, called OGRE-Multimedia. The integrated presentation of multimedia content inside VEs relied on the proposal of an XML-based representation to describe all the media objects to be presented and their synchronization relations, the *meta-multimedia document*. OGRE-Multimedia can be applied straightforward in different OGRE Virtual Reality applications since it is the result of an open-architecture where different APIs were applied in conjunction to provide the presentation of different kinds of media objects including the traditional images, audio, video, animations, etc., and also multimedia documents such as FLASH, and web-browsers as well. Indeed, the combination of Multimedia and Virtual Reality can be successfully applied to the design of robust applications where users feel more comfortable and have their focus inside the VE, definitely improving their feeling of immersion.

References

1. Sorden, S.D.: A cognitive approach to instructional design for multimedia learning. Informing Science Journal 8(1), 263–279 (2005)
2. VRML: Virtual Reality Modeling Language, http://www.w3.org/MarkUp/VRML/
3. ISO/IEC 19775:2004 — Extensible 3D (X3D), http://www.web3d.org/x3d/specifications/ISO-IEC-19775-X3DAbstractSpecification/
4. Java3D API, http://java.sun.com/products/java-media/3D/
5. ISO/IEC JTC 1/SC 29/WG 11N7608 — MPEG-4 BIFS white paper, http://www.chiariglione.org/MPEG/technologies/mp04-bifs/index.htm
6. Chee, Y.S.: Network Virtual Environments for Collaborative Learning. In: Proceedings of ICCE/SchoolNet 2001—Ninth International Conference on Computers in Education, Seoul, S. Korea, ICCE/SchoolNet, pp. 3–11 (2001)
7. McArdle, G., Monahan, T., Bertolotto, M., Mangina, E.: A Web-Based Multimedia Virtual Reality Environment for E-Learning. In: Proceedings Eurographics 2004, Grenoble, France (July 2004)
8. Halvorsrud, R., Hagen, S.: Designing a Collaborative Virtual Environment for Introducing Pupils to Complex Subject Matter. In: NordiCHI 2004: Proceedings of the third Nordic conference on Human-computer interaction. Tampere, Finland, pp. 121–130 (2004) ISBN: 1581138571
9. Bochenek, G.M., Ragusa, J.M.: Virtual (3D) collaborative environments: an improved environment for integrated product team interaction? In: Proceedings of the 36th Annual Hawaii International Conference on System Sciences, 2003, p. 10 (2003)

10. Pingali, G., Sukaviriya, N.: Augmented Collaborative Spaces. In: Proceedings of the 2003 ACM SIGMM workshop on Experiential telepresence, International Multimedia Conference, Berkeley, California (2003)
11. Carrozino, M., Tecchia, F., Bacinelli, S., Cappelletti, C., Bergamasco, M.: Lowering the development time of multimodal interactive application: The real-life experience of the XVR Project. In: ACM SIGCHI International Conference on Advances in Computer Entertainment Technology (2005)
12. Rodrigues, S.G., Oliveira, J.C.: ADVICe - um Ambiente para o Desenvolvimento de ambientes VIrtuais Colaborativos. In: XI Simpósio Brasileiro de Sistemas Mutimídia e Web - WebMedia2005, Poços de Caldas, MB, Brasil (2005)
13. Osawa, N., Asai, K., Saito, F.: An interactive toolkit library for 3D applications: it3d. In: Proceedings of the workshop on Virtual environments 2002, ACM International Conference Proceeding Series, Barcelona, Spain, vol. 23, pp. 149–157 (2002)
14. Garcia, P., Montalà, O., Pairot, C., Skarmeta, A.G.: MOVE: Component Groupware Foundations for Collaborative Virtual Environments. In: Proceedings of the 4th international conference on Collaborative virtual environments, Bonn, Germany, pp. 55–62 (2002)
15. Walczak, K., Chmielewski, J., Stawniak, M., Strykowski, S.: Extensible Metadata Framework for Describing Virtual Reality and Multimedia Contents. In: Proceedings of the 7th IASTED International Conference on Databases and Applications DBA 2006, Innsbruck, Austria, pp. 168–175 (2006)
16. OGRE 3D: Open source graphics engine, `http://www.ogre3d.org/`
17. RealPlayer - Real Networks (2004), `http://www.real.com/player/`
18. GRiNS – SMIL 2.0 Player Home Page, `http://www.oratrix.com/GRiNS/SMIL2.0/`
19. Adobe – Flash Player, `http://www.macromedia.com/software/flash/about/`
20. SMIL 2.0 - Synchronized Multimedia Integrated Language, `http://www.w3.org/AudioVideo/`
21. TinyXML – XML Parser, `http://www.grinninglizard.com/tinyxml/`
22. OIS – Object Oriented Input System, `http://sourceforge.net/projects/wgois/`
23. OpenAL – Cross-Platform 3D Audio, `http://www.openal.org/`
24. DevIL Development Team. DevIL – A full featured cross-platform Image Library, `http://openil.sourceforge.net/`
25. Simmons, A.: Main Page – NaviWiki, `http://navi.agelessanime.com/wiki/index.php/Main_Page`

Eighty-Color-Channel Lighting

Zoltán Márton[1] and Cecília Sik Lányi[2]

[1] Grandpace, József Attila u. 9., 8200 Veszprém, Hungary
[2] University of Pannonia, Egyetem u. 10., 8200 Veszprém, Hungary
marton.zoltan@grandpace.com, lanyi@almos.uni-pannon.hu

Abstract. Lighting simulation programs nowadays still use the CPU using global illumination. On the other hand computer games still use three channel color calculations. The modern GPUs (Graphics Processing Units) have enough performance to use more channels. Using more color (spectral) channels can create a more realistic virtual world. In this paper, we introduce a method to provide a better lighting for future's computer games and a possible replacement in some cases for those software render programs.

1 Introduction

Developing photorealistic virtual environments (VE), for example 3D animations or games or VE used in rehabilitation or therapy [1], we need to focus not only on the principle of ecological validity: modeling objects and environments, using good texture etc…, similar like real objects and environments, which fit to the users needs, but using correct lighting too.

What kind of light is correct in a scene? What is the quality of light? The answer would be: softness, intensity, color, attenuation [2].

What is photorealistic lighting? Most people don't realize that every light source has distinct color based on its color temperature [3]. We don't actually see this color because our eyes automatically adapt to changes in color temperature. Our eyes automatically correct the color of the light so that white objects look white under a large variety of yellowish white to bluish white lights [3]. This has to be included in the rendering program. On the other hand if we would like to emphasize the situation in which the observer sees the scene, as e.g. whether we render a street lighting scene in fog or dawn, or a Disco surrounding, some further considerations are necessary, and the present techniques are not always optimal to do such renderings. The problem becomes even more critical if not the color of the light, but of objects, illuminated with the lights of different spectral power distribution are considered.

In real world scenes specularity of objects and their reflection properties are important. Normally in computer graphics these are not rendered correctly. This is one of the first things to be learned in comparing photography and computer-rendering techniques. Phong and Gouraud specularity renderings are fake, they substitute perfectly round light sources with various controls to assimilate the way light interacts in the real world [4].

Lighting a scene in 3D is much like lighting a scene for photography, film or theatre. As an element of design, light must be considered a basic influencing factor at the beginning of the creative process and not something to be added later [5].

I. Lovrek, R.J. Howlett, and L.C. Jain (Eds.): KES 2008, Part III, LNAI 5179, pp. 473–480, 2008.
© Springer-Verlag Berlin Heidelberg 2008

Color and lighting is the most critical question and also one of the more complicated things in developing 3D environments. We have to understand relevant aspects of optics, physics, computer science, human perception and arts too. In the real world an artist does not matter if his/her hand, clothes are dirty with the paints or a sculptor sees the final statue in a rough stone with his/her mind's eye, a 3D animator or software engineer has to know everything about the renderer, what he/she uses. 3D rendering software and algorithms are behind the scenes, what they developed.

Nowadays VR systems use for lighting:

- software solution methods or
- hardware based methods with the R,G,B channels.

But with 3 channels the differences produced by sources of similar color temperature but different spectral power distribution can not be correctly rendered. We developed a new method for better simulating lighting. More than 40 color channels are used, this is already nearly as good as if the continuous spectrum of the light would be used. The method is based on the principles of colorimetry and radiometry.

2 Eighty-Color-Channel Lighting

The method is based on colorimetry, and Phong shading. The colorimetry base consists of using the CIE standard $\overline{x}, \overline{y}, \overline{z}$ spectral responsivity functions to provide a more realistic lighting. The Phong shading is required to do realistic shading.

2.1 Colorimetry Basics

The previously mentioned $\overline{x}, \overline{y}, \overline{z}$ spectral responsivity functions have been determined using additive color matching experiments. First the test subjects had been shown a light of a known wavelength and intensity, and had to match it with the mix of three basic color stimuli (red, green, blue). This test provided the so called $\overline{r}, \overline{g}, \overline{b}$ responsivity functions, but there was a problem with them. The test subject could not always mix exactly the same color from those basic colors. So to match the colors they had to add one color to the test light. This means that the $\overline{r}, \overline{g}, \overline{b}$ functions have negative lobes. To solve this problem the scientists have chosen to do a linear transformation on them. The $\overline{x}, \overline{y}, \overline{z}$ color matching functions are the result from that transformation. [6]

A color can be described with three coordinates X, Y, Z, (tristimulus values), which can be calculated from the sensitivity functions and the spectral distribution of the light to be evaluated, as shown in Equation 1

$$\begin{bmatrix} X \\ Y \\ Z \end{bmatrix} = \int_{380\,\text{nm}}^{780\,\text{nm}} I(\lambda) \begin{bmatrix} \overline{x}(\lambda) \\ \overline{y}(\lambda) \\ \overline{z}(\lambda) \end{bmatrix} \mathrm{d}\lambda \tag{1}$$

To do this on a digital system the integration is substituted by summation, usually for practical purposes on a 5 nm wavelength step size and bandwidth.

$$\begin{bmatrix} X \\ Y \\ Z \end{bmatrix} = \sum_{380\,\text{nm}}^{780\,\text{nm}} I(\lambda) \begin{bmatrix} \overline{x}(\lambda) \\ \overline{y}(\lambda) \\ \overline{z}(\lambda) \end{bmatrix} \Delta\lambda \tag{2}$$

where $\Delta\lambda$ is the band width and step size with which the measurement is made, in our case it will be 5 nm.

We can modify the spectral distribution of the light so that it contains the 1/5nm constant. So we get equation (3), where the I' is the modified spectral distribution. This equation will be used later in the demo program.

$$\begin{bmatrix} X \\ Y \\ Z \end{bmatrix} = \sum_{i=0}^{80} \begin{bmatrix} X(i \cdot 5nm + 380nm) \\ Y(i \cdot 5nm + 380nm) \\ Z((i \cdot 5nm + 380nm) \end{bmatrix} I'(i \cdot 5nm + 380nm) \tag{3}$$

To get computer display RGB values from the XYZ tristimulus values we need to transform the results. This requires us to know the display's characteristics: the XYZ tristimulus values of the display primaries, the white point and the gamma characteristic of each channel.

2.2 Phong Shading

The Phong shading is based on BFRD models. The simplest way to approximate the intensity (I) of a point is to use the surface normal vector (N) and the normalized light vector (L), it is known as flat shading.

$$I = N \cdot L \tag{4}$$

This is only for diffuse lighting, specular highlight and ambient (or indirect) lighting is more difficult.

A better way is to use the normal of the vertices instead surface normal, and interpolate the intensity between them. This method is called ground shading. The best approach is to interpolate the normal for each point and calculate equation (4) with that normal. If the interpolation is done by a look-up table it is known as bump-mapping. [7]

The demo program does not do bump-mapping; it does a simple sphere interpolation with the normal of the vertices. Because light has quadratic attenuation we need to calculate that too. Every light in the program is a point light.

2.3 Shader Model

A shader is a small program running on the GPU itself. There are two and some times three different kind of shaders: vertex shader, pixel shader and geometry shader. The last one is only present in DX10 capable video cards. The capabilities of the shaders depend on the version numbers they have. There are vertex shader 1.0, 1.1, 2.0, 2.0+, 3.0, 4.0; pixel shader 1.0, 1.1, 1.2, 1.3, 1.4, 2.0, 2.0a, 2.0b, 2.0x, 2.0+, 3.0, 4.0 and geometry shader 4.0.

The collection of those shaders are called shader model. As the shaders have their own versions so have the shader models. Shaders and shader models were developed by the video card vendors.

3 Development

Some decisions had to be done at the development of the program. Firstly there was need to choose the 3D API (Application Programming Interface). Nowadays there are two mainstreams 3D APIs available for Windows: Direct3D and OpenGL. OpenGL is more flexible and extendable than Direct3D, but Direct3D has better hardware abstractions. Direct3D does not depend so much on the hardware vendors than OpenGL does. Secondly the program had to work on most of the ATI and nVidia cards, and not just on the latest ones. The program needs at least minimal HDR (High Dynamic Range) support. This means that the card has to support float render target (float buffer) and Shader Model 2.0.

3.1 OpenGL

Although OpenGL has serious compatibility problems which make the development more complicate, OpenGL has been chosen, because it is flexible, and there is no need to rewrite the application when a new version comes out. OpenGL is extendable, that means, it has a basic API and some on OS (Operating System) and hardware vendor depending extensions. It is better to have an intermediate level API which does all the abstractions that are required. [8]

On the other hand a framework has been used, which abstracted beside of the required OpenGL abstractions, also the basic file I/O, the memory management, the thread management, the console and the windowing sub-system. Practically the program utilizes this framework.

3.2 The Shader Model 2.0 Pathway

The real development of the program started from here. Actually this Shader Model 2.0 pathway consists of two different pathways, because GeforceFX and Radeon 9500 and above cards support HDR in different ways. Basically both Shader Model 2.0 pathways work in the same way. Two floating point textures are created at start-up. The lighting is done in two passes. At the first pass the calculations are done from 465 nm to 780 nm (this means sixteen textures are utilized in the first pass from the required twenty per surface). The result is stored in the first floating point texture. The second pass calculations are done from 385 nm to 460 nm. The result are stored in the second floating point texture. These calculations are done by one shared vertex shader (vertex program) and by two different pixel shaders (fragment programs). The two textures are finally combined by another vertex shader, pixel shader pair. The pixel shader does the transformation into RGB color space and detects if the given intensity is too high to be displayed. The difference between the two pathways is on ATI card, the float texture must be a 2D texture and has to be bound to the float format pixel buffer, while on nVidia GeforceFX cards the float texture has to be a rectangular texture (texturerect) and there is no need to be bound to the float pixel buffer. The 2D textures of the Shader Model 2.0 have to be in both resolutions power of two. So the ATI way is more memory consuming than the nVidia way.

The texture coordinates of a 2D texture have to be normalized, which means that if it would access a texel to the matching pixel at coordinates (x, y) one would need to divide them by the matching texture resolution.

$$\begin{bmatrix} s \\ t \end{bmatrix} = \begin{bmatrix} 1/\text{texture width} & 0 \\ 0 & 1/\text{texture height} \end{bmatrix} \begin{bmatrix} x \\ y \end{bmatrix} \tag{5}$$

With rectangular textures one can simply access the matching texel of the pixel. So these differences require to have two different shader pairs. The work of this pathway is shown in Fig. 1.

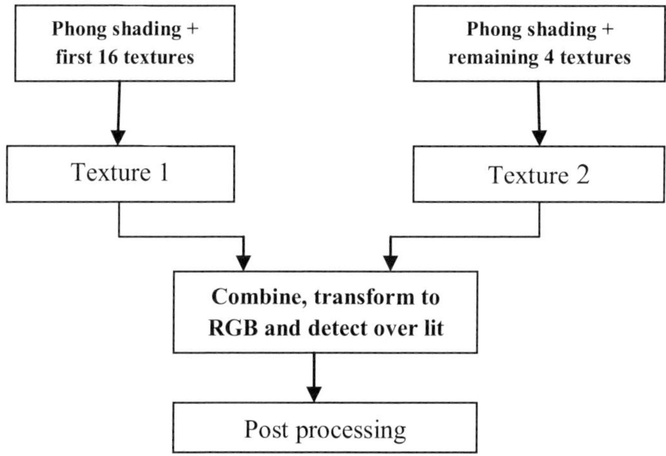

Fig. 1. Dataflow in Shader Model 2.0 path

3.3 The Shader Model 3.0 Pathway

The Shader Model 3.0 pathway works like the Shader Model 2.0 version, but there is no need for two different sub-pathways. Video cards supporting this shader model have the capability to do hardware floating point blending, and those support non power of two texture dimensions. So there is no need to create two textures and the texture can have exactly the same resolution as the client area of the window.

As previously mentioned in this case the HDR blending feature can be used to lower the overhead of the Shader Model 2.0 path, but the rendering must be still done in three passes. The first pass consist of drawing, using the first pass pixel shader for Phong shading with Z Test (Depth Test) set to less or equal test. At the second pass the drawing is done with the second pass shader, the writing to the Z-Buffer (Depth Buffer) is disabled and the additive blending is turned on. The third pass contains this time just the transformation and the over lit detection, and this operates this times just with one to float pixel buffer bind, non power of two, 2D texture. This process is shown on Fig. 2.

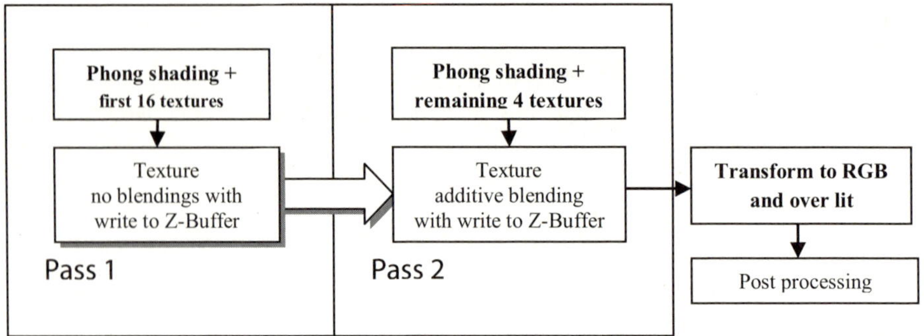

Fig. 2. The Working of the Shader Model 3.0 path

As it can be seen it is a bit simpler than the previous path. The shaders concerned in Phong shading are the same as in the previous pathway.

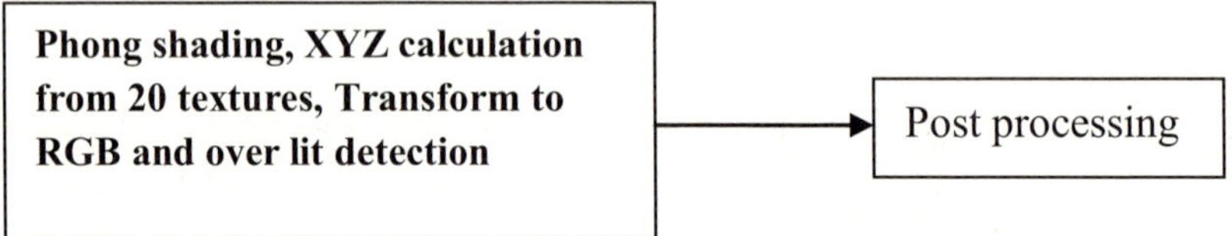

Fig. 3. The dataflow in Shader Model 4.0 case

This is the simplest pathway, because all three passes are done in a single shader. In previous shader models there was a limit which said sixteen textures could be utilized by a pixel shader. This limit has been increased in Shader Model 4.0 to thirty-two. So in this pathway there is no need to create float pixel buffers and float textures bind to them. Since it runes in one pass the Phong shading has to be done just once. That all means we can avoid a lot of overhead, but there will be a complex and long pixel shader, which has to do the Phong shading, calculate equation (3), transform *XYZ* to *RGB* and detect over lit. Fig. 3 shows the data flow of this process. Since the program uses OpenGL this pathway can run on pre Vista versions of Windows, but currently only on nVidia cards.

3.4 Results

The best frame per second performance, which was measured by the program it self, on the same hardware is by using the shader model 4.0 path. This path can have up to 50% better performance than the shader model 3.0 path. The lowest performance has been shown, when the ATI shader model 2.0 path has been used.[1] These tests were made on Geforce 8600 GT video card, AMD Athlon64 3500+ CPU, 3GB of RAM under Windows Vista Ultimate x64. The test results are shown in Fig. 4.

---

[1] The program automatically chooses the best pathway available. To do these tests the program has been modified for these occasions.

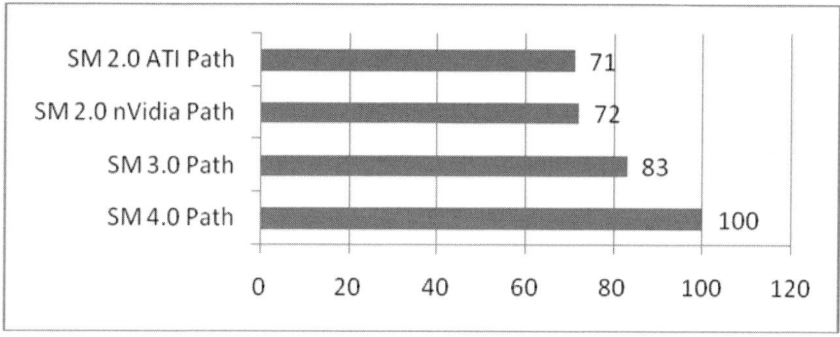

Fig. 4. The results of the different SM (shader model) pathway benchmark

Some other tests were made with the spectral coefficients of existing materials and spectral distributions of existing light sources. There are six light sources at the same position, and same intensity. Fig. 5. shows the result pictures. The program does not contain any chromatic adaptation model yet.

Fig. 5. Results of the rendering. Spectral distributions of existing light sources and a non-existing equi energetic light source are being used. (From left to the right, from top to down, in the first picture a HP1 standard light is being used, in the second one HP2 and so on. In the last one is the equi energetic lightsource being used).

As one can see on Fig. 5. the differences between to different colored surfaces are depending on the spectral distribution of the light source and on the reflection coefficients of the surfaces as it can be expected in real life.

4 Current State

The program currently does not support specular and ambient lighting. The post processing consists of the simple checking of the alpha channel and if it finds a value above 128 then it displays a lamp icon on the upper right corner.

As previously mentioned the shader model 4.0 pathway works only on nVidia cards, this may be changed by OpenGL 3.0.

5 Future Work

Extending the program with a chromatic adaptation model to simulate better the real life. Bringing shader model 4.0 path to work under ATI video cards. Doing more test to proof the current approximation model.

6 Conclusion

In this paper a method to create a more realistic lighting, has been shown and with a small demo program and its results presented the capabilities of the program have been demonstrated. The program utilizes s radiometry based method. In its present form this method runs, but it requires a lot of resources, mainly video memory. In the future this method could be used in computer games for example.

On the other hand this method needs just a bit more extension and it could be used to replace nowadays global illumination based lighting simulation software. Because that software would use the GPU it could create real-time experience.

References

1. Sik Lányi, C.: Virtual Reality in Healthcare. In: Ichalkaranje, A., et al. (eds.) Intelligent Paradigms for Assistive and Preventive Healthcare, pp. 92–121. Springer, Heidelberg (2006)
2. Birn, J.: Lighting & Rendering. New Riders Publishing (2000)
3. Fleming, B.: 3D Photorealism Toolkit. John Wiley & Sons, Inc., Chichester (1998)
4. Kundert-Gibs, J., Derakhshani, D.: Maya Secrets of the Pros. SYBEX Inc. (2005)
5. Berndt, C., Gheorghian, P., Harrington, J., Harris, A., McGinnis, C.: Leraning Maya 6, Rendering, Alias Systems (2004)
6. CIE (2003) Technical report: Colorimetry third edition, CIE Publ. 15 (2004)
7. Phong, B.T.: Illumination of Computer-Generated Images, UTEC-CSs-73-129 (1973)
8. Lengyel, E.: The OpenGL Extensions Guide. Charles River Media (2003) ISBN 1584502940

Virtual Reality Supporting Environmental Planning Processes: A Case Study of the City Library in Gothenburg

Kaj Sunesson[1], Carl Martin Allwood[2], Ilona Heldal[1,*], Dan Paulin[1], Mattias Roupé[3], Mikael Johansson[3], and Börje Westerdahl[3]

[1] Dept. of Technology Management and Economics,
Chalmers University of Technology, SE - 412 96 Göteborg, Sweden
[2] Dept. of Psychology, Lund University, SE-221 00 Lund, Sweden
[3] Dept. of Civil and Environmental Engineering, Chalmers University of Technology,
SE-412 96 Göteborg, Sweden

Abstract. The introduction of Virtual Reality (VR) models in the environmental planning process can change the traditional roles of the parties involved. This paper examines the influence of VR-models on the role of professionals in an architectural competition concerning proposals for rebuilding the city library in Gothenburg. The study shows that VR was experienced as useful by the jury, while troublesome by the architects. In the future architects are likely to have to produce the VR-models for their proposals by themselves. Therefore smaller firms may experience it as difficult to enter the market or to have full control over their VR contributions. Further results show how VR can be more efficiently used in the evaluation process by improving communication between different stakeholders.

Keywords: Virtual reality, VR-models, Architectural competition, Building planning, Visualization, Environmental planning.

1 Introduction

New technology often causes disruption and change in established practices. In the future the use of Virtual Reality (VR) models is likely to increase in the city planning process. This will lead to changes also in how the activities and roles in the early parts of urban building processes are conducted. The purpose of this paper is to portray the effects of the use of VR-models in an architectural competition evaluation process from the standpoint of different professions, and to illustrate and discuss how VR can influence the roles of important professions with focus on architecture. It is known that different groups of professionals use graphical models differently depending on their background [1]. For example, sketches, used in the early phases of a planning process, can show simulated effects deviating from the effects that the users will experience with the new building. A sketch may also mean different things for different individuals [2]. Experiencing graphical representations can also vary depending on

---

* Contact author: Email: ilona.heldal@chalmers.se

I. Lovrek, R.J. Howlett, and L.C. Jain (Eds.): KES 2008, Part III, LNAI 5179, pp. 481–490, 2008.

the culture [3], physical properties of the artefacts [4], light conditions [5], the used technology [6], or colours [7].

In an early work, Hall 1996 (reviewed in [8]) assumed four relevant interest categories in the urban planning process: 1) city planners, 2) architects working for clients, 3) the clients themselves and 4) the general public. In this context, modern VR-models provide many benefits. For example, although not yet optimal, VR-models can offer a representation that is closer to reality than other methods of presenting a concept of a planned building or urban area. Furthermore, VR-models can provide a communication medium that makes it possible for all interested parties to have access to a common representational medium, which makes it easier for them to achieve an understanding of the planned building object on the basis of their own common sense [9]. These benefits are already shown, for example, by VR-models supporting communication between politicians and engineers [10], informing the public with visualization of virtual heritage [11], communication between experts and the public in road planning [12], or involving different public segments, e.g. school children in evaluating urban changes [13], or disabled people evaluating their home [14].These studies show that the impact of VR-models "on cognition and preference might be significant" ([11], p. 1, 2007). Still, the issue of *how* VR functions as a means of communication in ongoing urban development processes for evaluations is insufficiently researched [16]. Although exploratory usability-oriented studies involving VR programs have been carried out (e.g. [16]), very few studies have been reported on the role that VR plays, and could play, in ongoing urban planning contexts (e.g. [17]). The present study analyzes the conditions and events for an instance when VR was used in an architectural competition in one urban planning process. A more focused examination is presented by Sunesson & al [18].

Traditionally and currently, architects belong to a specialized profession and, in many of their activities, tend to work in some isolation from other parties in the construction process. The production of sketches and plans for new buildings in specific environments is usually done according to profession-specific criteria, concerning how best to combine aesthetics, function and form, how the features of the building interact with the properties of the environment, etc. [2, 19]. Taking a more pessimistic approach, Bourdakis [9] suggested that architects "have a reputation for submitting non-realistic perspectives, omitting parts of schemes that are not fully designed and even hiding areas of schemes behind carefully placed trees and other features" (p. 8).

1.1 The Present Case Study

In this study we analyzed a real event involving the use of presentations of VR-models in an architectural competition concerning the best model for a large rebuilding of the city library in the second biggest city of Sweden, Göteborg. The library is placed at a square with great symbolic status for the city. Thus, the remaking of the library is an issue of great importance to the city.

In 1999, for the first time in Göteborg, it was decided that VR technology should be used in a large urban planning area. The VR-model for the project would be produced at the Visualization Studio at Chalmers University of Technology. The resulting model was successfully used to aid communication mainly with politicians and the general public. At present, the Visualization Studio works in close collaboration with the City Planning Authority in Göteborg (CPA) to generate a large VR-model, covering the major parts of the city, that will be used in future urban planning and building design projects. So far, two specific projects have resulted. The first of these is associated with the redevelopment of the city library, further described in this study. To our knowledge, this is the first time that VR technology has been used in this way for a public building in Sweden. Based on the suggestions from the four architectural firms that the Visualization Studio elaborated, the VR-models were used in the current evaluation. As we have noted, this is unlikely in the future, when architects themselves will have to make their own models. Considering the opinions from an independent party, e.g. the Visualization Studio, we can relate the work of the four different architectural firms to each other. Thus, this process created a unique opportunity to examine differences and similarities in the attitudes of the architectural firms in relation to using VR-models in their future projects.

To produce the VR-models, MRViZ – open source software, developed by Visualization Studio (see http://www.chl.chalmers.se/~roupe/MRVIZ.htm)- was used. This software program can import models from Archicad (see http://www. graphisoft.com), Autodesk® (see: http://www.autodesk.com), and Sketchup (see http://www. sketchup.com) products. These are typical software programs used by architectural firms. For this project two architectural contributions were made in Sketchup and one in Archicad, while one firm sent its DWG files (Architectural Desktop from Autodesk) to the Visualization Studio. The VR-model was formed as a surrounding environment in which each of the models of the proposed buildings was placed (see Figure 1). Each of the contributions could easily be presented while keeping the surrounding environment constant. The system supported on-demand switching between the different proposals when reviewed by the participants. The surroundings simulated an area which is approximately 500x500 m^2 around the city library. The surroundings were imported from the virtual city model of Göteborg, a project mentioned above. In the constructed VR-model of the city library the users can select areas, choose predefined views, or look around from certain points of the model. They can also fly around, walk around, or follow volume changes. MRViZ allows populating the area with avatars, or using 3D sound, if so desired.

2 Method

In total, ten interviews with twelve informants from the different professions involved in the competition were conducted. The interviewers were: three from the CPA, three from the jury, four from the four architectural firms, and two 3D-artists. One of the three informants from the CPA had a central role in the evaluation process as a secretary for the jury group. Another informant was an architect who participated in the jury. The third was a GIS coordinator and worked with IT-strategy questions. All three were males, aged between 35 and 62. One of the three informants of the jury represented the library and her expertise was how to operate libraries. The second, a woman, represented a political party in the city's cultural committee. The third, a male, represented the building committee. Their ages were between 50 and 60. Four informants were architects who had worked with the proposals that were part of the

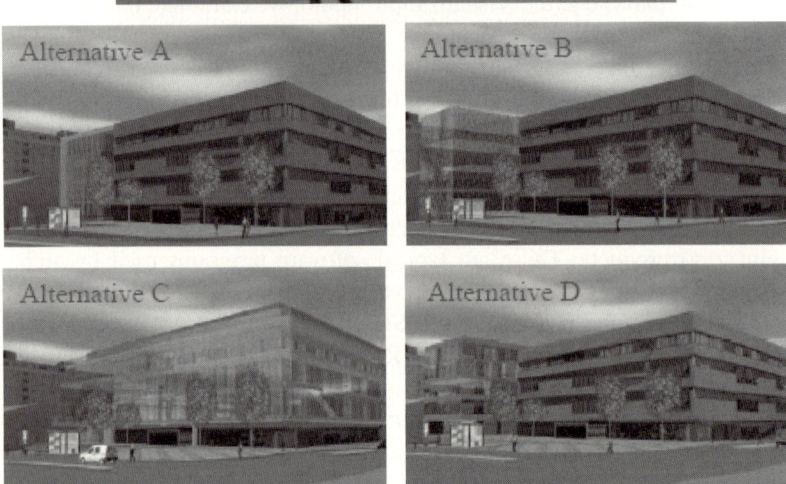

Fig. 1. The VR-model of the actual library and the four suggestions: A, B, and D exploit the place on the left of the actual library and C around and above it

evaluation. They came from three of the four firms that had contributed proposals to the competition. In one interview four architects took part concurrently. Two were women and two men. Three were approximately 30 years old and one was 45 years old. The two 3D-artist who made the VR-models were approximately 30 years old males and were interviewed at the same time.

2.1 Study Design and Interviews

Semi-structured interviews, between 50 and 90 minutes long, were conducted at the informants' workplaces by two of the authors (except two interviews) and were tape-recorded. The common sections of the interviews included the following:

- The informants' prior experience of 3D modelling and VR. (For the architectural firms the experience of the whole firm was also asked for.)
- A description of the process and communication when the VR-models were produced and used.
- Evaluation of pros and cons of using VR.
- Implications of the future use of VR for professions and evaluation processes.

3 Results

We first briefly describe the main events in the architectural competition as they were reported by one or more of the informants. Next the views of each participating profession as stated in the interviews are presented as aggregated for each profession.

3.1 Main Events in the Evaluation Process

In 2003 the cultural committee and the city library of Göteborg described the need for bigger premises, and in January 2006 the politically appointed building committee gave the CPA the commission to develop a new plan for the city library and the area around it. Four architectural firms were invited for parallel submissions of proposals. Each of the firms would develop a proposal (below also called contribution) for how a rebuilding of the city library could be designed. A jury was appointed to evaluate the proposals. This jury consisted of eight members from different interest groups: two came from the town council's cultural committee, two from the town council's building committee, two from the CPA, one from the company that administrates official buildings in Göteborg and one from the city library.

Initially the architectural firms were not asked to create any 3D or VR-model, but in May 2006 an e-mail was sent out to the four firms asking for digital material, including the CAD models that the firms had developed. All the firms agreed to send their material. It was decided that the models should be prepared for use in the VR-model by the Visualization Studio. The 3D-artists' task, given to them by the city planning office, was to make VR-models with all contributions equal *in quality*. This way of handling the issue was at least partly because of the late inquiry about the digital material. The VR-model was formed as a surrounding environment in which each of the models of the proposed buildings was put. The system supported on-demand switching between the different proposals when reviewed by the participants.

The four contributions were presented to the jury in August 2006. The task for the jury was to evaluate the proposals and to recommend one of them for further elaboration. To fulfill their task the jury had nine meetings, and in one of these a VR-model that showed the contributions was presented and discussed. At this meeting, which took place at the Visualization Studio, the VR-model was shown on a power wall, and a stereoscopic effect was created by the use of crystal eyeglasses. Later the jury had two or three further meetings that did not involve VR presentations.

3.2 The City Planning Authority Informants

This was the first time when CPA used VR in this manner in an evaluation process. The CPA has no widespread knowledge about VR (even 3D-modelling is not always used) but some parties/persons have good experience of VR. All three informants agreed that the initiative for using VR in the evaluation process came from the CPA and was put as a suggestion to the jury. The informants thought that the use of the VR-models made a big difference for how the jury understood the four contributions. The possibility to understand the interaction with the surroundings and to actually see the volumes was said to have meant a great deal for many of the jury members.

All informants pointed out the importance of being able to show the planned buildings in VR. One informant focused on the pedagogical aspects of using VR, and noted that it is important to find a quality level in the VR-rendering which is good enough to really depict the content of a proposal, but not so good that the general public understands the proposal as already finished. The informants thought that VR will be used much more in the future.

3.3 The Jury Informants

Two of the informants had some experience of tasks similar to this one in judging new buildings. The representative from the building committee also stated that he was used to reading blueprints and being able to see how these will appear in reality. All three informants use computers regularly, but none of them work with (or had much if any experience of) VR or 3D-graphics. No informant reported exerting any influence on how the VR-model was to be shaped or was actually developed.

Some of the informants thought that it was somewhat inconvenient to have to travel to another part of town to look at the model. The informants reported not having any particular expectations when going to this meeting. They all also reported being satisfied with not having to manage the model themselves during the show. One informant thought that it was enough to see the VR-model on one occasion, and no one seemed to think that it would have been very important to have seen the model more than this time. The VR-model gave much information but the informants felt that it would have been too much effort to have used it in more meetings. Here the informants referred to having to use the crystal eyes when viewing the model, and having to pay attention to people pointing to the power wall instead of just discussing. One of the informants mentioned that the model did not include the interior of the buildings and that this was an important part of the evaluation.

The credibility of the VR-models was rated high. One statement was that the VR-model was seen as giving a good presentation of how the contributions would appear in reality. It was also reported that the jury referred back to the VR-model in later meetings when they discussed different matters. All the jury informants criticized the aesthetic aspect of the model. One saw a risk that even the foundations of the architect's idea of the building could be missed, because the model seemed finished in a sense but was not good enough to "explain" what the architect had in mind when designing the building. The informants all confirmed the 3D-artists' view (see below) that the VR-model did explain the size, the volume of the building and its interaction with the surroundings. All informants thought that VR will be a part of future evaluation processes. One of them thought that VR is an important tool because it can explain the content of the blueprints for persons not used to these. Another informant thought that VR use will place more demands on future contributions. One informant stated that for the architects VR will imply one more interest group (the general public) to take into account.

3.4 The Architects

All three architects said that their knowledge about VR was "small". They did not know what possibilities the technique provides or what features they could ask for.

All three architectural firms used 3D-CAD more or less on a daily basis for presenting their jobs. One of the firms had special personnel for making most of its 3D illustrations. The second firm made some of the smaller illustrations in-house but otherwise hired help. The third firm hired help for most of its illustrations, although it did much of its work sketching in 3D (Sketchup). The architectural firms had some criticism about the process for deciding which proposal would be developed further, but none of the firms criticized the use of VR as a way to visualize the proposals.

All the firms had experienced that information about the use of VR as a tool for presenting their proposals came abruptly, and that the request to send in the material had come late in the development process. They had all made the sketches and drawings that originally were asked for when the inquiry about their 3D material came. Before sending in the material, one of the firms "cleaned" the model so that less information could be misinterpreted. All three firms stated that it took them about 8 hours of work to prepare the material for sending it off. Two of the firms recalled a feeling of worry at this stage. The basis of this worry was the lack of control over how the 3D-artists would render their proposals. The communication between the firms and the 3D-artists during the time the VR-models were produced consisted of one or two e-mails with screenshots that the architect firms received and could give a response to. In addition, some phone calls were made to clarify particular questions. None of the architectural firms saw the model before the VR-models were shown to the jury. Two of the firms felt that the screenshots did not do justice to their contribution, and one of them also stated that it did not know what to express in its response due to its lack of experience of VR.

Architectural firms did not think that the use of VR was important for the outcome of the competition. They thought that the decision was made according to other types of criteria. All the firms thought that the screenshots did not really show their proposals in an aesthetic manner. All the firms thought that VR will affect their work in the future and that VR will be a common tool. Two of the firms explained that they saw it as a new tool and a better possible way to express their proposals. The third firm was more sceptical about what value it will provide. All three firms said that an important purpose of the sketches that are used today is to give an initial feeling and to market the proposal. The representative of the third firm felt that VR just means "more of the same" of what is going on today, but he thought that it would be necessary to use VR in order to be seen in the "media stream".

If all presentation materials have their own role, including VR as a new presentation material with new values, and these all are needed in the planning process, according to Al-Kodmany [2], te architects have to face greater expectations – which also means a greater work load at the present stage. Here it has to be noted that the status and duties of the architects can vary from one country to another.

3.5 The 3D-Artists

The quality of the material from the architectural firms differed, ranging from work in Sketchup with a somewhat low level of detail to work made in Architectural Desktop of somewhat better quality, and work in Archicad with high quality. Due to the difference in quality, the contributions had to be worked with at different levels. The Archicad model could be put straight into the VR environment. Most work was done on the Sketchup models. All the four contributions were rendered in VR in parallel. When making the models, the 3D-artists sent screen dumps to the architectural firms and asked for feedback, and in a few phone calls specific questions to the firms were asked. The communication mostly concerned questions on material representations. At this stage every part of the building was already decided upon. The 3D-artists had 60 hours to finish all the VR-models. The time was sufficient for making the models, but additional time had to be spent on implementing a better representation for glass reflections. One of the 3D-artists worked with the external model (representing the surrounding environment) and the other worked with adapting the buildings to the surrounding environment. Overall, the 3D-artists thought that the making of the models and the communication with the architectural firms worked well. During the VR presentation one of the 3D-artists controlled the showing of the model. The members of the jury group could direct what perspectives and which parts of the model were seen by instructing this person, and the jury used this possibility.

The 3D-artists concluded, from the jury members' communication during the meeting, that the model had given a broader understanding about the size and volume of the contributions. Also the interaction with the building's surroundings seemed to be clarified. The jury did not appear to have felt that the aesthetic aspects of the models had been better clarified by experiencing the VR-models. The 3D-artists thought that VR will be much more used in future city planning projects and that architects will have to learn to make VR-models themselves in the future to pursue their profession. However, one big threshold for widespread use of VR may be that, for a simple use of VR, a common model of the surroundings is needed. Without such a common environmental model, the cost will be high for making VR-models of every proposal with surroundings and terrain.

4 Discussion

The notion that VR will influence the architectural profession is supported by our data, but exactly how is not equally clear. The discussion by [2], reviewed in the introduction, about the future role of different professions in the urban construction process, was fairly well taken. However, although VR will influence the construction planning process, this does not imply that the architectural profession has to be "downgraded" as suggested by [2]. The architect informants in this study tended to think of VR more as a new tool that gives a possibility to express their ideas than as a threat to their profession. However, the comment from one of them that VR mostly might be a necessary demand for being seen in the stream of media, without its adding any substantial new value, is an indication that more may be at stake.

Another conclusion from this study is that there were many indications that VR could have been more efficiently used in the evaluation process. The 3D-artists thought that the communication worked well – and in one sense it certainly did – but whether the small amount of input to the production process from the architects was an effect of their satisfaction, or was due to lack of experience with VR, is not clear. Not knowing what to ask for or to demand in the context of VR-rendering of a proposal can be seen as a problem. As reported above, it was not obvious to the architectural firms what response they should give to the screenshots they were sent from the

3D-artists. The meeting of different stakeholder interests in the construction process is also evidenced by statements by the informants from the jury and city planning office that the VR-model clarified how it "really will look". Since neither the jury representatives nor the architects knew what to ask for or what to expect from the showing, the quality (and thus the value) of the VR-models may not have reached full potential.

The present study shows that it is important for all parties in the construction process to be clear about how VR will be used in the evaluation process. Moreover, clear information should be given to all parties early in the process about how the VR-models are to be produced and, on the basis of how the models are produced, how they should best be interpreted. In this context it is significant that the architects expressed worry that it was someone else who would transform their proposals into VR, and that the proposals could be distorted in the process. Given that VR will be used much more in the future, architects will have to learn to produce VR-models by themselves. This could be a disadvantage for small architectural firms because of the extra investments in knowledge, time and money that are needed to enter the market.

References

[1] Whyte, J., Bouchlaghem, D., Thorpe, T.: Visualization and Information: A Building Design Perspective. In: Proceedings of International Conference on Information Visualization, IV, London, pp. 104–109 (1999)

[2] Al-Kodmany, K.: Visualization Tools and Methods in Community Planning: From Freehand Sketches to Virtual Reality. Journal of Planning Literature 17(2), 189–211 (2002)

[3] Sik Lányi, C., Sik, A., Sik, G.: Virtual world –A brave new world" – but what kind of colours are used in these virtual environments? Poster, 26TH Session of the CIE, Beijing, China, vol. 1, pp. D1-99–102 (2007)

[4] Sik Lányi, C., Lányi, Z., Tilinger, Á.: Using Virtual Reality to Improve Space and Depth Perception. Information Technology Education, 2003, 291–304 (2002), http://www.jite.org/documents/Vol2/v2p291-303-27.pdf

[5] Fekete, J., Sik Lányi, C., Schanda, J.: Night-time driving – new light sources in car headlamps. In: CIE Midterm Meeting and International Lighting Congress, León, Spain, May 12-21, pp. 58–65 (2005)

[6] Billger, M., Heldal, I.: Virtual Environments Versus a Full-Scale Model for Examining Colours and Space. In: Proceedings of the 1st International Conference on Virtual Concepts, pp. 249–256 (2003)

[7] Mátrai, R., Sik Lányi, C., Zs., K., Schanda, J.: The importance of colours and contours in navigation. Lux Junior (2005)

[8] Bourdakis, V.: Virtual Reality: A Communication Tool for Urban Planning. In: Asanowicz, A., Jakimowitz, A. (eds.) CAAD – Towards New Design Conventions, pp. 45–59 (1997)

[9] Westerdahl, B., et al.: Users' evaluation of a virtual reality architectural model compared with the experience of the completed building. Automation in Construction 15, 150–165 (2006)

[10] Drettakis, G., et al.: Design and Evaluation of a Real-World Virtual Environment for Architecture and Urban Planning. Presence, Teleoperators and Virtual Environments 16(3), 318–332 (2007)

[11] Laing, R., et al.: Design and use of a virtual heritage model to enable a comparison of active navigation of buildings and spaces with passive observation. Automation in Construction 16(6), 830–841 (2007)
[12] Heldal, I.: Supporting participation for planning new roads by using virtual reality systems. Virtual Reality, Special Issue for E-society 11(2), 145–159 (2007)
[13] Buchecker, M., Hunziker, M., Kienast, F.: Participatory landscape development: Overcoming social barriers to public involvement. Landscape and Urban Planning, 29–46 (2003)
[14] Sik Lányi, C., Hajnal, Z.: Design and developing a virtual home for aphasia patients. In: Assistive technology from virtuality to reality, 8th European conference for the advancement of assistive technology in Europe, pp. 158–162 (2005)
[15] Lange, E.: Issues and Questions for Research in Communicating with Public through Visualizations. In: Buhmann, E., Paar, P., Bishop, I., Lange, E. (eds.) Proceedings at Anhalt University of Applied Sciences: Trends in Real-Time Landscape Visualization and Participation. Wichmann Verlag, Heidelberg (2005), http://www.masterla.de/conf/pdf/conf2005/11lange_c.pdf
[16] Roussou, M., Drettakis, G.: Can VR be Useful and Usable in Real-World Contexts? Observations from the Application and Evaluation of VR in Realistic Usage Conditions. In: Proc. HCI International (2005)
[17] Caneparo, L.: Shared Virtual Reality for Design and Management: The Porta Susa Project. Automation in Construction 10, 217–228 (2001)
[18] Sunesson, K., Allwood, C.M., Paulin, D., Heldal, I., Roupé, M., Johansson, M., Westerdahl, B.: Virtual Reality as a New Tool in the City Planning Process. Journal of Tsinghua University-Science and Technology (to appear, 2008)
[19] Svensson, C., Tornberg, E., Rönn, M.: Arkitekttävlingar, Gestaltningsprogram och Arkitektonisk Kvalitet. Working Paper, TRITA-ARK-Forskningspublikationer, Skolan för arkitektur och samhällsbyggnad, KTH, Stockholm, Sweden (2006)

A Prototype for Human-Like e-Learning System

Kenji Yoshida[1], Yuji Shinoda[2], Koji Miyazaki[3],
Kayoko Nakagami[4], and Hirotaka Nakayama[5]

[1] Konan Boys' High School
[2] Institute of Intelligent Information and Communications Technology, Konan University
8-9-1 Okamoto Higashinadaku, Kobe 658-8501 Japan
shinoda@center.konan-u.ac.jp
[3] Nirvana Technology Inc.
[4] Kamiminami Osaka City Junior High School
[5] Dept. of Information Science and Systems Engineering, Faculty of Science and Engineering,
Konan University

Abstract. There are two types of e-Learning systems, namely, a synchronous type system and an asynchronous type system. Interactive communications with a teacher are important for education. However, an asynchronous, lecture type e-Learning system has some difficulty for keeping students' motivation and encouraging active learning. In this paper, we introduce a human-like e-Learning system based on machine learning techniques. This system has 3 kinds of subsystem. The fist is the detection of students' situation by patterning. The second is the automatic question answering. The third is to generate facial expressions of animations synchronized with pre-recorded audio data. Our human-like e-Learning system provides experimental contents for active learning and communicative functions such as chat. As a result, students can learn with a pleasant feeling.

Keywords: e-Learning, Edutainment, Automatic Answering System, Human-like System.

1 Introduction and Purpose of the Research

E-Learning systems are spreading over companies and universities rapidly. Synchronous type systems which use TV conference systems have a sense of reality and interactivity with the teacher. This system also has the demerit of limitation of time and place. On the other hand, an asynchronous e-Learning system has no limitation on time and place. Students are able to have a lecture anytime or everywhere. However, there is no interaction between students and the teacher mostly in existing e-Learning systems where students listen to the lecture in a passive manner only. Many existing asynchronous e-Learning systems becomes simply a portal to send any contents files through the internet. As a basic idea, however, we emphasize that e-Learning systems should assist students to learn by discovery through trial-and-error experiments. Moreover, e-learning systems should be human-like in order to keep the interest of students.

I. Lovrek, R.J. Howlett, and L.C. Jain (Eds.): KES 2008, Part III, LNAI 5179, pp. 491–498, 2008.

To solve this issue, we suggest a human-like e-Learning system which has the ability to teach according to students' learning situation. To human-like e-Learning system, we require 3 important functions: 1) The system must have function to distribute contents to students depending on his learning situation. 2) The system must accept students' questions at any time and answer automatically. 3) The system must have a function of entertainment that is known as edutainment. Edutainment is a word which combines education and entertainment. It is desirable for e-Learning systems to have a function of entertainment so that students keep interest in the contents.

In this paper, we construct a prototype of a human-like e-Learning system by using machine learning techniques. The key of our research is to integrate the three required functions into one system. Students access the learning site, where they learn the contents which include a page that lets student be active (through experiment and trial-and-error). The system watches the students' activity, and then the system judges the students' situation. Further contents are served based on the judgment of the system. This system serves contents according to each student. This is similar to a role-playing game. It is important for the e-Learning system to serve the contents that matches situation of each student.

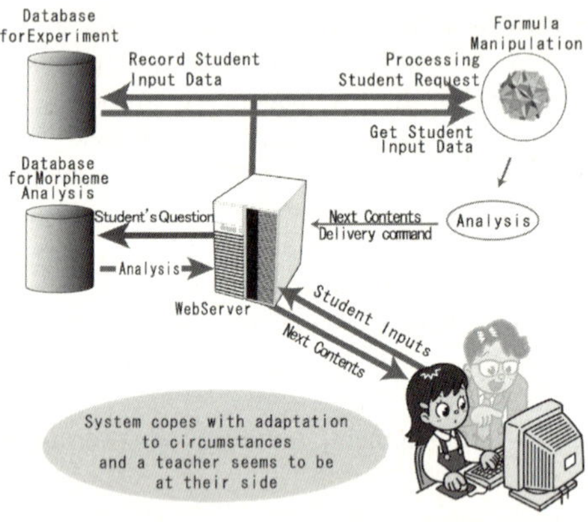

Fig. 1. Image of human-like e-Learning system

2 Constructions of the System

We made the contents of high school mathematics syllabus unit "Quadrics and their graphs". The contents are a lecture and some experiments of drawing graphs. Fig. 2 shows the image of the contents on our system. This system has 3 kinds of subsystem to provide human-like e-Learning system. The first is the detection of students' situation by patterning. The second is the automatic question answering. The third is the facial expressions of animations that synchronized with audio data.

In the following, we explain those three functions along an example of graphs of quadratic functions in a course of mathematics in high school. In this course, students are supposed to learn the relation between the parameters (a,b,c) and the graph by inputting the value of (a,b,c) in a manner of trial-and-error. Students have a chance to ask any questions at will. The teacher in the system observes their action, and gives them an advice. If students do not listen to the teacher's comment, she may get angry. Through this e-Learning system, students can learn with a pleasant feeling by discovery through trial-and-error experiments.

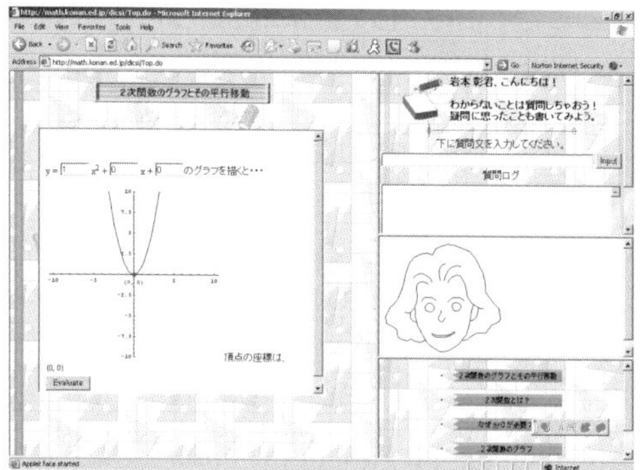

Fig. 2. Contents Image

	Student's Input	Generate Pattern		
		Change	a	b	c
1st	$y=\boxed{1}x^2+\boxed{2}x+\boxed{3}$	1st -> 2nd	1	0	0
2nd	$y=\boxed{2}x^2+\boxed{2}x+\boxed{3}$	2nd -> 3rd	0	1	1
3rd	$y=\boxed{2}x^2+\boxed{3}x+\boxed{4}$	3rd -> 4th	1	1	1
4th	$y=\boxed{1}x^2+\boxed{2}x+\boxed{3}$	4th -> 5th	0	1	1
5th	$y=\boxed{1}x^2+\boxed{0}x+\boxed{0}$	Pattern	1	1	1

Fig. 3. Patterning of a Student's Experiment Data

2.1 Adaptive Distribution of Learning Contents

Usually, there are differences in the degree of progress of students. To distribute contents depending on student's situation is one of the essential functions for human-like e-Learning systems. The system observes input data of students and chooses appropriate instruction contents according to students' situation [1].

We create learning contents that include experiments. In the contents, students read explanations and have to operate graphs under instructions by inputting mathematic expressions for the graphs. These contents are a kind of e-Learning drill. Intelligent mathematics drills are known in the field of CAI [2]. In these works, the model of the students' are used to describe how the students understand the targeted field. In our paper, however, we construct a model of the students for finding out the objective of the students' experiment.

For making teaching contents according to students' situation, we patternize student's input data [3]. For example, a student have experiments of drawing "y = ax^2 + bx + c" graphs. Fig. 3 is an example of patterning of student's experimental input data. From observation result the system generate Patterns form Student's input as (a, b, c). If the pattern is (1,0,0), the students is judged to note parameter a. (0,1,0) is judged as the student notes parameter b. (0,0,1) is judged as the student notes parameter c. (1,1,1) is judged as the student notes nothing. (0,0,0) is judged as the student could not operate. By this patterning, the system provides the next suitable content for the student.

To provide learning environment for students to learn through trial and error, we define typical series of patterns of students' experimental input data. These patterns describes students' situation of their experiments. For example, if a student continuously changing parameter a, he might consider the effect of this parameter. We generated 634 kinds of students' situations, and add 5 types of tags that describe the meaning of the situations.

To find a students' situation from his experiments, we use support vector machines [4] [5] [6]. We trained the support vector machines in advance, and try to find the meaning of the students' experiments. We constructed a system to give advices that matches with the students' situation. If he did not listen to our lecture, we give him negative feedback. Usually, teacher scolds a student who does not listen to the lecture in real class. So it is a necessary function for the system.

2.2 Automatic Question Answering

If a student has question, the system must answer quickly. Otherwise, the students may lose interest to the content. In this section we introduce automatic question answering.

2.2.1 Template of Questions

It is difficult to answer questions immediately in an e-Learning system without answer models. We prepared some template questions as follows [7].

1. How does a graph change when the parameter a is changed ?
2. How does a graph move when I change parameter b ?
3. How does a graph move when I change parameter c ?
4. Please teach me relation between the y-intercept and parameter a, b, c.
5. What is the relation between apex and parameter a, b, c?
6. How can I get the coordinate of apex of parabola?
7. Is the graph of quadratic function symmetrical?
8. Please tell me how to do right-left parallel translation of the graph?
9. Please tell me how to do up-down parallel translation of the graph?
10. Can you draw the graph of an side opened parabola?

A student inputs a question; the system selects a most similar template question to the inputted question by using TF-IDF. The system displays the selected template question, and asks the student whether the meaning of the template question is the same as that of the student's question. If the question of the student has a different meaning, the system displays a list of other candidates of template questions. Then the student can select a question from the list. This procedure is described in more detail as follows.

2.2.2 Morpheme Analysis

It is necessary for the system to understand students' question correctly. To deal Japanese sentences, we use the morphological analysis system 'Sen' [8]. The template questions and students' question are decomposed by Sen into indexing words. The system computes the precision by using these words lists. The system uses necessary word class (part of speech) and measures distance of the student's question and template questions.

2.2.3 Vectorization of Question Word

Let T_i (i=1, 2, 3, ..., n) denote template questions. Then we get the word list (WT_i) of T_i. And let S_j (j=1, 2, 3, ..., N) denote students' questions. Then we get the word list (WS_j) of T_i. For example,

T_1 : "a の値を変化させるとグラフはどのように変化しますか？"

    (How does a graph change when the parameter a is changed ?)

WT_1 ={a, 値, 変化, する, グラフ}

    (WT_1 ={a, value, change, do, graph})

S_1 : "a を変化させるとグラフはどのようになりますか？"

WS_1 ={a, 変化, する, グラフ, なる}

We introduce the following notations:

w_i	: i-th word in the question W
m	: Number of w
n	: Number of template questions
N	: Number of student question of the past
$f_{i,k}$	: Frequency of w_i in the k-th question
n_i	: Number of questions which includes w_i

$$W = \left\{ \bigcup_{i=1}^{n} WT_i \right\} \bigcup \left\{ \bigcup_{j=1}^{N} WS_j \right\} \tag{1}$$

$$m = |W| \tag{2}$$

Logarithmic TF(Term Frequency) weight for k-th term $l_{k,j}$ is given by

$$l_{i,k} = \log(1 + f_{i,k}) \quad (i = 1,2,..., m) \tag{3}$$
$$(k=0,1,2,...,n)$$

where $k = 0$ shows student's question and $k = 1, 2, ..., n$ shows template questions. Here $l_{i,0}$ is the weight for a term of student's question. Probabilistic IDF (Inverse Document Frequency) weight for i-th term, g_i is given by

$$g_i = \log \frac{n - n_i}{n_i} \quad (i = 1,2,..., m) \tag{4}$$

$\boldsymbol{q_0}$: Weight vector for terms of student question

$$q_0 = (l_{1,0} g_1, l_{2,0} g_2, ..., l_{m,0} g_m) \tag{5}$$

$\boldsymbol{q_k}$: Weight vector for terms of template question

$$q_k = (l_{1,k} g_1, l_{2,k} g_2, ..., l_{m,k} g_m) \quad (k = 1,2,..., n) \tag{6}$$

We compute the similarity between student's question q_0 and templates q_k ($k = 1,2, \ldots, n$) by cosine measure:

$$\cos\theta = \frac{q_0 \cdot q_k}{\|q_0\| \|q_k\|} \tag{7}$$

The system selects the template question which has the maximal cosine measure.

TF-IDF is often used in text mining. However, a student's question is sometimes too short to identify the template questions. Therefore we update weights whenever a student asks the system a question.

2.3 Facial Expressions of Animations That Synchronized with Audio Data

Human teacher's voice is more warmhearted than voices generated by speech synthesis software. In this section we introduce our development for an expression of animations synchronized with the pre-recorded audio data.

2.3.1 Automatic Drawing of Facial Expression

From a view point from edutainment, we constructed an automatic drawing system of emotional animation for given vowed data. This animation is useful from two reasons. The first reason is the contents become more attractive by the animation's actions and voices. The second reason is that the contents are constructed with less effort. Fig. 4 shows example of a character in happiness and anger including information of vowel (Japanese hiragana).

In our method, the system displays the animated teacher depending on the students' learning behavior. If a bad behavior of student is observed, the system orders the animated teacher to display an angry face and to speak some comment or advice. These activities are based on rules that are stored student action database. Inputs of the rules are based on the detection of students' situations that are mentioned in the section 3.1.

This automatic facial expression drawing system reduces our task a lot. There are some systems that generate computer graphic speaking characters. TVML [9] is one of a works to generate computer graphic of speaking characters' face and voice from a given text. Our method focused to add facial expressions to recorded teacher's voice. We need only to prepare the audio data for each explanation. Usually, without our proposed automatic drawing method, it is time consuming to generate animations synchronized with pre-recorded voice data. Synchronizing the mouth movement and the voice is complex. Besides, it is difficult for us to modify the animation in case that voice data are changed. In the case of e-Learning system using video capture, we need skill to edit the movie. On the other hand, the proposed real-time drawing system for automatic facial expression overcomes those difficulties.

2.3.2 Data for Facial Drawing

Facial drawing data consists of five items.

1. Audio file name
2. Length of audio play time （sec）

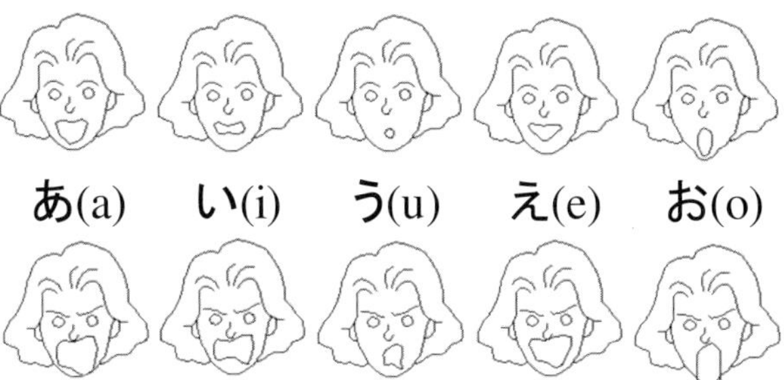

Fig. 4. Normal Facial Expression and Angry Facial Expression for Each Vowel

   3. Japanese hiragana character data
   4. Rate of pronouncing time for each character in a sentence
（Short: 1 <---> 5 <---> 9 :Long）

For example, the following two sentences are included in a Web page according to the order of the above four items;

./wave/0_1.wav

5.38

にじかんすうのぐらふと、そのへいこういどうについてべんきょうしましょう

555555 5 555 5 9 55 5555 555 5 558 55555 55555

Here we comment on the calculation way of play time of character. A value r_i is the relative time length of voice of vowel of i-th character. T is the length of audio play time. The play time for the character t_i is computed as follows.

$$R = \sum_{i=1}^{n} r_i \tag{8}$$

$$t_i = \frac{r_i T}{R} \tag{9}$$

The shape of a mouth of a vowel keeps for t_i seconds. In this way we use the audio data and vowel sound synchronization file, and the system generates a real time animation.

3 Conclusions and Future Study Directions

In this paper, we propose a prototype of a human-like e-Learning system by using machine learning techniques. We integrated three functions and construct as a private lesson system. This system gives advices to students, and receives questions in any time as if a teacher is standing by them. On the other hand, our current question and answer system cannot reply to questions which do not include the template question. In this case, the present system sends an e-mail to the teacher. We are able to change teacher's expression easily by using real-time facial expression drawing. In the proposed system, we are able to make the educational contents without video capture system or any special skills. These are an important merit for creating educational content in practice.

Acknowledgement. This work is supported by ORC (Open Research Center) project (2004-2009) of MEXT (Ministry of Education, Culture, Sports, Science and Technology), Japan.

References

1. Andoh, M., Ueno, M.: Contents estimate for e-Learning using an eye-mark recorder. JSiSE Research Report, vol. 19, pp. 11–18 (2004)
2. Anderson, J.R., Corbett, A.T., Koedinger, K.R., Pelletier, R.: Cognitive tutors: Lessons learned. Journal of the Learning Sciences 4(2), 167–207 (1995)
3. Yoshida, K., Nakayama, H.: Construction of an e-Learning System Controlled by Student Input. Mem.Konan Univ., Sci. & Eng. Ser. 51(2), 165–174 (2004)
4. Vapnik, V.N.: Statistical Learning Theory. John Wiley & Sons, Chichester (1998)
5. Nakagami., K., Nakayama, H.: On Multi Classifivation by Fuzzy Support Vector Machines. In: 23th Conference of The Society of Instrument and Control Engineers in Kyusyu, pp. 129–130 (2004)
6. Nakagami, K., Yoshida, K., Nakayama, H.: Optimal Contents Selection in E-Learning using Support Vector Machine. In: 37th Conference of The Society of Instrument and Control Engineers in Hokkaido, pp. 111–114 (2005)
7. Yoshida, K., Nakayama, H.: Auto-answering to learners' questions in an e-Learning system. Mem.Konan Univ., Sci.& Eng. Ser. 52(2), 95–104 (2005)
8. Sen Project, http://ultimania.org/sen/
9. Hayashi, Douke, Hamaguchi: Automatic TV Program Production with APEs. In: Proceedings of the Second International Conference on C5 Creating, Connecting and Collaborating through Computing Conference, pp. 18–23 (2004)

A Cross-LMS Communication Environment with Web Services*

Takashi Yukawa and Yoshimi Fukumura

Nagaoka University of Technology
1603-1 Kamitomioka-cho, Nagaoka-shi, Niigata 940-2188 Japan
`yukawa@vos.nagaokaut.ac.jp`

Abstract. The present paper proposes a cross-LMS communication system for distributed, autonomous, and collaborative learning environments. The authors have been conducting the eHELP project, which promotes the sharing of e-learning contents and the transfer of credits between academic institutions in Japan. A problem in achieving collaboration between students at different institutions is revealed, whereby different LMSs are used at the participating institutions. To solve this problem, a cross-LMS bulletin board system (cBBS) and a cross-LMS chat system (cChat) using web services are proposed. The present paper describes the requirements, designs, and implementations of these systems. Working examples of the cBBS are also presented.

1   Introduction

Based on recent progress in e-learning, a Credit Transfer System (CTS) has been promoted among a number of higher education institutions for the purpose of promoting flexible learning in Japan. In European countries, the European Credits Transfer System (ECTS) has been implemented in order to achieve cross-boarder education.

Nagaoka University of Technology (NUT) concluded a CTS agreement using e-learning among four national universities and six (at present 13) national technical colleges in Spring 2004 and started to deliver lectures through e-learning. The project is called eHELP, which stands for e-learning Higher Education Linkage Project [1].

NUT has been using the OpenSourceLMS [2] as a Learning Management System (LMS) to deliver lectures for eHELP on campus. In addition, NUT

- has been developing learning materials based on the international standard Sharable Content Object Reference Model (SCORM) 2004 [3].
- has a policy to use open-source systems for e-learning,
- has developed an original Bulletin Board System (BBS), a chat system, and a report submission system to fulfill the requirements for communications among learners and instructors.

---

* This research was supported by the Ministry of Education, Science, Sports and Culture of Japan through a Grant-in-Aid for Scientific Research (B), 19300276, 2007.

I. Lovrek, R.J. Howlett, and L.C. Jain (Eds.): KES 2008, Part III, LNAI 5179, pp. 499–506, 2008.

These facts are the reasons for employing OpenSourceLMS, which is SCORM2004 compliant, and are decisive factors in providing effective e-learning lectures on the NUT campus.

From the viewpoints of consistent operation, economical efficiency, and ease of maintenance, the unified LMS environment should be constructed over eHELP participants. However, since the institutions have different policies and have already constructed their own e-learning environments, it is very difficult for the members to switch to a unified LMS. Consequently, the students are required to master the operation of various LMSs.

Since the scale of the CTS will be increased, some technological measures are needed in order to provide consistent operation. These can be

- developing e-learning contents to be compliant with the SCORM standard, and distributing e-learning contents to the LMSs of the member institutions,
- providing communication interfaces that have a look and feel that is consistent with the local LMS, and sharing communication contents over LMSs using Web Services.

The first approach allows a student view any learning contents with his/her local LMS, and the second approach allows the student to communicate with students at other institutions using local user interfaces.

The present paper proposes a distributed communication environment which provides each learner a consistent operation even if the system consists different types of LMSs. The requirements for collaboration of heterogeneous e-learning institutions are clarified, then a design of cross-LMS communication system using the Web Service Architecture is proposed. Finally, the system is implemented and the feasibility of the cross-LMS collaborative communication is demonstrated.

2   Background

This section presents a brief introduction to the eHELP project and its environment and the LMSs used in the project.

2.1   eHELP Project and Learning Environments

The eHELP project is a project that constructs co-operative frameworks for higher education, verifies the effectiveness of e-learning by applying large-scale e-learning programs, and clarifies domains of applicability of various e-learning models (delivery configuration) including a cohort-based model. The objectives of the project are sharing educational resources between universities and colleges, providing mutual knowledge, and providing flexible learning for students.

During the initial phase of the project, each institution simply delivered its educational contents to the other participants using its own LMS. Each institution issued user accounts of its LMS for the students from other institutions and the students logged onto the system to take a course. The students have one user-ID for each of the institutions at which classes are taken. The students also have to acquire operation skills for each of LMSs used in the institutions. This is burdensome for the students because if they want to take courses at various institutions, they have to learn multiple LMSs.

The project is approaching the commencement of next phase, which focuses on collaborative learning between the institutions. Communication between students who log on to different LMSs becomes important in this phase.

2.2 Learning Management Systems

A Learning Management System (LMS) is a server system that delivers learning contents, manages user accounts of learners, and accumulates the learning histories of learners. In addition, the LMS provides functions for communication among learners and teachers. The communication functions are significant for collaborative learning. For example, if a teacher assigns a problem and wants the students to solve it through discussion, a BBS is the best solution in the context of an e-learning environment.

Institutions participating in eHELP use various types of LMS, including custom LMSs, commercial LMSs, and freely available LMSs. Some institutions use Moodle [4], which is the most popular free LMS, while others use Open-SourceLMS.

3 A Cross-LMS Communication Environment

As described in the previous section, it is cumbersome for students to manage numerous user-IDs and learn the operations of various LMSs. In order to reduce this redundancy, students should be able to log-on to the LMS with a single user-ID and should be able to access LMSs with a consistent look and feel.

In the eHELP project, learning contents can be built as SCORM compliant, and most LMSs are compatible with SCORM. Therefore, each participating institution simply copies the contents to its own LMS and is able to deliver the contents locally. However, this manner of delivery cannot provide traversal communication between students at different institutions. It is desirable for LMSs to add functions that will enable students to communicate with each other, regardless of their home institution.

The requirements for cross-LMS communication functions, and the design and implementation of these functions are described below.

3.1 Requirements for the Cross-LMS Communication Environment

When a student wishes to use a single LMS, in parallel, he/she should communicate with students in different institutions using BBS or Chat. Accordingly, the requirements for the cross-LMS communication functions are:

- A student accesses his/her local LMS and can read and post articles to/from BBS/Chat on the local LMS;

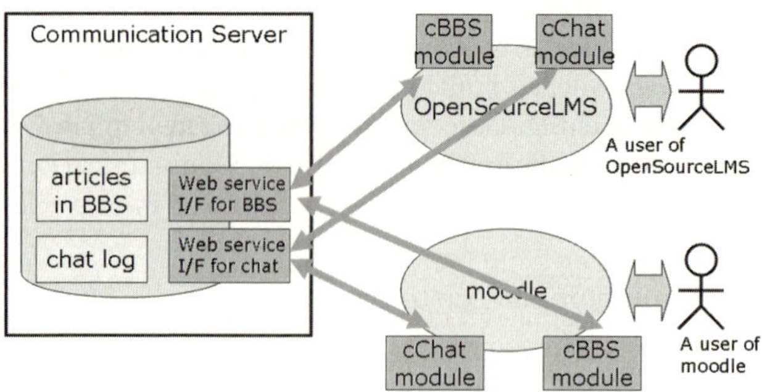

Fig. 1. System Structure

- Students in the same class can read every article posted to the LMSs, regardless of which LMS the students use to log-on;
- The articles can include a message text, message linkages (thread information), and attached files containing office documents, images, videos, sounds and/or Squeak animations [5].

3.2   Designs of cBBS and cChat

This subsection proposes designs for a cBBS and a cChat, which are communication systems that satisfy the above-mentioned requirements.

The decision of a model of the system, i.e., a server-client model or a peer-to-peer model, must be made. The authors have chosen a server-client model because a peer-to-peer system requires bothersome management tasks. That is, articles should be accumulated in a central server (communications server, CS), and a user interface is located in each LMS. To avoid the complicated configuration of a firewall at each institution, only the normal Web access port should be used for communicating between the CS and LMSs. Consequently, a Web Services Architecture is employed for developing the system.

Figure 1 shows a structural overview of the system. A collaborative-BBS (cBBS) and a collaborative-Chat (cChat) are provided. The CS stores every article of cBBS and all utterances in cChat. The CS also provides Web Service interfaces to access articles and utterances. The client-side modules in this system work with the LMSs and provide user interfaces that have a similar look and feel to their host LMS.

Most LMSs have their own BBS (called Forum in some LMSs) module and provide various functions for the convenience of teachers and students. Implementing every function supported by LMSs would not be wise because users would be confused by the functions from other LMSs. The authors studied the BBS functions of various LMSs and extracted the following functions as the essential functions:

- Reading and posting articles;
- Replying to any article;

- Thread display of a title list;
- File attachment for use in the integrated risk management education program, eSAFE [5];
- An important posting notification (IPN), which automatically alerts a BBS user that someone posts an article relating to his/her interest [6].

For cChat, the following functions are also extracted:

- Reading and writing sentences;
- An automated topic display (ATD) function that extracts focuses of interest of chat-rooms and provides them to users who are looking for the appropriate chat-room where he/she should join [7].

With implementing above functions, the client modules of cBBS and cChat can act as optional modules of each LMS. Therefore, a teacher may use them in his/her course in a similar fashion to the built-in BBS and Chat modules. The difference from the built-in modules is that the proposed modules can share messages among participating LMSs.

3.3 Implementation of cBBS and cChat

The LMS implementation issues are identified as follows:

- What type of storage system should be used for message store?
- What type of protocol framework should be used for CS–Client interaction?
- How are IPN and ATD modules incorporated to the client modules?

The authors chose a Relational Database Management System (RDBMS), specifically PostgreSQL, for message store because it satisfies both the versatility and performance requirements. The CS–Client interface framework is chosen as Web Services, as mentioned in the previous sections.

In order to achieve the IPN and ATD functions, natural language processing, which consumes a lot of processing resources, is employed. These functions should be implemented as independent programs and invoked as asynchronous processes to avoid interfering with the basic functions in cBBS and cChat. The IPN process reads messages through the CS server process, processes them according to the methods proposed in [6], and then writes the result to the CS server.

As an example of CS-Client interaction, Figure 2 shows the system sequence when a user is reading an article. The steps in the sequence are:

1. A user invokes the BBS interface;
2. The cBBS module sends a request for a title list;
3. The CS process accesses the data store and returns the title list;
4. The cBBS module displays the title list on the screen;
5. The user views the title list and chooses a message to view;
6. The cBBS module sends a request for the article.
7. The CS process retrieves the article from the message store and returns it.
8. The cBBS module displays the article.

Every interaction in this sequence uses the Web Service APIs.

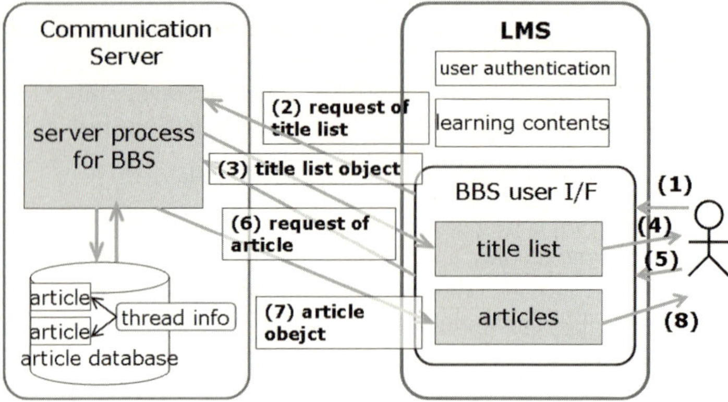

Fig. 2. System Sequence for Reading an Article

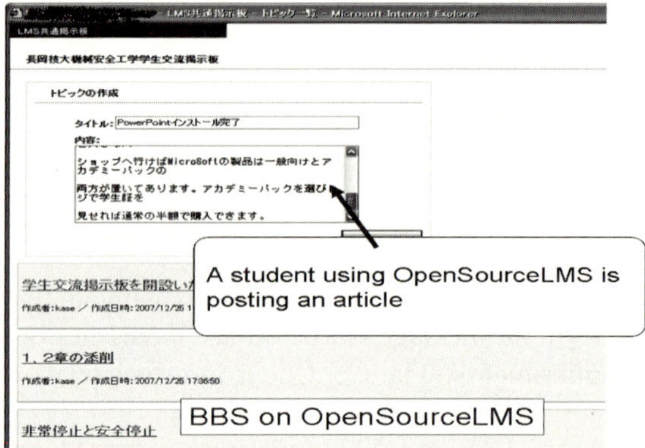

Fig. 3. Screen-shot of a Working Example (1)

3.4   A Working Example

Screen-shots of a working example of cBBS are shown in Figures 3 through 5. The articles in the figures are written in Japanese because the system is used in Japanese institutions. However, the contents of the articles are not significant in the example showing the behavior of the system.

In Figure 3, a student (Student A) using OpenSourceLMS is posting an article. After the article is posted, another student (Student B) who is using Moodle can read the article, as shown in Figure 4. Note that Students A and B are studying at different institutions and are using the LMSs operated by their local institutions. The user interface modules for both students also work with their local LMSs. Therefore, the look and feel of each of the interfaces is consistent with that of the local LMS. Student B then attempts to reply to the article, as

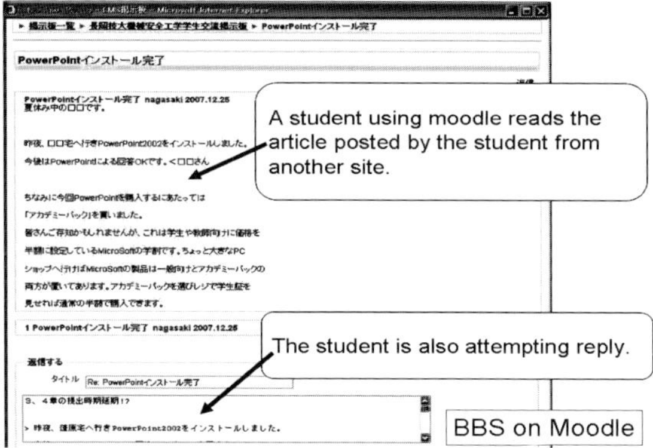

Fig. 4. Screen-shot of a Working Example (2)

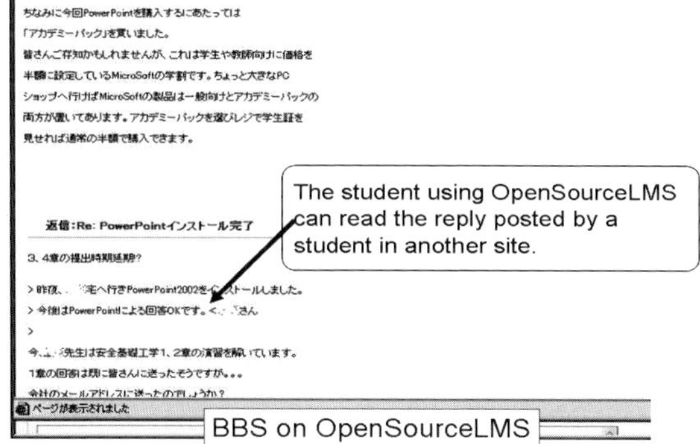

Fig. 5. Screen-shot of a Working Example (3)

shown in Figure 4. When Student B posts a response, it can be read by Student A, as shown in Figure 5.

4   Discussion

Achieving the cross-LMS communication environment would have some impacts on the management and design of e-learning courses.

In the eHELP project, the teacher attempts to construct a community of learners in an e-learning course using a BBS. This attempt is currently limited to a group of students at an identical institution. The cBBS enables this attempt to be extended to students at different institutions. It is known that the community helps to keep learners motivated, and therefore the number of students who drop out of courses is expected to be reduced. This would have an impact on the management of e-learning course.

Another teacher may plan to have students solve mathematical problems collaboratively in an e-learning class. Collaborative work would induce better understanding of the subject. If the students are distributed over several institutions, the cChat can be used for this collaboration. This would have an impact on the design of e-learning course.

The cross-LMS communication environment and the learning designs mentioned in this section allow students to feel mental experiences (including affinity and sens of community) as if they belong to a single physical classroom, although they are studying at different institutions. The students are then able to collaborate with other students smoothly. This would contribute to assuring the educational quality of the CTS with e-learning.

5   Summary and Future Research

In the present paper, the requirements and design of a cross-LMS communication system were addressed. In order to popularize the Credit Transfer System using e-learning, a distributed learning environment is developed and cross-LMS communication functions are required for seamless interactions between students at different institutions. Collaborative BBS and Chat systems with the Web Services Architecture were proposed herein. The system consists of a centralized communication server and the client modules, each of which corresponds to a different function and LMS. The client module accesses the message storage through the Web Services API provided by the communication server. The design and management of e-learning courses in a distributed environment were also discussed.

The proposed system has just recently been implemented. The quantitative evaluation of the concepts mentioned in the previous section are areas for future study.

References

1. Fukumura, Y., Maruyama, K.: e-Learning program for supporting to create safety and secure society. Journal of Multimedia Education Research 12(2), 37–48 (2006) (in Japanese)
2. http://www.elc.or.jp/cgi-bin/scorm_engin/lms/index-scorm.html (2004) (in Japanese)
3. http://www.adlnet.org/index.cfm?fuseaction=scormabt (2004)
4. http://moodle.org/
5. Yukawa, T., Amarume, H., Tochio, M., Kimura, T., Nakahira, K.T., Fukumura, Y.: An intelligent bulletin board system for an e-Learning program for safety engineering. In: Proc. ED-MEDIA 2006, pp. 2493–2500 (2006)
6. Yukawa, T., Amarume, H., Fukumura, Y.: An Important Posting Notification Function in an Intelligent Bulletin Board System for e-Learning. In: Apolloni, B., Howlett, R.J., Jain, L. (eds.) KES 2007, Part III. LNCS (LNAI), vol. 4694, pp. 761–768. Springer, Heidelberg (2007)
7. Yukawa, T., Kawano, K., Suzuki, Y., Tansuriyavong, S., Fukumura, Y.: Implementing a Sense of Connectedness in e-learning. In: Proc. ED-MEDIA 2008, pp. 1198–1207 (2008)

A Web-Based Asynchronous Discussion System and Its Evaluation

Hisayoshi Kunimune[1], Kenzou Yokoyama[2], Takeshi Takizawa[2],
Takuya Hiramatsu[1], and Yasushi Fuwa[2]

[1] Faculty of Engineering, Shinshu University
[2] Graduate School of Science and Technology, Shinshu University
4-17-1 Wakasato, Nagano City, Nagano, 380–8553 Japan
kunimune@cs.shinshu-u.ac.jp

Abstract. This paper describes the evaluation of an asynchronous discussion supporting system for web-based materials. The system, which is named "Writable Web," provides the features assisting asynchronous discussion by sharing annotations. We improve the features in order to support the annotation sharing of the system on the basis of the first trial experiment conducted in an experimental class. Then, we carry out the second experiment in an actual class. The system is evaluated from two aspects. One aspect is the effect of the asynchronous discussion, which is measured from the scores obtained in a written assignment, and other is the functionality of the features based on the result of a questionnaire survey carried out after the experiment.

Keywords: web-based training, e-Learning, collaborative learning, asynchronous discussion.

1 Introduction

Recently, web-based training courses have gained wide-spread popularity in educational facilities and companies. In these courses, the course materials are provided as web pages. The advantages of web-based materials over paper-based materials are as follows:

- Multimedia contents such as movies, audios, and images can be included.
- The contents can be updated anytime by lecturers.

On the other hand, learners prefer to write annotation texts and marks on paper-based materials for better understanding of the subject[6]. However, learners cannot benefit from the abovementioned features of the web-based materials if they take printouts of the same to write annotations, This is because the printouts do not include no multimedia contents, thus becoming outdated when the web pages are updated. Many learners who study by distance e-learning face several problems related to printed web-based materials[5].

I. Lovrek, R.J. Howlett, and L.C. Jain (Eds.): KES 2008, Part III, LNAI 5179, pp. 507–514, 2008.

In addition, online discussions during web-based training foster the learners' knowledge construction. In order to provide an online discussion environment in asynchronous distance-learning courses, many web-based learning materials include online forums[1,4]. In discussions regarding the learning materials, learners often need to quote related parts of the material because online forums are separated from the learning materials.

To solve these problems, several systems that allow the user to incorporate annotations into the web-based contents and to share these annotations have been proposed. Koivunen has proposed a system "Annotea," which requires each user to install the client software[3]. Cadis has offered an annotation sharing system that needs to prepare proprietary servers such as Microsoft SQL Servers and Office Server Extensions: this system works only on with Windows[2].

We have proposed an annotation sharing system called "Writable Web," which is implemented as a web server application: thus, users need not install any special client software. The result of the first trial experiment showed that learners mainly used this system for understanding the important parts of materials clearly and for organizing and improving their knowledge. Moreover, the experiment showed that learners are dissatisfied with the usability of this system. On the basis of these results, we improve the annotation sharing functions and user interface of this system to offer usefulness to help share annotations. Sharing annotations directly on learning materials will facilitate online asynchronous discussions.

In this paper, we introduce an overview of "Writable Web" and discuss the evaluation of the improved functions and usability from the results of the second trial experimental conducted with the improved system.

2   Overview of "Writable Web"

2.1   Main Features of the System

"Writable Web" has the following improved features over other annotation systems mentioned in the previous section[5]. The system works as a server-side web application between the users' web browser and web servers that provide web-based learning materials. Therefore, "Writable Web" works on commonly used web browsers, without the need for a particular client software. Figures 1 and 3 show the systems that run on Firefox and Internet Explorer, respectively.

Users can directly attach two types of annotations to a given part of the text on the web pages, namely, highlighting (called as marker) and attaching text (called as memo) by using the system.

The system allows multiple users to write and share annotations after authentication by their IDs and passwords. The annotations written by each user are stored in a database on the server: therefore, users can share their annotations online. The usability of this system is improved according to the result of the first trial experiment, and the system has an organized structure of frames and usable floating menu, which enables one to choose the desired type of annotation.

2.2   Functions for Supporting Asynchronous Discussion

"Writable Web" has the following features that support asynchronous online discussions and communications:

– Making comments to shared memos.
– Organizing comments to thread structures.
– Notifying newly arriving comments to a user's memo or comment.
– Showing the rank of the web pages that are bookmarked and annotated by most of the users.

Though users can view the ideas of other users by sharing annotations, they cannot express or share their own ideas or opinions with other users. This system allows communications among users by making comments to shared memos. The existing web-based training courses have forums that enable communication among lecturers and learners. However, these forums are separated from the learning material texts, and users often need to quote a part of the text. On the other hand, our system facilitates communication directly on the texts of the learning materials. Therefore, users can clearly recognize that the discussion is related to the given part of the material. Figure 1 shows the form used to write a comment to a memo shared by another user. This system organizes the comments attached to a memo as a thread structure, as shown in Fig. 2.

This system provides a portal page to each user after the user logs into the system (Fig. 3). The portal page offers the users a form to input the URL and a bookmarking function to select web-based materials. The history of a user's annotations is also provided to the user on this page. Moreover, the portal page shows the users newly arrived comments attached to their memos or comments (Fig. 4) and displays the rank the web pages that are bookmarked and annotated by most of the users (Fig. 5). These functions show the user the summary of the memos and comments and the activities of the discussions on each page. Therefore, by using these functions, the users can easily follow the asynchronous discussions on many materials.

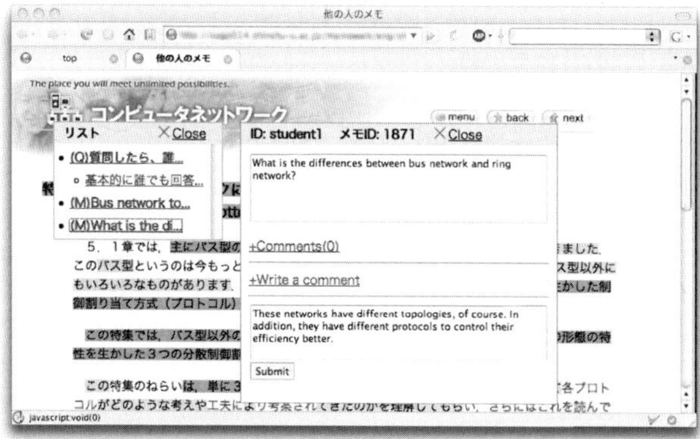

Fig. 1. Writing a comment to a memo

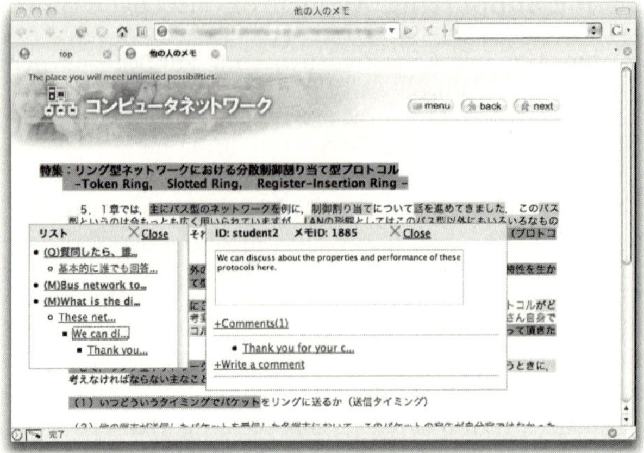

Fig. 2. The thread structure of comments

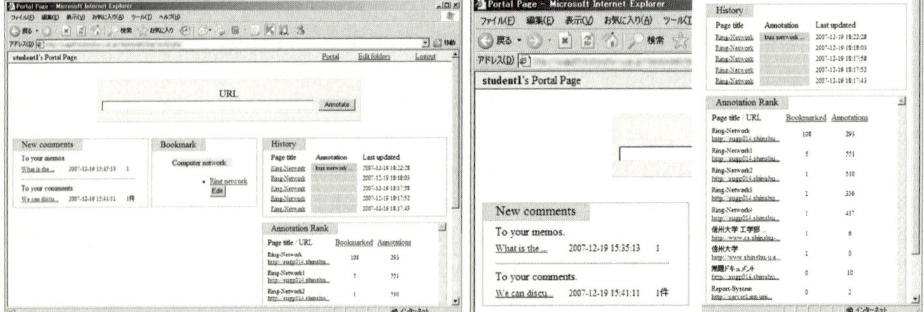

Fig. 3. An example of portal page

Fig. 4. The list of newly arrived comments

Fig. 5. The ranks of bookmarks and annotations

3   Trial and Evaluation

We conducted a week-long second trial experiment in order to evaluate the abovementioned features and user interface of the system with 88 undergraduate students. The students participated in a blended learning program called "Computer network," where they used "Writable Web" to annotate on the web pages in four rounds of face-to-face lectures and to have online asynchronous discussions on the web material during the lectures. The material consists of five web pages. The first page describes the introduction of the material. The second, third, and fourth pages describe first page describes the characteristics of the three types of network topologies, and the last pages describes the comparison between them. Therefore, students can discuss the overview of these network topologies on the first page, the features of each network topology from the second page to the fourth page, and the differences between these topologies on the fifth page. We believe that in order to participate in the discussions, each student must understand the contents of the material: furthermore, viewing other students' opinions helps him/her understand the concepts in a better manner.

3.1   Evaluation of Asynchronous Discussions

All the annotations, including memos and markers, are shared among all the students in this experiment. Thus, students can attach comments to other students' memos and comments. In this experiment, the students annotated 735 markers (avg. 8.5 markers for each student), 1030 memos (avg. 12.0 memos), and 133 comments (avg. 1.5 comments) to 113 memos (avg. 1.2 comments for each memo).

After this experiment, the lecturer gave an assignment to the students. In this assignment, the students had to summarize the differences in the characteristics between several network topologies with valid reasons. The same assignment was given to 84 undergraduate students from the previous year's class who had not participated in the asynchronous discussions. We discuss the difference in the assignment scores between these two sets of students for the evaluation of the asynchronous discussions. The lecturer of the class grades the assignment papers on the basis of the following nine viewpoints on a scale of zero (not understood) to two (understood well).

(1) states that the throughputs of token-ring and slotted-ring networks are almost the same.
(2) states that the throughputs of token-ring and slotted-ring networks always are higher than those of register-insertion ring networks.
(3) describes the reasons for (1) and (2).
(4) states that the highest throughput of token-ring networks is less than 100%.
(5) describes the reasons for (4).
(6) states that the delays in communication encountered in token-ring and slotted-ring networks are shorter than those in register-insertion ring networks under low load conditions.
(7) describes the reasons for (6).
(8) states that the delay in register-insertion ring networks is shorter than that in the token-ring and slotted-ring networks under high load conditions.
(9) describes the reasons for (8).

A paper that includes the summary of the web-based materials is highly graded from the viewpoints of (1), (2), (4), (6), and (8) because the materials are based on the same. On the other hand, the other viewpoints serve to confirm a student's understanding of the mechanism of these networks. Therefore, we can separate these viewpoints into two groups —"viewpoints A" and "viewpoints B"— depending on their properties mentioned above.

Table 1 shows the scores obtained in the assignment held in 2006 and 2007. The mean scores based on viewpoints A are 7.42 (standard deviation (S.D.)= 1.51) in 2006 and 7.76 (S.D.= 2.26) in 2007. The mean scores are based on

Table 1. The scores of the assignments.

	2006 ($N = 84$) Mean (S.D.)	2007 ($N = 88$) Mean (S.D.)	t-test for equality of means
Viewpoints A	7.42 (1.51)	7.76 (2.26)	$t(152.8) = 1.17$ $p = 0.121$
Viewpoints B	4.05 (2.33)	4.89 (2.57)	$t(170) = 2.24^*$ $p = 0.014$
Total score	11.5 (3.29)	12.6 (4.42)	$t(160.6) = 1.99^*$ $p = 0.024$

Note: N implies the number of students in the given year's class. $* : p < .05$.

viewpoints B are 4.05 (S.D.= 2.33) in 2006 and 4.89 (S.D.= 2.57) in 2007. The means of the total scores based on viewpoints A and B are 11.5 (S.D.= 3.29) and 12.6 (S.D.= 4.42), respectively.

To determine where there are differences in the mean score based on viewpoints A and B in 2006 and 2007, the t-test is performed with the null hypothesis (H_0), which indicates that the "mean scores are the same", and the alternative hypothesis (H_1), which implies that the "mean score in 2007 is greater than that in 2006." The results of the t-test show that there is no significant statistical difference between the mean scores based on viewpoints A ($t(152.8) = 1.17$, n.s.). However, the mean score based on viewpoints B in 2007 is significantly greater than that in 2006 ($t(170) = 2.24, p < .05$). As a result, the mean of the total scores in 2007 is also greater than that in 2006 ($t(160.6) = 1.99, p < .05$).

This result shows that asynchronous discussions among students help improve raises their scores. In particular, these discussions improve the student's understanding, which is not possible by simply reading the learning materials.

3.2 Evaluation of Functionality

After the experiment, we carried out a questionnaire survey. The students were questioned about the usefulness and intended purpose of the features provided by the system, and 69 students answered these questions.

Figure 6 shows the results of the questions regarding the usefulness of the annotation features in the learning process. The first question is based on writing annotations directly on the web-based materials. The second question inquires about sharing annotations among users, and the third question is based on writing comments to others' memos. The last question is based on the usefulness of sharing annotations and writing comments to communicate with other users. From these results, we can observe that the features of writing and sharing annotations and writing comments are highly valued by the students. However, students opined that these features are not sufficient for supporting smooth communications.

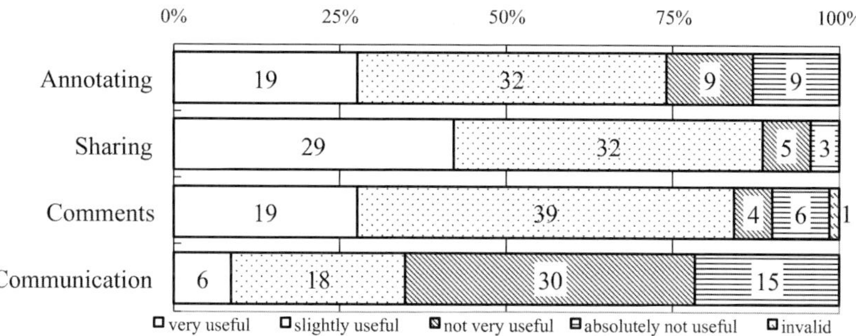

Fig. 6. Usefulness of annotations

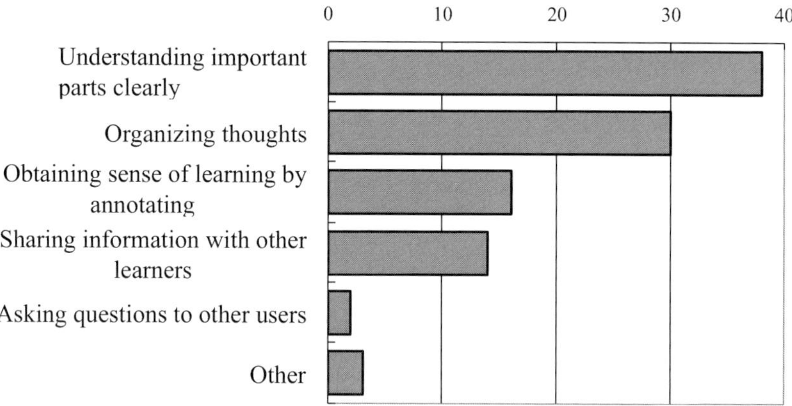

Fig. 7. Intended purposes of annotations

Figure 7 shows the intended purposes of the annotations. The annotations are mainly used to clarify important parts of the learning material and to organize the users' opinions. Therefore, the students mainly used the annotations to improve their understanding of the material, and approximately 20% of the students were of the opinion that one of the intended purposes of annotation is information sharing. However, when we asked the same question regarding the usefulness of annotations in improving their knowledge to 25 other students who had used the previous version of "Writable Web" [5], only 3 (12%) answered in the affirmative.

The results of the questions regarding the usefulness of the portal page features that support asynchronous discussions. According to these results, though the students evaluated the overall features of the portal page well, they felt that the ranking of the annotations and bookmarks is not particularly useful in asynchronous discussions.

On the basis of these results, we will improve and review the features of the system in order to foster asynchronous discussions and communications.

4   Conclusions

In this paper, we have discussed the evaluation of an asynchronous discussion system. We improved the main features and usability of this system on the basis of the results of the first experiment conducted on the previous version of the system. Thus, we implemented several new features in the system in order to foster asynchronous discussions by information sharing.

To evaluate the improved system, we conducted an experiment in the actual class with 88 students. The students used the system to organize their thoughts and highly evaluated the features of the system regarding annotating and annotation sharing. Moreover, the students could participate in asynchronous discussions on the web-based materials, thus obtaining a understanding of the subject than the students who did not participate in these asynchronous discussions.

Acknowledgement

This research was partially supported by the Ministry of Education, Science, Sports and Culture, Grant-in-Aid for Young Scientists (B), 18700642, 2006–2008.

References

1. Barolli, L., Koyama, A., Durresi, A., Marco, G.: A Web-based e-Learning System for Increasing Study Efficiency by Stimulating Learner's Motivation. Information Systems Frontiers 8(4), 297–306 (2006)
2. Cadis, J., Gupta, A., Gruding, J.: Using Web Annotations for Asynchronous Collaboration Around Documents. In: Proceedings of the 2000 ACM Conference on Computer Supported Cooperative Work, pp. 309–318 (2000)
3. Koivunen, M.: Annotea and Semantic Web Supported Collaboration. In: Proceedings of UserSWeb, 1st Workshop on End User Aspects of the Semantic Web, vol. 137(1), pp. 5–17 (2005)
4. Kunimune, H., Niimura, M., Wasaki, K., Fuwa, Y., Shidama, Y., Nakamura, Y.: The Current Activities and the Evaluations of Graduate School of Science and Technology on the Internet, Shinshu University. Transactions of Japanese Society for Information and Systems in Education. 22(4), 264–271 (2005)
5. Kunimune, H., Yokoyama, K., Niimura, M., Fuwa, Y.: An Annotation Sharing System for e-Learning Materials. In: Proceedings of ED-MEDIA 2007, pp. 3366–3371. Association for the Advancement of Computing in Education, Norfolk (2007)
6. O'Hara, K., Sellen, A.: A Comparison of Reading Paper and On-line Documents. In: Proceedings of the SIGCHI Conference on Human Factors in Computing Systems, pp. 335–342 (1997)

CSCL Data Structurization and Inter-LMS Sharing with Use of Web Services

Yasuhisa Tamura[1], Kazuya Sumi[1], Takeshi Yamamuro[2], and Masashi Maejima[2]

[1] Department of Information and Communication Sciences, Sophia University
Kioi-cho 7-1, Chiyoda-ku Tokyo 1028554 Japan
[2] Graduate School of Informatics, Sophia University
Kioi-cho 7-1, Chiyoda-ku Tokyo 1028554 Japan
`{ytamura,k-sumi,t-yamamu,m-maejim}@sophia.ac.jp`

Abstract. The authors specified XML bindings for CSCL data interchange, and developed an automatic CSCL data exchange function among LMSs. Conventional CSCL activities are processed without exchanging any information with other learner groups, even they are discussing the same topic. With use of the proposing function, a learner group is able to acquire basic information, related references and various standpoints for the given topic during the discussion or in advance. In order to execute XML based inter-LMS data transmission, XML bindings named COL-COX was specified. Also, an open source LMS of Moodle was utilized and expanded to install the proposing function to transmit CSCL data. With use of the proposing function, a learner group is able to get scaffolding information and knowledge on CSCL activities.

Keywords: Collaborative Learning, CSCL, Web services, XML bindings, automatic exchange.

1 Introduction

Based on theory of Social Constructivism [1], collaborative learning and CSCL (Computer Supported Collaborative Learning) have established specific research topics in the learning science and e-learning areas. In the CSCL environment, many learners exchange their opinions, cooperatively solve the given problem, and process the given duty in collaborative manner on the computer connected by a network or Internet. This means that many learners are connected virtually without restriction of learning places. This CSCL is expected to increase learners' cognitive workload, to facilitate deeper understanding of the target learning domain, and to train meta-cognitive and self-reflection skills through the communication between learners.

Researches on CSCL have been active from late 1990's. International conferences of CSCL [2] sponsored by ISLS (The International Society of the Learning Sciences) started in 1995 and are held in every two years. Japanese Society for Information and Systems in Education (JSiSE) started a task force for CSCL and Supporting Technology in 2003. The result of investigation and discussion has been published in [3]-[6]. The summary of this activity was published in the JSiSE journal as [7]. Also in the

I. Lovrek, R.J. Howlett, and L.C. Jain (Eds.): KES 2008, Part III, LNAI 5179, pp. 515–522, 2008.

standardization activity of ISO/IEC JTC1/SC36 [8], WG2 is working for standardization of data model for the collaborative learning and CSCL.

In these research and standardization activities, many problems have been pointed out for the CSCL. In this paper the authors focus on the narrowness of the collaborative learning space. The collaborative learning space means a physical or virtual learning environment where a learner group shares various types of information. From the viewpoint of system installation, it is an electronic bulletin board or online chat that the learner group does a discussion or cooperative work. Each learner group has the own and closed learning space. It means that the group members are unaware the existence of other learner groups and the information exchanged in these groups. The problem is that any learner groups should start their discussions from scratch, without any reference of other groups' discussion, even the same topic. In other words, the group has little scaffolding information for "Legitimate peripheral participation" [9] in the preceding discussions.

There are pros and cons that a learner group starts the discussion from scratch. One supporting opinion is that the process of collecting information and discussion itself is the essential collaborative learning process. Another supporting is that preceding information is harmful for thinking and developing his own opinion, if a learner is given a conclusion or major opinion for the given topic. It is also somewhat risky from the viewpoint of cognitive workload. An opposing opinion is that learning activity of preceding discussion is effectively used as scaffolding information to participate the discussion. Another opposing is that the learning activity of preceding discussion is enough for his cognitive workload, because he is unable to participate the discussion without the information shared in the community. This paper supports the latter standpoint.

To achieve this function, the authors developed a CSCL support environment that utilizes the Web services. The Web services are a collection of technical standard specification advocated by W3C [10], which provide the data format and the procedure to exchanging information automatically among the server on the network. With use of Web services, a couple of CSCL systems provide the function to share various discussions and reference information of CSCL activities in these LMSs. These LMSs forms one virtual learning community even if they are physically distributed.

There are some preceding researches. First, there are some to utilize Web services for general learning support functions. These majority objects are to share Learning Objects LOM (Learning Object Metadata) or to compose these collections: Learning Repository. This is classified into three (3) categories: an e-learning project that incorporates Web service technology, the combination of grid technology and e-learning, and the enhancement of SCORM specification. Elena Project [11] proposes the information exchange framework between servers with use of Web services for an intelligent and open e-learning environment. LeGE-WG [12] proposes a decentralized teaching material repository with use of grid technology. Neumann [13] proposes a function to request information from SCORM client PC grid servers with use of Web services.

Also, there are varieties of preceding researches to apply Web service to CSCL. IEEE LTTC [14] (Learning Technology Technical Committee) compiles a special issue of this approach. Hoppe [15] proposes a function that LORs (Learning Object Repositories) to use the Web service as a means to exchange information that may share the object to which the theme is the similar between two or more Learning Community. Ingram [16] practices CSCL in higher education that uses SharePoint Team Web Services. Dugenie [17] proposes CSCL activity in EleGI project (study support that uses the GRID technology).

In this paper, two major items are proposed. First, hierarchical information structure of CSCL is proposed. In the information hierarchy of CSCL, an organization ID is set to be top. Below this, a class, a CSCL activity, a topic, an utterance, a phrase, and finally a word form a hierarchical structure with their attached attribute. The authors provide an XML binding of this hierarchy as COLCOX (COllaborative Learning Contents Open eXchange).

The second is to provide a function that LMSs exchanges COLCOX data. In order to decrease network bandwidth and archived information, the COLCOX data is classified into two levels. In the first level, organization ID, class name, and CSCL activity (called PrimaryQuery) are communicated between LMSs at the first stage. The PrimaryQuery information enables to search the CSCL group with the same agenda. Only when the same agenda is discussed in other learner groups, details of the discussion (the rest of COLCOX data, called SecondaryQuery) are forwarded to the target LMS. This second level information transmission reduces the amount of communication, and accelerates to search CSCL topics in other LMSs. Learners are able to retrieve and view the content of the other groups' discussion with use of the proposed function.

As a result, it becomes possible to refer to the opinion that in other groups, and to extend range of the discussion rather than the discussion in one closed group. Moreover, a learner uses his discussion time effectively for the discussion itself to check reference and related material pointed out by other groups' learners, rather than searching these materials by themselves. In addition, he is able to participate in the discussion community after tracing preceding discussion of the same theme smoothly. In this case, the tracing activity means to acquire scaffolding information for the participation.

The proposal in this paper is one sub-function of a research project to support CSCL activity. In this project, Japanese natural language input is checked first. When the target input is difficult to analyze with use of natural language analysis scheme, a learner is suggested to correct his input sentence (refer to Horikoshi [18]). Next, the morphological analysis, parsing, and deep case analysis are applied to input Japanese sentence. As a result, various attributes of role, relation, etc. are attached into the target sentences and phrases (refer to Maejima [19]). COLCOX data is the result of analysis and attribute attachment of the preceding functions above. In this paper, the transmitting data itself is assumed to be prepared and arranged.

2 COLCOX: XML Bindings of CSCL Information

Web services technology by W3C [10] is applied in the proposal of CSCL information transmission. In the conventional WWW (World Wide Web), a WWW server transmits multimedia information to the client through the network by the trigger of clients' request which is manually generated. On the contrary, the Web services transmission is triggered automatically by a Web service client, not manually. Moreover, information transmitted with Web services is tagged by XML so that the server may interpret and process information easily. In order to utilize Web services for CSCL data transmission, an XML binding should be defined. The authors defined an XML binding named COLCOX (COllaborative Learning Contents Open eXchange) for CSCL data exchange.

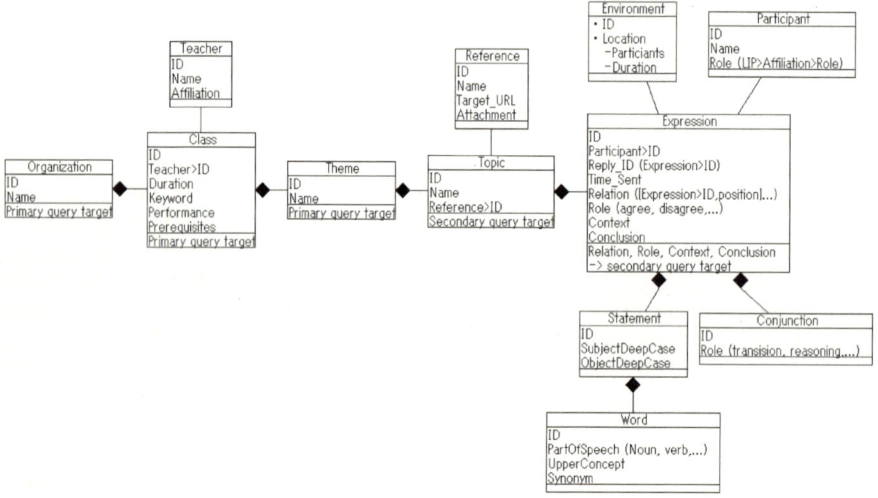

Fig. 1. UML description of COLCOX

There are some preceding proposals for this topic. IMS Learning Design [20] proposed a comprehensive data model and its XML bindings including collaborative learning activity. Also, Friesen [21] proposed a CSCL description framework in the standardization activity of ISO/IEC JTC1/SC36. To compare with these proposals, COLCOX contains more detailed parts of phrase role and sentence relation, to reflect the result of natural language analysis. Figure 1 shows the class chart of COLCOX with use of UML (Unified Modeling Language) language.

- Organization class: name, ID, and attached information of the organization where CSCL environment is set.
- Class class: name and ID of the target class are described.
- Also, keyword of the class description and learner's ability/ performance/ prior knowledge/ prerequisites are disclosed.
- Teacher class: show teacher's or facilitator's name, ID, and affiliation.
- Theme class: ID, theme name, and time period of the discussion are described. This "theme" means the one defined by a teacher or a facilitator.
- Topic class: ID, name, and attached information of the present discussion. The authors divided "Theme" and "Topic" classes, because some topics of the target discussion may vary from the given theme. This division allows the variation of discussed topics.

- Reference class: references and documents which are used while the discussion. Target URL of quoted Web pages or attached file are described as attributes.
- Expression class: single utterance that may include some sentences. ID, time set to the server, role, relation to another utterance, context (body text), and conclusions are described as an attribute.
- Environment class: ID and the place (location) of the target CSCL are described.
- Participant class: learners' IDs, names, and roles are described.
- Statement class: single sentence of a learner's utterance. Sentence ID and deep case of each object and subject are described.
- Conjunction class: conjunction ID, body, and role to connect sentences are described.
- Word class: each word text, ID, part of speech, higher-ranking concept, and synonym are described.

3 CSCL Data Exchange between LMSs

The data exchange function to retrieve and to reuse the CSCL data between LMS was developed. An existing LMS of Moodle [22], an open source LMS was enhanced for the development. There are following reasons for the enhancement. (1) Functions of the Forum (bulletin board), Online Chat and Vote functions are provided in Moodle. (2) The entire LMS is described with PHP, and the function enhancement is easy, including the functional addition of Web services. (3) Because Moodle is an open source software, function enhancement under certain condition is possible. (4) It is easy to install, so the delivery of the developed environment is also easy. (5) The developed function is able to deliver by the module unit, so trial environment delivery is also easy.

For the proposing function, three PHP modules are necessary to add officially distributed Moodle. (1) SimpleXML module: an enhancing module to operate the XML file with PHP. (2) MySQL module: an enhancing module to operate MySQL (database) with PHP. (3) SOAP module: an enhancing module to execute SOAP communication with PHP.

After installing these modules, the authors developed the function to request and transmit CSCL information with use of Web services between Moodle LMSs. The communication between LMS is divided into two levels as mentioned above. The first level of communication is called "Primary Query", while the second level is called "Secondary Query". Primary Query communication is automatically done when a learner group starts the discussion. Secondary Query is done when a learner requests to view another group's discussion. Figure 2 shows the outline of these communications and processing. Details of Primary and Secondary Query are follows.

(1) Primary Query
Top level information of CSCL activity is exchanged among LMSs. A learner is able to search if there is a discussion of same topic name in another LMS. In order to achieve this function, SOAP (Web services communication) module is built in Moodle function. The target data of Primary Query is stored in "P" file at the right. There are group-wise "P" files, so the same topic name of different groups can be identified.

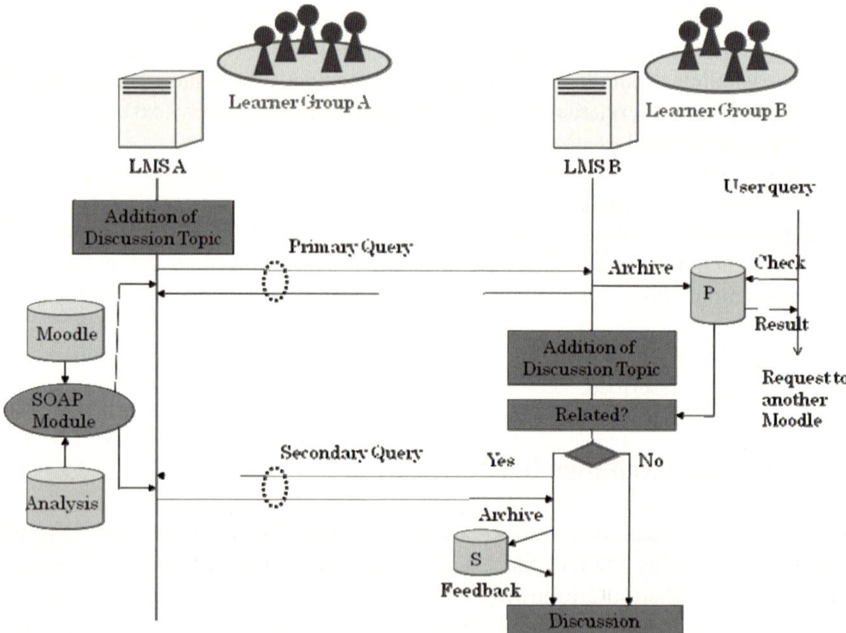

Fig. 2. Communications and Processing Outline

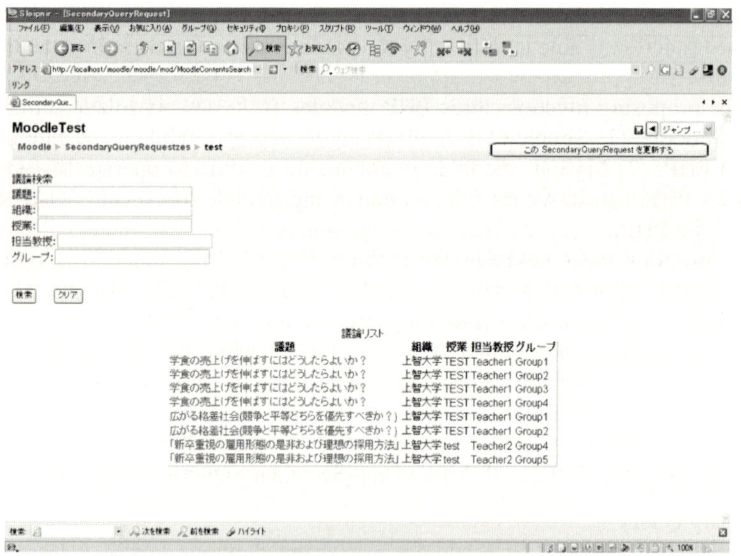

Fig. 3. Screen Snapshot

(2) Secondary Query

It is executed when there is a discussion on the same topic name on another Moodle. The detailed contents of the discussion and the attached result of the analysis are transmitted by this Query. It also utilizes SOAP module, and query result is archived in "S" file at the bottom.

Moreover, we developed a user interface for these functions. Figure 3 shows an interface snapshot for Secondary Query. Eight (8) discussions at the bottom are already picked up with use of Primary Query. More precise query is available to input "topic name", "organization name", "class name", "teacher name", and "group name" in the text boxes at the top.

4 Conclusion

As shown above, this paper discloses the details of CSCL data exchange between LMSs. XML bindings named COLCOX is proposed to exchange CSCL data automatically with use of the Web services. Moreover, the function is installed to exchange COLCOX data with use of Web services. It is done by enhancing Moodle forum module. This exchanging function is installed more than two Moodle systems in the laboratory for field test.

There are mainly four kinds of future tasks below. First, a directory server named UDDI should be provided. UDDI is one of standard specification provided in the Web services. In conventional environment, the proposing LMS should know the target IP address to exchange CSCL information in advance. It means that the number of LMS is limited and predefined. When the UDDI is available, any LMS are able to exchange the CSCL data. Before the communication, an authentication of each LMS should be clarified. The second is development of the communication protocol when the other LMS is off-line. Both Primary Query and Secondary Query are executed by the automatic operation, but currently no assumption for off-line situation. The third is to develop a friendly user interface to retrieve and reuse the shared CSCL information. Currently the interface only provides just the raw data of the discussion. COLCOX provides various additional attributes for each utterance, phrase and word. Searching function should utilize its useful information in order to execute more intelligent search. The last one to design a whole CSCL activity to utilize the proposing function. It is not still clear how human-side CSCL activity is accelerated. We should verify it through experiments.

Acknowledgement

This research was partially supported by the Ministry of Education, Science, Sports and Culture, Grant-in-Aid for Scientific Research (B), 19300284, 2007.

References

[1] Vygotsky, L.: Mind in Society: The Development of Higher Psychological Processes. Harvard University Press (1978)
[2] ISLS, http://www.isls.org/conferences.html

[3] JSiSE task force on CSCL. In: Proc. Workshop on Basic technological of CSCL, JSiSE 29th annual conference (2004) (in Japanese)

[4] JSiSE. In: Proc. Symposium on CSCL and support technology (2004) (in Japanese)

[5] Tamura: Study on information technology standardization of CSCL functions and data items on commercial LMS, pp. 56–59 (2004) (in Japanese)

[6] JSiSE task force on CSCL. In: Proc. Workshop on technological support of CSCL, JSiSE 30th annual conference (2005) (in Japanese)

[7] Kojiri, et al.: CSCL and support technology. JSiSE Journal 23(4), 209–221 (2006) (in Japanese)

[8] ISO/IEC JTC1/SC36, `http://jtc1sc36.org/`

[9] Lave, J., Wenger, E.: Situated Learning: Legitimate Peripheral Participation. Cambridge University Press, Cambridge (1991)

[10] World Wide Web Consortium (W3C), `http://www.w3.org/`

[11] Elena Project, `http://www.elena-project.org/`

[12] LeGE-WG: Learning Grid of Excellence Working Group, `http://www.lege-wg.org/`

[13] Neumann, F., Geys, R.: SCORM and the Learning Grid. In: Proc. Towards a European Learning Grid Infrastructure: Progressing with a European Grid (2004)

[14] Learning Technology newsletter on Learning Communities & Web Service Technologies, vol. 6(1) (2004), `http://lttf.ieee.org/learn_tech/issues/January2004/learn_tech_january2004.pdf`

[15] Hoppe, U., et al.: Building Bridges within Learning Communities through Ontologies and "Thematic Objects". In: Proc. CSCL (2005)

[16] Ingram, A.L., Parker, R.E.: Collaboration and Technology for Teaching and Learning. In: Proc. Ohio Learning Network (2003)

[17] Dugenie, P., Lemoisson, P.: A bootstrapping scenario for eliciting CSCL services within a GRID virtual community. In: Proc. First International ELeGI Conference on Advanced Technology for Enhanced Learning (2005)

[18] Horikoshi, Tamura: Mistake checking of CSCL text chat based on Japanese analysis. In: Japanese Society Mechanical Engineering (JSME) 47th Kanto Student Association Meeting (2008)

[19] Maejima, M., Tamura, Y.: Metadata Extraction and Description with use of RDF for e-Learning Context. In: Proc. WBE (Web-Based Education) (2008)

[20] IMS Learning Design specification, `http://www.imsglobal.org/learningdesign/`

[21] Friesen, N., Tamura, Y., Arnaud, M., Sohn, H.: Metadata for Synchronous and Asynchronous Collaborative Learning Environments: Scoping and Data Model Refinement, ISO/IEC JTC1 SC36 WG2 N0082, `http://collabtech.jtc1sc36.org/doc/SC36_WG2_N0082.pdf`

[22] Moodle, `http://moodle.org/`

Adventures in the Boundary between Domain-Independent Ontologies and Domain Content for CSCL

Seiji Isotani and Riichiro Mizoguchi

The Institute of Scientific and Industrial Research, Osaka University,
8-1 Mihogaoka, Ibaraki, Osaka, 567-0047, Japan
isotani@acm.org, miz@ei.sanken.osaka-u.ac.jp

Abstract. One of the main problems facing the development of ontology-aware authoring systems (OAS) is to link well-designed domain-independent knowledge (ontologies) with domain content. Such a problem comes from the fact that all OAS developed to date require end-users (non-experts) to create their own domain ontologies to run the system in real scenarios. In collaborative learning (CL), this problem hinders the development of OAS that aid the design of pedagogically sound CL sessions with strong technological support. In this paper, we propose a framework that connects an ontology for CL (CL ontology) with domain content without the use of domain ontologies. To check its usability, we present an example to model a geometry drawing course demonstrating that it is feasible to instantiate the CL ontology to represent a specific domain and connect it with adequate learning objects.

Keywords: ontological engineering, CSCL, ontology-aware authoring system.

1   Introduction

Ontologies have been employed in many intelligent educational systems with some degree of success [3]. Although their use covers a vast field, ontologies are especially important to allow for a more explicit representation of knowledge in intelligent authoring systems (IAS) for education where the knowledge is based on various instructional/learning theories. One of the main problems facing the development of IAS for collaborative learning (CL) is to link well-designed domain-independent knowledge (ontologies) with the contents of a specific domain. On one hand, we have a very powerful and sharable knowledge that can be used in many different situations to support the authoring of CL sessions with theoretical justifications. On the other hand, we have domain content that needs to be adequately connected with theoretical foundations to support a well-designed CL session through the use of technology.

To solve the problem of connecting ontologies with specific domains and learning objects (LO), related researches [6;7] ask end-users to create their own ontologies for the specific domain. This approach show many benefits and good results for semantic annotation; however, both are excessively time-consuming due to the fact that they require end-users or non-experts to create their own ontologies from scratch.

This research proposes a framework to deal with this problem that connects domain-independent ontologies, specifically the CL ontology [9], with domain-specific

I. Lovrek, R.J. Howlett, and L.C. Jain (Eds.): KES 2008, Part III, LNAI 5179, pp. 523–532, 2008.

content and LOs without the necessity of asking end-users to create new ontologies. This approach promotes a user-friendly way to implement the CL ontology by offering a framework along with templates that help users to link adequate LOs with the instantiated concepts in the ontology. In this paper we first show an overview of the prototype of an ontology-aware system that uses the CL ontology to support the design of CL sessions with theoretical justifications. Then we propose a framework for linking the CL ontology with domain-specific LOs. This framework is strongly based on our model GMIP that is presented in previous works. Finally, we demonstrate how to use this framework in a domain-dependent context.

2   Overview of CHOCOLATO

There are many learning theories that support group activities (*e.g.* [2;10;13]). Thus, to create group activities we can select an appropriate set of theories, considering necessary pre-conditions and desired benefits for learners. This flexibility of choosing suggests the difficulty of users (*e.g.* teachers) in selecting an appropriate set of learning theories to ensure learners' benefits. Therefore, to help these users, we need an elaborate system that considers different learning theories to support the CL design.

In CSCL, ontologies have been successfully applied to solve the problem of representing learning theories to support CL [9;11]. Furthermore, our previous research analyzed seven different theories (e.g. [2;4;10]) used to support CL activities to propose the Growth Model Improved by Interaction Patterns (GMIP) [8]. The GMIP is a graph model based on an ontological structure that describes an excerpt of learning theory. It represents, in a simplified way, the learner's knowledge acquisition and skill development processes, and explains the relationships between learning strategies, educational benefits and the interactions used to achieve these benefits.

The GMIP has twenty nodes that represent the levels of the learner's development at certain stages of learning. Each node is composed of two triangles. The upper-right triangle represents the stage of knowledge acquisition, while the lower-left triangle represents the stage of skill development. The nodes are linked with edges that show possible transitions between nodes in compliance with [1] and [12]. Using the GMIP graph, we show the benefits of a learning strategy by highlighting its path on the graph and associating each edge with interactions activities (top-right of Figure 1).

In order to develop a system to support the design of CL activities based on ontologies and our model, we have been developing CHOCOLATO (Concrete and Helpful Ontology-aware Collaborative Learning Authoring Tool) an ontology-aware system that uses ontologies developed in the Hozo Editor (http://www.hozo.jp) to provide its theoretical knowledge. And the sub-system called MARI allows the visualization of learning theories on the screen using the GMIP (Figure 1).

Using ontologies and the GMIP, MARI can select appropriate learning theories and strategies and suggest a consistent sequence of activities for learners in a group. The suggestions given by our system are guidelines that can be used to propose CL activities based on theories which (a) preserve the consistency of the CL process and (b) guarantee a suitable path to achieve desired benefits.

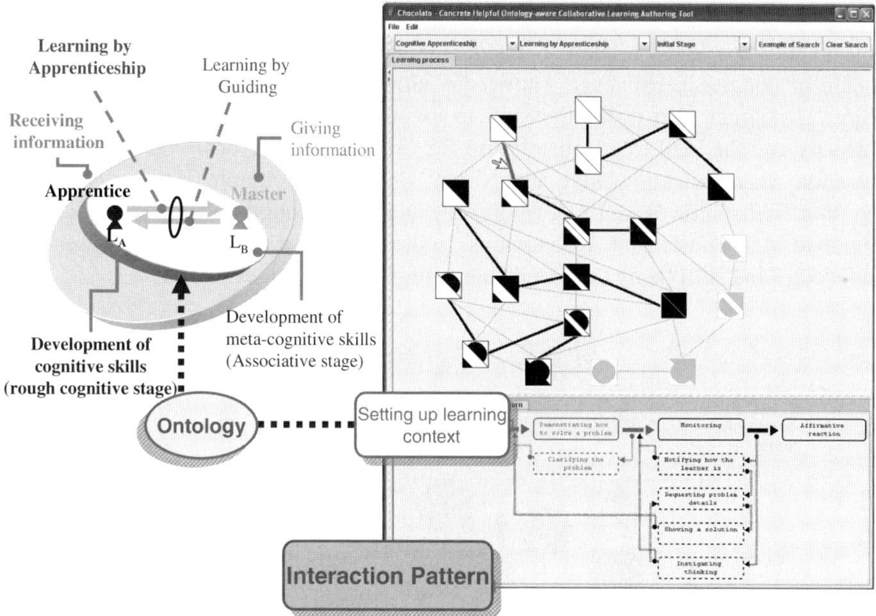

Fig. 1. A screenshot of MARI showing the Cognitive Apprenticeship theory using the strategy *Learning by Apprenticeship*. The top shows the path on the GMIP. The bottom shows the sequence of activities in compliance with the theory. The left side shows other information that can be extracted from our ontology.

3   A Framework to Support Ontologies, Domain Content and LOs

MARI prototype is strongly based on domain-independent ontologies and our model GMIP. This means that MARI can provide domain-independent recommendations that can be used in different situations and are justified by theories, but it does not consider the actual domain (*e.g.* mathematics) in which the recommendations will be applied. Because of this, some colleagues/researchers have pointed out that although our approach is theoretically valid, it can hardly be applied in real environments. Thus, to augment our research and show that a theoretically valid approach can be applied in real environments, we propose a framework to link domain-specific content into our model GMIP and our ontologies. This framework is shown in Figure 2. The proposed framework has four linked layers. The top two layers are completely domain-independent, representing the knowledge about CL, learning theories and the learning state of a learner. The two bottom layers are related to domain-dependent content. One is related to the knowledge and skills of the domain-specific content and the other is related to the LOs connected with this content.

We define the learning state layer (top layer in Figure 2) as a set of nodes of different GMIPs. As summarized in section 2, each node in GMIP represents the stages of knowledge acquisition and skill development. Furthermore, each GMIP represents a different piece of knowledge and a different skill acquired by a learner. For example, to draw a geometric object a learner needs knowledge about the properties of the object and knowledge related to the manipulation of available drawing tools (*e.g.* square or compass). However, knowledge alone is not enough to draw a geometric object. A learner also needs the skill to use the properties of the geometric object correctly and the skill to adequately use the drawing tools. According to [1;9;12] we have four stages of knowledge acquisition: Nothing, Accretion, Tuning, Restructuring; and five stages of skills development: Nothing, Rough-cognitive, Explanatory-cognitive, Associative and Autonomous. In the learning state layer we represent each knowledge and skill being developed according to these stages.

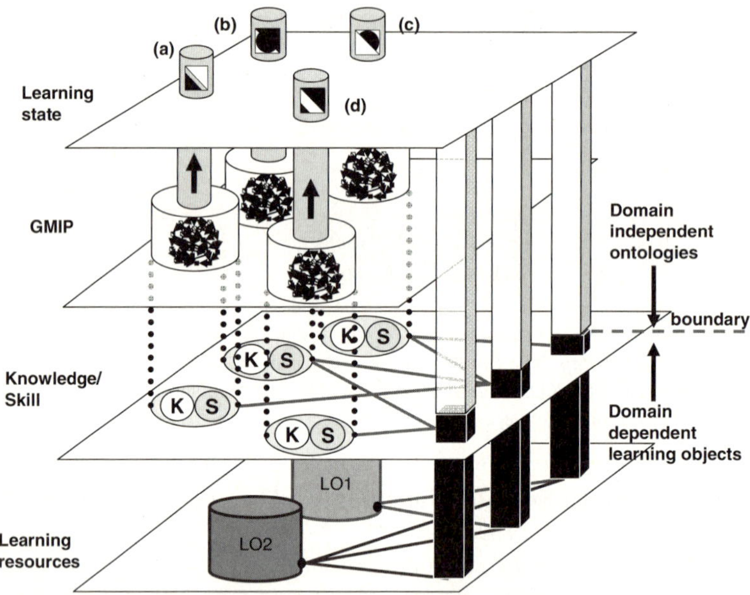

Fig. 2. Framework to link domain independent ontologies, domain specific content and LOs

The second layer is the GMIP. In this layer we show how learners can develop their knowledge/skills as transitions between nodes. In previous works we presented how this model was created and used to design CL activities. Now, the main difficulty is how to link the second and third layer in order to map the GMIPs with domain-specific knowledge and skills.

To define the third layer and link it with the second layer, first of all, given a domain-specific content and a learning goal, we must separate the knowledge from the skills necessary to achieve this goal in the specified domain. The knowledge to achieve the learning goal should be decomposed into different sub-knowledge pieces to be acquired. Similarly, the skills should be decomposed into sub-skills to be developed. The final structure will be a decomposition tree as shown in Figure 3 that identifies the minimum knowledge and skills necessary to construct a geometric object. The granularity of the decomposition tree depends on the learning goals and the expertise of the user who creates the tree. Observe that this tree is different from those proposed by [5, 11]. While those works provide decomposition trees that represent instructional design plans, our trees represent the knowledge and skills to be developed without any reference to how it will be developed. The design of CL activities can occur after linking this tree with the GMIP.

Using this approach, we can separate information about the content from information about how to learn the content. Such differentiation (boundary) is important when we think about learner-centered environments where the environments adapt the way to provide information or the way to teach the same content according to learning/teaching preferences.

To complete the mapping of knowledge and skills into our model GMIP it is necessary to explicitly identify the relationship of the knowledge and the skills in the tree as shown by the blue doted line in Figure 3. Each skill can be related to one or more pieces of knowledge and vice-versa. For each relationship between knowledge and skill we can create an instantiated GMIP, which will then be able to support the development of this knowledge and skill in the specific domain.

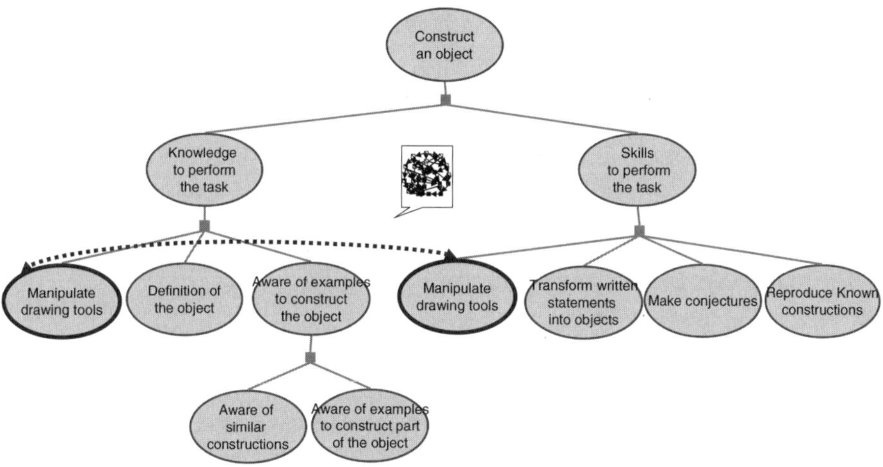

Fig. 3. Example of a decomposition tree to separate knowledge and skills for a specific goal

Finally, it is necessary to identify how each linked knowledge/skill fits into the stages proposed by [1; 9; 12]. To facilitate such a task, we provide templates that use the proposed definitions of stages of knowledge and skills. Such templates help users to adequately understand the knowledge and skill development process. Furthermore, it helps us to create a support system that semi-automatically maps specific skills/knowledge in our model GMIP. An example of the template for skill development instantiated for the example in Figure 3 (manipulate drawing tools) is shown in Table 1.

Table 1. Example of the template of skill development instantiated to represent the skill to manipulate a drawing tool. The phrases in bold are variables where specific skills can fit; all the others phrases come from the definitions provided by [1]. The GMIP graphical representation is shown in the middle column.

Stage Name	GMIP	Template definition
Nothing		Does not have the desired skills.
Rough-cognitive stage		Involves an initial encoding of **the skill to manipulate a drawing tool** into a form that is not sufficient for us to generate the desired behavior, usually by observing a process in which another person **manipulates a drawing tool**.
Explanatory-cognitive stage		Encodes **the skill to manipulate a drawing tool** into a form sufficient to permit a learner to **somewhat manipulate the drawing tool**.
Associative stage		Tunes the **skill to manipulate a drawing tool** through practice. Errors in the initial understanding of the skill are gradually detected and eliminated.
Autonomous stage		Demonstrates gradual continued improvements in the performance of the **skill to manipulate a drawing tool**.

The last layer in our framework is the learning resources layer. Each resource is a learning object (such as a tool, text, video, or activity) that can be used to improve a domain-specific knowledge or skill. Each learning object (LO) can be linked with different knowledge or skills. Thus, with the domain-specific goals provided in the third layer, an end user can add/remove LOs to satisfy specific conditions in the environment where the learning will occur.

This framework provides a high degree of flexibility to change the system. Thus, maintaining the same representation of knowledge and skills, changes in one layer do not affect the other layers. For example, a user can provide a new way to design a CL session, include this way as a path on GMIP, and the system can use it immediately without any change in the other layers. Similarly, a user can re-define the decomposition tree or add/remove LOs without any modification in the top layers. Because of this flexibility, it is possible to instantiate domain-independent CL ontologies to support the use of domain-specific content and LOs. Thus, we can offer much better support for users during the design of CL activities, taking into account the specific conditions of the domain and using the LOs available in the environment.

4 An Example of Application

To provide an example of application, we tested the usability of our framework to model a geometry drawing course. This course is one of the required courses for graduation in Mathematics at the University of Sao Paulo. It is comprised of two teachers, three teaching assistants and more than 150 students. In the geometry drawing course the flexibility of learning resources is very high. Thus, although the main learning goals remain the same, according to teachers' intentions, new exercises are included or removed from the curriculum while students are taking the course.

One of the main goals of this course is to provide learners with the knowledge and skills to (1) construct objects (such as an angle, a triangle, and a circle); (2) define objects (e.g., what is a point, an angle, a triangle); (3) classify objects according to some properties (e.g., triangle: equilateral or isosceles); (4) check properties (distances, angles, sides); (5) propose conjectures (e.g., this object is equal to another one because ...); and (6) generalize and make inferences about properties (e.g., any square is a rectangle, but not vice-versa).

Using our framework, we created decomposition trees that explicitly identify the knowledge and skills that need to be acquired to achieve the objectives listed above, as shown in Figure 3 for objective (1). Then, using the templates we provided (Table 1), we mapped the knowledge and skills into our model GMIP. Finally, about 70 exercises were created and linked with the respective knowledge and skills to serve as learning objects for this trial.

To give an example of how our system for CL design will work based on this framework, let us propose a CL session that helps a learner to acquire the skills to construct a geometric object. To prepare such a CL session, using the MARI interface the user can select the initial stages (pre-conditions) of learners. This step corresponds to identifying the stages of the learners in the top layer. Because each stage is instantiated for a domain specific goal (Table 1), the system will be able to support the user by offering explanations of each stage of the learning development considering the specific domain. Thus, it can facilitate a more accurate selection of initial stages for learners. Furthermore, this process can be done semi-automatically.

Next, the user can select the desired final stage (desired goal) for learners. Through the formal structure based on ontologies which allows MARI to evaluate theories and their features, it will check the learners' conditions, learning goals and special requirements related to the domain to offer a sophisticated group formation and design CL activities with theoretical justifications. If more than one theory can help a group of learners, the system will present all of them and the user can select the preferable one. Our approach uses theory-driven group formation with suggestions of role assignment and sequence of interactions to offer fundamental settings for an effective CL session and essential conditions to predict the impact of interactions in the learning process. Furthermore, MARI can suggest different interaction patterns (sequences of activities) to achieve the desired goal. One example of the pattern for the Cognitive Apprenticeship Theory is shown in the bottom-right of Figure 1. This process is done on the second top layer of our framework. Due to the limitation of space we cannot explain how the group is formed or how the role assignment is done.

In the third layer, the system will take the recommendations of the second layer and use the decomposition tree with the information of the specific domain to support the user in implementing the CL session. For example, for the first three interactions of the interaction pattern in the bottom-right of Figure 1, we show on Table 2 the differences between domain-independent events and domain-specific events. *Master* and *Apprentice* are the roles that learners will play during the CL session. These interactions are proposed to help the apprentice to develop the skill to manipulate a drawing tool (from *Nothing* to the *Rough-cognitive stage*) and to help the master to develop his cognitive and meta-cognitive skills (from the *Associative stage* to the *Autonomous stage*).

Table 2. Three interactions from the Cognitive Apprenticeship interaction pattern showing the differences between domain-independent and domain-dependent activities to develop the skill to manipulate a drawing tool

Interaction	Domain-independent activity		Domain-dependent activity	
	Master	Apprentice	Master	Apprentice
Set up learning context	Giving Information	Receiving Information	Showing the drawing tools that will be used and explaining their functionality	Becoming familiar with the drawing tools
Demonstrate how to solve a problem	Demonstration	Observing demonstration	Drawing one object using adequate tools and explaining the process	Observing the steps to draw the object using the tools
Clarify the problem	Identifying misconceptions	Externalization of misconceptions	Answering questions to identify weak points in the explanation and in the apprentice's skill in using the tools.	Asking questions about the correct way to use the tools.

Finally, each interaction of a pattern is linked with specific knowledge/skills in the decomposition tree. The knowledge/skills are then linked with specific learning objects that will support carrying out the CL process in the specific domain (fourth layer). In our example, in which the domain is geometric drawing, each interaction is associated with geometry exercises or texts explaining geometric concepts. Thus, the system can help the user to select adequate material to support the CL session and run group activities. For example, to support learners to manipulate drawing tools, learning objects related with the simple drawing of triangles, parallels, and circumferences are connected. In this case, drawing simple geometric objects is fundamental to familiarize learners with the given drawing tools.

This example demonstrates that it is possible to connect domain-independent ontologies with domain-dependent LOs in CSCL environments without asking end-users to create ontologies by themselves. Furthermore, because our ontologies are based on theories, our framework also gives some hints about how to use a theory to support real environments in a CL context. The next version of CHOCOLATO will have the functionalities presented in this section.

5 Conclusions

To create intelligent educational systems based on well-grounded theoretical knowledge and to apply them in real environments are two important challenges that research in the development of ontology-aware systems are facing nowadays. In order to solve these problems in the context of CL, we propose a framework that intends to connect CL ontology [9] with LOs intermediated by our model GMIP. By providing this connection, we can offer a more user-friendly way to design pedagogically sound CL sessions in a specific domain with strong technological support.

The proposed framework is divided into four layers interconnected by the concept of knowledge acquisition and skill development proposed by [1;9;12]. Thus, each layer has the flexibility to change any of its content or characteristics without affecting the other layers or system functionality. The top two layers (Figure 2) represent the domain-independent knowledge and the bottom two layers represent the domain-dependent knowledge that instantiates the top layers. To create the domain-dependent knowledge, a user needs to create a decomposition tree for the specific domain and map it into our model GMIP. This process is partially supported by our templates, which can be generated semi-automatically during the process of mapping. Such an approach seems to be more reliable than other approaches, especially because it removes the burden of creating domain ontologies for each domain of application.

To exemplify the use of our framework, we mapped a geometric drawing course (together with its LOs) into our model GMIP. This example demonstrated that it was feasible to instantiate the GMIP (and thus, the CL ontology) to represent the specific domain and connect it with domain-dependent LOs. Our future research intends to complete the implementation of CHOCOLATO using this framework in order to help users to (a) form groups and design CL activities in compliance with theories by suggesting adequate LOs during this process; and (b) analyze interaction among learners to identify the educational benefits acquired during the CL process.

References

1. Anderson, J.R.: Acquisition of Cognitive Skill. Psychological Review 89(4), 369–406 (1982)
2. Collins, A.: Cognitive apprenticeship and instructional technology. In: Idol, L., Jones, B.F. (eds.) Educational values and cognitive instruction. LEA (1991)
3. Devedzic, V.: Semantic Web and Education. Springer, Heidelberg (2006)
4. Endlsey, W.R.: Peer tutorial instruction. Educational Technology (1980)
5. Gagne, R., Briggs, L., Wager, W.: Principles of Instructional Design, 4th edn. HBJ College Publishers, Fort Worth, TX (1992)
6. Gasevic, D., Jovanovic, J., Devedzic, V.: Ontologies for Reusing Learning Object Content. In: Proceedings of the Int. Workshop on Ontologies and Semantic Web for E-Learning, pp. 650–659 (2005)
7. Knight, C., Gasevic, D., Richards, G.: Ontologies to integrate learning design and learning content. Journal of Interactive Media in Education (07), 1–24 (2005)
8. Isotani, S., Mizoguchi, R.: Deployment of Ontologies for an Effective Design of Collaborative Learning Scenarios. In: Haake, J.M., Ochoa, S.F., Cechich, A. (eds.) CRIWG 2007. LNCS, vol. 4715, pp. 223–238. Springer, Heidelberg (2007)
9. Inaba, A., Ikeda, M., Mizoguchi, R.: What Learning Patterns are Effective for a Learner's Growth? In: Proc. of International conference on Artificial Intelligence in Education, pp. 219–226 (2003)
10. Lave, J., Wenger, E.: Situated Learning: Legitimate peripheral participation. Cambridge University Press, Cambridge (1991)

11. Mizoguchi, R., Hayashi, Y., Bourdeau, J.: Inside Theory-Aware and Standards-Compliant Authoring System. In: Proceedings of the Int. Workshop on Ontologies and Semantic Web for E-Learning, pp. 1–18 (2007)
12. Rumelhart, D.E., Norman, D.A.: Accretion, Tuning, and Restructuring: Modes of Learning. In: Semantic factors in cognition. LEA, pp. 37–53 (1978)
13. Salomon, G.: Distributed Cognitions: Psychological and Educational Considerations. Cambridge University Press, Cambridge (1996)

Mammographic Image Contrast Enhancement through the Use of Moving Contrast Sweep

Zailani Mohd Nordin, Nor Ashidi Mat Isa, Umi Kalthum Ngah,
and Kamal Zuhairi Zamli

School of Electrical and Electronic Engineering,
Universiti Sains Malaysia,
Seri Ampangan, 14300 Nibong Tebal,
Pulau Pinang, Malaysia
zailani_mohd_nordin@hotmail.com, ashidi@eng.usm.my,
umi@eng.usm.my, eekamal@eng.usm.my

Abstract. Low contrast in mammographic image has always made detection of subtle signs such as the presence of micro calcification within dense tissue a challenge. Therefore, numerous researches have been conducted on contrast enhancement technique for mammographic image. However, most of these methods focus on enhancing specific range of intensity level producing an enhanced version of the original image. This paper proposes a new approach in dealing with contrast enhancement for mammographic images. Instead of producing just a single image, the approach proposed here will generate multiple different images where each of them corresponds to different range of intensity level. These images will then be combined together to form frames of a moving video which would display seamless transition from one intensity level to another. As a result, the radiologist who is performing visual inspection of the image can sweep through the entire range of intensity level of the image which has been contrast enhanced.

Keywords: Mammogram, Mammography, Mammographic, Contrast Enhancement, Contrast Sweep.

1 Introduction

Breast cancer is one of the most common types of cancer among women reported all over the world. It is also the most frequent type of cancer reported among Malaysian women. Based on a report by National Cancer Registry of Malaysia in 2003[1], around 30% of reported cases of cancer for women in Malaysia is attributed to breast cancer. One of the best tools used to detect early signs of breast cancer is mammography. The low doses of X-rays used in mammography are capable of detecting microcalcifications (MCC), one of the earliest signs of breast cancer as well as the presence of other abnormalities. However, one of the biggest challenges with mammography is that the produced image tends to be low in contrast making the detection of MCC difficult especially in regions with dense tissue. Quite a number of studies have been conducted on contrast enhancement techniques especially in the area MCC detection [2].

I. Lovrek, R.J. Howlett, and L.C. Jain (Eds.): KES 2008, Part III, LNAI 5179, pp. 533–540, 2008.

High contrast in an image means that the difference between one intensity level to its adjacent value is high. The ability to project the difference between intensity levels is important because it determines the amount of details which can be perceived by the observer. The differences in these intensity levels must be big enough before they are distinguishable by human eye [3].

Traditionally, contrast enhancement is done on an image by taking the image and running it through the contrast enhancement process to produce an enhanced version of the image. Morrow et al. [4] proposed the use of region base contrast enhancement technique by firstly identifying the region for feature of interest through the use of the seed growing technique. The contrast between the region of interest and its background was then enhanced to create greater separation between their intensity levels which would make the region of interest more visible to the observer. Laine et al. [5] proposed the use of dyadic wavelet analysis for image enhancement. The image was decomposed into multiresolution representation and processed through linear and non-linear operators which would enhance significant features in the image. Veldkamp et al. [6] proposed the used of local contrast enhancement with adaptive noise equalization which was proven to be more effective than using a fixed noised equalization alone. Contrast enhancement steps performed in the method proposed by Nunes et al. [7] was done by increasing the intensity level of pixels which have high probability of being related to MCC while the intensity level of pixels which most likely represent tissue was decreased. Cheng et al.[8] employed the use of fuzzy logic in contrast enhancement to enhance selected feature such as MCC while keeping the noise amplification to a minimum. Scharcanski et al. [9] proposed the used of adaptive local contrast enhancement technique. The enhancement was done adaptively, based on the gray level and noise value.

All contrast enhancement methods mentioned above have the same common goal i.e. to enhance regions having features of interest. Their advantages lie in their capability to target specific range of intensity levels to be enhanced which represent features of interest, a feat that may not be possible with a simple contrast enhancement technique. The simple contrast enhancement technique need to be performed globally or the range of intensity level need to be specified by the user and then enhanced manually, making it less effective compared to the methods above.

The method proposed here demonstrates that the effectiveness of a simple contrast enhancement technique can be increased through the use of a moving contrast sweep. Instead of producing the result in the form of a single contrast enhanced image, the method proposed here will produce multiple images which has been enhanced at different intensity levels. These images can then be combined to form frames for a moving video which will show seamless transition between intensity levels which have been contrast enhanced.

2 Methods

Sample images are obtained from Hospital University Sains Malaysia (HUSM), Kubang Kerian. They are actual mammographic images which have been digitized for our analysis. The programming routine which is described here is done through MATLAB.

The basic idea behind this technique is to develop a programming routine which opens up an image file, processed it through contrast enhancement techniques which could be sourced from an existing technique or a new one and then generate multiple enhanced images. Each image corresponds to a specific range of intensity level enhancement. The enhancement process is then repeated over the entire range of intensitities. The resulting images can then be layered on top of each other, forming moving video frames displaying seamless transition between intensity levels. This study focuses on mammographic images which are in gray scale format. If the image file uses color format, it must first be converted to gray scale before it can be processed through this methods.

The programming routine in this method is executed with the following parameters:

a. The file name of the input image,
b. The type of contrast enhancement to be used and
c. The sweep range of the intensity level where contrast enhancement will be performed.

There are two options for usage i.e. the full range of 0-255 or specifying a range of interest. Whatever the case may be, the lower and the upper limit of the intensity level range need to be specified. The steps of the sweep need to be ascertained too. This means that we can sweep every intensity levels within the range will be swept or by specifying the level value, it may be skipped. Finally, the spread of the intensity level which will determine the strength of the contrast need to be stated. A few more parameters may be added to the routine if a more complex contrast enhancement technique is desired.

2.1 Normalization

The first step to be executed after the image loading would be to normalizing the histogram so that the intensity level is evenly distributed across the histogram. These steps would help to simplify the contrast enhancement technique to be used. Performing normalization prior to performing contrast enhancement is not uncommon in this field. Cheng et al. [6] used histogram normalization with a predefined intensity level range prior to performing their contrast enhancement technique. Our histogram normalization technique is similar to the above method except that the intensity level range is determined adaptively in our approach. The normalization equation would be as follows:

$$j = \frac{i - i_{min}}{i_{max} - i_{min}} \times T \tag{1}$$

In equation 1, i represents the value of original intensity level and j represents the new intensity after the normalization process. i_{min} represents the minimum intensity value while i_{max} the maximum intensity value; both having significant amounts of pixels. T represents the total number of intensity level which can be represented in an image. For an 8-bit grayscale image this value would be 255.

The maximum and minimum intensity value in our approach is determined by counting whether the intensity level has a significant amount of data. In this case, the value which has been chosen to determine the significance of the intensity level is the number of pixel represented by the intensity level must be at least 0.00005% of the total number of pixels in the image. In our approach, we start measuring the amount of data represented by each intensity level starting from the highest possible level which is 255. We then start to work our way down the intensity level until we find a level which has a significant amount of data. This similar process is repeated from the lowest possible level in determining the minimum value for the intensity level.

The normalized image is then processed through contrast enhancement algorithm starting from the lowest intensity level in the range. For every single iteration, the intensity level is increased by the number of steps as specified at the beginning of the routine. For each iteration, the processed image is saved onto a file which records the intensity level as its name. For example if a full range (0-255) of the intensity level is being used with increment steps of 1, a total of 255 files will be generated where each one of them corresponds to an individual intensity level.

2.2 Contrast Stretching

For this paper, a simple non-linear contrast enhancement technique is used to redistribute the pixel intensity. The contrast enhancement technique used here is given by the equation 2:

$$j = \begin{cases} i & \text{if } i = i_x \\ \dfrac{i^s}{(i_x)^s} \times i_x & \text{if } i < i_x \\ 1 - \left(\dfrac{(1-i)^s}{(1-i_x)^s} \right) \times (1-i_x) & \text{if } i > i_x \end{cases} \tag{2}$$

In equation 2, i represents the value of the original intensity level and j represents the new intensity after the execution of the contrast enhancement process. i_x represents the targeted intensity level where the contrast enhancement is focused upon and s represents the spread magnitude of the intensity level which determines the strength of the contrast. This variable allows configurable contrast strength in the routine.

On a fairly fast workstation such as a quad core workstation, the time taken for this iterative process may take a few seconds or up to a few minutes depending upon the complexity of the contrast enhancement algorithm used and the size of the image.

2.3 Moving Video

The image files are then layered on top of each other to form frames of a moving video. This would provide a seamless transition between intensity levels making it visibly easier to see. Currently, an off the shelf movie editing software is being used for this task. However, further development is being planned in future to develop a customized tool for this task. This would allow greater freedom to the user in having

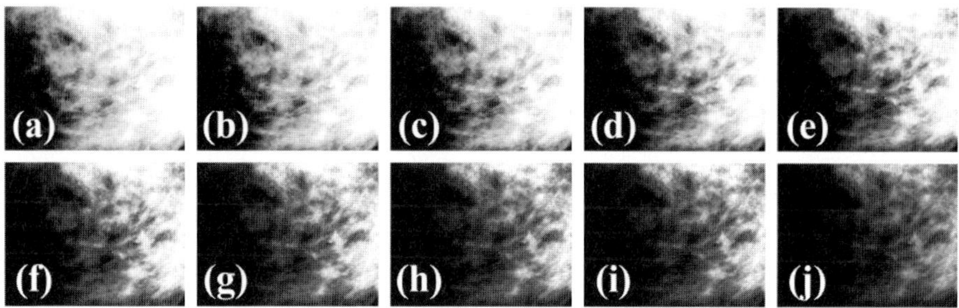

Fig. 1. Multiple images generated at multiple different intensity levels: (a)70 (b)90 (c)110 (d)130 (e)150 (f)170 (g)190 (h)210 (i)230 (j)250

greater flexibility and control to browse through the moving video frame. Figure 1 shows the transition of multiple frames of a sample image which has been contrast enhanced at different intensity level.

3 Results and Discussions

This method shows that MCC within mammographic image residing in different areas within the tissue can be exposed by performing contrast enhancement at different intensity levels.

Figure 2(a) shows the normalized version of the sample image 1. Clusters of MCC at the bottom left region of the image can reasonably be observed but clusters of MCC at the top right region are vague. Figure 2(b) shows contrast enhancement at intensity level of 150. Clusters of MCC at the bottom left region can be observed much clearly but clusters of MCC at the top right region cannot be detected at all. Figure 2(c) shows contrast enhancement at the intensity level of 255. Clusters of MCC at the top right region can now be clearly observed.

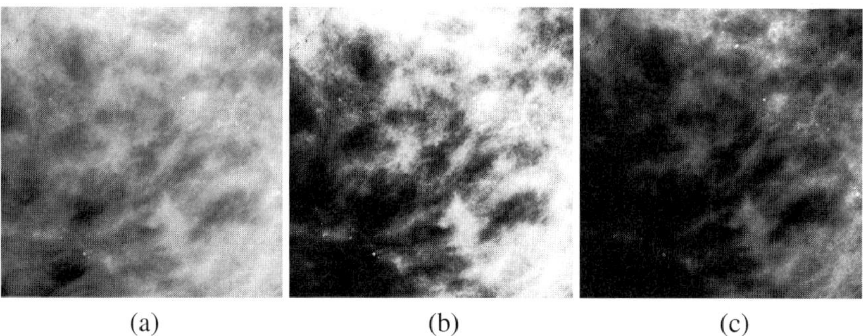

(a)   (b)   (c)

Fig. 2. (a)Image sample 1 which has been normalized (b) Image sample 1 which has been contrast enhanced at intensity 150 (c) Image sample 1 which has been contrast enhanced at intensity 255

538   Z.M. Nordin et al.

Figure 3(a) shows the normalized version of the sample image 2. Clusters of MCC at the top left region of the image can vaguely be observed. Figure 3(b) shows contrast enhancement at the intensity level of 242. The clusters of MCC have become much more evident.

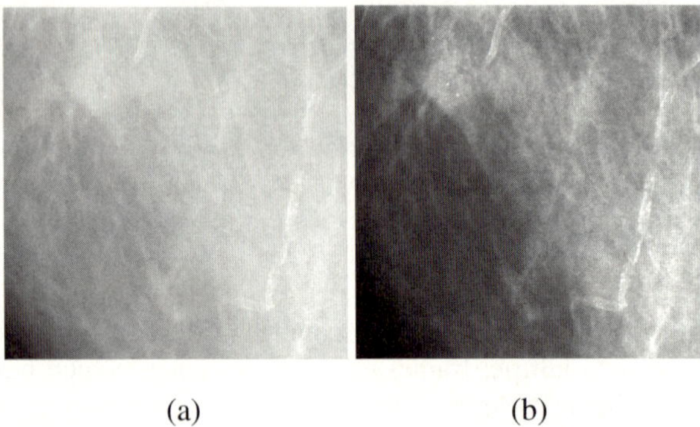

(a)   (b)

Fig. 3. (a)Image sample 2 which has been normalized (b) Image sample 2 which has been contrast enhanced at the intensity of 242

Figure 4(a) shows the normalized version of the sample image 3. Clusters of MCC at the top right region of the image are reasonably visible but those at the bottom left region are quite vague. Figure 4(b) shows contrast enhancement at the intensity level of 195. The cluster of MCC at the top right-hand region has become even more evident but not the ones at the bottom left-hand region. Figure 4(c) shows contrast enhancement at the intensity level of 250. The clusters of MCC at the bottom left region have become more defined.

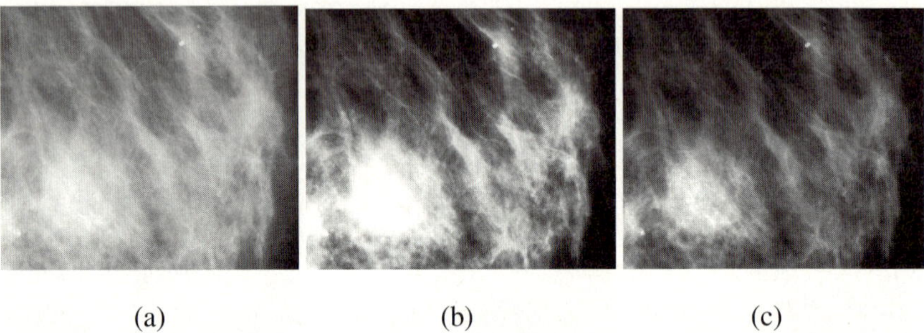

(a)   (b)   (c)

Fig. 4. (a) Image sample 3 which has been normalized (b) Image sample 3 which has been contrast enhanced at intensity 195 (c) Image sample 3 which has been contrast enhanced at intensity 250

The sample image shows that MCC residing in different regions within the breast tissue would require contrast enhancement to be performed at different intensity levels. Through this approach we can sweep through different intensity level seamlessly while at the same time, easily exposing clusters of MCC within different regions of the image. This approach can be combined with other existing contrast enhancement, denoising techniques and other image sharpening techniques, making them even more effective in exposing the MCC. However, further development is still required in developing a more refined software taking into consideration the need for a user friendly and flexible environment necessary for users while allowing a much greater degree of control over the image enhancement process.

4 Conclusion

Based on the method proposed in this paper, it has been demonstrated that simple contrast enhancement technique which were previously not very effective in exposing MCC within mammographic images may now be put to a greater use. Previously, the detection of MCC in the cases displayed shown would require a more complex contrast enhancement technique. However, through the method proposed in this study, a simple contrast enhancement technique would be just as effective. Furthermore, it is also simple to use. The radiologist who performs the visual inspection does not need to configure the various contrast enhancement parameters while using it. Instead all that is needed is to sweep through the entire range of intensity levels to find any suspicious signs within the mammographic image.

Acknowledgement

This work is supported by the Ministry of Higher Education (MOHE) Malaysia, under the Fundamental Research Grant Scheme entitled 'Investigation of Mammogram Image Characteristics for Medical Imaging Application'. This project is partially supported by Ministry of Science Technology and Innovation (MOSTI). The image processing technique in this work, are all applied and developed to be used in the aforementioned grant.

References

1. Lim, G.C.C., Yahaya, H.: Second Report of the National Cancer Registry - Cancer Incidence in Malaysia. National Cancer Registry (Retrieved January 5, 2008) (2003), http://www.acrm.org.my/publications.htm
2. Rangayyan, R.M., Ayres, F.J., Desautels, J.E.: A Review of Computer Aided Diagnosis of Breast Cancer – Towards the Detection of Subtle Signs. Journal of the Franklin Institute 344, 312–348 (2007)
3. Gonzalez, R.C., Woods, R.E.: Digital Image Processing, 2nd edn. Prentice Hall, New Jersey (2002)
4. Morrow, W.M., Paranjape, R.B., Rangayyan, R.M., Desautels, J.E.: Region Based Contrast Enhancement of Mammograms. IEEE Transactions on Medical Imaging 11(3), 392–406 (1992)
5. Laine, A., Fan, J., Schuler, S.: Contrast Enhancement by Dyadic Wavelet Analysis. In: 16th Annual International Conference of the IEEE, pp. 10a–11a (1994)

6. Veldkamp, W.J.H., Karssemeijer, N.: Normalization of Local Contrast in Mammograms. IEEE Transactions on Medical Imaging 19(7), 731–738 (2000)
7. Nunes, F.L.S., Schiabel, H., Benatti, R.H., Stamato, R.C., Escarpinati, M.C., Goes, C.E.: A Method to Contrast Enhancement of Digital Dense Breast Images Aimed to Detect Clustered Microcalcifications. In: IEEE International Conference on Image Processing, vol. 1, pp. 305–308 (2001)
8. Cheng, H.D., Xu, H.: A Novel Fuzzy Logic Approach to Mammogram Contrast Enhancement. Information Sciences 148, 167–184 (2002)
9. Scharcanski, J., Jung, C.R.: Denoising and Enhancing Digital Mammographic Images for Visual Screening. Computerized Medical Imaging and Graphics 30, 243–254 (2006)

Detection of Sprague Dawley Sperm
Using Matching Method

Mohd Fauzi Alias[1], Nor Ashidi Mat Isa[2], Siti Amrah Sulaiman[3],
and Kamal Zuhairi Zamli[4]

[1,2,4] School of Electrical & Electronic Engineering,
University Science Malaysia, Engineering Campus,
14300 Nibong Tebal, Penang Malaysia
kaka_shi2003@yahoo.com, ashidi@eng.usm.my, eekamal@eng.usm.my
[3] School of Medical Science, University Science Malaysia,
16150 Kubang Kerian, Kelantan Malaysia
sbsamrah@kb.usm.my

Abstract. Cross correlation algorithm is the common method for image matching technique. In this paper, an expert system based on cross correlation is designed to reduce the workload of pathologist and improve the screening technology in medical field. The system is proposed to detect, locate and classify between normal and abnormal sperm head by evaluating the similarity between two images. Using cross correlation, the highest value near to one is defined as the best matching value. When this condition is obeyed, the system shows the matched image at the output together with the matching indicator. Such result indicates this system is very helpful to pathologist.

Keywords: Template matching, correlation, normal and abnormal head sperm.

1 Introduction

During the last decades, there were a lot of errors occurred in pathology field reported especially on the detection method [1,3]. One of the reasons is the excessive workload of pathologist. Pathologist needs to determine certain criteria from samples for a limited time and eventually could produce errors from their careless or in other words, the increasing pathologist workload can be correlated with the frequency of errors [1,4].

The workload of a pathologist is considered high and a lot of energy and time is needed. For example, pathologists who work with sperm samples taken from Sprague Dawley rat need to determine and classify between the normal and abnormal sperms from the screening process. Each sperm appeared on the slide is classified and counted into different classes. This screening process covers the whole area on the slide. In getting the result, pathologists do this screening process including counting and classifying sperm manually. The work consumes lots of focus, energy and consistency in order to get the accurate results. Therefore, an expert system on Sprague Dawley sperm detection is designed to replace and reduce the workload of pathologist. The system will automatically classify and count the total number of normal and abnormal sperms.

I. Lovrek, R.J. Howlett, and L.C. Jain (Eds.): KES 2008, Part III, LNAI 5179, pp. 541–547, 2008.

Template matching often used in image analysis, computer vision and pattern recognition [5-7]. In image processing field, template matching is one of the most basic techniques and has also been used for many years and proposed to be one of the easiest methods in image matching research [8]. In template matching method, at least 2 images would be needed. One image is defined as a template or reference image and the other one as an input image. The operation concept of template matching is the input image is matched over the reference image [9-11]. Therefore, the reference image always has same or smaller size compared to the input image [9-12]. In the matching process, the reference image is shifted and overlapped on the input image until the whole area on the image is covered and processed.

Recently, template matching for non rigid image had received more attention and popular for research study [13,14]. In some cases, the shape and orientation of the researched image is not in great condition such as our image sample where few criterias need to be considered before the matching process can be applied.

2 Rat Sperm Morphology

Sperm samples, randomly taken from the Sprague Dawley rat had been used as research samples. The head of sperm is approximately 2.5μm long [2]. For a normal sperm, the head resembles a hook, therefore any other shapes are classified as abnormal sperm. The head contains nucleus operates as main controller of each sperm. The morphology of a normal rat sperm is given in Figure 1 [2].

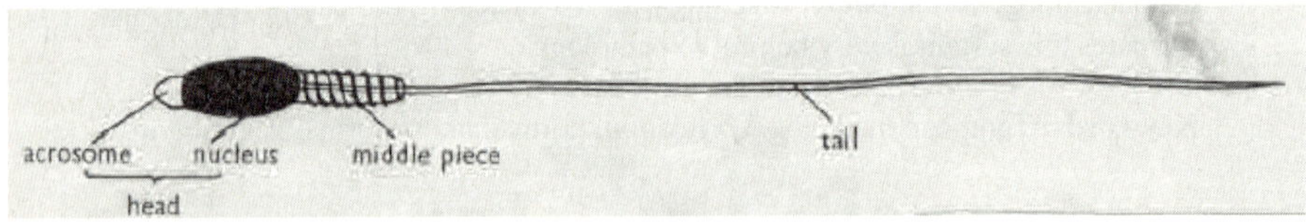

Fig. 1. The morphology of rat sperm

The system is designed to operate under magnification of 40x. During observation accurate result could be achieved when the processed image is magnified at 40x. Therefore, each image sample had been captured under magnification 40x correlated with the system designed. The normal spermatozoon after magnification at 40x is shown in Figure 2.

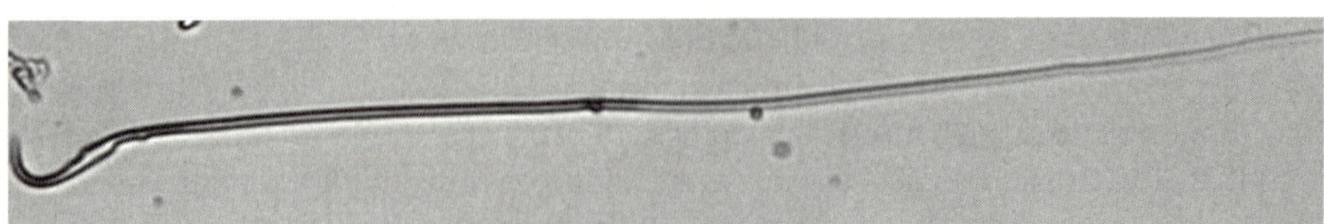

Fig. 2. The normal spermatozoon magnification at 40x

3   Template Matching Using Correlation

Template matching is one of the most common methods for similarity finding. To use this method, some of the important criterias need to be analyzed prior to get the best matching result. The important criteria need to be considered are the size of the image, orientation and the pixel intensity [5, 7, 8, 14, and 16]. In this research, matching is used as the main method to find the similarity between two images. Some other important methods have also been used in getting the best matching result such as segmentation, filtering and counting.

For segmentation method, grey-level value of data image was used in separating process [15-17]. Then it will differentiate the sperm and its background automatically. Segmentation method is used to eliminate the useless information pixel by setting them into certain pixel intensity [7,18].

For template matching technique, correlation method was being used. The result of the matching process lies in range 0 to 1 which indicates 0% and 100% matching percentage respectively. The result is in integer. When the result is near to 0, the matching percentage is near to 0% which proving no matching between both images. Using the same concept, when the result shows any integer near to 1, both images processed is almost similar or 100% match [7-9, 15, 18]. In statistics, correlation terms sometimes used to refer the covariance between two factors, such as to find the correlation between these 2 factors in their relationship. The higher correlation between these factors, the stronger relation between them.

For the research implementation, we used 2 dimensional cross correlation in template matching technique. The 2 dimensional cross correlation equation is shown in equation 1[9-11,19].

$$\text{Out}(i, j) = \sum_{m=0}^{M-1} \sum_{n=0}^{N-1} \text{In1}(m, n)\, \text{In2}(i+m, j+n). \tag{1}$$

Where In1 and In2 are the input images and Out is the result of the cross correlation between In1 and In2. In1, In2 and Out are in 2 dimensional image. In1 operates as the main image and In2 as the template images. The size of template image is usually smaller than the main image [9-12]. In the matching process, template image can be described as reference image. Template image is matched and shifted with the main image until it finds the highest matching percentage between both images. This process is repeated until all area on the main image is overlapped by the template image.

4   Result and Discussions

The system is designed to evaluate the input image and classify into two main classes which are known as normal and abnormal sperms. Template image is saved as the reference data. This reference data is chosen from the finest shape of normal sperm samples. Firstly, the system runs and saves the template data into one data array and then it becomes reference image for all input images. Secondly, the input images are segmented into binary form automatically and the resultant image displayed only in white and black pixels. Segmentation method is used to separate the sperm image from its background. Black pixel is the sperm image and white pixel is its background. The reference image is shown in Figure 3.

Fig. 3. The reference image for the system

The size of the reference image is 69 x 71 pixels. The black pixel shows the head of normal sperm and it cut from its tail to result accurate matching calculation.

For pre processing, the input image must undergo segmentation process to differentiate between the head sperm and its background. Filtering is then applied to decrease the amount of noises in the input image. Some of the resultant images are shown in Figure 4, 5 and 6 for Image01, Image02 and Image03 respectively.

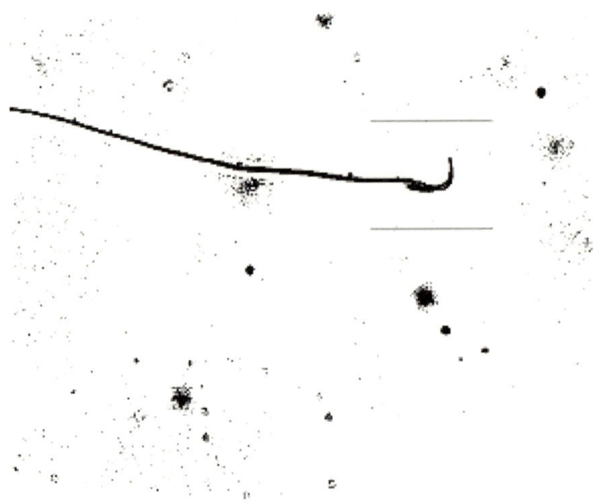

Fig. 4. The output Image01 with indicator

Figures 4, 5 and 6 show 2 lines at the top and the bottom of the head sperm to indicate the matching image (i.e. the sperm head) in the system. In this study, 50 random head sperm images had been tested and the result is tabulated in Table 1.

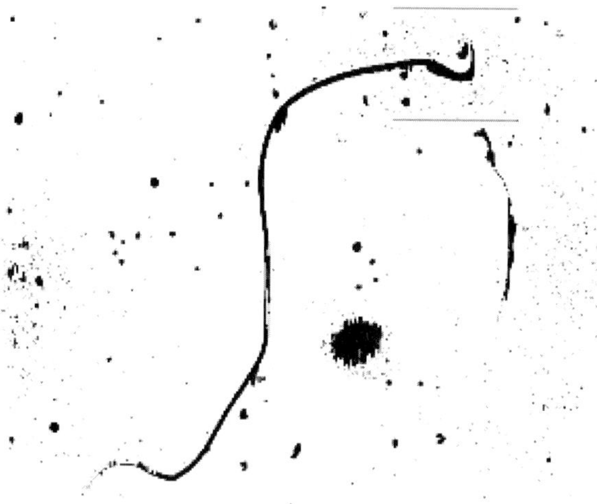

Fig. 5. The output Image02 with indicator

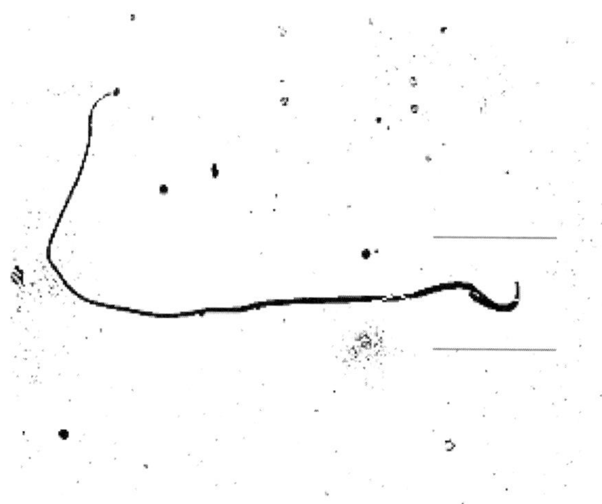

Fig. 6. The output Image03 with indicator

Table 1. Matching percentage of 50 input images

	Matching Percentage (%)					
	Above 70	Above 75	Above 80	Above85	Above90	Above95
Total number of correct detection	50	48	40	33	24	16
Percentage of correct detection	100%	96%	80%	66%	48%	32%

From Table 1, 100% of accuracy is achieved when the system considers the sperm head is correctly detected if 70% of area of the tested sperm image is matched with the reference sperm head image. 96% of tested images are correctly determined if the matching percentage is set to 75% and above. The system has also shown a high capability of detecting the sperm head by producing high accuracy although the matching percentage is increased. The system is capable to achieve the accuracy up to 80%, 66%, 48% and 32% for the matching percentage set to above 80, above 85, above 90 and above 95 respectively. The results obtained prove that the proposed system is suitable and high capability to be used as detection system for Sprague Dawley Sperm.

5   Conclusions

In this research, a system is designed to reduce the workload of the pathology. By using the best and suitable method, the system can be proven to reduce the workload of pathologist thus save more time. In this study, an expert Detection of Sprague Dawley Sperm system has been proposed to detect and classify the normal and abnormal sperm. The system requires 2 input images in applying the cross correlation method. Each image operates as main and template image respectively. The template image is set to be the reference image. The matching process is repeated until the whole area on the main image is covered. Some of pre processing methods are used in helping to get the best matching output such as segmentation, region growing and filtering. When the matching condition is obeyed, the system shows an indicator for the matched head sperm on the input image as the output display. As conclusion, this project has successfully carried out its objectives. The proposed template matching method performs promising assistant for pathologist and the matching accuracy is totally reliable for this project. More research needs to be done in this field to make the system more robust and be trusted 100% by consumer.

Acknowledgement

This project is partially supported by Ministry of Science, Technology and Innovation (MOSTI), Malaysia under Science Fund "Development of an Automatic Real-Time Intelligent Aggregate Classification System Based on Image Processing and Neural Network". The image processing techniques in this work are all applied and developed to be used in the aforementioned grant.

References

1. Stephen, S.R., Dana, M.G.: Anatomic Pathology Workload and Error, Pittsburgh, PA (2006)
2. Inveresk Research, Huntingdon Life Sciences, Sequani, Glaxo Wellcome.: Rat Sperm Morphological Assessment (October 2000)
3. Stephen, S.R., Dana, M.G., Richard, J.Z., Frederick, A.M., Stanley, J.G., Chris, J.: Anatomic Pathology Databases and Patient Safety, Pittsburgh (October 2005)

4. Mark, S.R.: The Use of Decision Analysis for Understanding The Impact of Diagnostic Testing Errors in Pathology, Pittsburgh (2006)
5. Yasuo, W., Katsutoshi, T.: A Fast Structural Matching and its Application to Pattern Analysis of 2D Electrophoresis Images. Japan (1998)
6. Khosravi, M., Schafer, R.W.: Template Matching Based on a Grayscale Hit or Miss Transform, Atlanta (1996)
7. Ghafoor, A., Iqbal, R.N., Shoab, K.: Robust Image Matching Algorithm, Pakistan (2003)
8. Xiong, X., Chen, Y., Li, T.: A Remote Sensing Image Subpixel Matching Algorithm Combined Edge with Grey, China (1997)
9. Gonzalez, R.C., Woods, R.E.: Digital Image Processing, 2nd edn. Prentice Hall, New Jersey (2002)
10. Myler, H.R., Weeks, A.R.: The Pocket Handbook of Imaging Processing Algorithms in C. PTR Prentice Hall, Englewood Cliffs (1993)
11. Sulaiman, S.N., Alias, M.F., Mat Isa, N.A., Abd Rahman, M.F.: An Expert Image Processing System on Template Matching, Penang, Malaysia (2007)
12. Yoneyama, S., Morimoto, Y.: Accurate Displacement Measurement by Correlation of Coloured Random Patterns, Japan (2003)
13. Zheng, Y., Doermann, D.: Robust Point Matching for Two-Dimensional Nonrigid Shapes. University of Maryland, College Park (2005)
14. Cheng, X.Y., Han, L.J., Ma, S.M.: Design and Realization of Medical Image Nonrigid Matching Algorithm, China (2006)
15. Banon, G.J.F., Faria, S.D.: Morphological Approach for Template Matching, Brazil (1997)
16. Coelho, L.D.S., Campos, M.F.M.: Similarity Based Versus Template Matching Based Methodologies for Image Alignment of Polyhedral-like Objects Under Noisy Conditions, Belo Horizonte, Brazil (1997)
17. Farooq, H., Salam, R.A.: Automatic Fish Recognition and Classification: A Framework, Pulau Pinang, Malaysia
18. Rahhal, J., Wang, Y.L., Atkin, G.E.: Template Matching for a Local Guidance System (1997)
19. Kagawa, K., Ogura, Y., Tanida, J., Ichioka, Y.: Prototype Demonstration of Discrete Correlation Processor-2 Based on High-speed Optical Image Steering for Large-fan-out Reconfigurable Optical Interconnections, Japan (2000)

Application of the Fuzzy Min-Max Neural Networks to Medical Diagnosis

Anas Quteishat and Chee Peng Lim[*]

School of Electrical & Electronic Engineering
University Science Malaysia, Malaysia
`cplim@eng.usm.my`

Abstract. In this paper, the Fuzzy Min-Max (FMM) neural network along with two modified FMM models are used for tackling medical diagnostic problems. The original FMM network establishes hyperboxes with fuzzy sets in its structure for classifying input patterns into different output categories. While the first modified FMM model uses the membership function and the Euclidian distance to classify the input patterns, the second modified FMM model employs only the Euclidian distance for the same process. Unlike the original FMM network, the two modified FMM models undergo a pruning process, after network training, to remove hyperboxes with low confidence factors. To assess the effectiveness of the three FMM networks in medical diagnosis, a set of real medical records from suspected Acute Coronary Syndrome (ACS) patients is collected and used for experimentation. The bootstrap method is used to analyze the results statistically. Implications of the experimental outcomes are discussed, and the potential of using the FMM networks a decision support tool for medical prognostic and diagnostic problems is demonstrated.

1   Introduction

In general, medical diagnosis is the process where physicians attempt to identify the disease a patient is suffering from based on the patient's physical symptoms, clinical tests data, and other related signs and information. In some cases, physicians may require additional opinion for disease diagnosis since some diseases have common symptoms and some are related to others. However, it is not always easy to obtain a second opinion or to statistically analyze the symptoms. This problem has inspired researchers to develop computerized medical diagnosis tools. Since medical diagnosis can be viewed as a pattern classification problem, artificial neural networks (ANN) have shown potentials as a tool to diagnose (classify) patients based on their symptoms. Indeed, ANNs have the ability to integrate data-based analytical techniques such as decision and classification theory, and to use knowledge-based approaches to provide useful information to support the decision making process [1]. From the literature, a lot of successful ANN applications to medical problems can be found. In [1], autonomously learning ANN models were used to classify real coronary care unit (CCU) patient records and trauma patient records into different categories. In [2] a pulmonary disease diagnostic system was proposed. The system used real clinical data to attempt to treat a whole category of distressed body organs. A model selection

---

[*] Corresponding author.

I. Lovrek, R.J. Howlett, and L.C. Jain (Eds.): KES 2008, Part III, LNAI 5179, pp. 548–555, 2008.

method that used the self-organizing map (SOM) for breast cancer diagnosis was demonstrated in [3]. In [4], the performances of a number of ANN models for breast cancer diagnosis were compared and analyzed. A neuromuscular disorder diagnosis system that employed two different ANN models (supervised and unsupervised) for analyzing features selected from electromyography (EMG) was presented in [5].

In this paper, use of ANN models to support the decision making process of suspected Acute Coronary Syndrome (ACS) patients is presented. A rapid and timely diagnosis of patients with suspected ACS is necessary as it could potentially lead to mortality if proper medical treatment is not provided to ACS patients as soon as possible. Specifically, the Fuzzy Min-Max (FMM) network [6], and two of its modified versions (MFMM1 and MFMM2) [7] were employed to classify in this work. The results obtained are analyzed statistically using the bootstrap method.

The paper is organized as follows. Section 2 describes the original FMM network. The modified versions of FMM are explained in section 3. The experiments using suspected ACS patient records along with the analysis for the results are presented section 4. A summary of this work is presented in section 5.

2   The Fuzzy Min-Max (FMM) Neural Network

FMM is an incremental learning system. It learns incrementally in a single pass through the data set. It refines the existing pattern classes as new information is received. It also has the ability to add new pattern classes online. In FMM, hyperboxes with fuzzy sets are formed to represent knowledge learned by the network. Learning in FMM comprises a series of expansion and contraction processes that fine-tune its hyperboxes to establish boundaries between classes. If overlapping hyperboxes of different classes occur in the input space, contraction is performed to eliminate the overlapping regions. The membership function is defined with respect to the minimum and maximum points of a hyperbox. It describes the degree to which a pattern fits in the hyperbox. The hyperboxes have a range from 0 to 1 along each dimension; hence the pattern space is an n-dimensional unit cube I^n. A pattern which is contained in the hyperbox has a unity membership function.

Mathematically, the definition of each hyperbox fuzzy set B_j [6] is defined by

$$B_j = \left\{ X, V_j, W_j, f\left(X, V_j, W_j\right) \right\} \qquad \forall X \in I^n \tag{1}$$

where $X = (x_1, x_2, \ldots\ldots, x_n)$ is   the   input   pattern, $V_j = (v_{j1}, v_{j2}, \ldots\ldots, v_{jn})$, $W_j = (w_{j1}, w_{j2}, \ldots\ldots, w_{jn})$ are the minimum and maximum points of B_j, respectively and $f\left(X, V_j, W_j\right)$ is the membership function. Figure 1 illustrates an example of the minimum and maximum points for a two-class problem where the universe of discourse is $\Re^3$.

Applying the definition of a hyperbox fuzzy set, the combined fuzzy set that classifies the Kth pattern class, C_k, is defined as

$$C_k = \bigcup_{j \in K} B_j \tag{2}$$

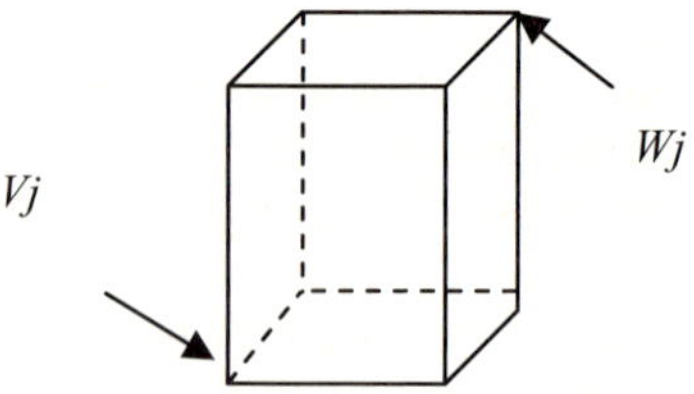

Fig. 1. A min-max hyperbox $B_j = \{V_j, W_j\}$ in $\Re^3$

where K is the index set of those hyperboxes associated with class k. One important property of this approach is that the majority of the processing is concerned with finding and fine-tuning the class boundaries. The learning algorithm of FMM allows overlapping of hyperboxes of the same class but eliminates overlapping of hyperboxes from different classes. The membership function for the jth hyperbox $b_j(A_h)$, $0 \le b_j(A_j) \le 1$, measures the degree to which the hth input pattern, A_h, falls outside hyperbox B_j [6]. As $b_j(A_h)$ approaches 1, the pattern is said to be more "contained" by the hyperbox. The resulting membership function is [6]:

$$b_j(A_h) = \frac{1}{2n} \sum_{i=1}^{n} \left[max\left(0, 1 - max\left(0, \gamma min\left(1, a_{hi} - w_{ji}\right)\right)\right) \right.$$
$$\left. + max\left(0, 1 - max\left(0, \gamma min\left(1, v_{ji} - a_{hi}\right)\right)\right) \right] \tag{3}$$

where, $A_h = (a_{h1}, a_{h2}, ..., a_{hn}) \in I^n$ is the hth input pattern, and γ is a sensitivity parameter that regulates how fast the membership value decreases as the distance between A_h and B_j increases.

The architecture of FMM consists of three layers of nodes, as shown in Figure 2. The first layer, F_A, is the input layer that contains input nodes equal in number to the number of dimensions of the input pattern. Layer F_C is the output layer. It contains

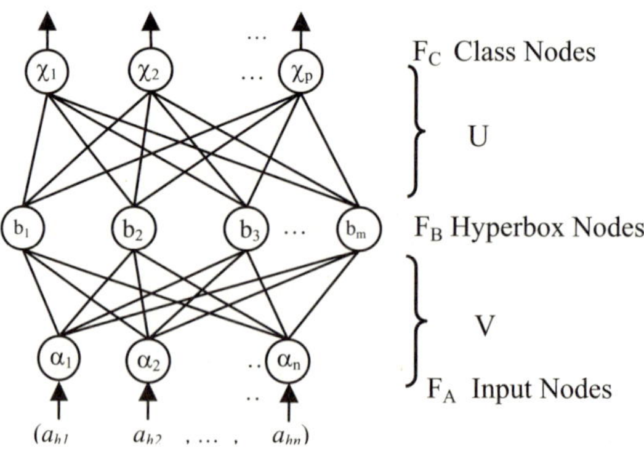

Fig. 2. A three layer FMM network

nodes equal in number to the number of classes. The middle or hidden layer, F_B, is called the hyperbox layer. Each F_B node represents a hyperbox fuzzy set, where F_A to F_B connections are the min-max points. The F_B transfer function is the hyperbox membership function defined by (3). The minimum and maximum points are stored in matrices V and W, respectively.

Note that the above provides an overview of the FMM network. Details of the FMM dynamics can be found in [6].

3 Modified Fuzzy Min-Max Neural Networks

In FMM, the size of a hyperbox is controlled by θ, which is varied between 0 and 1. When θ is small, more hyperboxes are created. When θ is large, the number of hyperboxes is small, and the size of hyperboxes is large. Therefore, when θ is large, a lot of hyperboxes may assume high membership function values for a new input pattern, in which the winner-take-all approach in determining the winning hyperbox may lead to inaccurate predictions by FMM.

To enhance the effectiveness of FMM, useful modifications are proposed in [7]. After the training stage, a pruning procedure is incorporated in FMM to reduce the number of hyperboxes, hence the network complexity. To assist pruning, a confidence factor is calculated using the method in [8]. A data set is divided into three: training set (for learning), prediction set (for pruning), and test set (for evaluation). The confidence factor for each hyperbox is based on its usage frequency and its predictive accuracy on the prediction sct, as follows

$$CF_j = (1 - \gamma)U_j + \gamma A_j \qquad (4)$$

where U_j is the usage of hyperbox j, A_j is the accuracy of hyperbox j, and $\gamma \in [0,1]$ is a weighing factor. The value of U_j is defined as the number of patterns in the prediction set classified by any hyperbox j, divided by the maximum number of patterns in the prediction set classified by any hyperbox with the same classification class. On the other hand, the value of A_j is defined as the number of correctly predicted set of patterns classified by any hyperbox j, divided by the maximum correctly classified patterns with the same classification class. The confidence factor identifies hyperboxes that are frequently used and generally give high classification accuracy, as well as hyperboxes that are rarely used and, yet highly accurate. Hyperboxes with a confidence factor lower than a user-defined threshold is pruned.

Modification to the prediction process of FMM is also suggested. In original FMM, the input patterns are classified based on the membership function calculated using equation (3). The degree of membership for the input pattern is calculated against all hyperboxes created in the learning process. The pattern is classified to the class associated with the hyperbox that has the highest membership function value.

Based on the above modifications, two modified FMM (MFMM) models are proposed [7]. The first modified FMM (MFMM1) model uses both the membership function and the Euclidean distance to predict its target output. The second modified FMM (MFMM2) model uses the Euclidean distance alone to predict the target output.

For MFMM1, given a new input pattern, the membership function values of all hyperboxes are first calculated. A pool of hyperboxes that have high membership function values is then selected. The number of hyperboxes selected can be based on a user-defined threshold. After that, the Euclidean distances between the selected hyperboxes and the input pattern are computed. The hyperbox with the shortest Euclidean distance is chosen as the winner. Figure 3 shows the classification procedure of a two-dimensional input pattern using the proposed method. Suppose hyperboxes 1 and 2 are selected, and E_1 and E_2 are the distances between the input pattern and the centroids of hyperboxes 1 and 2, respectively. Since $E_2 < E_1$, the input pattern is predicted to be in Class 2, even though it is contained in hyperbox 1. This modified FMM network, as shown in [7], is able to improve the classification results when the network has a small number of hyperboxes, and each with a large size (i.e. in situations where θ is large).

As for MFMM2, the Euclidian distance is calculated between the input pattern and all the hyperboxes. The hyperbox with the shortest distance is used to classify the input pattern. Details of the modifications can be found in [7].

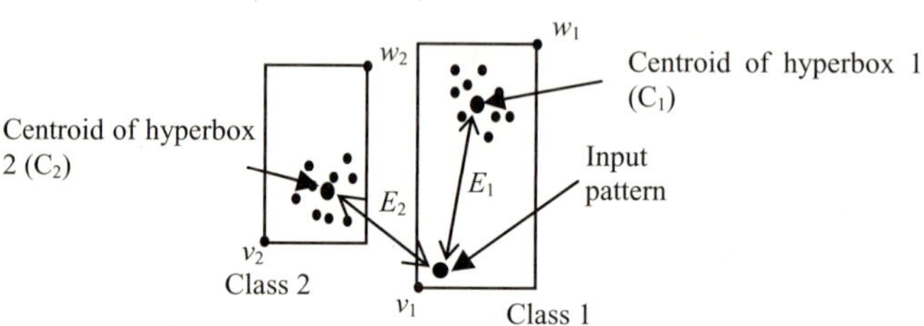

Fig. 3. The classification process of modified FMM models

4   Diagnosis of Acute Coronary Syndrome (ACS)

The ACS data set contained 118 real records of suspected heart attack patients admitted to Penang Hospital, Malaysia. After consulting with medical specialists, 16 features comprising clinical historical data, physical examination, ECG, and cardiac enzymes, and Troponin-T tests, were extracted from each record for this study. Out of the 16 features, 12 binary features were coded as 1, 0.25, and 0 for the presence, absence and missing values, respectively. Four real-valued features were age, degree of pain, duration of pain, and cardiac enzymes data. They were normalized between 0.25 and 1, and 0 was used to represent missing values. Among all the records, 103 patients were with ACS and 15 patients without ACS.

The experimental procedure was as follows. The data set was divided into three sub-sets: 60 samples for training, 23 samples for prediction, and 35 samples for test. Since the prediction data set was not needed for FMM, 83 samples were used for FMM training. The pruning threshold was set to 0.7 for both the modified FMM networks. Based on the 16 features, the FMM networks were employed to determine whether the patient suffered from ACS or otherwise. The experiments were repeated 10 times, each time the training data set was presented in a different sequence, and the average classification accuracy rates were computed. To further assess the results statistically, the bootstrap methods [9] was employed. The 95% confidence intervals of the results were estimated using 5000 bootstrap samples, and the *p*-value was used to quantify the performance statistically.

4.1 Results and Discussion

Figure 4 shows the average accuracy rates versus the hyperbox size, θ. The performance of the three networks was almost similar for small hyperbox sizes ($\theta <$ 0.1). However, as the hyperbox size increased, the FMM performance started to degrade. Compared with FMM, both modified FMM networks degraded in a slower manner. In general, MFMM1 yielded the best test accuracy rates, as shown in Figure 4. The highest result of MFMM1 was 85.43%.

The error bars in Figure 4 are the 95% confidence intervals of the average accuracy rates. Notice that the error bars of FMM did not overlap with those of MFMM1 and MFMM2, indicating that the FMM performances were statistically inferior to those of modified FMM networks.

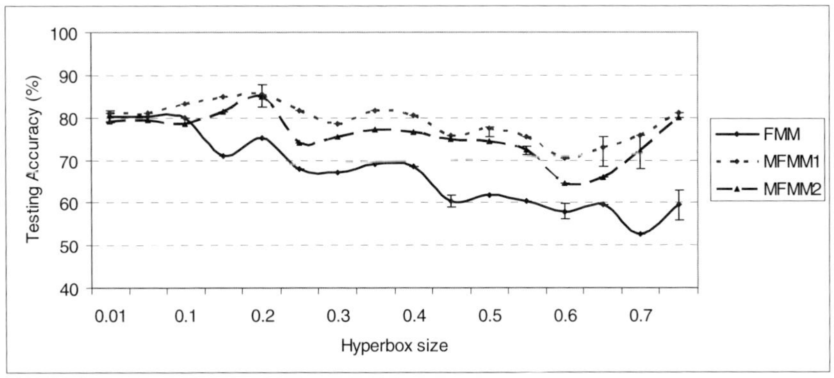

Fig. 4. Average test accuracy rate versus hyperbox size for the ACS data set

To further compare the performances of different FMM networks, the *p*-value hypothesis test was conducted. The null hypothesis stated that the average accuracy rates between FMM versus MFMM1, FMM versus MFMM2, and MFMM1 versus MFMM2 were equal (at the 95% confidence level). As shown in of Table 2 (column 2), the *p*-values of FMM versus MFMM1 are lower than the significance level ($\alpha = 0.05$) for $\theta \geq 0.15$. This means that the null hypothesis is rejected. In other words, there is a significant difference in terms of the test accuracy rates between FMM and MFMM1 for $\theta \geq 0.15$. The same trend (except for $0.55 \leq \theta \leq 0.65$) can be observed in Table 2 (column 3) for performance comparison between FMM and MFMM2. However, all the *p*-values between MFMM1 and MFMM2 are higher than 0.05, as shown in Table 2 (column 4), indicating that the performances of the two modified FMM networks are statistically equal at the 95% confidence level.

Table 1. The p-values for performance comparison between different FMM networks

Hyperbox size, θ	FMM vs. MFMM1	FMM vs. MFMM2	MFMM1 vs. MFMM2
0.05	0.8026	0.6016	0.562
0.1	0.0594	0.4322	0.0538
0.15	0	0.0004	0.266
0.2	0	0.0002	0.8034
0.25	0	0.0416	0.0278
0.3	0.004	0.011	0.43
0.35	0.0006	0.0256	0.2398
0.4	0.0138	0.0398	0.645
0.45	0.0094	0.005	0.9158
0.5	0.0082	0.0074	0.7446
0.55	0.0224	0.0962	0.5724
0.6	0.048	0.2962	0.4606
0.65	0.0298	0.3032	0.2598
0.7	0.001	0.0104	0.6104
0.75	0.027	0.0174	0.7416

5 Summary

This paper has presented an application of the FMM neural networks as a medical diagnosis tool. Two useful modifications have been incorporated into the FMM network: use of a pruning procedure to remove hyperboxes with low confidence factor from the network, and use of the Euclidian distance in the prediction process. Based on the modifications, two modified FMM networks have been introduced: MFMM1 uses both the membership value and the Euclidian distance for classifying the input pattern, while MFMM2 uses only the Euclidian distance for classification.

To assess the applicability of the FMM networks to medical diagnosis, a set of real patient records has been collected, and the FMM networks have been used to classify the patients into ACS and non-ACS categories. Stability of the results has been assessed and quantified using the bootstrap method. The results obtained have demonstrated that the modifications on the FMM network have improved the performances when the hyperbox size is high.

Future work will focus on using more data sets from different medical domains to further ascertain the capability and effectiveness of the modified FMM networks, so that they can be employed as a useful and usable decision support tool for medical prognostic and diagnostic tasks.

References

[1] Lim, C.P., Harrison, R.F., Kennedy, R.L.: Application of autonomous neural network systems to medical pattern classification tasks. Artificial Intelligence in Medicine 11, 215–239 (1997)
[2] Economou, G.P.K., Spiropoulos, C., Economopoulos, N.M., Charokopos, N., Lymberopoulos, D., Spiliopoulou, M., Haralambopulu, E., Goutis, C.E.: Medical diagnosis and artificial neural networks: a medical expert system applied to pulmonary diseases. In: Proceedings of the 1994 IEEE Workshop Neural Networks for Signal Processing, pp. 482–489 (1994)

[3] West, D., West, V.: Model selection for a Medical Diagnosis decision support system: A Breast Cancer Detection Case. Artificial Intelligence in Medicine 20, 183–204 (2000)

[4] Kıyan, T., Yıldırım, T.: Breast Cancer Diagnosis Using Statistical Neural Networks. In: International XII. Turkish Symposium on Artificial Intelligence and Neural Networks, pp. 754–760 (2003)

[5] Pattichis, C.S., Schizas, C.N., Middleton, L.T.: Neural network models in EMG diagnosis. IEEE Transactions on Biomedical Engineering 42, 486–496 (1995)

[6] Simpson, P.K.: Fuzzy Min-Max neural networks-Part 1: Classification. IEEE Transactions on Neural Networks 3, 776–786 (1992)

[7] Quteishat, A., Lim, C.P.: A modified fuzzy min–max neural network with rule extraction and its application to fault detection and classification. Applied Soft Computing 8, 985–995 (2008)

[8] Carpenter, G., Tan, A.: Rule extraction: From neural architecture to symbolic representation. Connection Science 7, 3–27 (1995)

[9] Efron, B.: Bootstrap methods: Another look at the jackknife. Annals of Statistics 7, 1–26 (1979)

SVD Based Feature Selection and Sample Classification of Proteomic Data

Annarita D'Addabbo[1], Massimo Papale[2], Salvatore Di Paolo[3],
Simona Magaldi[2], Roberto Colella[1], Valentina d'Onofrio[4],
Annamaria Di Palma[2], Elena Ranieri[5], Loreto Gesualdo[2], and Nicola Ancona[1]

[1] Istituto di Studi sui Sistemi Intelligenti per l'Automazione, CNR,
Via Amendola 122/D-I, 70126 Bari, Italy
daddabbo@ba.issia.cnr.it
[2] Molecular Medicine Center, Sect. of Nephrology, Dept. of Biomedical
Sciences and Bioagromed, Faculty of Medicine, University of Foggia
[3] Division of Nephrology and Dialysis, Hospital "Dimiccoli", ASL BAT, Barletta
[4] Dept. of Surgical Sciences , Faculty of Medicine, University of Foggia, Italy
[5] Dept. of Biomedical Sciences, Sect. of Clinical Pathology,
Faculty of Medicine, University of Foggia, Italy

Abstract. Feature selection becomes a central task when 'signature' profiles specific to a pathological status have to be extracted from high dimensional gene expression or proteomic data. In the present paper, we propose a feature selection method based on Singular Value Decomposition (SVD) and apply it to SELDI-TOF/MS proteomic data from a cohort of Type 2 Diabetics affected by Glomerulosclerosis and Membranous Nephropathy. We have selected a profile composed of 24 proteins that seems to be an effective signature for the pathology at hand, allowing to efficiently discriminate between the considered subtype of diabetes.

Keywords: Feature selection, SVD, proteomic data.

1 Introduction

Proteomic techniques based on mass spectrometry have made possible to investigate proteins over a wide range of molecular weights, as DNA microarray technologies have allowed to profile the expressions of thousand of genes simultaneously. These data promise to enhance understanding of life from a molecular point of view and seems to be particularly useful in medical diagnosis, treatment and drug design, but their analysis requires well founded mathematical tools that are adaptable to the large quantity of information and noise they carry on [1,2]. Actually, some learning machines, such as Support Vector Machines (SVM) and Regularized Least Square (RLS) Classifiers, have been demonstrated to be able to manage a great quantity of data and to be particularly robust to noise [2]. In particular, in the classification of gene expression data, SVM and RLS classifiers showed poor sensibility to the noise, in fact their error rate remained unchanged when the number of gene expressions, considered in input, ranged in a quite large interval [3]. So, from a diagnostic point of view, i.e. if the clinical status

I. Lovrek, R.J. Howlett, and L.C. Jain (Eds.): KES 2008, Part III, LNAI 5179, pp. 556–563, 2008.

has to be predicted, searching for a subset of features could not be relevant. On the other hand, from an etiological point of view or when the problem of best treatment is pursued, a central task becomes to discover 'signature' profiles specific to a pathological status or specific to different subtypes of the same disease. Several feature selection methods have been applied to DNA microarray and proteomic data [4,5,6,7] in order to select an opportune and significant subset of gene-expressions or proteins more correlated with the pathology at hand.

In the present paper, we propose a feature selection method based on Singular Value Decomposition (SVD) and apply it to proteomic data from a cohort of Type 2 Diabetics affected by glomerulosclerosis (class 1 DN) and Membranous Nephropathy. Renal injury in Type 2 diabetics shows a variable pattern of damage, ranging from typical glomerulosclerosis to prevailing chronic vascular damage, to glomerulonephritis with or without the presence of diabetic glomerulosclerosis [8]. On the other hand, Membranous Nephropathy is a glomerulonephritis characterized by diffuse thickening of glomerular capillary basement membranes and show damage pattern comparable to class 1 DN. Among proteomic approaches, SELDI-TOF/MS is a high throughput technique based on the chromatographic separation of proteins according to their physical characteristics. The coupling of these chromatographic surfaces to a laser desorption time of flight mass spectrometer allows to generate an accurate protein profile of a biological sample requiring minimal amounts of sample [9]. We chose to analyse urine protein profiles since urine contains proteins and peptides which are derived from a variety of sources including glomerular filtration of blood plasma, cell sloughing, apoptosis, proteolytic cleavage of cell surface glycosylphosphatidylinositol-linked proteins, and secretion of exosomes by epithelial cells [10,11,12]. Thus, a change in the amount of a given soluble protein that reaches the final urine can result from a change in its concentration in the blood plasma, a change in the function of the glomerular filter, or an alteration in the proximal tubule scavenging system. Based on these mechanisms, changes in excretion rate of specific urinary proteins can be indicative of systemic disease, glomerular disease, or diseases affecting the proximal tubule, respectively [13,14,15,16,17,18].

In section 2, we describe SVD, the RLS classifiers used to asses the selected protein profiles and our experimental set up. In section 3, data sets are described together with methods used to perform biomarker detection in the routine practice. Experimental results are shown in section 4 and conclusions follow.

2   Methods

2.1   Features Selection by SVD

Let the matrix X of size $n \times \ell$ represent the full data set, where n is the number of features and ℓ is the number of examples. In the present case, we have $\ell \ll n$. SVD [20] is a linear algebra method that compute the following matrix decomposition:

$$X = U \Lambda V^{\top} \tag{1}$$

where U is a $n \times \ell$ orthogonal matrix with columns u_i corresponding to eigen-examples, V is a $\ell \times \ell$ orthogonal matrix with columns v_i corresponding to eigen-features, and Λ is a diagonal $\ell \times \ell$ matrix, having the elements on the main diagonal arranged in decreasing order, such that $\lambda_1 \geq \lambda_2 \geq ... \geq \lambda_\ell \geq 0$.

We can associate to the first k eigen-features (eigen-examples) the following score

$$\sigma_k^2 = \frac{\sum_{i=1}^k \lambda_i^2}{\sum_{i=1}^\ell \lambda_i^2} \tag{2}$$

that indicates the fraction of the overall variance they capture. For this reason, fixed a value of σ_k^2, we can use the SVD decomposition to perform a space dimensionality reduction, i.e. we can project our data X into the space spanned by the first k eigen-examples u_i, obtaining a new data matrix X^*. Each example in X^* is composed by k features that are linear combination of the actual n features, i.e. they are mathematical abstractions and do not correspond to actual features in X. In many practical applications, it may be interesting to identify a subset of $n^* < n$ actual features capturing a portion of variance equal to σ_k^2, i.e. it is preferable to perform feature selection instead of feature construction [21].

For selecting features we perform the following procedure:

1. Let us be $x_t^\top$ the t-th row of X, i.e. the t-th actual feature. We compute, for each actual feature, the square of its projection into the space spanned by the first eigen-feature v_1, into the space spanned by v_1 and v_2, and so on. So, we construct the $n \times \ell$ matrix S, with the generic element

$$S_{tz} = \sum_{i=1}^z (x_t^\top v_i)^2 = \sum_{i=1}^z (\lambda_i u_{ti})^2 \tag{3}$$

   that indicates how much the t-th actual feature lies in the space spanned by the first z eigen-features.
2. We sort in descending order each column of S to know what actual features more lies in each subspace. We denote with P the matrix obtained by S sorting its columns.
3. We construct another $n \times \ell$ matrix T, with the generic element

$$T_{pz} = \sum_{i=1}^p P_{iz} \tag{4}$$

where the last row of T is the vector $(\lambda_1^2, \lambda_1^2 + \lambda_2^2, ..., \sum_{i=1}^\ell \lambda_i^2)^\top$.

It is worth noting that the generic element T_{pz} is equal to the variance captured by the p actual features that most lie in the space spanned by the first z eigen features. It is clear that the whole set of actual features is needed to capture λ_1^2 in the space spanned by v_1, to capture $\lambda_1^2 + \lambda_2^2$ in the space spanned by v_1 and v_2, and so on. Actually, the same portion σ_k^2 of variance can be captured in different manner (see figure 1). For example, $\sigma_k^2 = 0.9$ is captured from the whole set of actual features in the space determined by the first 5 eigen features. But $\sigma_k^2 = 0.9$ is also captured by 43 actual features in the space of the first 6 eigen features, by 34 in the space of the first 7 eigen features, and so on. In particular, fixed a value of σ_k^2, we search the element of the corresponding iso-variance curve that is closer to origin. In this way, the best compromise between number n^* of actual features and dimension k^* of eigen feature space is considered.

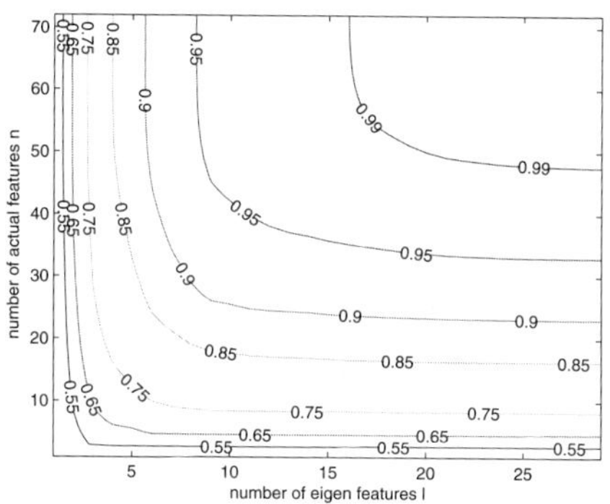

Fig. 1. Iso-variance curves of T matrix

2.2   Regularized Least Squares Classifiers

We assess the performances of our feature selection method by considering the prediction accuracies obtained by using RLS classifiers. In a two class classification problem, we have a set S composed of N independent and identically distributed data $(\mathbf{x_i}, \mathbf{y_i})$, where $\mathbf{x_i} \in \mathbb{R}^\mathbf{d}$ and $y_i \in \{-1, 1\}$ for $i = 1, 2, ..., N$, drawn from a fixed but unknown probability density function. In our setting, the vector $\mathbf{x_i}$ represents the proteic profile and the label y_i indicates the phenotype of the i-th subject. It is well known that classification is an ill-posed problem [22], i.e. there is not a unique solution to it. To make the problem well-posed, some form of constraint has to be imposed [23]. In a regularized approach [24], the solution is required to be smooth and constrained to belong to the reproducing kernel Hilbert space $\mathcal{H}_K$ induced by a positive definite kernel function K. In practice, this is equivalent to solve the following minimization problem

$$\min_{f \in \mathcal{H}_K} \frac{1}{\ell} \sum_{i=1}^{\ell} (y_i - f(\mathbf{x_i}))^2 + \lambda \|f\|_K^2, \tag{5}$$

where $\lambda > 0$ is called *regularization parameter* and $\| \cdot \|_K$ is the norm of the function f in $\mathcal{H}_K$. Independently of the adopted K, the solution of (5) is [24]:

$$f(\mathbf{x}) = \sum_{i=1}^{\ell} \mathbf{c_i} \mathbf{K}(\mathbf{x_i}, \mathbf{x}), \tag{6}$$

composed of a superposition of kernel functions centered on the available data, weighted by the coefficients c_i. The label y of a new input data $\mathbf{x}$ is given by:

$$y = \text{sign}\left(\sum_{i=1}^{\ell} c_i^* K(\mathbf{x_i}, \mathbf{x})\right). \tag{7}$$

2.3 Estimate of Prediction Accuracy and Statistical Significance

We use the leave-k-out cross validation (LKOCV) procedure [25] to provide a significant estimate of the generalization error. It consists of building T_1 pairs of training and test sets composed of r and $N - r$ examples respectively by random sampling without replacement the data set S. For each random split, we evaluate the error rate e_{n_i} of the classifier trained on r examples, testing it on $N - r$ examples. So, the LKOCV error e_n is given by $e_n = \frac{1}{T_1}\sum_{i=1}^{T_1} e_{n_i}$. Then, we apply the method of hypothesis testing to evaluate the statistical significance of the error rate e_n. For evaluating the *p-value* corresponding to e_n, we need to know the probability density function of e_n under the null hypothesis. Since it is unknown, we invoke non-parametric permutation tests [26]. In particular, we perform T_2 random permutations of the labels in the data set S. For each permutation, we split S in a training and a test set composed of r and $N - r$ examples respectively, by random sampling without replacement the data set S. Finally, we evaluate the error rate $e_{n_i}^0$ of the classifier trained on the r examples. Then the empirical probability density function of the error rate is constructed. The *p-value* of the error rate e_n is given by the percentage of $e_{n_i}^0 < e_n$.

3 Data Set Description

The urine protein profiles of 10 class 1 DN and 19 patients affected by Membranous Nephropathy were obtained by SELDI-TOF/MS analysis. For each sample we identified a list of mass peaks whose intensity values correlated with the protein amount in the urine sample. The Ciphergen Express software was than used to match the protein profiles and to identify a set of mass peaks shared by class1 DN and Membranous Nephropathy patients. By these analysis a cluster list was created. A cluster was defined as a group of peaks with similar masses but different intensities across multiple spectra. Once the peaks were clustered, the peak intensities from each spectrum were used to measure expression levels for that substance. To create the cluster list we manage two parameters: the signal to noise ratio (S/N) of each peak and the percentage of the spectra across multiple spectra in which a peak had to be present (minimum peak threshold). When setting 5 as S/N and 20% as minimum peak threshold, we identified 54 clusters; when the S/N was reduced to 4 and minimum peak threshold was set to 20%, we identified 58 clusters. Finally, when the S/N was 4 and minimum peak threshold was 15%, the number of clusters was increased to 72.

4   Experimental Results

In the first set of experiments, RLS classifiers were applied to the before mentioned proteomic data sets. Classification results, obtained by using linear RLS classifiers, are shown in table 1. In each experiment, the classification accuracy has been computed performing $T_1 = 200$ LKOCV, its *p-value* has been assessed by $T_2 = 10000$ random permutations of labels. In each LKOCV step, $n = 19$ examples have been used to constitute the training set. It is worth noting that when 72 actual features have been considered, the best prediction accuracy is obtained. When the number of features is reduced the capability to predict the pathological status by the protein profile is slightly reduced, confirming that feature reduction can not be a crucial step when the diagnostic purpose is addressed.

In the second set of experiments, we use SVD to automatically determine a protein signature characterizing the pathology at hand. We have applied SVD to the data set pre-processed by selecting peak threshold equal to 15% and S/N equal to 4. Different values of percentage of variance captured σ_k^2 have been considered. The prediction accuracies, obtained by linear RLS classifiers to reduced data sets, are shown in figure 2. In particular, with $\sigma_k^2 = 0.9$ a prediction accuracy equal to $69.3\%(p\text{-}value = 0.206)$ is obtained, selecting 24 actual features, as reported in table 2. It is important to note that the prediction accuracy obtained with the 24 actual features selected by SVD is comparable with the ones reported in table 1, where the number of considered features is higher: it is reasonable to think that a selection method based on data variance analysis may be more efficient that other selection criteria.

Table 1. Accuracy, specificity, sensitivity estimated by using a linear RLS classifier for data set with feature selection performed by Ciphergen Express software, as performed in routine practice by biologists and physicians

Protein number	Peak threshold	S/N ratio	Accuracy (p-value)	Sensitivity (p-value)	Specificity (p-value)
72	15	4	74.0%(0.075)	78.6%(0.144)	63.3%(0.425)
58	20	4	71.9%(0.074)	76.9%(0.142)	60.3%(0.426)
54	20	5	69.8%(0.208)	74.4%(0.139)	59.2%(0.414)

Table 2. Protein selected by SVD with $\sigma_k^2 = 0.9$

Protein mass/charge values

4062.2	4123.2	4167.7	4284.6	4352.3	4762.7	4850.9	5463.4	5556.7	5694.3	5863.6	7054.8
7916.6	8611.8	8788.1	8855.0	9089.7	9208.2	10052	10547	10800	11569	11716	11939

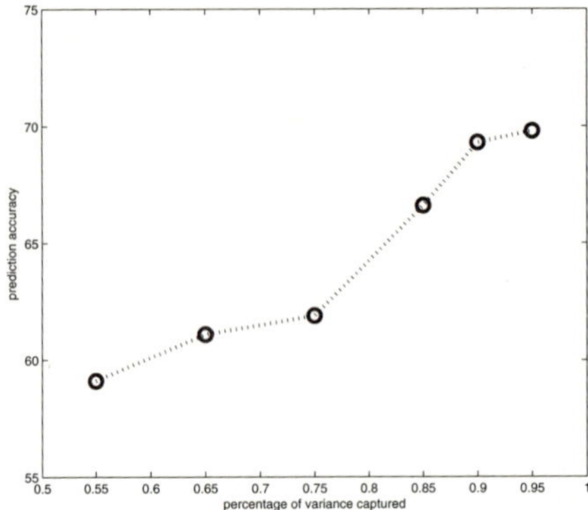

Fig. 2. Prediction accuracy of RLS classifier varying the number of input actual features, selected by SVD setting different values of percentage of variance captured σ_k^2

5   Discussion

In order to assess the performances of any feature selection method, we know that a proper approach is to include the feature selection procedure in a cross validation scheme, leaving the test sample out from the training set before undergoing any feature selection, as a pointed out in [19]. In this way, an honest predictive assessment is obtained, but different subset of features are selected. These subsets could be highly overlapped but also show up some features different case by case, reflecting sample variability and inherent heterogeneity [7]. However, in the present work the feature selection step is external to the LKOCV procedures, i.e., the feature selection was done with all the samples and the LKOCV was only done for the classification procedure. We chose this experimental set up because we want to mime with our method, the feature extraction performed on mass spectrometry data by biologists and physicians. In this way, we obtain a unique subset of feature, but we are aware that our results could be biased toward a potentially over-optimistic discrimination.

6   Conclusion

Our study shows that if the goal is to predict the phenotypic state of new subjects by analyzing their proteic profile then also proteins weakly associated with the phenotype have to be considered because they increase the prediction accuracy. On the contrary, if the objective is to determine proteic signatures associated to the trait of interest then protein selection schemes based on SVD provide profiles composed of a small number of proteins which accurately describe the pathology. Future work will focus on the biological interpretation of these signatures.

Acknowledgments. A.D. is PhD student of Dip. di Fisica, Universitá di Bari, Italy.

References

1. Golub, T.R., Slonim, D.K., et al.: Molecular Classification of Cancer: Class Discovery and Class Prediction by Gene Expression Monitoring. Science 286, 531–537 (1999)
2. Ancona, N., Maglietta, R., D'Addabbo, A., et al.: Regularized Least Squares Cancer Classifiers from DNA microarray data. BMC-Bioinformatics 6 (Suppl 4):S2 (2005)
3. Ancona, N., Maglietta, R., Piepoli, A.: D'Addabbo, et al: On the statistical assessment of classifiers using DNA microarray data. BMC-Bioinformatics 7, 387 (2006)
4. Guyon, I., Weston, J., Barnhill, S., Vapnik, V.: Gene selection for cancer classification using support vector machines. Machine Learning 46(1), 389–422 (2002)
5. Furlanello, C., Serafini, M., et al.: Entropy-based gene ranking without selection bias for the predictive classification of microarray data. BMC Bioinf. 4, 54–73 (2003)
6. Yasui, Y., et al.: A data-analytic strategy for protein biomarker discovery: profiling of high-dimensional proteomic data for cancer detection. Biostatistics 4(3), 449–463 (2003)
7. West, M., Blanchette, C.: Dressman, et al: Predicting the clinical status of human breast cancer by using gene expression profiles. PNAS 98(20), 11462–11467 (2001)
8. Mazzucco, G., et al.: Am. J. Kidney Dis., vol. 39, p. 713 (2002)
9. Vorderwulbecke, S., Cleverley, S., et al.: Protein quantification by the SELDI-TOF-MS based ProteinChip System. Nature Methods 2, 393–395 (2005)
10. Pisitkun, T., Shen, R.F., Knepper, A.: PNAS, vol. 101 (36), pp. 13368–13373 (2004)
11. Pisitkun, T., et al.: Molecular and Cellular Proteomics, vol. 5(10), pp. 1760–1771 (2006)
12. Rindler, M.J., et al.: J. Biol. Chem., vol. 265(34), pp. 20784–20789 (1990)
13. Fels, L.M., Bundschuh, I., Gwinner, W., et al.: Kidney Int. Suppl., vol. 47, pp. S81–S88 (1994)
14. Usuda, K., Kono, K., Dote, T., et al.: Arch Toxicol, vol. 72, pp. 104–109 (1998)
15. Nortier, J.L., Deschodt-Lanckman, M.M., et al.: Kidney Int., vol. 51, pp. 288–293 (1997)
16. Jungers, P., Hannedouche, T., et al.: Nephrol Dial Transplant, vol. 10, pp. 1353–1360 (1995)
17. Donaldio, C., Tramonti, G., Lucchesi, A., et al.: Ren Fail, vol. 20, pp. 319–324 (1998)
18. Lynn, K.L., Marshall, R.D.: Clin Nephrol, vol. 22, pp. 253–257 (1984)
19. Ambroise, C., McLachlan, G.J.: Selection bias in gene extraction on the basis of microarray gene-expression data. PNAS 99, 6562–6566 (2002)
20. Golub, G.H., Van Loan, C.F.: Matrix Computation. Johns Hopkins Univ. Press, Baltimore (1996)
21. Guyon, I., Elisseeff, A.: An introduction to Variable and Feature Selection. Journal of Machine Learning Research 3, 1157–1182 (2003)
22. Tikhonov, A.N., Arsenin, V.Y.: Solutions of ill-posed problems. W. H. Winston, Washington DC (1977)
23. Poggio, T., Girosi, F.: A Theory of Networks for Approximation and Learning. A. I. Laboratory, MIT, Cambridge (1989) A.I. Memo No. 1140
24. Girosi, F.: An Equivalence Between Sparse Approximation And Support Vector Machines. Neural Comp. 10(6), 1455–1480 (1998)
25. Mukherjee, S., Tamayo, P., Rogers, S., et al.: Estimating dataset size requirements for classifying dna microarray data. J. Comp. Biol. 10, 119–142 (2003)
26. Good, P.: Permutation tests: a practical guide to resampling methods for testing hypothesis. Springer, Heidelberg (1994)

Prediction of Crohn's Disease by Profiles of Single Nucleotide Polymorphisms

Roberto Colella[1], Annarita D'Addabbo[1], Anna Latiano[2],
Orazio Palmieri[2], Vito Annese[2], and Nicola Ancona[1]

[1] Istituto di Studi sui Sistemi Intelligenti per l'Automazione, CNR,
Via Amendola 122/D-I, 70126 Bari, Italy
colella@ba.issia.cnr.it
[2] U.O. Gastroenterologia, Ospedale CSS-IRCCS,
San Giovanni Rotondo, Italy

Abstract. This paper focuses on the comparison of two different approaches to the analysis of Single Nucleotide Polymorphism (SNP) profiles data regarding Crohn's Disease; the first one is based on a single SNP analysis, conducted by means of classical statistical tools, to assess the correlation existing between SNP's profile and phenotype; the second one makes use of classifiers based on Regularized Logistic Regression. The findings of the study show that the machine learning techniques adopted are able to provide statistically significant prediction accuracy of the phenotypic status of the subjects analyzed by SNP data. Moreover, they are poorly influenced by the noise embedded in the data and are suitable for genome-wide analysis.

Keywords: Logistic regression, SNP data.

1   Introduction

Crohn's Disease (CD) is an inflammatory bowel disease which affects about one million people in the world. It has been recently recognized that CD is a complex genetic disorder where susceptibility alleles and environmental exposure are required for disease development. Association studies have proposed several loci and genes correlated to CD, being *CARD15* the first validated gene [1]. More recently, genome-wide association studies have recognized other susceptibility loci, including OCTN[2], DLG5[3], IL23R[4], TNFSF15[5], and ATG16L1[6]. Unfortunately, for some of these genes different levels of significance have been found in different studies. These conflicting results could be explained by the approach adopted for data analysis in which single SNP analysis has been carried out.

This procedure, although provides interpretable insights on risk alleles, suffers of major limitations.

Single SNP analysis provides a limited view of complex diseases since it does not take into account epistasis and is unable to uncover the correlation between groups of SNPs and phenotype. Moreover, due to the multiple testing limitation, very conservative cut off values for significant association should be accounted,

I. Lovrek, R.J. Howlett, and L.C. Jain (Eds.): KES 2008, Part III, LNAI 5179, pp. 564–571, 2008.

thus possibly missing less powerful but true associations. When many different genes contribute to a given disorder like complex diseases, no particular gene has a remarkably large effect. Thus a specific phenotype may result from the combination of effects by a large number of moderately contributing genes [7]. In general, to capture effects of gene-gene interaction one needs an extremely large sample, and specific statistical methods and designs [8].

To overcome limitations of single SNP association studies, other experimental designs based on predictive studies have been proposed [9], evaluating the correlation existing between genetic markers and the phenotype by using well-founded methods and procedures developed in the field of statistical learning theory [10]. The strength of this methodology is to evaluate at the same time different candidate genes and polymorphisms and their possible interactions. Moreover, it also estimates whether the sample size is adequate for the prediction accuracy. The basic idea is that experimental results obtained by the analysis of SNP profiles may carry more interesting biological insights on the pathology at hand and may be more stable and reproducible than evidences drawn by single marker analysis, as similarly stated for gene list analysis of DNA microarray data [11].

In this paper we focus on the comparison between two different experimental designs: associative studies based on single SNP data and predictive studies in which a panel of SNPs is analyzed simultaneously. First of all, we performed single SNP association studies on two different data sets composed of gene polymorphisms previously confirmed or suggested for Crohn's disease. Successively, we analyzed these data sets by using Regularized Logistic Regression (RLR) classifiers. In this case, the whole SNP profiles were simultaneously considered. In section 2, a detailed description of single SNP analysis and RLR classifiers used in predictive design is reported. In section 3, data sets are described. Experimental results are shown in section 4 and conclusions follow.

2   Methods

2.1   Regularized Logistic Regression Classifiers

Predictive studies have been performed by using Regularized Logistic Regression (RLR) classifiers. In a two class classification problem, we have a set $S = \{(\mathbf{x}_1, y_1), ..., (\mathbf{x}_\ell, y_\ell)\}$ composed of ℓ independent and identically distributed data (examples) $(\mathbf{x}_i, y_i)$, where $\mathbf{x}_i \in \mathrm{I\!R}^d$ and $y_i \in \{0, 1\}$ for $i = 1, 2, ..., \ell$, drawn from a fixed but unknown probability density function $p(\mathbf{x}, y)$. In our setting, the vector $\mathbf{x}_i$ represents the SNP profile and the label y_i indicates the phenotype of the i-th subject.

The logistic regression [12] is based on the assumption that the posterior probabilities of K classes can be modelled as linear function in $\mathbf{x}$, with the constrain that they sum to 1. In a two-class problem, this position is written as

$$p_1(\mathbf{x}; \mathbf{w}) = \frac{e^{<\mathbf{w},\mathbf{x}>}}{1 + e^{<\mathbf{w},\mathbf{x}>}} \tag{1}$$

$$p_2(\mathbf{x}; \mathbf{w}) = 1 - p_1(\mathbf{x}; \mathbf{w}) = \frac{1}{1 + e^{<\mathbf{w},\mathbf{x}>}} \tag{2}$$

where $\mathbf{x} \equiv (1, \mathbf{x}_1, \mathbf{x}_2, ..., \mathbf{x}_d)$. The best parameter vector $\mathbf{w}^*$ is the one maximizing the log-likelihood function and is obtained by an iterative procedure known as the Newton-Raphson algorithm. In the present paper, we consider an opportunely constrained log-likelihood function

$$\mathcal{L}(\mathbf{w}) = \sum_{i=1}^{l} \left[y_i < \mathbf{w}, \mathbf{x}_i > - \log(1 + e^{<\mathbf{w},\mathbf{x}_i>}) \right] - \lambda \left(\|\mathbf{w}\|^2 - \alpha \right) \tag{3}$$

By applying Newton-Raphson's method, the best $\mathbf{w}^*$ is obtained through the recurrence equation:

$$\mathbf{w}_{n+1} = \mathbf{w}_n + (XWX^T + 2\lambda I)^{-1} \left(X(\mathbf{y} - \mathbf{p}) - 2\lambda \mathbf{w}_n \right) \tag{4}$$

Moreover, if X has full rank, $\mathbf{w}$ can be expressed as linear combination of the vectors $\mathbf{x}_i$ for $i = 1, ..., \ell$. This means that there exists a vector $\mathbf{c} \in \mathbb{R}^l$ so that

$$\mathbf{w} = X\mathbf{c} \tag{5}$$

and equation (4) becomes

$$\mathbf{c}_{n+1} = \mathbf{c}_n + (WK + 2\lambda I)^{-1}(\mathbf{y} - \mathbf{p} - 2\lambda \mathbf{c}_n) \tag{6}$$

with $K = X^T X$. A regularized kernel version of the logistic regression is so obtained, where many choices of the kernel function are possible.
Moreover, the problem of unbalanced data set has been addressed introducing an opportune diagonal matrix A, with elements

$$A_{ii} = \frac{(n_2 y_i + n_1(1 - y_i))}{l} \tag{7}$$

So, equation (6) is written as:

$$\mathbf{c}_{n+1} = \mathbf{c}_n + (WAK + 2\lambda I)^{-1}(A(\mathbf{y} - \mathbf{p}) - 2\lambda \mathbf{c}_n) \tag{8}$$

The probability for a new vector to belong to class $\mathcal{C}_1$ can be computed: $\mathbf{x}_i$ will be labelled as belonging to $\mathcal{C}_1$ if $p(\mathbf{x}_i; \mathbf{w}) \leq 0.5$, as belonging to class $\mathcal{C}_2$ otherwise.

2.2 Estimate of Prediction Accuracy and Statistical Significance

We use the leave-k-out cross validation (LKOCV) procedure [13] to provide a significant estimate of the generalization error. It consists of building T_1 pairs of training and test sets composed of r and $N - r$ examples respectively by random sampling without replacement the data set S.

At each random split, the optimal value for λ, varying in a pre-determined range $[\lambda_{min}, \lambda_{max}]$ is computed by using the k-fold cross validation technique; this consists in randomly separating each r-sized training data set into k subsets and evaluating the accuracy $Ac(\lambda)$ as the mean of the k values obtained training the RLR with the examples of the k-1 subsets and testing it on the examples belonging to the k-th subset; once the optimal value for λ has been computed, we evaluate the error rate e_{n_i} of the classifier trained on r examples, testing it on $N - r$ examples.

So, the LKOCV error e_n is given by $e_n = \frac{1}{T_1} \sum_{i=1}^{T_1} e_{n_i}$. Then,we apply the method of hypothesis testing to evaluate the statistical significance of the error rate e_n. For evaluating the *p-value* corresponding to e_n, we need to know the probability density function of e_n under the null hypothesis. Since it is unknown, we invoke non-parametric permutation tests [14]. In particular, we perform T_2 random permutations of the labels in the data set S. For each permutation, we split S in a training and a test set composed of r and $N - r$ examples respectively, by random sampling without replacement the data set S. Finally, we evaluate the error rate $e_{n_i}^0$ of the classifier trained on the r examples. Then the empirical probability density function of the error rate is constructed. The *p-value* of the error rate e_n is given by the percentage of $e_{n_i}^0 < e_n$.

3   Data Set Description

Two data sets have been analyzed in the present study. The first one, named D_1, contains 16 SNPs data, extracted from 178 healthy controls (HC) and 127 patients affected by Crohn's Disease (CD). The second data set, named D_2, consists of 28 SNPs data extracted from 122 HC and 108 CD patients.

It is worth noting that 15 SNPs are present in both D_1 and D_2; this has allowed us to extract two more samples, S_1 and S_2, from the two initial data sets, each including only the 15 common SNPs.

4   Experimental Results

4.1   Single SNP Analysis

The association of each single SNP with Crohn's Disease has been evaluated by using χ^2 test. It is worth noting that, for each SNP, the dominant genetic model holds. SNPs that best correlate with the phenotype (p-value ≤ 0.05) are shown in Table 1 for the data set D_1 and in Table 2 for D_2.

By considering the Bonferroni correction, only markers on MCP-1 and TNF-α genes are significant in D_1. By applying the Bonferroni correction to SNPs in D_2, only markers with χ^2 p-value < 0.018 are significant. Note that, after Bonferroni correction, only MCP-1 is significant in both samples. For the CARD15 SNPs, conflicting results are obtained. The marker on gene XBP1 that shows the higher level of significance is only present in D_2.

Table 1. Sample S_1: subset of SNP correlated with CD

Gene	SNP	χ^2 p-value
MCP-1	-A2518G	0.001
TNF-α	-C857T	0.027
CARD15	G908R	0.032
CARD15	L1007fsinsC	0.044

Table 2. Sample S_2: subset of SNP correlated with CD

Gene	SNP	χ^2 p-value
XBP1	rs6005893	3×10^{-6}
MCP-1	-A2518G	0.001
BSN	rs9858542	0.007
MYO9B	rs1545620	0.007
CARD15	R702W	0.022
FcGIIIA	G559T	0.024

4.2   Multiple SNP Analysis

In order to verify that experimental results obtained by the analysis of SNP profiles may be more stable and reproducible than evidences drawn by single marker analysis, we have considered the classification performances obtained by applying RLR classifiers to S_1 and S_2.

Figure (1) and figure (2) depict the accuracy of the RLR classifier trained with the whole panels S_1 and S_2, at varying the training data set size. The prediction accuracy estimates reach values of 61% for S_1 and 60% for S_2.

Successively, RLR classifier has been trained with the 28-SNP dataset, initially named D_2. In Fig.3 it can be observed that prediction accuracy increases up to 72%. Gaussian kernels have also been considered in this experiment, but no enhancement of the performance of the classifier has been obtained. This experimental evidence shows that linear interaction among features is sufficient to explain the correlation between the analyzed profile and phenotype.

5   Discussion

Single SNP analysis provides a limited view of complex diseases and produces, in different studies, conflicting results that can be explained by different genetic background and insufficient power of the sample size when investigating variants with low relative risks. The findings of this study suggest that

- The adopted classification technique, despite it does not aim at the search for susceptibility loci, provides information that seem to be more stable with respect to the variation of the data set and, above all, does not deteriorate when uncorrelated features, that can be considered as noise, are added;

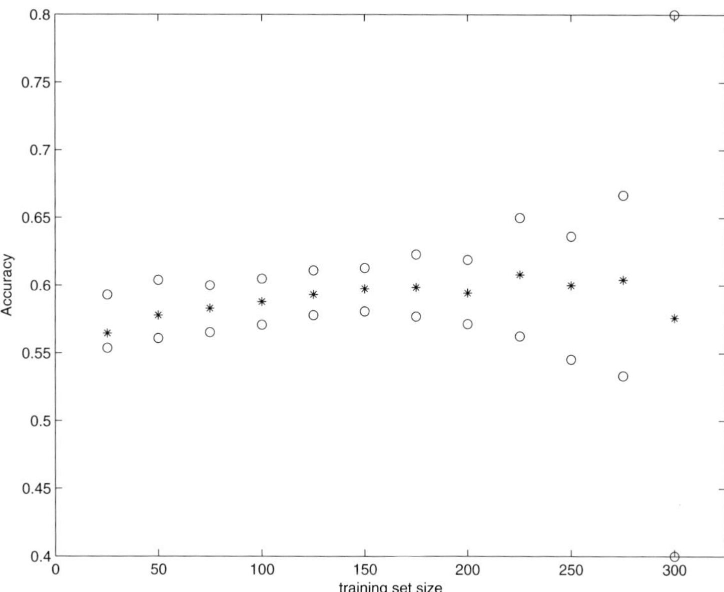

Fig. 1. Accuracy of the RLR classifier trained with the whole sample S_1. For each value of training data set size, the 25-th quantile,the mean and the 75-th quantile of the accuracy estimated in cross validation are depicted.

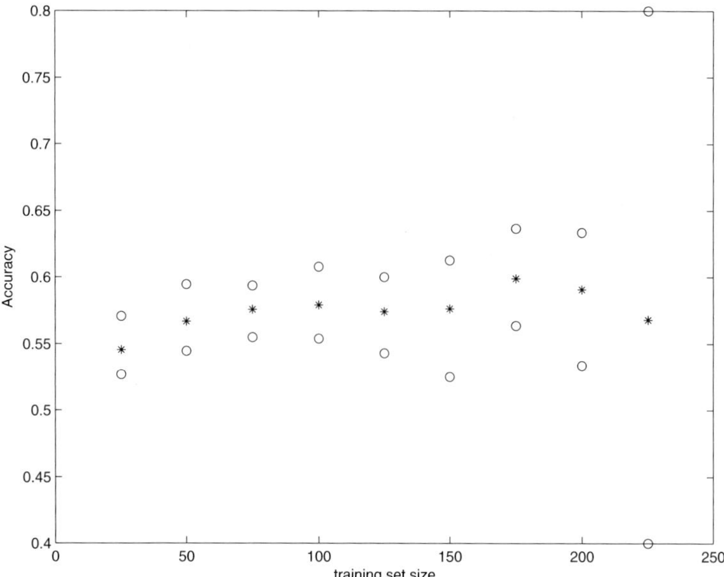

Fig. 2. Accuracy of the RLR classifier trained with S_2 data sample. For each value of training data set size, the 25-th quantile, the mean and the 75-th quantile of the accuracy estimated in cross validation are depicted.

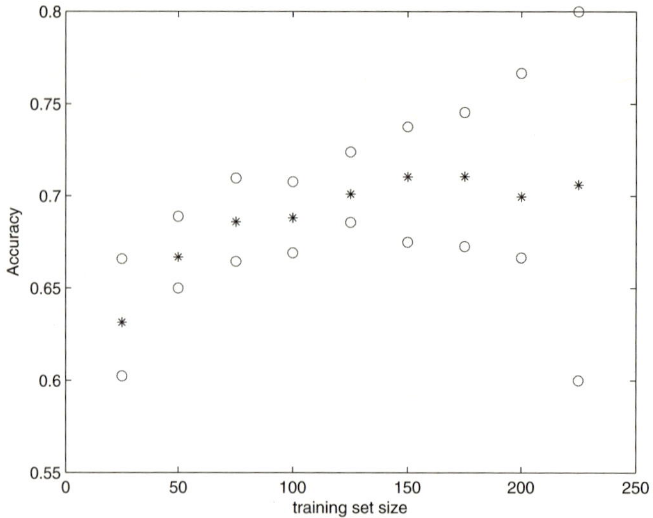

Fig. 3. Accuracy of the RLR classifier trained with D_2 data sample, composed of 28 SNPs. For each value of the training data set size, the 25-th quantile, the mean and the 75-th quantile of the accuracy values estimated in cross validation are depicted.

- The introduction of kernel in Eqn. 5 turns the search for the optimal vector w^* to the search for c^*, which requires the solution of a linear system with size equal to the number of examples; this implies a strong reduction in computational time, and makes this kind of machine learning approach widely suitable for genome-wide analysis, where data sets are often characterized by a huge number of features and a smaller number of examples;
- Linear models seems to be sufficient to account for the interactions among loci that are suspected to determine the susceptibility of a subject to Crohn's Disease.

6   Conclusion

Our study suggests that multivariate statistics which take into account gene-gene and gene-phenotype interactions provide better insights on complex diseases and make clear the usefulness of considering also loci weakly correlated to the phenotype in predictive studies. Moreover, our results need to be confirmed on dataset composed on greater number of markers.

Acknowledgments. A.D. is PhD student of Dip. di Fisica, Universitá di Bari, Italy.

References

1. Hugot, J.P., Chamaillard, M., Zouali, H., Lesage, S., Cezard, J.P., et al.: Association of NOD2 leucine-rich repeat variants with susceptibility to Crohn disease. Nature 411, 599–603 (2001)
2. Peltekova, V.D., Wintle, R.F., Rubin, L.A., Amos, C.I., Huang, Q., et al.: Functional variants of OCTN cation transporter genes are associated with Crohn disease. Nat. Genet. 36, 471–475 (2004)
3. Stoll, M., Corneliussen, B., Costello, C.M., Waetzig, G.H., Mellgard, B., et al.: Genetic variation in DLG5 is associated with inflammatory bowel disease. Nat. Genet. 36, 476–480 (2004)
4. Duerr, R.H., Taylor, K.D., Brant, S.R., Rioux, J.D., Silverberg, M.S., et al.: A genome-wide association study identifies IL23R as an inflammatory bowel disease gene. Science 314, 1461–1463 (2006)
5. Yamazaki, K., McGovern, D., Ragoussis, J., Paolucci, M., Butler, H., et al.: Single nucleotide polymorphisms in TNFSF15 confer susceptibility to Crohn disease. Hum. Mol. Genet. 14, 3499–3506 (2005)
6. Hampe, J., Franke, A., Rosenstiel, P., Till, A., Teuber, M., et al.: A genomewide association scan of non-synonymous SNPs identifies a susceptibility variant for Crohn disease in ATG16L1. Nat. Genet. 39, 207–211 (2007)
7. Risch, N.J.: Searcing for genetic determinants in the new millennium. Nature 405(6788), 847–856 (2000)
8. Sillanpää, M.J., Auranen, K.: Replication in genetic studies of complex traits. Ann. Human. Genet. 68, 646–657 (2004)
9. D'Addabbo, A., Latiano, A., Palmieri, O., Maglietta, R., Annese, V., Ancona, N.: Regularized Least Squares Cancer Classifiers may predict Crohn's disease from profiles of Single Nucleotide Polymorphisms. Ann. Hum. Genet. 71(4), 537–549 (2007)
10. Vapnik, V.: Statistical Learning Theory. John Wiley & Sons, Inc., Chichester (1998)
11. Subramanian, A., Tamayo, P., et al.: Gene set enrichment analysis: A knowledge-based approach for interpreting genome-wide expression profiles. PNAS 102(43), 15545–15550 (2005)
12. Hastie, T., Tibshirani, R., Friedman, J.: The Elements of Statistical Learning - Data mining, Inference and Prediction. Springer Series in Statistics, pp. 95–99 (2001)
13. Mukherjee, S., Tamayo, P., Rogers, S., et al.: Estimating dataset size requirements for classifying dna microarray data. J. Comp. Biol. 10, 119–142 (2003)
14. Good, P.: Permutation tests: a practical guide to resampling methods for testing hypothesis. Springer, Heidelberg (1994)

Decision Analysis of Fuzzy Partition Tree Applying AIC and Fuzzy Decision

Kimiaki Shinkai[1], Shuya Kanagawa[2], Takenobu Takizawa[1],
and Hajime Yamashita[1]

[1] Department of Mathematics, Graduate School of Education, Waseda University
`building-bridges@ruri.waseda.jp`
[2] Faculty of Engineering, Musashi Institute of Technology

Abstract. We often use fuzzy graph to analyze inexact information such as sociogram structure ([1] and [2]). Concerning the hierarchical cluster analysis of a fuzzy graph ([3], [4] and [5]), the number of clusters may have to be decided in the actual cluster analysis. In other word, we woud like to decide the optimal level with a partition tree. Concerning this problem, while AIC method in statistical analysis has been designed by us ([6] and [10]), we will now propose a fuzzy decision method which is based on the evaluation function paying attention to the size and number of clusters at each level.

Keywords: Fuzzy Graph, Partition Tree, AIC (Akaike's information criterion), Fuzzy Decision, Optimal level.

1 Introduction

A fuzzy graph G is defined by

$$G = (V, Y), \ V = \{v_i; 1 \le i \le n\}, Y = \{y_{ij}; 0 \le y_{ij} \le 1\} \tag{1}$$

where y_{ij} is the fuzziness of the arc from the node v_i to the node v_j.

If $f_{ij} = f_{ji}, f_{ii} = 1$ then G signifies the similarity relation among nodes. Here, by analyzing the fuzzy graph G, we have a partition tree P. In the cluster analysis of fuzzy graph, it is important and difficult problem how to decide the optimal cutting level as to a partition tree. By now, as example, the steepest descent method in the multivariate analysis and AIC method in the statistical analysis have been designed. Here, we shall consider the fuzzy decision method by constructing the evaluation functions that pay attention to each level of the partition tree.

2 Steepest Decent Method

We would define

$$R_z = \{G_1, G_2, \cdots G_m\}, (1 \le m \le n, 0 \le z \le 1) \tag{2}$$

as the division of $G, (|G| = n)$. For example, in Fig. 2,

I. Lovrek, R.J. Howlett, and L.C. Jain (Eds.): KES 2008, Part III, LNAI 5179, pp. 572–579, 2008.

$$R_{0.62} = \{\{1, 4, 7\}, \{2, 3\}, \{11, 6\}, \{9, 10, 5, 8\}\}.$$

The number of clusters of R_z could be defined as $x_{(z)}$ concerning the partition tree. In general, the more level z approaches 1, the more the number of clusters $x_{(z)}$ increases.

So, evaluation function $p(z)$ of the number of clusters and the derivative $p'(z)$ could be defined as the followings;

$$p(z) = \frac{x_{(z)} - x_{(0)}}{x_{(1)} - x_{(0)}} \in [0.1], \quad p'(z) = \left(\frac{x_{(z)} - x_{(0)}}{x_{(1)} - x_{(0)}}\right)' \in [0.1] \tag{3}$$

Here, the optimal level $z = z_0$ is decided as follows;

$$z_0 = \min\{z \,|\, \max(p'(z))\} \tag{4}$$

and we could obtain the optimal R_{z_0}

But this method might not be used now because this method could produce local minumum problem.

3   AIC Method

3.1   AIC (Akaike's Information Criterion)

Let $X_1, X_2, \cdots, X_n$ be independent and identically distributed random variables with the probability density function $f(x; \theta^*)$, where $\theta^* = (\theta_1^*, \theta_2^*, \cdots, \theta_K^*)$ is an unknown parameter of the distribution. Furthermore, for n random sample points $(x_1, x2, \cdots, x_n)$,

$$l(\theta) = \sum_{i=1}^{n} \log f(x; \theta) \tag{5}$$

is the log likelihood and $\hat{\theta}_K$ is the maximal likelihood estimator for $\theta^* = (\theta_1^*, \theta_2^*, \cdots, \theta_K^*)$ defined by

$$\hat{\theta}_K = \max_{\theta \in \Theta_K} l(\theta) \tag{6}$$

where Θ_K is the space of all candidates $\theta = (\theta_1, \theta_2, \cdots, \theta_K)$ for the real parameter θ^* .

AIC is an identification method using Kullback-Leibler information numbers as well as the consistency and the asymptotic normality of maximum likelihood estimators.

Now, we shall define AIC as following. The Kullback-Leibler information number $KLI(\theta^*, \theta)$ for the samples $(x_1, x_2, \cdots . x_n)$ is defined by

$$KLI(\theta^*, \theta) = \sum_{i=1}^{n} f(x; \theta^*) \log f(x; \theta^*) - \sum_{i=1}^{n} f(x; \theta^*) \log f(x; \theta) \tag{7}$$

which metrizes the distance between $f(x; \theta^*)$ and $f(x; \theta_i)$ using entropy. It is easy to see that $\hat{\theta}_K$ minimize the Kullback-Leibler information number $KLI(\theta^*, \theta)$ such that

$$KLI(\theta^*, \theta) = \min_{\theta \in \theta_K} KLI(\theta^*, \theta) \tag{8}$$

Here, AIC (Akaike's information criteria) is defined by

$$AIC(K) = -2l(\hat{\theta}_K) + 2K \tag{9}$$

For comparing several models of $\theta^* = (\theta_1^*, \theta_2^*, \cdots, \theta_K^*)$ for known K, we can find the optimal $\theta = (\theta_1, \theta_2, \cdots, \theta_K)$ in Θ_K with large probability when it attains the smallest $AIC(K)$ for $(x_1, x_2, \cdots . x_n)$. In general, when K is unknown, $-2l(\hat{\theta}_K)$ decreases as the number of parameters K increases. Therefore $AIC(K)$ also gives the optimal K from the definition.

3.2   Clustering Analysis Using AIC Method

Actually, we could obtain the optimal level $z = z_0$ and the optimal R_{z_0} as following steps.

Step. 1 Find the configuration of clusters on the optimal interval

Step. 2 Make partitions of the interval

Step. 3 Make histograms and calculate AIC for each histogram depending on cut level z.

Step.4 Find the optimal cut level z_0 which the value AIC is the smallest.

which is detailed in [6] and [10].

Here, AIC is obtained by the asymptotic normality for the maximal likelihood estimator $\hat{\theta}_K$. Therefor, when sample size is small, the error of the asymptotic normality is large and the degree of accuracy of AIC is low. AIC Method is reasonable based on the statistical inference, but it needs a lot of samples (over 100 samples).

For these reasons, we would propose the evaluation method based on Fuzzy Decision which could obtain the optimal R_{Z_0} even if there are not so many samples.

4   Fuzzy Decision Method

4.1   Evaluation Method

Definition 1. The number and the size of clusters of R_z could be defined as $x_{(z)}$ and $y_{(z)}$ concerning the partition tree. In general, the more level z rises, the more the number of clusters x increases, and the more the size of clusters y decreases.

Then, if we wish the condition that the clusters disperse and gather reasonably, evaluation function $p_{(z)}$ of the number of clusters and evaluation function $q_{(z)}$ of the size of clusters could be defined as follows.

$$p(z) = \frac{x_{(z)} - x_{(0)}}{x_{(1)} - x_{(0)}} \in [0.1], \qquad q(z) = \frac{y_{(z)} - y_{(1)}}{y_{(0)} - y_{(1)}} \in [0, 1] \qquad (10)$$

Here, the value of $y_{(z)}$ is defined by the size of the maximum cluster or the size of the mean cluster (arithmetic mean, power mean) of R_z.

Definition 2. The optimal level $z = z_0$ is decided as follows.

$$z_0 = \max\{z| \max(p(z) \wedge q(z))\} \qquad (11)$$

Definition 3. The power mean τ_n of $y_i > 0, 1 \leq i \leq n$ is defined as follows.

$$\tau_n = \left(\frac{1}{n} \sum_{i=1}^{n} y_i^n \right)^{\frac{1}{n}} \qquad (12)$$

Lemma 1. Concerning the arithmetic mean, the power mean, and the maximum value, the following inequalities hold.

$$\frac{1}{n} \sum_{i=1}^{n} y_i \leq \tau_n \leq \max_{i}\{y_i\} \qquad (13)$$

Lemma 2. If n is large enough, the following approximate equality holds.

$$\tau_n \sim \max_{i}\{y_i\} \qquad (14)$$

From Lemma1 and Lemma2, when the value of n is small, the power mean receives the influences other than max, and when the value of n goes large, the influence of the max grows.

Lemma 3. $q_1(z), q_2(z), q_3(z)$ are defined as the evaluation function that uses the size of the maximum cluster, the power mean cluster, the arithmetic mean cluster respectively.

If the fuzzy maximal decision interval by each method is $(a_1, b_1], (a_2, b_2], (a_3, b_3]$, then the following relation holds.

$$a_3 \leq a_2 \leq a_1, \quad b_3 \leq b_2 \leq b_1 \qquad (15)$$

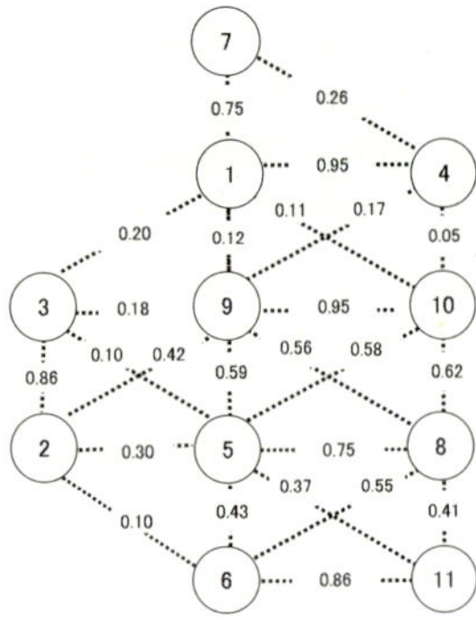

Fig. 1. Fuzzy Graph G

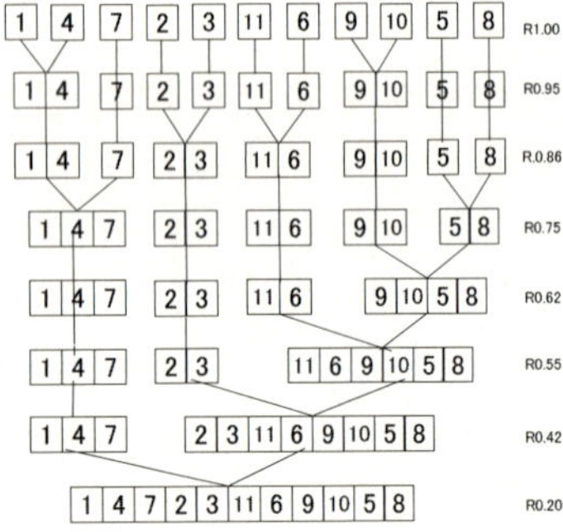

Fig. 2. Partition Tree P

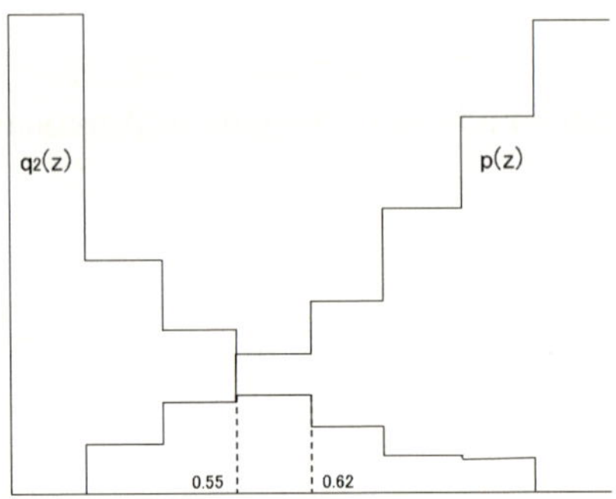

Fig. 3. Fuzzy Decision $p(z) \wedge q_2(z)$

4.2   Example of Level Analysis

By analyzing the fuzzy graph G in Fig. 1, we have a partition tree P in Fig. 2.

Consequently, we have obtained the following optimal levels. (see in Fig. 3.)

$$p(z) \wedge q_1(z)\text{method} \ (0.55, 0.62] \to 0.62 \tag{16}$$

$$p(z) \wedge q_2(z)\text{method} \ (0.55, 0.62] \to 0.62 \tag{17}$$

$$p(z) \wedge q_3(z)\text{method} \ (0.42, 0.55] \to 0.55 \tag{18}$$

Now, it would be notable that the decision interval by the power-mean method is placed between the maximum cluster method and the arithmetic mean method (4.1 Lemma 3). This is because we can obtain a more refined decision range.

5 Conditional Fuzzy Decision Method

5.1 Evaluation Method

Definition 4. If we consider the conditions about the number and the size of clusters wilfully, then evaluation function $p(z)$ of the number of clusters and evaluation function $q(z)$ of the size of clusters could be defined as follows.

$$p(z) = \begin{cases} \dfrac{x(z) - x_m + 1}{x_s - x_m + 1} & (x_m - 1 \le x(z) \le x_s) \\[2mm] \dfrac{x_M - x(z) + 1}{x_M - x_s + 1} & (x_s \le x(z) \le x_M + 1) \\[2mm] 0 & \begin{array}{l}(x(z) \le x_m - 1, \\ x(z) \ge x_M + 1)\end{array} \end{cases} \quad q(z) = \begin{cases} \dfrac{y_M - y(z) + 1}{y_M - y_s + 1} & (y_s \le y(z) \le y_M + 1) \\[2mm] \dfrac{y(z) - y_m + 1}{y_s - y_m + 1} & (y_m - 1 \le y(z) \le y_s) \\[2mm] 0 & \begin{array}{l}(y(z) \le y_m - 1, \\ y(z) \ge y_M + 1)\end{array} \end{cases} \tag{19}$$

Here, the condition of the number of clusters is [minmun value, optimal value, maximum value]=$[x_m, x_s, x_M]$, and of the size of clusters is [minmum value, optimal value, maximum value]=$[y_m, y_s, y_M]$.

Definition 5. The optimal level $z = z_0$ is decided as follows.

$$z_0 = \max\{z \mid \max(p(z) \wedge q(z))\} \tag{20}$$

Definition 6. If we consider the conditions and weights about the number and the size of clusters wilfully, then evaluation function $p(z)$ of the number of clusters and evaluation function $q(z)$ of the size of clusters could be defined as follows.

$$\{p(z)\}^a = \begin{cases} \left\{\dfrac{x(z) - x_m + 1}{x_s - x_m + 1}\right\}^a & (x_m - 1 \le x(z) \le x_s) \\[3mm] \left\{\dfrac{x_M - x(z) + 1}{x_M - x_s + 1}\right\}^a & (x_s \le x(z) \le x_M + 1) \\[3mm] 0 & \begin{array}{l}(x(z) \le x_m - 1, \\ x(z) \ge x_M + 1)\end{array} \end{cases} \quad \{q(z)\}^b = \begin{cases} \left\{\dfrac{y_M - y(z) + 1}{y_M - y_s + 1}\right\}^b & (y_s \le y(z) \le y_M + 1) \\[3mm] \left\{\dfrac{y(z) - y_m + 1}{y_s - y_m + 1}\right\}^b & (y_m - 1 \le y(z) \le y_s) \\[3mm] 0 & \begin{array}{l}(y(z) \le y_m - 1, \\ y(z) \ge y_M + 1)\end{array} \end{cases} \tag{21}$$

Here, the condition of the number of clusters is [minmun value, optimal value, maximum value]=$[x_m, x_s, x_M]$, and of the size of clusters is [minmum value, optimal value, maximum value]=$[y_m, y_s, y_M]$. And, if $a, b \ge 1$ then the evaluation of the conditions would be turned up, and if $0 \le a, b \le 1$ then the evaluation of the conditions would be turned down.

Definition 7. The optimal level $z = z_0$ is decided as follows.

$$z_0 = \max\{z \mid \max(\{p(z)\}^a \wedge \{q(z)\}^b)\} \tag{22}$$

Lemma 4. If $x_m = 2, x_s = x_M = x(1), y_m = 2, y_s = y_M = y(0), a = 1, b = 1$ then the evaluation function in 5.1 Definition 6 corresponds to the function in 4.1 Definition 1.

5.2   Example of Level Analysis

By analyzing the fuzzy parttition tree P in Fig. 2, we have obtained the following optimal levels.

5.2.1 Decision with Condition

If $[x_m, x_s, x_M] = [3, 5, 7], [y_m, y_s, y_M] = [1, 2, 4]$ then we have obtained the following optimal levels. (see in Fig. 4.)

$$p(z) \wedge q_1(z)\text{method}\ (0.62, 0.75] \to 0.75 \tag{23}$$

$$p(z) \wedge q_2(z)\text{method}\ (0.62, 0.75] \to 0.75 \tag{24}$$

$$p(z) \wedge q_3(z)\text{method}\ (0.62, 0.75] \to 0.75 \tag{25}$$

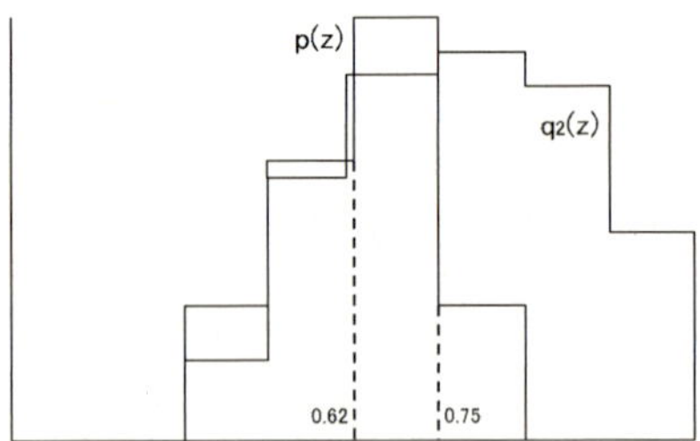

Fig. 4. Fuzzy Decision $p(z) \wedge q_2(z)$

5.2.2 Decision with Weighted Condition

If $[x_m, x_s, x_M] = [3, 5, 7], [y_m, y_s, y_M] = [1, 2, 4], a = \frac{1}{3}, b = 4$ then we have obtained the following optimal levels. (see in Fig. 5.)

$$p(z) \wedge q_1(z)\text{method}\ (0.75, 0.86] \to 0.86 \tag{26}$$

$$p(z) \wedge q_2(z)\text{method}\ (0.75, 0.86] \to 0.86 \tag{27}$$

$$p(z) \wedge q_3(z)\text{method}\ (0.62, 0.75] \to 0.75 \tag{28}$$

Here, the cutting level in 4.2 should be natural , and the cutting level in 5.2.1 could reasonably reflect the condition that the optimal size of clusters is 2 and the optimal number of clusters is 5. Moreover, the cutting level in 5.2.2 could reflect the condition that the optimal size of clusters is 2 should be turned up.

For a plactical application, we could effectively analyze the sociometry analysis which is detailed in [9] and [11].

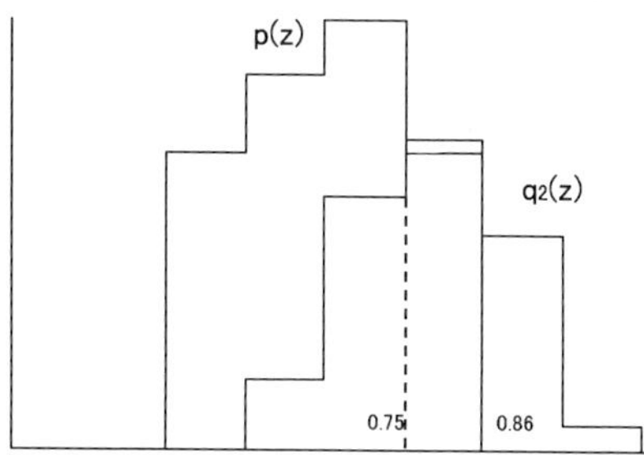

Fig. 5. Fuzzy Decision $\{p(z)\}^{\frac{1}{3}} \wedge \{q_2(z)\}^4$

6   Conclusion and Future Work

Our evaluation method could be effectively applied to general human society models which has not so many samples which is detailed in [7] , [8] and [11]. On the other hand, we must compare the result of AIC Method with that of Fuzzy Decision Method for a lot of samples in near future.

References

1. Moreno, J.: The Sociometry Reader. The Free Press (1960)
2. Zadeh, L.: Fuzzy Sets. Information and Control VIII, 338–353 (1965)
3. Lance, G., Williams, W.: A general theory of classificatory sorting strategies I. Hierarchical systems. Computer Journal 9, 380–383 (1967)
4. Romsburg, C.: Cluster Analysis for Researchers Lifetime Learning Publication (1984)
5. Miyamoto, S.: Introduction to Cluster Analysis:Theory and Applications of Fuzzy Clustering, Morikita-Shuppan, Tokyo (1999) (in Japanese)
6. Kanagawa, S., Tsuda, E., Shinkai, K., Yamashita, H.: Fuzzy Clustering Level Analysis Applying AIC. In: Proceedings of Joint 2nd International Conference on Soft Computing and Intelligent Systems and 5th International Symposium on Advanced Intelligent Systems (2004)
7. Shinkai, K., Yamashita, H.: Level Analysis of Fuzzy Partition Tree and its Application. In: Int'l Workshop on Soft Computing Applications (2005)
8. Shinkai, K., Tsuda, E., Yanai, A.: Decision Analysis of optimal Level in Fuzzy Partition Tree, SA-C3-3. In: 3rd Int'l Conference of Soft Computing and Intelligent Systems (2006)
9. Shinkai, K., Yamashita, H.: Level Analysis in Fuzzy PartitionTree and its Application. Mathematical Society of Japan, Division of Applied Mathematics 14 (2006)
10. Kanagawa, S., Uesu, H., Shinkai, K., Tsuda, E., Yamashita, H.: Fuzzy Clustering Level Analysis Using AIC Method for Large Size Samples. In: 2nd Intl. Conference on Innovative Computing Information and Control (2007)
11. Shinkai, K.: Journal of Intelligent and Fuzzy Sustems 19(1), 19–28 (2008)

Learning Environment for Improving Critical Thinking Skills Based on New Synthesis Theory

Hiroki Satake[1], Haruna Marumoto[1], Kazuhisa Seta[1],
Motohide Umano[1], and Mitsuru Ikeda[2]

[1] Osaka Pref. University, 1-1, Gakuen-cho, Naka-ku, Sakai, Osaka, 599-8531, Japan
[2] JAIST, 1-1, Asahi-dai, Nomi, Ishikawa, 923-1292, Japan
{seta,umano}@mi.s.osakafu-u.ac.jp,
satake@kbs.cias.osakafu-u.ac.jp, ikeda@jaist.ac.jp

Abstract. Critical thinking, thinking about one's own thinking processes, is a kind of meta-cognition. For advance of the quality of medical care, nurses must perform critical thinking processes to clarify the evidence of one's own decision-making explicitly and to validate it. To train critical thinking skills, we adopt a new synthesis theory as a base of our learning scheme. In this paper we propose a new synthesis theory based learning system for developing critical thinking skills in a nursing domain. Furthermore, we illustrate a scenario of a learner's system utilization to explain what functionaries are embedded in our system to encourage Critical Thinking learning.

Keywords: new synthesis, guidance information, critical thinking.

1 Introduction

Most of the serious decision-making processes on nursing assessments of a patient include risks in a situation where they should be performed from wide range of clinical viewpoints. In such a situation, evidence based decision-making plays an important role for advance of the quality of medical care. Based on this shared recognition, the concept of Evidence Based Nursing (EBN) has attracted attention [1].

Nurses have to perform Critical Thinking (CT) processes to clarify and validate the evidence of one's own decision-making explicitly for putting EBN into practice. Our research aims to build a learning system whereby a novice nurse can be aware of important evidence in the nursing care processes so that the nurse can learn CT methods.

To train CT skills, thinking skills with recognition of self-knowledge of one's own decision-making, we adopt a "new synthesis theory" based learning method.

New synthesis theory is proposed by Perkins and Salomon [2]. It argues that meta-cognitive ability is trained in a problem-solving context by synthesizing general know-how of performing meta-cognitive activity and context-dependent one.

2 What Is Critical Thinking?

Critical thinking (CT), thinking about one's own thinking processes, is a kind of meta-cognition. Meta-cognitive activities are tacit, latent and context-dependent:

I. Lovrek, R.J. Howlett, and L.C. Jain (Eds.): KES 2008, Part III, LNAI 5179, pp. 580–587, 2008.

abilities for performing meta-cognitive activities also depend on individual situations. This is a main cause of difficulties to train meta-cognitive skills. In this section, we consider the characteristics of CT to realize a learning environment for training CT skills effectively.

CT is a judgment activity based on calm, objective, and logical thinking: someone performing CT must shake free from the pitfalls of thinking based on an awareness of the influence of preoccupations [3].

CT activity that monitors one's own thinking processes plays an important role, especially in a nursing domain, because decision-making in the domain can directly affect a patient's condition.

We take an example of CT quoted from a precedent study in the nursing domain [4]. The scenario is as follows.

A nurse walking on the road witnessed a car accident; a man riding a bicycle was hurled about 20 m by the impact. She rushed to the man.

The problem confronted by the nurse in this situation is to perform objective and valid assessment of the accident victim's condition. We take two examples of thinking processes for illustrating CT processes in this situation.

Example 1 (Including no CT processes). The nurse makes a decision to perform artificial respiration immediately on the patient with respiratory failure caused by being hurled about 20m. In this case, the nurse plans to perform artificial respiration with NO consideration of the other nursing care plans. This means that the nurse tries to perform the nursing care plan without validating one's own decision-making processes, i.e., no CT processes.

Example 2 (Including CT Processes). The nurse considers one's own thinking and confirms both it is not preoccupied and something is not overlooked or not. The nurse should consider not only artificial respiration, but also watching and waiting for ambulance in this case. Intuitively, waiting and watching with doing nothing for the nurse confronting the patient seems irrational. But the nurse should assume the possibility of spinal damage because the patient was hurled 20m and crashed against the ground. If he has spinal damage, the nurse has to consider a risk that airway control for artificial respiration further aggravates it. The important point of the decision-making in this case is performing evidence based nursing processes including CT processes: the nurse has to perform a decision making whether she tries to perform artificial respiration or just watch and wait for ambulance with explicit recognition of their respective fatal risks. Consequently, CT plays an important role to clarify evidence for rational decision-making.

In general, well-organized domain knowledge (nursing domain knowledge in this situation) is necessary to perform planning processes at the cognitive level. No marked difference is apparent between the domain knowledge of a nurse described in Example 1 and the one in Example 2 to perform planning in this situation.

The differences are their meta-level knowledge related to how to perform critical thinking from appropriate viewpoints to make valid nursing plans (feasibility, risk factor, human life priority, and so forth), in addition to their awareness of why it is necessary to perform critical thinking.

Benner pointed out that expert nurses judge consciously, whereas inexperienced ones do unconsciously [5]. One reason for this dichotomy is the death of opportunities for training CT skills because well-formed learning schemes are not built. Expert nurses accumulate and develop CT skills in their daily work.

Our research goal is developing a learning environment that can aid learners in composing meta-level knowledge with appropriate viewpoints to perform critical thinking adequately.

3   Design Principles for Learning System

Most of the serious decision-making processes on the hospital wards include risks in the situation where they are required to perform rapidly for lack of time. In such situations, evidence based decision-making plays an important role for advance of the quality of medical care. Based on this shared recognition, the concept of Evidence Based Nursing (EBN) has attracted attention. There are three levels of evidence in the nursing domain as follows.

1. Evidence at data level: vital signs of a patient and their interpretation
2. Evidence at cognitive level: interpretation of a causal relationship that is clarified based on the nursing domain knowledge
3. Evidence at meta-cognitive level: criteria of a choice of nursing care plans

Nurses have to perform rational CT processes to validate ones' own decision-making by clarifying these three kinds of evidence explicitly for advance of the quality of medical care. Our research aims to build a learning system whereby a novice nurse can be aware of these three kinds of evidence in the nursing care processes so that the nurse can learn CT methods.

To train CT skills, thinking skills with recognition of self-knowledge of one's own decision-making, we adopt a "new synthesis theory" based learning method [2]. According to the new synthesis, J. T. Bruer says "we should combine the learning of domain-specific subject matter with the learning of general thinking skills, while also making sure that children learn to monitor and control their thinking and learning" [7].

In general, it is difficult for a novice nurse to perform CT in decision-making processes. A reason is that they cannot adapt principles of CT to individual situations even they know general level CT methods.

Design principle of our learning system is to embed guidance information which prompts the learner to combine general know-how of CT with problem-solving context dependent one of CT.

To realize this, we develop the learning system shown as Fig. 1 which is embedded the following five kinds of knowledge into the system shown as Fig. 1: (1) paper patient as Fig. 1(i) that sets an individual situation (problem-solving context), (2) the vocabulary for nursing assessment as Fig. 1(ii), (3) correct evidence structure as Fig. 1(iv), (4) CT knowledge at general level (Alfaro's key questions) quoted from [4] as Fig. 1(v), (5) problem-solving context dependent guidance as Fig. 1(v).

The knowledge of 1, 3 and 5 are described by expert nurse (E.N.) in our authoring system. The vocabulary is quoted from standardized specification called NANDA (North American Nursing Diagnosis Association).

In general, the more serious condition the patient is, the more rapidly his condition changes. A nurse should perform CT processes to deal with the changes, i.e., to monitor

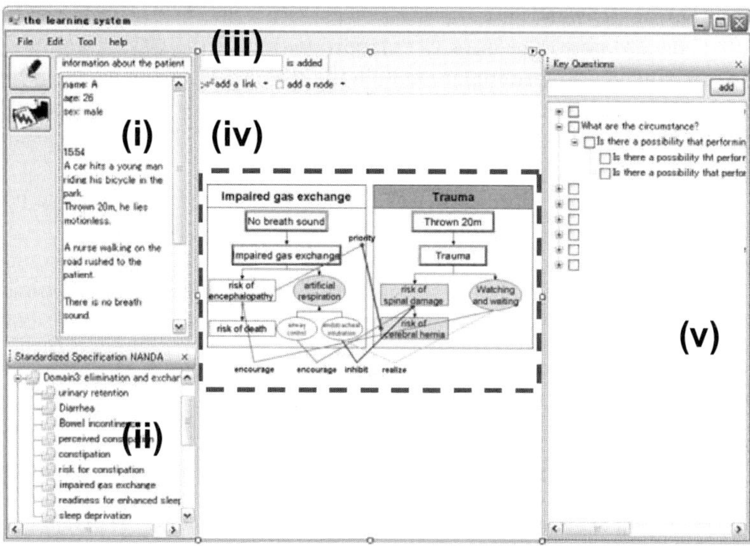

Fig. 1. Learning System for Training CT Skills

and reconstruct one's own decision-making processes and to explain the reason why they should be reconstructed to others. It is quite important ability to be an E.N. [6]. In order to perform them appropriately, the nurse, with explicit recognition of evidence, should verify the changes of the patient's condition properly which may revise one's own decision-making results. Thus, we recognize a paper patient including serious vital changes as important to train CT skills.

The vital signs shown as Fig. 1(i) is the example quoted in section 2. E.N. edited this learning material in the authoring environment we prepared for producing appropriate learning materials for CT learning. E.N. in the authoring environment describes vital condition of a patient and its evidence structures to perform appropriate CT processes in the situation. In this example, for instance, E.N. describes that: "At first, I chose "watching and waiting" based on the priority that recognition of the risk of cerebral hernia caused spinal damage as more important than the risk of life maintenance. Afterward, I changed my decision-making by observing the changes of the situation that the delay of the ambulance got increased the value of life maintenance risk more than the one of cerebral hernia risk."

The learner in this environment is required to describe evidence structures shown as Fig. 1(iv) so that the learner can consciously perform evidence based nursing care processes.

We'll explain the detail of the evidence structure in section 4.

Based on the learner's description, the system provides guidance information on CT that should be performed for this particular patient. Consequently, the learner has to perform nursing care processes including CT processes.

In the learning system, after the learner made a decision of "watching and waiting", the system shows the information about the changes of the condition and guidance information. For instance, the system adds the information of "the delay of the ambulance" in Fig. 1(i) and gives a question of "You (the learner) chose "watching and waiting" according to the priority that recognition of the risk of cerebral hernia caused spinal damage as more important than the risk of life maintenance in Fig. 1(v). Now, is there any changes of the patient's condition that revise your decision-making?". The guidance information is provided based on E.N.'s evidence structures described in the authoring environment.

This guidance stimulates the learner to detect the critical changes of the condition and to recognize the evidence of one's own decision-making explicitly.

By providing such guidance information to the learner, the system encourages the learner's awareness of importance of self-justification of one's own decision-making to reconstruct one's own decision making according to the changes of patient's condition.

4 Learning Support for Critical Thinking

Supporting a learner to acquire know-how of decision-making according to a change of a patient's condition, it is required to grow up an ability of interpreting the meaning of the change based on accurate understanding of three kinds of evidence described above. In our research, we think it is important for the learner to construct following three kinds of knowledge that facilitate valid decision-making processes based on the evidence:

1. Knowledge for problem-solving context dependent data interpretation
2. Domain knowledge in a problem-solving context
3. Meta-cognitive knowledge in a problem-solving context

In this section, we'll explain an overview of our learning system based on a scenario of a learner's utilization for illustrating how to support a learner to construct above knowledge.

4.1 Constructing Knowledge for Context Dependent Data Interpretation

At first, the learner in our system has to assess the patient shown as Fig. 1(i) and choose names of nursing assessments from standardized vocabulary in Fig. 1(ii). Then, rectangles corresponding to respective nursing assessments that the learner chose are shown in Fig. 1(iv). Then, the learner has to describe causal relations (an evidence structure) for each nursing assessment in respective rectangles as shown in Fig. 1(iv).

In this phase, for instance, the learner describes the evidence of the nursing assessment "impaired gas exchange" by marking a vital sign of "no breath sound" described in Fig. 1(i) with a mouse and associating it with the nursing assessment "impaired gas exchange" in Fig. 1(ii). This association means the nursing assessment of "impaired gas exchanges" is interpreted based on the vital evidence of "no breath sound."

In Fig. 1(iv), the rectangle with double line represents evidence or nursing diagnosis of a patient. The rectangle with single line represents a risk. The ellipse represents nursing activity.

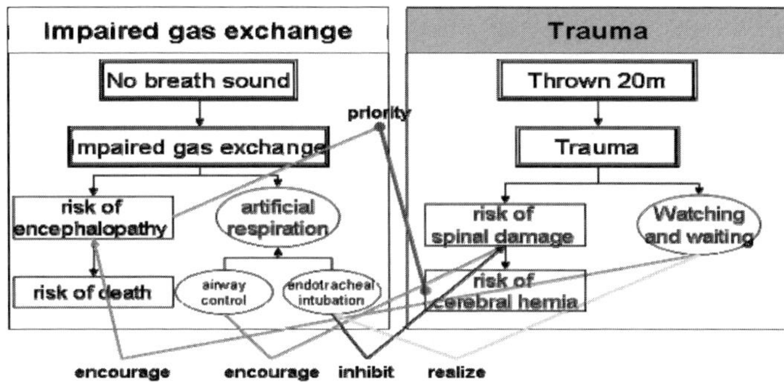

Fig. 2. An example of evidence structures (magnified image of the dotted rectangle in Fig. 1(iv))

By requesting the learner to describe the evidence structures, the system intends to make the learner be conscious of the data that should be focused on in the problem-solving context and the correspondences among data and respective their interpretations for supporting the learner to compose the knowledge for problem-solving context dependent data interpretation know-how.

The information that system provides for the learner to achieve the learning goal is the differences among descriptions by an expert nurse to recognize the gap between the expert nurse's context-dependent data interpretations results and the learner's ones. We think it is the important information for the learner to construct the criteria for selecting vital signs which should be focused on and context-dependent data interpretation methods in a problem-solving context.

4.2   Constructing Domain-Knowledge in a Problem-Solving Context

The learner, then, has to describe evidence structures that supports respective nursing assessments based on own domain knowledge (knowledge for nursing) to make appropriate nursing care plans.

Fig. 2 shows an example of evidence structures that is a magnified image of the dotted rectangle in Fig. 1(iv). The learner describes that the "impaired gas exchange" encourages the "risk of encephalopathy" and further it does the "risk of death". Furthermore, the learner describes that performing "artificial respiration" can inhibit the progress of "encephalopathy" and "artificial respiration" is achieved by "airway control" or "endotracheal intubation".

After describing evidence structures in each rectangle for each nursing assessment (local evidence structures), the learner has to describe evidence structures among nursing assessments (global evidence structures): causal relations among two evidences each of which is described in another rectangle.

In the figure, for instance, a causal relation is represented: performing "airway control" for "artificial respiration" encourages the "risk of spinal damage" considered as a risk of "trauma". Furthermore, another causal relations are also described: "artificial respiration" by "endotracheal intubation" that is realized by "watching and waiting for an ambulance" can inhibit the "spinal damage".

By requesting the learner to describe global evidence structures, the system intends to encourage: the learner's appropriate use of own domain knowledge in the problem-solving context, one's awareness of the importance of capturing causal relations from wider viewpoint and one's spontaneous attitude to perform such kinds of activities.

The information that system provides for the learner to achieve the learning goal is the differences among the associations that an expert nurse described and the ones the learner did, e.g. difference of a causal relation type, direction of a causal relation and so on: we think differential information among evidence structures is the important information for the learner to construct the problem-solving context dependent domain knowledge that could be a basis of capturing causal relations from wider viewpoint to perform appropriate nursing assessment.

4.3 Constructing Meta-cognitive Knowledge in a Problem-Solving Context

Then, according to guidance information from the system in fig. 1(v), the learner has to discover causal relations that the learner couldn't do by oneself.

In general, domain knowledge in a learner's head is not always well-organized for adapting any problem-solving contexts. By requesting the learner to perform the activity, the system intends to encourage: the learner's meta-cognitive awareness to use own domain knowledge, i.e. critical thinking methods in the problem-solving context.

The guidance information for the learner to achieve the learning goal is displayed, for instance, "Is there a possibility that performing some activities changes the value of something risk?" with highlighting related nursing activities when the learner couldn't describe a global evidence (causal relation), in this case performing "airway control" for "artificial respiration" encourages the "cerebral hernia" considered as a risk of "trauma".

The guidance information is organized from the know-how of CT at general level to the one at particular patient level. Above question is general level. The system can highlight nodes in the evidence structures which should be focused according to each general level indicated by the learner in Fig. 1(v). Thus, when the learner indicates more specific question "Is there a possibility that performing "artificial respiration" changes the value of something risks?", the system highlights more limited portion that "airway control" that is one of the methods of performing "artificial respiration".

In this way, the system gradually provides the context dependent guidance information for the learner to combine general (problem-solving context in-dependent) CT know-how with problem-solving context dependent one. We believe this function, that encourages to combine general level know-how with domain specific level one, plays a quite important role to enhance CT skills because it is difficult for them to combine general level know-how to specific level one directly.

This guidance support is developed based on the new synthesis theory

Finally, the learner has to decide the priority by clarifying the justification of one's own decision-making, for instance, what risk is the most important and why it is more important than other risks and so on. It is described in graphical representation and natural language.

By requesting the learner to perform this activity, the system intends to encourage the learner's awareness of the importance of self-justification of one's own decision-making. It is quite important to catch a meaningful vital changes based on which the nurse should revise one's own decision-making.

5 Conclusion

The concept of Evidence Based Nursing (EBN) has attracted attention. For advance of the quality of medical care, nurses must perform critical thinking processes to validate ones' own decision-making by clarifying the evidence of one's own decision-making explicitly.

In this paper, we firstly proposed a learning system which facilitates valid decision-making processes by providing guidance information based on the new synthesis theory. And then, we explained what kind of guidance information is provided for the learner to synthesize the general level know-how of CT and the domain specific one.

References

1. Robert, M.W.: Making Care Safer – A Critical Analysis of Patient Safety Practices. Agency for Healthcare research and Quality (2001)
2. Perkins, D.N., Salomon, G.: Are cognitive skills contest-bound? Educational Researcher 18, 16–25 (1989)
3. Eugene, B.Z., James, E.J.: Critical Thinking – A Functional Approach. Wadsworth (1992)
4. Rosalinda, A.L.: Critical Thinking in Nursing: A Practical Approach. W. B. Saunders Company, Philadelphia (1996)
5. Benner, P.: From Novice to Expert: Excellence and Power in Clinical Nursing Practice. Addison-Wesley Publishing Company (1984)
6. Ryoko, N.: Expert Nurse. Herusu Shuppan Company (2003)
7. Jhon, T.B.: Schools for thought – a science of learning in the classroom. MIT Press (1993)

English Grammar Learning System Based on Knowledge Network of Fill-in-the-Blank Exercises

Takuya Goto[1], Tomoko Kojiri[1], Toyohide Watanabe[1],
Takeshi Yamada[2], and Tomoharu Iwata[2]

[1] Graduate School of Information Science, Nagoya University
Furo-cho, Chikusa-ku, Nagoya, 464-0803, Japan
{tgoto,kojiri,watanabe}@watanabe.ss.is.nagoya-u.ac.jp
[2] NTT Communication Science Laboratories, Japan

Abstract. To understand English grammar is essential to write/speak/ read English appropriately. Fill-in-the-blank exercise of English grammar is one of the popular types of exercises, which is introduced to check acquired/in-acquired grammatical knowledge by evaluating the word selected by a learner for each sentence. Since such exercise-based learning is effective to acquire practical knowledge, our objective is to construct English grammar learning system using fill-in-the-blank exercises. In order to make learners study with the exercises effectively, the system should grasp the grammatical knowledge that learners need to obtain and give series of appropriate exercises. In our system, the knowledge network, which represents relations among exercises according to the differences in their knowledge, is introduced. In the knowledge network, exercises are linked by the inclusive relation of their knowledge. Based on the acquired/in-acquired grammatical knowledge determined by the learner's answer, the system traverses knowledge network and poses the exercise which contains the knowledge that the learner needs to study.

Keywords: Fill-in-the-blank exercise, English grammar, individual learning support, knowledge network.

1   Introduction

To understand English grammar is essential to write/speak/read English. The appropriate words are different for each situation, so it is necessary to study English grammar with various sentences. Fill-in-the-blank exercise of English grammar is one of the popular types of exercises, which is adopted in many tests for evaluating grammatical knowledge, such as TOEIC and TOEFL. In the exercises, learners select appropriate words/phrases with the correct grammar for the blank in the sentence from several choices. By tackling many exercises, learners acquire grammatical knowledge. Therefore, the objective of our research is to construct the English grammar learning system using fill-in-the-blank exercises. In order to make learners acquire grammatical knowledge effectively, the

I. Lovrek, R.J. Howlett, and L.C. Jain (Eds.): KES 2008, Part III, LNAI 5179, pp. 588–595, 2008.

system should grasp the knowledge which learners do not obtain and give a series of exercises corresponding to such knowledge. The effective learning is that the learner is able to notice his/her dis-understanding knowledge automatically. So, it is important to grasp learner's dis-understanding knowledge and give exercises based on it. However, acquired/in-acquired knowledge is different even when learners chose the same incorrect answer for the same exercise, since there is various grammatical knowledge to grasp in the sentence. So, if learners cannot answer a exercise, the system is not able to judge which grammatical knowledge is understood incorrectly. In our research, understanding degree is set for individual knowledge which represents ratio of correctness of exercises that include the grammatical knowledge. By defining the understanding degree, grammatical knowledge whose understanding degree is low is considered as in-acquired knowledge.

In learning support systems based on exercises, exercises are classified by their knowledge used to derive the answer. Kojiri, et al. proposed a mechanism for automatically generating answers to mathematical exercises and constructed network structure which represented the relation between mathematical exercises based on the steps of learners' solving process[1]. Ishima, et al. developed an intelligent tutoring system for high school chemistry and classified chemistry exercises into five types based on solving algorithms of exercsies[2]. In the solving process of English fill-in-the-blank exercises, since learners analyze grammatical structure of the whole sentence, it is necessary to classify exercises by grammatical knowledge that is acquired to solve the exercise.

In our system, exercises are categorized based on all grammatical knowledge which are included in the sentence of the exercise. Based on the inclusive relations of such grammatical knowledge of exercises, the knowledge network is introduced. The knowledge network represents relations between exercises according to the difference of grammatical knowledge. By traversing the knowledge network toward exercises that contain in-acquired grammatical knowledge, the system can give exercises that contain similar grammatical knowledge to the current exercise which holds in-acquired knowledge.

2   Learning by Fill-in-the-Blank Exercises

Fill-in-the-blank exercises of English consist of three parts: sentence, blank, and choices. When the learner solves the exercise, he/she firstly decides the main structure of the whole sentence by analyzing types of every word in it, such as verb or noun. The grammatical knowledge used to determine the structure of the whole sentence, i.e. sentence patterns, is called a *Sentence Structure Knowledge(SS-Knowledge)*. If choices consist of different types of words, the correct choice can be selected successfully by the SS-Knowledge. If there are choices of the same type of words, the learner has to consider not only the sentence pattern but also the appropriate declension or conjugation of particular words or phrases. The grammatical knowledge which decides such declension or conjugation is defined as *Partial Structure Knowledge(PS-Knowledge)*. Not only

PS-Knowledge of the blank in a sentence, but also that of related words to the blank are necessary.

In adaptive learning using exercises, the method of posing exercises is very important and various posing algorithms based on the understanding degrees of individual learners were developed [3,4,5]. Those algorithms change the difficulty of exercises from basic ones to practical ones step by step. Basic exercises consist of simple SS-Knowledge and contain a little PS-Knowledge. On the other hand, practical exercises are composed of complicated SS-Knowledge with a lot of different PS-Knowledge. Since it is necessary for learners to grasp all grammatical knowledge included in the exercise globally, such as PS-Knowledge and SS-Knowledge, difficulties of exercises are different for individual learners according to the combination of grammatical knowledge. So, for the purpose of making learners understand the new grammatical knowledge, exercises by which learners can focus on the target grammatical knowledge should be provided appropriately. That is, exercises that hold a lot of acquired grammatical knowledge and a little in-acquired grammatical knowledge may be effective to notify learners the new grammatical knowledge. Therefore, exercises to provide learners should be managed based on the acquired/in-acquired grammatical knowledge which they contain. For example, if a learner answers the exercise correctly, another exercise with one more PS-Knowledge is effective to acquire new knowledge. When a learner cannot answer the exercise, another exercise with one grammatical knowledge eliminated may help the learner to understand the in-acquired grammatical knowledge.

For the purpose of giving exercises along grammatical knowledge, the database of fill-in-the-blank exercises needs to be structured based on grammatical knowledge. In our approach, knowledge network is introduced, which indicates all combinations of grammatical knowledge existing in the database and their inclusive relations. Figure 1 shows the conceptual imagination of the knowledge network. In the knowledge network, the nodes that are composed of the smallest number of the least knowledge correspond to the most basic exercises. Paths from the root nodes to the goal nodes represent series of exercises for particular grammatical knowledge from basic ones to practical ones. By selecting appropriate nodes on the knowledge network based on the learner's answer and providing exercises that correspond the nodes, it is possible to support individual learners to study specific grammatical knowledge efficiently.

3   Knowledge Network

Knowledge network represents the relations between grammatical knowledge included in the database. Knowledge network is composed of nodes and links. Nodes correspond to the set of grammatical knowledge which consists of fill-in-the-blank exercises. Links indicate inclusive relations between nodes.

By analyzing the English grammar textbook[6], 5 types of SS-Knowledge and 24 types of PS-Knowledge are defined. Figure 2 is the example of attaching grammatical knowledge. "Subject(S) + Verb(V) + Direct Object(O)" is added to this

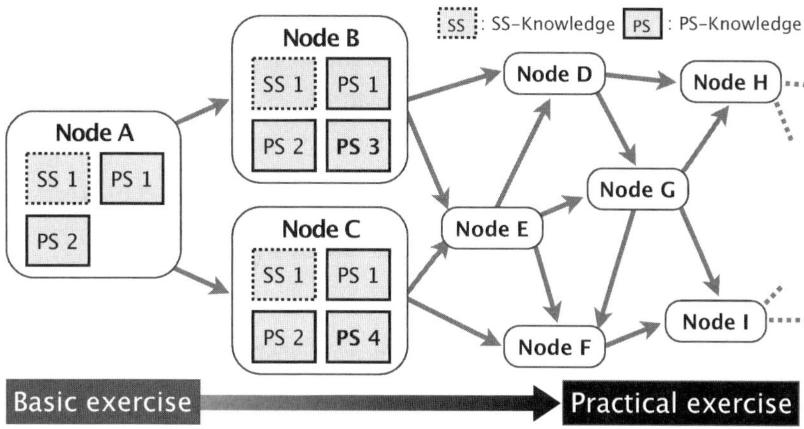

Fig. 1. Conceptual imagination of knowledge network

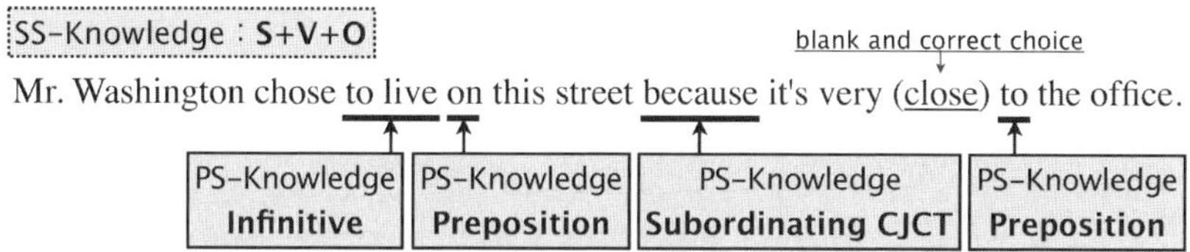

Fig. 2. Example of attaching grammatical knowledge

sentence as SS-Knowledge and "Infinitive", "Preposition", and "Subordinating conjunction(Subordinating CJCT)" are attached as PS-Knowledge.

Links between nodes indicate inclusive relations of grammatical knowledge. There are two types of inclusive relations: one is the inclusive relation of sentences and the other is the inclusive relation of words. The inclusive relation of words is defined by the embracement of existing PS-Knowledge. Since words included in exercises are different according to the given sentences, the amount and the types of PS-Knowledge differ according to them. If the amount of PS-Knowledge increases, more grammatical knowledge is needed to be understood. Therefore, links are attached to the nodes that have one more PS-Knowledge than the current node. On the other hand, the inclusive relation of sentences is derived from the relations of SS-Knowledge. SS-Knowledge represents the complicacies of sentences and some of one type of SS-Knowledge are extended by another one. For example, "S+V" needs to be understood before studying "S+V+O". Thus, links are also attached to nodes that contain the same PS-Knowledge and different SS-Knowledge of inclusive relations.

Figure 3 is the example of knowledge network. This network consists of four nodes. Nodes A and B, and nodes C and D have the same PS-Knowledge but different SS-Knowledge, such as "S+V+O" and "S+V+O+O". Since their SS-Knowledge have inclusive relations of sentence, nodes A and B, and nodes C and

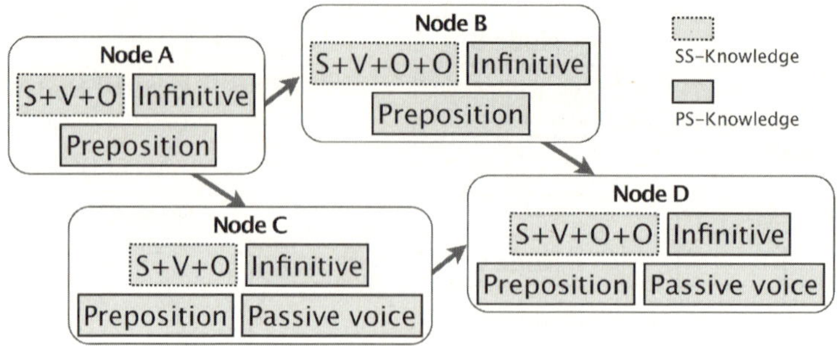

Fig. 3. Example of knowledge network

D are connected by links, respectively. On the contrary, nodes A and C contain the same SS-Knowledge, and PS-Knowledge of node C consist of that of node A and "passive voice". So, the link is attached from node A to node C. Similarly, another link is added from node B to node D.

4   Mechanism for Selecting Exercises

Based on the learner's answer, exercises which help the learner acquire dis-understanding knowledge are estimated. When the learner selects a correct answer, he may understand all grammatical knowledge included in the exercise. However, understanding of it can be inferred according to the word class of the choice that the learner selected, when he could not answer correctly. Table 1 shows the understanding/dis-understanding of knowledge based on the difference in word class of the learner's choice with the correct one. A learner tends to grasp the SS-Knowledge by observing the structure of the whole sentence and to find appropriate declension or conjugation from choices. Therefore, if the learner selected the wrong choices which have the different word class from the correct one, he/she may not understand the SS-Knowledge. On the other hand, when the learner selected the wrong choices with the same word class, he/she is expected to understand the SS-Knowledge, but may not be able to decide appropriate declension or conjugation. So, the PS-Knowledge in the blank is estimated as dis-understanding knowledge.

Table 1. Understanding/dis-understanding knowledge based on learner's choice

Learner's choice	Understanding knowledge	Dis-understanding knowledge
Different word class with right choice	-	SS-Knowledge
Same word class with right choice	SS-Knowledge	PS-Knowledge

According to the understanding/dis-understanding knowledge estimated by the learner's answer, understanding degree for each grammatical knowledge is determined. Understanding degree for the grammatical knowledge i, which is represented as c_i, is calculated as follows:

$$c_i = (s_i - f_i)/(s_i + f_i) \quad (-1 \leq c_i \leq 1) \tag{1}$$

c_i is defined so as to take the value between -1 to 1. When c_i is 1, the learner may understand the knowledge i, but if c_i takes -1, he/she does not understand the knowledge i at all. c_i is calculated based on s_i and f_i. s_i indicates the number of exercises by which the learner is estimated to understand the knowledge i. f_i corresponds to the number of exercises by which the learner is estimated no to understand it. For example, if the learner answered eight exercises that contain "Infinitive" correctly, $s_{\mathrm{Infinitive}}$ becomes 8. When the learner selected incorrect choices of the PS-Knowledge for two exercises that also have "Infinitive", $f_{\mathrm{Infinitive}}$ is set to 2. In this example, $c_{\mathrm{Infinitive}}$ is calculated as 0.6.

The exercise to give to the learner is determined based on the understanding degrees for individual grammatical knowledge. If a learner selected a right choice and the understanding degree of all grammatical knowledge in the current exercise is higher, the system grasps that the learner understands all the knowledge successfully and tries to teach new grammatical knowledge. So, it follows the link from the corresponding node to the node in which one new type of grammatical knowledge whose understanding degree is the lowest is added to the current one. If the understanding degree of grammatical knowledge of the current exercise is lower, the system regards that the learner does not understand the knowledge and tries to teach it again by giving another exercise of the dis-understanding knowledge in the other node. On the other hand, when the learner answers wrongly and he/she has not answered all exercises of the current node, the system tries to teach the knowledge again and gives another exercise of the current node. If he/she has already answered all exercises, the system follows the link from the corresponding node to the node in which one type of grammatical knowledge whose understanding degree is the highest is removed. By traversing the knowledge network according to the learner's answer and understanding degrees, the learner can tackle exercises that contain grammatical knowledge that he/she does not understand deeply.

5   Evaluation

355 fill-in-the-blank exercises were prepared in the database and 29 grammatical knowledge is applied. Since links based on inclusive relation of PS-Knowledge are attached from nodes to nodes which contain one additional type of PS-Knowledge, 104 knowledge networks were generated. Of all knowledge networks, 103 were smaller-scale networks that have less than 10 nodes. The last one is a larger-scale network which consists of 54 nodes. In order to make learners study acquired/in-acquired grammatical knowledge, it is necessary to construct knowledge network using all grammatical knowledge. By connecting nodes that

have inclusive relation of more than two PS-Knowledge, all exercises can be structured as one knowledge network.

We implemented an web-based English grammar learning support system based on our mechanism for selecting exercises. The knowledge network of larger-scale was applied. After a learner selects a choice through the web page, the system displays the result of learner's answer and the explanation to derive the correct answer. The system, then, determines the next exercise. If the system cannot select appropriate exercise because current node is the root node, the leaf nodes or the node that does not contain enough exercises, it finds the new node that contains the knowledge whose understanding degree is the lowest.

The experiment to evaluate the effectiveness of our selection mechanism is executed. Sixteen students in our laboratory were asked to participate in the experiment. The experiment consisted of three phases: pretest phase, learning phase, and posttest phase. In the pretest phase, participants were asked to answer twenty exercises. During the pretest, neither the results of their answers nor the right answer were displayed. In the learning phase, participants were divided into two groups: namely eight participants for each group, and participants of each group were given twenty exercises. Participants in one group studied using our system (*group 1*), and those of the other group used the system which randomly selects the exercise from the database (*group 2*). For the group 1, the system grasped the understanding degrees of participants based on results of their pretests and selected the start node for individual participants. In the learning phase, both results of their answers and the explanation to solve each exercise were shown. In the posttest phase, all participants tackled the same exercises provided with the pretest again, but the orders of posing exercises were different.

Table 2. Average scores in pretest and posttest

	Group 1	Group 2
Pretest	15.9	15
Posttest	16.1	15.9

Table 3. Average values of increasing understanding degrees for knowledge that answered incorrectly in pretest

	Group 1	Group 2
Increase of understanding degrees	42.3%	21.8%

Table 2 indicates the average scores in the pretest and the posttest. The changes of the total scores did not show significant differences between both groups. On the other hand, Table 3 indicates the average values of increasing understanding degrees for all grammatical knowledge that participants were not answered correctly in the pretest phase. For the group 1, understanding degrees for dis-understanding knowledge were increased about 42.3%, while 21.8% for the group 2. For both groups, almost all grammatical knowledge which were incorrectly answered were provided in the learning phases. Therefore, from the result, our system can give appropriate fill-in-the-blank exercises, which helps learners to study in-acquired grammatical knowledge effectively.

During the learning, the system could not select exercises from the corresponding node for five times because of the lack of the number of exercises and links. In order to provide appropriate exercises, more exercises should be prepared in our database.

6   Conclusion

In this paper, we proposed a framework of learning English grammar using fill-in-the-blank exercises. We classified exercises based on their grammatical knowledge and represented knowledge of the database as the knowledge network. Based on the experimental result, our selection mechanism and knowledge network was able to give appropriate exercises and was effective for studying dis-understanding knowledge step by step.

Currently, the selection methods of exercises are common to all learners. However, the progresses of understanding the knowledge are different from each learner. If learners can understand the knowledge with a few explanations, they may need a few exercises for one grammatical knowledge. On the other hand, if learners tend to understand the knowledge slowly, they may prefer to tackle the exercises of the same knowledge repeatedly. Thus, the mechanism to navigate the knowledge network according to the learner's learning tendencies should be developed.

References

1. Kojiri, T., Hosono, S., Watanabe, T.: Automatic Generation of Answers for Complex Mathematical Exercise. In: Proceedings of ICCE 2006, Workshop, pp. 25–32 (2006)
2. Ishima, N., Ueda, T., Konishi, T., Itoh, Y.: Developing Problem Representation, Knowledge Representation and Problem Solving System for Intelligent Educational System of High School Chemistry. In: Proceedings of ICCE 2006, Workshop, pp. 41–48 (2006)
3. Fischer, S.: Course and Exercise Sequencing Using Metadata in Adaptive Hypermedia Learning Systems. ACM Journal of Educational Resources in Computing 1(1), 5 (2001)
4. Scheiter, K., Gerjets, P.: The Impact of Problem Order: Sequencing Problems as a Strategy for Improving One s Performance. In: Proc. of 24th Annual Conference of the Cognitive Science Society, pp. 798–803 (2002)
5. Cooley, R.E., Abdullah, S.C.: Controlling Problem Progression in Adaptive Testing. In: Proc. of ICCE 2000, vol. 1, pp. 635–642 (2000)
6. Quackenbush, E.: New Direction of English. Translation by Okada, N., Biseisya, Japan (in Japanese) (2000)

The Planetary Simulator for Generalized Understanding of Astronomical Phenomena from Various Viewpoints

Masato Soga, Yasunori Nakanishi, and Kohe Tokoi

Faculty of Systems Engineering, Wakayama University,
930 Sakaedani, Wakayama 640-8510, Japan
soga@sys.wakayama-u.ac.jp

Abstract. We developed the planetary simulator that is able to simulate astronomical phenomena in the solar system. The planetary simulator can set not only target planet but also the planet of viewpoint. It can simulate planets' revolutions around the sun fixing viewpoint on another planet. Therefore it can help generalized understanding of wax and wane of inner planets. The planetary simulator also supports viewpoint guiding function for self-learning. Moreover, the planetary simulator shows a small dynamic guide map that shows current learner's viewpoint, target and scope. Finally, We developed the function by which a learner can perform the virtual experiment in the simulator to understand Kepler's laws of planets' revolutions.

Keywords: Learning environment, Education, Astronomy, Simulator, Simulation, Viewpoint, Astronomical phenomena.

1 Introduction

Astronomy is a domain where high technology can assist learners understanding, since learners cannot touch the real celestial bodies in hand. Several learning environments were developed by using web, Internet, computer graphics and so on.

As a learning environment for constellations, a planisphere type astronomy learning environment based on a structured constellation database were developed on web using Ajax [1]. Also, a multi-platform contents management system for online constellation learning is developed [2].

As a learning environment by using telescope, [3] is the first remote telescope in the world that supplies live picture of celestial bodies by Internet. Learners were able to control remote telescope from web user interface. Learners in Germany controlled remote telescope located in Japan, and they watched live picture of the moon by using time difference between Japan and Germany [4].

On the other hand, simulator by computer graphics can assist understanding of astronomy, since the object is too large and too far away to see in the real world. Several simulators were developed for astronomical education as follows.

Astronomical VR Contents [5] can simulate traveling from the earth to the universe like traveling by spaceship. A novel tangible user interface consisting of a glove and a

I. Lovrek, R.J. Howlett, and L.C. Jain (Eds.): KES 2008, Part III, LNAI 5179, pp. 596–603, 2008.

physical avatar attached on its surface are developed for the Astronomical VR Contents, since the capability to support spatial perception plays an important role in developing material for astronomy education [6]. Virtual reality graphics of the solar system can be controlled manually with the glove and avatar.

Stella Navigator is an astronomical simulator produced by Astroarts co. ltd [7]. It simulates night sky in any time, any place. It can simulate astronomical phenomena only from the earth viewpoint.

As research on a method of expression of space phenomena with CG animation, the collision of the comet Shoemaker-Levy 9 with Jupiter in July 1994 is simulated [8].

Mitaka is a simulator developed in 4D2U project by National Astronomical Observatory of Japan [9]. It simulates seamlessly from the earth to the end of the universe at 13.7 billion light years away like traveling in the universe. A learner can select a target celestial body from menu, then, he/she can travel virtually between the target celestial body and the end of universe. Planets, moon, and some famous stars are prepared for the target celestial bodies. Mitaka is very excellent software to understand the universe from the solar system to the end of the universe. It shows distance from the target celestial body, so that the learner can know the size of every structure in the universe.

However, those pre-developed simulators are not appropriate software to understand generally astronomical phenomena in the solar system, such as antecedence, inner planets' wax and wane, inner planets' transit in front of the sun, apparent declination change of Saturn's ring, and so on. Most astronomical phenomena in the solar system are caused by specific relative positions between the targets and observers. Those pre-developed simulators are not designed to set observers' various viewpoint.

Our purpose is to develop an appropriate learning environment for astronomical phenomena that occurs in the solar system. We suppose that learners are high school students. The learning environment should enable the learner to understand the principle of the phenomena. Furthermore, the learner should be able to understand the phenomena more generally.

We think that the following two functions are necessary to assist learners understanding for astronomical phenomena in the solar system. One is the function by which a learner can select a viewpoint and a target to observe the astronomical phenomena. This function enables the learner to understand the principle of the phenomena in the 3D space. Moreover, the function enables the learner to generalize the understanding of the phenomena. The other is the function by which a learner can perform the virtual experiment in the simulator to understand law of planets' revolutions.

For the former function, the planetary simulator supports viewpoint and target selection function. For the latter function, the planetary simulator support virtual experiment environment.

2 Viewpoint Setting and Guiding

We developed the planetary simulator that is able to simulate astronomical phenomena in the solar system. The planetary simulator can select not only target planet but also the planet of viewpoint. It can simulate planets' revolution around the sun fixing viewpoint on another planet. Therefore it can help generalized understanding of wax and wane of inner planets. If the viewpoint is on the Earth, wax and wane occur on the Venus and the Mercury. If the viewpoint is on the Mars, wax and wane occur not only on the Venus and the Mercury but also on the Earth. Through these observations, a learner can know that wax and wane can occur on any inner planets from the planet's viewpoint. The learner can get generalized understanding of wax and wane of planets.

It is difficult for a novice learner to set the viewpoint and target appropriately to observe astronomical phenomena. Therefore, blending lessons is suitable for learning astronomical phenomena by the planetary simulator.

2.1 Viewpoint Guiding Function

The planetary simulator also supports viewpoint guiding function for self-learning. The simulator shows eye vector that indicates viewpoint for observing the astronomical phenomena. If the learner put his/her viewpoint on the eye vector, he/she can observe the astronomical phenomena.

Figure 1 shows eye vectors to assist the learner to understand the principle of wax and wane of the earth. In the figure 1, five eye vectors are shown simultaneously. However, actually every eye vector is shown in turn for the learner. When the learner sets his/her viewpoint on the first eye vector, then the planetary simulator shows next eye vector. By continuing this process, the learner can learn the principle of wax and wane of the earth.

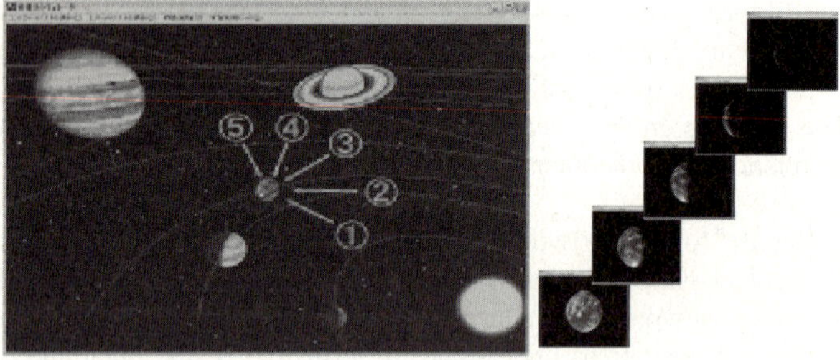

Fig. 1. Viewpoint guidance function for earth's wax and wane. Actually, every eye vector is shown in turn for the learner. When the learner sets his/her viewpoint on the first eye vector, then the planetary simulator shows next eye vector. Every right picture shows every scene from every viewpoint on the every eye vector. By continuing this process, the learner can learn the principle of wax and wane of the earth.

For another example, Figure 2 shows Viewing guidance function for Mercury's transit in front of the sun. In the Figure 2, red orbit is Mercury's and blue orbit is earth's orbit. At first, the simulator shows yellow line (Fig. 2, left picture), and makes the learner set viewpoint on the yellow line (Fig. 2, right picture). Then, the learner can find Mercury's transit in front of the sun. Also, the learner can find both Mercury's orbit and earth's orbit are viewed horizontally (Fig. 2, right picture). The explanation and operation instruction is shown in the window at upper left corner.

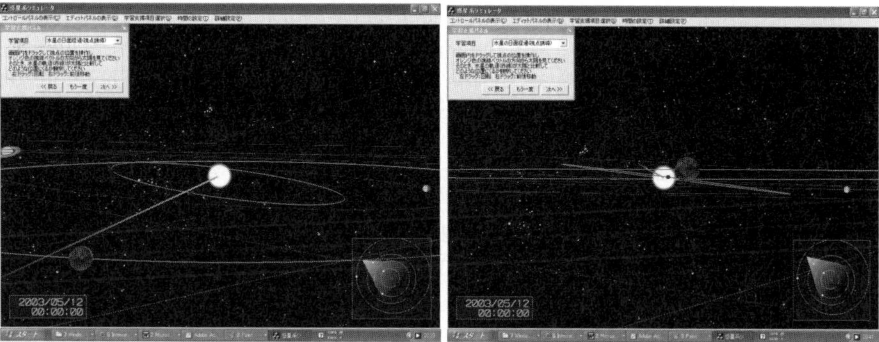

Fig. 2. Viewing guidance function for Mercury's transit in front of the sun. At first, the simulator shows yellow line (left picture), and a learner set viewpoint on the yellow line (right picture). Then, the learner can find Mercury's transit in front of the sun.

Figure 3 shows the example in which Mercury's transit in front of the sun cannot be seen. At first, the simulator shows yellow line (Fig. 3, left picture), and makes the learner set viewpoint on the yellow line (Fig. 3, right picture). Then, the learner can find that Mercury's transit in front of the sun does not occur this time. Also, the learner can find Mercury's orbit cannot be viewed horizontally this time (Fig. 3, right picture).

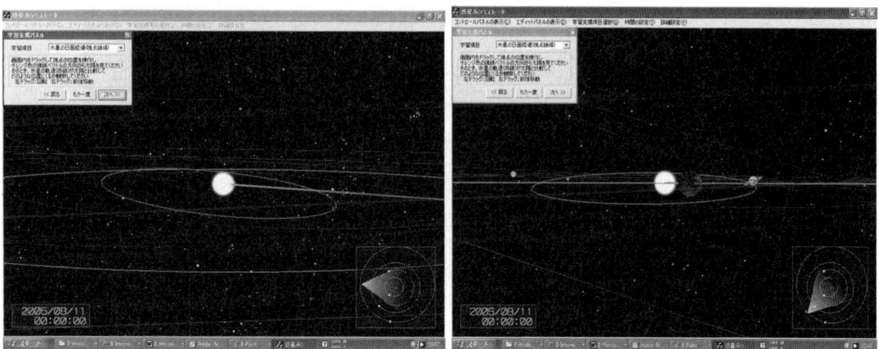

Fig. 3. Viewing guidance function for Mercury's passing under the sun. At first, the simulator shows yellow line (left picture), and a learner set viewpoint on the yellow line (right picture). Then, the learner can find Mercury is passing under the sun. Also the learner can find Mercury's orbit cannot be viewed horizontally this time.

2.2 Dynamic Guide Map

When a learner set his/her viewpoint in the solar system, he/she sometimes cannot find where he/she is located in the solar system. To avoid this situation, the planetary simulator shows a small dynamic guide map that shows current learner's viewpoint, target and scope. The small dynamic guide map is shown in the lower right corner of the simulator window. Actually, the dynamic guide map is minified planetary simulator from the viewpoint of the north ecliptic pole.

When a learner observes motion, wax and wane of a target planet from another planet's viewpoint such as the earth, the learner can observe relative position between target planet and viewpoint planet by the small dynamic guide map. Therefore the learner can understand relation between planet's relative position and apparent motion, waxing, waning simultaneously.

This function is very useful and important. For instance, suppose that a learner wants to understand the Venus's motion, wax and wane from the earth viewpoint. Venus looks like full moon, when it is located behind the sun. After that, Venus wanes and apparent disc size become larger. When Venus is catching up with the earth during its revolution around the sun, Venus looks like new moon and apparent disc size become maximum. The learner can understand relation between planets' relative positions by the small dynamic guide map and he/she can observe apparent motion, wax and wane simultaneously (Fig. 4).

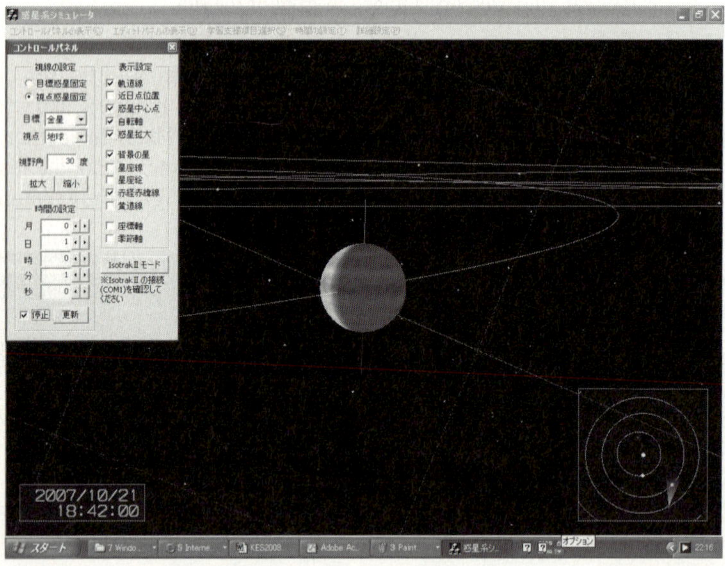

Fig. 4. Simulation of Venus revolution observed from the earth viewpoint and small dynamic guide map that shows positions of Venus and the earth

3 Virtual Experiment Environment

Every planet revolves around the sun by following Kepler's laws. The planetary simulator simulates eight planets' revolution, but it is not enough to learn Kepler's laws, since planets' orbits are almost circle. We developed the function by which a learner can perform the virtual experiment in the simulator to understand Kepler's laws of planets' revolutions (Fig. 5).

At the beginning, the simulator simulates 8 planets' and Pluto's orbits around the sun. Then, a learner can select a planet and change the orbit's parameters of the planet, such as eccentricity, inclination, longitude of the ascending node, and argument of periapsis.

Eccentricity may be interpreted as a measure of how much this shape deviates from a circle. Inclination is the angular distance of the orbital plane from the plane of the ecliptic, normally stated in degrees. Longitude of the ascending node is the angle from the vernal equinox as the origin of longitude, to the direction of the ascending node, measured in the ecliptic. The angle is measured counterclockwise (as seen from north of the ecliptic) from the vernal equinox the to the node. Argument of periapsis is the angle of an orbiting body's periapsis (the point of closest approach to the sun), relative to its ascending node (the point where the planet crosses the ecliptic from South to North). The angle is measured in the orbital plane and in the direction of motion.

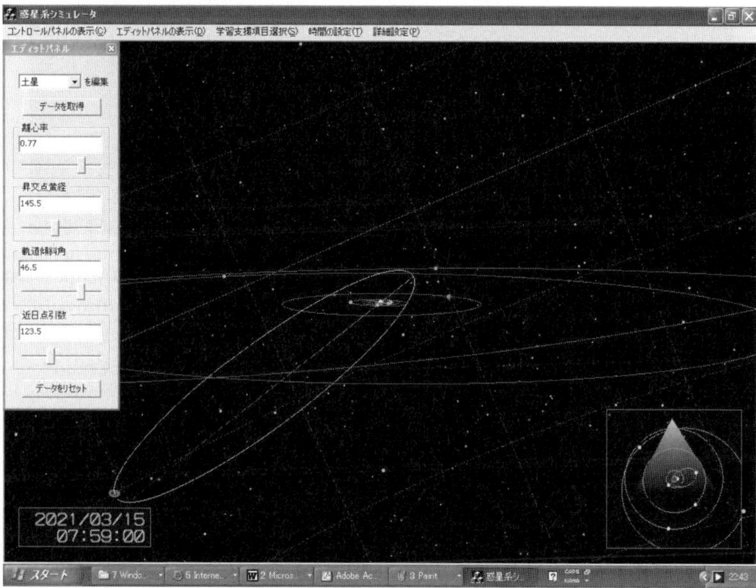

Fig. 5. Virtual experiment environment: Saturn's orbit parameters are changed and the orbit became like comet's orbit. Learners can simulate Saturn's revolution following Kepler's second law.

The simulator shows sliding bars for changing the parameters. When a learner moves the sliding bars, then the planet's orbit also changes interactively in the virtual world. For example, if a learner changes eccentricity of Saturn's orbit into larger number, then the orbit becomes long ellipse like comet's orbit. Saturn moves on the long ellipse orbit depending on Kepler's laws. Saturn moves fast near the sun, and moves slow far from the sun. If a learner sets the sun as the target, and sets Saturn as viewpoint, then the learner can observe the solar system by traveling on the Saturn's orbit. The learner can experience the Kepler's second law. Moreover, if the learner increases inclination value, then the Saturn's orbit inclines against the ecliptic. Also, the learner can observe other planets revolve around the sun away from ecliptic. This fact gives the learner an opportunity to think about the reason why other planets move along ecliptic from the viewpoint of the earth.

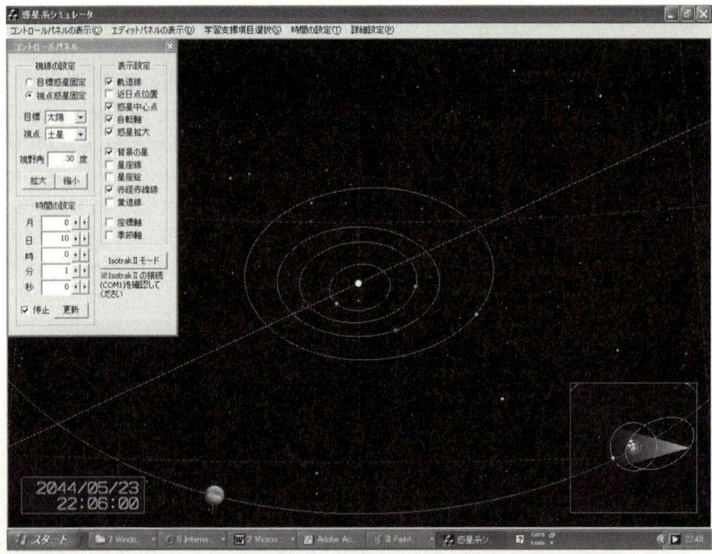

Fig. 6. If the learner increases inclination value, then the Saturn's orbit inclines against the ecliptic. Also, the learner can observe other planets revolve around the sun away from ecliptic. This fact gives the learner an opportunity to think about the reason why other planets move along ecliptic from the viewpoint of the earth.

Thus, learners can understand phenomena more generally by using our planetary simulator.

4 Conclusion

In this paper, we describe the planetary simulator that enables to simulate astronomical phenomena in the solar system. The planetary simulator can set not only target planet but also the planet of viewpoint. It can simulate planets' revolutions around the sun fixing viewpoint on another planet. Therefore it can help generalized understanding of wax and wane of inner planets.

The planetary simulator also supports viewpoint guiding function for self-learning. The simulator shows eye vector that indicates viewpoint for observing the astronomical phenomena. If the learner put his/her viewpoint on the eye vector, he/she can observe the astronomical phenomena.

Moreover, the planetary simulator shows a small dynamic guide map that shows current learner's viewpoint, target and scope. The small dynamic guide map is shown in the lower right corner of the simulator window.

Finally, We developed the function by which a learner can perform the virtual experiment in the simulator to understand Kepler's laws of planets' revolutions. A learner can select a planet and change the orbit's parameters, such as eccentricity, inclination, longitude of the ascending node, and argument of periapsis. After changing the planet's orbit into long ellipse, the planet moves fast near the sun, and moves slow far from the sun. If a learner sets the sun as the target, and sets the planet as viewpoint, then the learner can observe the solar system by traveling on the planet's orbit. The learner can experience the Kepler's second law.

Thus, learners can understand phenomena more generally by using our planetary simulator.

References

1. Mouri, K., Endo, M., Iwazaki, K., Noda, M., Yasuda, T., Yokoi, S.: Development of a Planisphere Type Astronomy Education Web System Based on a Constellation Database using Ajax. In: Gabrys, B., Howlett, R.J., Jain, L.C. (eds.) KES 2006. LNCS (LNAI), vol. 4252, pp. 1045–1052. Springer, Heidelberg (2006)
2. Kondo, M., Yasuda, T., Yokoi, S., Ura, M., Endo, M., Iwazaki, K., Mouri, K., Noda, M.: A Multi-platform Contents Management System for Online Constellation Learning. In: Proc. of Communication, Internet and Information Technology (CIIT 2007), pp. 73–78 (2007)
3. Soga, M., Watanabe, K., Tanaka, T., Tanaka, H., Sakamoto, M., Toyomasu, S., Okyudo, M.: A Remote Control System of a Telescope by WWW and Real Time Communication for Astronomical Education. In: Proc. World Conf. on Educational Multimedia and Hypermedia / World Conf. on Educational Telecommunications (ED-MEDIA/ED-TELECOM 1998), vol. 2, pp. 1305–1310 (1998)
4. Soga, M., Okyudo, M., Toyomasu, S., Tanaka, T., et al.: Astronomical Experimental Class and Evaluation of Interactive Remote Telescope System by Time Difference between Germany and Japan. In: Proc. of International Conference on Computers in Education 2001 / School Net 2001, pp. 702–705 (2001)
5. Ando, M., Yoshida, K., Tanikawa, T., Kato, H., Kuzuoka, H., Hirose, M.: Development of Astronomical VR Contents for Group Study. In: Proceedings of the Virtual Reality Society of Japan Annual Conference (CD-ROM), vol. 9(2A3-1) (in Japanese) (2004)
6. Yamashita, J., Kuzuoka, H., Fujimon, C., Hirose, M.: Tangible Avatar and Tangible Earth: A Novel Interface for Astronomy Education. In: CHI 2007 extended abstracts on Human factors in computing systems, pp. 2777–2782 (2007)
7. Stella Navigator, http://www.astroarts.co.jp/shop/showcase/stlnav8/index-j.shtml
8. Tokai, S., Yasuda, T., Yokoi, S., Toriwaki, J.: A Method of Expression of Space Phenomena with CG Animation: The Collision of the Comet Shoemaker-Levy 9 with Jupiter in July 1994. In: Proc. Computer Graphics International 1995, pp. 267–280 (1995)
9. 4D2U Project, http://4d2u.nao.ac.jp/english/index.html

A Discussion Model for System Design Novices

Shoich Nakamura[1], Miho Watanabe[2], Atsuo Hazeyama[2], Setsuo Yokoyama[2]
and Youzou Miyadera[2]

[1] Fukushima University, Kanayagawa 1, Fukushima 960-1296, Japan
`nakamura@sss.fukushima-u.ac.jp`
[2] Tokyo Gakugei University, 4-1-1, Nukui-Kita, Koganei, Tokyo 148-8501, Japan
`{yokoyama, miyadera}@u-gakugei.ac.jp`

Abstract. The opportunities to discuss system design as part of system development through a network continue to increase. In our university, system development education for novices is also done through a network. However, the usefulness of discussion on system development for novices is limited by various unresolved problems left from the discussion and the inability of novices to discuss these matters in a comprehensive way. We propose a discussion model designed for novices that will control such discussion smoothly, and then evaluate the model's usefulness. This model is based on state transition using discussion states and discussion comment categories.

Keywords: discussion model, discussion state transition, discussion support, system design, system development education.

1 Introduction

Recent advances in network technologies has enabled more distributed cooperative development of software systems [1]. For such cooperative development to succeed, effective support and an appropriate environment is needed to assist communication through discussions and so on. Although chat systems and bulletin board systems (BBSs) are often used for these purposes, engaging in this kind of communication skillfully can be difficult since problems regarding discussion initiation and conversational gaps often occur. Moreover, beginners (such as students of a system development class in a university) often do not know how to proceed with such discussions. To solve these problems, this research is aimed at developing a method to support smooth discussion in such contexts.

Previous efforts to develop means of discussion support have been reported. Inaba et al. developed a system [2] to identify slumps in a discussion and absent learners based on a conversation model constructed through analysis of discussions among learners in cooperative problem solving over a distributed computer environment. Conklin et al. developed groupware [3] to support discussions in the upper process of software design based on the IBIS model devised through a structured approach. Winograd et al. tried to extract the common structures in general conversations to help users grasp discussion state transmissions based on these structures [4]. Kotani et al. developed a system which observes the

I. Lovrek, R.J. Howlett, and L.C. Jain (Eds.): KES 2008, Part III, LNAI 5179, pp. 604–615, 2008.

progress of discussions via Web chat. This system displays in real-time the roles of learners in a discussion based on favorable utterances. Kaiya et al. analyzed the discussions in a cooperative software specification process and proposed a model to express discussions by assigning roles to topics within them.

These methods are not applicable for complex discussions, such as regarding system development, since they do not consider practical discussion situations and are too abstract. Moreover, utterance types in these systems are unclear and a clear decision regarding them is difficult for discussion members. Thus, it is difficult to control a discussion as the relevant situation changes. A few systems have been specialized for system development, but members participating in the development still have to assign the parts in those systems. Unfortunately, beginners usually find it difficult to play a specific role in such discussions. Analysis of discussion contents through natural language methods is an alternative approach, but such methods are subject to many restrictions and they are generally not very practical. Thus, our goal in this research has been to develop a more applicable method.

In this research, we have limited the target to system development based on the waterfall model. We propose a discussion support model that controls discussions as smoothly as possible by focusing on discussion transitions defined by clear utterance types and discussion states.

As a first step, two analyses were done.

(1) An analysis of the required utterance types for each system development process.
(2) An analysis of discussion transitions in each process.

Based on these analyses, we developed a model that identifies discussion states from utterance types and their transitions to enhance smooth discussion. This approach, which uses support from the discussion form side, shares a concept with some of the methods just mentioned. Our approach has practical merits in that it is specialized for system development and controls related discussions based on particular utterances and their transitions.

2 Problems in System Development

2.1 Analysis of BBS Use in System Development

To categorize the problems that typically hinder discussion concerning system development, we analyzed the BBS used in a class "system development exercise", which is assigned to third-year undergraduate students in the authors' university. In this class, students attempt exercises related to cooperative system development based on the waterfall model. Specifically, all BBS utterances through one semester, which consisted of 3180 utterances related to 271 topics by 26 members, were examined one by one. We focused on discussion failure points and diagnosed the reasons for failures from neighboring utterances. We defined a discussion failure point as a case where almost no utterance appeared and/or a new question was uttered before the previous question had been answered.

The analysis detected 149 discussion failure points. Three main reasons for these failures were identified:

1) abandonment of an unresolved discussion (49.6%)
2) emission of discussion (23.4%)
3) confused discussion (20.1%)

These three causes of failure are particularly serious for novices since novices accounted for more than 90 percent of the failures. Through this research, we aim to solve the discussion problems caused by these three failures.

2.2 Related Research

Most research on support for system development discussion [2][3][4][5][6] has not considered the third problem stated above. Moreover, the scope of the support target has usually not been precisely defined and the discussion models are too abstract. Therefore, no means of effectively controlling complex discussions, such those concerning system development, is currently available. Existing discussion models cannot satisfactorily cover actual utterances since they are not based on a wide enough range of situations. It's also impossible for many models to support beginners' discussions on system development since they aim to enhance discussion through differences in experience and knowledge among members. Thus, there is a great need for research into ways to support beginners' discussions concerning system development.

3 Discussion Support Model Design

Our intention has been to develop a model for supporting discussion based on the waterfall model. The waterfall model consists of requirements analysis, design, implementation, testing, and integration/maintenance. System development progresses in this order. After completion of a process, it is not permissible to return to a previous process. That is, operations progress to the next process as soon as a process is satisfactorily completed. Our research has targeted the processes from requirements analysis to testing because exercises for beginners seldom consider the processes of integration and maintenance.

In this paper, we propose a model that expresses discussion transitions in terms of discussion states and utterance types for each process. To realize this model, we did the two analyses described in this section:

(1) analysis of essential utterance types for each process
(2) analysis of discussion transitions in each process

3.1 Analysis of Essential Utterance Types for Each Process

To build a discussion support model specialized for system development, we extracted essential utterance types for each process. Specifically, we first classified

Table 1. Extracted utterance types for each process

Process	The extracted utterance types		
Requirements analysis	Question	Contact	Proposal
	Agreement	Disagreement	Confirmation request
	Reply to question	Question to question	Explanation
Specifications drawing	Question	Contact	Proposal
	Agreement	Disagreement	Confirmation request
	Reply to question	Question to question	Explanation
	Acceptance	Rejection	
Model diagram drawing	Question	Contact	Proposal
	Agreement	Disagreement	Confirmation request
	Reply to question	Question to question	Explanation
	Acceptance	Rejection	
Program design	Question	Contact	Proposal
	Agreement	Disagreement	Confirmation request
	Reply to question	Question to question	Explanation
	Acceptance	Rejection	
Implementation	Question	Contact	Proposal
	Agreement	Disagreement	Confirmation request
	Reply to question	Question to question	Completion report
Testing	Question	Proposal	Explanation
	Agreement	Disagreement	Error information
	Reply to question	Question to question	Correction
	Completion	Completion report	

all utterances for each process based on the BBS analysis described in Sect.2.1.
Then we manually examined the contents of each utterance one by one and
investigated the typical types of utterance for each process. From the results of
these analyses, we grouped similar utterance types and supplemented insufficient
types. Consequently, the essential utterance types were extracted

Table 1 shows the extracted utterance types for each process. The meanings of
each utterance type are shown in Table 2. In this research, the design process is
divided into three sub-processes - specifications drawing, model diagram drawing
and program design - since these are the primary tasks studied in the target
exercise class. We expect the display of these utterance types to help users grasp
discussion states.

3.2   Analysis of Discussion Transitions in Each Process

Analyzing BBS use as in the previous case allowed us to identify the parts where
discussion progressed smoothly. The utterance types from Table 1 were assigned
to each utterance in the smooth parts one by one. Through this task, we extracted tendencies in the transitions of utterance types in those smooth parts.
For example, in the requirements analysis process, the discussion transition

Table 2. Meanings of the essential utterance types

Utterance type	Meaning
Question	Question to the utterance
Contact	A message to group members
Suspension	Difficulty in immediate answer
Reply to question	Reply to question
Question to question	Further question to the question
Agreement	Agree to the utterance
Disagreement	Disagree to the utterance
Explanation	Supplement the utterance
Proposal	Indication of new idea
Confirmation request	Confirmation request of the completed work before submission
Acceptance	Acceptance of the work
Rejection	Rejection of the work
Completion report	Completion report of coding or correction of a part
Completion	Completion of all test items and progress to next process
Correction	Incompletion of any test items or existing of any errors, and continuation the current operation
Error information	Discovery of errors by testing

"Contact'Question'Reply to question'Question'Reply to question'Agreement' Reply to question'Agreement'Agreement" has been extracted.

After that, we designed discussion processes by grouping the extracted discussion transmissions. It is known that relations of the discussion process illustrate each system development process. Consequently, typical discussion processes for each development process were extracted (Table 3).

4   Development of a Discussion Model

This section describes a discussion model which expresses the transitions by discussion state and utterance type. Based on the classification of discussion processes described in Sect.3, we created discussion models for each development process. This paper describes the development process of requirement analysis, and Fig.1 shows a discussion model for requirement analysis.

First, discussions based on the question discussion process and contact discussion process take place. When these are completed, the discussion proceeds to the proposal discussion process. In the same way, after agreement is reached in the proposal discussion process, the discussion moves to the confirmation request process. When all discussions in the requirements analysis process have been completed, the discussion moves forward to the specification drawing process. Otherwise, the discussion returns to the proposal discussion process.

As described, there are small discussion processes in each development stage. Moreover, transitions occur among these discussion processes. Overall, the transitions in each development stage make up the total development process.

Table 3. Discussion processes for each development process

Process	Discussion process	
Requirements analysis	Question discussion process	Contact discussion process
	Proposal discussion process	Confirmation request process (1)
Specifications drawing	Question discussion process	Contact discussion process
	Proposal discussion process	Confirmation request process (2)
	Confirmation request process (3)	
Model diagram drawing	Question discussion process	Contact discussion process
	Proposal discussion process	Confirmation request process (3)
Program design	Question discussion process	Contact discussion process
	Proposal discussion process	Confirmation request process (3)
Implementation	Question discussion process	Contact discussion process
	Proposal discussion process	Completion report process (1)
Testing	Error information process	Completion report process (2)

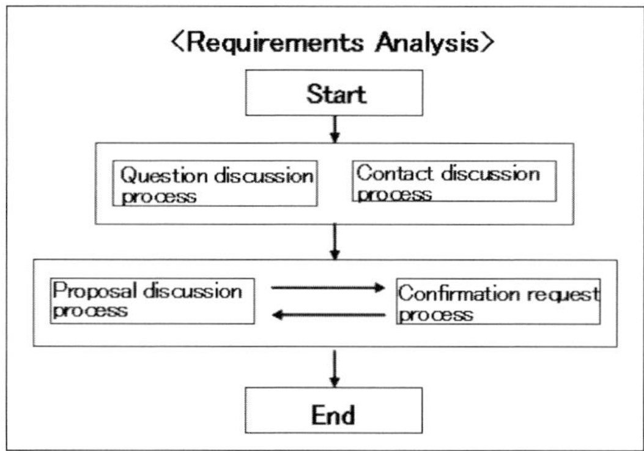

Fig. 1. Discussion model for requirement analysis

Here, the discussion model is explained in detail for the proposal discussion model as an example. Figure 2 shows a model for the proposal discussion process. As a proposal arises, the state moves from 0 to 1. In state 1, if all members express "agreement" regarding the proposal, the proposal is acknowledged and the state moves to 6. Conversely, if there is any "disagreement", the state returns to 0. Similarly, if any members contribute an "explanation" to the proposal, the state remains in 1. If a "question" is received in state 1, there is movement to state 3. If a "question to question" is received in state 3, the state changes to 4. State 3 also changes to 5 when a "reply to question" is received. If either "suspension", "reply to question", or "question to question" is received in state 4, the state does not change. Moreover, if both "suspension" and "reply to question" are received in state 5, the state remains unchanged, but it returns to state 3 when "agreement" is received.

Since it is possible to determine that a discussion has been finished successfully by examining whether the state has reached 6, this model enables to avoid the neglectfulness of the discussions.

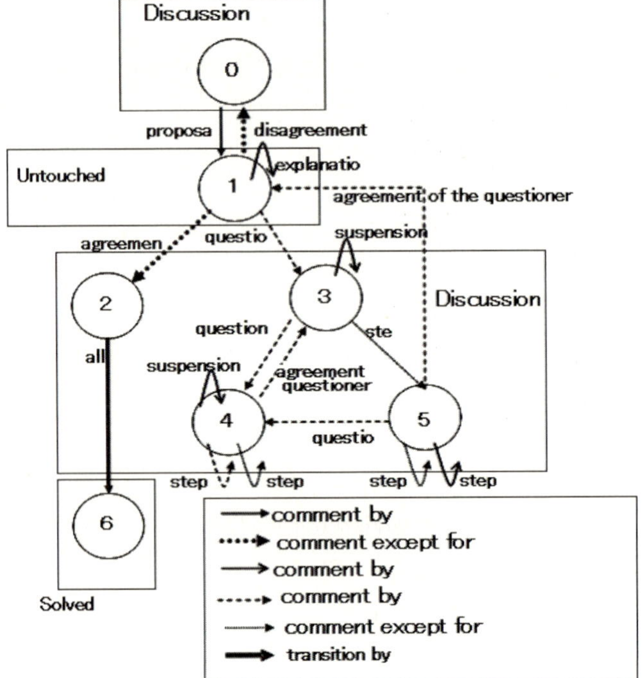

Fig. 2. Model for proposal discussion process

In a discussion model such as that shown in Fig.2, state 1, states 2 to 5 and state 6 respectively express "untouched situation", "discussion in progress" and "finished discussion".

Consequently, discussion models for each development process have been developed, and it is now possible to control discussions by developing a support system that uses these discussion models.

5   Design and Development of the Evaluation System

A discussion based on the proposed model proceeds by grasping the discussion states for an individual process through the flow of discussions and the kinds of comment made. Accordingly, discussion support for novices can be achieved. To evaluate the proposed model, we developed a discussion support system that incorporates it.

A user uttering a comment in the development system BBS has to select a comment type (from a choice of 15) as mentioned regarding analysis 1. This method restricts the comment types used. The miscellaneous comment type "others" was added to evaluate the appropriateness of this model. Users can select "others" when they cannot select an appropriate comment type for their utterances, and this means that the discussion has deviated from the discussion model. Therefore, after selecting "others" once, users can only select "others" for subsequent comments. When the discussion is completed, the users can select "discussion completed in the other case", and then the discussion is terminated.

For discussion completion that does not follow the discussion transition like this, the selection label of "discussion completion as unresolved" is also available in the system. The label "discussion completion as unresolved" is selected when the discussion has to be completed with unsolved states occurring. By using this, users can terminate the discussion even if they argue along other than the discussion transition model. The discussion states are visually presented to users by color-coding the background of each discussion topic. "Untouched discussion", "discussion middle", "solved discussion" and "other" are colored pink, yellow, light blue, and white, respectively. Although "discussion completion as unresolved" and "discussion completed in the other case" are shown in the same light blue as "solved discussion", their borders are changed from unbroken to broken lines.

Furthermore, the discussion states of each process can be easily grasped. If no discussion of the process has begun, the discussion stat0s color is pink, meaning "untouched discussion". If there is some unsolved problem in the process, the discussion states color is yellow, meaning "discussion middle". If not all discussions have been completed, the discussion states color is light blue, meaning "solved discussion".

The BBS screen that we developed is shown in Fig.3. When a process is selected from area A, areas B and C are displayed on the right side. Users can input the discussion contents through B. In this case, the user needs to input his or her name, destination, discussion solution term, title, text and the kinds of comment. Using such meta-information, the discussion structure is displayed in the tree on C, and each discussion is colored according to its states. When the title of a comment is selected, the comment content is displayed and a reply is permitted.

6   Evaluation

6.1   Outline

In this section, we examine the effectiveness of the proposed discussion model. Especially, we examine the efficacy of system development by focusing on the three problems mentioned earlier.

Seventeen students from the class described in Sect.2.1 participated in this evaluation. The class consisted of four groups, and each group consisted of four or five students. The number of persons, the subject and the BBS type for each group are shown in Table 4. Groups A and B were applied to the discussion support system described in Sect.5. Groups C and D were applied to the normal BBS which has no restrictions on comment type, discussion transition navigation or discussion states, and has no classification of the processes. We will call this BBS the "normal system". The evaluation lasted from October 2005

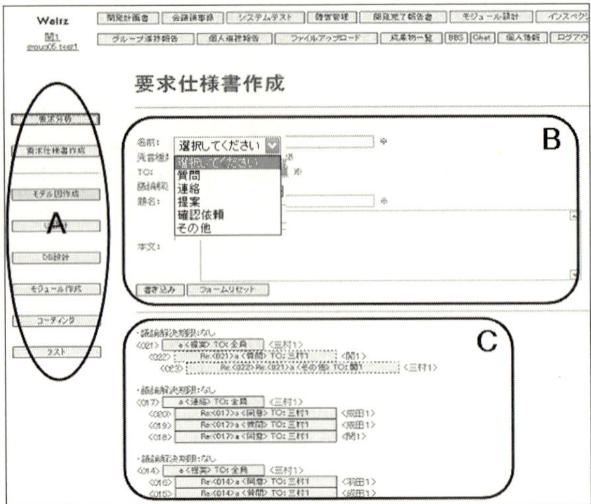

Fig. 3. BBS screen

Table 4. Evaluation groups

Group	Subject	BBS
A	Employment information management system	Our developed system
B	Total school affairs system	Our developed system
C	Employment information management system	Normal BBS
D	Employment information management system	Normal BBS

to November 2005, and covered the "requirement analysis" and "specifications preparation" processes.

6.2 Effectiveness of the Model

The effectiveness of the discussion model was evaluated with groups A and B, both of which applied the developed discussion system, as evaluation targets. The evaluation method was as follows:

(1) Measure the proportion of all comments that were classified as the "Others" comment type from BBS data
(2) Analyze why the comment type "Others" was selected by using a questionnaire.

The results for (1) are shown in Table 5 for each subject.

Apparently, "Others" is generally used when the discussion deviates from the discussion model. The proportion of "others" is 10% at most, and in many cases the proportion of "others" in one subject is low. For subjects where the proportion

Table 5. Proportion of the comment kind "Others"

Group	Process	Proportion of "Others"
A	Requirements Analysis	0.0%
	Preparation of Requirements Specification	10.2%
B	Requirements Analysis	10.0%
	Preparation of Requirements Specification	3.9%

Table 6. Questionnaire results

Id	Question item	A+B	C+D
I	Smooth discussion was achieved	100.0%	16.6%
II(a)	The discussion states could be grasped	88.8%	37.5%
II(b)	The button "discussion completion as unresolved" was used	12.5%	—
III(a)	The selection of discussion types was useful	100.0%	—
III(b)	Restriction of the comment types was useful	66.6%	—
IV	The classification of the colored process was useful	88.8%	—

of "others" was high, if "Others" was selected once, the other selections were not permitted. The reasons that "Others" was selected were as follows:

1. When the discussion is too long, users want to start a new discussion as a way to continue
2. The discussion stopped for long time.

Furthermore, by analyzing each comment carefully, we found there were appropriate comments that could be used instead of "Others". Therefore, the discussion model proposed in this paper is appropriate, because there were no problems with the discussion model, such as stagnant discussion.

Next we evaluated the efficacy of the model. The evaluation target was again groups A - D of Table 4. The questionnaire was completed after the evaluation period. Results are shown in Table 6. The participants could select yes or no for each question, and the values in Table 6 are the proportion of respondents who selected "yes". Regarding smooth discussion support of the purpose of this study, the groups using our system obtained good results for subject I (Table 6).

We then evaluated the particular problems of system design: (1) unresolved problems left from the discussion, (2) discussion emission and (3) not understanding the direction of the discussion.

The main cause of the first problem is that students cannot easily grasp the discussion states. The results related to this are shown in II(a) of Table 6. Members of groups A+B commented that the oversight of discussion was gone, and it was easy to decide for which discussion to make a comment. Members of groups C+D commented that it was difficult to grasp and to see the whole BBS. The states of discussion should be clarified. Moreover, as show in II(b) of Table 6, the usage of "discussion completion as unresolved" was low. Therefore, the discussion model seems to have worked effectively.

A possible solution to the second problem is to control the discussion transitions. Questionnaire results regarding this are shown in III(a) and III(b) of Table 6. The following comments were obtained.

- It is shown how the user corresponds to each comment.
- There are no deviations.
- It is good to unify the types of comment.

Therefore, the emission of discussion seems to have been controlled.

The third problem can also be solved by controlling the discussion transitions. The question results for the color classification of the discussion states are shown in IV of Table 6. The following comments were obtained.

- It is easy to understand which discussion is the current one
- It is easy to notice changes in the discussion states

These results indicate that users could more easily grasp the discussion progress and better understand the discussion method and contents. The evaluation supports our belief that this discussion model can solve the three main problems of system design that we have considered here.

7   Conclusion

This paper describes a discussion model that can be used to control the discussion of novices. We initially analyzed the actual BBS use in use in system development. Next we extracted the essential utterance types and discussion transitions for each system development process. Then, we designed the discussion processes for each development process. Based on these analyses, we finally developed the discussion model which focuses on discussion states and their transitions. After that, we developed the evaluation system and examined the discussion model through the experiments. Consequently, the results of experiments have shown that the model enables smooth discussion control.

Future works will include further evaluation of the model by introducing it into practical system design discussions and its improvement.

Acknowledgments

The authors would like to thank Ms. Harumi Seki and Ms. Mikako Miura for their contributions which include careful analyses in the initial step of this study. The authors would also like to thank members of YokoMiya-Lab. at Tokyo Gakugei University for their cooperation and valuable discussions.

This work was supported in part by the Ministry of Education, Culture, Sports, Science and Technology, Japan and by Japan Society for the Promotion of Science under Grant-in-Aid for Young Scientists (B) (No.20700632), Grant-in-Aid for Exploratory Research (No.19650239) and Grant-in-Aid for Scientific Research (B) (No.19300272) respectively.

References

1. Matsuhita, M., Iida, H., Inoue, K.: An Interaction Support Mechanism in Software Development. IEICE Trans. J81-D1(9), 1072–1081 (1998)
2. Inaba, A., Hasaba, Y., Okamoto, T.: An Intelligent Supporting of Discussion for the Distributed Cooperative Learning Environment. IEICE Trans. J79-A(2), 207–215 (1996)
3. Conklin, J., Begeman, M.L.: gIBIS: A Hypertext Tool for Exploratory Policy Discussion. In: Proc. of the Conference on Computer-Supported Cooperative Work (CSCW 1988), pp. 140–152. ACM Press, New York (1988)
4. Winograd, T.: Where the Action Is. BYTE 13(3), 256–259 (1988)
5. Kotani, T., Seki, K., Matsui, T., Okamoto, T.: Development of Discussion Supporting System Based on the Value of Favorable Words' Influence. Trans. JSAI 19(2), 95–104 (2004)
6. Saeki, H.K.M.: VDP: Framework for Analyzing Cooperative Software Process. IPSJ Technical Reports, 91-HI-37 91(52), 9–16 (1991)
7. Kawamura, K.: Fountain of the intellect (accessed March1, 2008), `http://www.st.rim.or.jp/~k-kazuma/SD/SD.html`

Information Fusion for Uncertainty Determination in Video and Infrared Cameras System

Cornel Barna

Aurel Vlaicu University Arad, Romania
`barnac@rdslink.ro`

Abstract. In the paper I present a new method for measuring uncertainty based on multiple information fusion. A system composed of video and infrared cameras is used. From this system, the features of the observed objects are extracted, separate for the video and infrared cameras. Based on these characteristics, the uncertainty value for video and infrared observations are determined. Three information fusion methods are applied, and for each of them the aggregated uncertainty measure is determined. Finally, a global uncertainty estimation is obtained from the previous results, which can be used for the labeling decision process of recognizing an object.

Keywords: uncertainty, information fusion, video and infrared cameras.

1 Introduction

The complexity of the captured images, as well as the numerous parameters that can influence the localization and recognition in images triggered the appearance of numerous articles, which propose numerous methods [3]. One issue related to this subject is to measure the confidence in the obtained results [7]. In order to assign the best label to an unknown object, according to the obtained information, we must know how reliable the resulted data are, especially if we obtain contradictory information from different chains of acquisitions. In this paper a new view is presented, analyzing the uncertainty of the observation based on fusion data provided by video and infrared cameras, and basing the final decision process on uncertainty estimation obtained from multiple fusion methods. The computation is based on soft computing theory and fusing information from fuzzy type similarity measurements.

2 Video and Infrared Uncertainty Measurement

To determine the degree of confidence for the estimated characteristics is one of the most important issues in feature discovery and labeling. It has often been underlined [1], [2] that crisp interpretation of the classification results is causing the loss of an important part of the initial information, and for this reason it is preferable to either use a fuzzy form for out coming data or to establish a continence measure. In these cases, an uncertainty mesurement can be used.

Equivalence classes imply three main proprieties: reflexivity, symmetry and transitivity. In numerous applications, transitivity doesn't hold (see Luce paradox), fact

I. Lovrek, R.J. Howlett, and L.C. Jain (Eds.): KES 2008, Part III, LNAI 5179, pp. 616–621, 2008.

determined by the approximation of the observation (two characteristic measurements seem to be equal, but in reality there is little difference between them). For this reason, the use of a more general relationship, such as similarity, in which the transitive propriety is not required, is more adequate. This is especially true for granular computing and fuzzy logic, where the observations are rough. For this reason, the fuzzy similarity relationship proposed in [6] will be used. This relationship is obtained by overlaying two fuzzy relationships: that of inclusion and that of dominance. The first one implies a fuzzy implication relationship and establishes the degree of inclusion of a variable x in an interval boarded by a constant value a. The second one implies the fuzzy implications between a boarded constant b and the variable x. Thus, the similarity relationship is:

$$sim(x,a,b) = (a \rightarrow x) \otimes (x \rightarrow b) \tag{1}$$

For example, Fig.1 represents the similarity function for the product t-norm and Fig.2 represents the similarity function for the Lukasiewschi type t-norms:

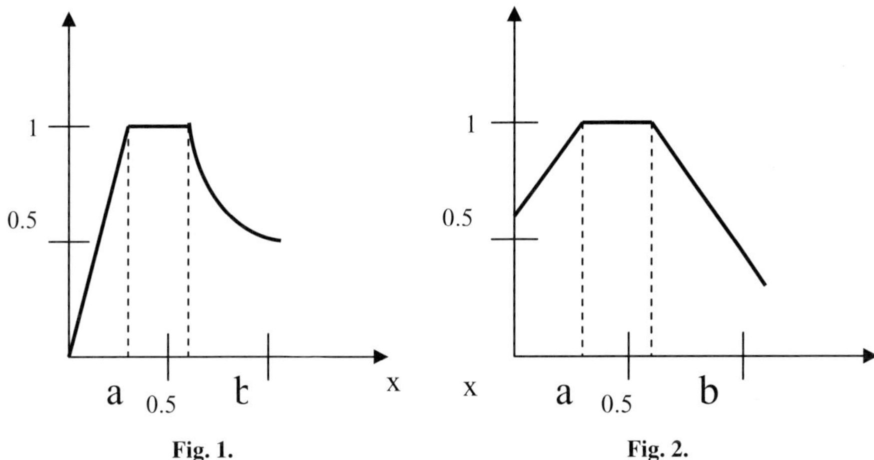

Fig. 1.   Fig. 2.

Thus, the label of the unknown feature is determined by the nearest prototype, which is also the most similar to it:

$$S(x,a,b)=\max(sim(x,a,b)) \tag{2}$$

The feature is also compared to the next nearest prototype. The dissimilarity is then determined from this reference point. Dissimilarity is considered the complementary measure of similarity, and its function is determined by using the negation of the latter one. In the present paper, the classical negation function will be used. The resulting relationship for this measure is:

$$D(x,a,b)= \min_{i=1,N} (1\text{-}sim(y,a_i,b_i)) \tag{3}$$

From the relationship between these two measurements the measure of uncertainty, expressed as a difference between the similarity with the closest prototype and the dissimilarity from the second closest prototype can be determined. This type of me-surement supposes an optimist point of view, in the sense that all the object proto-types are known from the scene. But the assumption is reasonable, because if the object is unknown and has no prototype it can be viewed as an indistinctiveness case, in the sense defined by Pawlak. The proposed relation for uncertainty mesurement has one of these two forms presented in relations 4 and 5:

$$U(x,a,b) = S(x,a,b) - D(x,a,b) \tag{4}$$

If the result of this measurement is positive, then small values indicate small uncer-tainty, because the labeled feature is similar to the designed prototype, and is weakly similar to the next closest one. On the other hand, if the result is large, it can be con-cluded that uncertainty is important because the difference of similarity between the feature and the next nearest prototypes is small, and the possibility of it belonging to each of these two classes is high, resulting that the probability of making a mistake is considerable.

If the difference has a negative value, it can be concluded that the feature does not belong to the investigated class because it is more similar to another prototype, or it can be labeled on the information provided by the existent prototypes.

Or, it can be used as a ratio relation, as presented:

$$U(x,a,b) = \frac{S_1(x,a,b)}{S_2(x,a,b)} \tag{5}$$

The relation expresses that the uncertainty is growing if the nearest prototype S_1 is more similar to the second nearest prototype S_2. This means that the possibility of mislabeling the features is higher if the two prototypes have the same similarities.

3 Information Fusion

For the final decision process three type of information fusion are selected, based on the experimental conclusion that the more different theoretical bases it has, the more independent results it generates.

The first fusion is made using a heuristic method. This is obtained using relation 6

$$U_E = a \cdot U_v + b \cdot U_i \tag{6}$$

Where a, b are constants which must satisfy the conditions $a+b=1$, a, $b \in$ [0, 1], and U_v, U_i represent the results provided by the video and the infrared uncertainty measurements (see relations 4 and 5). The performance of this information fusion depends largely on the chosen constants values, which must be established from experiments.

The second information fusion between the results obtained from the two types of captors is of fuzzy type. Because the information obtain from the video cameras is more reliable when we have good light conditions, in daytime the image is better that in nighttime. Complementary, the infrared images are more accurate when the object temperatures have a larger gradient, meaning that night images are better than daytime ones. Therefore, the reliability of the result obtained from the combined video and infrared cameras system is obtain with the following relation:

$$1 - U_G = \frac{\mu_v \cdot (1 - U_v) + \mu_i \cdot (1 - U_i)}{\mu_v + \mu_i} \tag{7}$$

where:

U_v, U_i are the video and the infrared cameras' uncertainty, and U_G is the system uncertainty and μ_v, μ_i represent technical characteristics of video and infrared cameras' sensibility to light and heat respectively.

It thus results that the global uncertainty is defined by the relation:

$$U_G = \frac{\mu_v \cdot U_v + \mu_i \cdot U_i}{\mu_v + \mu_i} \tag{8}$$

It can be easily observed that the global uncertainty is improving the system performance, because in day conditions the video information is dominant, having in the same time lower uncertainty, while in night conditions the infrared information has a bigger influence on the result, because it's uncertainty is better.

The last fusion process implied an information fusion between the results obtained by a reinforcement type method [4]. Because the information obtained from the video cameras is generally more significant than the information from the infrared cameras, the latter will only reinforce the conclusion of the former. Thus, one will operate on the different features obtained from both methods in order to increase or decrease the uncertainty obtained from the video cameras. The following relationship will be operated:

$$U_F = \min(1, U^1) \tag{9}$$

where

$$U^1 = U_V + \frac{1}{K} \left(\sum_{i=1,n} \alpha_i \frac{|f_{Vi} - f_{Ii}|}{|f_{Vi}|} \right) U_I \tag{10}$$

and f_V, f_I are the estimated surfaces obtained by video and infrared cameras respectively, $\alpha_i \in [0,1]$ are weigh coefficients, and K is a normalization factor.

Finally we can derive a global result from all the three methods using a weighted sum:

$$U = a \cdot U_E + b \cdot U_G + c \cdot U_F \tag{11}$$

The constants can be obtained using an adaptive process [5], which increases the importance of that information fusion method with the best fitting characteristic for the real experimental condition.

4 Experiments

A set of 3x3x8 experiments with features extracted from images was performed. This was done using an elongated, a round and a elliptic object, for each of which three illumination condition were created. Two MDV213 type video cameras were used, and two TEM-412 type cameras with infrared projectors were used for obtaining the infrared images. For obtaining a 3D information, a triangulation process was performed. Seven sets of attributes were used. By extracting the area and the second order moment from the images the shape extraction was performed. The segmentation was made using the background extraction.

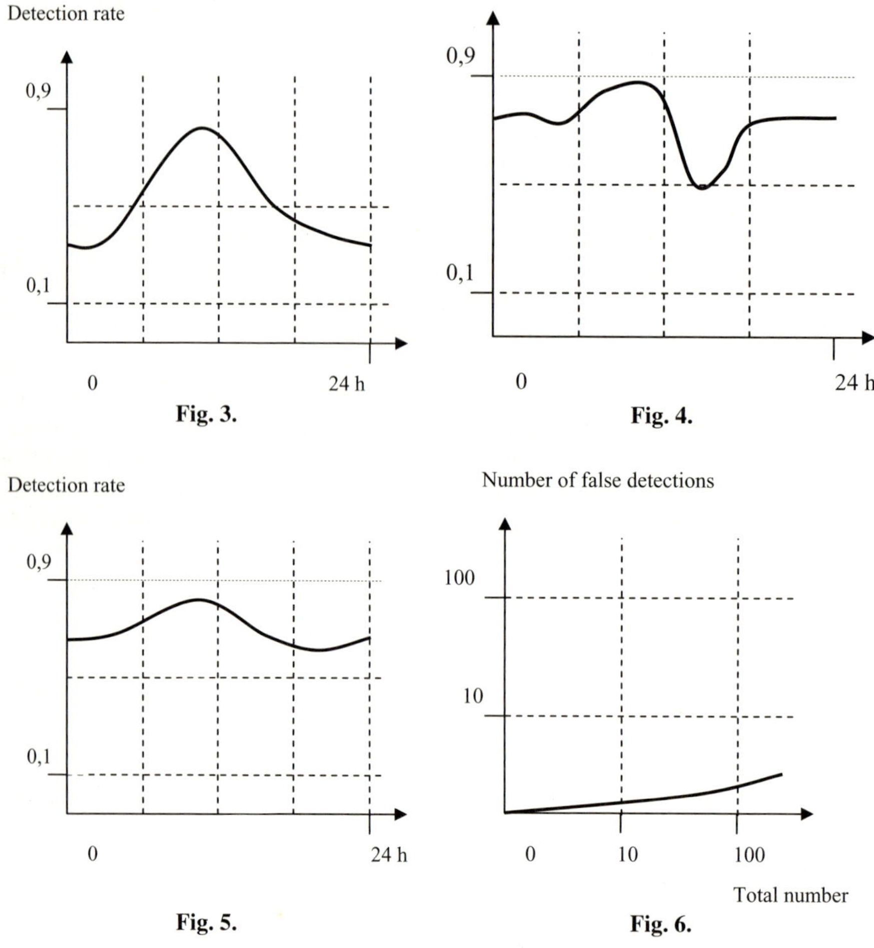

The observations were made for 24 hours cycles. As is presented in figure 3, the fuzzy fusion method was more accurate in day conditions. Figure 4 presents the observations for the heuristic fusion method and the variations of the results with the changing of external conditions can be observed. That occurred because this method is a rigid one. The results for the reinforced method have generated quite good results, as it can be seen in figure 5, but it was observed that this method is sensitive to the parameters values. Finally, figure 6 presents the global results of the three methods, and the low rate of false detections, with underline the effectiveness of these fusion methods, can be observed.

5   Conclusions

Recognition techniques are mostly based on information provided by video cameras. In this article, a dual video and infrared camera system is used. The result is obtained by applying three fusion procedures, and it computes an uncertainty mesurement for each one.

For image recognition we have proposed a system composed of two video cameras and two infrared cameras. This system is used for computing three information fusion results, based on: a heuristic method, a fuzzy method and a reinforcement method, which were selected because they have different theoretical bases. First, for each of video and infrared camera systems, geometric characteristics are obtained which in a second stage are used in three types of information fusion.

Finally, an aggregation operation is performed using the three information fusion, by using adaptively determined weighting constants.

References

1. Alkoot, F., Kittler, J.: Modified product fusion. Pattern Recognition Letters 23(8) (2002)
2. Bargiera, A., Pedrycz, W., Hiroto, K.: Logic-based granular prototyping. In: 26-th Annual International Computer Software COMPSAC (2002)
3. Barna, C.: An Alarm System Design with PIR using Fuzzy Sensor Fusion and Genetic Algoritm Program. In: IEEE SOFA 2005, Szeged-Arad, pp. 195–199 (2005)
4. Barna, C.: Alarm Systems. In: Vlaicu Arad, A. (ed.) Design using Soft Computing (2007) ISBN: 978-973-752-153-8
5. Negnevitsky, M.: Artificial Intelligence A guide to intelligent systems. Addison-Wesley, London (2002)
6. Pedrycz, W.: Knowledge-Based Clustering From Data to Information Granules. Willey-Interscience. ISBN: 0-471-46966-1
7. Yager, R.: A General approach to the Fusion of Imprecise Information. Inter'l Journal of Intelligent system 12 (1997)

Various Speaker Recognition Techniques Using a Special Nonlinear Metric

Tudor Barbu and Mihaela Costin

Institute for Computer Sciences of the Romanian Academy, Iasi branch, Romania
tudbar@iit.iit.tuiasi.ro, mccostin@iit.iit.tuiasi.ro

Abstract. In this paper we compare voice recognition techniques using a special nonlinear metric. We describe text-dependent and text-independent speaker recognition methods based on a mel-cepstral analysis in the feature extraction stage and a supervised classification. The Hausdorff-based metric proposed by us is able to measure the distance between different sized speech feature vectors resulted from the mel-cepstral featuring process. Then, a minimum mean distance classifier uses this new distance to identify the speakers.

Keywords: text-dependent speaker recognition, text-independent speaker recognition, mel-cepstral analysis, supervised classification.

1 Introduction

Voice recognition represents the process of automatically recognizing who is speaking on the basis of individual features included in the speech waves ([1], [2]). The voice recognition methods can be divided into text-dependent and text-independent techniques. The former approaches discriminate the users based on the same spoken utterance ([3]-[5]), while the latter do not rely on a specific speech ([5]-[8]).

Also, speaker (voice) recognition consists of voice identification and voice verification. Speaker identification is the process of determining which registered speaker provides a given spoken utterance, while speaker verification represents the procedure of accepting or rejecting the identity claim of an input speaker ([1]).

The Mel Frequency Cepstral Coefficients (MFCCs) are the dominant features used for speech and speaker recognition. So, we propose a melodic cepstral analysis for vocal feature extraction ([4]-[6], [8]). In our previous works we approached other feature extraction methods, such as LPC (cepstral) analysis, (auto)regression (AR) coefficients and the pitch based techniques ([5]).

The identification process consists of two stages: speech feature extraction and classification. Often, we obtain different sized feature vectors from the featuring process, therefore the most used Euclidian distance cannot be utilized anymore. So, a new special metric, based on Hausdorff distance for sets, is proposed in the next section ([4],[5],[8]).

In the third section some speech-dependent ([3]-[5]) voice recognition systems are provided. In the fourth section a speech-independent ([5]-[8]) voice recognition technique is described. They utilize the *DDMFCC* analysis in the feature extraction stage and a supervised classifier, using the Hausdorff-derived metric, at the classification

I. Lovrek, R.J. Howlett, and L.C. Jain (Eds.): KES 2008, Part III, LNAI 5179, pp. 622–629, 2008.

level. Some experiments performed with the proposed recognition technologies are also described. The paper ends with a conclusions section.

2   A Hausdorff-Derived Distance

In this section we propose a special nonlinear metric which is able to compute the distance between different sized matrices having a single common dimension. It derives from the Hausdorff metric for sets ([4],[5],[8]), described as:

$$H(A,B) = \{\max_{a \in A} \inf_{b \in B} dist(a,b), \max_{b \in B} \inf_{a \in A} dist(a,b)\} \tag{1}$$

Let us consider now matrices having a single common dimension (the number of rows), instead of sets. Thus, $A = (a_{ij})_{n \times m}$ and $B = (b_{ij})_{n \times p}$, $dist$ being the Euclidean metric. Let us introduce two more helping vectors, $y = (y_i)_{p \times 1}$ and $z = (z_i)_{m \times 1}$, then compute $\|y\|_p = \max_{0 \le i \le p} |y_i|$ and $\|z\|_p = \max_{0 \le i \le m} |z_i|$. With these notations, we get:

$$d(A,B) = \max \left\{ \sup_{\|y\|_p \le 1} \inf_{\|z\|_m \le 1} \|By - Az\|, \sup_{\|z\|_m \le 1} \inf_{\|y\|_p \le 1} \|By - Az\| \right\} \tag{2}$$

This restriction based metric represents the Hausdorff distance between the sets $B(y : \|y\|_p \le 1)$ and $A(z : \|z\|_m \le 1)$ in the metric space R^n, so it can be written as:

$$d(A,B) = H(B(y : \|y\|_p \le 1), A(z : \|z\|_m \le 1)) \tag{3}$$

Next, after eliminating the terms y and z from the above formula, we finally obtain the following Hausdorff-based distance:

$$d(A,B) = \max \left\{ \sup_{1 \le k \le p} \inf_{1 \le i \le m} \sup_{1 \le i \le n} |b_{ik} - a_{ij}|, \sup_{1 \le i \le m} \inf_{1 \le k \le p} \sup_{1 \le i \le n} |b_{ik} - a_{ij}| \right\} \tag{4}$$

The resulted nonlinear function d verifies main distance properties: the positivity, the symmetry and the triangle inequality. While not representing a Hausdorff metric anymore, the Hausdorff-based distance d constitutes a satisfactory discriminator between the feature vectors, therefore it could be successfully used in the next classification processes.

3   Text-Dependent Speaker Recognition Approach

Let us now consider a vocal signal S to be featured. Our feature vector extraction approach uses a short-time analysis of the involved audio signal ([4],[5],[8]). Thus, the signal is divided in overlapping frames having the length 256 and overlaps of 128 samples. Then, each resulted segment is windowed, by multiplying it with a Hamming window of length 256. The spectrum of each windowed sequence is then computed, by applying *FFT* (*Fast Fourier Transform*) to it. The cepstrum of each windowed frame $s[n]$ is then computed as:

$$C[n] = IFFT(\log|FFT(s[n])|)) \tag{5}$$

where *IFFT* represents the Inverse FFT. Next we use the mel-scale, which translates regular frequencies to a scale that is more appropriate for speech. It is described as:

$$mel\,(f) = 2595 \cdot \log_{10}(1 + f / 700) \tag{6}$$

Thus, a sequence of mel-frequency cepstral coefficients (*MFCCs*) is obtained for each frame. Each such MFCC set represents a melodic cepstral acoustic vector. Next, a derivation process is performed on these MFCC acoustic vectors.

Delta mel cepstral coefficients (*DMFCC*) are computed as the first order derivatives of mel cepstral coefficients. Then, the *delta delta mel frequency cepstral coefficients* (*DDMFCC*) are obtained as the second order derivatives of MFCCs. These derivative processes are used because of the intra-speaker variability, telling us how fast the voice of the speaker is changing in time ([8]).

Thus, a set of DDMFCC acoustic vectors result for the initial voice signal. Each of them is composed of 256 samples, but the speech information is codified mainly in the first 12 coefficients. So, each acoustic vector is truncated at its first 12 samples and then it is positioned as a column of a matrix.

The resulted DDMFCC acoustic matrix constitutes a powerful speech discriminator which works successfully as a feature vector for the processed vocal sound. Let us note $V(S)$ the feature vector of the speech signal S.

Each feature vector has 12 rows and a number of columns depending on the length of the speech signal. Therefore, because of their different dimensions, these speech feature vectors can be compared only using a nonlinear metric like that introduced in the previous section ([4]-[6],[8]).

We propose a supervised classifier for our voice (speaker) recognition system ([1],[10]), providing a *minimum mean distance classification* approach ([4],[5],[8]). The training set of our classifier contains a collection of spoken utterances, generated by the registered (advised) speakers. Each vocal utterance from the training set constitutes a vocal prototype and represents the same speech.

If we consider N advised speakers, then the resulted training set receives the form $M = \{M_1,..., M_N\}$, where each $M_i = \{s_1^i,..., s_{n(i)}^i\}$ represents the set of signal prototypes corresponding to the ith speaker. For each s_j^i, where $i = \overline{1, N}$, $j = \overline{1, n(i)}$, the previously described vocal feature extraction is then performed, and $\{\{V(s_1^1),..., V(s_{n(1)}^1)\}, ..., \{V(s_1^N),..., V(s_{n(N)}^N)\}\}$ represents the feature training set of our classifier.

Then we consider a sequence of input vocal utterances having the same speech, to be classified by the speaker criteria, using these training feature vectors (*prototype vectors*). Let it be $\{S_1,..., S_n\}$, each signal S_i, for $i = \overline{1, n}$, corresponding to the same spoken text (speech). The feature extraction process is performed on them, the feature set $\{V(S_1),..., V(S_n)\}$ thus being obtained.

The classification approach provided by us, the *minimum mean distance classifier*, represents an extended variant of the *minimum distance classifier* ([1],[10]). Thus, there must be N classes, each of them corresponding to a different registered speaker.

Our procedure inserts each input vocal sequence in the class of the closest registered speaker, that is the speaker corresponding to the smallest mean distance between the feature vector of the input signal and the prototype vectors of the speaker.

Thus, the mean distance between the input S_i and the training subset M_j, related to the jth speaker, is computed as $\dfrac{\sum_{k=1}^{n(j)} d(V(S_i), V(s_k^j))}{n(j)}$. Therefore, the closest speaker is identified as the p_i^{th} registered speaker, where:

$$p_i = \arg\min_j \frac{\sum_{k=1}^{n(j)} d(V(S_i), V(s_k^j))}{n(j)}, \ \forall i \in [1, n] \tag{7}$$

where d is the Hausdorff - based metric given by relation (4). Obviously, signal S_i has to be inserted in the p_i^{th} class, where $p_i \in [1, N]$

This classification result, consisting of N classes of vocal utterances, represents also the result of the speaker identification process. For each input speech, the closest advised speaker is thus identified. The next stage of the recognition process, the speaker verification, must decide if that identified speaker is the one who really produced it.

Therefore, a verification operation should be performed for each previously obtained speaker class. Let these classes be $C_1,..., C_N$. There are various techniques which could be used for this purpose. The most used of them are the *threshold-based* methods and we provide such an approach too ([4],[8]). Thus, we set a threshold value T and then compare the resulted minimum mean distance values with it, as in:

$$\forall i \in [1, N], \forall S \in C_i : \frac{\sum_{k=1}^{n(i)} d(V(S), V(s_k^i))}{n(i)} \leq T \tag{8}$$

If the condition (8) becomes true for a voice sequence S and a class C_i, then the utterance S is accepted by the recognition system as an advised vocal input generated by the ith registered speaker. Otherwise, S is rejected by our system, as being provided by an unadvised speaker. The speaker verification procedure ends when all spoken utterances produced by unregistered system users are rejected.

Any threshold-based recognition approach implies the task of choosing a proper threshold value. We propose an automatic threshold detection method, considering the overall maximum distance between any two prototype vectors belonging to the same training feature subset, as a threshold value ([4]). Consequently, a satisfactory threshold is obtained from the following relation:

$$T = \max_{i \leq N} \max_{k \neq t} \sum_{k=1}^{n(i)} d(V(s_k^i), V(s_t^i)) .$$

(9)

4 Text-Independent Speaker Recognition Technique

Text-independent recognition systems involve impressing volumes of training data ensuring that the entire vocal range is captured. It is useful for not cooperative subjects (for example in surveillance systems).

The most successful speech-independent recognition methods are based on *Vector Quantization* (*VQ*) or *Gaussian Mixture Model* (*GMM*). The *VQ*-based methods are parametric approaches which use *VQ* codebooks consisting of a small number of representative feature vectors [6], while the GMM-based methods represent non-parametric techniques using K Gaussian distributions [7].

In our previous work [4,5,9], our research concentrated also on speaker recognition methods based on *vowel characteristics*. We used the AR coefficients together with an ART classifier for the speaker recognition task ([5]).

The main steps of an autoregressive schema we designed as automatic speaker recognition tool are as follows: compute the signal spectrum, by a classical Fourier Transform (FFT), then plot the temporal evolution of spectral energy, detect the maxima and the inflection points by inverse descent procedure; select a number of 4-7 frequency bands, having the limits bounded by the inflexion points. On each band is practiced a zero phase filtering. In fact the selected bands correspond this way to the formantic areas of main vowels most often presented in a certain language, and the segments that interest us more are situated in high frequencies, that proved to be more selective in speaker recognition. The AR coefficients are computed on each band, their values constituting the training set for a clustering structure. To them, we added, in more trials different complementary information on the respective band, as we previously discussed, with various results. When the classification results from all available bands are combined, the performance is comparable to conventional recognizers. We obtained results ranging from 80% up to 90%, depending on the recognition structure and on the number of coefficients used as features on each channel. Also, we could obtain a satisfactory text-independent speaker recognition approach by modifying the text-dependent recognition method described in the last subsection [8]. Thus, we use the same DDMFCC analysis in the feature extraction stage. Other speech-independent voice recognition methods, such as those based on Vector Quantization, utilizes MFCC-based unidimensional feature vectors ([6]). The speech feature extraction approach proposed by us creates bidimensional vocal feature vectors, as *DDMFCC* – based truncated acoustic matrices.

A supervised minimum-distance classifier is also used in the next stage of the recognition process. The difference is that this time we use another kind of training set. Thus, we consider a large spoken text, containing all the English language phonemes. Each registered speaker should provide this speech several times ([8]).

The formula (7) is also used in the classification process, then the voice verification procedure is performed. We propose a threshold-based verification technique, but not an automatic one, like in the previous case. The threshold value T, from the relation (8) is chosen empirically ([8]).

5 Experiments

We performed a lot of numerical experiments on various speech datasets, successfully testing the work of our provided speaker recognition systems. A high recognition rate, approximately 85%, has been obtained for both DDMFCC – based technologies. Let us describe a simple text-independent voice recognition experiment. We consider 3 registered speakers and a long sequence of words, containing all the English phonemes, to be spoken by each of them.

The training set contains 4 vocal utterances having that sequence of words as text. We have one recording for the first advised user, two recordings for the second user and one recording for the last one. The training set of the system, and the corresponding training feature vectors represented as color images, are displayed in Fig. 1.

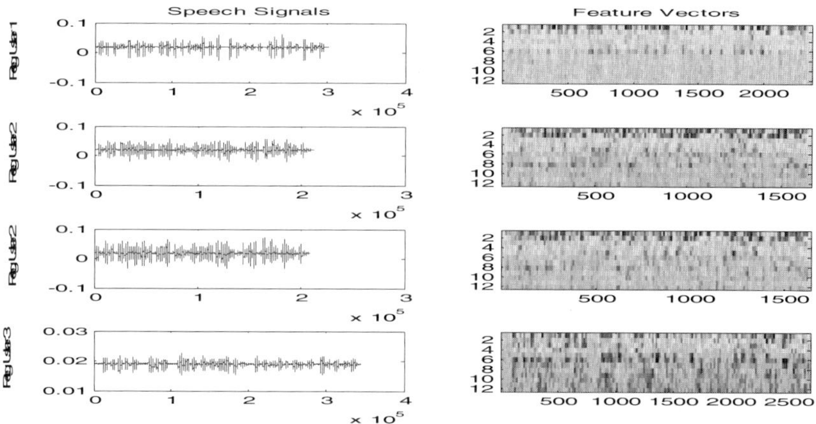

Fig. 1. The prototype speech signals and their feature vectors

Then, we consider a sequence of 9 input speech utterances to be recognized, each of them having a different spoken text. Their signals S_i and the feature vectors $V(S_i)$ are represented in the next two figures.

From Table 1, where the computed mean distance values are registered, we obtain the identification result: Speaker 1 $=>\{S_2, S_6, S_9\}$, Speaker 2 $=>\{S_1, S_3, S_7\}$ and Speaker 3 $=>\{S_4, S_5, S_8\}$. After setting the threshold $T = 7.5$ and applying the relation (8), we get the final recognition: Speaker 1 $=>\{S_2, S_6, S_9\}$, Speaker 2 $=>\{S_1, S_3, S_7\}$, Speaker 3 $=>\{S_4, S_5, S_8\}$ and Unregistered Speaker $=>\{S_5\}$.

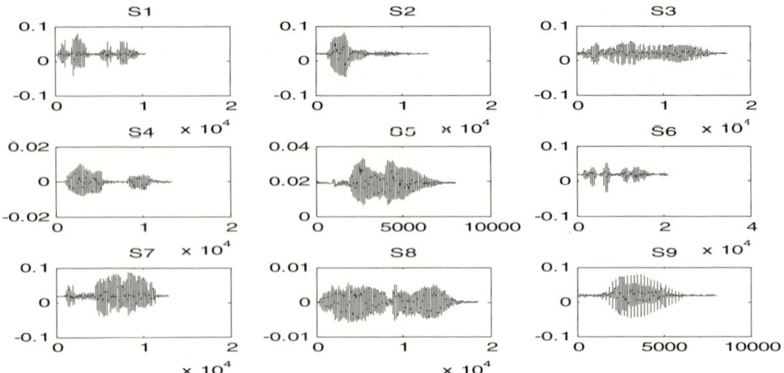

Fig. 2. The input speech signals

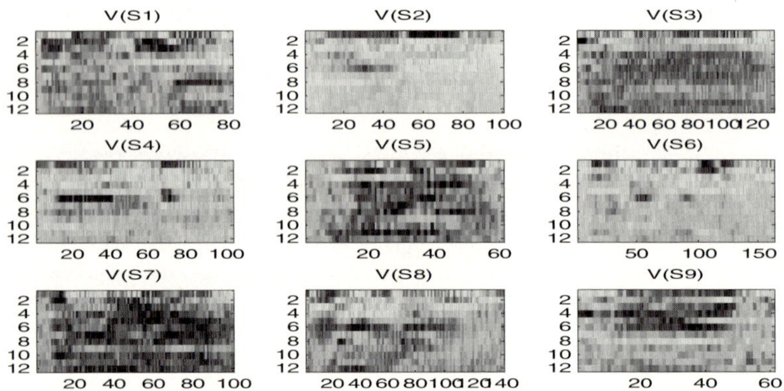

Fig. 3. The speech feature vectors

Table 1. Mean Distance Values

	Speaker 1	Speaker 2	Speaker 3
Input 1	5.6678	3.4676	6.2476
Input 2	4.1606	7.2581	6.4327
Input 3	5.9543	3.8976	4.3671
Input 4	6.0946	4.7342	2.9857
Input 5	10.6853	9.7366	8.6545
Input 6	5.1522	5.7855	6.3879
Input 7	7.5031	4.8871	10.8775
Input 8	8.7624	7.9964	5.8976
Input 9	3.2465	8.0245	4.9082

6   Conclusions

We have focused on the speaker recognition domain in this paper. A voice recognition system makes it possible to use one or more identifiers to verify the person

identity and control access to various security services such as voice dialing, banking by telephone, database access services, information services, voice mail, security control for confidential information areas or remote access to computers.

We have approached two speaker recognition types in this paper, speech-dependent and speech-independent respectively, and obtained satisfactory results. The main contributions of our work are as follows: the Hausdorff derived metric, the bidimensional DDMFCC feature vectors and the minimum mean distance classifier.

References

1. Cole, R.A., Mariani, J., Uszkoreit, H., Zaenen, A., Zue, V.: Survey of the State of the Art in Human Language Technology. Cambridge University Press, Cambridge (1997)
2. Squires, B., Sammut, C.: Automatic speaker recognition: An application of machine learning. In: Proceeding of the 12th International Conference on Machine Learning (1995)
3. Toutios, A., Margaritis, K.G.: Development of a Text-Dependent Speaker Identification System with the OGI Toolkit. In: Vlahavas, I.P., Spyropoulos, C.D. (eds.) SETN 2002. LNCS (LNAI), vol. 2308, pp. 525–530. Springer, Heidelberg (2002)
4. Barbu, T.: Speech-dependent voice recognition system using a nonlinear metric. International Journal of Applied Mathematics 18(4), 501–514 (2005)
5. Barbu, T., Costin, M.: Comparing Various Automatic Speaker Recognition Approaches. In: Proceedings of Symposium of Electronics and Telecommunications, ETC 2004, 6th edn. (2004)
6. Bagge, N., Donica, C.: ELEC 301: Final Project Text Independent Speaker Recognition. ELEC 301 Signals and Systems Group Projects (2001)
7. Reynolds, D.A., Rose, R.C.: Robust text-independent speaker identification using Gaussian mixture speaker models. IEEE Trans. Speech Audio Processing 3(1), 72–83 (1995)
8. Barbu, T.: A supervised text-independent speaker recognition approach. In: Proceedings of the 12th International Conference on Computer, Electrical and Systems Science, and Engineering, CESSE 2007, vol. 22, pp. 444–448 (July 2007)
9. Bălaş, V., Costin, M., Bălaş-Timar, D.: Automatic Voice Emotional Evaluation. In: Proceedings of the 6th International Conference on Recent Advances in Soft Computing, Canterbury, United Kingdom, p. 80 (July 9-10, 2006)
10. Duda, R., Hart, P., Stork, D.G.: Pattern Classification. John Wiley & Sons, Chichester (2000)

A Collaborative Information System Architecture for Process-Based Crisis Management

Omar Tahir[1], Eric Andonoff[1], Chihab Hanachi[1], Christophe Sibertin-Blanc[1],
Frédérick Benaben[2], Vincent Chapurlat[3], and Thomas Lambolais[3]

[1] University of Toulouse, IRIT Laboratory, 2 rue du Doyen Gabriel Marty,
31042 Toulouse Cedex, France
{tahir,andonoff,hanachi,sibertin}@univ-tlse1.fr
[2] DRGI, Ecole des Mines d'Albi-Carmaux, Route de Teillet, 81013 ALBI Cedex 9, France
{Frederick.Benaben}@enstimac.fr
[3] Laboratoire de Génie Informatique et d'Ingénierie de Production (LGI2P)
Parc Scientifique G. Besse, site EERIE-EMA, 30035 Nîmes cedex 1, France
Vincent.Chapurlat@ema.fr, Thomas.Lambolais@ema.fr

Abastract. In computer-based crisis management, participating actors have to put in common their information systems (IS). One of the main issues they have to face is to interoperate and coordinate these ISs in order to ease their collaborations and take the right decisions at the right moment. For that, they have to set up a Collaborative Information System (CIS) for orchestrating a Collaborative Process (CP) whose activities are performed by the participating ISs. The aim of this paper is to define the functional requirements of such a CIS and then to propose a conceptual architecture meeting these requirements. This architecture supports a perception-decision-action iteration for crisis reduction. Indeed, it provides means to capture information about the impacted real world, to support the definition and the adaptation of the CP managing the crisis reduction, and to coordinate and assign to each participant the actions to be undertaken. These actions modify the real world and lead to perform a new iteration until the crisis resolution.

Keywords: Architecture, Collaborative Information System, Collaborative Process, Adaptiveness, Crisis Management.

1   Introduction

In crisis situations, participating actors/organizations have to act simultaneously and in emergency for reducing the crisis and its impacts on the real world. To achieve this common goal efficiently, these actors must collaborate or at least act in a coordinated way in order to make their activities as efficient as possible.

Crisis coordination is a difficult task since it requires to take into account activities' distribution and actors' autonomy (they decide by themselves how to perform their missions and services) and to combine their competences in a flexible way in order to meet the evolving requirements of the crisis. In order to support such complex coordination, organizations have developed several computer-based tools such as GIS [9] or Collaborative Tools [5].

I. Lovrek, R.J. Howlett, and L.C. Jain (Eds.): KES 2008, Part III, LNAI 5179, pp. 630–641, 2008.

For instance, in the framework of the IsyCri ANR[1] (French Research Agency) project, we collaborate with the "Préfecture" of the Tarn[2] in order to develop a Collaborative Information System helping them to manage crisis. Within this framework, as soon as a crisis occurs, a crisis cell is created and leaded by the "Préfet"[3]. This crisis cell is made up of the representatives of the different actors implied in the crisis resolution [6]. In such a context, the coordination role of the crisis cell is to elaborate a collaborative process, for crisis response, synthesizing different plans of actions proposed by the participating partners (police, military forces, medical organizations...). Doing so, the crisis cell avoids activities' inconsistency and redundancy, and ensures the convergence of the different activities towards the common goal. This need for coordination is even important in cases of crisis that occur in an ill-structured context. For instance, in humanitarian crisis, the actors' coordination helps to maximize the services they offer to the injured persons, refugees, and so on, but also to reduce time of their interventions and to increase their reactivity. Besides, Non Governmental Organizations (NGO) or organizations in charge of coordination such as OCHA (Office for the Coordination of Humanitarian Affairs), systematically have been using platforms for information exchange and sharing (e.g. http://www.reliefweb.int). Then, the participating actors can better position their missions both geographically but also with regard to the activities they undertake

This work is developed in the framework of the IsyCri ANR project [6]. It adopts a computer-based approach to support coordination and it is based on three assumptions: (i) a Collaborative Information System (CIS) should be set up to support crisis resolution, (ii) the partners' Information Systems should be integrated to the CIS, and (iii) crisis resolution is process-oriented: a Collaborative Process (CP) is defined for specifying the coordination of partners' activities.

The coordination of the actions of the different partners involved in a crisis resolution is supported by the CIS which can be seen as an Inter-Organizational Information System [8] acting as a middleware component between the ISs of the involved partners.

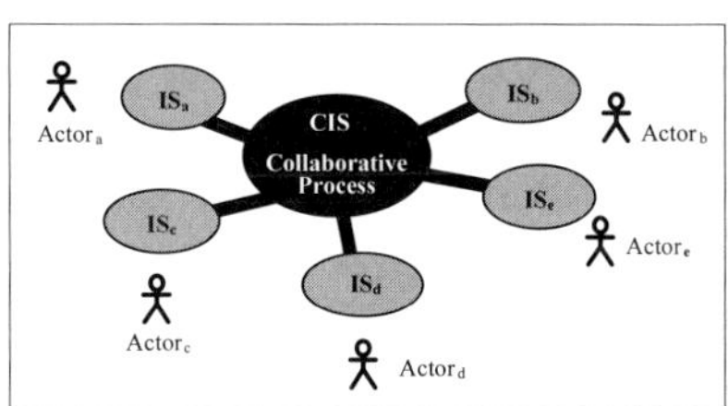

Fig. 1. Collaborative Information System

The CIS (see figure 1) is responsible for the management (definition, adaptation and orchestration) of the CP and it requires the partners' ISs to perform the missions or services they have in order to reduce the crisis. Doing so, the different partners master their missions and services and consequently the autonomy of their ISs is preserved.

The aim of the paper is to propose a CIS architecture for crisis management. More precisely, the main research questions addressed in this paper are: which are the requirements that a CIS must meet? And how does a CIS architecture for crisis management look like?

The proposed architecture supports a perception-decision-action iteration for crisis reduction. Indeed, it provides means to capture information about the impacted real world, to support the definition and the adaptation of the collaborative process managing the crisis reduction, and to coordinate and assign to each participant the actions to be undertaken. These actions modify the real world and lead to a new iteration, where these tasks are performed again, taking into account the new crisis context.

The paper is organized as follows. Section 2 presents the requirements of a collaborative process-based information system. In section 3, we give the CIS conceptual architecture. First, we provide an overview of how the partners' ISs can be integrated with the CIS. Then, we detail its internal architecture by giving its functional components, its kinds of users (roles) and its databases. Section 4 presents the dynamics of our architecture, i.e. how its different components interact, through several UML sequence diagrams. Section 5 discusses our proposition and concludes the paper.

2   Requirements for a Collaborative Information System

The IsyCri project aims at developing a framework and a software prototype namely the CIS for crisis management. From the study of historical information about crisis, the way they have been managed, and on the basis of meetings, interviews and workshops with actors involved in crisis cells (the potential CIS' users), we have identified four major objectives to be met by the CIS:

– Providing an overview on the crisis and its evolution.
– Providing reliable information about the means (plans, processes, resources …) that are available to resolve the crisis.
– Ensuring communication interoperability: increasing interaction capability between the information systems of the implied partners.
– Allowing the coordination of the activities of the actors that intervene in the crisis resolution: supporting the orchestration of the dynamic Collaborative Process followed for managing the crisis.

Hereafter, we present the main functional requirements of the CIS. However, for the real system (and not for the prototype), non functional requirements are of paramount importance: in particular, reliability, safety and real-time constraints purposes should be taken into account, since the system is designed to run in a very disrupted environment (with unreliable data connections in particular). In particular, the CIS supports the Collaborative Process, helps organization partners in the crisis management, but it cannot make any decision on their behalf. Actors remain the pilots, in any circumstances.

Let us precise some vocabulary and the kinds of users involved in a CIS in order to clarify the functional requirements. The *Impacted Real World* is the observed system. The *Treatment System* is the added-on system set up to resolve the crisis. It includes Actors and their ISs, the CIS, and the Collaborative Process. A *Crisis Model* is a representation of a crisis in accordance to a Crisis meta-model given in [6]. A *Crisis Cell* is a group of people in charge of the crisis management by using the CIS. Three kinds of users (roles) of a CIS have been identified. A crisis cell generally includes one person who is globally and ultimately responsible for the proper management of the crisis (e.g. in our project, the "Préfet"); his activity corresponds to the *Pilot* role, that can be fulfilled by any person delegated for this. A crisis cell also includes one person for each of the civil institutions involved in the crisis management (e. g. the police, firemen, water supply, hospital...); this person is the representative of his institution and stands as the *Partner* role. Finally, we also consider the *Crisis Model Manager* role that handles the observed information about the situation coming from the partners' organizations and is responsible for updating the Crisis Model according to these information.

From the previous objectives assigned to the CIS, we have derived four main functional requirements to be met. The CIS should give means to:

– *Manipulate and share a dynamic crisis representation*. Four services should be provided for that purpose. First, the CIS should give access to the crisis model and views on it to Partners. Second, it should help the partners, via simulation or analysis, to assess the situation of the Impacted Real World. Third, the CIS should give means to update the Crisis Model. Finally, the CIS should offer tools for validating and verifying the crisis model (notably its consistency).

– *Define, adapt, configure and enact the crisis collaborative process*. All these functions are to be performed by the Pilot of the CIS. The definition consists in creating the collaborative process from the Crisis meta-model which contains both descriptions of past crisis and the way they have been resolved. This Collaborative process is obviously adapted according to the evolution of the crisis. The configuration consists in granting rights to the different partners in order to allow them to perform some activities of the Collaborative Process. Finally, the enactment corresponds to the execution of the Collaborative Process, where the CIS should have means to access to the partners' Information Systems in order to extract information or submit activities to be performed. Except for the adaptation function, the other ones could be supported by a conventional Workflow Management System [1].

– *Manage data*. This function is offered to each Partner. It should give means to retrieve data about the impacted real world, to build scoreboards in order to provide synthetic information for risks evaluations, and to simulate the consequences of possible actions.

– *Administrate the CIS*. This function allows the Pilot to define, configure and modify the parameters required to connect the partners' Information Systems to the CIS and give them means to communicate with one another (chat, forum...).

3 Conceptual Architecture of a Collaborative Information System

The purpose of this section is to present the conceptual architecture of a CIS for crisis management. This architecture is independent from any implementation consideration. A CIS connects the different ISs of partners involved in a crisis resolution. This architecture is presented in top-down way. First, we give an overview of how the partners' ISs can be integrated with the CIS. Then, we give a more detailed view of the CIS focusing on its internal architecture and presenting its functional components, and the involved databases. The dynamic aspects will be presented latter.

3.1 Integration of the Partners' ISs with a CIS

Figure 2 shows the three components involved during a crisis: the Impacted Real World, a set of ISs belonging to the different partners involved in the crisis resolution and a CIS which coordinates the actions undertaken by these partners.

More precisely, the CIS manages a Collaborative Process (CP) which describes the orchestration of the actions to be undertaken in order to reduce the crisis. Moreover, the CIS does not directly access or act on the impacted real world but rather uses the partners' ISs to do that. In fact, the partners' IS are the "eyes and hands" of the CIS on the impacted real world. Indeed, in a crisis situation, the impacted real world is initially observed. This observation, which is done via the partners' ISs, is modelled and represented in the CIS by the *Crisis Model*. From this Crisis Model and using what we call the *Treatment Model*, which models a set of actions (and their coordination) to be undertaken when a given situation or a given event is detected, a CP is then built in response to the crisis. This CP is then orchestrated by the CIS which orders the different participating partners to perform actions. These actions modify the real world which is observed at any time. Then, the CP can possibly be adapted or totally changed according to the new perception of the crisis, as mentioned, it is an iterative process.

Figure 2 also shows the three activities of a CIS: *Observation* of the Impacted Real World (Observe and get Observations arrows), *Deliberation* in order to build a CP which is convenient to reduce the crisis (Deliberate arrow) and *Act* on the Impacted Real World (Ask for Performing Actions and Perform Actions arrows). These three activities are repeated until the resolution of the crisis.

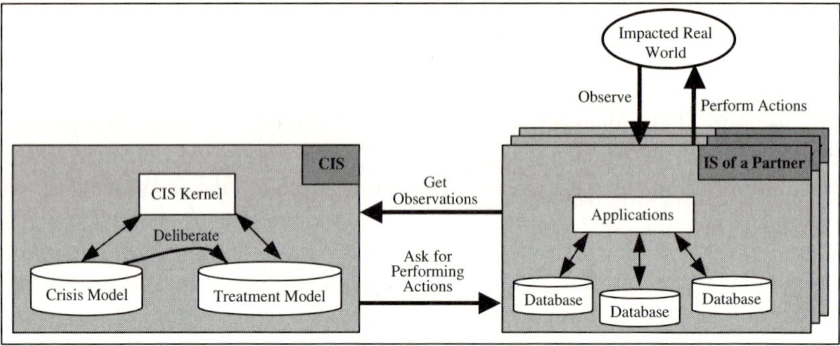

Fig. 2. Global View of a CIS

The CIS kernel is composed of a set of functional components which manage the CP, the Crisis Model and the Treatment System and support the Deliberation activity. These components will be detailed in the next section.

3.2   Internal View of a CIS

We now focus on the internal architecture of a CIS presenting the involved databases, the functional components and the interfaces that make it up. Figure 3 below presents this internal architecture, which meets the functional requirements presented in section 2 and is organized around:

– 3 databases, each one represented by a cylinder.
– 7 functional components, each one represented by a rectangle.
– 10 interfaces supporting the interaction between functional components and kind of users. Interfaces are represented by arrows.

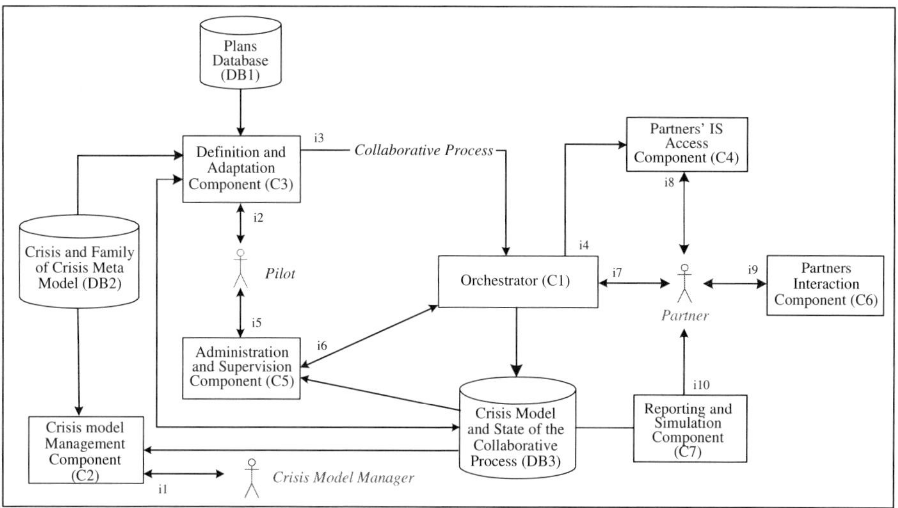

Fig. 3. Internal View of a CIS

We distinguish three databases, DB1, DB2, DB3, storing respectively:

– A set of processes, each one defining a set of coordinated activities to be undertaken when a given situation or event occurs (Treatment Model of figure 2).
– A Crisis and Crisis Family Meta Model.
– The actual representation of the impacted real world (Crisis Model of figure 2) and the current state of the CP.

DB3 is used to have an accurate representation of both impacted real world and undertaken actions, while DB2 and DB1 are used for the Deliberation activity in order to build a convenient CP for crisis reduction. In fact, DB2 and DB1 are built from crisis ontologies [6] defining respectively the crisis concepts, their links and a crisis classification (civil, humanitarian, social…) regarding DB2, and, regarding DB1, the plans and actions to undertake in order to resolve each type of crisis. Concerning DB3, the crisis model corresponds to an instance of these ontologies. These latters have been implemented in OWL (http://www.w3.org/TR/owl-features/) using Protégé software (http://protege.stanford.edu/).

Regarding the functional components, the *Orchestrator* component (C1) is the main CIS component. Its role is to orchestrate the CP. It consults the partners' ISs via interface I4, interacts with a partner to ask for the execution of services via interfaces I7 and I8, and receives reports on the execution state of the requested services via interface I7. The Orchestrator remains under the control of the Pilot through the *Administration and Supervision* component (C5), at the same time as it stores the state of the CP in the DB3 database. The *Crisis Model Management* component (C2) serves to build and maintain a model of the crisis, as reliable and up-to-date as possible in the DB3 database. The *Definition and Adaptation* component (C3) permits the Pilot to build the CP at the beginning of the crisis, but also to check this CP and adapt it according to the evolution of the Crisis Model (DB3) and the Treatment Model (DB2 and DB1).

The *Administration and Supervision* component (C5) aims at giving rights to partners, monitors the orchestration of the CP and makes the guidance of CP execution by the pilot possible (alternative, choice, …). The *Partners' IS Access* component (C4) allows the CIS to be connected with the partners' ISs. It transmits to the partners the requests for missions and services execution. The *Partners Interaction* component (C6) provides communications services (chat, forum…) helping to take the appropriate decision on the processing system including the CP. Finally, the *Reporting and Simulation* component (C7) can be used to visualize both the impacted real world and the CP state. It can also be used to simulate the crisis evolution and its consequences.

4   The Dynamics of the CIS: How to Use the CIS to Manage the Crisis

Figure 3 shows the functional architecture of a CIS: the main services it offers to the crisis cell members and the information it needs to manage. In this section we present its dynamics, that is how the components of this architecture are used and interact to manage a crisis. First of all, we have to notice that the controlled management of a crisis includes two sub-processes:

– A *Collaborative Process* (CP) performed by the *operational* agents of the Partners' organizations, which act on the field to secure population and goods, prevent the occurrence of new disastrous events, repair the damages, rescue people in danger and so on.
– A *Control Process* performed by the members of the crisis cell, that consists in collecting information about the actual situation of the crisis and giving appropriate orders to the operational agents of the partners' organizations in charge of fighting the crisis concretely; in fact, the control process is a kind of meta-process since it aims at defining the Collaborative Process (CP), adapting or reshaping it according to the evolution of the situation and supervising its execution.

The CIS does not directly intervene in the crisis field. Instead, it is the tool used by the crisis cell members to manage the crisis control process. Thus, the main part of the SIC's dynamics is related to the way it supports the crisis control process.

A description of this control process is shown in Figure 4 as a UML2 interaction overview diagram [12]. It is an abstract view of the process that stands at the same level as the functional architecture in Figure 3. Once the crisis cell for the crisis management is set up, the control process includes five tasks: two for managing the crisis model (a representation of the impacted real world) shared by the crisis cell members, two for defining and adapting the structure of the CP according to the crisis model, and one for supervising the CP.

The *Crisis_Model_Initialization* task collects information about the crisis from the partners and, among the known crisis models stored in the DB2 database (Crisis and Family of Crisis meta-model), selects the one that best matches to the situation. The *Crisis_Model_Management* task aims at maintaining the crisis model in conformance with information about the impacted real world; in case of an expected evolution, it is just a matter of updating the values of some parameters or variables, while in case of an evolution that changes the nature of the crisis, a new crisis model has to be built in the same way than in the initialization task. In both cases, the Orchestration task presented below is stopped.

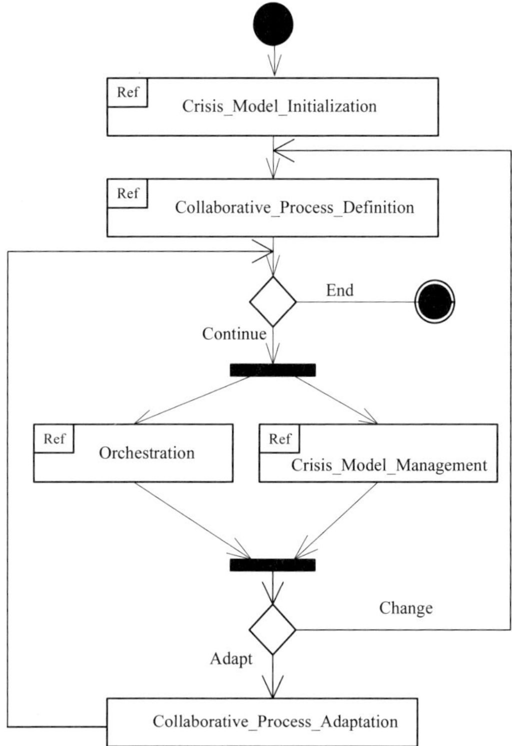

Fig. 4. UML overview diagram of the Control Process

The *Collaborative_Process_Definition* task consists in finding the most appropriate actions to reduce the crisis; the crisis model gives the expected results of these actions, while the DB1 database (Plan database) contains the action plans and means that can be enacted by the partners according to their respective responsibilities. Thus, defining the CP is a matter of finding the best match between what should be done to reduce the crisis and what can be done by the partners, and to coordinate the activities the partners are requested to perform.

The *Collaborative_Process_Adaptation* task accounts for the updates of the crisis models that require only little changes in the structure of the CP such as removing or adding actions, using information to select the most relevant alternative or reordering some actions in a more positive way.

Finally, the *Orchestration* task is in charge of supervising the execution of the CP. It is performed by the Orchestrator component, a workflow engine that sends to the partners (or partners' Information Systems) the actions they have to perform and maintains the current state of the CP thanks to the execution reports supplied by the partners.

The structure of the control process described in Figure 4 should be read as follows. Once the Crisis_Model_Initialization is achieved by the Crisis Model Manager, the Pilot defines the initial CP. Then, the process *Continues* and, concurrently, the Orchestrator orchestrates the execution of this CP while the Crisis Model Manager deals with the available information to update the Crisis Model. In the *Adapt* case of a simple update, the orchestration of the CP is stopped in order to be updated and then a new < orchestration // Crisis_Model_Management > loop is started. In case of a *Change* in the Crisis Model, the orchestration of the CP is also stopped, and the Pilot enters a new performance of the Collaborative_Process_Definition task. If the resulting new CP is empty then the *End* signal provokes the termination of the control process since the crisis is considered as being solved, otherwise the *Continue* signal makes the control process enter in the loop again.

We have specified UML Sequence Diagrams that define standard scenarios of how the members of the crisis cell and the components of the SIC interact to perform each of these tasks.

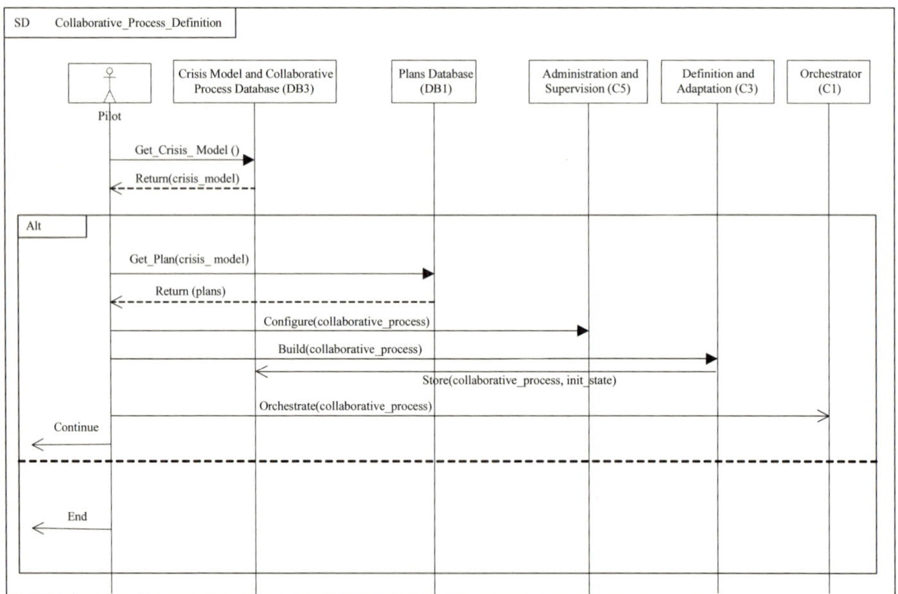

Fig. 5. Sequence Diagram of the Collaborative_Process_Definition task

Because of space limitation, we only present here two of these sequence diagrams for the Collaborative_Process_Definition and the Crisis_Model_Management tasks. These sequence diagrams are respectively visualized in figures 5 and 6. They are self-explanatory and we only comment on the included frames that appear in them. In figure 5, the *Alt* indication in the upper left corner of the included frame introduces an alternative between the two parts separated by the horizontal dashed line; that means that, according to the result returned by the DB3 database, either the Pilot builds and stores an effective collaborative process and emits a *Continue* signal, or he just emits an *End* signal that produces the termination of the control process. The same holds in the sequence diagram of figure 6, where the task ends either with a *Change* or with an *Adapt* signal.

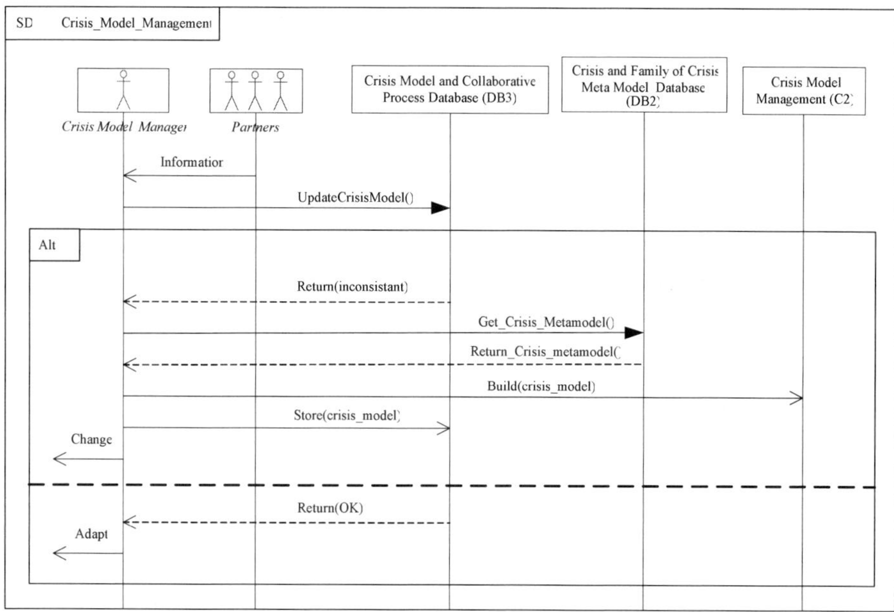

Fig. 6. Sequence Diagram of the Crisis_Model_Management task

5   Discussion and Conclusion

This paper has described a Collaborative Information System architecture for process-based crisis management. This architecture features three main advantages. First, it supports the collaborative-process design, enactment and adaptation. *Adaptation* is an innovative aspect regarding traditional workflow management systems [1] and very few works address this aspect [2]. In our architecture, adaptation is supported at two different levels: the crisis model may be changed and the collaborative process can be transformed. The second advantage is that *Context-awareness* is made possible thanks to databases able to record and structure information about the crisis, its evolution and possible reactions plans. Third, interfaces are defined to connect partners' information systems to the CIS. Regarding this aspect, interoperability issues in the context of Services Oriented Architecture are discussed in [6].

Regarding the state of the art, in the Crisis Information System field, even if a lot of information systems have been developed to support crisis management, most of them are specific to a type of crisis [3], focus on the influence of innovative tools [4], or are limited to simulation or geographic information management and visualization [7]. Collaborative tools (such as CSCW) have also been developed in the Crisis field [5]. Most often they are used specially to support interactions among partners and document sharing among them. Nevertheless, such tools do not adopt a process perspective. In this way, the closest work to ours is [11], developed in the context of the WORKPAD project (http://www.workpad-project.eu). In this work, a P2P architecture is designed from users requirements, and it includes data storage and communication, middleware and user layers. This architecture also manages processes and workflow patterns mining is made possible. What makes our architecture different from the one proposed in [11] is that our collaborative process is a first class citizen and also our architecture remains conceptual and does not impose any technological choice (P2P, Web Services, Grid, etc). Although our architecture is independent from any implementation consideration, a SOA approach fits in it [6]. The collaborative process may invoke activities implemented as web services registered in an UDDI and coordinated by an orchestration engine. An implementation on top of the open source PETALS [14] Platform is in progress. PETALS is an infrastructure that implements a SOA architecture with the Enterprise Service Bus (ESB) technology. On the one hand, PETALS supports mediation and interoperability between heterogeneous software components (thanks to adapters), and, on the other hand, it provides service orchestration engine based on standards like Business Process Execution Language (BPEL). In our context, PETALS supports the interaction between the partners' information systems, and also orchestrates the collaborative process. At design time the crisis collaborative process is defined with BPMN (Business Process Modeling Notation), which is a high level description language, while at run time it is transformed onto BPEL [13] which is an operational language and moreover supported by PETALS.

Although our conceptual framework is a solid basis for an implementation, there are some open issues, notably crisis process adaptation. This issue assumes a very deep study of the parameters to be adapted in a process. It may concern the process control structure, the organizational model inherent to the process, allocations policies of tasks to actors, among other aspects. The complexity of this issue is due to the fact that all these adaptations may be investigated in a dynamic context, i.e. at run time during the process execution.

Acknowledgments

This work has been supported by the ANR (French Research Agency) through the IsyCri project. The authors would like to thank the other project partners namely Thales Communication and EBM Websourcing.

References

1. van der Aalst, W., van Hee, K.: Workflow Management, Models, Methods, and Systems. MIT Press, Cambridge (2004)
2. Adams, M., ter Hofstede, A., Edmond, D., van der Aalst, W.: A Service-Oriented Implementation of Dynamic Flexibility in Workflows. In: 14th Int. Conf. on Cooperative Information System, Montpellier, France, pp. 291–308 (2006)
3. de Addams-Moring, R.: Tsunami Self-evacuation of a Group of Western Travelers and Resulting Requirements for Multi-Hazard Early Warning. In: 4th Int. Conf. on Information System for Crisis Response and Management, Delft, The Netherlands (2007)
4. Avery-Gomez, E., Turoff, M.: Interoperable Communication: an Analysis of SMS Text-Message Exchange. In: 4th Int. Conf. on Information System for Crisis Response and Management, Delft, The Netherlands (2007)
5. Cai, G.: Extending Distributed GIS to Support Geo-Collaborative Crisis Management. Geographic Information Science 11(1), 4–14 (2005)
6. Couget, P., Benaben, F., Lauras, M., Hanachi, C., Chapurlat, V.: Information Systems Interoperability as a way to Partners Integration in a Crisis Context. In: 20th Int. Conf. on Software and Systems Engineering and their Applications, Paris, France (2007)
7. Dilekli, N., Rashed, T.: Towards a GIS Data Model for Improving the Emergency Response in the Least Developing Countries: Challenges and Opportunities. In: 4th Int. Conf. on Information System for Crisis Response and Management, Delft, The Netherlands (2007)
8. Divitini, M., Hanachi, C., Sibertin-Blanc, C.: Inter Organizational Workflows for Enterprise Coordination. In: Omicini, A., Zambonelli, F. (eds.) Coordination of Internet Agents, pp. 46–77. Springer, Heidelberg (2001)
9. Erl, T.: Service-Oriented Architecture: Concepts, Technology and Design. Prentice-Hall, Englewood Cliffs (2005)
10. Kumar, V., Bugacov, A., Coutinho, M., Neches, R.: Integrating Geographic Information Systems, Spatial Digital Libraries and Information Spaces for conducting Humanitarian Assistance and Disaster relief Operations in Urban Environments. In: 7th International Symposium on Advances on Geographic Information Systems, Kansas City, Missouri, USA (1999)
11. de Leoni, M., de Rosa, F., Marrella, A., Mecella, M., Poggi, A., Krek, A., Manti, F.: Emergency Management: from User Requirements to a Flexible P2P Architecture. In: 4th Int. Conf. on Information System for Crisis Response and Management, Delft, The Netherlands (2007)
12. Object Management Group UML 2.0 Specification (2005), http://www.omg.org
13. Ouyang, C., van der Aalst, W., Dumas, M., ter Hofstede, A.: From BPMN Process Models to BPEL Web Services. In: 4th Int. Conf. on Web Services, Chicago, USA, pp. 285–292 (2006)
14. PETALS (2008), http://petals.objectweb.org/

A Role-Based Framework
for Multi-agent Teaming

Jinsong Leng[1], Jiuyong Li[2], and Lakhmi C. Jain[1]

[1] School of Electrical and Information Engineering,
Knowledge Based Intelligent Engineering Systems Centre,
University of South Australia, Mawson Lakes SA 5095, Australia
Jinsong.Leng@postgrads.unisa.edu.au,
Lakhmi.Jain@unisa.edu.au
[2] School of Computer and Information Science,
University of South Australia, Mawson Lakes SA, 5095, Australia
Jiuyong.Li@unisa.edu.au

Abstract. Multi-agent teaming is a key research field of multi-agent systems. BDI (Belief, Desire, and Intension) architecture has been widely used to solve complex problems. The theory of joint behavior has been widely used to solve the team level optimisation problems. Due to the inherent complexity of real-time and dynamic environments, it is often extremely complex and difficult to formally specify the joint behavior of the team a *priori*. This paper presents a role-based BDI framework to facilitate cooperation and coordination problems. This BDI framework is extended and based on the commercial agent software development environment known as JACK Teams. A real-time 2D simulation environment known as soccerbots has been used to investigate the difficulties of multi-agent teaming. The layered architecture has been used to group the agents' competitive and cooperative behaviors, which can be learned through experience by using the reinforcement learning techniques.

Keywords: Agents, BDI Architecture, Cognitive Architecture, Role, Multi-agent Teaming, Decision Making.

1   Introduction

An intelligent agent is a hardware and/or software-based computer system displaying the properties of autonomy, social adeptness, reactivity, and proactivity [17]. Multi-agent systems (MASs) are composed of several agents to solve the complex problems, and usually operate in real-time, stochastic and dynamic environments. A team is a set of agents having a shared objective and a shared mental state [5,6]. Many theoretical and applied techniques have been applied to the investigation of agent architecture with respect to coordination, cooperation, adaptation, and communication, in order to provide a framework for implementing artificial intelligence and computing techniques. In addition, the issues of complexity, uncertainty, and incomplete knowledge should be considered when designing agent-based systems.

I. Lovrek, R.J. Howlett, and L.C. Jain (Eds.): KES 2008, Part III, LNAI 5179, pp. 642–649, 2008.

Agent-oriented Software Engineering (AOSE) is an active research area, which aims to simplify the construction of complex systems by using abstractions of nature [8]. For example, Open Agent Architecture (OAA) [4] is a team-oriented agent architecture for integrating a community of heterogeneous software agents in a distributed environment. STEAM (a Shell for TEAMwork) [16] is a hybrid teamwork model using the joint intension theory formalism, which is built on top of the Soar architecture [9].

A real-time strategic teamwork is described using the concepts of formations in [15]. Homogeneous agents are able to switch roles within formations at runtime. This flexibility increases the performance of the overall team. JACK [2] is an agent development environment with agent-oriented concepts, including: Agents, Capabilities, Events, Plans, Knowledge Bases (Databases), Resource and Concurrency Management. The JACK Teams [3] provides a hierarchical team-oriented framework using some unique concepts: teams, roles, and team plans. MadKit platform [7] provides general agent facilities in terms of organization (agent, group, role). An organization oriented MAS emphasises on capabilities and constraints, on organizational concepts such as roles (or function, or position), groups, tasks and interaction protocols.

This paper addresses the role-based BDI architecture, which is the extension to the commercial agent software development environment known as JACK Teams. The team tasks are specified by a set of roles to guide the joint behaviors coherently at the team level. Due to the dynamic nature of MASs, it is difficult to develop an agents' teamwork model applicable for various application domains in general. The agents' teamwork research aims to improve concept understanding, to develop some reusable algorithms, and to build high-performance teams in dynamic and possibly hostile environments. In this paper, a role-based BDI architecture is proposed for addressing the issues of multi-agent teaming. The extension to JACK Teams architecture has been made to form an agent team capable of dealing with incomplete information, adaption and learning. Finally, a role-based soccer team is implemented in a real-time 2D simulation environment known as soccerbots [1].

The rest of the paper is organised as follows: Section 2 introduces the major properties of roles and team formation. Section 3 details the JACK and JACK Teams; the extension to JACK Teams has then been presented. A soccer agents' team is implemented using this framework in Section 4. Finally, we discuss future work and conclude the paper.

2   Roles and Team Formation

2.1   Roles

There are several ways to form a agents' team. One option is to construct a team by assigning tasks and responsibilities on the fly. Another option is to specify the tasks by a set of roles. If the team structure is known in advance, we are able to assign the responsibilities and capabilities to different roles that may be performed in the team. The role-based approach can make the team to perform more efficiently with the minimum communication and negotiation cost.

The concept of roles has been widely used for enterprise modelling. The role is a set of responsibilities to be taken by team members. Lind [12] describes a role as a logical grouping of atomic activities according to the physical constraints of the operational environment of the target system. The major characteristics of a role are reported in [12].

Form agents' teaming perceptive point of view, a role is an extension of the agents current knowledge, and plays in a (joint) process like problem solving, planning or learning [12].

2.2   Roles Assignment and Team Formation

For a specific agent-based system, one needs to specify the tasks and responsibilities and then group them into roles. A set of roles is defined to perform the joint behaviors coherently at the team level. Each agent is designed capable of playing one or more roles in a team.

Each agent is also capable of playing several roles at the same time and each role may be performed by different agent. Team formation is the process to map roles onto agents using either static assignment or dynamic assignment. The static assignment implies that the relationship between roles and agents is fixed when a team is formed. The dynamic role assignment reallocates the roles to agents during running time. Static role assignment does not consider the role changes when the agents internal state changes, while dynamic role assignment needs some interaction protocol to guide the dynamic assignment process.

3   The JACK Teams Based BDI Framework

We summarise the major properties of the BDI framework for constructing a coherent agents' team, as follows:

- A coordinated cooperation by explicitly designing team plan;
- A hierarchical team structure for constructing team;
- Roles for defining the relationship and dependency between agents;
- Team formation by assigning roles statically or dynamically;
- Beliefs transformation for sharing knowledge.

3.1   JACK Teams Development Environment

JACK [2] is commercial software tool with some embedded functions and building blocks to help construct BDI agents. Within this platform, each agent is triggered by an event, resulting in updating its beliefs about the world and desires to satisfy, driving it to form intentions to act. An agent is an event-driven deliberation entity. An intention is a commitment to perform a plan. A plan is an explanation of how to achieve a particular goal, and it is possible to have several plans for same goal.

BDI agents exhibit reasoning behaviour under both pro-active (goal directed) and reactive (event driven) stimuli. Adaptive execution is introduced by flexible plan choice, in which the current situation in the environment is taken into account. A BDI agent has the ability to react flexibly to failure in plan execution. Agents cooperate by forming teams in order to achieve a common goal.

Some reasoning and deliberation functions are provided in JACK, for instance, alternative approaches can be applied in case a particular strategy (plan) fails. For example, JACK agents can be employed to control the players' actions in a simulation environment known as Unreal Tournament (UT) [13]. The execution process is illustrated in Fig. 1.

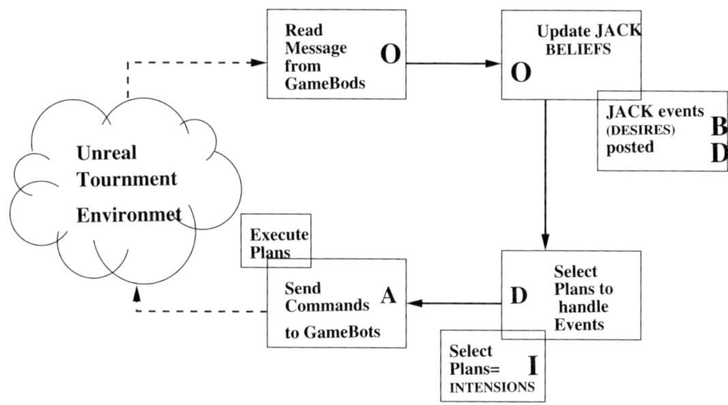

Fig. 1. JACK BDI Execution Model [13]

The JACK Teams [3] provides a hierarchical team-oriented framework using some unique concepts: teams, roles, and team plans. The focus of this paper is on the team programming model provided by JACK teams as a core technology. The JACK Teams model provides a natural way to specify team structure and coordinated behavior related to team structure in terms of BDI model entities. Roles are used to describe the requirements and capabilities.

The JACK Teams can provide the basic pro-active and reactive behaviors by using the integrated reasoning and deliberation functions. Joint behaviors are available by choosing the 'appropriate' plans. As discussed above, uncertainty is a common property and makes it infeasible to predict all possible situation a priori in dynamic and non-deterministic systems. Automation and learning are two significant challenges, especially for human-agent interactions or intra-agent team.

3.2 Extension to JACK Teams

The JACK Teams does not address the issues of adaption and learning. Furthermore, scaling up large state space is a challenge problem as well. The extension to JACK Teams with regard to learning ability and partial observability over large state space is of paramount importance for solving control problems in the complex and dynamic agent-based systems.

The team entity is able to coordinate the individual members' actions by find the optimal joint behavior. Some extensions to JACK Teams have been made so as to enhance the autonomy and adaption by learning through experience, outlined as follows:

- The on-line self-learning or self-evolving is particularly useful when designing a complex and dynamic agent-based system. A abstract class has been designed to enhance the on-line learning abilities. The decision theory can be applied to solve the optimisation control problems. In our previous work, we use temporal difference learning techniques to learn the control process in the game of soccer [10,11]. In this BDI framework, the team entity or sub-team entity can extend the abstract class LearningAgent.java, which includes some methods for on-line learning.
- The major difficulty for joint behavior is to deal with large state space. An abstract class State.java is used to extend the Belifset. Within JACK Teams platform, the Beliefset is a rational database to save the information and knowledge. The purpose of the abstract class State.java is to provide some function approximation techniques to deal with large state space.
- JACK Teams provides the functions to synthesize the beliefs of sub-teams. In the noisy environment, individual member may have incomplete or inaccurate information. We need some other techniques to deal with partial observation and information sharing problems.

4   Soccer Agents Team Implementation

4.1   Simulation System

SoccerBots is part of TeamBots(tm) software package [1], which provides a simulated environment for studying team-related issues. Each team can have 1–5 players. This system can be modified by adding some soccer control and learning strategies. The soccer game is a real-time, noisy, adversarial domain, which has been regarded as a suitable environment for multi-agent system simulations [14]:

- Real-time: it acts in a dynamically changing environment.
- Noisy: which affects both sensors and effectors.
- Collaborative: the agents must cooperate (play as a team) to achieve the jointly desired goal.
- Adversarial: it involves competing with opponents.

As indicated above, the aim of this paper is to provide a role-based BDI framework capable of investigating the following issues:

- Competitive skills: can the individual agents learn to intercept, shoot, dribble, clear;
- Communication and cooperative skills: when and how to pass; when and where to move to receive a pass.

- Coordination: forecasting where and when other individual players are liable to move.
- Learning and agent team architecture: is there an interaction between the individual learning and that necessary for the team to perform optimally.

4.2   Implementation

The JACK team entity acts like a coach who is designed to control the actions of a soccer player performed in the SoccerBots simulation environment. A team reasoning entity (team agent) is another JACK agent with its own beliefs, desires, and intensions. Every agent player can be regarded as a sub-team and performs a certain role that defines the relationship between team and sub-teams. The relation between soccer team and sub-team is illustrated in Fig. 2.

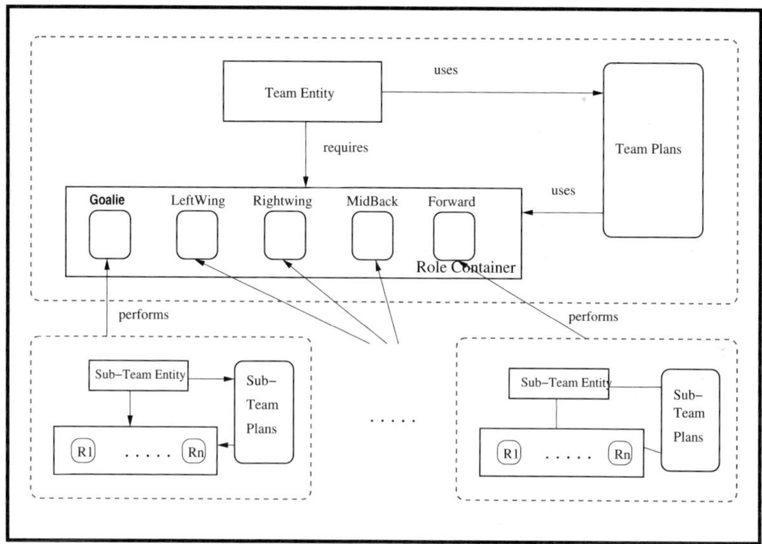

Fig. 2. JACK Soccer Team Framework

The JACK Teams forms a team by using the method @establish(). The team plan is used to control the actions of the soccer players performed in the SoccerBots simulation environment. The team controls the sub-teams in parallel by using the method @parallel().

In the real-time situation environment, performance is a high priority to be considered to follow the changes of motion of the ball and the players. The layered BDI agents are considered to operate the soccer players. Each sub-team entity plays a role in the soccer team. The plans in a sub-team entity are able to execute both reactive and proactive. For example, the forward can have the reactive actions to shoot, intercept, and pass the ball. Meanwhile, it has some plans to proceed the proactive actions, i.e., learning competitive skills through experience. The team entity provides a power coordination means to control the sub-team entities actions. The strategies of team can be specified in the team plans. The cooperative strategies can also be learned by using the learning algorithms.

5   Conclusion and Future Work

This paper presents our initial research on a role-based BDI framework to facilitate the coordination and cooperation in MASs. The multi-agent team can be constructed in a layered architecture: individual agent behaviors (competitive skills) are specified in the low level, and team behaviors (cooperative skills) are addressed in the high level. However, this framework needs an additional team entity that has the whole knowledge of team members and the ability for coordinating individual member. Such assumption may not be suitable for some application domains, i.e., negotiation in ad-hoc manner.

Even this framework provides an abstract way to model the multi-agent teaming , the difficulty remains due to the lack of effective techniques to solve the multi-agent behavior optimisation problems. Another practical problem is that an explicit model for role assignment (either static or dynamic) is necessary to build an efficiency team. Further work includes to compare team level performance by using different approaches, i.e., hand-coded programs versus learning techniques.

We are in the process of developing an effective agent-teaming architecture with competitive and cooperative learning algorithms. Dynamic agents role assignment will be considered for agents' teaming. Due to the stochastic nature of MASs, it is required to use some optimisation techniques, in order to find the optimal team behaviors and build robust multi-agent team.

Acknowledgment

The authors gratefully acknowledge the valuable comments and suggestions of the reviewers, which have improved the quality and presentation of this paper.

References

1. Teambots (2000), http://www.cs.cmu.edu/~trb/Teambots/Domains/SoccerBots
2. Agent Oriented Software Pty Ltd. JACK Intelligent Agents User Guide (2002)
3. Agent Oriented Software Pty Ltd. JACK Teams User Guide (2002)
4. Cheyer, A., Martin, D.: The Open Agent Architecture. Autonomous Agents and Multi-Agent Systems 4(1-2), 143–148 (2001)
5. Cohen, P., Levesque, H., Smith, I.: On Team Formation. Contemporary Action Theory (1998)
6. Cohen, P.R., Levesque, H.J.: Teamwork. Nous 25(4), 487–512 (1991)

7. Gutknecht, O., Michel, F., Ferber, J.: Integrating Tools and Infrastructure for Generic Multi-Agent Systems, Autonomous Agents 2001. ACM Press, New York (2001)
8. Jennings, N.R.: On Agent-based Software Engineering. Artificial Intelligence 117, 277–296 (2000)
9. Laird, J.E., Newell, A., Rosenbloom, P.S.: Soar: an architecture for general intelligence. Artificial Intelligence 33(1), 1–64 (1987)
10. Leng, J., Fyfe, C., Jain, L.: Reinforcement Learning of Competitive Skills with Soccer Agents. In: Apolloni, B., Howlett, R.J., Jain, L.C. (eds.) KES 2007, Part I. LNCS (LNAI), vol. 4692, pp. 572–579. Springer, Heidelberg (2007)
11. Leng, J., Jain, L., Fyfe, C.: Simulation and Reinforcement Learning with Soccer Agents. International Journal of Multiagent and Grid Systems 4(4) (to be published, 2008)
12. Lind, J.: Agents, Multiagent Systems and Software Engineering. In: Lind, J. (ed.) Iterative Software Engineering for Multiagent Systems. LNCS (LNAI), vol. 1994, pp. 9–33. Springer, Heidelberg (2001)
13. Sioutis, C.: Reasoning and Learning for Intelligent Agents. PhD thesis, School of Electrical and Information Engineering, University of South Australia (2005)
14. Stone, P.: Layered Learning in Multiagent Systems: A Winning Approach to Robotic Soccer. MIT Press, Cambridge (2000)
15. Stone, P., Veloso, M.: Task Decomposition, Dynamic Role Assignment, and Low-Bandwidth Communication for Real-Time Strategic Teamwork. Artificial Intelligence 110(2), 241–273 (1999)
16. Tambe, M.: Towards Flexible Teamwork. Journal of Artificial Intelligence Research 7, 83–124 (1997)
17. Wooldridge, M., Jennings, N.: Intelligent Agents: Theory and Practice. Knowledge Engineering Review 10(2), 115–152 (1995)

About the Buildings' Danger Control Systems Vulnerabilities

Daniela E. Popescu[1] and Marcela F. Prada[2]

[1] Computer Science Department, University of Oradea, Electrical Engineering and Information Technologies Faculty, Romania
Tel.: (40) 259 408180; Fax: (40) 259 408408
depopescu@uoradea.ro
[2] Construction Department, University of Oradea, Construction and Architecture Faculty, Romania
Tel.: (40) 0723 287 766; Fax: (40) 259 408408
mprada@uoradea.ro

Abstract. Analysis is an important security component, because it provides a systematic way to approach computer and communications security. There is no single context within which to address the subject. There are distinct environments, each of which appears in a different light and must be considered with some degree of independence. The present paper give a potential solution for reducing the vulnerabilities for danger control systems. A danger control system is designed for managing different building dangers and any specific building activity. So, the danger control systems are an essential prerequisite for the reliable and efficient functioning of the life-safety measures and security in a building. The present paper gives the fault and recovery tress for a potential solution that enhances the quality for danger control systems.

Keywords: security system, testability, vulnerabilities, programmable outputs.

1 Basic Security Concepts

There are two basic security concepts that are important and easily visualized. These concepts are a "building block" security support structure and a "multiple barrier" configuration of security measures.

A secure overall operating environment is attained, as shown in Fig.1, by building a support structure based on the six environments we have outlined. We can view these environments as containing building blocks, each of which contributes to the support structure. The arrangement of the blocks in columns illustrates the sense of independence we infer for the environments. The blocks contain a rough outline of the subjects to be addressed. The more blocks used, the more solid the structure. It is appropriate to keep the analogy in mind.

The physical security is frequently the first barrier with which an adversary is confronted. There are seven main aspects to physical security: intrusion prevention, intrusion detection, proper information destruction, power protection, fire protection, water protection and contingency planning. Fig.2. is showing the level architecture for security systems [2].

I. Lovrek, R.J. Howlett, and L.C. Jain (Eds.): KES 2008, Part III, LNAI 5179, pp. 650–657, 2008.

SECURE OPERATING ENVIRONMENT					
Intrusion Prevention	Interviews	Laws	Access control	Access control	Encryption
Intrusion Detection	Backgroun d Screening	Policies	Reliability	Multilevel security	Dialup Control
Environm. protection	Training	Procedures	Electrical Protection	Structured developm.	Network controllers
Disaster recovery	Monitoring	Respons. actions	Hardware logic	Auditing	Fiber optics

Fig. 1. Security building blocks

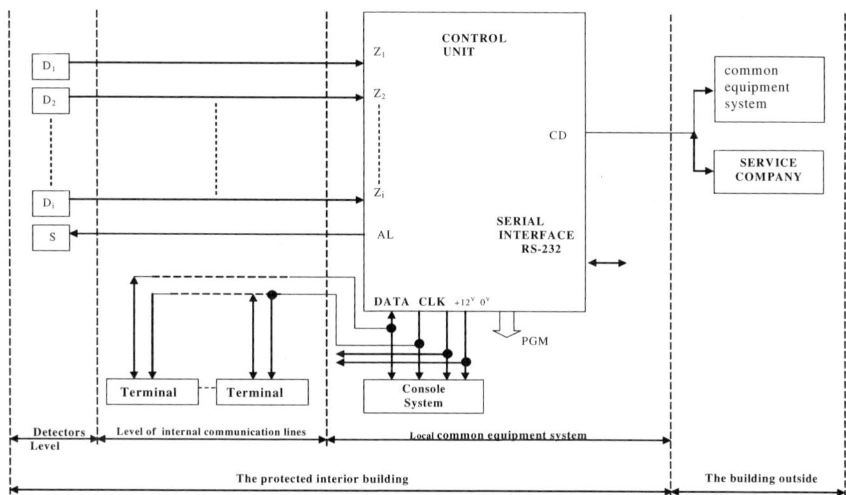

Fig. 2. The level architecture for security systems

2 Intrusion Prevention

Many of the same measures we may have considered for our own property are also appropriate in a computer or communication environment. These include fences, guards, windows barriers, lights and locks. The differences when applying these concepts to computers and communications may be slight if the facilities involved are relatively inexpensive or have relatively relaxed security requirements. However, most facilities do not meet this description. The result is that more technical sophistication, more complexity and more reliability are frequently appropriate.

In thinking about security for our personal possessions, most of us realize that the threat is real because of personal experience or because of the incidents reported in the news media. As reported incidents repeatedly illustrate, computers and communication media are subject to real threats also.

The threats of physical penetration lead to several vulnerabilities. Among these are theft of hardware; theft of software; compromising data (copying or reading); damage to or alteration of hardware, software of data; and misappropriation of resources (theft of computer or communication time or capabilities).

However, intrusion protection is almost totally a delaying measure. This means that an adversary, given enough time and resources, can penetrate any physical barrier. The effectiveness of the measure is therefore strongly dependent on the time delay and adversary will incur in overcoming the barrier. The only caveat is that some barriers are easier to couple to other security measures, such as intrusion alarms and response actions.

3 Fences and Multiple Barriers Vulnerabilities

Fences of various types are commonly used as barriers.

The common use of multiple security barriers is an important and useful strategy, but should not be accepted without examination. One reason for using multiple barriers is that some measures have inherent weakness that cannot be overcome within practical expense constrains.

Cost is often a reason for using multiple barriers. It might appear that since it is difficult to make any barrier completely effective, multiple barriers are most cost-effective than single barriers. This is not always true, as will be demonstrated below.

A numeric illustration of the performance of a multiple-barrier strategy can be provided by the following model. Assume that there are n possible security barriers, each having a successful protection probability of $1-e^{-c_1}$, where c_i represents the cost spent on the i th security barrier. This model is chosen to approximate the case we intuitively expect, where in increasing expenditures result in decreasing increments of improvement.

Now compare using one barrier to achieve a probability goal, p_g, with the strategy of using n barriers to achieve the same goal.

Using one barrier (the i th), $p_g = 1 - e^{c_i}$. Using multiple barriers,

$$p_g = 1 - (1 - p_1) \cdot (1 - p_2) \cdots (1 - p_n)$$
$$= 1 - e^{-(c_1 + c_2 + \ldots + c_n)}$$

Equating the two expressions, we find:

$$c_i = c_1 + c_2 + \ldots + c_n$$

This means (for the model we used) that one barrier at a given total expenditure can provide the same protection as multiple barriers combining to the same total expenditure.

This calculation is not intended to discredit the multiple barrier concept, but rather to reinforce the contention that intuition may mislead one as to the benefits. A model with a more precipitous decline in the security provided by additional expenses might have yielded a more favorable payoff for multiple barriers.

The vulnerabilities of fences include the ease with which destructive devices can be thrown over, the possibility that credentials can be passed back through the barriers ("passback attack") and their inability to protect against airborne penetration.

4   The PGM Outputs Used as an Enhancement Testability Solution for Danger Control Systems

The control units have some special outputs called Programmable Output Module [3][4][5][6]. There are many possibilities for programming them (24), but our option was to program them with the definition 10 - Latched System Event (Strobe Output). This programming mode permits to program PGM outputs to activate when special events occur. These events can be programmed in the following manner:

[1]-- *Burglary (Delay, Instant, Interior)*
[2]-- *Fire (Fire Keys, Fire Zones)*
[3]-- *Panic (Panic Zones)*
[4]-- *Medical (Medical and Emergency Zones)*
[5]-- *Supervisory (Supervisory, Freezer and Water Zones)*
[6]-- *Priority (Gas, Heat, Sprinkler)*
[7]-- *Holdup (Holdup Zones)*

Practically, the events received at the zone inputs, are registered in the CU event buffer, and then, they can be transmitted (by an appropriate programming) to the programmable output, that can be used in the testing process. It is important to note that a major problem for the above mentioned method is the fact that after the event occurring, the PGM output so defined will remain stuck-at 1 until we reset it with a reset cod from the console [3] [4] [6]. This is because the CU producers did not provide this definition mode (10) for PGM output for testability reasons; it was introduce for producing some interactions inside control system. On the other hand is important to maintain the current control unit state after reset.

Globally, the control unit may have 2 states:

- Armed
- Disarmed

We tried to find another way than that provided by the CU producers, for resetting PGM outputs defined as *Latched System Event;* the two majors requests for this new way are:

- Automatic reset for a PGM output defined as *Latched System Event,* without the operator need
- Keeping the current CU state even after the reset

So, we considered the (23) PGM output definition - Maintained Keyswitch Arm Zone, that provides successively CU arm / disarm states when this zone is momentary violated / secured. The definition permits us to conclude that at the moment when a zone defined with (23) definition is violated / secured, the arm cod is automatically transmitted on the data internal line of the controller (CU).

For testing the solution, we used 2 CU zones and one PGM output. Z1 zone is defined as 24 hours zone; this meaning a zone when the produced events are recorded to the PGMx output, defined as Latched System Event (10). Z8 zone is defined by "Maintained Keyswitch" definition and it has the role to reset the event and to bring it

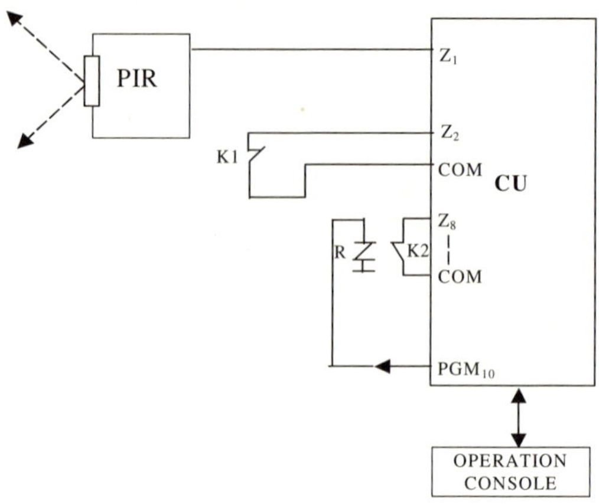

Fig. 3. The Implementation for the Automatic Reset

in the initial state. The implementation for the automatic reset is given in Fig.3: the events are produced on Z1 zone by a motion sensor (passive infrared – PIR).

A PC 5010 [4][5] was programmed for testing the solution and the obtained results were good.

5   Testability Development Possibilities Based on the Proposed Solution

To be noted that this solution permits to obtain test vectors automatically, without the need of an extra hardware [2]. It is sufficient the existing hardware. Moreover, it easily to be seen the special flexibility provided by these control units in tracing the various system events. As an example, defining a PGM output by (09) definition - System Trouble Output, this will be activated / deactivated on occurrence / loss of fault conditions mentioned in [4].

Based on the arguments pointed out, on the PGM output flexibility, we assume that there are enough conditions to conclude that based on different test criteria (CTi , i =1,....k) and on the needed testability degree for a given application, we can make k groups of test vectors CTi.

Fig.4 illustrates a potential use of the adopted solution by a redundant structure, static, global with voting and self-test. In this manner, the danger control system's testability degree may be spectacularly extended based on a large number of criteria. By example we can make up test vector groups referring to the risk sectors of a security system: intrusion, fire extinguishing, gas lacks, flood, panic, medical, holdup [8]. There are also some other criteria that may take part in making up test vectors according the testability imposed requirements.

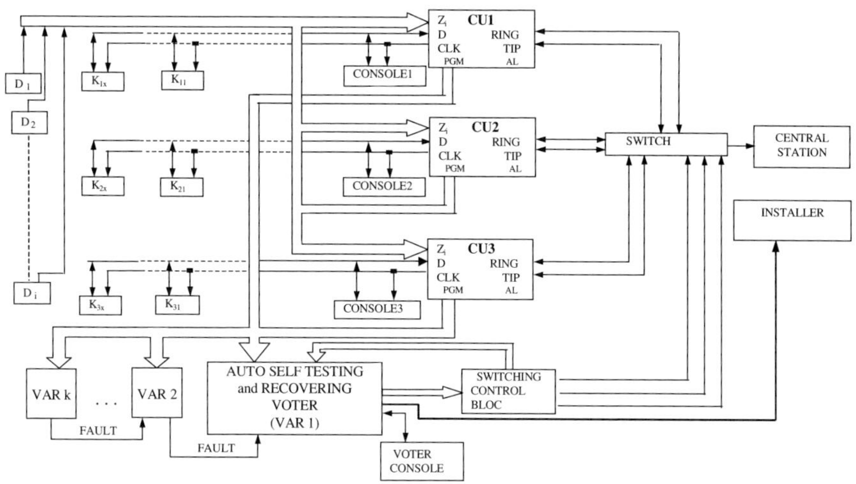

Fig. 4. A Redundant Structure, Static, Global With Voting and Self-Test

Fig.5 gives the fault and recovery tree structure for the situation of case A – where UC1 is the active unit and UC2 and UC2 are units in standby. The B and C cases are similar. For the diagnose process there are used three programmable outputs (one for each control unit): 1PGM, 2PGM and 3PGM.

These programmable outputs were defined with the definition (10) *"Latched System Event"*, and the these three outputs were setted to "on" for all the events in the system: *burglary, fire, panic, medical, supervisory, priority, holdup.*

So, the PGM output programming will activate PGM to all the events mentioned above. These outputs were used in the self-testing process. Initially, the system is no faulty, and UC1 ensure messages transmission. The ordinograms consider all the possible fault combination.

The first fault in DEF1 may have 3 causes: fault at UC1 (def_1), fault at UC2 (def_2) or fault at UC3 (def_3). If the fault affects UC1 an automated switch must be made, and if the fault affects UC2 (def_2) or UC3 (def_3) the switch must not be made (because UC1 is without fault), it must be made only a reconfiguration.

- The adopted structure ensures the system functioning after the occurrence of DEF1 and there also are two possibilities: the faulty unit is repaired and is replaced in the functional system.
- the faulty unit is not repaired so that the functioning will be ensured by the other two units.

The system will continue to function in this incomplete way; it can detect a second fault that can appear in the system: DEF2, but he can not localize it, because it has no valid criteria for voting – there are only two units that are compared. Only the operator can take a decision based on the information displayed on the system panel. So, in this point, the reconfiguration must invalidate the voting process and permit only the comparison for the unfaulty control units PGM outputs. If the switching must be done, this is realized by introducing a special allocated code.

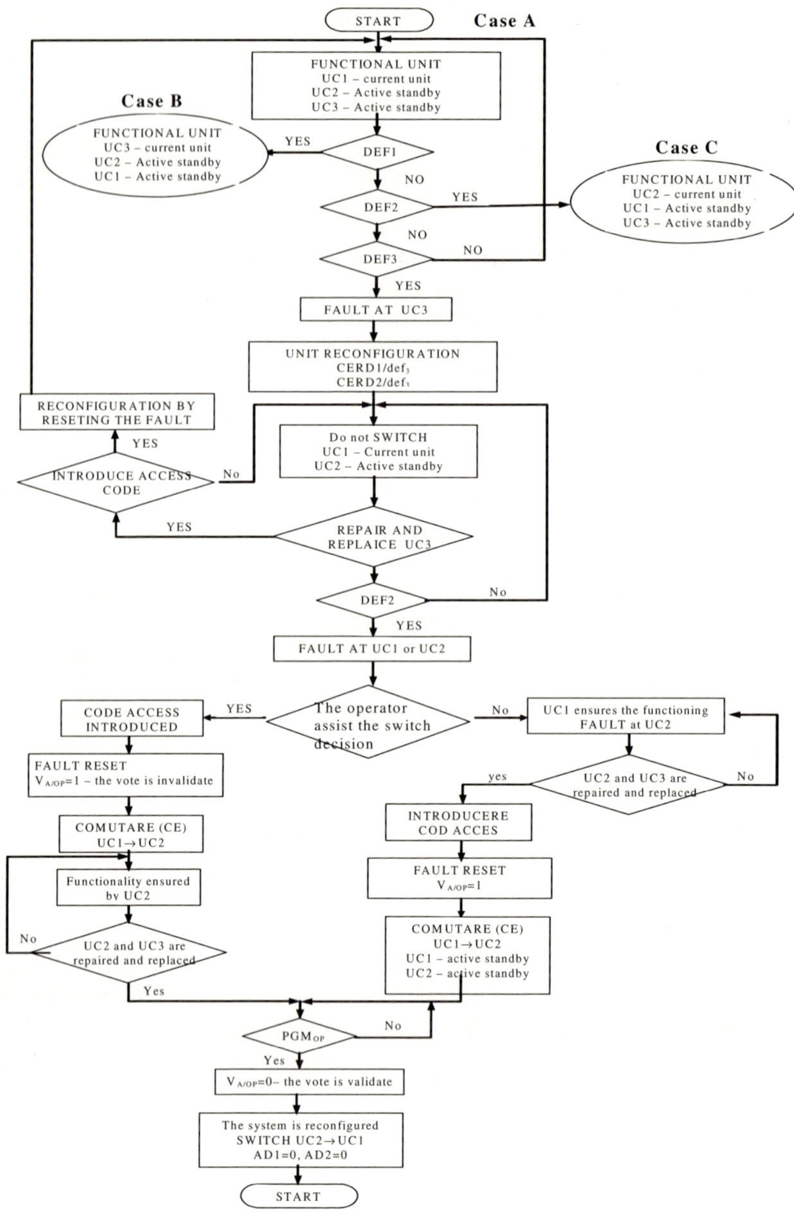

Fig. 5.

It is important to point that after this manual switching decision both the voting process and the comparing process must be invalidated by a conforming reconfiguration (otherwise, incorrect decisions can ocuure).

The faulty and recovery tree takes account on all the possibilities and it is important to mention that this kind of reservation offers enough time for repairing process without interrupting the security system functioning.

6 Conclusions

By applying the voting approach both at the UC level and at the sensors level the reliability and testability of a security system is much increased. This is very important for avoiding the false alarms and for signaling a fault at the UC level. The proposed structure from Fig.5 was realized and verified in practice.

So, a new solution is promoted by this paper (contrasting with [3] [4] [6]), that allows the automatic reset of the control units PGM outputs, offering the possibility for tracing the system events based on programmable outputs. So, the condition of maximizing the system observability which determines the enhancing of the control system testability. This solution doesn't require an extra hardware for generating test vectors, these being generated by danger control system detection elements themselves, during normal functioning. To be noted the modularity and flexibility of the solution that permits its expansion by simple repeatability according to high testability requirements, based on different criteria directly imposed by the implemented application, with direct involvement in danger control system enhancing quality.

References

[1] Anderson, R.: Security Engineering – A Guide to Building Dependable Distributed Systems, 2nd edn. Wiley Computer Publishing, Chichester (2008)
[2] Cerberus, A.G.: Danger Management System, Switzerland, white Papers, pp. III.3-III.25 (2004)
[3] Napco Security Systems, Inc., Gem-P3200, Control Panel – Programming Instructions, New York (2000)
[4] Digital Security Controls Ltd., Panel Control PC5010 – Installation Manual, Canada (2002)
[5] Digital Security Controls Ltd, Panel Control PC5010 – Programming Worksheets, Canada (2001)
[6] Digital Security Controls Ltd., Panel Control PC5020 – Installation Manual, Canada (2000)

Traffic Management by Constant Time to Collision

Valentina E. Balas and Marius M. Balas

Aurel Vlaicu University of Arad, Romania
`balas@inext.ro`

Abstract. The paper is presenting a new method for the management of the traffic flow on highways, based on the constant time to collision criterion. The criterion is applied at two levels, for each car implied in traffic, and for the whole highway. Each car is provided with a constant time to collision cruise controller, that is maintaining optimal distance-gaps between cars, in accordance to the technical data of each car, the actual speed, and an imposed time to collision. The highway's traffic management center has the possibility to impose the same optimal time to collision to all the cars in the traffic. This way the traffic is permanently organizing itself, by distributing the cars such way that the collision risk is uniformly distributed. The traffic intensity is controlled by the imposed time to collision.

Keywords: knowledge embedding by computer models, constant time to collision, fuzzy-interpolative cruise controllers.

1 Introduction

Automate driving is enhancing the driving performance and reducing the crash risks. The Advanced Driver Assistance Systems ADAS are such systems. When designed with a safe Human-Machine Interface they are able to increase the car's and the traffic' safety [1], [2], [3]. Examples of such systems are: the In-Vehicle Navigation System usually with GPS and Traffic Message Channel TMC for providing up-to-date traffic information, the Adaptive Cruise Control ACC, the Lane/Road Departure Detection/warning system, the Collision warning system, the Intelligent speed adaptation or intelligent speed advice ISA, the Night Vision, the Adaptive Light Control, the Pedestrian Protection System, the Automatic parking, etc. Other systems are the Autonomous Intelligent Cruise Control AICC and the Collision Avoidance Systems CAS. Some systems also feature Forward Collision Warning Systems FCWS or Collision Mitigation Avoidance System CMAS, which warns the driver and/or provides brake support if there is a high risk of a rear-end collision. Some of these systems can be linked to a car's cruise control system, allowing the vehicle to slow when catching up the vehicle in front and accelerate again to the preset speed when traffic allows. A key problem in this issue is the measurement and the control of the distance gap between two following cars. In some previous papers [4], [5], [6], we introduced a fuzzy interpolative distance-gap control method that is using a Constant Time to Collision Planning CTCP, in the sense of the Planning System concept [7]. This approach was also discussed in [8]. This work is continuing the investigation of CTCP in the domain of the traffic management.

I. Lovrek, R.J. Howlett, and L.C. Jain (Eds.): KES 2008, Part III, LNAI 5179, pp. 658–663, 2008.

2 The Constant Time to Collision Criterion

Several indicators measure the characteristics of the traffic flow: the Time-to-Collision TTC, the Time-to-Accident, the Post-Encroachment-Time, the Deceleration-to-Safety-Time, the Number of Shockwaves, etc. TTC is the time before two following cars (Car2 is following Car1) are colliding, assuming unchanged speeds of both vehicles:

$$\text{TTC} = \frac{d}{v_2 - v_1} \tag{1}$$

The TTC is linked to the longitudinal driving task. Negative TTC implies that Car1 drives faster, i.e. there is no danger, while positive TTC is leading to unsafe situations. By assessing TTC values at regular time steps or in continuous time, a TTC trajectory of a vehicle can be determined. Doing this for all vehicles present on a road segment one can determine the frequency of the occurrence of certain TTC values, and by comparing these distributions for different scenarios, one can appreciate the traffic safety [2]. We introduced in ref. [9] also the *Inverse Time to Collision*.

The central issue in cars' safety is to impose an appropriate distance between cars, d_i. AICC is imposing a particular polynomial $d_i(v_2)$ law:

$$d_i(v_2) = z_0 + z_1 \cdot v_2 + z_2 \cdot v_2^{\,2} = 3 + z_1 \cdot v_2 + 0.01 \cdot v_2^{\,2} \tag{2}$$

Several settings are recommended, for example $z_1 = 0.8s$ or $z_1 = 0.6s$. Two objections can be drawn against this polynomial $d_i(v_2)$ law:

- no adaptation to the traffic intensity is offered: if (3) is tuned for the highest possible traffic, when the traffic is decreasing, the following cars will continue to maintain the same short distance-gaps between them;
- $-z_1$ and z_2 are artificially introduced parameters, they have no significance for humans - highway operators or drivers - and they are not linked to the physical features of the system.

The *Constant Time to Collision* criterion CTTC consists in imposing *stabilized* TTCs by means of the Car2 cruise controller. Applying CTTC brings two obvious advantages:

- a constant collision risk for each vehicle involved;
- the possibility to control the traffic flow on extended road sections, if each vehicle will apply the same TTC that is currently recommended by the Traffic Management Center [12]: long TTC means *low traffic flow* and *higher safety* while short TTC means *high traffic flow* and *higher risk*.

The on-line TTC control is not convenient because when the two cars have the same speed $v_2 - v_1 = 0$, the denominator of TTC is turning null. That is why CTTC must be implemented off-line, with the help of $d_i(v_2)$ mappings (fig. 1). The CTTC implementation by $d_i(v_2)$ distance-gap planners is possible because *a distance gap planner using* TTC *will produce* CTTC. We studied this method by computer simulations, using a Matlab-Simulink model of the tandem Car1-Car2, introduced in other previous papers [4], [5], [6], [10].

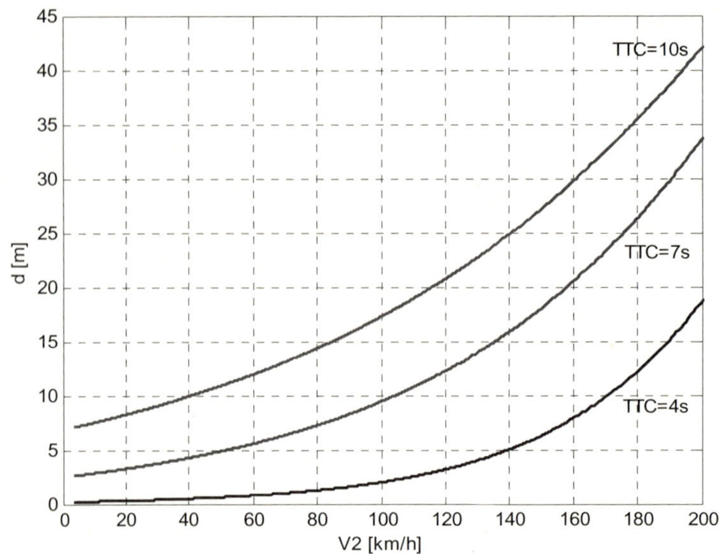

Fig. 1. The recorded $d_i(v_2)$ mappings for three different TTC

Since the design of the planners is performed with the help of functional models of the cars, accurate knowledge about the specific behavior and parameters of each car (traction and braking forces, weight, aerodynamic coefficient, etc.) can be taken into account, which is not possible to the simplified and leveling analytic model (2).

The application of this method is imposing to the car manufacturers to provide each type of automobile with a computer model.

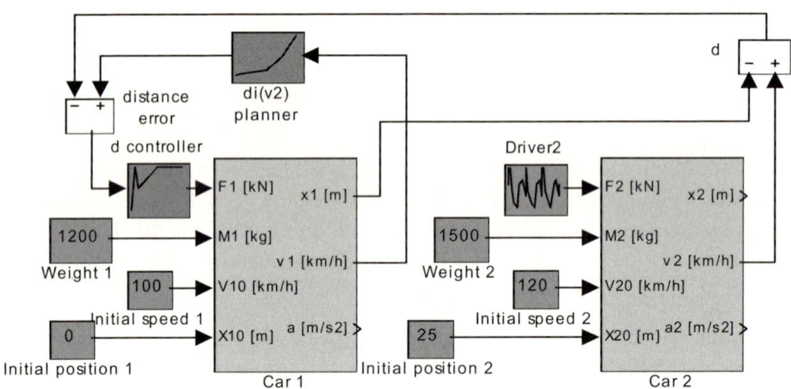

Fig. 2. A cruise control system with distance controller and CTTC $d_i(v_2)$ plan

The distance-gap planners are designed as follows. The simulation scenario consists in braking Car1 until the car is immobilized, starting from a high initial speed. A TTC controller is driving the Car2 traction/braking force such way that during the whole simulation TTC is stabilized to a desired constant value. The continuous braking allow us to avoid the v_2-v_1=0 case. We will use the recorded d mapping as the desired $d_i(v_2)$ planner for the given TTC. The figure 1 planners are determined for three TTC values: 4s, 7s and 10s. These planners can be easily implemented with the help of the look-up tables with linear interpolation.

The use of the CTTC planning technique is essentially facilitating the task of the distance controller that is actually driving the traction/braking force of a real car during the cruise regime, as shown in fig. 2. Very simple fuzzy-interpolative PD controllers or even linear controllers [4], [10], [11], etc. can such way control successfully the car following task. The implementations can be basically achieved by look-up-table techniques [8], [11], etc.

3   The Traffic Management by Constant Time to Collision

The superior application level for the CTTC criterion is the management of the traffic over extended highways' segments. Assuming that each car that is participating at the traffic is provided with a cruise controller with CTTC planner, the Traffic Management Center TMC has the possibility to impose the same TTC to all the cars. This way the highway system becomes a distributed one. Each car is trying to reach and to maintain the position that respects the imposed TTC to the previous car. This trend has as major advantage a constant distribution of the collision risk for each car.

Lets consider that TMC is imposing a 7s TTC. If the traffic is not too intense, the tendency of the cars will be to form platoons that are able to maintain TTC=7s. It is to remark that the distance between the cars belonging to the same platoon are not necessarily identical, even for constant speeds, because each type of cars has its particular $d_i(v_2)$ planner, in accordance to its technical parameters (weight, aerodynamics, engine power, brakes, etc.)

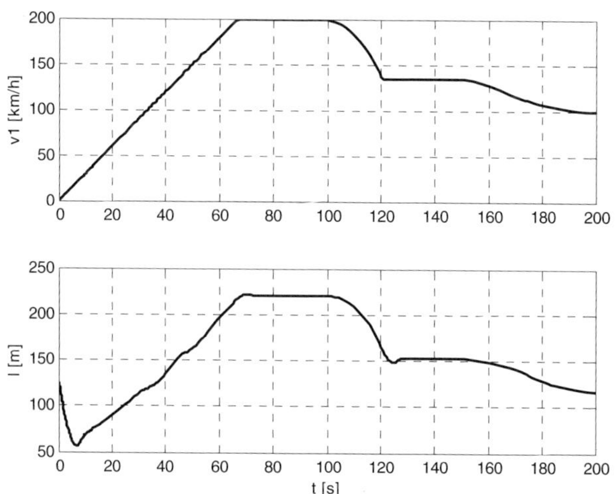

Fig. 3. The speed of the first car v1 and the resulting length of the 5 cars platoon l

If the traffic is beginning to decrease the number of the cars that are included into platoons will decrease too, and empty zones will develop on the highway. In this case TMC should increase the imposed TMC value, either continuously or by discrete values, say TMC=10s. This way the disposable space of the highway will be better covered and the collision risk will decrease for each car. In the opposite case, if the traffic is increasing, the cars will not be able to maintain the desired TTC and the afferent distance-gaps. TMC will be forced to reduce the imposed TTC, either continuously or by discrete values, say TMC=4s. This way the density of the traffic will increase and the collision risk will increase too, but this will happen in a smooth and controlled manner, the risk continuing to be equally distributed over each car.

Our research is only at the initial stage, but the preliminary simulations are confirming that CTTC criterion is potentially able to cope with the highway traffic. In the following figures Fig. 3 and Fig. 4 one can observe the behavior of a 5 cars CTTC platoon that is starting from random initial positions and is aggregating in less than 10s, for TTC=15s.

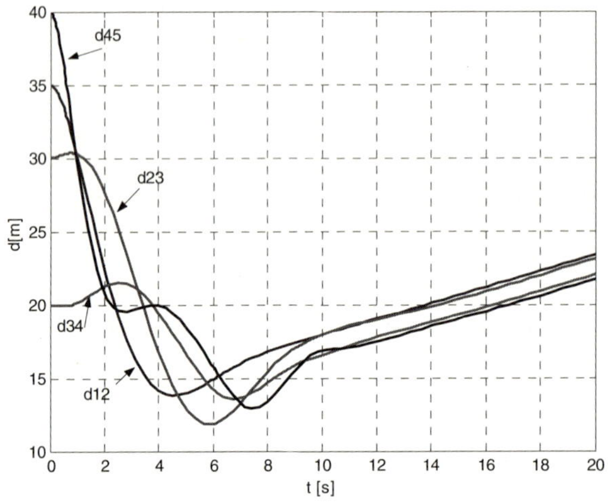

Fig. 4. The aggregation of the 5 cars platoon, with the distance-gaps between cars

4   Conclusions

The time to collision index was used in previous papers for determining the optimal distance-gap between following cars. The criterion can also be used in the highway traffic management. Assuming that each car is provided with a constant time to collision cruise controller, the traffic management center can impose the same time to collision to all the cars. This way the highway system becomes a distributed one, each car trying to reach and to maintain the position that respects the imposed time to collision to the previous car. This trend has as major advantage a constant distribution of the collision risk for over all the cars of the highway.

Besides the simplicity and the advantageous interpolative implementation, all the time to collision based tools have a common feature: they are embedding precise knowledge about the technical data of the automobiles on which they are installed, thanks to the functional computer model that stands behind their design. This adaptive capability is promising to contribute to the improvement of the future highway traffic that is presenting so many elements of uncertainty.

References

1. Girard, A.R., Borges de Sousa, J., Misener, J.A., Hendrick, J.K.: A Control Architecture for Integrated Cooperative Cruise Control and Collision Warning Systems, Berkeley, University of California, `http://path.berkeley.edu/~anouck/papers/cdc01inv3102.pdf`
2. Minderhoud, M.M., Hoogendoorn, S.P.: Extended Time-to-Collision Safety Measures for ADAS Safety Assessment, Delft University of Technology, `http://www.Delft2001.tudelft.nl/ paper%20files/paper1145.doc`
3. Zhang, Y., Kosmatopoulos, E.B., Ioannou, P.A., Chien, C.C.: Autonomous Intelligent Cruise Control Using Front and Back Information for Tight Vehicle Following Maneuvers. IEEE Trans. on Vehicular Technology 48(1), 319–328 (1999)
4. Balas, M., Barna, C.: Using CCD cameras for the car following algorithms. In: Proc. of IEEE International Symposium on Industrial Electronics ISIE 2005, Dubrovnik, June 20-23, 2004, pp. 57–62 (2004)
5. Balas, M., Balas, V.: Optimizing the Distance-Gap between Cars by Fuzzy-Interpolative Control with Time to Collision Planning. In: Proc. of The IEEE International Conference on Mechatronics, Budapest, Hungary ICM 2006, Budapest, July 3-5, 2006, pp. 215–218 (2006)
6. Balas, M., Balas, V., Duplaix, J.: Optimizing the Distance-Gap between Cars by Constant Time to Collision Planning. In: Proc. of IEEE International Symposium on Industrial Electronics ISIE 2007, Vigo, pp. 304–309 (June 2007)
7. Passino, K.M., Antsaklis, P.J.: Modeling and Analysis of Artificially Intelligent Planning Systems. In: Antsaklis, P.J., Passino, K.M. (eds.) Introduction to Intelligent and Autonomous Control, pp. 191–214. Kluwer, Dordrecht (1993)
8. Balas, M.: Le flou-interpolatif, present et perspectives. In: Seminaire LSIS St.-Jerome, Marseille, France (September 21, 2006)
9. Balas, V., Balas, M.: Driver assisting by Inverse Time to Collision. In: Proc. of The IEEE International Conference on Mechatronics, Budapest, Hungary WAC 2006, Budapest, July 24–27(2006)
10. Precup, R.E., Preitl, S., Balas, M., Balas, V.: Fuzzy Controllers for Tire Slip Control in Anti-Lock Braking Systems. In: Proc. of The IEEE International Conference on Fuzzy Systems FUZZ-IEEE 2004, Budapest, July 25-29, 2004, pp. 1317–1322 (2004)
11. Kóczy, L.T., Balas, M.M., Ciugudean, M., Balas, V.E., Botzheim, J.: On the Interpolative Side of the Fuzzy Sets. In: Proc. of the IEEE International Workshop on Soft Computing Applications SOFA 2005, Szeged-Arad, August 27-30, 2005, pp. 17–23 (2005)
12. ITS Decision. Traffic Management Centers, `http://www.calccit.org/itsdecision/serv_and_tech/Traffic_management/TMC/tmc_summary.html`

Structuring Spatial Knowledge and Fail-Safe Expression for Directive Voice Navigation

Taizo Miyachi[1], Jens J. Balvig[2], Ipei Kuroda[3], and Takeshi Suzuki[4]

[1] School of Information Science and Technology, Tokai University
1117 Kitakaname, Hiratsuka, Kanagawa 259-1292, Japan
`miyachi@keyaki.cc.u-tokai.ac.jp`
[2] Export Japan Co, Ltd.
[3] IBM Japan e-Communications Co., Ltd.
[4] Kanagawa Prefecture Hiratsuka School for the Visually Impaired

Abstract. Navigation systems in mobile terminals are often useful for finding a way to the destination. However it is not easy for pedestrians to consult small maps with a sense of direction in a target area. Voice navigation is easy and useful for everybody to easily know spatial information in a map. However pedestrians should build mental maps based on the spatial information in voice guide. We propose and examine a quiet route navigation system with directive voice beams for pedestrians in order to safely reach to the destination. A method for making a fail-safe voice guide that allows pedestrians to easily understand and remember spatial information avoiding misunderstanding is also discussed.

1   Introduction

Safe and imageable [2] walking anywhere a pedestrian expects makes a delightful space in a city and improves city life. Such walking not only reduces up to 70 % of death in traffic accidents between pedestrians and cars [3] but also makes a healthy aged city with small medical expenses. Lots of visitors can easily walk around sight-seeing spots and the visitor business might become prosperous. Pedestrians with a sense of direction can sometimes walk anywhere using navigation systems by cell phones [4, 5] or by the Ubiquitous Communicator (UC) of FMA Project under the ubiquitous environments in Japan [1]. However most pedestrians can not do so as they have to not only consult a small map in which most roads have no name but also have a sense of direction in a target area. Aged people and a person with low vision can not also see a map in a small display of a mobile terminal and can not easily operate the terminal. Voice navigation is easy and useful for everybody. However it is not easy for pedestrians to build mental maps with spatial structures in voice guides.

In this paper we propose a quiet route navigation system with directive voice beams that allows visitors including aged people and pedestrians with low vision to safely walk around a delightful town by themselves. We discuss a method for making a voice guide for pedestrians to easily understand and remember detail guides after pedestrians listen to them once or twice. Duplicate expressions, a pair of landmarks, and guide equivalence for a behavior to shorten a voice guide are also discussed.

I. Lovrek, R.J. Howlett, and L.C. Jain (Eds.): KES 2008, Part III, LNAI 5179, pp. 664–672, 2008.

2 Tasks of Mobility and the Navigation Systems

Pedestrian can not often reach to the destination consulting a small map in a mobile terminal. The pedestrian's tasks on the way to the destination should be clarified.

(1) Pedestrian's tasks to safely reach to the destination are classified into T1 to T7.

(T1) Find the destination on the map and ensure the direction to the destination
(T2) Find a safe route to the destination
(T3) Find the start point and chose a pavement and a crosswalk in the right direction
(T4) Pay attention to the embedded dangers and avoid obstacles.
(T5) Look out moving bicycles and cars on the pavement
(T6) Modify a line of walking or change the route to avoid dangers
(T7) Find a way in the last several ten meters to the destination

(T1), (T2) and (T7) are difficult for strangers as most roads have no name in Japan. From (T3) to (T6) are important for aged pedestrians, pedestrians with low vision, etc. The weak needs useful predictions for the safe and delightful moving.

(2) Navigation systems and safe mobility
Navigation systems in both mobile terminals and mobile phones have already been used all over the world. They are mainly classified into four categories.

(i)   Car navigation systems and portable navigation systems
(ii) Internet services of map and navigation (ex. Google map / earth, NAVITIME)
(iii) Navigation by mobile phones with GPS (ex. EZ NAVIWALK, NAVITIME)
(iv) Accurate navigation systems with sensors (ex. FMA system in Japan, etc.)

  Navigation systems have useful functions but they have mainly seven problems to reach to the destination. Users of the systems can not often reach to the destination.

(a) **Consulting a map is hard.** Human acquires more than 70 % of information by a sense of sight. However it is hard for a pedestrian with little sense of direction to consult a map. The pedestrian can not reach to the destination with the map as most roads have no name in Japan. It is not easy for visitors to notice the corner to turn in the immediate space consulting a map.

(b) **Entrance of the destination.** Navigation systems do not usually show users the entrance of the destination. They show the users only a point such as a grocery store near the destination instead of the entrance. The users can not sometimes reach to the destination and returns without accomplishing their purposes. A user must find a way to the entrance of destination in the last several ten meters by himself / herself.

(c) **Getting lost and wrong way.** A pedestrian usually gets lost or walks the wrong way by many turns in a complicated route in down town where many houses stand close together and narrow paths have no name.

(d) **Tough route.** The system usually merely shows the shortest route without considering toughness of the route, steep slopes and too low stairs.

(e) **Wrong direction.** The system sometimes shows a wrong direction by voice navigation because of close parallel roads, a dead end, and wrong sensing of the direction.

(f) **A map is too small** for an aged person and a visually impaired person to look it in a small display of mobile terminal.

(g) **Sudden stop and staying for a while** cause collisions on the pavement.

3 An Easy Voice Navigation with Directional Voice Beams

A city consists of three kinds of areas and paths: (a) safe, (b) watch, and (c) dangerous. There are lots of static and dynamic dangers in the areas. It is basically hard for a visitor only to consult a map and to safely reach to the destination avoiding tough ways and dangers. Voice navigation is useful for the pedestrian to easily understand a safe and delight route. However it is hard for human not only to remember spatial objects [8] in voice guides but also to structure them in a mental map. Moreover the memory size of pedestrian becomes small while walking. A simple structure of spatial information is indispensable in such situations. We discuss a method for making a voice guide for pedestrians to easily understand and remember the spatial objects after listening to the voice guide once or twice. Two points are important for a safe mobility. (1) How to easily make the spatial information that a pedestrian can easily understand and perform appropriate actions. (2) How to avoid misunderstanding.

3.1 A Method for Structuring Navigation with Priority of Spatial Objects

A pedestrian familiar with the target area can easily walk to the destination but can not often explain a stranger how to reach to the destination.

Human mainly performs three tasks to understand voice navigation. (a) Listen to the navigations and understand them, (b) Build a mental map of routes in brain, (c) Remember information for a safe mobility. Easy listening and understanding without misunderstanding are important for task (a). A simple structure of spatial objects and relationships among them so as to perform important actions are important for (b). A suitable structure for the characteristics of human memory is important for (c).

We discuss the characteristics and limit of capability of human memory.

(1) Familiar objects in a mental map. Familiar objects become anchors in a mental map and make human easily remember cognitive information for a long time.

(i)   Human can remember many familiar objects for longer time than unfamiliar ones.
(ii) Human can easily listen to familiar objects in a short time.
(iii) A pedestrian can easily remember a familiar object and a relationship among familiar objects. He/she can also easily find them as he/she merely listens to one of them. It is sometimes hard for the pedestrian to find unfamiliar objects.

Familiar objects have already been remembered in long-term memory without additional tasks. They are useful anchors in the long-term memory so as to utilize the knowledge and situations remembered in mental maps in the long-term memory.

(2) Understanding voice navigation and the magical number. Human has the magical number 7 ± 2 in short-term memory.

Example 1. Across the crosswalk at the intersection and turn in the direction of three o'clock. TSUTAYA locates 70m ahead on your left. There is a parking in front of TSUTAYA. Then LAWSON locates 25m ahead on your left. Cars often enter into the paths on the both side of LAWSON. Give attention to the sound of moving car and walk carefully. Go straight at the intersection 5m ahead of LAWSON. Example 1 includes eleven spatial objects such as TSUTAYA. The pedestrian basically forgets this information when he/she acquires additional information. The number of objects in voice guides easily exceeds the magical number. As it is impossible for a pedestrian to hear and remember all of them in short-term memory, structuring spatial information in long-term memory is indispensable to remember them.

(3) Perceptual information and cognitive information

A pedestrian acquires and uses useful perceptional information such as a pole of signal in immediate space and remember it in short-term memory. On the other hand, in moving to an invisible space and a remote space, he/she acquires useful cognitive information from a map or a voice guide, and builds a mental map with actions to be performed. The cognitive information should be easily matched with objects and a set of useful actions in immediate space to avoid dangers and keep safety.

(4) The whole route and a set of parts of the route

It is easier for a pedestrian to understand spatial information in each sub-route than in the whole route. A pedestrian can avoid the dangers and enjoy delightful surroundings and events in the sub-route.

(5) Human easily forgets qualitative value in a shorter time than the other objects.

We propose a method for structuring spatial information based on (1) to (5).

  (i) Classify spatial information into ten categories: area, node, sign, direction, obstacle, action, road condition, distance and values. A list of signs that correspond to nodes shows a whole route with the direction. A list of signs shows a sub-route. Obstacles and dangers embed in the sub-route. A pedestrian take care of the road conditions, and performs actions to cope with the obstacles. The last category is the exact value such as distance since human easily forgets the value in a short time.

 (ii) Choose familiar spatial objects such as big signs and famous landmarks. Make a list of relationships among the familiar spatial objects that shows a whole route or a sub-route with the direction (See Fig. 1).

(iii) Make a relationship from an obstacle to spatial objects.

(iv) Make a relationship between obstacles and actions to cope with the obstacle.

 (v) Make a relationship from a value to the familiar objects in the upper level.

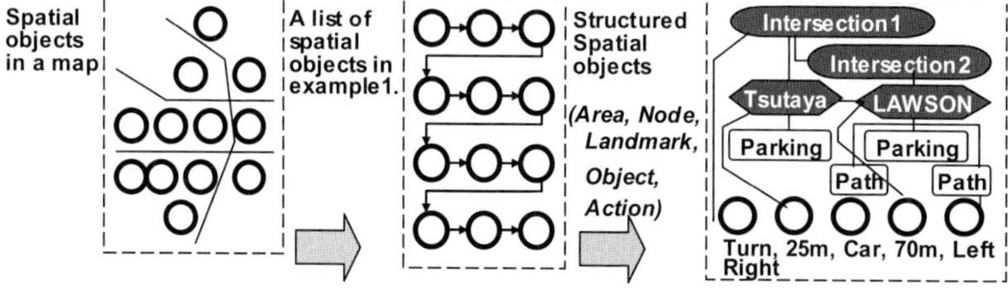

Fig. 1. Spatial objects in example 1 and structuring the spatial objects

3.2   Misunderstanding by a Pedestrian and a Fail-Safe Explanation

Recognitions of spatial objects by pedestrians changes depending on their interesting and the situations. The pedestrian misses a spatial object or misunderstands direction when he/she encounters the danger such as turning cars. A fail-safe expression of navigation should be prepared for the safe mobility. The right direction at the start point and a way in the last several ten meters to the destination are very important for pedestrians.

(1) Misunderstanding at the start point and duplicate expression. The right direction to proceed at the start point is the most important. The wrong choice of the direction usually makes a movement to the destination very difficult or sometimes makes it impossible in a hilly area. For example, a pedestrian misunderstands "turn in the direction of 3 o'clock" as a meaning of "turn left" in a hilly area. He/she might walk down to a point far away from the destination as the roads spread in each direction. "Duplicate expression" is basically wrong in regular sentences but is necessary for the pedestrians in order to avoid such misunderstanding. The navigation system should make the pedestrian ensure the right-turn in an expression "Turn right in the direction of three o'clock" with a directive voice beam by a parametric speaker (PS). The duplicate expression is also useful at a start point such as the exit of subway.

(2) Hard discovery of a way in the last several ten meters. Navigation systems usually stop the guide at a point near the destination such as a grocery store. It is hard for the user to discover a way in the last several ten meters to the entrance of destination on the ground or underground. There are few signs close to the destinations such as signs of department stores both underground and on the ground. Such destinations are mainly classified into three categories (a) station: bus, subway, (b) public building: city office, school, hospital, and (c) private enterprise: department store, bank. Pedestrians also hardly notice unfamiliar signs even in front of them. The navigation system should show a pair of objects at a distance such as the sign of drag store and an object close to the pedestrian at the same time by the directional voice beam.

Example 3. "Landmarks are drug store, Japanese bar, and McDonald's. An obstacle is a pipe fence in front of the drug store where you should turn right."
A pipe fence is a kind of obstacle but it becomes a very useful landmark.

(3) Sudden obstruction. A pedestrian sometimes misses the chance to discover the sign by an obstruction such as sudden appearance of a car entering a car parking. The system should show both next landmark and the landmark that he/she has already passed by. The pedestrian can ensure the right route where he/she is walking. A navigation system should also predict typical patterns of the movement and the stereotype of dangerous objects in order to cope with upcoming situations. The pedestrian can have inner reserve for next actions by such information.

(4) Guide equivalence for a behavior. Two guides are called "guide equivalent" when two behaviors guided in different expressions with different spatial objects are the same. One of the guides can be omitted in order to shorten the voice guide as human has the magical number in short-term memory.

Example 4. "Landmarks are Tsutaya and LAWSON" is guide equivalent to "walk 70 meters forwards from Tsutaya" on the safe pavement.

We propose additional rules. (vi) "Duplicate expressions with a far landmark and a close landmark," and "prediction based on (1), (2), and (3)" should be expressed in the navigation, (vii) A guide equivalent to the other guide for a behavior can be omitted, and (viii)The exact values are explained in the end. The example of the navigation in the first area is shown in Example 5 (See Fig.2). The navigation system tells a whole route to the station by a list of signs with white characters at first. The system tells a sub-route by a list of signs. Then spatial objects are explained following the route. The exact values such as 5m are explained in the end.

Example 5. We will go to Tokai University Mae station. Land marks are Tsutaya, the big intersection, Daiei, and Drug store. In the first area Landmarks are Tsutaya and LAWSON. Obstacles are in and out of cars at parking in front of Tsutaya and in and out of cars on the roads both sides of the LAWSON. The surface of road is good. Go straight at the intersection 5m away from the LAWSON (See the first area in Fig.5).

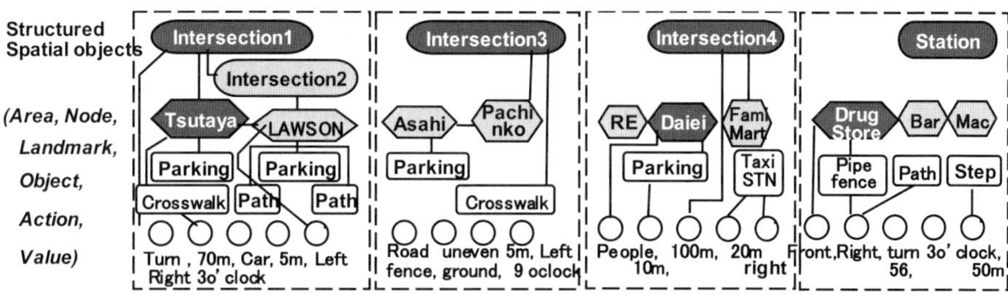

Fig. 2. An image of structure of spatial objects in a route

4   A Quiet Route Navigation System and Experiment

We propose a quiet navigation system by parametric speakers (PS) [6] that guides pedestrians to reach to the destination by themselves. The system supplies pedestrians with suitable information at each node such as an intersection by voice beams of ultrasonic wave with high directivity, which is anchored in the immediate space (Refer to Fig.3). The control system finds an appropriate voice guide which is received from the navigation center through the Pod casting server corresponding to the requests which are sent from a cell phone using Bluetooth or RFID tags. The voice guides are made by a text to speech system: SmartVoice. The pedestrian can also know actions to cope with danger with accurate direction and an artificial space

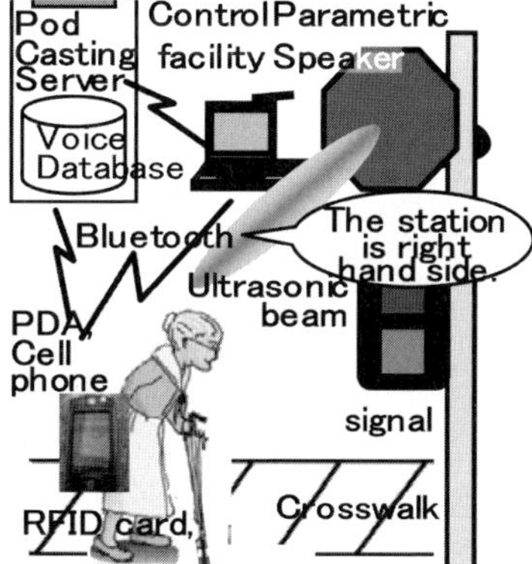

Fig. 3. A quiet route navigation system with a mobile terminal

affordance both on the ground and under ground. The pedestrian can listen to the explanation again by the terminal.

■ **The set up of a navigation system**

We examined navigation by the PS and a mobile terminal. The start point was the north gate of Tokai University that is 52 meters above

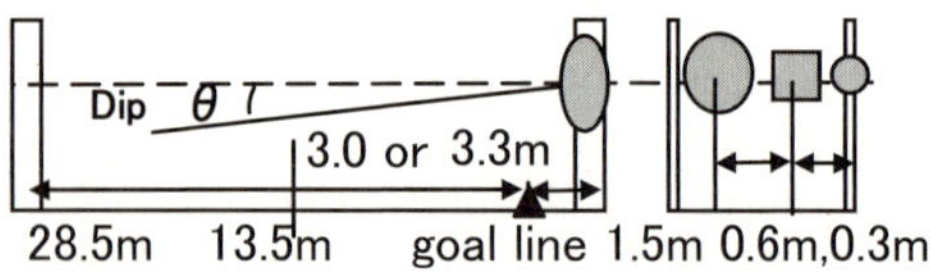

Fig. 4. The set up of parmetric speakers

sea level. The destination was the railway station. The dips (θ) of speaker is 16°. Sound pressure was between 75 and 64 dBA measured at 1.6 m high.

(1) Pre-test 1. Easy catch of a voice explanation at the start point and in the park. Subjects are eight sighted men (7 in their 20s one in his 50s) and five visually impaired persons. The contents of the explanation are a guide for walking to the railway station. All pedestrians could understand long explanations transmitted by the PS. The voice of the guide was low but clear like a voice heard by an open-air type of ear phone since the system reproduced the voice close to pedestrian's ears.

(2) Pre-test 2. Guide in the right direction near Hiratsuka station and in the Univ.

 – All pedestrians were guided in the right direction and walked inside of the crosswalk. The difference between tracks is less than 1.5 m at each point.
 – Six sighted pedestrians could hear a voice navigation by PS at the point 55 meters away in an under ground path. Four sighted pedestrians could do so at the point 70 meters away on the ground. They could find a way to the station.

We could ensure that the navigation system guides a pedestrian both from the front and from the left/right sides. The pedestrian could also modify the direction.

(3) Test. Orientation from Tokai University to Tokai University Mae station

■ **Subject.** A subject is a woman of about thirty with low vision.
■ **Course.** Start point is the north gate of Tokai University and destination is Tokai-University Mae station in a hilly area. (See Fig. 5)

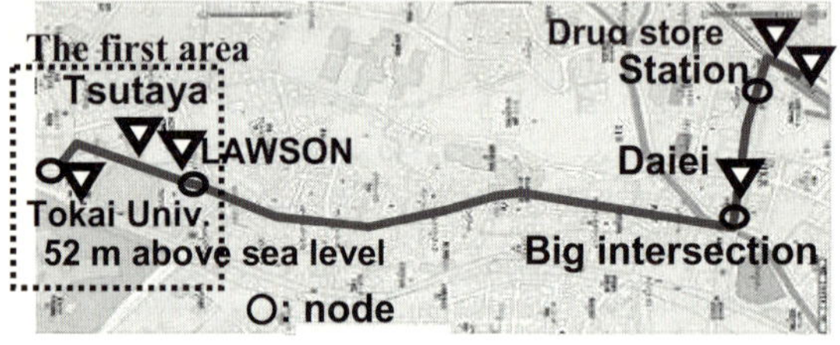

Fig. 5. A safe route to the train station

Fig. 6. A pair of landmarks

■ **Testing method.** (1) The pedestrian took pre-test2 before the test and could hear a long explanation by the PS. (2) The pedestrian listened to the voice navigation once or twice at each node such as an intersection and walked to a target point. She repeated this task until she reached to the destination. She used a mobile terminal and heard a voice guide in front of her instead of the voice by the PS as it rained.

■ **Success on the trip.** She could safely reach to the station by herself through four sub-routes using the navigation system in spite of obstructions and a tough intersection.

Sub-route1 (from Tokai Univ. to the second intersection). The pedestrian with low vision easily listened to the guide and remember a list of landmarks at the first intersection. She safely crossed a street and discovered big signs of Tsutaya and LAWSON. When she passed the parking of LAWSON, suddenly a car turned right before her and entered into the path as the navigation system warned. She stopped and watched the car passing, then reached to the second intersection with no trouble.

Sub-route2 (from the second intersection to Daiei store). The pedestrian easily built mental map of the second sub-route hearing the second guide. A steep slope and a bumpy road started in the route. She unconsciously kept paying attention to both a car upcoming from the right hand side and the shape of bumping road. So she missed the sign of AHAHI news paper on the left hand side. An unpredictable situation waited for her at the third intersection. She could not look for a signal for pedestrians since there was no such signal. She could not also find signals for car drivers because of her narrow visual field. The signal for car drivers was extremely tall because the driver driving down could watch it from a distance. She noticed a man passing her and crossed the street following him at the intersection.

Sub-route 3 and 4(from Daiei store to Tokai University Mae station). When she reached to the station building she could not find an unfamiliar sign of drug store. The sign was not big and similar that of next store. Fortunately she could notice pipe fences close to her since she could remember them in her mental map as a pair of landmarks (See Fig. 6). If a common pedestrian could not remember such small objects the directive voice guide by the PS would be very helpful for the pedestrians to find a right way in the last several ten meters to the entrance of destination.

■ **Hearing from the subject A.** In the area where more than two famous signs locate the other spatial objects can be omitted. The road near the signs is usually safe. She could have inner reserve and cope with a car suddenly entering into the path as she had heard the prediction. A cold shiver ran down her spine when a bike passed beside her at the intersection with no signal for pedestrians. A pair of landmarks with pipe fences was very helpful to find a way to station in the last several ten meters.

■ **Hearing from the subjects B.** Five visually impaired persons told that the directed voice navigation was very useful to find the right direction and a right way. They strongly expected that the systems were settled at important nodes in the town.

5 Conclusions

We proposed a method for structuring spatial objects corresponding to the characteristics and limit of capability of human memory. We also examined a route navigation system for a pedestrian to safely reach to the entrance of destination by herself. She could easily remember main spatial information in her mental map and could safely reach to the destination in spite of sudden obstruction and tough intersection. We found that the system was effective for this purpose and the voice guide with accurate direction by PS could solve the problems (a) to (g) except (d) mentioned in section two. The system chose an easy way for problem (d) and assisted the pedestrian by predictions concerning condition of road and upcoming dangers. We could ensure that the duplicate expression was effective to avoid misunderstanding at each intersection. The pair of spatial objects was indispensable for finding a way to the entrance of destination. We found the pedestrian used two famous signs based on the guide equivalence.

References

1. Ministry of Land, Infrastructure and Transport (2008), `http://www.jiritsu-project.jp/`
2. Lynch, K.: The Image of the City. MIT press, Cambridge (1960)
3. Police in Japan (2007),
   `http://www.keishicho.metro.tokyo.jp/kotu/kourei/koureijiko.htm`
4. NAVITIME JAPAN: NAVITIME (2008), `http://www.navitime.co.jp/`
5. KDDI: EZ NAVI Walk (2008), `http://www.au.kddi.com/ezweb/service/ez_naviwalk/`
6. N. Kyouno: Technology Trends on Parametric Loudspeaker. JSME (2004-7)
7. Miyachi, T., Balvig, J., Kisada, W., et al.: A Quiet Navigation for Safe Crosswalk by Ultrasonic Beams. In: Apolloni, B., Howlett, R.J., Jain, L. (eds.) KES 2007, Part III. LNCS (LNAI), vol. 4694, pp. 1049–1057. Springer, Heidelberg (2007)
8. Seki, Y., Sato, T.: Development of auditory orientation training system for the blind by using 3-D sound. In: Proc. of CVHI 2006, CD-ROM (2006)

Ubiquitous Earthquake Observation System Using Wireless Sensor Devices

Hirokazu Miura, Yosuke Shimazaki, Noriyuki Matsuda, Fumitaka Uchio,
Koji Tsukada, and Hirokazu Taki

Faculty of Systems Engineering, Wakayama University

Abstract. Japan suffers from a high number of earthquakes, so the technology for an earthquake observation is highly demanded. One of information on an earthquake is seismic intensity information which is measured with seismic intensity meters and is announced by the media such as TV. The seismic intensity is a value observed at a location where a seismic intensity meter is placed. This means that the seismic intensity shows wide area information such as administrative districts, e.g., city, town or village. However sites where equal seismic intensity was observed would not necessarily suffer the same damage, because damage depends on the type of construction and on the nature of the seismic motion. Therefore, it is difficult to estimate the individual damage in personal area. In order to measure all the locations, it is necessary to place huge number of seismic intensity meters. In the paper, we develop the system which can observe an earthquake by using wireless sensor network technology. We use wireless nodes equipped with acceleration sensors. The system can provide useful information to predict the individual damage.

1   Introduction

In Japan, the technology for an earthquake observation is highly demanded, because an earthquake happens frequently. One of information on an earthquake is seismic intensity information which is measured with seismic intensity meters and is announced by the media such as TV. This information includes magnitude of an earthquake and location where larger earthquake than intensity 1 is observed. Seismic meters are placed by strict constraint to measure ground motion. The seismic intensity is a value observed at a location where a seismic intensity meter is placed. This means that the seismic intensity shows wide area information such as administrative districts, e.g., city, town or village, and it depends on location of seismic intensity meters. The seismic intensity occasionally varies even within a city. In general, the damages caused by earthquakes are different according to the location or architecture of buildings. The seismic intensity is usually measured on the ground, but the shaking on upper stories of buildings may be amplified greatly. On the other hand, sites where equal seismic intensity was observed would not necessarily suffer the same damage, because damage depends on the type of construction and on the nature of the seismic motion. Therefore, it is difficult to estimate the individual damage in personal area. In order to measure all the locations, it is necessary to place huge number of seismic

I. Lovrek, R.J. Howlett, and L.C. Jain (Eds.): KES 2008, Part III, LNAI 5179, pp. 673–679, 2008.

intensity meters. There are about 600 JMA (Japan Meteorological Agency) [1] seismic intensity observation stations throughout Japan as of April 1996. However, those meters can not cover all the locations.

In the paper, we develop the system which can observe an earthquake by using wireless sensor network technology [2][3][4]. We use wireless nodes equipped with acceleration sensors. The system can provide useful information to predict the individual damage. Sensor nodes are arranged at observing point. The sensor node measures acceleration and send it to the system periodically. Earthquake observation system recognizes the time when an earthquake occurs and finishes from this acceleration data. Then the system outputs seismic information which indicates the situation the acceleration varies by seismic motion during earthquake. We believe that the user can intuitively estimate damage cased by earthquakes from our system.

2 Ubiquitous Earthquake Observation System

2.1 System Overview

This system recognize earthquake by acceleration from sensor nodes. The system continues to obtain the acceleration during earthquake.In the end of the earthquake, the system provides the observation results.

This system consists of sensor nodes equipped with accelerometer, base station and earthquake observation program. Fig. 1 shows system overview.

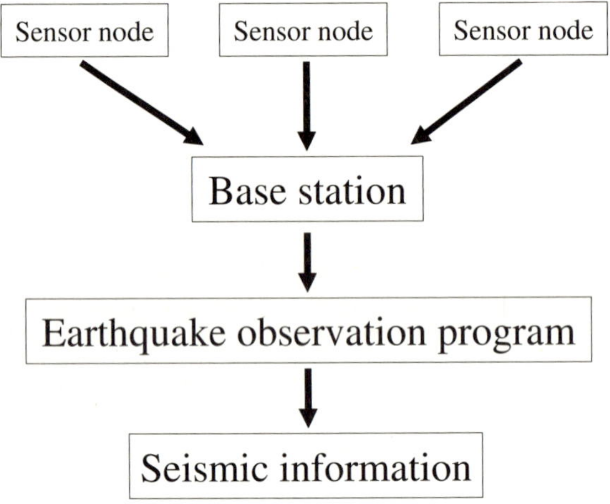

Fig. 1. System overview

The sensor nodes always measure the acceleration and send it to the base station. The earthquake observation program obtains the acceleration datum through the base station. The program detect earthquake from obtained acceleration data and outputs seismic information (Fig.2). In Fig.2, the horizontal axis shows observed time, and the vertical axis on the left side and on the right side shows acceleration (gal) and the

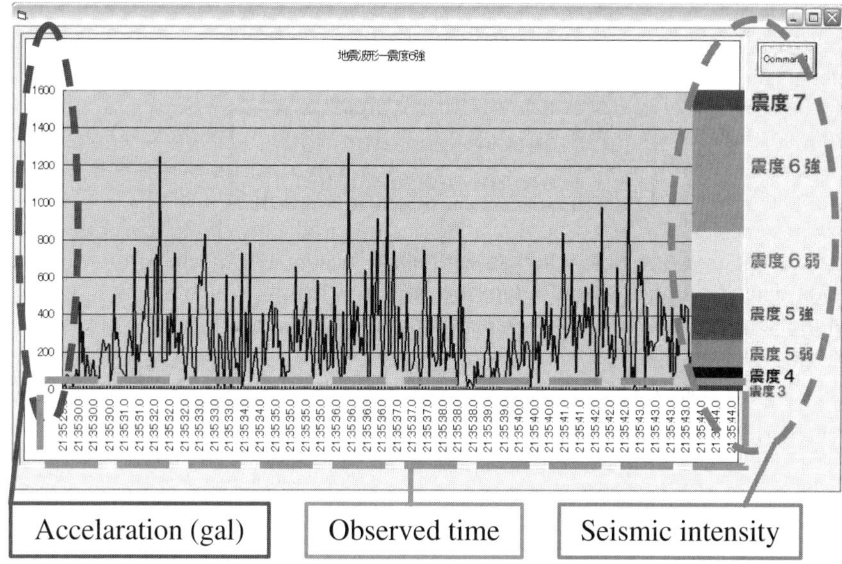

Fig. 2. Output image

seismic intensity, respectively. Figure 2 indicate the situation the acceleration varies by seismic motion during earthquake. The seismic intensity is now divided into 10 scales in Japan. It is difficult to calculate an accurate seismic intensity. For the simplicity, Table 1 is used for indication of the seismic intensity.

Table 1. Maximum acceleration corresponding to each seismic intensity

Seismic intensity scale	Maximum acceleration (gal)
Intensity 4	40 – 110
Intensity 5 Lower	110 – 240
Intensity 5 Upper	240 – 520
Intensity 6 Lower	520 – 830
Intensity 6 Upper	830 – 1500
Intensity 7	Larger than 1500

2.2 Earthquake Observation Program

This section describes the earthquake observation program in detail. Fig.3 shows flow chart of the earthquake observation. When the obtained acceleration data is higher than a threshold, the program calculates the resultant acceleration of x and y axis. The

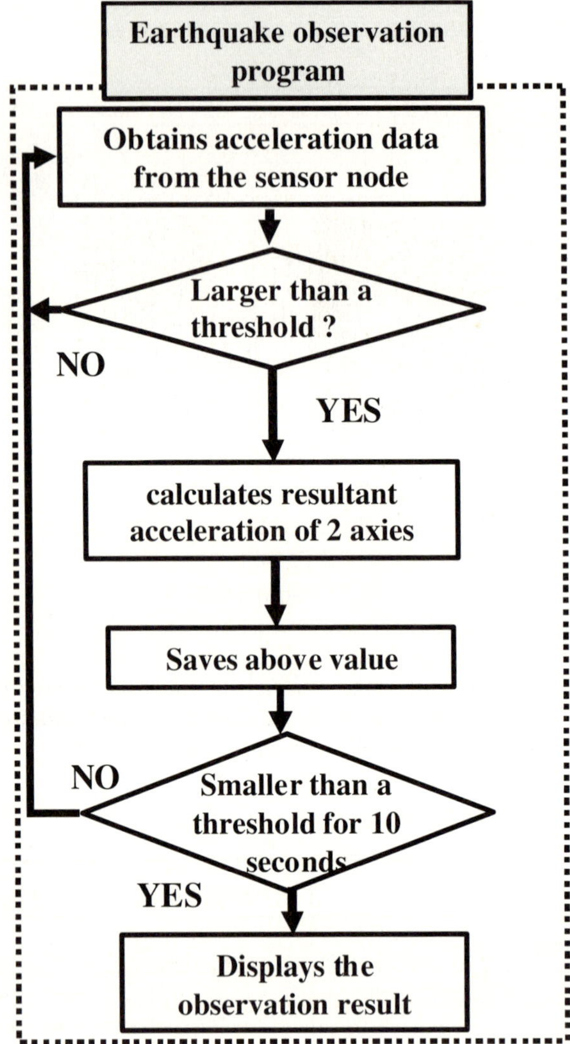

Fig. 3. Flow chart of the earthquake observation

initial setting of the threshold is 30 gal. The program reserves the resultant acceleration and that time. After the system observes the lower acceleration than 50 gal for 10
seconds, the observation is finished.

During earthquake observation, the sensor nodes can be broken by an earthquake
and communication failure between the node and base station can occur. In that case,
we cannot obtain the seismic information from that node. However, information of
node error is also useful information for estimation of damage. In the case where there
is no updating of information during a certain period, the system detect node error and
inform it to a user.

3   Evaluation

We implement our localization system using an ad-hoc wireless sensor device, Mote [4] in Fig.5. MICA2 and MIB510 in Fig.5 are used for the sensor node and the base station in our system, respectively.

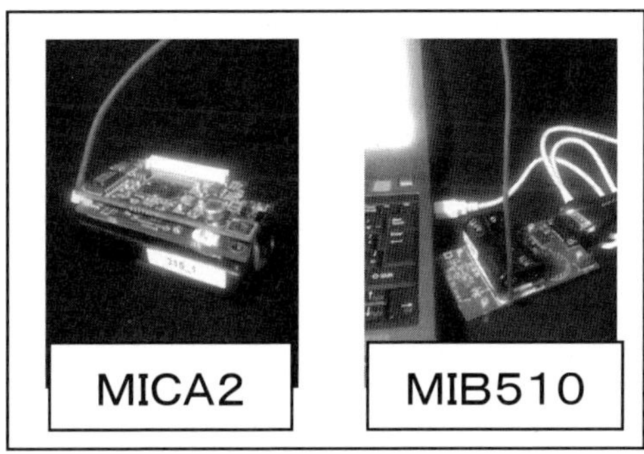

Fig. 4. Mote

3.1   Evaluation of Basic Characteristics

We investigated the basic performance of our system experimentally. We arranged the sensor nodes at 3 locations (A, B, C). Then, we generate dummy earthquake by shaking the locations for 10 seconds. We compare the generation time, duration time and finish time of the earthquake in the observation results with the actual shaking. Furthermore, we investigate whether the system can detect the error when breakdown of sensor node and communication failure occurs. Dummy earthquake for 15 seconds is generated and one sensor node among 3 nodes was stopped after 5 seconds past. In this case, we confirmed that the system showed the message which indicates node error.

Table 2 shows the experimental results. Fig.5 shows the output image in location B from the system. From these results, the system can detect 10 seconds earthquake. However 10 seconds delay can be seen in the results of generation and finish time.

Table 2. Generation time and finish time of shake

	start	finish
Actual motion	16:35:30	16:35:40
Location A	16:35:41	16:35:50
Location B	16:35:40	16:35:50
Location C	16:35:40	16:35:51

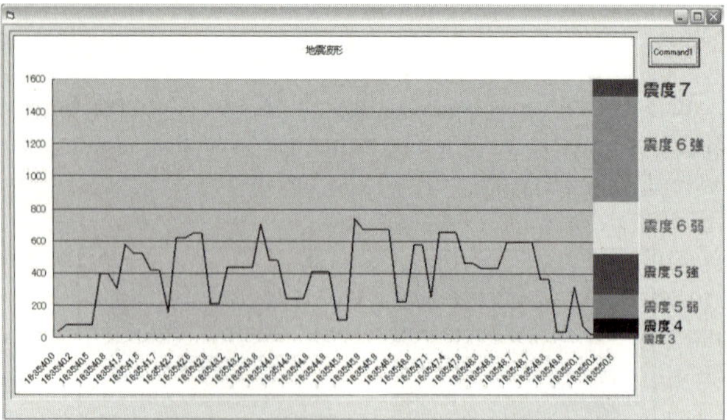

Fig. 5. Results in location B

3.2   Evaluation in Different Seismic Intensity

We investigated the performance of our system in different seismic intensity. We install our system in an earthquake experience car which has mobile earthquake simulation room. Earthquakes from intensity 4 to intensity 7 are generated in this experiment. Fig.6 shows the typical results in each seismic intensity. The observed acceleration varies intensely as the intensity become larger, as shown in Fig.6. The user can intuitively estimate damage cased by earthquakes from these results.

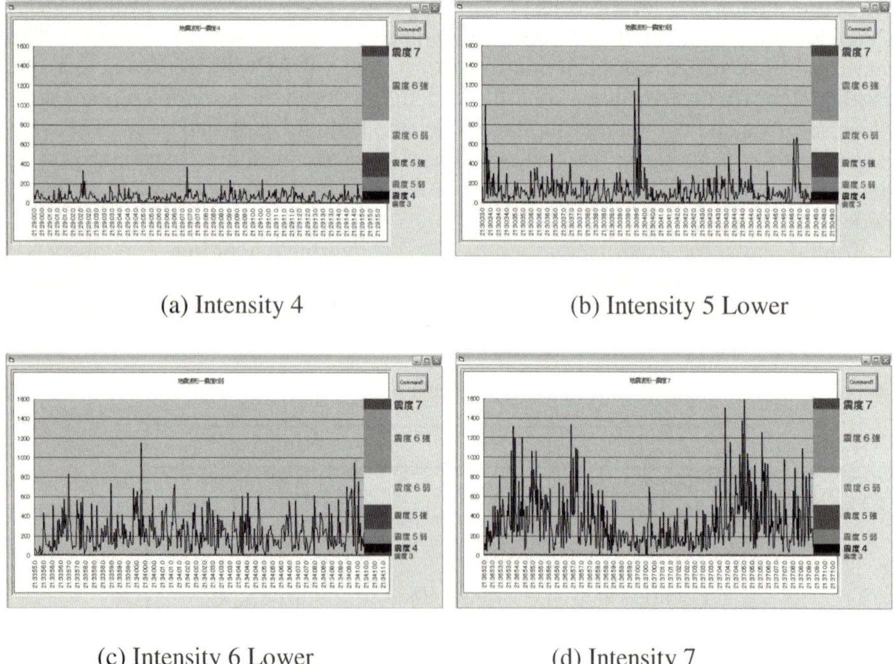

    (a) Intensity 4   (b) Intensity 5 Lower

    (c) Intensity 6 Lower   (d) Intensity 7

Fig. 6. Results in each seismic intensity

4 Conclusions

In the paper, we develop the system which can observe an earthquake by using wireless sensor network technology. We use wireless nodes equipped with acceleration sensors. The system output the graph which shows observed time and acceleration value. From our experimental results, the system can provide useful information to predict the individual damage and the user can intuitively estimate damage cased by earthquakes from our system. Future work will integrate the localization system [5][6] and this earthquake observation system to create seismic intensity map.

References

1. Japan Meteorological Agency (JMA), `http://www.jma.go.jp/jma/indexe.html`
2. Ganesan, D., Estrin, D., Woo, A., Culler, D.: Complex Behavior at Scale: An Experimental Study of Low-Power Wireless Sensor Networks, Technical Report UCLA/CSD-TR 02-0013 (2002)
3. He, T., Huang, C., Blum, B.M., Stankovic, J.A., Abdelzaher, T.F.: Range-Free Localization Schemes in Large Scale Sensor Networks. In: The Ninth Annual International Conference on Mobile Computing and Networking (MobiCom 2003), San Diego, CA (September 2003)
4. Mote, `http://www.xbow.com/`
5. Taki, H., Hori, S., Miura, H., Matsuda, N., Abe, N., Sakamoto, J.: Indoor Location Determination using a Topological Model. In: Khosla, R., Howlett, R.J., Jain, L.C. (eds.) KES 2005. LNCS (LNAI), vol. 3684, pp. 143–149. Springer, Heidelberg (2005)
6. Miura, H., Sakamoto, J., Matsuda, N., Taki, H.: Adequate RSSI Determination Method by Making Use of SVM for Indoor Localization. In: Gabrys, B., Howlett, R.J., Jain, L.C. (eds.) KES 2006. LNCS (LNAI), vol. 4252, pp. 628–636. Springer, Heidelberg (2006)

An Effectiveness Study of Pictogram Elements for Steps in Manufacturing Procedures

Atsuko K. Yamazaki[1], Kenji Goto[2], Hirokazu Taki[3], and Satoshi Hori[4]

[1] Faculty of Engineering, Shibaura Institute of Technology, 307 Fukasaku, Minuma-ku,
Saitama-shi, Saitama, Japan
atsuko@sic.shibaura-it.ac.jp
[2] Mitsubishi Electric Corporation
Pge510@yahoo.co.jp
[3] Faculty of Systems Engineering, Wakayama University, 930 Sakaedani, Wakayama-shi,
Wakayama, Japan
taki@sys.wakayama-u.ac.jp
[4] Department of Manufacturing Technologists, Institute of Technologists,
333 Maeya, Gyoda-shi, Saitama, Japan
hori@iot.ac.jp

Abstract. In this study, the authors investigated the effectiveness of pictograms for steps in manufacturing procedures. Pictograms for manufacturing steps, such as "cut", "push" and "measure," were designed for the study. Their effectiveness was examined by means of both questionnaires and a comprehension experiment related to the subjects' response actions to the pictograms. The questionnaires were presented to 72 Japanese and 40 Portuguese-speaking subjects. The comprehension experiment was conducted with a different group of 69 Japanese subjects. Results from the questionnaires and the experiment suggest that the meaning of a pictogram containing an object can be comprehended better than one without an object. In particular, the pictograms for procedures performed by a body part, such as pushing by a finger and stepping by a foot, induced erroneous responses when they were presented without an object. On the other hand, the pictograms for procedures using a tool, such as cutting with scissors and measuring with a scale, tended to be comprehended more correctly and spontaneously than the procedural pictograms containing body parts.

Keywords: pictogram, comprehensibility, manufacturing instructions, communication.

1 Introduction

Pictograms are often used in situations where the meaning of a message needs to be comprehended quickly, since well-designed pictograms are intended to convey meanings perceptively. They are also useful for situations where verbal communication is difficult or takes time [1] [2]. For example, a traffic sign with no written words is often observed in the countries and regions where more than one language is used [3][4]. Pictograms included in medicine instructions have been

I. Lovrek, R.J. Howlett, and L.C. Jain (Eds.): KES 2008, Part III, LNAI 5179, pp. 680–686, 2008.

regarded as a way to help low-literate patients [5]. Many studies point out that pictorial symbols and signs can be a useful means of delivering information to people with limited verbal skills or to people who have difficulty communicating in the dominant language [2] [5] [6] [7]. In this sense, proper pictograms can be effectively used in situations such as factories where workers and supervisors do not share the same language and cultural background.

Some studies have been conducted to evaluate the effectiveness of public information pictograms [2][5][8]. Mackett-Stout and Dewar [7] found significant positive correlations between the legibility distance and comprehensibility of the signs in their evaluation tests. Warning that many public signs and pictograms have been placed in work sites and public places without undergoing any evaluation in spite of international standards for evaluating the comprehensibility of pictorial symbols, Wolff and Wogalter [2] point out the importance of conducting evaluation tests for safety-related signs. They also mention the needs for analyses on factors that influence comprehension tests. Dowse and Ehlers studied pictograms for medical instructions to patients from eight different South African language groups. They pointed out that pictograms developed locally were more successful than American pictograms [5].

Many of the pictograms used in public places are ones representing the meaning of a noun, but pictograms used to convey the meaning of a verb or an action are not as common. In this study, we focused on pictograms for verbs or phrases to express actions in manufacturing processes. We designed instructional pictograms with sings depicting processing actions, since manufacturing steps in factory settings often contain verbs that represent important procedures. In addition, adequate pictograms can be expected to be very useful in manufacturing settings, where worker reactions are often critical for ensuring productivity and a safe working environment.

Yamazaki *et al* examined the effectiveness of pictograms representing actions in lathe procedures [9]. The results of the questionnaire experiment showed that the pictograms developed for lathe instructions can convey their meanings as effectively as common public signs. In particular, the pictograms with the symbols emphasizing actions for "turn" and "fix" displayed a significant adaptability for use in real manufacturing settings due to their high comprehensibility scores compared to that of an emergency exit sign.

In order to test whether pictograms can be effectively employed as instructions for procedures in actual manufacturing settings, experiments should be conducted with manufacturing workers as subjects. It is also important to test the same pictogram on subjects who speak different languages and have different cultural backgrounds. In this study, pictograms for the manufacturing steps used for making septic tanks were created and their effectiveness was examined through questionnaires and a comprehension experiment. The effects of pictogram elements were also investigated in this study.

2 Pictograms for Steps in Manufacturing Procedures

Manufacturing fiberglass reinforced plastic (FRP) septic tanks usually includes the process of forming an FRP tank unit from raw materials by means of a molding press. This molding process consists of cutting a sheet of raw materials into a piece of a certain length, measuring the piece, placing the piece on a metal mold in the press machine and pushing a button to start the pressing process. These are the steps performed by a worker in an actual operation at a septic tank factory of Nikko Company in Japan. At the factory, if an emergency situation occurs during the procedure, the worker is required to inform nearby co-workers of the emergency by calling out and to push a button to stop the machine. In this study we chose three steps from the molding procedure and the actions related to an emergency situation for use in our pictogram studies. The procedural steps selected for the questionnaires and the comprehension experiments were: 1) "cut" the sheet into a piece with scissors; 2) "measure" the piece with a scale; and 3) "push" the button to start the machine. The emergency action, "call out" to co-workers, was also chosen for this study. Two other manufacturing steps, "step on" the pedal and "hit" the needle were also adopted for the purpose of our studies since they are very common steps in the factory, even though the molding procedure does not include these actions.

Table 1. The correctness of subjects' interpretation of pictograms for all subjects

Pictogram	Cut	Measure	Push	Call	Hit	Step
(a) With the object						
(b) Without the object						

Two types of pictograms were created for each action. One type has a pictorial sign representing an action only. The other type consists of the action sign and a picture element representing the object of the action. For example, one of the two pictograms for "cut" the sheet was the picture of scissors only, but the other one was composed of the pictures of scissors and a sheet. Before conducting the questionnaires and the comprehension experiments, we did a preliminary test to determine the designs of the pictograms to be investigated in the experiments. Eleven subjects who were Japanese male college students were asked to evaluate the clarity of pictogram designs in terms of their meanings. The pictograms evaluated most positively by the subjects in the preliminary test were then used in the questionnaires and the comprehension experiment. The pictograms were designed to be monotonic in color in order to minimize the effects that colors might have on subject perception. These pictograms are shown in Table.1.

3   Comprehension Questionnaires

In order to examine how effectively the pictograms convey their meanings and to see if language backgrounds can affect the comprehensibility of the pictograms, an experiment with questionnaires was conducted with two groups: one group of Japanese-speaking subjects and another group of Portuguese-speaking subjects. The respondents to the questionnaires were 72 Japanese-speaking males and 32 Portuguese-speaking subjects from Brazil, only three of whom were female. All Japanese subjects were working at the factory manufacturing septic tanks. Twenty-two of the Brazilian subjects were working at the same factory, but the rest of them were employed at another factory producing car parts..

Two questionnaires (Questionnaire A and B) were constructed. Questionnaire A contained the pictograms with the objects and Questionnaire B had the pictograms without the objects. In both, the pictograms were shown with instructions regarding the questionnaire and an example answer. All subjects were asked to write the meaning of each pictogram in their native languages. In addition, the first section of both questionnaires asked for each subject's personal information, such as sex, age, job title and the length of employment. Questionnaire A contained the pictograms shown in (a) of Table 1, and Questionnaire B used the pictograms shown in (b) of Table 1.

4 Comprehension Experiment

The comprehensibility of each pictogram was also evaluated in terms of response actions to the pictograms in a comprehension experiment conducted in a simulated work environment. Sixty-nine Japanese subjects who were 1st and 2nd year male college students participated in the experiment. The subjects were divided into two groups, Group A and Group B. Subjects in Group A were to respond to the pictograms with the objects that are shown in (a) of Figure 1 through 6, and Group B were to do the same for the pictograms without the objects in (b) of Figure 1 through 6. The numbers of the subjects in Group A and Group B were 33 and 36 respectively. Each subject was to enter a room without being given any prior instructions or information. Once in the room, he was asked to stand in front of a desk upon which were placed a roll of plastic sheet, a pair of scissors, a scale and a mock up of a button (i.e. as if for activating a device). Also, he was told that there was a foot pedal under the desk. Then, he was instructed to take a spontaneous action appropriate to the procedural step of any picture shown to him. Finally, he was shown the pictograms one by one. The response time of each subject's action to each pictogram was measured and the correctness of the subject's response was evaluated.

5 Questionnaire and Comprehension Experimental Results

The comprehensibility of each pictogram presented in the questionnaires was evaluated according to a scoring system applied to the answers written by the subjects. The subject's interpretations were assigned scores of 0, 1, 2 or 3. When an answer for a pictogram perfectly corresponded to the meaning of the pictogram, a score of 3 was given to the answer. The correctness of each subject's interpretation for each pictogram was calculated as a percentage of the total responses from both the Japanese subject group and Brazilian subject group. The score and percentage results

Table 2. Average scores (scale of 0 to 3) and percentages of correct interpretations by all subjects (Japanese and Portuguese-speaking) in the questionnaire experiment

Pictogram	Cut	Measure	Push	Call	Hit	Step
With the object (N=59)	3.0 (100%)	3.0 (100%)	92 (97.3%)	2.67 (88.9%)	3.0 (100%)	3.0 (100%)
Without the object (N =54)	2.92 (97.3%)	2.92 (97.3%)	0.65 (21.6%)	2.38 (79.3%)	2.86 (95.2%)	1.43 (47.6%)

Table 3. Average scores (scale of 0 to 3) and percentages of correct interpretations by the Japanese subjects in the questionnaire experiment

Pictogram	Cut	Measure	Push	Call	Hit	Step
With the object (N=36)	3.0 (100%)	3.0 (100%)	2.31 (76.9%)	2.67 (88.9%)	3.0 (100%)	3.0 (100%)
Without the object (N =36)	2.92 (97.3%)	2.92 (97.3%)	0.65 (21.6%)	2.37 (79.3%)	2.86 (95.2%)	1.43 (47.6%)

are listed in Table 2 for all subjects. Table 3 shows the results for the Japanese subjects and the results from the Portuguese-speaking subjects are listed in Table 4.

The comprehensibility of each pictogram evaluated by the two groups showed that there was not a big difference between the Japanese group and the Portuguese-speaking group. Both groups interpreted the meanings of the pictograms with the objects better than the ones without objects. This tendency was clearer for the pictograms representing an action using a body parts, such as "push" and "step." The pictograms showing a procedural step with a tool, such as "cut" and "measure" obtained higher percentages for meaning correctness (Table 2, 3 and 4).

Table 4. Average scores (scale of 0 to 3) and percentages of correct interpretations by the Portuguese-speaking subjects in the questionnaire experiment

Pictogram	Cut	Measure	Push	Call	Hit	Step
With the object (N=15)	3.0 (100%)	3.0 (100%)	2.0 (66.7%)	3.0 (100%)	3.0 (100%)	3.0 (100%)
Without the object (N =17)	2.18 (72.7%)	2.0 (66.7%)	0.55 (18.2%)	2.73 (90.9%)	2.45 (81.8%)	0.27 (9.1%)

The results from the comprehension experiment displayed the same tendency. The correctness of each subject's action for each pictogram was also calculated in percentage. The response time for each pictogram obtained from all subjects were averaged. The results obtained in the comprehension experiment are shown in Table 5.

Table 5. The percentages of correct response by the subjects and the average response time

Pictogram	Cut	Measure	Push	Step
With the object (N=35)	82.9% 7.81sec	82.9% 4.71sec	97.1% 2.7sec	94.3% 3.57sec
Without the object (N =34)	47.1% 9.49sec	50% 6.75sec	47.1% 4.41sec	61.8% 4.78sec

6 Discussion

The results of the questionnaire experiment showed that the pictograms developed for manufacturing steps can convey their meanings effectively. In particular, the pictograms with the objects for "cut", "hit" and "measure" displayed significant adaptability for use in real manufacturing settings because of their high comprehensibility scores.

The comparison of pictograms (a) and (b) for the same procedural step indicated that the adequacy of pictogram designs affects the comprehensibility of a pictogram. The results for pictograms with an object and without an object for a procedural step were compared in terms of their comprehensibility. The questionnaire results obtained from both groups showed that the pictograms with an object could be comprehended more effectively than ones without an object. This tendency was more pronounced for the pictograms for procedural steps performed by body parts, such "push" and "step. The pictograms including body parts without an object induced erroneous responses from many subjects in the comprehension experiment. On the other hand, the experiment results indicated that the pictograms for the procedural steps using tools, such as "cut" and "measure", tended to be comprehended correctly more often than the ones containing the body parts.

Since the correlation test results showed that a subject's individual level of procedural skills did not affect the scores in the comprehension experiment, we suspect that the illustrated actions have different levels of comprehensibility when a pictogram expressing the procedure instruction. In other words, it is important to consider whether a movement is sufficiently abstracted and embedded in a pictogram or whether the design of a sign corresponds well with the image of an action that the targeted audience may already have. Yamazaki *et al*'s study indicates that pictograms symbolizing movements should be emphasized rather than those illustrating an object subjected to the action. The analysis of the results from this study and the results from Yamazaki *et al*'s study suggest that an emphasized and appropriate illustration of an action in a pictogram with the clear picture of an object can elicit better comprehensibility of the pictogram for a procedural step.

7 Conclusion and Future Work

In this study, we examined the effectiveness of pictograms for manufacturing operations in terms of their comprehensibility. Although the results showed the significant adaptability for the use of pictograms in giving instructions, more detailed analyses of factors affecting their comprehensibility should be performed. As Wolff and Wogalter [2] mention, it is important to evaluate what determines the effectiveness or clarity of a symbol. Further studies should be conducted to examine what influences a person's comprehension of a pictogram's meaning when it is used in manufacturing settings.

In order to test whether pictograms can be effectively employed to instruct workers in procedures in an actual manufacturing setting, we plan to conduct experiments with workers of different cultural backgrounds. Morinaga *et. al.* [10] suggest that natural language sentences can be automatically converted to pictograms to represent a series of procedures, the meaning of which can be conveyed to an operator. Using such a conversion system, pictograms can be produced to instruct the next operation or to remind a cautionary action to a worker in manufacturing settings.

Finally, we would like to thank students at the Institute of Technologists who participated in this study. We also would like to extend our gratitude to Mr. Hideyuki Tagami at Nikko Company., who provided us with cooperation and thoughtful advice.

References

1. Barnard, P., Marcel, T.: Representation and Understanding in the Use of Symbols and Pictograms. In: Easterby, R.S., Zwaga, H.J.G. (eds.) Information Design, pp. 37–75. John Wiley and Sons Ltd, Chichester (1984)
2. Wolf, J.S., Wogalter, M.S.: Comprehension of pictorial symbols: Effects of context and test method. Human Factors 40, 173–186 (1986)
3. Dewar, R.: Criteria for the design and evaluation of traffic sign symbols. Transportation Research Record 1160, 1–6 (1988)
4. Zakowska, L.: Perception and recognition of traffic signs in relation to drivers characteristics and safety – a case study in Poland. Safety Science Journal 19, 227–234 (1995)
5. Dowse, R., Ehlers, M.: Pictograms for conveying medicine instructions: comprehension in various South African language groups. South African Journal of Science 100(11/12), 687–693 (2004)
6. Foster, J.J.: Evaluating the effectiveness of public information symbols. Information Design Journal 7, 183–202 (1994)
7. Mansoor, E.L., Dowse, R.: Effect of pictograms on readability of patient information materials. Pharmacother 37, 1003–1009 (2003)
8. Mackett-Stout, J., Dewar, R.L.: Evaluation of public information signs. Human Factors 23, 139–151 (1981)
9. Yamazaki, A.K.: Study of meaning comprehensibility of pictograms for lathe procedural instructions. In: The Proceeding of KES 2007, Italy, pp. 1058-1064 (2007)
10. Ito, K., Hashida, K.: Ontology mapping to promote making and understanding pictograms. DBSJ Letters 5(2), 93–96 (2006)

Analysis of Continuous Tracks of Online Aerial Handwritten Character Recognition

Kazutaka Ogura[1], Yoshihiro Nishida[1], Hirokazu Miura[2], Noriyuki Matsuda[2], Hirokazu Taki[2], and Norihiro Abe[3]

[1] Graduate School of Systems Engineering, Wakayama University,
Wakayama 640-8510, Japan
[2] Wakayama University, Wakayama 640-8510, Japan
[3] Kyushu Institute of Technology, Fukuoka, 820-8502, Japan

Abstract. This paper describes the on-line handwritten character recognition method in an arbitrary place without the information of up and down moving of the pen lifting. This method can be applied to the character interface (aerial handwritten character recognition device) with which the user moves the hand in the air. To realize the system, we selected the gyro sensor for the aerial handwritten character recognition device. This device can detect the moving direction. The advantage of this device is able to be used anytime and anywhere like a pen. It is not necessary to preparer the table in order to detect absolute position. We developed the blackboard model recognition system that includes 2 types DP matching recognition programs. The system is able to recognize the Japanese hiragana characters in 84% accuracy.

Keywords: black board model, dynamic programming, handwritten character, on-line character recognition, gyro sensor, motion vector.

1 Introduction

In present character recognition, it is main current to write the character in a specific memo object with the input device. For instance, the input device has the tablet and the display, etc. These are the best when using in the house or using it on the information display at the station. To realize the mobile input device, we must develop the tool that does not use written object but uses writing trajectory.

There is a voice recognition device for mobile information input. But, sometimes, there are other limitations, for example, we must be quite in the public train. So, the voice interface is not useful in the public train and noisy environment.

We selected a gyro sensor (a gyro mouse) for the input device that is able to be used in the air without the tablet. Our character recognition system analyzes the trajectory of the device movement as the handwriting.

In the preceding researches, these character recognition systems use the pen lifting information (pen up and down information) [1] [2]. Therefore, the system was prepared the basic information including the state of the pen up and down. The recognition bias is also based on the pen lifting. The pen up and down is natural operation to write the character on the paper. When we write the character in the mobile environment in which

I. Lovrek, R.J. Howlett, and L.C. Jain (Eds.): KES 2008, Part III, LNAI 5179, pp. 687–694, 2008.

there is not a tablet or a paper, the pen up and down operation can not be done. Therefore, the aerial handwritten character recognition system has not to use the pen up and down information.

Alphabet and number characters are able to be recognized in Palm OS without pen lifting, though it is necessary for the user to write the character on the tablet. Our proposal method can recognize the hiragana characters using aerial handwriting device in the mobile environment. A wearable input device is one of our target like FMRID[3] and DigiTrack[4].

We are planning to use our system for sentence or keyword input interface of mail, the DVD recorder and the cellular phone.

2   Character Recognition with Gyro Sensor

2.1   Gyro Sensor

In the early study, we have developed the another character recognition system using a CCD camera [5]. In this technique, the direction element of the camera movement is extracted from the frame difference. The direction data are converted into the moving vector information. The system compares observed data and reference character data in the prepared character dictionary. In this research, we use the gyro sensor that can detect the moving direction in three-dimensional space instead of the CCD camera. The following items are compared with the CCD camera and the gyro sensor usage in our method.

- It is not necessary to do heavy image processing of the usage of the CCD camera.
- This gyro information processing is lighter than one of the CCD camera
- The calibration operation for the gyro sensor is unnecessary.

So, it is easy for low performance computer to use this device at any mobile environment.

2.2   Sensor Information

The shape and area of the aerial handwritten character are free form and size. So, our system can not treat this information. Information on the direction vector is extracted from the movement of two dimension coordinates obtained from the sensor. It is encoded into a direction symbol. The system selects a character candidate according to the symbol.

2.3   Method of Writing Character

There are several issues to implement this system. They are "how to detect the start and the end position of a character" and "how to select one candidate of a kanji character made from a series of kana characters". Therefore, to avoid these issues, the user gives the initial point and termination of a character to the system for each character.

I explain the process of the character input using this system. To begin with, the user sends the starting sign to the system by the start key. After that, the user draws the character without lifting the gyro mouse in the air. At the finish of the character writing, the user tells the termination to the system by the end key. Then, the proposal algorithm recognizes the character automatically, and the result of the selected character is shown on the screen.

3 Direction Code

3.1 Detection of Direction Vector

When the user moves the input device, the direction encoded data (called "direction code data") are accumulated. Similarly, terminal points are accumulated. At the points, drawing motions were stop or changed their direction. These terminal points are identified in two dimension coordinates. The system judges the terminal points when the drawing is in the narrow area.

3.2 Decision of Direction Code Data

The information of the pen up and down is not obtained because the character is written in the air. It is not limited in the two dimensional area like a flat tablet. The one fragment of the drawing stroke is not a complete datum to judge the direction. A set of a stroke and a termination of the drawing is important to decide the direction code data. When judged that there is neither place nor movement that the initiation, the termination, and the direction element of the stroke change greatly, the direction code is added. The start point of the drawing is assumed to be 'S' shown in Fig.2 Fig.1 shows Japanese character "SE". The start point of "SE" is the left hand terminal of the character.

The direction vectors are identified as this code 'S' and eight directions which are expressed nine characters of 'a'-'h' shown in Figure 2.

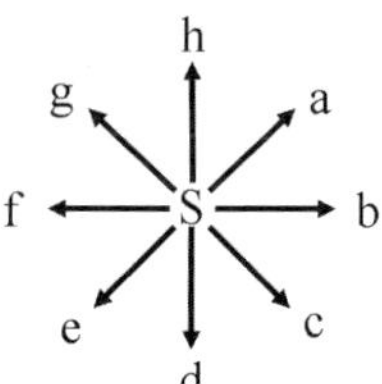

Fig. 1. Japanese Character "SE"   **Fig. 2.** Direction code data

3.3 Normalization of Personal Difference of Writing Speed

The character writing speed is different among the users. There are not only the differences of the stroke speed but also the difference of intervals between strokes. To normalize such a personal difference, the length of the stroke is adjusted. For example, the direction code data "bbbba" includes a continuous part of "bbbb", so the system converts it to short code "ba".

This system doesn't use the length of drawing stroke and the character size in aerial handwritten character. Therefore, this method uses the series of the stroke direction code

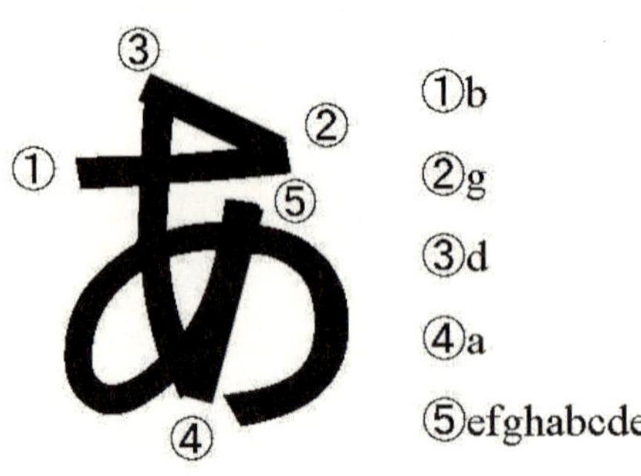

Code of '**あ**' : SbSgSdSaSefghabcde

Fig. 3. Sample direction code data

Table 1. A part of dictionary code

Character	Code
あ	SbSgSdSaSefghabcdeS
い	SdSaScdS
う	ScSfeSabcdeS
え	ScbSeSbaSeSabcdcbS
お	SbSgSdefghabcdefghabcS
か	SabcdeShSdeSabcdS
き	SbSfedcbSgSdcSfedcbS
く	SeScS
け	SdShabSgSdeS
こ	SbSedcbShS

"a"-"h". The sample code of Japanese character is shown in Figure 3. Some characters' code dictionary is shown in Table 1.

4 Evaluation of Input Data

The time series data are converted into the direction code data. They are compared with the code dictionary. In this research, the recognition technique is based on the DP matching which is used to analyze the time series data generally.

4.1 DP Matching

The DP matching is a pattern recognition technique to calculate the degree of similarity between the dictionary code and the observed object numerically. The matching rate is evaluated by this numerical value. The DP rate indicates minimum numerical value when the input pattern corresponds to the dictionary one.

At first, the recognition technique 1 uses the DP matching to decide the direction code data. In the second process, the series of the direction code data is compared to the dictionary.

If the input code is equal to the same code in a character reference data of the dictionary, the value 1 is added to the decision rate else the value 10 is added to the decision rate. This comparison process is done for each character in the dictionary is calculated. As a result, the character, which rate is the smallest value, is selected for the candidate of the input character. The penalty is added to the rate when the direction is difference from the reference code of the dictionary. The result of this technique is dependent on the feature of stroke data.

In this method, the input code is equal to the same code in a character reference data of the dictionary, the value 1 is added to the decision rate, too. This method uses other penalty calculation. So, we defined new penalty code rate shown in Fig.2.

- As the difference between the input code and the dictionary one is bigger, the penalty value is increased according to this difference.
- The period of the drawing termination is also calculated. If stopping time in the point is short, the penalty is added to the rate.

This method enhances the difference of the direction more than the recognition technique 1. This method decreases the penalty of code "S", so relatively the specification of strokes is stronger than the recognition technique 1.

Figure 4 shows the difference of the evaluation values of two methods when the system recognizes stroke directions of the Japanese character of Fig.3.

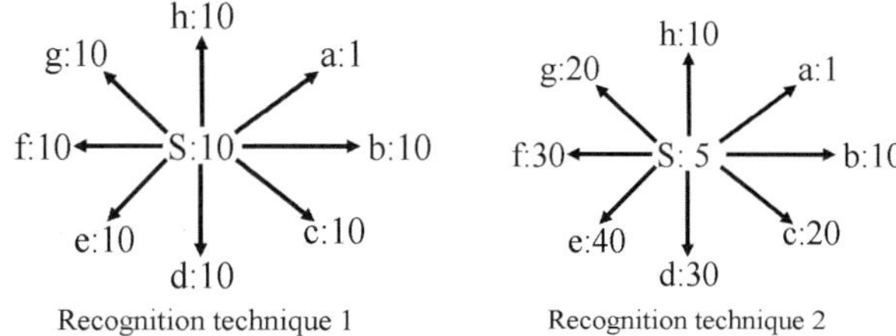

Fig. 4. Difference of direction evaluation rate

4.2 Recognition Rate of Recognition Technique

Each ratio of two recognition techniques was examined in correct recognition cases. We have examined the correctness to recognize a set of similar characters. To compare two methods, they recognized the same input data in the same dictionary. To study the recognition ratios, each hiragana character is judged three times. Smallest value of the recognition ratio shows the most suitable character. The dictionary data was made from the character written by us, another person made the test input data.

In the recognition technique 1, the recognition success ratio was 80.8% and the failure ratio was 19.2%. In the recognition technique 2, the success ratio was 82.2% and the failure ratio was 17.8%. The recognition technique 2 is better than technique 1.

4.3 Consideration of This Recognition Technique

Normally, the evaluation rate is from 20 to 40. When the rate is more than 100, there are repeated patterns like "ababab…". There is no big difference of the evaluation rates between two recognition techniques. However, the set of the candidate characters are not same. For example, cha-14 and cha-15, etc. are misunderstood to a candidate of cha-3. They are treated a similar character set in the recognition technique 1. On the other hand, cha-13 and cha-26, etc. are misunderstood to a same character in the recognition technique 2. In the recognition technique 1, the characters which have many strokes like cha-17 or cha-27 are recognized successfully. The recognition technique 2 recognizes cha-3 and cha-28 which have small circle successfully.

The high score candidates have small numerical value as the decision rate. The cases of these score differences between top two candidates which are smaller then 10 are 35%. There are important features in Japanese Hiragana characters. They are

"Tome", "Hane" and "Harai" which are represented by the pen up and down. For example, the stroke without the pen lifting almost can not find the difference among cha-29, cha-16, and cha-30. Then the recognition of them makes mistakes.

High recognition failure rate characters are shown in the following Tables 2.

Table 2. A part of character of recognition mistake

		cha-1 "き"	cha-2 "く"	cha-3 "す"	cha-4 "ち"	cha-5 "ぬ"	cha-6 "ほ"	cha-7 "ま"	cha-8 "わ"
	Recognition Character								
Recognition technique1	Failure Candidate	cha-9 "を"	cha-10 "い", cha-11 "と", cha-12 "リ", etc.	cha-13 "は", cha-14 "ゆ", cha-15 "る", etc.	cha-3 "す", cha-16 "ら"	cha-17 "ね"		cha-18 "か", cha-19 "せ", cha-20 "や", etc.	cha-21 "あ"
Recognition technique2	Failure Candidate	cha-22 "さ", cha-23 "そ", cha-24 "た"	cha-10 "い", cha-25 "へ"	cha-13 "は", cha-26 "よ"	cha-16 "ら"	cha-17 "ね"	cha-13 "は", cha-7 "ま"	cha-20 "せ", cha-13 "は", cha-26 "よ", etc.	

cha-27 "み", cha-28 "な", cha-29 "う", cha-30 "ろ", cha-31 "て", cha-32 "ん"

We developed the other solution to solve this misrecognition problem. The reshaped characters, which are easy to be recognized, are developed like "Graffiti". The other method is machine learning to tune decision parameters. We are also developing an advanced method which the recognition technique 1 and 2 are combined.

5 Compound Judgment Technique

We propose the compound judgment technique that combines two recognition techniques. The system uses the blackboard model that includes multiple knowledge resources or multi-decision resources.

5.1 Blackboard Model

The blackboard is a decision making inference system that can be accessed from two or more agents as knowledge resources. The stroke data and terminal data of the input character are written in its system. The blackboard can be accessed by all agents at any time. Each agent reads and writes data on the blackboard according its judgment. Or there is a use to exchange data through a blackboard between agents.

In this proposal technique, the numerical values of the input character are calculated by the two recognition techniques, and they are stored in the blackboard. These values are judged by an evaluation algorithm (i.e. our proposal decision resource).

They are maintained until all judgments are finished. This system lists up the candidates.

5.2 Judgment Process

At first, the time series data of two dimensions coordinates input is extracted using the gyro mouse. The system converts the data into the direction cord data. After that, it takes recognition technique 1 about the data that I converted. The recognition technique 1 compares the input data and the dictionary to calculate the evaluation values. These values for each character are written on the blackboard.

The system compares these values and makes the ordering of the candidates in ascending order. If the difference between top two characters score is more than 10, the system decides the best candidate. But if the difference is small, it fails to select the candidate. So, in the failure case, the system will consider the other way that is the recognition technique 2. It gives priority to the result of the recognition technique 2 more than the technique 1. But, if the differences between top and second score are similar, it is impossible for this way to judge the result.

Finally, these two recognition techniques select same character "A" for best character. The system selects this character "A". If the techniques select other character, the system summarizes the evaluation values for each character and select the best score one.

5.3 Evaluation of Proposal Technique

We evaluated the recognition rate of the Japanese hiragana each character three times in order to evaluate the blackboard system. As a result, before the final stage, i.e. the value summarizing method, the system selects the correct character in 68.8%. In the total process, the system selects the correct character in 84.4%.

5.4 Consideration of Proposal Technique

We improved the judgment success rate in the proposal method. The blackboard system makes better selection than stand-alone evaluation techniques. The system covers the miss-selection of stand-alone techniques. Some characters cha-13, 27, 31 are selected successfully, cha-32 is never succeeded to recognize. We must make a good dictionary to use machine-learning technology to recognize cha-32.

6 Conclusion

We proposed the compound recognition technique based on the blackboard model as a new recognition technique in aerial handwritten character. It was actually confirmed to be able to recognize the Japanese hiragana characters at the probability of about 84%.

We are planning to improve the recognition ratio using machine learning technology. In the future, we believe that the smaller aerial handwritten character device will be on the fingers. The small wearable device will be a general ubiquitous input device.

References

1. Sudou, T., Nakai, M., Shimodaira, H., Sagayama, S.: On-line Overlapped Handwriting RecognitionBased on Substroke HMM, IEISE, PRMU2001-65, pp.163–170 (2002-03)
2. Oki, T., Nakagawa, M.: Implementation and Evaluation of a Combined Recognition System using Normalization of Likelihood Values, IEISE, PRMU2002-196, pp.43–48 (2003-0110)
3. Sugikawa, A., Suzuki, K.: FMRID:Finger Motion Recognition Input Device. In: Interaction 2000, pp. 91–98 (2000)
4. Mine, K., Oobuchi, R.: DigiTrack: A pointing device for wearable information appliances, Japan Society for Software Science and Technology. In: WISS 2001, pp. 125–130 (2001)
5. Nishida, Y., Naemura, M.: Online Aerial Handwriting Interface by Using Video Camera, IEISE, ITS 2005-67, pp.119–124 (2005)

Interactive Learning Environment for Drawing Skill Based on Perspective

Yoshitake Shojiguchi, Masato Soga , Noriyuki Matsuda, and Hirokazu Taki

Faculty of Systems Engineering, Wakayama Universityu
930 Sakaedani, Wakayama, 640-8510 Japan
soga@sys.wakayama-u.ac.jp

Abstract. We developed an interactive learning environment for drawing skill based on perspective. The environment supports function for learners to diagnose error of drawn lines based on perspective by themselves. The environment does not fix motif for drawing, although motif is limited to 3D figure that consists of flat surfaces.

Keywords: Learning environment, Sketch, Perspective, Skill.

1 Introduction

In the research domain of supporting painting or sketch, Bill Baxter developed an interactive haptic painting system with 3D virtual brushes [1]. This system is an excellent system that combines virtual world with real world. However, the system focuses on virtual reality of haptic feeling during painting. It helps a user to paint something in virtual world, but it does not help the user to get painting skill, because it cannot diagnose the picture painted in the virtual world by the user.

On the other hand, we have been developing sketch learning environments that have diagnosis function based on cognitive process of interaction during sketch drawing. When a learner draws a sketch, at first he/she perceives a motif and recognizes it. Then he/she selects appropriate action in mind. After that he/she acts, such as to draw an outline of the motif, or repair the shape and so on. Repeating these processes, the sketch is completed.

To generalize these process, skill repeats (1)Recognition, (2)Selection of appropriate action, (3)Action. Repeating process (1)-(3), finally a work is produced. From this point of view, if the work is wrong, the cause is in one of process (1)-(3) or more. Therefore it is important to diagnose process (1)-(3) to learn the skill. In the domain of sketch drawing, if the cause is in recognition, the learner might have wrong model of the motif in his/her mind. If the cause is in action, the learner might not know how to use his arm when he/she tries to draw smooth curves. If the cause is in selection, he/she might not able to select appropriate method to draw dark shadow.

Based on the consideration, we developed sketch learning environment with automatic diagnosis and advice. The learning environment consists of four modules that diagnose or promote (1)Recognition, (2)Selection, (3)Action, and (4)Work. The learning environment that diagnose learners drawn sketch is developed first [2]. Then, the learning environments that supports learners recognition[3][5] and that diagnose

I. Lovrek, R.J. Howlett, and L.C. Jain (Eds.): KES 2008, Part III, LNAI 5179, pp. 695–700, 2008.

learners arm motion[4][5] are developed. However, these learning environments fix motif, such as plate and glass, because the environments need knowledge of the motif while diagnosing drawn sketches by learners.

To solve the problem, we are developing motif-independent learning environment. The motif-independent learning environment assists learners to diagnose drawn lines by perspective. The learning environment is different from CAD system. CAD system does not support learners' learning.

2 How to Draw Lines by Perspective

In this chapter, we explain how to draw lines by perspective. At first, we explain the theoretical method. After that, we explain practical method.

2.1 Theoretical Method for Drawing Lines by Perspective

Theoretical method for drawing lines by perspective is as follows. At first, decide position of a vanishing point (Fig.1). Then, draw lines from the vanishing point. After that, leave necessary lines and erase unnecessary lines.

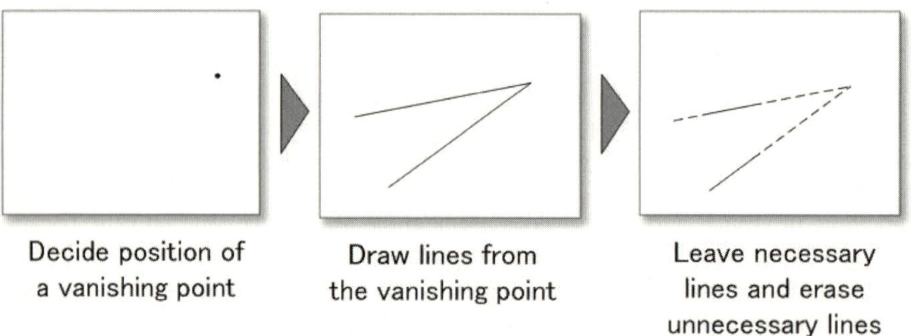

Fig. 1. Theoretical method for drawing lines by perspective

2.2 Practical Method for Drawing Lines by Perspective

However, sometimes the paper is not large enough to set position of a vanishing point inside of the paper. In such a case, a learner has to set the vanishing point outside of the paper, and the learner draws parallel lines on the estimated lines from the estimated vanishing point (Fig.2). Therefore, ability to estimate lines and vanishing point affects quality of drawn pictures by perspective.

Also, the learner cannot keep the estimated vanishing point on the paper. Therefore, the learner has to estimate again the position of the vanishing point from drawn lines, when the learner adds other parallel lines by perspective.

This method can be applied when vanishing point is located close to the paper. However, if vanishing point is far away from the paper edge, the method is not easy to apply, because position error of vanishing point becomes large.

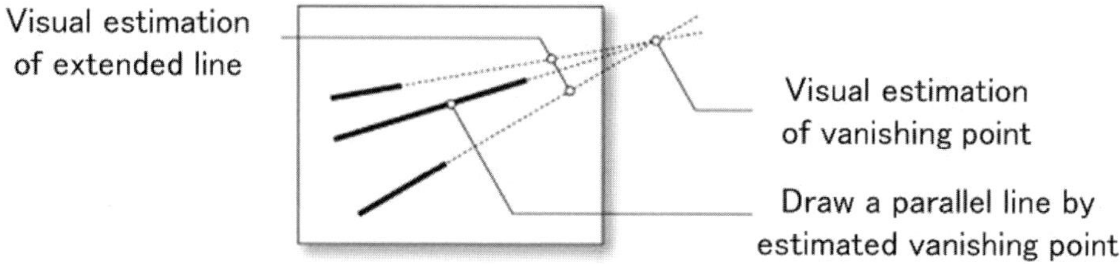

Fig. 2. Practical method to add a new parallel line by estimation of vanishing point

In this case, another practical method can be applied to add a new line. At first, a learner estimates a trapezoid that is formed by two drawn lines (Fig.3). After that, the learner estimates dividing points by equal ratio both on the bottom line and on the top line of the trapezoid. Finally, the learner draws the line by connecting two dividing points.

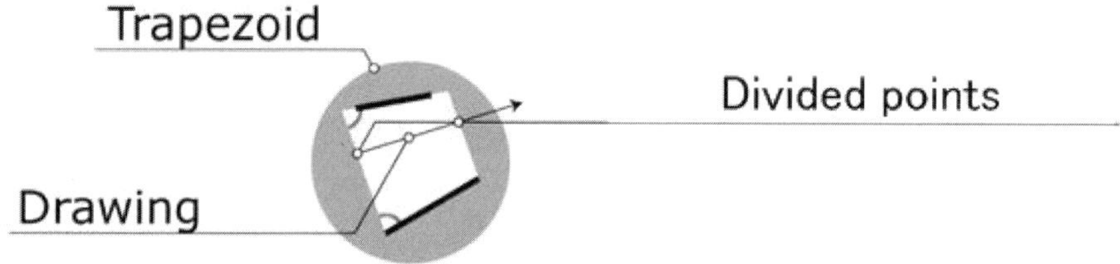

Fig. 3. Practical method to add a new line by estimation of trapezoid

Thus, to add a new line needs to estimate either vanishing point or trapezoid. It is difficult task for novice learners. Therefore, it is valuable to design and to develop the learning environment that trains and supports diagnosis of such task.

3 Design of Learning Environment

Based on consideration in chapter 2, our research goal is to design and to develop the learning environment that trains drawing skill and assists self-diagnosis of line drawing by perspective. Figure 4 shows basic design of the learning environment. The interaction between learners and the learning environment is based on repetition of action and recognition. In the figure 4, at first, a flat surface is given to a learner by perspective. The learner draws a line by perspective. In this stage, the learner has to estimate either vanishing point or trapezoid to draw the new line.

Then, the learning environment shows the flat surface with drawn line by orthogonal view. The learner can recognize error, if the drawn line is not parallel to edge of the flat surface. Then the learner corrects the error by the orthogonal view.

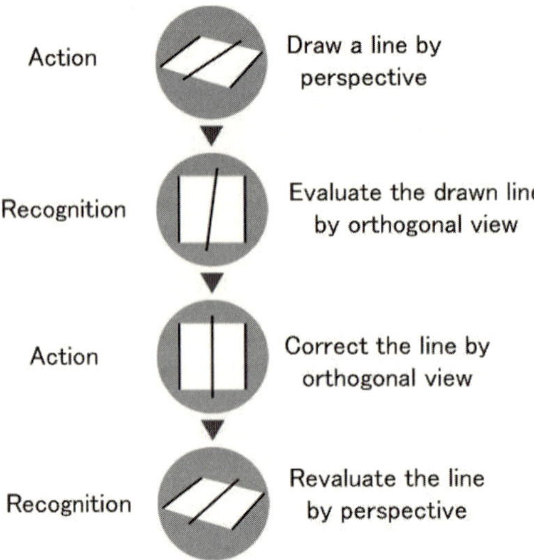

Fig. 4. Drawing and self-diagnosis process in the learning environment

After that, the learning environment shows the surface with the corrected line by perspective. The learner can recognize the corrected line by perspective. The learner can learn how to draw a line on the surface by perspective.

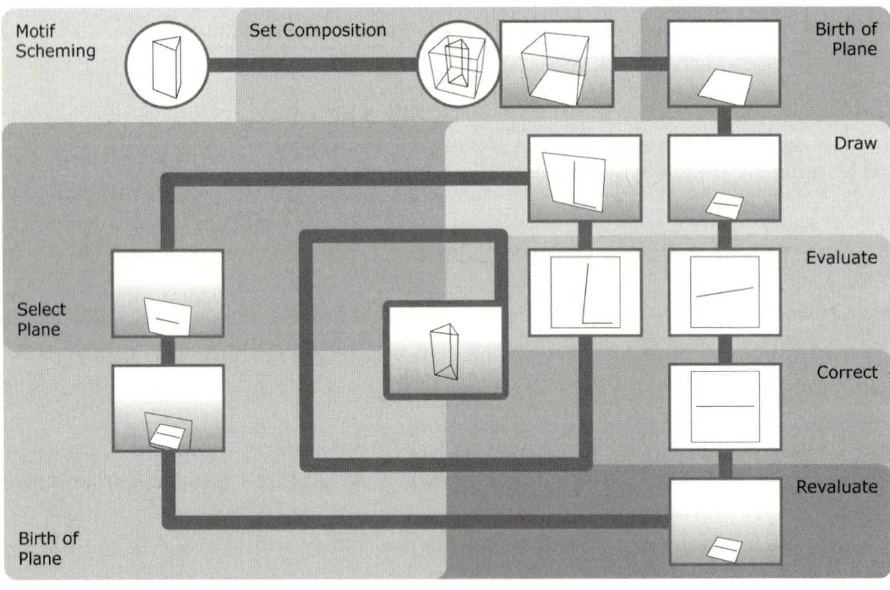

Fig. 5. Whole drawing and diagnosis process of a 3D figure by the learning environment

Repeating these processes, the learner can construct a 3D figure. Figure 5 shows whole interaction design of the learning environment. We limit 3D figures to the figures that consist of flat surfaces. 3-point perspective is used in the prototype system. Tablet PC is recommended to use the learning environment, since the learner can use pen instead of mouse. The learner draws lines on the display of the tablet by using the pen. Figure 6 shows a scene of using prototype system of the learning environment on a tablet PC.

In the figure 5, "Motif Scheming" (at upper left corner) and "Set Composition" (at upper middle) are the process in learner's mind. The learner decides what he/she constructs as a 3D figure, and learner commands the learning environment to generate a new flat surface. After that, "Draw", "Evaluate", "Correct", and "Revaluate" are already explained as figure 4. Repeating these interactions, every surface is constructed. Finally the 3D figure is completed (at the center of fig.5).

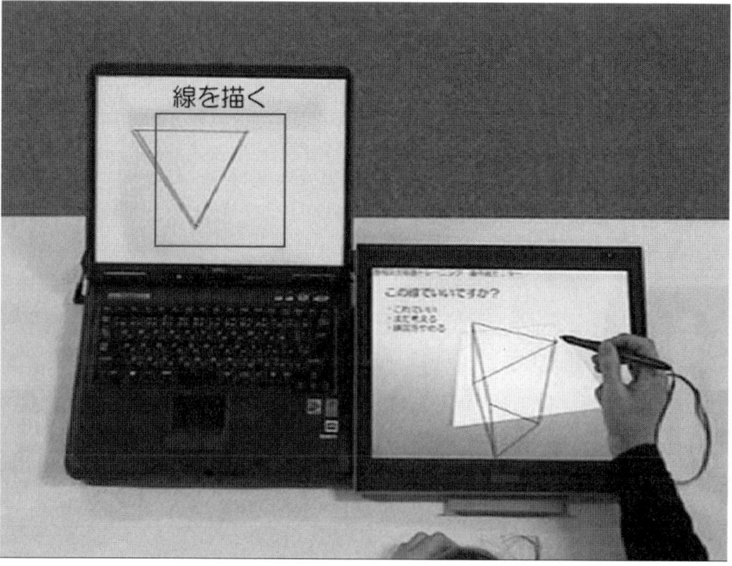

Fig. 6. A scene of using prototype system of the learning environment on a tablet PC

4 Conclusion

In this paper, we describe motif-independent learning environment by perspective. The learning environment trains drawing skill and assists self-diagnosis of line drawing by perspective. Drawing and self-diagnosis process in the learning environment are repetition of action and recognition. Repeating these processes, the learner can construct a 3D figure.

Currently, the learning environment assists learners' self-diagnosis. However, it does not support automatic diagnosis function. We are currently developing automatic diagnosis function.

Acknowledgments. This research was partially supported by the Ministry of Education, Science, Sports and Culture, Grant-in-Aid for Scientific Research   (B), 19300282, 2008.

References

1. Baxter, W., Scheib, V., Lin, C.M., Manocha, D.: DAB: Interactive Haptic Painting with 3D Virtual Brushes. In: Proc. of the 28th annual conference on Computer graphics and interactive techniques, pp. 461–468 (2001)
2. Takagi, S., Matsuda, N., Soga, M., Taki, H., Shima, T., Yoshimoto, F.: An educational tool for basic techniques in beginner's pencil drawing. In: Proc. of Computer Graphics International 2003, pp. 288–293 (2003)
3. Soga, M., Matsuda, N., Taki, H.: A Sketch Learning Support Environment that Gives Area-dependent Advice during Drawing the Sketch. Transactions of the Japanese Society for Artificial Intelligence, 96–104 (2008) (in Japanese)
4. Soga, M., Maeno, H., Koga, T., Wada, T., Matsuda, N., Takagi, S., Taki, H., Yoshimoto, F.: Development of the Sketch Skill Learning Support System with Real Time Diagnosis and Advice showing Learner's Arm Motion Animation. Transactions of Japanese Society for Information and Systems in Education 24(4), 311–322 (2008) (in Japanese)
5. Soga, M., Matsuda, N., Takagi, S., Taki, H., Yoshimoto, F.: Sketch Learning Environment based on Drawing Skill Analysis, KES2007. In: Apolloni, B., Howlett, R.J., Jain, L. (eds.) KES 2007, Part III. LNCS (LNAI), vol. 4694, pp. 1073–1080. Springer, Heidelberg (2007)

Intelligent GPS-Based Vehicle Control for Improved Fuel Consumption and Reduced Emissions

S.H. Lee, S.D. Walters, and R.J. Howlett

Centre for SMART systems
Engineering Research Centre, University of Brighton
Moulsecoomb, Brighton, BN2 4GJ, UK
S.H.Lee@Brighton.ac.uk, S.D.Walters@Brighton.ac.uk,
R.J.Howlett@Brighton.ac.uk

Abstract. The development of in-vehicle control systems such as Global Positioning Systems (GPS) provides static and dynamic road information. This widely accessible technology can be used to develop an auxiliary control subsystem to reduce vehicle fuel consumption as well as improve road safety and comfort. The raw GPS data from the receiver were processed and integrated with the past trajectory using a Neuro-fuzzy technique. The system essentially used a fuzzy logic derived relief map of the test route and this was further validated and corrected based on the past trajectory from the GPS sensor. The information was then processed and translated in order to estimate the future elevation of the vehicle. Experimental results demonstrated the feasibility and robustness of the system for potential application in vehicle control for reduced fuel consumption and emissions.

1 Introduction

The Global Positioning System (GPS) has been increasingly used in real-time tracking of vehicles, especially when GPS is integrated with ever increasingly powerful geographic information system (GIS) technologies. The accuracy and reliability of low-cost stand-alone GPS receivers can be significantly improved to meet the technical requirements of various transportation applications of GPS, such as vehicle navigation, fleet management, route tracking, vehicle arrival/schedule information systems (bus/train) and on-demand travel information. Systems that were previously only intended for fixed installation in vehicles are gradually being replaced on the market by portable systems that require no connection to the vehicle other than the power supply. To an increasing extent, GPS navigation is becoming a software product that can also be installed on handheld computers, laptops and mobile phones.

Global positioning determination is based primarily on using the GPS. Stand-alone systems such as handheld computers use this exclusively, whereas fixed installation systems also run 'dead reckoning' if they have additional in-vehicle sensors. Dead reckoning ensures exact position determination even if no GPS signals can be received, e.g. in tunnels. To measure the distance travelled, all that is needed is a speedometer output signal. The change of direction is ascertained by a rotation rate sensor or gyroscope. Hence, the absolute direction of travel can be determined by the Doppler effect of the GPS signals [1]. The levels of accuracy that can be achieved are in

I. Lovrek, R.J. Howlett, and L.C. Jain (Eds.): KES 2008, Part III, LNAI 5179, pp. 701–708, 2008.

the range of 3 to 5m, and 10 to 20m in the case of measuring altitude relative to sea level. With the autonomous European Satellite Navigation System Galileo, expected in 2008, an opportunity of a joint system 'GPS + Galileo' with more than 50 satellites will provide many advantages for civil users and vehicle systems, in terms of availability, reliability and accuracy [2].

Future GPS may not only be used to guide the vehicle but information from the system may also be used to control/influence the engine through given control parameters in a safe and cost-effective manner. A GPS receiver provides reliable reference position data which can be manipulated to provide more significant road information such as gradients or even road traffic congestion updates when it is combined with the vehicle telematics. It is a technology integrated with computers and mobile communications technology in vehicle navigation systems. This information can be used to not only inform the driver but also to enhance the control of several systems of the vehicle. The vehicle speed, gear selection and even the application of brakes could be appropriately chosen and strategically designed. The idea is to provide the control system with this essential information that the driver normally uses when driving. Good driving requires consideration of several inputs. This is a complex, exhausting and demanding task even for commercial vehicle drivers and thus supporting control functionality is of great interest. It is believed to be even valuable to obtain road information beyond the sight of the driver. Whilst all these driving decisions have to be made manually by the driver in the interest of comfort and fuel efficiency, the newly intelligent vehicle controller aims to address these tasks.

The applications of intelligent systems, i.e. software systems incorporating artificial intelligence, have shown many advantages in control and modelling of engineering systems. They have the ability to rapidly model and learn characteristics of multi-variant complex systems, exhibiting advantages in performance over more conventional mathematical techniques. This has led to their being applied in diverse applications in power systems, manufacturing, optimisation, medicine, signal processing, control, robotics, and social/psychological sciences [3, 4]. The Adaptive Neuro-Fuzzy Inference System (ANFIS), developed in the early 1990s by Jang [5], combines the concepts of fuzzy logic and neural networks to form a hybrid intelligent system that enhances the ability to automatically learn and adapt. Hybrid systems have been used by researchers for modelling and prediction in various engineering systems. The basic idea behind these neuro-adaptive learning techniques is to provide a method for the fuzzy modelling procedure to learn information about a data set, in order to automatically compute the membership function parameters that best allow the associated Fuzzy Inference System (FIS) to track the given input/output data.

Current trends towards sustainable transportation require dramatically reduced fuel consumption and emissions. This project will address the issue and challenges imposed by legislation and guidelines relating to fuel consumption and exhaust emissions with the aim to facilitate the reduction of exhaust emissions and fuel consumption through precise control of the vehicle. Techniques include the fusion of data from sources that are external as well as internal to the vehicle; also from analysis of these data using special intelligent systems techniques and tools. The system essentially used a fuzzy logic derived relief map of the test route, and this was further validated and corrected based on the past trajectory from the GPS sensor. The information was then processed and translated in order to estimate the future elevation of the vehicle. The following sections describe the project work-in-progress and reports on initial experimental results. The devised system can potentially be used in vehicle control for reduced fuel consumption and emissions.

2   Techniques

A challenge of the project was how to meet higher safety and obtain reduced fuel consumption by the use of live GPS road information for vehicle control. The approach was to use GPS to track the vehicle, and also to create the base map. At all other times GPS readings are used to validate or correct the base map when a reliable signal is available and of sufficient accuracy. The correct vehicle position was achieved by tracing this GPS signal received at a predetermined time interval.

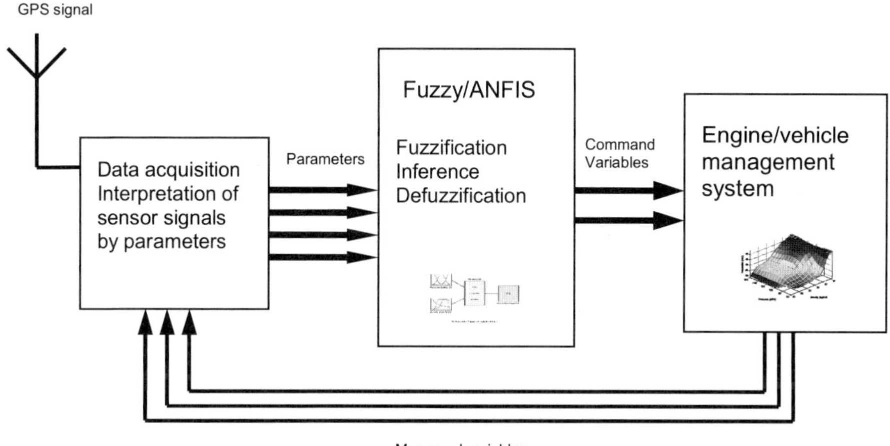

Fig. 1. Engine/vehicle control using internally and externally acquired data

The conceptual overview of this investigation is shown in Figure 1. The control system acquired the position data through the serial interface, so that it can be used to improve the operation of several sub-systems in the vehicle, e.g. controlling a series of actuators or settings. Distance travelled and vehicle speeds were recorded along the way, a 10m distance span was used initially and this was maintained by on-board timers. The ANFIS technique was used to derive a relief map of the test track, and here the position data was translated and represented by two input and one output membership functions together with twelve rules as part of the Neruo-fuzzy optimisation routine. The relief map that was devised under the scheme has been used for future gradient prediction. The chosen intelligent technique involved extraction of necessary representative features from a series of data points. Previous work using this Neuro-fuzzy derived technique for modelling fuel spray penetration was described in [6] and achieved good results.

Similar control techniques based on predictive parameters have been proven useful and achieved better results. Model Predictive Control (MPC) is an optimisation algorithm which has shown that a 2.5% reduction of fuel consumption can be achieved by controlling the speed of a vehicle. The control signals were; percentage of throttle opening, activation of brakes and gear selection. The control algorithm there has been tuned and optimised according to some criteria, e.g. the main issues are to minimise costs, time and fuel consumption [7]. Similar work described in another publication has been designed and simulated on cruise control [8]. The simulation showed a reduction of fuel consumption in the range of 1.5 to 3.4% was achieved. It used a dedicated logic in a finite number of simulated driving situations, given that the topography of the road such as gradient is a known input to the system. The Control of the vehicle powertrain has been carried by DaimlerChrysler, the research suggested a three-dimensional digital road map was used in order to let the cruise control replicate a skilled driver [9]. A lowered fuel consumption of 4.1 to 5.2% was attained. Furthermore, cruise control has now been incorporated with radar technology to record the distance and speed relative to the vehicle in front as well as additional data such as position of other vehicles in vicinity. The system used such information to regulate the time gap between vehicles. The interface was developed in the European project MAPS&ADAS to obtain the map data from the on-board data provider [10]. This is an advance convenience system which adapts the speed to the vehicles around and keeps a safe distance.

All in all, a number of approaches were researched. A substantial amount of work has been carried out on how the interface between vehicle control system and the GPS system should be designed. The investigation was focused on information retrieval and processing. Location data could be available to the vehicle control unit in a variety of formats, resolutions and temporal accuracies. Data processing and fusion forms the main part of this project. This information was made available and able to combine with other sensory data of the vehicle.

3   Experimental Setup

The experiments have been performed on a small passenger vehicle. A test route was established on the outskirt of Eastbourne in East Sussex. A stand-alone laptop with a handheld GPS device was used throughout the experiment. The devised system was not connected to the vehicle control system. Therefore, an external GPS receiver was used and that data were logged together with the time from the on-board clock through a serial Bluetooth interface. The aim of this investigation was to use a GPS receiver in conjunction with custom-written Matlab software to collect and store three-dimensional vehicle position data. The incoming stream of data was used to estimate the future elevation of the vehicle; this data was expected to be of use further for dynamically influencing the control of an engine. The programme flowchart in Figure 1 showed how these algorithms tied together to form a fuzzy predictive control system.

The main software was divided into several functional modules, each of which performed their own set of calculations and the optimisation was performed by the Neuro-fuzzy module, the FIS generated was stored in the computer memory, whilst the timing of all these activities was governed by the on-board clock and timers.

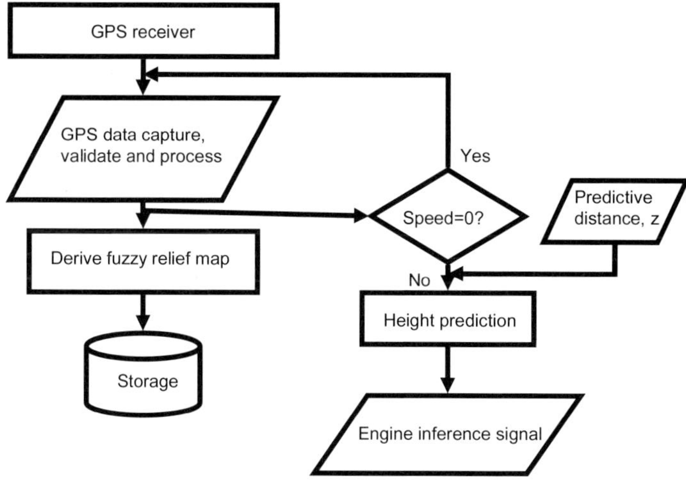

Fig. 2. Predictive fuzzy inference system

4 Road Gradient Estimation

The fuzzy predictive control scheme is shown in Figure 3. The operation is triggered by a start signal and the status of the GPS data, a few given set points are needed i.e. predictive distance and sampling rate. It begins with the first position data of the vehicle and this is registered and as a result a reference trajectory can be designed. From the reference trajectory, the next reference position is obtained according to the preset

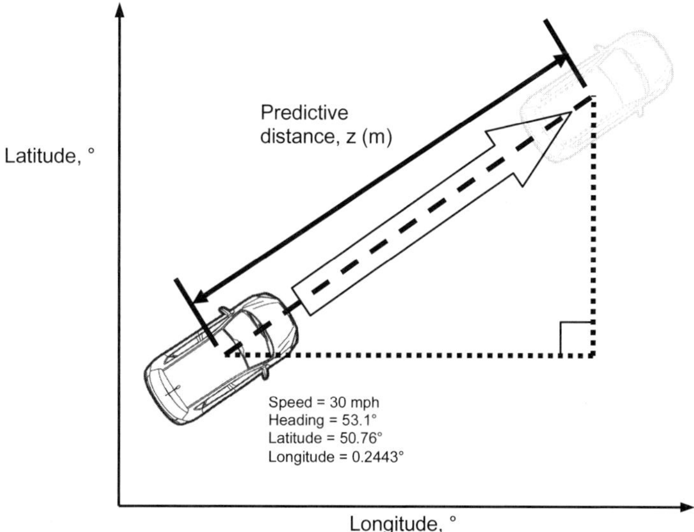

Fig. 3. Road gradient predictive scheme

distance span. Meanwhile, the predictive algorithm calculates the next position of the vehicle using the current speed gathered from the GPS receiver. Base on the difference between the current and the predicted position, the fuzzy controller deduces the height at a set distance ahead and subsequently calculates the gradient.

To calculate the distance between two points on the Earth requires the use of spherical geometry and trigonometric math functions otherwise known as the Great Circle Distance Formula.

$$Distance\ (m) = r \times \arccos\left[\begin{array}{l}\sin(lat1)\times\sin(lat2)+ \\ \cos(lat1)\times\cos(lat2)\times\cos(long2-long1)\end{array}\right] \quad (1)$$

where r is the radius of the earth, 6378.7km. The variables lat1, long1 and lat2, long2 are the current position and predicted position, respectively.

The software is capable of handling double-precision floating point as this formula requires a high level of floating point mathematical accuracy. The future location deduced using the described algorithms is of particular interest since it provides information about the condition of the road ahead, in order to realise the appropriate control signal.

5   Experimental Results

A first batch of measurements is shown in Figure 4, the solid line represented the height data, this was used to train the Nero-fuzzy network to produce a base relief map of the route, second run was performed at a variable sampling rate i.e. speed dependent sampling. These data were used as test data; it was done by the predictive

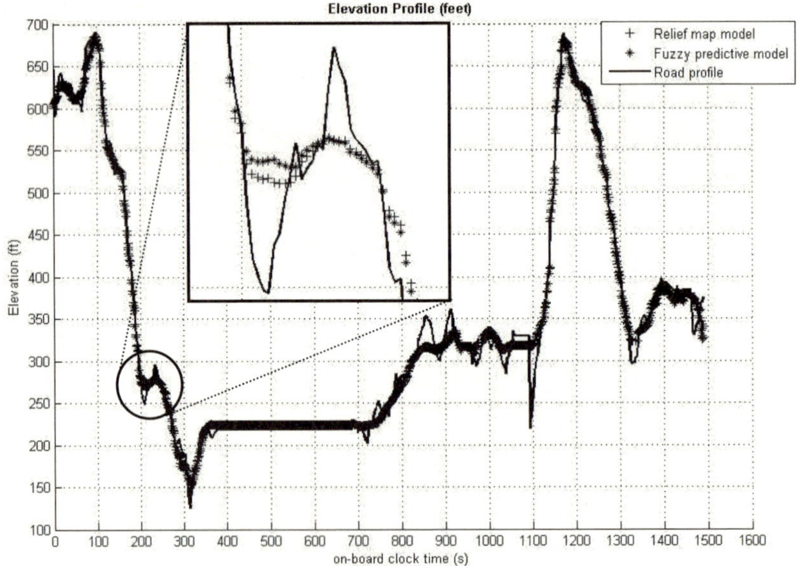

Fig. 4. Measurement of road elevations

algorithm where future gradient estimation was computed. The result was compared and automatically logged for off-line analysis.

6 Conclusions and Future Work

This paper has demonstrated that intelligent system can be used for predictive control of a vehicle. The technique represented a convenient and robust method of achieving road prediction, to form a fuzzy system that 'looks ahead' leading to improved fuel consumption and a consequent reduction in exhaust emissions. A new algorithm was demonstrated, which integrates live GPS data with the existing fuzzy logic devised relief map; matching software was developed and successfully implemented. This Neuro-fuzzy paradigm utilised simple map matching criteria, determined the gradient ahead based on current GPS position, and subsequently affect potential to influence the control of an engine. The GPS data observations are combined with fuzzy logic derived position to provide vehicle height information every two seconds.

Experimental results demonstrated the feasibility and advantages of this predictive fuzzy control on the trajectory tracking of a vehicle. Over 900 vehicle positions were generated and computed on each 9.8 mile test run using the newly devised algorithms. A similar number of test data were collected and compared to the height information generated by the predictive algorithm. The results showed that a good agreement was achieved between the predictive and the actual position data. The correlation coefficient of the elevation estimated by the Nero-fuzzy technique is 0.996, indicating good correlation.

The technique developed in road height estimation performs well and will be simulated using Simulink. This can be further improved with more low-cost GPS receiver technology and the integration with in-vehicle sensors as well as the engine operating parameters. Due to the fact that the method is tested and used on known and repeated routes, the system is intended and ideal to be used on buses or fleet vehicles. Future work will be focused on system integration that can be cost effectively developed and eventually made to influence the control of a test engine.

Acknowledgements

This work has been co-funded by the European Union under the Interreg IIIa Programme of the European Regional Development Fund, Intelligent Vehicle On-board Systems 'VBIS'.

References

1. Automotive Electrics Automotive Electronics, Robert Bosch GmbH, 5th edn. Wiley, Chichester (2007)
2. Directorate-General Energy and Transport, Galileo European Satellite Navigation System, `http://ec.europa.eu/dgs/energy_transport/galileo/index_en.htm`
3. Kalogirou, S.A.: Applications of artificial neural-networks for energy systems. Appl. Energy 67, 17–35 (2000)

4. Xu, K., Luxmoore, A.R., Jones, L.M., Deravi, F.: Integration of neural networks and expert systems for microscopic wear particle analysis. Knowledge-Based Systems 11, 213–227 (1998)
5. Jang, J.: ANFIS: Adaptive network-based fuzzy inference systems. IEEE Transactions on Systems, Man, and Cybernetics 23, 665–685 (1993)
6. Lee, S.H., Howlett, R.J., Walters, S.D.: An Adaptive Neuro-fuzzy Modelling of Diesel Spray Penetration, Paper No. SAE-NA 2005-24-64. In: Seventh International Conference on Engines for Automobile (ICE 2005), Capri, Napoli, Italy (September 2005) ISBN 88-900399-2-2
7. Hellstrom, E.: Explicit use of road topography for model predictive cruise control in heavy trucks. Master thesis. Linkoping University, Sweden (2005)
8. Wingren, A.: Look Ahead Cruise Control: Road Slope Estimation and Control, Linkoping University, Sweden. LiTH-ISY-EX-05/3644-SE (2005)
9. Lattemann, Frank, Neiss, K., Terwen, S., Connolly, T.: The predictive cruise control – a system to reduce fuel consumption of heavy duty trucks. SAE Technical paper 2004-01-2616 (2004)
10. Loewenau, J.P., Richter, W., Urbanczik, C., Beuk, L., Hendriks, T., Pichler, R., Artmann, K.: Real time optimization of active cruise control with map data using standardised interface. In: Proceedings of the 12th World congress on ITS, San Francisco, CA, USA, paper 2015 (2005)
11. Powers, W.F., Nicastri, P.R.: Automotive vehicle control challenges in the 21st century. Control Engineering Practice 8, 605–618 (2000)

Agent-Based Decision Making through Intelligent Knowledge Discovery

Marina V. Sokolova[1,2] and Antonio Fernández-Caballero[1]

[1] University of Castilla-La Mancha, Polytechnical Superior School of Albacete,
Campus Universitario s/n, 02071-Albacete, Spain
smv1999@mail.ru, caballer@dsi.uclm.es
[2] Kursk State Technical University, Kursk, ul.50 Let Oktyabrya, 305040, Russia

Abstract. Monitoring of negative effects of urban pollution and real-time decision making allow to clarify consequences upon human health. Large amounts of raw data information describe this situation, and to get knowledge from it, we apply intelligent agents. Further modeling and simulation gives the new knowledge about the tendencies of situation development and about its structure. Agent-based decision support system can help to foresee possible ways of situation development and contribute to effective management and planning.

1 Introduction

Humans, as the active element of environment, interact with it and receive the feedback, that is not always positive though, and sometimes appears to be unhealthy or even harmful. Negative environmental impact upon human health is a complicated one and can be, summed up from permanent effects, such as sun radiation, Earths geomagnetic field; from humans life and work conditions (characteristics of profession, quality of portable water, air,) which also include characteristics of the place they live in - transport and industrial activity, use of vehicles; and from instable characteristics like meteorological conditions.

Humans adapt to natural unhealthy factors, often, at the cost of the organisms reversible and irreversible disfunctions, which can be then inherited and came to light as birth malfunctions and defects.

During a long time an expert-system paradigm remained a standard solution for environmental assessment issues, as this approach supplied researches with all the necessary instrumentation: sets of expert information and analytical support data. Later, appeared multi-agent system based applications, called by the requirements to real-time decision making and intelligent data mining.

Some authors apply agent-approach to emergency assessment system development [1] related to traffic accidents response by allowing fast situation description, generation possible solutions and selection the decision maker for that case. MAS approach was utilized to increase the efficiency of hospital processes and to foresee possible effects of changes that can happen when expenses are reduced and the health-care budgets are restricted and stabilized. Different research institutes and organizations confirm the ecological contribution to illness

I. Lovrek, R.J. Howlett, and L.C. Jain (Eds.): KES 2008, Part III, LNAI 5179, pp. 709–715, 2008.
© Springer-Verlag Berlin Heidelberg 2008

and death in adults and children [3]. The World Health Organization (WHO) declared the necessity of broader analysis of environmental risk factors over public health, with a strong emphasis on mortality [4].

2   The DSS Structure

Large amounts of raw data information describe the "environment - human health" system, but not all the information can be of use though. For the situation modeling we orient to factual and context information, presented in data sets and we use intelligent agents to extract it.

So, the information transforms from the initial "raw" state to the "information" state, which suggests organized data sets, models and dependencies, and, finally, to the "new information" which has a form of recommendations, risk assessment values and forecasts. The way the information changes is given on the figure below:

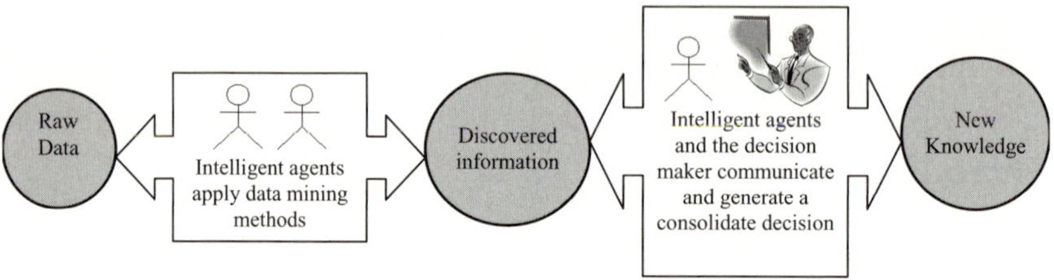

Fig. 1. The Information Change in ADSS

The hidden information is discovered by agents, but for new information construction not only intelligent agents, but knowledge of decision maker or expert are involved. The agent-based decision support system (ADSS) we are creating, provides these information changes [10].

The process of information change, shown on the Fig.1, corresponds to the MAS life cycle flow, which, in case of MAS counts the following steps: (1) domain and system requirements analysis; (2) design; (3) implementation; (4) verification; (5) maintenance [5].

The life cycle stages are supported as with theoretical methodologies, so with software tools. To determine the system and domain requirements, we suggest using an OWL-language based toolkit, as OWL has become a standard for ontologies description [6]. The Protege Ontology Editor and Knowledge-Base Framework [7] complies a set of procedures for ontology creation and analysis, offering a set of plug-ins covering viewers, problem-solving methods, knowledge trees, converters, etc. According to our proposal, ontologies can be represented by means of Protege and later may be incorporated into MAS.

In order to provide the system design to use the Prometheus Development Tool (PDT) [8], which provides a wide range of possibilities for MAS planning and implementation: the system architecture, the system entities, their internals and communications within the system and with outer entities. The most important advantages of PDT are an easy understandable visual interface and the possibility to generate code for JACK^{TM} Intelligent Agents [9], which is used for MAS implementation, verification and maintenance.

3   Agent-Based Decision Making

As it comes from the agent nature, the agents autonomy means decision making [11]. Actually, the agents we use in the ADSS, belong to BDI agent model are intelligent by their nature, and have to make decisions every time while executing. Being acting within three logical levels of the ADSS, agents solve a set of problems. So, data search, fusion and pre-processing is being delivered by two agents, which do a number of tasks, following the work flow:

`Information Search` $\longrightarrow$ `Data Source Classification` $\longrightarrow$
`Data Fusion` $\longrightarrow$ `Data Preprocessing` $\longrightarrow$ `Believes Creation`

`Information Search` obliges agents to search for data storages which can obtain the necessary information, and then classify the found sources in accordance with their type, presence of ontology concepts and the file structure organization.

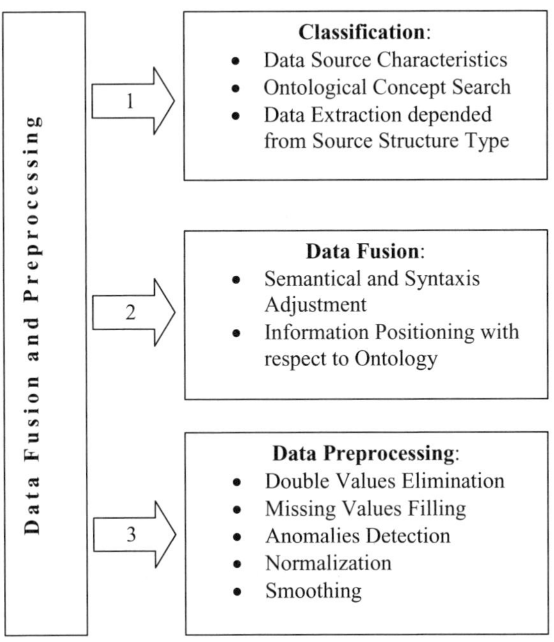

Fig. 2. The Tasks solved at the First Logical Level

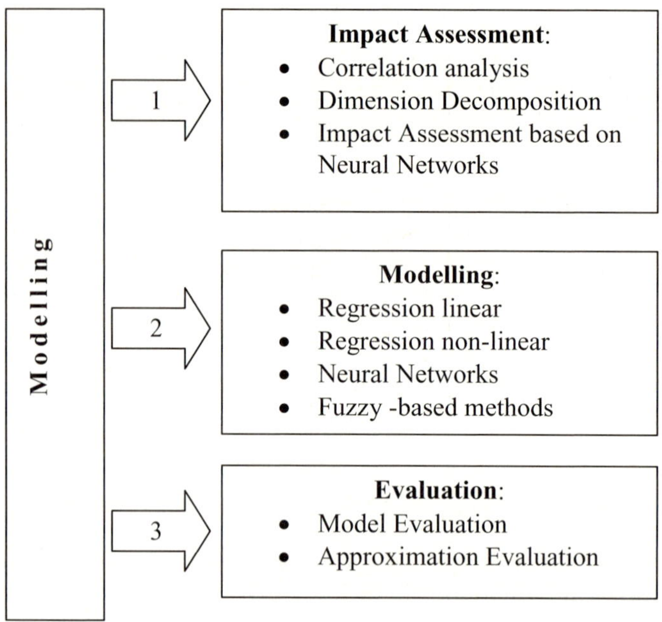

Fig. 3. The Tasks solved at the Second Logical Level

After these tasks have been solved, the next work to do is to search the necessary values and their characteristics,agree with the domain ontology. The crucial task here is to provide the semantical and syntaxis identity of the retrieved values, saying they have to be preprocessed before being placed into ontology and agents belief. The properties for the "pollutant" concept include scale, period of measurement, region, value, pollutant class and "disease" include age, gender, scale, period of measurement, region,value, disease class.

If recovered values satisfy all the requirements imposed or have been adjusted properly, they are placed in ontology. Agent then pre-process the newly created data sets (they are checked for anomalies, double and missing values, normalized and smoothed) and create global belief, utilized for further calculations.

The second logical level is completely based on autonomous agents, which decide how to analyze data and use their abilities to do it. The principal tasks to be solved here are:

- to state the environmental pollutants which impacts on every age and gender group and determine if they are associated with examined diseases groups;
- to create the models which explain dependencies between diseases, pollutants and groups of pollutants.

Here we are aimed to discover the knowledge in form of models, dependencies and associations from the pre-processed information which comes from the previous logical layer. The work flow of this level includes the following tasks:

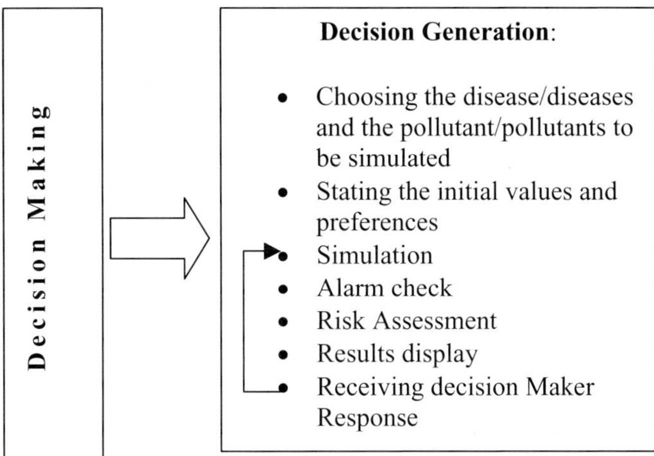

Fig. 4. The Tasks solved at the Third Logical Level

```
State Input and Output Information Flows⟶ Create models ⟶
Assess Impact⟶ Evaluate models ⟶ Select Models⟶ Display The
                         Results
```

Preliminary, we state the inputs (the pollutants) and the outputs (the diseases) for every model. The principal errors to be avoided here are to include in the model input variables which are highly correlated to each other, and to include the variables which correlate with the dependent output variable, as in this case, we would not receive independent components and the model would not to be adequate. These difficulties are anticipated and warned by correlation analysis and factor dimension decomposition, which is based on neural-network approach [12].

The set of created models is wide and contain linear and non-linear regression models, neural-networks based and fuzzy-models and their modifications. The models are validated. The selection of the best models for every disease is realized by statistical estimators, which validate the approximation abilities of the models.

The third level of the system is dedicated to decision generation, so, both the decision making mechanisms and human-computer interaction are both important here. The system works in a cooperative mode, and it allows decision maker to modify, refine or complete the decision suggestions, provided be the system and validate them. This process of decision improvement repeats until the consolidated solution is generated. The work flow is represented below:

```
State Factors for Simulation⟶ State the Values of Factors ⟶
    Simulate⟶ Evaluate Results ⟶ Check Possible Risk ⟶
  Display The Results ⟶ Receive Decision Maker Response⟶
    Simulate⟶ Evaluate Results ⟶ Check Possible Risk⟶
                Display The Results
```

First, the decision maker selects the factor or factor, then states pollutants and the initial values for simulation. Agents simulate models, calculate risk assessment values and check for alarm. It is more than one recommendation generated.

Then the agents calculate the decision making criterion, select the best decision and recommends it to the system user. If he accepts it, his session of work with this factor is finished, and the recommendation is accepted. If not, the sequence of steps from `Receive Decision Maker Response` till `Display The Results` repeats until the consolidated decision is generated. As a result created models and dependencies become the foundation to receive new knowledge in form of variants of situation development and possible advantages and losses.

4   Conclusion

The agent-based Decision Making problem is a complicated one, especially for a general issue as environmental impact upon human health. Although supposing it to be a tractable problem, we should note some essential advantages we have reached, and some directions for future research. First, the ADSS supports decision makers in chousing the behavior line (set of actions) in a such a general case, which is potentially difficult to analyze and foresee. As for any complex system, ADSS allows pattern predictions, and the decision maker's choice is to be decisive. Second, in spite of our time consuming modeling work, we are looking forward to both revise and improve the system and deepen our research. Third, we consider to make more experiments varying as with data structure, trying to apply the system to the other, but similar application field. Consequently, our future work can be drawn on various levels.

Acknowledgements

Marina V. Sokolova is the recipient of a Postdoctoral Scholarship (Becas MAE) awarded by the Agencia Española de Cooperación Internacional of the Spanish Ministerio de Asuntos Exteriores y de Cooperación.

References

1. Smirnov, A., Pashkin, M., Levashova, T., Shilov, N., Kashevnik, A.: Role-Based Decision Mining for Multiagent Emergency Response Management. In: Gorodetsky, V., Zhang, C., Skormin, V.A., Cao, L. (eds.) AIS-ADM 2007. LNCS (LNAI), vol. 4476, pp. 178–191. Springer, Heidelberg (2007)
2. Herrler, R., Heine, C.: Approaches and Tools to Optimize and Manage Clinical Processes. In: Application of Agents and Intelligent Information Technologies, pp. 39–65. IGI Publishing, USA (2007)
3. Kaiser, R., Romieu, I., Medina, S., Schwartz, J., Krzyzanowski, M., Knzli, N.: Air pollution attributable postneonatal infant mortality in U.S. metropolitan areas: A risk assessment study. Environmental Health: A Global Access Science Source 3(1), 4 (2004)

4. World Health Organization, `http://www.who.int/en/`
5. ISO/IEC 12207 home page, `http://www.iso.org/iso/`
6. OWL Web Ontology Language home page (2004),
   `http://www.w3.org/TR/owl-features/`
7. Prometheus Design Tool home page, `http://www.cs.rmit.edu.au/agents/pdt/`
8. Padgham, L., Winikoff, M.: Prometheus: A pragmatic methodology for engineering intelligent agents. In: Proceedings of the Workshop on Agent Oriented Methodologies (Object-Oriented Programming, Systems, Languages, and Applications), pp. 97–108 (2002)
9. Jack^{TM} Intelligent Agents home page,
   `http://www.agent-software.com/shared/home/`
10. Sokolova, M.V., Fernández-Caballero, A.: A multi-agent architecture for environmental impact assessment: Information fusion, data mining and decision making. In: 9th International Conference on Enterprise Information Systems, ICEIS 2007 (2007)
11. Weiss, G.: Multi-agent Systems: A Modern Approach to Distributed Artificial Intelligence. The MIT Press, Cambridge (2000)
12. Sokolova, M.: Thesis for Doctorate Degree. Kursk, Kursk State Technical University press (2004)

Conceptual Modeling in a Meta-model of Sustainability Indicators

Alfredo Tolón-Becerra[1] and Fernando Bienvenido[2]

[1] Dept. of Rural Engineering,
[2] Dept. of Computer Science
University of Almeria
Ctra. Sacramento s/n, La Canada ,04120 Almeria, Spain
atolon@ual.es, fbienven@ual.es

Abstract. During last 30 years, due to the interest in sustainable development and ecology, different systems of sustainability indicators have been developed. These systems are increasingly complex, varying their scope both for territorial and thematic areas. A key point developing sustainable policies is the accessibility of good indicators. Developing a knowledge model about them can facilitate their management and interpretation. Due to the complexity of diverse set of indicators developed for 29 areas in 9 countries, we were taken to develop a conceptual model of these indicators, which associate to the different indicators their validity scope, specific interpretation and use. We have integrated the different indicators using conceptual maps (using CMap-Tool) and associating to them semantic labels. The web developed to manage the different projects includes these labels (it is a semantic web) and all the referenced texts have been labeled using them.

Keywords: Sustainability indicators, sustainable development, enviromatics, sustainability conceptual maps, semantic web, rural development.

1 Introduction

The concept of sustainable development, born in the 80's, has matured worldwide with the publication of the Brundtland Report in 1980 [1], where it is defined as *"development that meets the needs of the present without compromising the ability of future generations to meet their own needs"* [2] [3] [4]. Starting from this definition, several international and national organisms adapt the concept of sustainable development to their specific objectives, making their own definition. In this sense, the Spanish Ministry of Environment, in its first report "Sustainability in Spain 2005", points the objective of evolving to a more sustainable development; which must be equilibrated and harmonious in three dimensions: economic, social and environmental ones. We advance to the sustainable development when it presents a dissociation between the economic grow and the environmental degradation and the use of the natural resources.

Independently of the type/level of sustainability to implement, extra strong, strong, with critic natural capital constant and weak sustainability; always, it is required to

I. Lovrek, R.J. Howlett, and L.C. Jain (Eds.): KES 2008, Part III, LNAI 5179, pp. 716–723, 2008.

evaluate the state of the system, measuring the natural and human capitals, previously to the definition of the specific policies. Indicator systems are the ways this capitals and their evolution area presented.

Next section presents the definitions of indicator and indicator systems we are using and the evolution of these later systems. In section 3, a meta-model of sustainability indicators, that includes the way they are generated and managed, is presented. This model, developed by our team, is been used in different projects (see table 1), been tested in 27 areas of environmental interest (see table 2) with extremely diverse conditions as their dispersion shown.

Table 1. Projects where have been applied our model and tools

Projects	Countries
Development of a network of sustainability indicators for a dynamic monitoring of the development processes in LEADER and PRODER rural areas. (AGL2003-04540)	Spain
Construction of a network of Sustainability Indicators for dynamic monitoring of processes of Tourism and Rural Development in Europe and Latin America. Application to Costa Rica (A/0632/03)	Costa Rica / Spain
Creating a network of sustainability indicators in rural areas from Havana (AI1/04)	Cuba / Spain
Construction of a network of sustainability indicators into three Cuban rural areas (A/1899/04)	Cuba / Spain
Network on Sustainability indicators of Rural Development in "Frailesca Region" and "Selva de Chiapas" – Mexico (A/3622/05)	Mexico / Spain
Iberoamerican network on Sustainability indicators in Rural Spaces (South American Block) (A/4987/06)	Argentina / Brasil / Spain / Uruguay
Iberoamerican network on Sustainability indicators in Rural Spaces (Andean Block) (A/5698/06)	Ecuador / Spain / Venezuela
Iberoamerican network on Sustainability indicators in Rural Spaces (Mesoamerican Block) (A/5696/06)	Costa Rica / Cuba / Mexico / Spain
Participative network on Sustainability indicators in Iberoamerican Rural Spaces (South American Block) (A/7735/07)	Argentina / Brasil / Spain / Uruguay
Participative network on Sustainability indicators in Iberoamerican Rural Spaces (Mesoamerican Block) (A/7733/07)	Costa Rica / Cuba / Mexico / Spain

From these works, we saw the possible use of a really wide set of indicators, while usually in specific conditions no all of them are required. So we realized the necessity of order, qualify and combine these indicators in order to get a really applicable model. All the indicators used in these projects have been integrated in a set of conceptual maps, as shown in section 4, associating to them different labels, that let us to give semantic value to the different parts of our web and to all the documents that are referenced. An semi-automatic document labelling system is a future objective. We wish to thank all the researchers of the different projects that have collaborate in the evaluation of the meta-model and the assembling of the knowledge model and label systems.

Table 2. Natural areas where we have applied our model and tools

Countries	Areas	Projects
Argentina	- San Carlos/Mendoza –Apóstoles/Misiones	A/4987/06, A/7735/07
Brasil	- Projeto Onca Taperoá - Colonia Ituberá	A/4987/06, A/7735/07 A/4987/06, A/7735/07
Costa Rica	- Monteverde - Puerto Viejo	A/0632/03, A/5696/06, A/7733/07
Cuba	- San José de las Lajas - Guines - Consejo Popular Jamaica - Consejo Popular Norte - Consejo Popular Nazareno	AI1/04, A/5696/06, A/7733/07 A/1899/04 A/1899/04 A/1899/04
Ecuador	-Caseiches-Guaranda - Nabón-Azuay .	A/5698/06
Mexico	- Selva de Chiapas - Frailesca	A/3622/05, A/5696/06, A/7733/07
Spain	- Alpujarras (Almeria) - Alto y Bajo Almanzora - Serranía de Ronda (Málaga) - Bajo Guadalquivir (Seville) - Sierra norte de Madrid (Madrid) - Sierra del Jarama (Madrid) - Alto Vinalopó (Alicante) - Vega Baja del Segura (Alicante)	All All AGL2003-04540 AGL2003-04540 AGL2003-04540 AGL2003-04540 AGL2003-04540 AGL2003-04540
Uruguay	- Bañados del Este - Cuenca lechera del Sur	A/4987/06, A/7735/07
Venezuela	-Juan Germán Roscio (Park) -José Tadeo (Park)	A/5698/06 A/5698/06

2   Sustainability Indicators and Systems of Indicators

Now, starting from the definition of indicator and index, we are going to define the concept of system of indicators, and their frameworks; analyzing how these later have evolved along last 30 years.

2.1   Definition of Indicator and Index

One indicator, according to the OECD, is a parameter, or derived value of other parameters, which provides information about the status of a phenomenon, and that has a significance that extends beyond that one directly associated with the parameter value. This organism also defines the concept of index as a single value composed aggregating or weighting a set of parameters or indicators [5].

The European Environment Agency defines an indicator as "the representative value of an observed phenomenon to be studied, in general, indicators quantify information by aggregating many different data, the result is synthesized information, simplifying the information that can help to reveal complex phenomena".

The definition given by the Spanish Ministry of Environment is "a variable that has been socially endowed with a meaning added to that one derived from its own scientific settings, in order to reflect synthetically a social concern with respect to the environment, being inserted coherently in the decision-making process ".

2.2 Indicator Systems and Their Framework

In the report about the sustainability in Spain of 2006, a system of indicators is defined as "a set of tools for monitoring and evaluating the compliance with a common goal to all of them, the economic development, environmental improvement and the quality of life, essential to the effective application of the concept of sustainable development."

A system of Sustainability Indicators can also be defined as an ordered and cohesive set of indicators which allows, in a continuous way, the evaluation, monitoring and control of the environmental, social and economic sustainability of the development processes in a specific geographical area or set of them.

Indicator systems require an organizing framework that allows the interaction between the different dimensions of sustainable development, giving them a better usefulness and an adequate order [5]. These indicators organizing frameworks offer alternative organizations of the possible indicators, whose usefulness depends on the final end use to which the information is intended. They provide a synergistic effect to the individual information contained in each indicator, so far they are assigned particular attributes, and are not mutually exclusive. Frameworks help:

- Organizing indicators in a consistent manner.
- Doing compatible indicators.
- Guiding the collection of information.
- Communicating synthesised information to managers for decision making.
- Suggesting logical groupings to integrate related information.
- Identifying information gaps.
- Distributing the work load when generating reports.

2.3 Evolution of the Indicator Systems

Historically, the systems of indicators can be classified in three generations, taking account of their process of development and application [5], as shown in figure 1. The first generation systems, originated in the 80's with the work done by the OECD, are characterized by being highly theoretical and exclusively environmental. The main frameworks used by the systems of indicators of the first generation were : PSR (Pressure-State-Response) –mainly used OECD-, DSR (Driving Force – State – Response), DPSR (Driving force-Pressure-State-Response) and SPSIR (Driving Force-Pressure-State-Impact-Response) –mainly used by EEA-.

The use of second-generation systems began in the 90's, through the development of systems at national level; we can highlight the efforts made by Mexico, Chile, United States, United Kingdom, Spain, etc. They incorporated a multidimensional approach (economic, environmental and social) of sustainable development; in recent years have been included a fourth dimension, the institutional one, due to the relevance and influence of the policies dictated by the different control bodies (local and national governments, international agencies, etc…). The development of these systems has been led by the Commission on Sustainable Development of the United Nations, with indicators that fall into each of the dimensions of development, but unlinked between them.

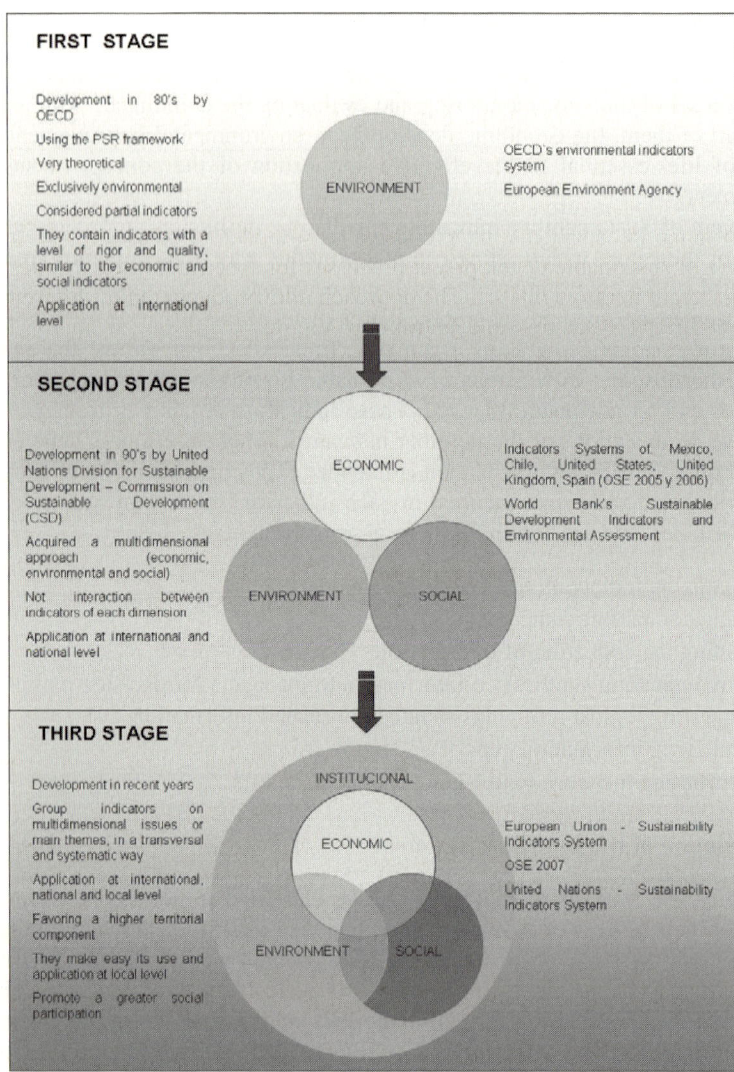

Fig. 1. Evolution of the systems of indicators

In recent years, the need of linking the three development dimensions and their indicators among themselves generated the third generation of indicators frameworks; which offers quick access to a much larger world of meanings, grouping indicators under themes or multidimensional areas, in a transversal and systematic way. We can highlight the efforts made by the European Union, through the Working Group on Sustainable Development Indicators. While they were generated at an international level, the new systems of indicators include a greater territorial component that facilitates its use and implementation at the local level and promote greater social participation. This is a key point of our use of this framework schema, because our detailed work in specific areas, as shown in our meta-model.

3 Meta-model of Sustainability Indicators

Reed et al. [6] describe the process of developing and implementing a system of sustainability indicators through a mixed top down-bottom up approximation. It consists of four phases, including twelve stages that close a recurring cycle of sustainability evaluation; it is an empirical frame of reference, which should be added several tools for its implementation, according to the context of the study areas.

Based on the Reed schema, we propose a meta-model, to be used a base to assemble models of sustainable development processes for specific territorial spaces, which consists of eight iterative phases. The approach allows adaptation to different territorial scales, different sectors, and bottom-up, top-down and mixed approaches. The phases of this general meta-model are grouped according to the structure shown in figure 2. This meta-model has been used in 8 LEADER and PRODER areas (counties) in Spain, and it is actually been tested in 19 work areas in America.

<table>
<tr><td>1.</td><td colspan="2">CONTEXT OF THE INDICATORS SYSTEM</td></tr>
<tr><td></td><td>1.1.</td><td>Geographic scope of the Indicators System</td></tr>
<tr><td></td><td>1.2.</td><td>Sectorial approach of the Indicators System</td></tr>
<tr><td></td><td>1.3.</td><td>Political-Administrative Context</td></tr>
<tr><td></td><td>1.4.</td><td>Selecting key time periods</td></tr>
<tr><td>2.</td><td colspan="2">PREVIOUS RESEARCH WORKS</td></tr>
<tr><td></td><td>2.1.</td><td>Characterization of the Indicators System</td></tr>
<tr><td></td><td>2.2.</td><td>Characterization of useful Databases</td></tr>
<tr><td>3.</td><td colspan="2">GENERATION OF INDICATORS</td></tr>
<tr><td></td><td>3.1.</td><td>Initial Structure of the Indicators System. Classification. Priority main themes</td></tr>
<tr><td></td><td>3.2.</td><td>Generation of simple indicators by thematic areas</td></tr>
<tr><td></td><td>3.3.</td><td>Generation of complex indicators (indexes). Final structure of the Indicators System.</td></tr>
<tr><td>4.</td><td colspan="2">SELECTION PROCESS INDICATORS</td></tr>
<tr><td></td><td>4.1.</td><td>Characterization and evaluation of Indicators</td></tr>
<tr><td></td><td>4.2.</td><td>Selection of indicators</td></tr>
<tr><td>5.</td><td colspan="2">CONSTRUCTION AND APPLICATION OF SELECTED INDICATORS</td></tr>
<tr><td></td><td>5.1.</td><td>Characteristics, construction techniques and selection of indicators</td></tr>
<tr><td></td><td>5.2.</td><td>Data needed to calculate indicators</td></tr>
<tr><td></td><td>5.3.</td><td>Calculation and application of the indicators in the areas.</td></tr>
<tr><td>6.</td><td colspan="2">DEVELOPMENT OF THE INDICATORS SYSTEM</td></tr>
<tr><td></td><td>6.1.</td><td>Estimated Thresholds Values, critics and desirable. Reference Values.</td></tr>
<tr><td></td><td>6.2.</td><td>Estimated Objectives Values: approximation to the Threshold Value Desirable</td></tr>
<tr><td>7.</td><td colspan="2">DISCUSSION AND INTERPRETATION</td></tr>
<tr><td></td><td>7.1.</td><td>Thematic Analysis: analysis for main themes in the network</td></tr>
<tr><td></td><td>7.2.</td><td>Geographical Analysis: analysis of the entire indicators system in each geographic area</td></tr>
<tr><td>8.</td><td colspan="2">COMMUNICATION OF INDICATORS SYSTEM</td></tr>
<tr><td></td><td>8.1.</td><td>Structure of each indicator information and diffusion</td></tr>
<tr><td></td><td>8.2.</td><td>Use of indicators system.</td></tr>
</table>

Fig. 2. Meta-model of sustainability indicators. A base for the development of specific systems

4 Conceptual Modeling of Indicators

The general set of indicators included in the meta-model, must be adapted to each study area taking account of the existing knowledge, development agents (different groups of persons, with different backgrounds and interests), social conditions, policies, etc, in a double top-down / botton-up strategy. This means that the set of indicators, their interpretation and importance, evolve along the development of the projects and further. In this case, the use of the information technologies is critics at the different stages of application of the meta- model, as shown in figure 3.

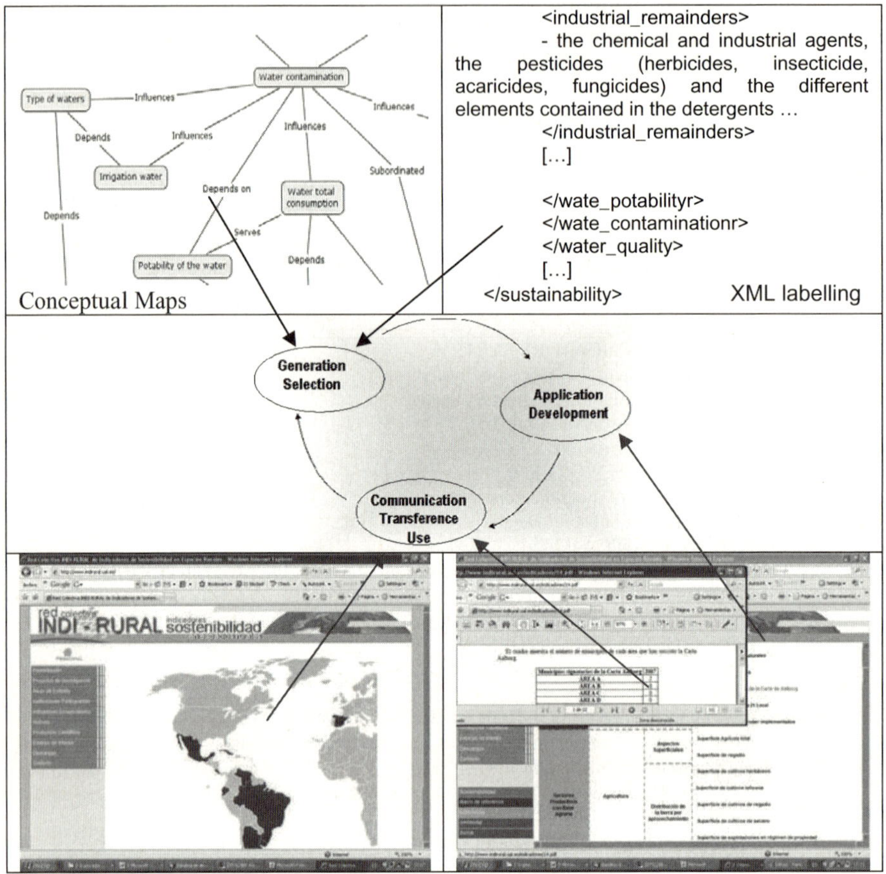

Fig. 3. Application of the information technologies in the use of our meta-model

Main tools used have been:

- An Internet Web page (http://www.indirural.ual.es/index.html) for diffusion.
- An Intranet for collective work (more than 14 teams working together along a wide area in the Americas and Europe (with the manager, researcher, evaluator and policy maker roles).
- A set of Conceptual Maps [7], that let us to order all the indicators. They are easily understandable by the different participants (no only sustainability or IT experts). We have used the CMap-Tool by IHMC to develop them and create the internet pages added to our web page (Indirural).

- A labeling system for the different indicators; all the texts included are labeled actually. This set of labels have been developed using the XML structure. Actually we are working in using the indicator labels structure to classify important texts on sustainability.

5 Conclusions and Further Works

In order to facilitate and adequate application of sustainable development policies, we must change our decision making scope to more manageable areas and fields, at the local level. This means that the indicators must be adapted to their specific characteristics and interpretations. After working with a wide variety or areas (see table 2) in different projects (see table 1), we realized the necessity of ordering the indicators in a simple way, accessible for all the agents in a top-down / botton-up development strategy (taking account of the ideas, requirements and vision of all the people affected). This have been done using conceptual maps, tool that let us not only to order the different concepts, but to express them easily for all the agents. From this order, we have extracted a set of labels, linked to the languages used (Portuguese and Spanish), and the meaning of the indicators (definition, scope, validity, ...). This set of labels is been used as a XML labeling system for our web page, and it is being used to label pre-existent texts, that can be automatically analyzed.

Future works, include 1/ the analysis of the practical use of the labeled texts by the decision-making agents in actual participative projects, 2/ developing a tool for semi-automatic labeling of sustainability texts (they would require by the moment some supervision), and 3/ including in our public web the analyzer for labeled text (actually restricted to the extended work team just testing it).

References

1. United Nations World Commission on Environment and Development (WCED): Our Common Future. Oxford University Press, Oxford (1987)
2. Hueting, R., Reijnders, L.: Broad sustainability contra sustainability: the proper construction of sustainability indicators. Ecological Economics 50, 249–260 (2004)
3. Pearce, D., Atkinson, G.: The concept of sustainable development: An evaluation of its usefulness ten years after Brundtland. Centre for Social and Economic Research on the Global Environment, University College of London and Working Paper, pp. 98–102 (1998)
4. International Union for Conservation of Nature (IUCN): World conservation strategy: Living resource conservation for sustainable development. IUCN, Switzerland (1980)
5. Gallopín, G.: Indicators and Their Use: Information for Decision-making. Part One-Introduction. In: Moldan, B., Bilharz, S. (eds.) Sustainability Indicators. A Report on the Project on Indicators of Sustainable Development. SCOPE 58, pp. 13–27. Wiley, Chichester (1997)
6. Reed, M., Frases, E., Dougill, A.: An adaptative learning process for developing and applying sustainability indicators with local communities. Ecological Economics 59, 406–418 (2006)
7. Cañas, J., et al.: Concept Maps: Integrating Knowledge and Information Visualization. In: Tergan, S., Keller, T. (eds.) Knowledge and Information Visualization. LNCS, vol. 3426, pp. 205–219. Springer, Heidelberg (2005)

Using Global Optimization
to Explore Multiple Solutions
of Clustering Problems

Ida Bifulco, Loredana Murino, Francesco Napolitano, Giancarlo Raiconi,
and Roberto Tagliaferri

NeuRoNe Lab, DMI, University of Salerno, via Ponte don Melillo, 84084 Fisciano
(SA) Italy
{ibifulco,lmurino,fnapolitano,gianni,rtagliaferri}@unisa.it
http://www.neuronelab.dmi.unisa.it

Abstract. A classical approach to clustering consists in running an algorithm aimed to minimize the distortion. Apart from very limited and simple cases such problem cannot be solved by a local search algorithm because of multiple local minima. In this paper a Global Optimization (GO) algorithm is used to overcome such difficulty. The proposed algorithm (Controlled Random Search) iterates by maintaining a population of solutions which tends to concentrate around the most "promising" areas. From Data Mining point of view such an approach enables to infer deep information about the underlying structure of data. Collecting and presenting such information in a human understandable manner can help the choice between several possible alternatives. Numerical experiments are carried out on a real dataset, showing that GO produces solutions with much better distortion values than the classical approach, while graphical representation of the whole solution set can be useful to data exploration.

Keywords: clustering, global optimization, k-means.

1 Introduction

Clustering can be considered as one of the most important unsupervised learning problems and it deals with finding a structure in a collection of unlabeled data. Clustering techniques have been applied to a wide variety of research problems. For example, in the field of medicine, clustering diseases and cures for diseases gives a key to understand relations between them. In the recent years, clustering was proved to be a useful technique to reveal the hidden structure of data provided by genomic experiments. As the amount of analyzed data grows, often also both noise intensity and the number of candidate clusters grows: in such a situation the weakness of some of the most used clustering approaches becomes more evident. Such weakness can be of different types, two of which are the instability of clustering with respect to data and the instability of clustering with respect to the algorithm initialization.

I. Lovrek, R.J. Howlett, and L.C. Jain (Eds.): KES 2008, Part III, LNAI 5179, pp. 724–731, 2008.
© Springer-Verlag Berlin Heidelberg 2008

The first type of instability consists in the fact that, if the clustering algorithm is applied to data slightly varied (i.e. adding or deleting a little amount of data, etc.), the resulting clustering is completely different from the original one; the second type of instability occurs when the same algorithm runs several times on the same data: varying only the initialization of the algorithm produces different clusterings. To analyze the effect of the first type of instability, a large amount of work was published in recent times under the topic of clustering assessment [14,12,15].

For what concerns the second type of instability, it can occur only on those algorithms on which the dynamic grouping of data in clusters starts from an arbitrary or random initial grouping and evolves following an updating rule. Consider as an example the well known approach of $k-$means:

Algorithm 1 (K-means)

1. *Initialize k centroids, $(c_1, ..., c_k)$.*
2. *Assign each pattern to its nearest centroid.*
3. *Update each centroid as the mean of all the patterns assigned to it in step 2.*
4. *Repeat from step 2 until no reassignment is possible.*

This is an optimization algorithm which attempts to find centroids $c_j, j = 1, 2, ..., k$ and the corresponding clusters $C_j, i = 1, 2, ..., k$ that solves the problem:

$$\arg \min_{C_j, c_j} \{ J = \sum_{j=1}^{k} \sum_{x_i \in C_j} \|x_i - c_j\|^2 \} \tag{1}$$

where $x_i \in \mathbb{R}^n$ are the input patterns. As all local iterative algorithms even Algorithm 1 converges to the global minimum on J only if some tight hypotheses on J are fulfilled. The only properties that can be proved are: Algorithm 1 stops in a finite number of iterations; any fixed point of Algorithm 1 is a stationary (minimum) point for J. Note that if the number of patterns is not too high and/or if the true structure of data is not complex (small number of clusters, well divided) then several runs of the algorithm randomly started are generally sufficient to find global minimum. If such assumptions are not assured, the only way to attempt to find the absolute minimum is to resort to an algorithm specifically designed for GO. This approach was followed by several authors using principally genetic algorithms [10].

The aim of our research is twofold: the former is to exploit a clever GO algorithm based on Controlled Random Search (CRS) [13] that, as proved in other applications, is much more efficient of a straight genetic one (section 2); the latter is to represent the structure of data using intelligent visualization techniques depicting different obtained clusterings, in order to introduce subjective judgement of an expert to choose the most "reasonable" clustering between several plausible ones (section 3,4). Several papers searched to find the "real" structure of data using cluster aggregation techniques [2,3,4] or following a "meta clustering" approach [5] . In this paper we are more interested in a method to define the population of plausible clusterings and in its visual representation.

2   Global Optimization: Price Algorithm

A GO algorithm aims at finding a global minimizer or its close approximation of a function $f : S \subset \mathbb{R}^n \to \mathbb{R}$. A point y^* is said to be a global minimizer of f if $f^* = f(y^*) \leq f(y) \; \forall y \in S$. CRS is a direct search technique based on a mix of random and heuristic steps. It is a kind of contraction process where an initial sample of N points is iteratively concentrated by replacing the worst point with a better point. The replacement point is either determined by a global or a local technique. The global technique is an iterative process in which a trial point, the new point, is defined in terms of $n + 1$ points selected from the whole current sample of points until a replacement point is found [13]. Some CRS algorithms also apply a local technique where the replacement point is searched for a subset of best points in the current sample. In particular, the key point of the Price's algorithm is the set S^l (l is the iteration counter): at the first iteration ($l = 0$) this set is constituted by m randomly chosen points (centroid) over a compact set D and then it collects the best points produced in the procedure. At each iteration a new trial point is produced along a direction which is randomly chosen over a finite number of vectors determined by the points belonging to S^l (Algorithm 2).

The rationale behind this approach is that, as the number of iterations increases, the set S^l should cluster points around the global minimum and the directions used should become more effective than directions chosen at random on $\mathbb{R}^n$. Modification of the original Price's algorithm was proposed in [6] and [7] (the version used in this paper) improving the local search phase without significant increase in computational cost.

Algorithm 2 (Price).

0. Set $l = 0$; determine the initial set $S^l = \{y_1^l, y_2^l, ..., y_m^l\}$, where the points y_i^l ($i = 1, ..., m$) are chosen at random over D; evaluate f at each point y_i^l.

1. Determine the points y_{min}^l, y_{max}^l and the values f_{min}^l, f_{max}^l such that

$$f_{max}^l = f(y_{max}^l) = \max_{y \in S^l} f(y)$$

$$f_{min}^l = f(y_{min}^l) = \min_{y \in S^l} f(y)$$

If the stopping criterion is satisfied, then stop.

2. Choose at random $n + 1$ points y_{ir}^l over S^l. Determine the centroid $centr^l$ of the n points $\{y_{i_1}^l, y_{i_2}^l, ..., y_{i_n}^l\}$ where

$centr^l = \frac{1}{n} \sum_{r=1}^{n} y_{ir}^l$

Determine the trial point $\tilde{y}^l$ given by

$\tilde{y}^l = centr^l - (y_{i_0} - centr^l)$;

If $\tilde{y}^l \notin D$ go to step 2; otherwise compute $f(\tilde{y}^l)$

3. If $f(\tilde{y}^l) > f_{max}^l$ than take

$S^{l+1} = S^l$. Set $l = l + 1$ and go to Step 2.

4. If $f(\tilde{y}^l) < f_{max}^l$ than take

$S^{l+1} = S^l \bigcup \{\tilde{y}^l\} - \{y_{max}^l\}$. Set $l = l + 1$ and go to step 1.

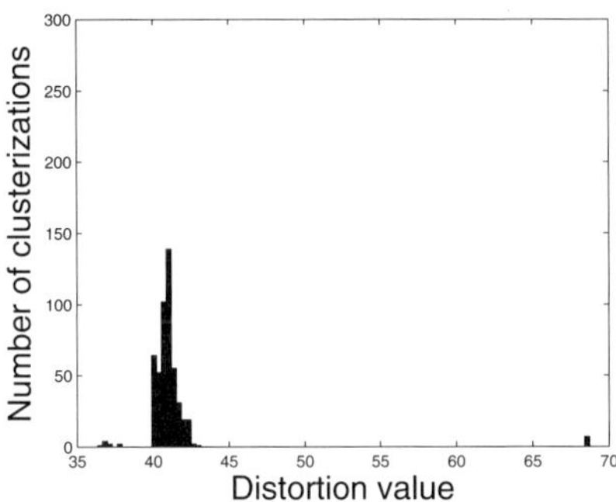

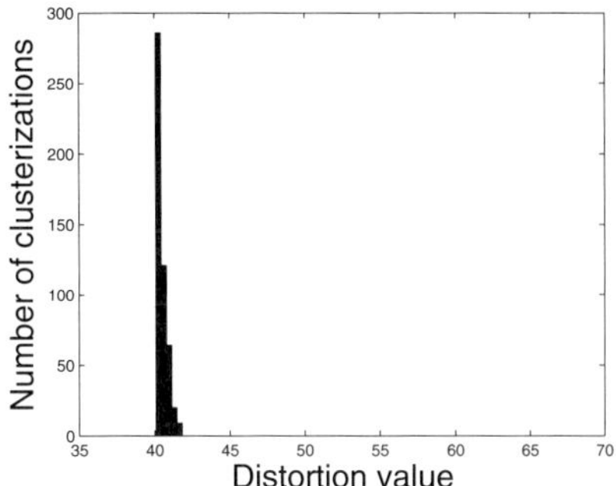

Fig. 1. Histogram of distortion values (*Price* algorithm)

Fig. 2. Histogram of distortion values (*K-means* algorithm with random start)

We carried out the cluster optimization using the following procedure. Let be $x_i \in \mathbb{R}^n, i = 1, ..., M$ the data to be clustered, $c_1^*, ..., c_k^* = y^*$ the unknown k-centroids that represent the global optimizer for J. We first construct the initial population of N k-centroids chosen at random in the solution space:

$$c_j^r \in \mathbb{R}^n, j = 1, 2, ..., k, \ r = 1, 2, ..., N$$

then the Price/Genetic algorithm [7] is started to minimize the figure of merit J as function of $c_j, \ j = 1, 2, ..., k$ only. So that at each iteration the function f is well defined as follows. Given a set of centroids $y = \{c_j\}$, data is grouped according to:

$$C_j^* = \{x_i| \ \|x_i - c_j\| \le \|x_i - c_p\| \forall p \in \{1, ..., k\}\} \ \ for \ j = 1, ..., k$$

then the objective function of GO is computed as

$$f(y) = J(c_1, ..., c_k, C_1^*, ..., C_k^*)$$

The Price algorithm is iterated and, at the last step, a new population is available, which is concentrated around the most pronounced minima of J and characterized by a much lower average value of J. Because the GO step has been performed considering only the centroid vectors as independent variables, therefore they do not necessarily correspond to effective local minima and so Algorithm 1 is iterated until convergence to the final population itself. As a consequence, at most we have a population of N local minima for problem (1). Since, as one can easily imagine, not all the so obtained solutions are distinct, then the next step is that of eliminating duplicated solutions, obtaining a population of $N' \le N$ different solutions of the clustering problem. The described algorithm is referred to the *k-means* approach but it can be extended to any other approach by substituting the appropriated figure of merit.

3   Analysis of Solutions Ensemble

As pointed out above, the clustering problem for large datasets characterized by a high level of noise is "ill posed" in the sense that, if the structure underlying data is complex and depending on how much the data sampling is adapted to capture this structure, the solution of the problem is not unique. In such situation the number of solutions obtained starting from different starting points is surprisingly high. As an example, consider the results obtained for yeast dataset [9]; it is constituted by 1484 patterns (genes), any of those characterized by 6 features (real numbers). Any record is also characterized by a label identifying the belongings to one of 10 classes and the dataset has been designed to test classification algorithms. Here we are interested only in unsupervised learning and we shall not consider the labels. By applying the GO with $k = 10$ and 500 random starting points, a final population of 371 distinct clusterings is obtained. In fig. 1 the histogram of corresponding distortion values is showed. It is interesting to compare these results with those obtained simply starting from the same number (500) of random started runs of the clustering algorithm (fig.2). It is clear that:

- GO finds an absolute minimum (36.25) less than that of random search (40.02).
- GO best explores the local minima range.

The effect of the aforementioned ill-posedness is that data itself can be grouped in several different ways giving distortion values (1) not too different. In such cases the blind acceptance of the absolute minimum of distortion cannot be the right answer, and a plausibility criterion, based on the judgement of the expert of domain, can be very useful to resolve the indeterminacy. In order to formulate such judgement the expert must be supplied by information about the "most attracting" alternatives in a clearly understandable manner. The use of a CRS approach gives the advantage to furnish not only an estimate of the absolute minimum but also the ability to locate areas of the search domain where critical points are located, by maintaining a population of candidate points (any of them representing a candidate clustering) that tend to cluster around most "interesting" areas, thus giving a lot of information about the global behavior of the function. On the other hand, if data points are high dimensional then it is not easy to elicit the meaning of the terminal population. Our approach consists in introducing a similarity measure between clusterings and using such measure to visualize relations between candidate clusterings. This can be obtained either grouping similar clusterings in *cluster of clustering* or projecting them in a low (2 or 3) dimensional space where interpretation of results is much easier [5].

4   Similarity Measures and Similarity Plots

Given a number of clustering solutions for a given dataset, a measure to assess the similarity between them must be defined. Well known similarity functions are found in literature, like Minkowski Index, Jaccard Coefficient, correlation and matching coefficients (all found in [1]). In our studies we used a measure based on the entropy of the confusion matrix between clustering solutions [8]. The measure is defined as follows.

Given two clustering solutions A and B, where A is made of n clusters and B is made of m clusters, the confusion matrix M between A and B is an $n \times m$ matrix, in which the entry (i, j) reports the number of objects in the $i - th$ cluster of the clustering A falling into the $j - th$ cluster of the clustering B. Let R_i be the i-th row of M and $H(R_i)$ the corresponding entropy, then we can compute the similarity between B and A as the *mean entropy* of the clusters of B versus A. The a-priori class probability $P(C_i)$, that is the probability of a randomly chosen pattern of belonging to C_i, can be approximated as $(number - of - objects - in - C_i)/(total - number - of - objects)$, giving rise to the formula:

$$S(M) = \sum_i P(C_i) \cdot H(R_i)$$

We finally define our similarity measure as:

$$S_a(M) = S(M) + a \cdot S(M^T)$$

which is symmetric for $a = 1$ and has interesting properties in discovering sub-clustering relationships when $a \neq 1$, as discussed in [8].

When L clusterings C_l, $l = 1, 2, ..., L$ are concerned, the measure $S_{l,t}$ can be computed for any pair of clusterings C_l and C_t and assembled in a *similarity matrix*.

On the basis of such a matrix, several graphical representations can be devised to explore the whole set of solutions. Solutions (clusterings) themselves can be viewed as points to group together in order to find a smaller number of solutions, where each one represents a particular alternative. Moreover, if a solution is found which is reasonable, solutions in its neighborhood can be easily explored to discover the most suitable.

The first graphical rapresentation that can be used is based on the agglomerative hierarchical clustering approach; in fig.3 the corresponding dendrogram is plotted for the 371 distinct solutions of the Yeast dataset. In the figure a threshold is set (horizontal dotted line) which puts into evidence several groups of clusterings that are *similar*, identified by the same color.

The other approach we used to show the set of solutions is the Multi Dimensional Scaling (MDS, [11]) projection employed to visualize clustering solutions as points in a low-dimensional space, in which near points represent similar solutions. Other information can be embedded in the MDS projection of the clustering solutions by adding colors to the points. Interpolation through such colors finally produces a *clustering map*, as shown in fig. 4. Figure represents location of solutions in the MDS plane where distance from two points tends to approximate $1 - S_{l,t}$: therefore two near points on the map represent similar partitions of the dataset. The other information is the distortion value corresponding to any point represented in the third coordinate by contour lines in

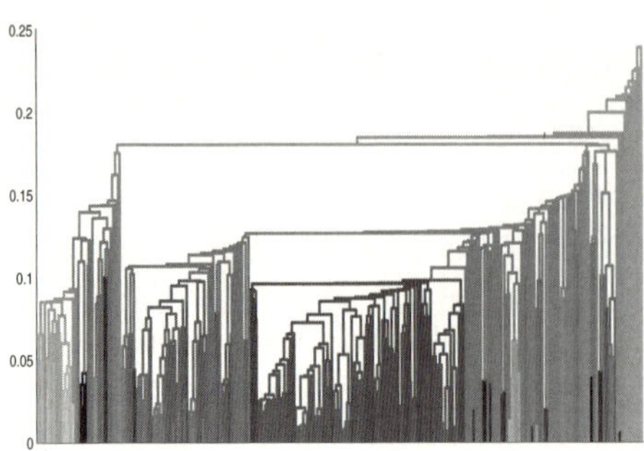

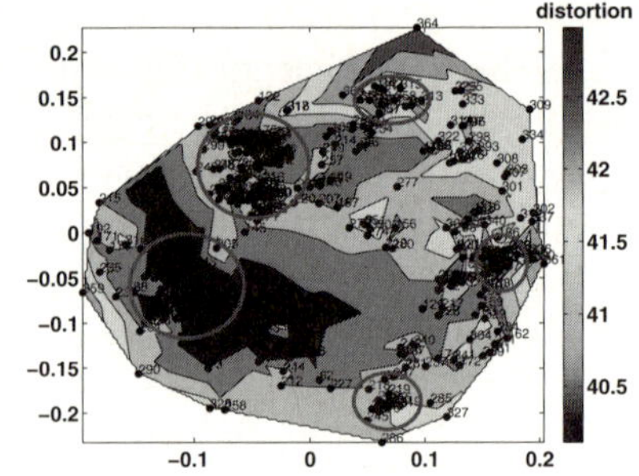

Fig. 3. Example of clustering map visualization (Yeast dataset)

Fig. 4. Example of clustering map visualization (Yeast dataset)

the map. Analyzing the map we can discover large areas of local minima, circled on the map, that can be interpreted as alternative subsets of solutions, two of them are characterized by distortion value near to global minimum.

5   Concluding Remarks

In this paper we have introduced a framework where we used the GO approach to find multiple solutions (i.e. different clustering solutions) to a clustering problem and a way to show the information regarding the space of solutions in terms of similarity distances, MDS and dendrogram. All of them can permit to the domain expert to interactively analyze the results and to infer useful information to make decisions.

References

1. Ben-Hur, A., Elisseeff, A., Guyon, I.: A stability based method for discovering structure in clustered data. Pacific Symposium on Biocomputing 7, 6–17 (2002)
2. Gionis, A., Mannila, H., Tsaparas, P.: Clustering aggregation. ACM Trans. Knowl. Discov. Data 1, 1 article 4 (2007)
3. Filkov, V., Skiena, S.: Integrating microarray data by consensus clustering. Int. J. AI Tools 13(4), 863–880 (2004)
4. Topchy, A.P., Jain, A.K., Punch, W.F.: Clustering Ensembles: Models of Consensus and Weak Partitions. IEEE Trans. Pattern Anal. Mach. Intell. 12, 1866–1881 (2007)
5. Caruana, R., Nam Nguyen, M.E., Smith, C.: Meta Clustering. In: Perner, P. (ed.) ICDM 2006. LNCS (LNAI), vol. 4065, pp. 107–118. Springer, Heidelberg (2006)
6. Brachetti, P., De Felice Ciccoli, M., Di Pillo, G., Lucidi, S.: A new version of the Price's algorithm for global optimization. Journal of Global Optimization 10, 165–184 (1997)
7. Bresco, M., Raiconi, G., Barone, F., De Rosa, R., Milano, L.: Genetic approach helps to speed classical Price algorithm for global optimization. Soft Computing Journal 9, 525–535 (2005)

8. Bishehsari, F., Mahdavinia, M., Malekzadeh, R., Mariani-Costantini, R., Miele, G., Napolitano, F., Raiconi, G., Tagliaferri, R., Verginelli, F. (eds.): PCA based feature selection applied to the analysis of the international variation in diet. LNCS (LNAI), pp. 551–556 (2007) ISSN: 0302-9743

9. Nakai, K., Kanehisa, M.: A Knowledge Base for Predicting Protein Localization Sites in Eukaryotic Cells. Genomics 14, 897–911 (1992)

10. Di Gesú, V., Giancarlo, R., Lo Bosco, G., Raimondi, A., Scaturro, D.: GenClust: A genetic algorithm for clustering gene expression data. BMC Bioinformatics 6, 289–300 (2005)

11. Kruskal, J.B., Wish, M. (eds.): Multidimensional Scaling. Sage, Thousand Oaks (1978)

12. Kuncheva, L.I., Vetrov, D.P.: Evaluation of Stability of k-Means Cluster Ensembles with Respect to Random Initialization. IEEE Transactions on Pattern Analysis and Machine Intelligence 28, (11) 1798–1808 (2006)

13. Price, W.L.: Global optimization by controlled random search. Journal of Optimization Theory and Applications 55, 333–348 (1983)

14. Smith, S.P., Dubes, R.: Stability of a Hierarchical Clustering. Pattern Recognition 12, 177–187 (1980)

15. Valentini, G., Ruffino, F.: Characterization Of Lung Tumor Subtypes Through Gene Expression Cluster Validity Assessment. RAIRO-Inf. Theor. Appl. 40, 163–176 (2006)

Robust Clustering by Aggregation and Intersection Methods

Ida Bifulco, Carmine Fedullo, Francesco Napolitano, Giancarlo Raiconi,
and Roberto Tagliaferri

NeuRoNe Lab, DMI, University of Salerno, via Ponte don Melillo, 84084 Fisciano
(SA) Italy
{ibifulco,fnapolitano,gianni,rtagliaferri}@unisa.it
http://www.neuronelab.dmi.unisa.it

Abstract. When dealing with multiple clustering solutions, the problem of extrapolating a small number of good different solutions becomes crucial. This problem is faced by the so called Meta Clustering [12], that produces clusters of clustering solutions. Often such groups, called meta-clusters, represent alternative ways of grouping the original data. The next step is to construct a clustering which represents a chosen meta-cluster. In this work, starting from a population of solutions, we build meta-clusters by hierarchical agglomerative approach with respect to an entropy-based similarity measure. The selection of the threshold value is controlled by the user through interactive visualizations. When the meta-cluster is selected, the representative clustering is constructed following two different consensus approaches. The process is illustrated through a synthetic dataset.

Keywords: consensus clustering, meta clustering, mds visualization, dendrogram visualization.

1 Introduction

In the last years many papers have been published in literature regarding the use of clustering techniques in the Knowledge Discovery and Data Mining field, for data analysis in many applications and scientific areas [3,4,5].

The main idea of cluster analysis is to group together similar patterns depending on the chosen features. The most used algorithms start from a random or arbitrary initial configuration and then evolve to a local minimum of the objective function. In complex problems (in many real cases) there are several minima and more than one can explain in a convincing manner the data distribution. In this case, we need, at least, to run many times the algorithms to choose more reasonable solutions. This is due to the intrinsic ill-posedness of clustering where the existence of a global solution cannot be assured and small perturbations of data due to noise can lead to very different solutions. Some discussions about this point can be found in [6,7,8].

As a consequence of these facts, in some cases researchers try to assess the stability and reliability of the obtained clusterings [11,10,9]. Stability means that

I. Lovrek, R.J. Howlett, and L.C. Jain (Eds.): KES 2008, Part III, LNAI 5179, pp. 732–739, 2008.

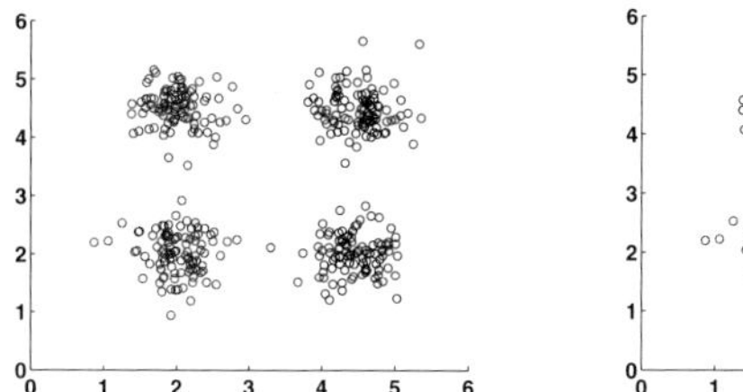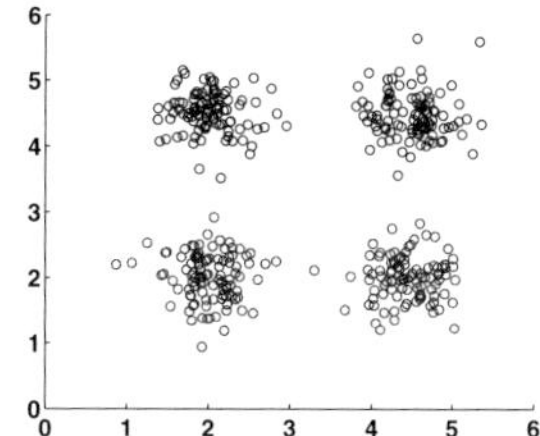

Fig. 1. Multiple minima example. The shown dataset has two equivalent and symmetric 2-clusters solutions.

slight perturbations on the inputs does not influence the obtained clusters too much. Reliability, instead is the tendency of a pattern to belong always to the same cluster with high probability or possibility.

Stability implies the concept of uniqueness of solution. For example, when using *k-means* the main problem is finding the parameter k assuring the best stability value. But k indicates also the zoom level we choose to analyze data. In some cases, once fixed k, we can have more than one stable solution. For example, suppose to have a mixture of four well separated Gaussians with the same number of patterns but different means and variance. It is obvious that in this case the best stability is obtained by solutions with four clusters. Let us suppose, instead, that we seek for a more rough grouping of data, for example with only two clusters. In this case there are two obvious different best solutions with respect to the distortion function that are equivalently reasonable. The two solutions cannot be merged together in a good way if we want to maintain $k=2$ or the complete dataset (Fig.1). The process of merging clustering solutions is called consensus and two approaches to this problem, one of which is introduced in this paper, are illustrated in Section 2. The other one was chosen for its ability to exclude points from the consensus, which is also a feature of our method. Meta clustering [12] is the process of clustering multiple solutions and is introduced in Section 1. Section 3 shows the application of such ideas to a synthetic dataset.

2   Meta Clustering

When dealing with multiple clustering solutions, the problem of extrapolating a small number of good different solutions becomes crucial. Grouping similar clustering solutions together is in turn a clustering problem. This process is called Meta Clustering [12]. Partly following the approach of [12], we divide the meta clustering process into 4 steps: 1) Generate many good different base-level clusterings (in our case local minima of the objective function generated by global optimization techniques [13]). 2) Measure the similarity between each couple of base clusterings. 3) Exploit the computed similarities to build a dendrogram and an MDS Euclidean embedding of the clustering solutions. 4) Use consensus algorithms to obtain a single solution from each meta-cluster.

Quality measures for the aggregations (such as accuracy, compactness etc.[12]) are used to decide the threshold of "mergeability" of the clusterings: a high decrease of quality for an aggregated solution suggests that the base solutions should not be merged.

The use of interactive tools derived from our previous work ([15,14]) permits to give the user information he needs to make decisions on the best meta clustering solutions.

3   Consensus Clustering

Consensus clustering, also known in literature as clustering ensembles and clustering aggregation, is the process of extrapolating a single clustering solution from a collection, in such a way to maximize a measure of agreement. This problem is NP complete [16]. In [17] some methods are introduced for consensus proving a 3-approximation algorithm. On the other hand, [18] suggests that the complexity of approximation algorithms can be unjustified, given the comparable performances of heuristic approaches. They also indicate the combined use of unsampling techniques and approximation algorithms as the best compromise. In [19] three EM-like methods are illustrated showing comparable performance with other 11 algorithms. EM approach is also exploited in [20], but combining weak clusterings. In [21] consensus clustering was improved by adaptive weighted subsampling of data. In the following we shall consider two kinds of approach: the consensus method of [17], and a novel method that obtains a subpartition of data using the intersection between corresponding clusters belonging to all solutions that are included in a metacluster.

3.1   Clustering Aggregation

The first algorithm we use in this paper is the consensus "Balls Algorithm" proposed in [17] for the correlation clustering problem. It first computes the matrix of pairwise distances between patterns. Then, it builds a graph whose vertices are the tuples of the dataset, and the edges are weighted by the distances. Aim of the algorithm is to find a set of vertices that are close to each other and far from other vertices. Given such a set, it can be considered a cluster and removed from the graph. The algorithm proceeds with the rest of the vertices. The algorithm is defined with an input parameter α that guarantees a constant approximation ratio.

3.2   Intersection Method

Let be $X = \{x_1, ..., x_N\}$, $x_i \in \mathbb{R}^n$ a set of N data points and $\mathcal{C} = \{\mathcal{C}^1, ..., \mathcal{C}^m\}$ a set of partitions (clustering) of such data, we want to infer from these a new solution (clustering) in agreement with all the collected information. To achieve this we must first state a model for comparing different partitions. This problem is defined as *m-partition clustering* in [1] as follows: let $\Delta(A, B) = |(A \setminus B) \cup (B \setminus A)|$ for two sets (partitions) A and B and let Σ be a collection of m partitions

$\mathcal{C}^1, \mathcal{C}^2, \ldots, \mathcal{C}^m$ of Σ with each partition $\mathcal{C}^i = \{C^i_1, C^i_2, ..., C^i_k\}$ containing exactly the same number k of subsets. Let a *valid solution* be a sequence of m permutations $\sigma = (\sigma_1, \sigma_2, ..., \sigma_m)$ of $\{1, 2, ..., k\}$ that *aligns* the partitions. For any permutation ρ of $\{1, 2, ..., k\}$, $\rho(i)$ is the $i-th$ element of ρ for $1 \leqslant i \leqslant k$, we want to find σ that minimize :

$$f(\sigma) = \sum_{i=1}^{k} \sum_{1 \leq j < r \leq m} \Delta(C_j^{\sigma_j(i)}, C_r^{\sigma_r(i)}) \tag{1}$$

this is an NP-hard problem for $m > 2$ [2], so we have developed a heuristic approach based on a greedy strategy for it. The main steps performed by the procedure to achieve this objective are:

1. Clustering population ordering
2. Similarity matrix for cluster computation
3. Cluster ordering
4. Intersection of solutions

Clustering population ordering. Let Σ be a collection of m partitions $\{\mathcal{C}^1, \mathcal{C}^2, \ldots, \mathcal{C}^m\}$ with each partition $\mathcal{C}^i = \{C^i_1, C^i_2, ..., C^i_k\}$ containing exactly the same number k of subsets. The idea behind the developed algorithm is to view the sequencing problem of such collection as a graph optimization one. Consider a complete graph on which each node correspond to a clustering. The distance matrix has elements $d_{i,j} = 1 - S(i, j)$ where similarities S are computed by a symmetric version of the measure defined in [14]. In such terms finding a chain of neighbors can be viewed as finding a 'minimum path' connecting all nodes (a slight variation of the TSP problem). To approximately solve such last problem we use the following greedy strategy: at the first step the procedure selects from the initial set the clustering pair having the maximum similarity value and we define it *(Left,Right)*. Then, at each iteration, it is selected the nearest (from a similarity point of view) to *Left* and the nearest to *Right* and are added to the left and to the right of the current 'best clustering pair' and so on, until no clustering can be added to the solution.

Compute similarity matrix for cluster. Let be $\mathcal{C} = \{\mathcal{C}^1, ..., \mathcal{C}^m\}$, a sorted clustering set, and $C^i_j = \{C^i_1, ..., C^i_k\}$ for $i = 1, ..., m$, the k clusters of $i - th$ clustering. A similarity measure between clusters of a sorted clustering pair can be defined as follow:

$$sim(C^i_j, C^{i+1}_l) = \frac{|C^i_j \cap C^{i+1}_l|}{|C^i_j|} \quad j, l = 1, ..., k \tag{2}$$

which summarizes the amount of common data between two clusters of lined up clustering solutions. This is a 'forward' measure because it takes into account only the similarity between clusters of successive clustering $(\mathcal{C}^i, \mathcal{C}^{i+1})$ for $i = 1, ..., m - 1$.

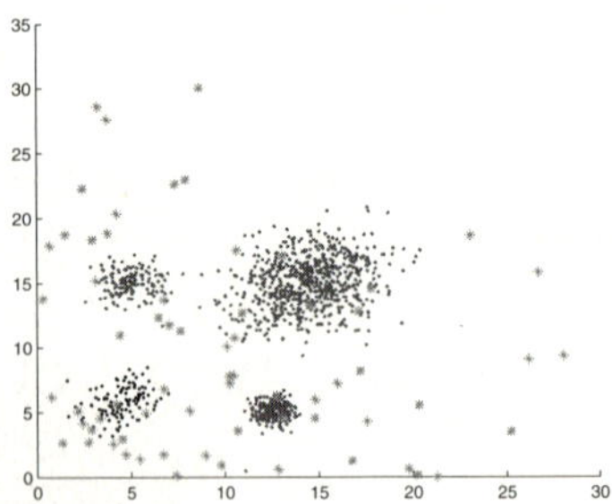

Fig. 2. Synthetic dataset with random noise

Cluster ordering. Let $\mathcal{C} = \{\mathcal{C}^1, ..., \mathcal{C}^m\}$ the set of clusterings sorted as described in the previous subsection, where each $\mathcal{C}^i = \{C_1^i, ..., C_k^i\} = \{set\ of\ k\ clusters\}$, and ***sim*** is a similarity matrix between clusters defined by (2). We can compute a greedy approximation for the cluster ordering problem as follows:

Step 1. Sort the clusters $\mathcal{C}^1$ such that[1]:
$$|C_1^1| \geq |C_2^1| \geq ... \geq |C_k^1|$$
Step 2. Set $\hat{C}_j^1 = C_j^1$ for $j = 1, ..., k$
Step 3. For each cluster in the current clustering and for each clustering in the solution set compute:

$$\hat{C}_j^{i+1} = \arg\max_{C_s^{i+1}}\{\boldsymbol{sim}(\hat{C}_j^i, C_s^{i+1})\} \quad C_s^{i+1} \neq \hat{C}_r^{i+1} \mid r < j \qquad (3)$$

note that the max value in (3) is computed taking into account only clusters not previously chosen.

Step 4. output the sequence of clusterings $\{\hat{C}^1, ..., \hat{C}^m\}$

Intersection of solutions. The final step of our procedure attempts to compute a consensus clustering among a set of ordered clusters by (3) as follows:

$$\tilde{C}_j^{intersect} = \{\hat{C}_j^1 \bigcap \hat{C}_j^2 \bigcap .... \bigcap \hat{C}_j^m\} \quad j = 1, ..., k \qquad (4)$$

4   Experimental Results

In order to test the proposed procedures we applied them to a synthetic dataset composed by a mixture of four Gaussians with different variances and additive noise (see Fig.2). We start by running the Optimization Algorithm (see [13]) to find local minima obtaining 36 different solutions. We build the hierarchical tree on the 36 solutions as shown in Fig.3, left, where leaves are labeled by numerals in ascending order of distortion value. The tree is built using the complete linkage method based on the similarity matrix as defined in [14]. We exploited interactive

---

[1] $|X|$ is the number of elements in set X.

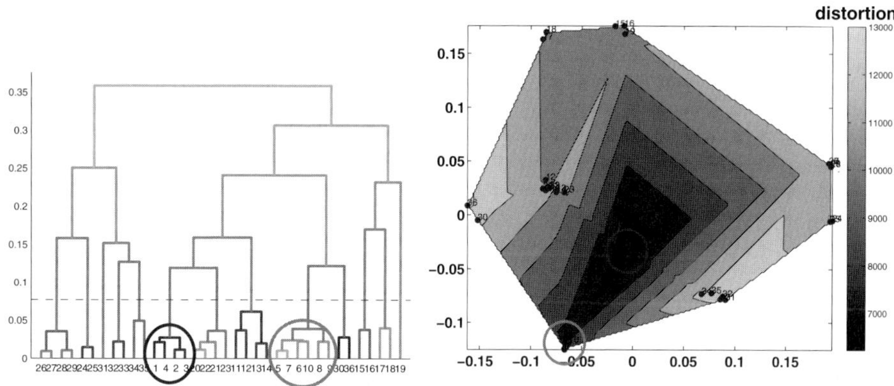

Fig. 3. Meta Clustering Dendrogram and MDS Clustering Map. Two subtrees have been chosen on the dendrogram and the solutions corresponding to their leaves are reported on the right map. They correspond to the two regions of minimum distorsion on the map.

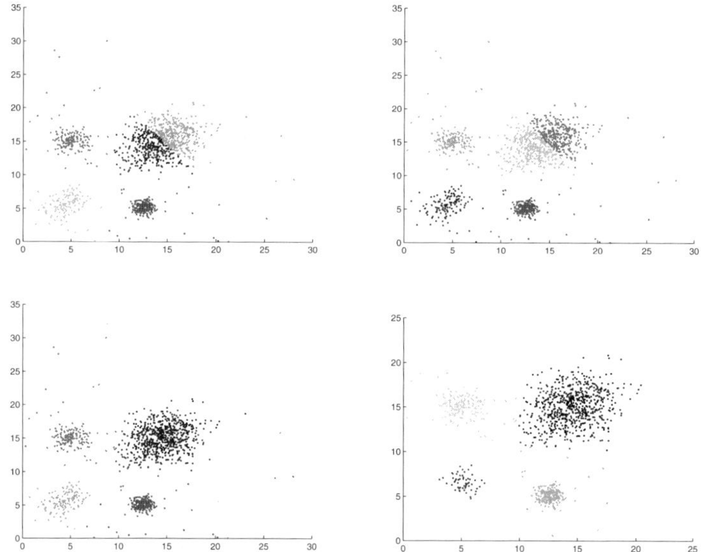

Fig. 4. Results of the two consensus approach performed on the two selected subtrees. First row, left to right: Ballclust on the blue subtrees (5 big clusters plus 2 singleton clusters), intersection method on the blue subtree (5 big clusters and 15 points discarded). Second row, left to right: Ballclust on the green subtree (4 big clusters plus 10 singleton clusters), intersection method on the green subtree (4 big clusters and 92 points discarded). It can be seen that the points discarded by intersection are mainly noise.

visualizations such as the hierarchical tree (see Fig.3, left), where the user can select any subtree; the MDS Clustering Map (see Fig.3, right), that put into evidence the solutions currently selected on the tree; the consensus solution performed on the leaves of the selected subtree (see Fig.4). Finally, some quality indicators are shown to the user, such as distortion values, number of clusters, number of points per cluster, number of points eliminated by the intersection method, mean dissimilarity between the solution and the leaves of the subtree, etc. Looking at figure 4 it is evident that both meta-clusters lead, using consensus techniques, to a final clustering that well explains the real data structure. The intersection method excludes more data from the final clustering, but results in a more robust solution.

5   Conclusions

In this paper we showed how Meta Clustering can be exploited to extract a small number of different and representative solutions, together with consensus algorithms. We used two different approaches to consensus clustering: the first method, known in literature, builds a new clustering by minimizing a disagreement function, while the second one, introduced in this paper, produces a subpartition by cluster intersection. All our tools are implemented in an interactive environment that simplifies the analysis of the results. Our future plans are to include the application of the most cited consensus algorithms in literature and the criteria to evaluate the quality of the proposed solutions in our interactive visualization tool.

References

1. Berman, P., DasGupta, B., Kao, M., Wang, J.: On constructing an optimal consensus clustering from multiple clusterings. Inf. Process. Lett. 104, (4) 137–145 (2007)
2. Gusfield, D.: Partition-distance: A problem and class of perfect graphs arising in clustering. Information Processing Letters 82, 159–164 (2002)
3. Amato, R., Ciaramella, A., Deniskina, N., et al.: A Multi-Step Approach to Time Series Analysis and Gene Expression Clustering. Bioinformatics 22, 589–596 (1995)
4. Jiang, D., Tang, C., Zhang, A.: Cluster Analysis for Gene Expression Data: A Survey. IEEE Transactions on Knowledge and Data Engineering 16, (11) 1370–1386 (2004)
5. Xui, R., Wunsch, D.: Survey of clustering algorithms. IEEE Transactions on Neural Networks 16(3), 645–678 (2005)
6. Hu, Y., Hu, Y.P.: Global optimization in clustering using hyperbolic cross points. Pattern Recognition 40(6), 1722–1733 (2007)
7. Agarwal, P.K., Mustafa, N.H.: k-means projective clustering. In: Proceedings of the Twenty-Third ACM SIGMOD-SIGACT-SIGART Symposium on Principles of Database Systems, pp. 155–165. ACM Press, New York (2004)
8. Kaukoranta, T., Franti, P., Nevalainen, O.: Reallocation of GLA codevectors for evading local minima. Electronics Letters 32(17), 1563–1564 (1996)

9. Valentini, G., Ruffino, F.: Characterization Of Lung Tumor Subtypes Through Gene Expression Cluster Validity Assessment. RAIRO-Inf. Theor. Appl. 40, 163–176 (2006)
10. Bertoni, A., Valentini, G.: Random projections for assessing gene expression cluster stability. In: Proceedings IEEE International Joint Conference on Neural Networks, vol. 1, pp. 149–154 (2005)
11. Kuncheva, L.I., Vetrov, D.P.: Evaluation of Stability of k-Means Cluster Ensembles with Respect to Random Initialization. PAMI 28(11), 1798–1808 (2006)
12. Caruana, R., Elhawary, M., Nguyen, N., Smith, C.: Meta Clustering. In: ICDM, pp. 107–118 (2006)
13. Bifulco, I., Murino, L., Napolitano, F., Raiconi, G., Tagliaferri, R.: Using Global Optimization to Explore Multiple Solutions of Clustering Problems. In: KES 2008 (2008)
14. Ciaramella, A., Cocozza, S., Iorio, F., Miele, G., Napolitano, F., Pinelli, M., Raiconi, G., Tagliaferri, R.: Interactive data analysis and clustering of genomic data. Neural Networks 21, 368–378 (2007)
15. Napolitano, F., Raiconi, G., Tagliaferri, R., Ciaramella, A., Staiano, A., Miele, A.: Clustering and visualization approaches for human cell cycle gene expression data analysis. International Journal Of Approximate Reasoning 47(1), 70–84 (2008)
16. Barthélemy, J.P., Leclerc, B.: The median procedure for partitions. In: Cox, I.J., Hansen, P., Julesz, B. (eds.) Partitioning Data Sets, American Mathematical Society, Providence, RI, pp. 3–34 (1995)
17. Gionis, A., Mannila, H., Tsaparas, P.: Clustering aggregation. ACM Trans. Knowl. Discov. Data 1(4), 1 (2007)
18. Bertolacci, M., Wirth, A.: Are approximation algorithms for consensus clustering worthwhile? In: 7th SIAM International Conference on Data Mining, pp. 437–442 (2007)
19. Nguyen, N., Caruana, R.: Consensus Clustering. In: Perner, P. (ed.) ICDM 2007. LNCS (LNAI), vol. 4597, pp. 607–612. Springer, Heidelberg (2007)
20. Topchy, A., Jain, A.K., Punch, W.: Clustering ensembles: models of consensus and weak partitions. IEEE Transactions on Pattern Analysis and Machine Intelligence 27(12), 1866–1881 (2005)
21. Topchy, A., Minaei-Bidgoli, B., Jain, A.K., Punch, W.F.: Adaptive clustering ensembles. Pattern Recognition. In: Proceedings of the 17th International Conference, vol. 1, pp. 272–275 (2004)

Comparison of Genomic Sequences Clustering Using Normalized Compression Distance and Evolutionary Distance

Massimo La Rosa[1,2], Riccardo Rizzo[2], Alfonso Urso[2],
and Salvatore Gaglio[1,2]

[1] Dipartimento di Ingegneria Informatica, Universitá di Palermo, Italy
[2] ICAR-CNR, Consiglio Nazionale delle Ricerche, Palermo, Italy

Abstract. Genomic sequences are usually compared using evolutionary distance, a procedure that implies the alignment of the sequences. Alignment of long sequences is a long procedure and the obtained dissimilarity results is not a metric. Recently the normalized compression distance was introduced as a method to calculate the distance between two generic digital objects, and it seems a suitable way to compare genomic strings. In this paper the clustering and the mapping, obtained using a SOM, with the traditional evolutionary distance and the compression distance are compared in order to understand if the two distances sets are similar. The first results indicate that the two distances catch different aspects of the genomic sequences and further investigations are needed to obtain a definitive result.

1 Introduction

In recent years, the growing availability of biological data has driven computer scientists to develop algorithms and methodologies able to support the analysis of this kind of information. Typical biological data are, for instance, genetic and protein sequences, molecular structures, chemical compounds, gene expressions data. Given this large amount of raw data, one of the most used approach to extract useful and functional information is to perform data mining techniques, such as unsupervised or supervised clustering.

Our work focuses, in particular, on biological data made up of bacteria DNA sequences, that can be easily found in very large databases like GenBank [1] or EMBL [2], and aims at finding a simple and reliable clustering tool that can help biologists in their studies concerning species identification and phylogenesis.

One of the major issue is related to the information content of nucleotide sequences: they are not directly representable using a feature space, so the only information we have is in terms of pairwise similarities/dissimilarities.

There exists several types of distances for comparing two protein or nucleotide sequences, the so called "evolutionary distances" [3], and all of these methods are based on the concept of sequence alignment [4,5]. This methodology has, however, at least two major drawbacks: first of all, the algorithms that perform

I. Lovrek, R.J. Howlett, and L.C. Jain (Eds.): KES 2008, Part III, LNAI 5179, pp. 740–746, 2008.
© Springer-Verlag Berlin Heidelberg 2008

the alignments are often computationally expensive, especially when aligning very long sequences, and depend on several parameters; secondly, evolutionary distances are not metrics, that is triangular inequality does not hold.

For all of these reasons, it is more advantageous the use of a distance metric, alignment-free, that overcomes the limitations remarked above. Recently, Li et al. [6] have developed a metric, based on Kolmogorov complexity [7]. This similarity metric, called "Universal Similarity Metric" (USM), can be generally applied to compare any two objects that can be represented as a binary string; applying USM to nucleotide sequences, it is possible to obtain a parameter-free similarity measure, having the properties of a metric, without any *apriori* alignment.

The aim of this work is to use the USM in order to compute a dissimilarity matrix of a bacteria dataset, made up of a single "housekeeping gene"; from this matrix we shall perform an unsupervised clustering, using an extension of the classical Self-Organizing Map algorithm [8], tuned up to cope with input data expressed only in terms of their pairwise dissimilarities. The choice of a dataset containing only one "housekeeping" gene, more specifically the 16S rRNA gene, is justified by recent trends in bacterial genomics that have shown how this gene is particularly suitable for clustering and classification purposes [9,10]. The obtained clustering results will be compared with the ones coming from a dissimilarity matrix computed with evolutionary distance on the same dataset. With this approach, we want to test the effectiveness of USM applied to short DNA sequences, about 1400 base pairs, rather than whole genomes, as a valid input data in order to perform a meaningful unsupervised clustering of bacteria.

2   Related Work

The use of USM for DNA sequences comparison has been done in [6] and [11]. In those papers, a dissimilarity matrix of complete mammalian mithocondrial genomes has been computed in order to build a phylogenetic tree or to visualize the data through the Multidimensional Scaling technique [20].

A deeper study of USM and its applications to biological datasets is presented in [23]. Authors of this paper tested three approximations of USM by using 25 different compressors (see next section) on six biological datasets. Then, from the dissimilarity matrices obtained, they built several phylogenetic trees and compared them with the corresponding trees created using the classical evolutionary distance. From all of these experimental results, they further validated the use of USM for biological sequences.

A clustering approach, using Median Som [18], based on protein sequences and evolutionary distances was carried out in [12]. Median Som was also adopted by [13] for clustering of human endogenous retrovirus sequences, given a distance matrix based on the FASTA similarity scores [14].

The generation of a topographic map was at the basis of the work of [15], in which the algorithm proposed by [16] was used to obtain a visualization of a well defined bacteria dataset, containing only the DNA sequences of *type strains*, that is sample species, of the Gammaproteobacteria class.

3 Universal Similarity Metric

Li et al. introduced the concept of Universal Similarity Metric in [6], using the theory of Kolmogorov complexity [7]. The USM represents a class of distance measures obtained as an approximation of Kolmogorov complexity, since this one is not computable. Among this class of measures, we will consider the one called "Normalized Compression Distance" (NCD), in which the Kolmogorov complexity of an object x is approximated with the size of its compressed version.

So, given a real-word reference compressor C, NCD is defined as:

$$\mathrm{NCD}(x, y) = \frac{C(xy) - \min\left\{C(x), C(y)\right\}}{\max\left\{C(x), C(y)\right\}} \tag{1}$$

where $C(xy)$ is the compressed size of the concatenation of x and y, and $C(x)$ and $C(y)$ are the compressed sizes respectively of x and y. Roughly speaking, NCD is a number between 0 and 1: the smaller the number, the more similar are the objects. In our system, the generic objects x and y are the nucleotide sequences, that can be seen as strings composed of the four letters of DNA alphabet: $\{a, c, g, t\}$.

Using a normal compressor C, according to the definitions in [11], NCD has the properties of a similarity metric. A large set of real-world compressors are normal compressors: we have decided to use *GenCompress* [17], a normal compression algorithm designed to work with DNA sequences and able to give the best compression ratios, providing this way the best approximation of Kolmogorov complexity.

4 Clustering Algorithms

One of the most used algorithms for data clustering is Kohonen's Self-Organizing Map (SOM) [8]. SOM algorithm has also the great advantage to provide a direct visualizations of clusters, because it defines a set of neurons, also called models, that are arranged in a rectangular lattice, creating a topographic map. During learning phase, each input pattern is mapped to its closest neuron, known as best matching unit (*bmu*): when the learning is complete, the coordinates of each pattern into the two-dimensional lattice are the same of its own *bmu*.

Unfortunately, SOM does not work with dataset expressed only in terms of pairwise dissimilarities, as in our situation. An adaptation of SOM to non-vectorial data is the so called Median SOM [18]. Median SOM, however, has a strong limitation: the neurons are selected among the set of patterns, providing this way a severe restriction especially for small datasets. For this reason, we used a novel technique, called Relational Topographic Map [19], that is able to work with nonvectorial data and, at the same time, overcomes the limitation introduced by Median SOM.

Relational Topographic Map [19], differently from Median SOM, represents the neurons as linear combinations of input patterns. If we indicate with $\boldsymbol{x}^j$ the generic pattern for which the distance $\left\|\boldsymbol{x}^i - \boldsymbol{x}^j\right\|^2$ exists, and and if we consider the generic neuron $\boldsymbol{w}^i$ as a linear combination of input patterns, that is $\boldsymbol{w}^i = \sum_j \alpha_{ij} \boldsymbol{x}^j$, with $\sum_j \alpha_{ij} = 1$, then $\left\| \boldsymbol{w}^i - \boldsymbol{x}^j \right\|^2$ is equal to:

$$\left\| \boldsymbol{w}^i - \boldsymbol{x}^j \right\|^2 = (D \cdot \alpha_i)_j - 1/2 \cdot \alpha_i^t \cdot D \cdot \alpha_i, \tag{2}$$

where D is the pairwise distance matrix among patterns (more details in [19]). Moreover, the distance between two neurons on the map can be computed as follows:

$$\left\| \boldsymbol{w}^i - \boldsymbol{w}^j \right\|^2 = \alpha_j^t \cdot D \cdot \alpha_i - 1/2 \cdot \alpha_j^t \cdot D \cdot \alpha_j - 1/2 \cdot \alpha_i^t \cdot D \cdot \alpha_i \tag{3}$$

Using this transformation, the distance between neurons and patterns will depend only on the pairwise dissimilarities among input patterns. During learning phase, the weights of these linear combinations are updated and, as usual, input patterns are mapped to their *bmu*. As demonstrated by its authors, Relational Topographic Map gives better clustering results than Median SOM because the neurons are not constrained to overlap to the patterns.

5   Experimental Results

5.1   Dataset Description

All of our experiments were conducted on a test dataset derived from the one used by Garrity and Lilburn in [24]. From the original dataset, consisting of 1436 small subunit ribosomal RNA sequences, we selected only the 16S rRNA sequences, whose length is about 1400 bp, obtaining a total of 1285 gene sequences. The sequences were labeled with their own order, using the same classification as in [24]. There is a total of 16 orders, and 18 bacteria sequences are unclassified. From this bacteria sequences dataset, two dissimilarity matrices were computed: the first matrix was computed using the evolutionary distance with the Jukes and Cantor method [21]; the second dissimilarity matrix was obtained using NCD, according to (1), and GenCompress compressor.

5.2   Algorithm Setup

We trained, with the Relational Topographic Map algorithm, and with both dissimilarity matrices, 40 square maps of 10×10 dimension, and for each configuration we obtained 40 maps. We chose to consider 10×10 maps, since having 16 preclassified orders, with 100 neurons there are about six neurons per order. Each pair of corresponding maps, trained with the two input matrices, had the same initialization and a learning process of 40 epochs.

5.3   Map Analysis

A first comparison between maps was done as follows: each cell in a map was given a unique label according to the Order of the most of the elements belonging

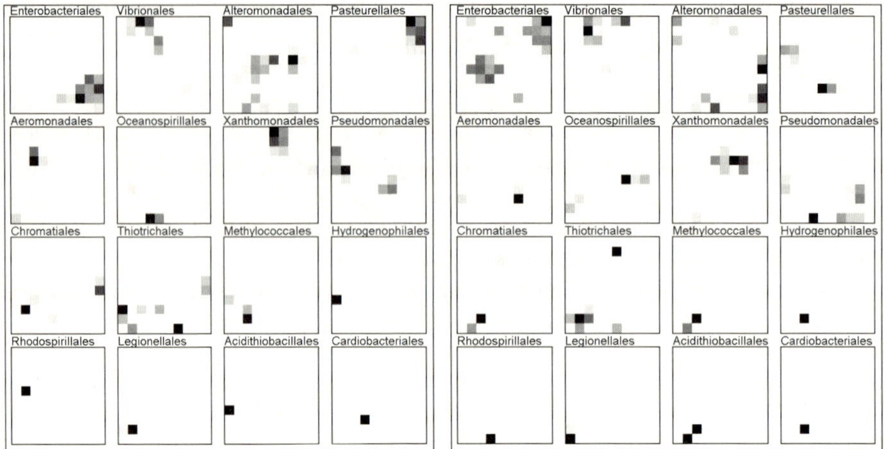

Fig. 1. Map trained with evolutionary distance, on the left, and with NCD, on the right. Both maps contains the lowest number of misplaced patterns. Considering the two dissimilarity matrices, we obtained a quite similar clustering, but different topological placement.

to that cell; then, for each cell, we counted up the number of misplaced patterns, according to the label previously assigned. We found that the maps trained with evolutionary distances and NCD have, on average, respectively 29.75 and 37.25 misplaced patterns, and considering the total number of 1285 elements of our dataset, that means we obtain an error rate, respectively, of 2.3% and 2.8%. This means the two distances alow to obtain a quite similar clustering results.

Furthermore, we applied the measure of dissimilarity defined in [22], which allows to compare two Self-Organizing Maps.

This measure has been adapted to work with Relational Topographic Map, using (2) to compute the distance between each input pattern and its own *bmu*, and (3) to compute the distance between neurons on the map; then it was computed for every pair of maps trained with the two input matrices: the result we obtained is that there is a mean similarity of 21.61% among the two sets of maps referred to the two dissimilarity matrices. This low value of similarity can be interpreted considering that pairwise matrix computed through NCD represents a very strong perturbation with respect to the pairwise matrix computed through evolutionary distance; so, even if the two clusterings are quite similar, according to the error rate decribed above, the two maps, from a topographic point of view, are very different.

In Fig. (1), we show the two maps, on the left the one trained with evolutionary distance and on the right the other one trained with NCD, with the lowest number of misplaced elements. The visualization is organized in submaps: each submap represents the distribution of one order, written at the top of the submap itself. The gray level of each cell is proportional to the number of sequences belonging to that node: in each submap, the maximum number of elements in a node is given the darkest shade of gray. It is possible to see that, for example, a large group like "Enterobacteriales", in the map obtained with NCD is split into two main clusters, whereas in the map obtained with evolutionary distance it preserves an evident compactness. Another difference is between "Thiotricales" orders: in the upper map it is spread through the map, while in the lower map it forms two clusters: one at the bottom left corner and another one at the top of the map. As for the other orders, in both maps they exhibit well defined clusters, although their positions upon maps do not coincide.

This situation reflects the fact that, considering the two dissimilarity matrices, we obtained a quite similar clustering, but different topological placement.

6   Conclusions

In previous studies, Normalized Compression Distance has been used to create phylogenetic trees of complex species based on whole genomes. In this work, we applied NCD and classical evolutionary distance in order to obtain two dissimilarity matrices of a dataset composed of the 16S rRNA gene sequences of bacteria. Our main goal was to use these two dissimilarity matrices as input data for an unsupervised clustering process, performed with an extension of Self-Organizing Map, and to compare the clusterings computed this way. We found that the two matrices provide a quite similar clustering result. On the other hand, using a distance measure to compare the maps, we found that, although with the same initialization, the maps, in terms of topological positioning, are different. From these two contrasting results we can state that further investigations are needed in order to understand the bind between Normalized Compression Distance and evolutionary distance and, subsequently, if NCD can substitute evolutionary distance in genome comparisons.

References

1. National Center for Biotechnology Information, Entrez Nucleotide query, `http://www.ncbi.nlm.nih.gov/entrez/query.fcgi?db=Nucleotide`
2. European Molecular Biology Laboratory, `http://www.ebi.ac.uk/embl/`
3. Nei, M., Kumar, S.: Molecular Evolution and Phylogenetics. Oxford University Press, New York (2000)
4. Needleman, S.B., Wunsch, C.D.: J. Mol. Biol. 48, 443–453 (1970)
5. Thompson, J.D., Higgins, D.G., Gibson, T.J.: CLUSTAL W: improving the sensitivity of progressive multiple sequence alignment through sequence weighting, position specific gap penalties and weight matrix choice. Nucleic Acids Research 22, 4673–4680 (1994)
6. Li, M., Chen, X., Li, X., Ma, B., Vityi, P.M.B.: The similarity metric. IEEE Trans. Inf. Theory 50(12), 3250–3264 (2004)
7. Li, M., Vitanyi, P.M.B.: An Introduction to Kolmogorov Complexity and its Applications, 2nd edn. Springer, New York (1997)
8. Kohonen, T.: Self-organizing maps. Springer, Heidelberg (1995)

9. Drancourt, M., Bollet, C., Carlioz, A., Martelin, R., Gayral, J., Raoult, D.: 16S Ribosomal DNA Sequence Analysis of a Large Collection of Environmental and Clinical Unidentifiable Bacterial Isolates. J. Clin. Microbiol. 38, 3623–3630 (2000)
10. Drancourt, M., Berger, P., Raoult, D.: Systematic 16S RNA Gene Sequencing of Atypical Clinical Isolates Identified 27 New Bacterial Species Associated with Humans. J. Clin. Microbiol. 42, 2197–2202 (2004)
11. Cilibrasi, R., Vitanyi, P.M.B.: Clustering by Compression. IEEE Trans. Inf. Theory 51(4), 1523–1545 (2005)
12. Somervuo, P., Kohonen, T.: Clustering and visualization of large protein sequence databases by means of an extension of the self-organizing map. In: Proceedings of the Third International Conference on Discovery Science, pp. 76–85 (2000)
13. Oja, M., Somervuo, P., Kaski, S., Kohonen, T.: Clustering of human endogenous retrovirus sequences with median self-organizing map. In: WSOM 2003 Workshop on Self-Organizing Maps, September 9-14, 2003 (2003)
14. Pearson, W., Lipman, D.: Improved tools for biological sequence comparison. Proc. Natl. Acad. Sci. USA 85, 2444–2448 (1988)
15. La Rosa, M., Di Fatta, G., Gaglio, S., Giammanco, G.M., Rizzo, R., Urso, A.: Soft Topographic Map for Clustering and Classification of Bacteria. In: R. Berthold, M., Shawe-Taylor, J., Lavrač, N. (eds.) IDA 2007. LNCS, vol. 4723, pp. 332–343. Springer, Heidelberg (2007)
16. Graepel, T., Burger, M., Obermayer, K.: Self-organizing maps: generalizations and new optimization techniques. Neurocomputing 21, 173–190 (1998)
17. Chen, X., Kwong, S., Li, M.: A compression algorithm for DNA sequences. Engineering in Medicine and Biology Magazine 20(4), 61–66 (2001)
18. Kohonen, T., Somervuo, P.: How to make large self-organizing maps for nonvectorial data. Neural Networks 15(8-9), 945–952 (2002)
19. Hasenfuss, A., Hammer, B.: Relational Topographic Maps. In: R. Berthold, M., Shawe-Taylor, J., Lavrač, N. (eds.) IDA 2007. LNCS, vol. 4723, pp. 93–105. Springer, Heidelberg (2007)
20. Torgerson, W.S.: Multidimensional scaling: I. Theory and method. Psychometrika 17, 401–419 (1952)
21. Jukes, T.H., Cantor, R.R.: Evolution of protein molecules. In: Munro, H.N. (ed.) Mammalian Protein Metabolism, pp. 21–132. Academic Press, New York (1969)
22. Kaski, S., Lagus, K.: Comparing Self-Organizing Maps. In: Proceedings of the 1996 International Conference on Artificial Neural Networks (1996)
23. Ferragina, P., Giancarlo, R., Greco, V., Manzini, G., Valiente, G.: Compression-based classification of biological sequences and structures via the Universal Similarity Metric: experimental assessment. BMC Bioinformatics 8, 252 (2007)
24. Garrity, G.M., Lilburn, T.G.: Self-organizing and self-correcting classifications of biological data. Bioinformatics 21(10), 2309–2314 (2005)

A One Class Classifier for Signal Identification: A Biological Case Study

Vito Di Gesù, Giosuè Lo Bosco, and Luca Pinello

DMA - Università di Palermo, Italy
{digesu,lobosco,lucapinello}@math.unipa.it

Abstract. The paper describes an application of a one-class KNN to identify different signal patterns embedded in a noise structured background. The problem become harder whenever only one pattern is well represented in the signal, in such cases one class classifier techniques are more indicated. The classification phase is applied after a preprocessing phase based on a Multi Layer Model (MLM) that provides a preliminary signal segmentation in an interval feature space. The one-class KNN has been tested on synthetic data that simulate microarray data for the identification of nucleosomes and linker regions across DNA. Results have shown a good recognition rate on synthetic data for nucleosome and linker regions.

1 Introduction

One-class classifiers have been introduced in order to discriminate a target class from the rest of the feature space [1].

This approach is mandatory when only examples of the target class are available and it can reach an high performance if the cardinality of the target class is much greater than the other one so that too few training examples of the smallest class are available in order to properly train a classifier. Recently, Wang and Stolfo [2] have shown that a one class training algorithm can perform equally well as two class training approaches in the specific problem of masquerader intrusion detection.

Here, a one class KNN's is proposed in order to identify signal pattern slightly differentiated and embedded in a noise structured background. Examples of such kind of signal are those provided by microarray data of the *Saccharomyces cerevisiae* [3], where the goal, in this case, is the identification of nucleosomes and linker regions across DNA. Nucleosomes are the fundamental repeating units of eukaryotic chromatin and their position can be regulated in vivo by multi-subunit chromatin remodeling complexes and can influence gene expression in eukaryotic cells [4,5]. Alterations in chromatin structure, and hence in nucleosome organization, can result in a variety of diseases, including cancer, highlighting the need to achieve a better understanding of the molecular processes modulating chromatin dynamics [6].

To measure nucleosome positions on a genomic scale, a DNA microarray method has been recently developed. This new approach allows the identification of nucleosomal and linker DNA sequences on the basis of susceptibility of

I. Lovrek, R.J. Howlett, and L.C. Jain (Eds.): KES 2008, Part III, LNAI 5179, pp. 747–754, 2008.
© Springer-Verlag Berlin Heidelberg 2008

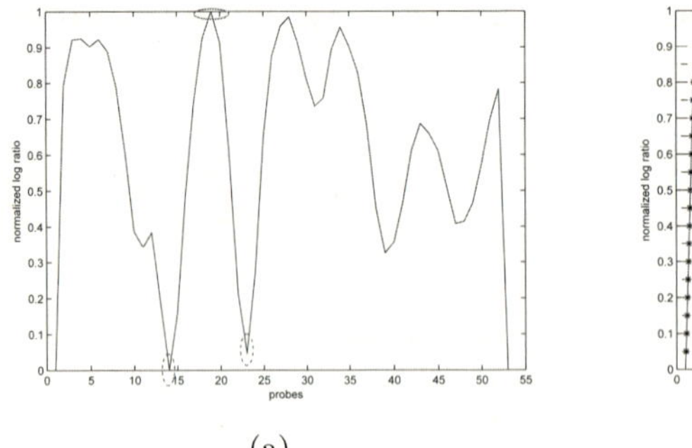 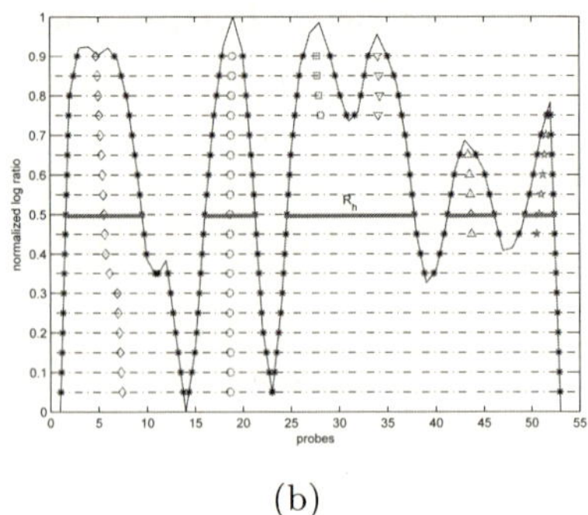

(a)   (b)

Fig. 1. *(a) Input Signal:* The input signal is the logarithmic ratio of the *green channel* to *red channel* value for each spot of the microarray. Nucleosomes correspond to peaks (marked by black circle), surrounded by lower ratio values corresponding to linker regions (marked by dashed circles) that are nucleosome free. *(b) Pattern identification and extraction:* In this example 6 patterns are retrieved, identified by rhombus, circle, square, triangle down, triangle up, star. Each pattern identifier is replicated for each of its feature values and pointed in each one of its middle point.

linker DNA to micrococcal nuclease [3]. This method allows the representation of microarray data as a signal of green/red ratio values that shows nucleosomes as peaks of about 150 base pairs long, surrounded by lower ratio values corresponding to linker regions. Note that, because of noise in the data signal and large-scale trends in mean hybridization values, a naive threshold-based approach for determining nucleosome positions is considered highly inaccurate. Therefore, a Hidden-Markov Model (HMM) approach has been used to discriminate regions representing nucleosomes and linkers. However, this approach suffers from the constrain imposed by its static topology, as a consequence lots of potential good input data are discarded from the analysis. Moreover, this approach does not take into account the shape information of the green/red ratio values.

In particular, in this paper, we aim to discriminate between two classes of patterns the *well positioned nucleosomes* (WPN) and the *linkers* (LN). For an example of an input signal which also shows the two pattern classes see Figure 1a.

The one-class KNN is applied after a preprocessing phase, based on a Multi Layer Model (MLM) [7] that provides a preliminary signal segmentation in an interval feature space. In this work the one class KNN has been tested on synthetic microarray data. Results have shown a good recognition rate on simulated data for nucleosome and linker regions. The paper is organized as follow: Section 2 describes the one class KNN classifier; the MLM preprocessing phase is outlined in Section 3; method validation is performed in Section 4; final remarks and discussion are given in Section 5.

2   One Class KNN

Here, we propose a one class training KNN which is a generalization of the classical KNN classifier [8]. Let T_p be the training set for a generic pattern p

representing a *positive* instance, and δ a dissimilarity function between patterns. Then the decision rule for an unknown pattern x is:

$$j = \begin{cases} 1 & \text{if } |\{y \in T_p \text{ such that } \delta(y, x) \leq \phi\}| \geq K \\ 0 & \text{otherwise} \end{cases} \tag{1}$$

where $j = 1$ means that x is positive, otherwise it is negative. Informally, the rule says that if there are at least K pattern in T_p dissimilar from x at most ϕ, then x is supposed to be a positive pattern, otherwise it is negative. Therefore, the one class training KNN depends on both parameters K and ϕ and their choice has to be decided by an optimization procedure.

In the particular case of nucleosome positioning, the positive patterns are the WPN while the LN are the negatives and the choice of the best K and ϕ for this specific problem will be described in detail in subsection 4.2.

3   Multi Layer Model

In this section an outline of the MLM for the analysis of mono-dimensional signal is provided. It can be considered a preliminary step that provides to the one-class KNN the proper feature space and input pattern to be classified. The MLM procedure is carried out as follows:

Preprocessing. A preprocessing is necessary in order to reduce the effect of the signal noise. Starting from the input signal, $\mathbf{S}$ each fragment S_t, $1 \leq t \leq T$, is convolved by an averaging kernel window $w = [\frac{1}{4}, \frac{1}{2}, \frac{1}{4}]$.

Training set and model construction. Since we know that WPN are shown as peaks of a bell shaped curve, in order to locate the position of a nucleosome, all local maxima of the input signal are automatically extracted from the convolved signal $\mathbf{X}$ of $\mathbf{S}$. Each convolved fragment X_t is processed in order to find $L(X_t)$ local maxima $M_t^{(l)}$ for $l = 1, \cdots, L(X_t)$. The extraction of each sub-fragment for each $M_t^{(l)}$ is performed by assigning all values in a window of radius os centered in $M_t^{(l)}$ to a vector, F_t^l of size $2 \times os + 1$: $F_t^l(j) = X_t(M_t^{(l)} - os + j - 1)$, for $j = 1, 2, ..., 2 \times os + 1$. The selection of the *significant* sub-fragments - to be used in the model definition - is performed by satisfying the following rule:

$$\begin{cases} F_t^l(j+1) - F_t^l(j) > 0 & j = 1, \cdots, os \\ F_t^l(j+1) - F_t^l(j) < 0 & j = os + 1, \cdots, 2 \times os \end{cases} \tag{2}$$

After the selection process $G(X_t)$ sub-fragments remain for each X_t. The model of the *interesting pattern* is then defined by considering the following average:

$$\overline{F}(j) = \frac{1}{T} \sum_{t=1}^{T} \frac{1}{G(X_t)} \sum_{h=1}^{G(X_t)} F_t^h(j) \quad j = 1, \cdots, 2 \times os + 1$$

That is, for each j, the average value of all the sub-fragments satisfying Eq. 2. The training set of the *interesting pattern* is $T_p = \{F_t^l | 1 \leq t \leq T \ , \ 1 \leq l \leq G(X_t)\}$.

<u>Interval identification</u>. The core of the method is the interval identification by considering H threshold levels t_h ($h = 1, ..., H$) of the convolved signal $\mathbf{X}$. For each t_h a set of intervals $R_h = \{I_h^1, I_h^2, \cdots, I_h^{n_h}\}$ is obtained; where, $I_h^i = [b_h^i, e_h^i]$ and $\mathbf{X}(b_h^i) = \mathbf{X}(e_h^i) = t_h$.

<u>Interval merging and pattern definition</u>. This step is performed by taking in account that bell shaped pattern must be extracted for the classification phase. Such kind of patterns are characterized by sequences of intervals $\left\{I_j^1, \cdots, I_{j+l}^n\right\}$ such that $I_j^i \supseteq I_{j+1}^{i+1}$; more formally a pattern P_i is defined as:

$$P_i = \{I_j^{i_j}, I_{j+1}^{i_{j+1}}, \cdots, I_{j+l}^{i_{j+l}} \mid \forall I_h^{i_h} \ \exists! I \in R_{h-1} : I = I_{h-1}^{i_{h-1}} \supseteq I_h^{i_h}\}$$

where, j defines the threshold, t_j, of the widest interval of the pattern. From the previous definition it follows that P_i is build by adding an interval $I_{h+1}^{i_{h+1}}$ only if it is the unique in R_{h+1} that includes $I_h^{i_h}$. Note that, this criterion is inspired by the consideration that a nucleosome is identified by bell shaped fragment of the signal, and the intersection of such fragment with horizontal threshold lines results on a sequence of nested intervals.

<u>Pattern selection</u>. In this step the *interesting patterns* $\mathbf{P^{(m)}}$ are selected following the criterium:

$$\mathbf{P^{(m)}} = \{P_i \ : \ |P_i| > m\}$$

i.e. patterns containing intervals that persists at least for m increasing thresholds. This further selection criterion is related to the height of the shaped bell fragment, in fact a small value of m could represents noise rather than nucleosomes. The value m is said the *minimum number of permanence*.

<u>Feature extraction</u>. Each pattern $P_i \in \mathbf{P^{(m)}}$ is identified by $I_j^{i_j}, I_{j+1}^{i_{j+1}}, \cdots,$ $I_{j+l}^{i_{j+l}}$, with $l \geq m$. Straightforwardly, the feature vector of P_i is a $2 \times l$ matrix where each column represents the lower and upper limits of each interval from the lower threshold j to the upper threshold $j + l$. The representation in this multi-dimensional feature space is used to characterize different types of patterns.

<u>Dissimilarity function</u>. A dissimilarity function between patterns is defined in order to characterize their shape:

$$\delta(P_r, P_s) = \sum_{h=1}^{H} |a_h^{r_h} - a_h^{s_h}| \tag{3}$$

where $a_h^{r_h} = e_h^{r_h} - b_h^{r_h}$, $a_h^{s_h} = e_h^{s_h} - b_h^{s_h}$.

Note that a generic pattern P could not be defined for all the $h = 1, \cdots, H$. The dissimilarity δ is then calculated after completing the feature vector of patterns P_r and P_s to H components by using the pattern model $\overline{F}$ in the case a component is not defined. In particular, this dissimilarity is a distance that takes into account the *shape* of a pattern. An example of signal and the relative interesting patterns is given in Figure 1b.

4   Experiments and Validation

Experiments have been carried out on synthetic data. The purpose is both to evaluate the expected performance of the one-class KNN for different signal to noise ratio and study the threshold, ϕ, and number of K, that optimize the one-class KNN.

4.1   Synthetic Signal Generation

A procedure to generate synthetic signal, resulting from the evolution of a first prototype introduced in [7], has been developed allowing us to assess the feasibility of our method on controlled data. Comparing to the previous version, by using more parameters, it allows a more reliable generation of synthetic microarray data. Generated signals emulate the one coming from a tiling microarray where each spot represents a *probe i* of resolution r base pairs overlapping o base pairs with probe $i + 1$. In particular, the chromosome is spanned by moving a window (probe) i of width r base pairs from left to right, measuring both the percentage of mononucleosomal DNA G_i (*green channel*) and whole genomic DNA R_i (*red channel*) within such window, respecting also that two consecutive windows (probes) have an overlap of o base pairs. The resulting signal $V(i)$ for each probe i is the logarithmic ratio of the *green channel* G_i to *red channel* R_i. Intuitively, nucleosomes presence is related to peaks of V which correspond to higher logarithmic ratio values, while lower ratio values shows nucleosome free regions called *linker regions*. This genomic tiling microarray approach takes inspiration from the work of Yuan et al. [3] where the authors have used the same methodology on the *Saccharomyces cerevisiae* DNA. Here we have defined a model able to generate such signals characterized by the following parameters:

- *nn:* The number of nucleosomes we want to add to the synthetic signal.
- *nl:* The length of a nucleosome (we know that in real case a nucleosome is 150 base pairs long)
- λ: Mean of the poisson distribution used to model the expected distances between adjacent nucleosomes;
- *r:* The resolution of a single microarray probe.
- *o:* The length in base pairs of the overlapping zone between two consecutive probes.
- *nr:* The number of spotted copies (replicates) of nucleosomal and genomic DNA on each probe of the microarray;
- *dp:* The percentage of the delocalized nucleosomes over the total number of nucleosomes;
- *dr:* It represents the range which limit the delocalization of a nucleosome in each copy of *nr*. It is defined in base pairs.

- *nsv:* The variance of the green channel in each probe, even in absence of nucleosomes due to the cross hybridization. This variance follows a normal distribution with mean 0.1.
- *pur:* The percentage of DNA purification, which is the probability that each single DNA fragment of the nr copies appears in the microarray hybridization.
- *ra:* Relative abundance between nucleosomal and genomic DNA.
- *SNR:* The linear signal to noise ratio of the synthetic signal to generate. Note that the noise is assumed to be gaussian.

Initially, a binary mask signal M is generated by considering as 1's all the base pairs representing a nucleosome (the *nucleosomal regions*) and as 0's the regions representing linkers (*the linker regions*). Note that, the beginning of each nucleosomal region is established by the Poisson distribution with mean λ. The mask signal M will be used in order to validate the classification results.

The red channel of the microarray (the genomic channel) results from the generation of nr replicates $I_1^R, \cdots, I_{nr}^R$ each one starting from an initial nucleosomal region of random size $b \sim U(0, r)$ (uniformly distributed in the range $[0, r]$), followed by continuous nucleosomic region of r base pairs.

Conversely, in order to simulate the green channel (the nucleosomic channel) nr replicates , $I_1^G, \cdots, I_{nr}^G$ are considered, each one initially equal to M and subsequently modified by perturbing each staring points x_D^i of the nucleosome to consider as delocalized such that $x_D^i = x_D^i + \mu$ with random $\mu \sim U(dr)$. Note that the percentage of nucleosomes to consider as delocalized is established by the parameter dp.

Afterwards, each nucleosomal region on the generic replicate I_i^R and I_i^G can be switched off depending on a the value of a random variable $\alpha \sim U(0, 1)$. Precisely, each nucleosomal region veryfing the test $\alpha < pur$ is considered and set to 1, otherwise it is not considered and set to 0. This results in new replicates T_i^R and T_i^G. Finally, the generated synthetic signal V for a probe i is so defined:

$$V(i) = \{log_2 \left(\sum_{j=1}^{nr} \frac{T_j^G(k)*ra}{T_j^R(k)} + \epsilon \right) \mid$$
$$(r - o)i - r + o + 1 \le k \le (r - o)i + o\} \tag{4}$$

where $\epsilon \sim N(0.1, nsv)$.

4.2 Derivation of Optimal K and ϕ from the ROC Curve

The one-class KNN performances depends on the threshold, ϕ, and the number of neighbors, K, that are used in the classification phase. Both of them can be determined from the *Receiver Operating Characteristic ROC* curve obtained in the validation phase. A *ROC* curve is a two dimensional visualization of the *true positive rate* (TPR) versus the *false positive rate* (FPR) in the case of a 2 class problem. In particular, it is obtained by calculating FPR and TPR for various values of a classification threshold parameter. The definition of FPR and TPR can be generally stated as follow:

$$FPR = \frac{FP}{FP+TN} \qquad TPR = \frac{TP}{TP+FN}$$

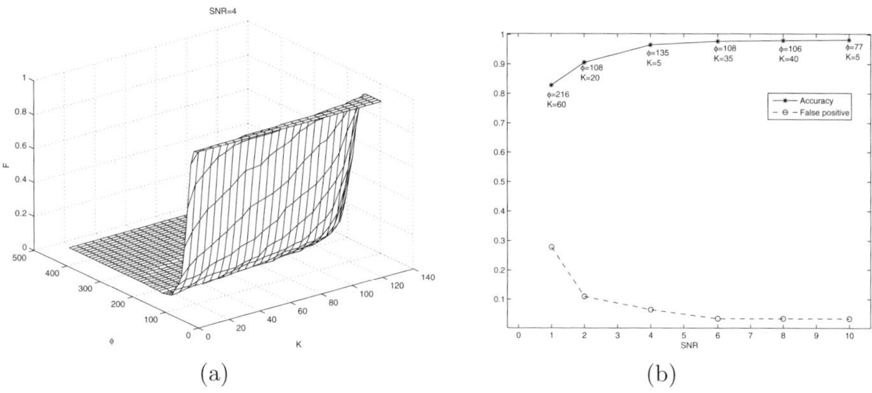

(a)   (b)

Fig. 2. *(a) One-class KNN calibration:* Plot of the function $F(\phi, K)$ for SNR=4; *(b) Accuracy and FPR values:* Best Accuracy and *FPR* values versus SNR, the couples (ϕ, K) causing such results are also reported.

Where, $FP = false\ positives$, $TP = true\ positives$, $FN = false\ negatives$, $TN = true\ negatives$. Here, we consider as positive (negative) examples the WPN's (LN). Since the One class KNN depends on the two values K and ϕ, rather than a two dimensional curve, we obtain a ROC surface. The goal is to find the couple of values (ϕ^*, K^*) that maximizes the accuracy of the classification on a validation set. Giving the function $F(\phi, K) = FPR(\phi, K) + 1 - TPR(\phi, K)$ where $FPR(\phi, K)$ and $TPR(\phi, K)$ represent the FPR and TPR of the one class KNN when using the parameters ϕ and K, it is easy to show that the pair $(\phi^*, K^*) = arg\min_{(\phi, K)}\{F(\phi, K)\}$ maximizes the accuracy since $min_{(\phi, K)}\{F(\phi, K)\} = min_{(\phi, K)}\{FPR(\phi, K)\} + max_{(\phi, K)}\{TPR(\phi, K)\}$.

4.3   Results

All parameters used in the generation of synthetic data have been inspired by biological considerations and are $nn = 200$, $nl = 250$, $\lambda = 200$, $r = 50$, $o = 20$, $nr = 100$, $dp = 0$, $dr = 0$, $pur = 0.8$, $nsv = 0.01$, $SNR = \{1, 2, 4, 6, 8, 10\}$ and $ra = 4$, resulting in 6 synthetic signals at different SNR. The training set Tp is represented by all WPN's that fit better the conditions in Eq. 2 with $os = 4$, because, by biological consideration, we know that a nucleosome is around 150 base pairs which corresponds to 8 probes. Thus, the training set T_p and consequently its size TL, are automatically selected by the MLM depending on the generated input signal, resulting that, for the specific experiments reported here, $TL = \{63, 98, 127, 142, 145, 147\}$ for $SNR = \{1, 2, 4, 6, 8, 10\}$ respectively. The optimal parameters for the MLM are derived as described in [7] and have resulted $H = 20$ and $m = 5$. The performances have been evaluated measuring the correspondence between the classified WPN or LN regions and the ones imposed in the generated signal. In particular, a new mask M' is obtained converting M into probe coordinates such that a probe value is set to 1 (e.g. shows a nulceosome portion) if the corresponding base pairs in M include at least a 1. The real nucleosomal (linker) regions RNR (RLR) are represented by M' as contiguous sequence of 1's or 0's respectively. Here we consider that a nulceosomal (linker) region CNR (CLR) has been classified correctly if there is a match of at least $l = 0.6 \times L$ contiguous 1's (0's) between CNR (CLR) and the corresponding RNR (RLR) in M' where L is the length of RNR (RLR). The value 0.6 has been chosen because it represents a 60% of regions overlap very unlikely to be due to chance. The optimization of the parameters (ϕ, K) has been performed for $SNR = \{1, 2, 4, 6, 8, 10\}$. Fig. 2a shows the trend of the function $F(\phi, K) = 1 + FPR(\phi, K) - TPR(\phi, K)$ as a function (ϕ, K) for $SNR = 4$ while Fig. 2b reports the best *Accuracy* and *FPR* values versus SNR, showing also, for each SNR signal, the estimated (ϕ, K) causing such values. From this study, it results that the average accuracy and FPR over the 6 experiments is 94% and 9% respectively.

5   Final Remarks

In this paper we have shown that the One class KNN classifier, by using the pre-processing of the Multi Layer method, is able to distinguish between nucleosome and linker patterns in the particular problem of the nucleosome positioning. At this stage of the work, we have validated our results on simulated data, using a method to generate synthetic microarray experiments. In particular, the results have shown an average *Accuracy* of 94% on 6 simulated signals generated at different signal to noise ratios. Future work will be devoted on the validation on real microarray data such as the *Saccharomyces cerevisiae* DNA.

References

1. D.M.J. Tax, One-class classification. Ph.D. thesis, Delft University of Technology, `http://www.ph.tn.tudelft.nl/davidt/thesis.pdf`
2. Wang, K., Stolfo, S.J.: One-Class Training for Masquerade Detection. In: Workshop on Data Mining for Computer Security, Melbourne, Florida, November 19-22, 2003, pp. 10–19 (2003)
3. Yuan, G.C., Liu, Y.J., Dion, M.F., Slack, M.D., Wu, L.F., Altschuler, S.J., Rando, O.J.: Genome-Scale Identification of Nucleosome Positions in S. cerevisiae. Science 309, 626–630 (2005)
4. Corona, D.F.V., Tamkun, J.W.: Multiple roles for ISWI in transcription, chromosome organization and DNA replication. Biochem. Biophys. Acta. 1677(1-3), 113–119 (2004)
5. Saha, A., Wittmeyer, J., Cairns, B.R.: Chromatin remodelling: The industrial revolution of DNA around histones. Nature Reviews Molecular Cell Biology 7, 437–447 (2006)
6. Jacobson, S., Pillus, L.: Modifying chromatin and concepts of cancer. Current Opinion in Genetics and Development 9, 175–184 (1999)
7. Corona, D.F.V., Di Gesù, V., Lo Bosco, G., Pinello, L., Yuan, G.: A New Multi-Layers Method to Analyze Gene Expression. In: Apolloni, B., Howlett, R.J., Jain, L. (eds.) KES 2007, Part III. LNCS (LNAI), vol. 4694, pp. 862–869. Springer, Heidelberg (2007)
8. Jain, A.K., Dubes, R.C.: Algorithms for clustering data. Prentice Hall, NY (1988)

A Fuzzy Extension of Some Classical Concordance Measures and an Efficient Algorithm for Their Computation

Michele Ceccarelli and Antonio Maratea

Research Center on Software Technology
University of Sannio, Via Traiano 11, Benevento, Italy
{ceccarelli,amaratea}@unisannio.it

Abstract. Many indexes have been proposed in literature for the comparison of two crisp data partitions, as resulting from two different classifications attempts, two different clustering solutions or the comparison of a predicted vs. a true labeling. Most of these indexes implementations have a computational cost of $O(N^2)$ (where N is the number of data points) and this fact may limit their usage in very big datasets or their integration in computational-intensive validation strategies. Furthermore, their extension to fuzzy partitions is not obvious. In this paper we analyze efficient algorithms to compute many classical indexes (most notably the Jaccard coefficient and the Rand index) in $O(d^2 + N)$ (where d is the number of different classes/clusters) and propose a straightforward procedure to extend their computation to fuzzy partitions. The fuzzy extension is based on a pseudo-count concept and provides a natural framework for including memberships in computation of binary similarity indexes, not limited to the ones here revised. Results on simulated data using the Jaccard coefficient highlight a higher consistence of its proposed fuzzy extension with respect to its crisp counterpart.

Keywords: Cluster Stability, Concordance Measure, Validity Index, Efficient Algorithm.

1 Introduction

Data partitions are generated in many different frameworks and can result from many different problems. How to evaluate their fitting to data or model structure, their agreement with a priori knowledge and by the end their meaningfulness is indeed a non trivial problem that has generated many different approaches. One approach is in the use of indexes that measure "similarity" among two partitions (see for example [11]). These indexes actually can be used to measure accuracy in a broad sense (when comparing a true solution vs. a predicted one) or they can be interpreted as generic concordance measures between two different partition. Proper evaluation of accuracy is essential to estimate the predictive power of a model, while having good measures of concordance is essential to evaluate i.e. the goodness of fit of a model, the

I. Lovrek, R.J. Howlett, and L.C. Jain (Eds.): KES 2008, Part III, LNAI 5179, pp. 755–763, 2008.
© Springer-Verlag Berlin Heidelberg 2008

consistence of a solution, the effect of a parameter or the effectiveness of a set of features [3]. Partitions can be "putative", that means the effect of a prediction algorithm, or "natural", that means a priori known. In classification, putative partitions result from the attempt to recognize, among given predefined classes, the right class of belonging of each point [2]. In clustering, putative partitions result from the attempt to group similar points in meaningful ways [10]. For all classes of algorithms that produce putative partitions there are hence at least two different uses of similarity indexes: as a measure of quality or as a measure of performance. The former falls within the so called *internal* validation problem and aims to quantify "predictions agreement" horizontally comparing partitions, while the latter falls within the so called the *external* validation problem and aims to quantify "prediction goodness", with one partition hierarchically dominant with respect to the others. However, we note that the latter is just a special case of the former, obtained when one of the two partitions is a natural partition, and hence concordance measures include as as special case the accuracy measures. From here on, we will generically talk about concordance measures omitting the specification of their usage.

In this paper we will also consider the extension of concordance measures to fuzzy partitions [4]. Indeed, some approaches for comparing fuzzy clustering solutions in the framework of clustering validation have already been proposed in literature (see for example [1]), but we do not want just to propose another validation index, rather we want to investigate, in a broad sense, the ways to compare fuzzy partitions, both as quality than as performance measures. This is actually an open question as the classical concordance measures are typically crisp and binary [7] and, when applied to fuzzy partitions, they require the defuzzifications of the memberships. Recently there have been some works to extend a concordance index in order to manage fuzzy partitions (i.e. [5,14]), however these works are based on the set–theoretic definition of the indexes and their extension is crudely based the notion of fuzzy set. Here, we take into account the fact that the concordance indexes for clustering should be based on some parameters which depend on particular labeling of the observed data without the need of a binary representation of the partitions. This produces two main advantages: we show that *(i)* the concordance parameters can be computed directly from a *concordance matrix* which is built from the labels given to each data point and *(ii)* the fuzzy extension of the concordance parameters directly follows from the way of computation of the concordance matrix, where we use the fuzzy membership of the data points to the various clusters instead of the crisp labels.

The paper is organized as follows: the next section defines the concordance measures that we consider in this paper; the following section reports a systematic approach for computing the concordance parameters from the concordance matrix; the fuzzy case is reported in the fourth section. The last section reports some preliminary results showing the stability of the fuzzy indices for clusters having a significant overlap.

2   Concordance Measures

Given N points in a finite dimensional space and two label vectors ℓ_1 and ℓ_2 corresponding to two different data partitions (not necessarily in the same number of parts), the issue of measuring their "concordance" is not a trivial issue and there are many possibilities and indexes to solve this task in literature [6].

$$\ell^1 = \ell_1^1, \ell_2^1, \ldots, \ell_p^1, \ldots, \ell_N^1 \; ; \; \ell_p^1 \in \{1, 2, \ldots, d_1\} \tag{1}$$

$$\ell^2 = \ell_1^2, \ell_2^2, \ldots, \ell_q^2, \ldots, \ell_N^2 \; ; \; \ell_q^2 \in \{1, 2, \ldots, d_2\} \tag{2}$$

Most classical and widely used indexes are computed on the base of the following four concordance parameters:

$$N_{11} = \sum_p^N \sum_q^N \mathbf{I}(\ell_p^1 = \ell_q^1 \wedge \ell_p^2 = \ell_q^2) \tag{3}$$

$$N_{10} = \sum_p^N \sum_q^N \mathbf{I}(\ell_p^1 = \ell_q^1 \wedge \ell_p^2 \neq \ell_q^2) \tag{4}$$

$$N_{01} = \sum_p^N \sum_q^N \mathbf{I}(\ell_p^1 \neq \ell_q^1 \wedge \ell_p^2 = \ell_q^2) \tag{5}$$

$$N_{00} = \sum_p^N \sum_q^N \mathbf{I}(\ell_p^1 \neq \ell_q^1 \wedge \ell_p^2 \neq \ell_q^2) \tag{6}$$

Where N_{11} is the number of points that are in the same group w.r.t. ℓ^1 and also w.r.t. ℓ^2; N_{10} is the number of points that are in the same group wrt ℓ^1 but in different groups wrt ℓ^2; N_{01} is the number of points that are in different groups with respect to ℓ^1 but are in the same group wrt ℓ^2; N_{00} is the number of points that are in different groups wrt ℓ^1 and also wrt ℓ^2 and $\mathbf{I}$ is the indicator function.

One of the most known and widely used measures of concordance is the Jaccard's coefficient $\mathcal{J}$ [9] (or equivalently the Jaccard's distance $1 - \mathcal{J}$). It is an index originally proposed for quantifying the overlap between two binary vectors and then adapted to measure the concordance of crisp partitions. One desirable property that it has is its independence from the ordering of vectors. Jaccard similarity coefficient is defined as follows:

$$\mathcal{J} = N_{11}/(N_{11} + N_{01} + N_{10}) \tag{7}$$

Other indexes based on the same four quantities are for example the Rand index [13], whose definition follows:

$$\mathcal{R} = (N_{11} + N_{00})/(N_{11} + N_{01} + N_{10} + N_{00}) \tag{8}$$

The Adjusted Rand index [8] is:

$$\mathcal{R}^{adj} = \frac{N_{11} - \frac{(N_{11}+N_{01})(N_{11}+N_{10})}{N_{00}}}{\frac{(N_{11}+N_{01})+(N_{11}+N_{10})}{2} - \frac{(N_{11}+N_{01})(N_{11}+N_{10})}{N_{00}}} \tag{9}$$

The Fowlkes-Mallows index [6] is:

$$\mathcal{FM} = N_{11}/\sqrt{(N_{11} + N_{01})(N_{11} + N_{10})} \tag{10}$$

The Minkowski measure [12] follows:

$$\mathcal{M} = \sqrt{(N_{01} + N_{10})/(N_{01} + N_{11})} \tag{11}$$

A review of the different properties and drawbacks can be found, i.e. in [5].

3   Computational Aspects

Here we want to extend the results reported in [6] by showing that all concordance parameter can be computed from the matrix $\mathbf{C}_{d_1 \times d_2} = [c_{ij}]_{d_1 \times d_2}$ built with the absolute frequencies of occurrence of labels pairs (that may be seen as a contingency table).

$$[c_{ij}] = \sum_p^N \mathbf{I}(\ell_p^1 = i \wedge \ell_p^2 = j) \tag{12}$$

Table 1. Contigency table for the partitions ℓ^1 and ℓ^2

$\ell^1\backslash\ell^2$	1	2	...	j	...	d_2
1	c_{11}	c_{12}	...	c_{1j}	...	c_{1d_2}
2	c_{21}	c_{22}	...	c_{2j}	...	c_{2d_2}
...	...	...	...	...	...	...
i	c_{i1}	c_{i2}	...	c_{ij}	...	c_{id_2}
...	...	...	...	...	...	...
d_1	c_{d_11}	c_{d_12}	...	c_{d_1j}	...	$c_{d_1d_2}$

Proposition 1. Quantities (3,4,5,6) can be equivalently computed from matrix $\mathbf{C}_{d_1 \times d_2} = [c_{ij}]_{d_1 \times d_2}$ as follows:

$$N_{11} = ||\mathbf{C}||_\mathcal{F}^2 - N \tag{13}$$

$$N_{10} = \sum_j^{d_2} (s_{.j})^2 - ||\mathbf{C}||_\mathcal{F}^2 \tag{14}$$

$$N_{01} = \sum_i^{d_1} (s_{i.})^2 - ||\mathbf{C}||_\mathcal{F}^2 \tag{15}$$

$$N_{00} = \sum_i^{d_1} \sum_j^{d_2} c_{ij} \dot{s}_{ij} \tag{16}$$

where $||\mathbf{C}||_\mathcal{F}^2$ is the square of the Frobenius norm of matrix $\mathbf{C}$, $||\mathbf{C}||_\mathcal{F}^2 = \sum_i^{d_1} \sum_j^{d_2} c_{ij}^2$, $s_{.j} = \sum_i^{d_1} c_{ij}$ is the sum of its j-th column, $s_{i.} = \sum_j^{d_2} c_{ij}$ is the sum of its i-th row and $\dot{s}_{ij} = \sum_{l\neq i}^{d_1} \sum_{m\neq j}^{d_2} c_{lm}$ is the sum of all elements of matrix $\mathbf{C}$ excluding row i and column j.

Proof. Let us consider N_{11}. If $c_{ij} = 0$ then there is no value of r for which $\ell_r^1 = i$ and $\ell_r^2 = j$, none of the points presents that specific combination of labels and there is no contribution to the N_{11}. If $c_{ij} = 1$ then there is just one value of r for which $\ell_r^1 = i$ and $\ell_r^2 = j$ and again there's no contribution to the N_{11} (remember that at least one *pair* is needed to verify condition defining N_{11}). If $c_{ij} > 1$ then the specific combination of labels $\ell_r^1 = i$ and $\ell_r^2 = j$ happens two times or more and the contribution to the N_{11} is given by the number of pairs that's possible to extract from c_{ij}, that is:

$$\binom{c_{ij}}{2} = \frac{c_{ij}(c_{ij} - 1)}{2} = \frac{c_{ij}^2 - c_{ij}}{2} \tag{17}$$

summing over all i and j, remembering that $\sum_i^{d_1} \sum_j^{d_2} c_{ij} = N$, we get:

$$N_{11} = 2\sum_i^{d_1}\sum_j^{d_2} \binom{c_{ij}}{2} = \sum_i^{d_1}\sum_j^{d_2} c_{ij}^2 - N = ||\mathbf{C}||_{\mathcal{F}}^2 - N \tag{18}$$

For what concerns N_{01} (the same holds for N_{10}), we must consider all pairs that have the same value of ℓ^2. This value can be obtained for each value of j simply summing the j-th column of $\mathbf{C}$. We call $s_{.j}$ the sum of the j-th column

$$s_{.j} = \sum_i^{d_1} c_{ij} \tag{19}$$

and the number of all pairs that's possible to extract is hence given by

$$\binom{s_{.j}}{2} = \frac{s_{.j}(s_{.j} - 1)}{2} = \frac{(s_{.j})^2 - s_{.j}}{2} \tag{20}$$

What happens in this way is that we count also the pairs with both labels equal, that is the N_{11}, so we must subtract this number from the above. Summing over all i and j, remembering that $\sum_j^{d_2} s_{.j} = N$, we get:

$$N_{01} = 2\sum_j^{d_2} \binom{s_{.j}}{2} - ||\mathbf{C}||_{\mathcal{F}}^2 + N = \sum_j^{d_2}(s_{.j})^2 - \sum_j^{d_2} s_{.j} - ||\mathbf{C}||_{\mathcal{F}}^2 + N = \sum_j^{d_2}(s_{.j})^2 - ||\mathbf{C}||_{\mathcal{F}}^2 \tag{21}$$

Finally, as far as for N_{00}, we must compute the times $\ell_r^1 \neq i$ and $\ell_r^2 \neq j$. For each pair (i, j), this number is the sum of all elements of matrix $\mathbf{C}$ excluding row i and column j.

$$\dot{s}_{ij} = \sum_{l\neq i}^{d_1}\sum_{m\neq j}^{d_2} c_{lm} \tag{22}$$

If $c_{ij} = 0$, then there is no contribution to the N_{00}. If $c_{ij} = 1$, then the contribution to the N_{00} is exactly given by the summation (22). If $c_{ij} > 1$, then we must count s_{ij} so many times as is c_{ij}. This happens because we cannot count pairs that both have $\ell_r^1 = i$ and $\ell_r^2 = j$, but we must consider just the pairs that s_{ij} form with each c_{ij} independently. The total contribution for each pair (i,j) is hence:

$$c_{ij}\dot{s}_{ij} = c_{ij}\sum_{l\neq i}^{d_1}\sum_{m\neq j}^{d_2} c_{lm} \tag{23}$$

and summing over all i and j, we get:

$$N_{00} = \sum_{i}^{d_1}\sum_{j}^{d_2} c_{ij}\dot{s}_{ij} \tag{24}$$

At this point, it is trivial to show the following proposition.

Proposition 2. quantities (13,14,15,16) can be computed in time $O(d_1 \cdot d_2 + N)$.

4   Fuzzy Extension

Once computation of concordance indexes has been reformulated in terms of matrix $\mathbf{C}$, it is legitimate to ask if this reformulation can be useful also for comparison of fuzzy partitions. The extension of crisp concordance measures to fuzzy partitions is non trivial because in the latter case each point has a membership value with respect to each cluster/class and as a consequence its contribution must be accounted with respect to all possible cluster/class assignments. Thus, in filling matrix $\mathbf{C}$, each point must contribute to all of its values. The approach we adopted is to consider memberships as *pseudo counts*, such that if a point p has membership μ_{pi}^a to cluster/class i in first partition (called a) and membership μ_{pj}^b to cluster/class j in the second partition (called b), its contribution to term c_{ij} is the sum of the two membership values.

$$[c_{ij}]_\mu = \sum_{p}^{N}(\mu_{pi}^a + \mu_{pj}^b) \tag{25}$$

The rationale can be clarified a follows: if i.e. two points p_1 and p_2 have both a membership value of 0.5, respectively to cluster/class i in first partition and to cluster/class j in second partition, they should be counted as a single point that has a complete assignment (membership 1 in just one of the two cluster/classes). If their memberships sum to 1.5 then they should be counted for "one complete assignment and half" and so on. This is a generalization of the crisp case that allows to account for all membership values and take full advantage of the expressive power of membership functions. Higher memberships are the most relevant and hopefully meaningful for the data partition, while lower membership are less relevant and more easily determined by noise and data artifacts. To increase robustness of the concordance measures' computation we boosted the influence of higher memberships (whose sum is greater than 1) and reduced the influence of lower memberships (whose sum is smaller than 1) through a power α applied to the second member of equation (25).

$$[c_{ij}]_\mu = \sum_p^N (\mu_{pi}^a + \mu_{pj}^b)^\alpha \tag{26}$$

The higher the power α, the most robust is the index computation. Given two different fuzzy partitions of the same N points and once matrix $\mathbf{C}$ has been filled according to formula (26), it is straightforward to compute indexes (7,8,9,10,11) in the fuzzy case. The only difference is in formula (13), for which does not hold anymore the simplification $\sum_i^{d_1} \sum_j^{d_2} c_{ij} = N$ and the sum must remain explicit.

$$N_{11} = ||\mathbf{C}||_{\mathcal{F}}^2 - \sum_i^{d_1} \sum_j^{d_2} c_{ij} \tag{27}$$

Fuzzy indexes computed in this way can be used to quantify similarity of fuzzy partitions both as "predictions agreement" than as "prediction goodness" as they allow even to compare a fuzzy partition with a crisp partition, considering the latter as a degenerate case of the former (in which membership are 1 for the only cluster of belonging of each point and 0 otherwise). Computational complexity for fuzzy partitions is $O(Nd_1d_2)$.

5   Application and Results

We used 10 datasets of simulated data and the classical fuzzy c-means algorithm to test our method. In each dataset 450 data points were generated from three

Table 2. Datasets of simulated data used in the experiments

Dataset	Instances	Features	Classes	μ_1, μ_2, μ_3	$\Sigma_{123} = \Sigma_1 = \Sigma_2 = \Sigma_3$
X_1	450	3	3	(1,0,0) (0,0,0) (-1,0,0)	$\Sigma_{123} = (0.1)I_3$
X_2	450	3	3	(1,0,0) (0,0,0) (-1,0,0)	$\Sigma_{123} = (0.2)I_3$
X_3	450	3	3	(1,0,0) (0,0,0) (-1,0,0)	$\Sigma_{123} = (0.3)I_3$
X_4	450	3	3	(1,0,0) (0,0,0) (-1,0,0)	$\Sigma_{123} = (0.4)I_3$
X_5	450	3	3	(1,0,0) (0,0,0) (-1,0,0)	$\Sigma_{123} = (0.5)I_3$
X_6	450	3	3	(1,0,0) (0,0,0) (-1,0,0)	$\Sigma_{123} = (0.6)I_3$
X_7	450	3	3	(1,0,0) (0,0,0) (-1,0,0)	$\Sigma_{123} = (0.7)I_3$
X_8	450	3	3	(1,0,0) (0,0,0) (-1,0,0)	$\Sigma_{123} = (0.8)I_3$
X_9	450	3	3	(1,0,0) (0,0,0) (-1,0,0)	$\Sigma_{123} = (0.9)I_3$
X_{10}	450	3	3	(1,0,0) (0,0,0) (-1,0,0)	$\Sigma_{123} = I_3$

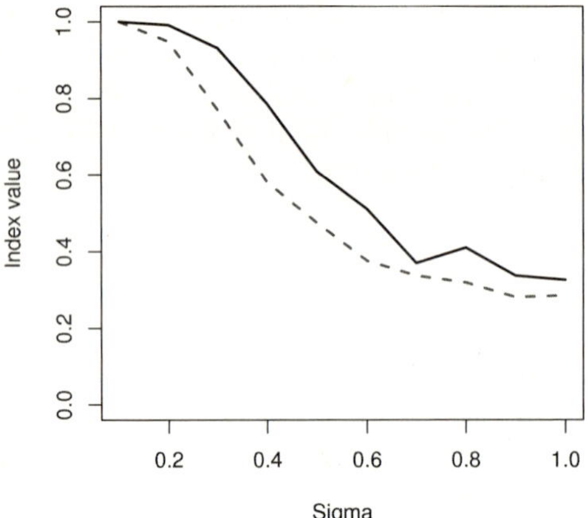

Fig. 1. Comparison of the classical $\mathcal{J}$ index computed on defuzzified fuzzy c-means solutions (dashed line) and fuzzy Jaccard coefficient $\mathcal{J}_\mu$ computed with membership functions produced by the fuzzy c-means algorithm (solid line). On x axis values of σ characterizing the ten datasets $X_1, \ldots, X_{10}$.

different gaussians (150 points from each one), with different mean and equal variance (see Table 2). True labeling ℓ^T is assumed to be given by the data generation procedure. What changes from one dataset to the other is essentially the amount of data overlap. Data go from good separation (dataset X_1) to massive overlap (dataset X_{10}).

For each dataset X we performed a fuzzy c-means clustering with constant parameters (fuzzifier $f = 1.5$ and the right number of clusters) and obtained both the predicted membership matrix $\mathbf{M}_X$ than the predicted crisp labeling consequent to defuzzification ℓ^X. We first computed the standard Jaccard coefficient $\mathcal{J}$ between the defuzzified solution ℓ^X and the true labeling vector ℓ^T and then we computed the fuzzy Jaccard coefficient $\mathcal{J}_\mu$ between the predicted membership matrix $\mathbf{M}_X$ and the degenerate membership matrix of the true solution $\mathbf{M}_T$ (in which membership are 1 for the only cluster of belonging of each point and 0 otherwise). In figure (1) it is shown the comparison of the two values for increasing amount of overlap. We can see that in all cases the fuzzy Jaccard coefficient $\mathcal{J}_\mu$ dominates standard coefficient $\mathcal{J}$ and is hence more consistent with true labeling. When overlap increases, both indexes lower their values, but the fact that $\mathcal{J}_\mu$ accounts for all memberships instead of just the highest allows a more gradual decrease of its values and a better relative performance with respect to $\mathcal{J}$ regardless of the amount of overlap (on tested datasets).

6   Conclusions

In this paper we analyzed the methods to compute many classical indexes (most notably the Jaccard coefficient and the Rand index), and showed that all the classical concordance parameters can be obtained from the contigency matrix. This in the crisp case produces an $O(d^2 + N)$ algorithm for the classical Rand,

Jaccard, Fowlkes and other indexes. The contigency matrix formulation led also to an extension of such indexes for fuzzy clustering. The fuzzy extension is based on a pseudo-count concept and provides a natural framework for including memberships in computation of binary similarity indexes. The reported results over simulated datasets having an increasing amount of overlap show that the proposed method manifest a major stability with respect to the overlap of the clusters.

References

1. Bezdek, J.C., Pal, N.R.: Some new indexes of cluster validity. IEEE Trans. Systems Man Cybernet-B 28(3), 301–315 (1998)
2. Bishop, C.M.: Neural Networks for Pattern Recognition. Clarendon Press, Oxford (1996)
3. Ceccarelli, M., Maratea, A.: Assessing Clustering Reliability and Features Informativeness by Random Permutations. In: Apolloni, B., Howlett, R.J., Jain, L. (eds.) KES 2007, Part III. LNCS (LNAI), vol. 4694, pp. 878–885. Springer, Heidelberg (2007)
4. Ceccarelli, M., Maratea, A.: Improving fuzzy clustering of biological data by metric learning with side information. International Journal of Approximate Reasoning 47, 45–57 (2008)
5. Campello, R.J.: A fuzzy extension of the Rand index and other related indexes for clustering and classification assessment. Pattern Recognition Letters 28, 833–841 (2007)
6. Fowlkes, E.B., Mallows, C.L.: A method for comparing two hierarchical clustering. Journal of the American Statistical Association 78(383), 553–569 (1983)
7. Handl, J., Knowles, J., Kell, D.B.: Computational cluster validation in post-genomic data analysis. Bioinformatics 21(15), 3201–3212 (2005)
8. Hubert, A.: Comparing partitions. Journal of Classifification 2, 193–198 (1985)
9. Jaccard, S.: Nouvelles recherches sur la distribution florale. Bull. Soc. Vaud. Sci. Nat. 44, 223–270 (1908)
10. Jain, A.K., Murtym, M.N., Flynn, P.J.: Data clustering: a review. ACM Computing Surveys 31, 264–323 (1999)
11. Jardine, N., Sibson, R.: Mathematical Taxonomy. John Wiley and Sons, Chichester (1971)
12. Jiang, D., Tang, C., Zhang, A.: Cluster analysis for gene-expression: a survey. IE Transactions on Knowledge and Data Engineering 16, 1370–1386 (2004)
13. Rand, W.M.: Objective criteria for the evaluation of clustering methods. J. Amer. Statist. Assoc., 846–850 (1971)
14. Setnes, M., Cross, V.: Compatibility-based ranking of fuzzy numbers. In: Annual Meeting of the North American Fuzzy Information Processing Society NAFIPS, pp. 305–310 (1997), doi:10.1109/NAFIPS.1997.624057

An Algorithm to Assess the Reliability of Hierarchical Clusters in Gene Expression Data

Roberto Avogadri[1], Matteo Brioschi[2], Francesca Ruffino[1], Fulvia Ferrazzi[3], Alessandro Beghini[2], and Giorgio Valentini[1]

[1] DSI - Dip. Scienze dell' Informazione, Università degli Studi di Milano, Italy
{avogadri,ruffino,valentini}@dsi.unimi.it
[2] DBioGen - Dip. Biologia e Genetica per le Scienze Mediche, Università degli Studi di Milano, Italy
{matteo.brioschi,alessandro.beghini}@unimi.it
[3] Dip. Informatica e Sistemistica, Università degli Studi di Pavia, Italy
fulvia.ferrazzi@unipv.it

Abstract. The validation of clusters discovered in bio-molecular data is a central issue in bioinformatics. Recently, stability-based methods have been successfully applied to the analysis of the reliability of clusterings characterized by a relatively low number of examples and clusters. Nevertheless, several problems in functional genomics are characterized by a very large number of examples and clusters. We present a stability-based algorithm to discover significant clusters in hierarchical clusterings with a large number of examples and clusters. Preliminary results on gene expression data of patients affected by Human Myeloid Leukemia, show how to apply the proposed method when thousands of gene clusters are involved.

1 Introduction

The unsupervised discovery and validation of clusters underlying data is a central issue in several branches of bioinformatics [1], as well as the proper visualization of clustering results [2]. Different clustering validation techniques (see [3] for a recent review), and software tools implementing classical validity indices (such as the *Dunn's index* and the *Silhouette index*) have been proposed [4].

Several recent methods to estimate the validity of the discovered clusterings are based on the concept of stability: multiple clusterings are obtained by introducing perturbations into the original data, and a clustering is considered reliable if it is approximately maintained across multiple perturbations [5, 6, 7, 8]. Despite their successful application in several bioinformatics domains, they are well-suited to unsupervised problems characterized by a relatively low number of clusters and/or examples [9, 10]. Indeed if we try to apply them to the analysis of a very high number of clusters, computational problems may arise. For instance, to assess the reliability of clusters of N genes using DNA microarray data, we usually deal with thousands of examples (genes) and with an exponential (2^N) number of potential clusters.

I. Lovrek, R.J. Howlett, and L.C. Jain (Eds.): KES 2008, Part III, LNAI 5179, pp. 764–770, 2008.

Considering that clusters of genes may show a hierarchical multi-level organization [11], we could reduce the computational complexity by examining a linear number of clusters, computed by a hierarchical clustering algorithm.

The main idea of this work consists in the assessment of the reliability of the clusters discovered by a hierarchical clustering algorithm, using a stability based measure mutuated from our previous work [10]. Differently from our previous approach, we do not need to know in advance the correct or the approximate number of clusters, but we can directly apply a stability measure that estimates the reliability of each individual cluster of the dendrogram computed by a hierarchical algorithm, thus reducing the complexity to a linear number of clusters with respect to the number of available examples.

In the next section we describe the proposed algorithm. In Sect. 3 we introduce an application of the algorithm to the discovery of significant gene clusters in patients affected by Human Myeloid Leukemia, by using DNA microarray gene expression data prepared and analyzed by our research group using the Affymetrix hgu133plus2 GeneChip. Then we discuss the advantages and the limitations of the proposed method. In the conclusions we propose some research lines for future work.

2   The Algorithm

Our algorithm is founded on a stability based approach to discover the significant clusters identified by a hierarchical clustering algorithm.

The main logical steps of the algorithm are the following:

1. **Hierarchical clustering of the original data.** A hierarchical clustering algorithm is applied to the original data to discover the clusters whose reliability will be evaluated through the steps listed below.
2. **Multiple perturbation of the original data.** The original data are perturbed by randomized projections [12], by subsampling or bootstrapping procedures [13], or by controlled noise injection.
3. **Multiple hierarchical clustering of the perturbed data.** Multiple clusterings are obtained by applying the same hierarchical clustering algorithm as in the step (1) to the perturbed data.
4. **Construction of the similarity matrix.** A similarity matrix that stores the frequency by which each pair of examples falls into the same cluster in the "perturbed" clustering is built [14].
5. **Computation of the stability indices.** For each cluster obtained through the hierarchical clustering of the original data (step 1), a stability index [10] is computed using the similarity matrix constructed at step 4. The stability index S (see line (11) of the pseudo-code of the algorithm) has values between 0 (low stability) and 1 (high stability).
6. **Selection of the most reliable clusters.** Using the stability indices computed in the previous step, the most reliable clusters are selected. Several approaches can be used; the easiest one consists in the selection of the clusters whose stability is above a given threshold.

The pseudo-code of the algorithm is reported below:

Cluster stability algorithm:

```
Input:
- A data set D = {x_i ∈ ℝ^r, 1 ≤ i ≤ N}.
- A hierarchical clustering algorithm C.
- A number n on perturbations of the data.
- A procedure that realizes a randomized map μ : ℝ^r → ℝ^m, m < r.
Begin algorithm
```

    (1) $\{A_1, \ldots, A_{2N-1}\} := \mathcal{C}(D);$

    (2) `C := {A_i|A_i is not a leaf or the root};`

    (3) $M := 0$;

    (4) $d := 0;$

```
Repeat for j = 1 to n
```

       (5) $D^j := \mu(D);$

       (6) $\{B_1^j, \ldots, B_{2N-1}^j\} := \mathcal{C}(D^j);$

       (7) $C^j := \{B_i^j | B_i^j$ `is not a leaf or the root` $\};$

       (8) $d := d + $ `depth` $(\mathcal{C}(D^j)) - 1;$

```
      For each B_k^j ∈ C^j
          For each (x_t, x_v) ∈ (B_k^j × B_k^j)
```

          (9) $M(t, v) := M(t, v) + 1;$

```
          end For
      end For
   end Repeat
```

    (10) $M := \frac{M}{d};$

```
   For each A_k ∈ C
```

       (11) $S(A_k) := \frac{1}{|A_k|(|A_k|-1)} \sum_{(x_t, x_v) \in A_k \times A_k} M(t, v);$

```
   end For
end algorithm.
Output:
```
- $S = \{s(A_i) | A_i \in C\}.$

In this algorithm a randomized map is applied to perturb the data. Note that with abuse of notation we represent clusters and nodes with the same symbols, as well as dendrograms and corresponding clusterings. At line (2), from the original hierarchical clustering composed by $2N - 1$ clusters (line (1)), only the internal $N - 2$ nodes are selected. Indeed it is easy to see that all the singleton clusters (the leaves of the dendrogram) and the "root" cluster are always present in any hierarchical clustering and as a consequence their stability is always 1 (maximum stability).

The "core" of the algorithm is represented by the `Repeat` loop. At each iteration we obtain an instance of the perturbed (projected) data (step 5); then a hierarchical clustering algorithm is applied to the perturbed data, considering only the internal nodes (steps $6 - 7$). After updating the cumulative depth of the n dendrograms (8), the following two nested iterative loops update the similarity matrix M, by adding 1 to the entry $M(t, v)$ if the examples x_t and

$\boldsymbol{x}_v$ are both present in the cluster B_k^j. To maintain the value of each entry of the matrix M between 0 and 1 we need to normalize it by d (step 10). Indeed each pair of examples may belong to a number of clusters equal at most to the depth minus one of the corresponding tree (step 8). The output of the algorithm consists in the set of stability indices computed for each node of the hierarchical clustering C.

3   Results and Discussion

As an example of application of the proposed algorithm, we analyzed gene expression data of eight samples, including seven patients affected by Human Myeloid Leukemia at diagnosis and one healthy donor as control. Samples were analyzed using the Affymetrix hgu133plus2 GeneChip. Each gene on this chip is represented by 11 oligonucleotides, termed a "probeset". The hgu133plus2 contains 54675 probe sets and it analyzes the expression level of 47400 transcripts and variants including 38500 UniGene clusters at the time of array design.

During the laboratory procedures biotin-labeled RNA fragments are hybridized to the probe array. The hybridized probe array is stained with streptavidin phycoerythrin conjugated and scanned by the GeneChip Scanner 3000 $Affymetrix$. From the image files .cel files containing a single intensity value for each probe cell delineated by the grid are obtained. We used Bioconductor [15] packages to assess the quality level of the data, using standard Affymetrix tests, as well as other quality check tests such as the Relative Log Expression (RLE) plot and Normalization Unscaled Standard Error (NUSE) [16]. Fig. 1 shows the MA plots of the expression levels of the seven samples using the healthy donor as reference. All quality checks assured the high quality of the gene expression data. Background correction, normalization and summarization have been performed using the Robust Multi-array Average (RMA) procedure that summarizes the probe level data to obtain gene expression levels [16] and the "expresso" method from the $Affy$ Bioconductor package [17].

To reduce the high number of examples (54613 probe sets with the exclusion of the Affymetrix chip control probes), we used a z-test to select the genes whose gene expression levels significantly differ from the healthy donor control patient. At a 0.1 significance level we selected 1007 genes. Using the algorithm described in Sect. 2 and the classical average-linkage algorithm to perform the hierarchical clusterings, we iterated 50 random projections from the original 7-dimensional space to a lower 5-dimensional space, using $Bernoulli$ random projections [11].

The results are showed in Table 1. Different thresholds $0 < \alpha < 1$ have been considered, in order to select the set R_α of reliable clusters, among those belonging to the clustering C in the original space:

$$R_\alpha = \{A_i \in C | S(A_i) > \alpha\}$$

The last column represents the ratio values with respect to the total number of clusters (1005), obtained excluding the singleton and the "root" clusters. From

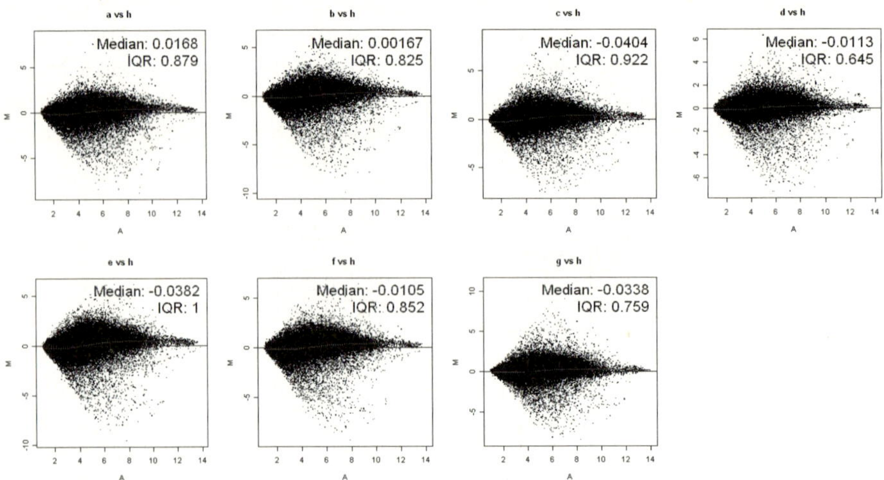

Fig. 1. MA plots of the seven patients affected by Human Myeloid Leukemia using the healthy donor as reference

these results we may observe that 43 clusters show a stability larger than 0.8, and only 5 clusters show a very high reliability (stability larger than 0.9).

The proposed approach shows several limitations that need to be carefully considered for future work.

For instance, the algorithm has a bias versus very low sized and very large sized clusters. Indeed it is easy to see that singleton clusters and the clusters that contains all the examples are always present in every hierarchical clustering algorithm, thus resulting in a stability equal to 1. All the other clusters lie in between: hence it is necessary to include a proper correction with respect to the cluster size.

Another relevant problem is the choice of the threshold α to select the significant clusters. In the proposed algorithm the choice is somehow arbitrary: we considered very reliable the 5 clusters selected with a threshold equal to 0.9, but there is no reason to consider this threshold as a warranty of reliability. Moreover this problem is related to the previous one, because the threshold should be related to the cardinality of the clusters.

The choice of classical hierarchical algorithms to discover the clusters of genes may represents another limitation. Even if clusters of genes may show a hierarchical structure [18], a gene may belong to multiple nodes in different non-nested subtrees of the hierarchical structure, and classical hierarchical clustering algorithms cannot capture these characteristics of the data. To this end a possible more consistent approach could be a fuzzy or probabilistic hierarchical clustering approach, in order to address the problem of "not-hierarchically-related" clusters.

From a bioinformatics standpoint we need also to biologically validate the clusters discovered as reliable by the proposed method. To this end we need a careful biological and bio-medical analysis of the clusters of genes individuated

Table 1. Number of clusters of the original hierarchical classification with a stability larger than α. The last row represents the ratio of the selected clusters with respect to the total number of clusters.

α	Number of clusters	Ratio
0.1	1004	0.999
0.2	919	0.914
0.3	680	0.677
0.4	392	0.390
0.5	227	0.226
0.6	138	0.137
0.7	76	0.076
0.8	43	0.043
0.9	5	0.005

as significant. To support this bio-medical task functional enrichment methods are often used to find if one or more of gene modules (e.g. Gene Ontology classes or KEGG pathways) are significantly over-represented among the relevant genes selected in the experiment [19, 20]. Over-representation of a given gene module means that genes with a particular property have been activated or deactivated in the experiment.

4   Conclusions

We presented an algorithm to discover reliable clusters in hierarchical clusterings characterized by a large number of examples and clusters. The method proposes a stability-based approach that uses multiple randomized projections of the original data and a stability measure constructed through a similarity matrix that summarizes multiple clusterings on the perturbed data. A preliminary application to patients affected by Human Myeloid Leukemia discovered a relatively small number of gene clusters that need to be biologically validated. The aim of this preliminary work consists in showing the applicability of a stability-based method to discover significant clusters when their number is relatively high and classical stability-based methods are not applicable for computational complexity reasons. Nevertheless in future works we need to address the problem of the bias of the stability measure and we need also a principled method to choose the threshold to select the set of significant clusters. We are working on a non-parametric statistical test to solve both these open problems.

References

[1] Datta, S.: Comparison and validation of statistical clustering techniques for microarray gene expression data. Bioinformatics 19, 459–466 (2003)
[2] Napolitano, F., Raiconi, G., Tagliaferri, R., Ciaramella, A., Staiano, A., Miele, G.: Clustering and visualization approaches for human cell cycle gene expression data analysis. Int. J. Approx. Reasoning 47, 70–84 (2008)

[3] Handl, J., Knowles, J., Kell, D.: Computational cluster validation in post-genomic data analysis. Bioinformatics 21, 3201–3215 (2005)

[4] Bolshakova, N., Azuaje, F., Cunningham, P.: An integrated tool for microarray data clustering and cluster validity assessment. Bioinformatics 21, 451–455 (2005)

[5] Kerr, M., Curchill, G.: Bootstrapping cluster analysis: assessing the reliability of conclusions from microarray experiments. PNAS 98, 8961–8965 (2001)

[6] Monti, S., Tamayo, P., Mesirov, J., Golub, T.: Consensus Clustering: A Resampling-based Method for Class Discovery and Visualization of Gene Expression Microarray Data. Machine Learning 52, 91–118 (2003)

[7] Ben-Hur, A., Ellisseeff, A., Guyon, I.: A stability based method for discovering structure in clustered data. In: Altman, R., Dunker, A., Hunter, L., Klein, T., Lauderdale, K. (eds.) Pacific Symposium on Biocomputing, Lihue, Hawaii, USA, vol. 7, pp. 6–17. World Scientific, Singapore (2002)

[8] McShane, L., Radmacher, D., Freidlin, B., Yu, R., Li, M., Simon, R.: Method for assessing reproducibility of clustering patterns observed in analyses of microarray data. Bioinformatics 18, 1462–1469 (2002)

[9] Smolkin, M., Gosh, D.: Cluster stability scores for microarray data in cancer studies. BMC Bioinformatics 36 (2003)

[10] Bertoni, A., Valentini, G.: Randomized maps for assessing the reliability of patients clusters in DNA microarray data analyses. Artificial Intelligence in Medicine 37, 85–109 (2006)

[11] Bertoni, A., Valentini, G.: Model order selection for bio-molecular data clustering. BMC Bioinformatics 8 (2007)

[12] Achlioptas, D.: Database-friendly random projections: Johnson-lindenstrauss with binary coins. Journal of Comp. & Sys. Sci. 66, 671–687 (2003)

[13] Efron, B., Tibshirani, R.: An introduction to the Bootstrap. Chapman and Hall, New York (1993)

[14] Dudoit, S., Fridlyand, J.: Bagging to improve the accuracy of a clustering procedure. Bioinformatics 19, 1090–1099 (2003)

[15] Gentleman, R., et al.: Bioconductor: open software development for computational biology and bioinformatics. Genome Biology 5 (2004)

[16] Irizarry, R., Hobbs, B., Collin, F., Beazer-Barclay, Y., Antonellis, K., Scherf, U., Speed, T.: Exploration, normalization, and summaries of high density oligonucleotide array probe level data. Biostatistics 2, 249–264 (2003)

[17] Gautier, L., Cope, L., Bolstad, B., Irizarry, R.: Affy–analysis of affymetrix genechip data at the probe level. Bioinformatics 20, 307–315 (2004)

[18] The Gene Ontology Consortium: Gene ontology: tool for the unification of biology. Nature Genet. 25, 25–29 (2000)

[19] Khatri, P., Draghici, S.: Ontological analysis of gene expression data: current tools, limitations, and open problems. Bioinformatics 21, 3587–3595 (2005)

[20] Dopazo, J.: Functional interpretation of microarray experiments. OMICS 3 (2006)

Integrating Web Services and Intelligent Agents in Supply Chain for Securing Sensitive Messages

Esmiralda Moradian

Department of Information Science, Computer Science,
Uppsala University,
Box 513, SE-751 20, Uppsala, Sweden
Esmiralda.Moradian@dis.uu.se

Abstract. Security is a global issue for today's businesses that operate at the crossroads on multiple e-supply chains. Organizations increasingly use agent technologies and web services that allow assembling unique business processes. Businesses share information and manage electronic transactions with trading partners throughout the e-supply chains. The key factor in business success is decisions made on the basis of correct information that needs to be protected. In this paper, we present an approach of integrating web services and intelligent agents in supply chain in order to protect secret business information from being disclosed, modified and lost. To prevent loss and disclosure of sensitive information but also provide information integrity and availability, we apply meta-agents to monitor the software agents' actions and then direct the software agents' work.

Keywords: Information security, confidentiality, integrity, availability, intelligent agents, meta-agents, monitoring.

1 Introduction

Businesses share information and manage electronic transactions with trading partners throughout the e-supply chains. The key factor to business success is decisions made on the basis of correct information that needs to be protected. Security is essential to supply chain management, since relationships between all involved parties i.e., customers, suppliers and intermediaries (supplier's suppliers), are based on the flow of information and transactions. Internet technologies help organizations to reduce time and cost by enabling integration of different value chain activities through increased flow of information. Electronic communications gives the opportunity to companies to outsource core value chain activities and support activities. Interoperability between supply-chains is essential to organization in order to interact within multiple supply chains.

Supply chain management includes overseeing of information and finances as they flow in a process from buyer to seller across supply networks. To achieve maximum efficiency, e-business systems must be secured and protected from cyber-fraud, denial-of-service attacks, and the like [4]. One of the problems in e-supply chain is protecting the business information from the outside and at the same time providing availability of

I. Lovrek, R.J. Howlett, and L.C. Jain (Eds.): KES 2008, Part III, LNAI 5179, pp. 771–778, 2008.

information to authorized parties. Uninvited parties are attracted by valuable information flowing through the electronic supply chain. If one of the links snaps or is temporarily inactive consequences can be enormous [4]. Another problem is sharing information without disclosing secret information to competitor organizations.

Currently, traditional mechanisms such as firewalls, routers and proxies, are used to send message through network [15]. In order to effectively meet the demands of a secure supply chain in today's environment, a comprehensive and integrated security focus is required. Enhanced security must be balanced with efficient and synchronized flow of information. Sensitive messages need to be protected both on application and transport level. The goal of message monitoring is to provide secure message exchange. Security considerations like unauthorized access, unauthorized alteration of messages and man in the middle attacks that have arisen with the introduction of Web Services are identified by the WS-I Basic Security Profile. To prevent disclosure and modification of information but also secure flow through e-supply chain we use agent technology, together with web services technology. For successful composition of processes among business partners, information about adoption of web services security by trading partners is not enough. Organizations should have knowledge about message level security. We use meta-agents on top of software agents to monitor message transport throughout supply chain.

2 Related Work

Intelligent agents and web services have been used to secure e-business. One solution is to adopt agents for e-commerce in order to handle trust, security and legal issues. Fasli [3] discusses the use of intelligent agents in e-commerce and highlights risks that emanate from stealing information.

Another solution proposed by Atallah *et al.* [1] is design of Secure Supply-Chain Collaboration (SSCC) protocols. The authors discuss enabling supply-chain partners to cooperatively achieve desired system-wide goals without revealing the private information of any of the parties.

3 Web Services

Service oriented architecture permits different solutions for sharing the business logic between the different applications. Web services technologies are increasingly deployed in business organizations to achieve their system collaboration. Incorporated into business networks, web services can support both internal and external business operations including e-commerce, finance, manufacturing, supply-chain management and customer-relationship management [13].

Web services are self-contained, self-describing, modular applications that have open, Internet-oriented, standards-based interfaces designed to be used by other programs or applications rather than by humans [14]. A primary purpose of web services architecture is to integrate different web services and to open up the corporate network to the external world. SOAP (Simple Object Access Protocol), WSDL (Web Services Description Language) and UDDI (Universal Description Discovery and

Integration) are nowadays the core of web services. These are built by using XML and various Internet protocols, for example HTTP (Hyper Text Transfer Protocol), see Figure 1. The use of too many protocols can create security holes and openings for unauthorized parties that can gain access to the data within transactions and alter or steel secret information. Web services rely on message security that focuses on confidentiality and integrity [15]. The Web services follow request/response model of interaction and are stateless [13]. The services rely on SOAP that exchange data over networks. Using the SOAP protocol over HTTP, application sends a request to a service at a given URL. When the service receives the request, it processes the request, and returns a response. A company can be the provider of web services and also the consumer of other web services [13].

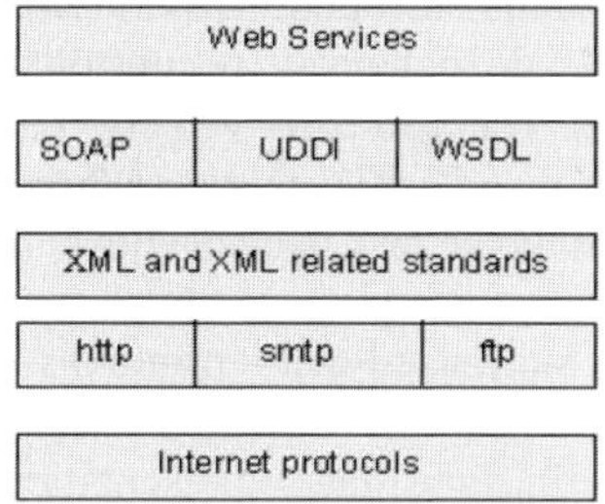

Fig. 1. Web services building blocks (Source: Hartman, B. et al., Mastering Web Services Security. pp. 27)

If any part of chain in network is compromised, the entire WS-security model breaks down [6]. Lack of understanding web services security risks prevent many organizations from implementing the services [2]. A shortcoming of Web services lies in its use of HTTP for message transport. The HTTP transport can use Secure Socket Layer (SSL) as a tunnel for securing message content. SSL is inadequate because of existing limitations [15]. To ensure the integrity of messages, XML encryption and XML digital signature are used today. Cryptography is used to protect sensitive data as it passes by on the wire. SOAP is designed to support more than one intermediary that can forward or reroute SOAP messages based upon information either in the SOAP header or in the HTTP header [15].

4 Software Agents

As an organization's supply chains become more complex, it becomes essential to control the flow of information of business partners. Using a multi-agent system is necessary due to complex environment of supply chains. Each software agent acts as a ground-level information collecting agent in order to monitor and control functions in supply chain in an efficient and ubiquitous way. The software agents work with one task at the time [18] in a straightforward manner where the environment is partially observable, deterministic, episodic, static and discrete.

In a partially observable environment, the agents cannot predict the behaviour of the environment, since the state is unclear [9]. In our system the agents can detect aspects that are relevant to the choice of action and have access to information that business partners agreed to share, like IP addresses, protocols, servers, routers and the all the components involved in the message passing. The agents follow the messages through supply chain, i.e., organisation's application server or router, where the route is determined by the protocol, at application-level and transport-level.

The agents work in an environment that is episodic, in which episodic defines the task [8]. The task environment is divided into atomic episodes, where each episode has an agent that performs a single task [18]. In our work, the agent is moving between a two nodes and when the agent has reached its end destination, it becomes a launching point for the next agent. Each agent "hands over" the message to the successor agent that can proceed the work. Handing over messages will continue until the message has reached its end destination, i.e. last node.

Static environment refers to the surrounding world that is unchanged during execution [8]. The software agents do not need to control the surrounding environment while determining the actions. The agents move between servers, routers and network components following the protocols. The protocol directs each package through the networks and the agent performs the task. The agents collect all the information available during its travel, like time, IP addresses and message digests. The information can easily be expanded or minimized with protocols, including other levels than transport and application [18].

A discrete environment defines the state representation of the environment, as well as, the way time is handled. The discrete environment also described perceptions and actions of the agents [8]. A discrete environment has a finite number of states and a discrete set of perceptions and actions. The actions of the agents follow the messages between the routers; meanwhile the agents record everything that happens. The agents perceive the environment in which the routers work and record the path between the routes [18]. This can ensure the confidentiality of information and avoid messages being wiretapped when information is transmitted on line.

By integrating agents and web services into supply chain, safe transmitting of valuable corporate information on Internet can be achieved, securing that the data transmitted on network is not stolen by hacker. The actions and the results from the software agents are monitored by an administrative meta-agent.

5 Meta-agents

Meta-agents act as meta-level information and collect ground-level software agents in a multi-agent system, which would be an efficient and ubiquitous way of monitoring and controlling the supply chain network. Meta-agents in our system can compare and combine information received from software agents, but the information and combinations will vary with time. Moreover meta-agents are capable to learning, negotiating and collaborating in proactive, social and adaptive manner. The main role of the meta-agents is to shift between strategies and make decisions. These strategies can be implemented in a single agent or a group of agents [9].

Meta-agents can be applied over software agents in a multi-agent system to keep track of multi-agents and control their behaviour while they move between nodes [10]. These meta-agents are not predefined but are created from the software agents and their actions. The meta-agents are built on the agents and used to inspect the behaviour of the agents when reaching a result, where the behaviour is due to the characteristics in the environment. The meta-agents can also perceive the reason for that result and show the status of the static and dynamic characteristics in the environment. The benefit of using meta-agents for multi-agent systems is the ability to find the fastest way between nodes under given circumstances, by taking several different approaches between the start and goal into account. [11]

In this paper, we use meta-agents on networks to secure information that passes through supply chain. While the software agents are moving around in the networks, the meta-agents need to consider the circumstances at a given time and act upon unexpected events. The meta-agent is superior to the software agents and is capable of monitoring and controlling the agents, and then evaluating them [18]. Each set of software agents that have performed and accomplished the task is encompassed by the meta-agents.

The meta-agents are autonomous, which means that the agents are to some extent capable of deciding what they need to do in order to satisfy their objectives [12]. Moreover, the meta-agents interact with software agents and other meta-agents to perform tasks. They cooperate with the software agents in the network by sending messages directly to the software agents but also via messages through other meta-agents. The communication with the software agents consists of gathering information from the software agents and giving commands that the software agents have to execute.

6 Securing Messages by Using Meta-agents

Enterprises and organisations use web services technology to manage transactions [13]. Web services adoption in e-Commerce can improve and simplify organizations' information interchange. The ability to solve complex problems increased the use of software agents in e-commerce. According to W3C [7] "…software agents are the running programs that *drive* web services — both to implement them and to access them" [7]. E-Commerce challenge is to improve information flow and provide correct information throughout supply chain [4]. Message integrity means that message did not change in transit, either by mistake or in purpose, i.e., transmitted information has not been altered. Integrating web services and software agents is necessary for securing e-supply chain. To prove the integrity of business information, to avoid that the data of trading partners being wiretapped or lost on its way of transmitting on Internet, and to establish a safe and reliable network we use meta-agents upon software agents.

To secure business messages and provide message integrity we propose the coordination of supply activities of an organization between trading partners by intelligent agents. We applied meta-agents over software agents to orchestrate the work of software agents and provide information symmetry in order to reduce inefficiency in managing supply chains network that can have impact on security. Software agents travel with SOAP message from buyer to the last supplier in the chain through intermediary suppliers and transport level routers. Software agents copy the routines automatically when the intermediary inserts a new URL [18]. Because of fear that secret information will leak to the competitor organizations (or between different departments within organization), which will take advantage of the information we propose the use of more than one meta-agent. These meta-agents can communicate with each other and handle over information, which is not classified as secret, like product description, price, time and amount as well as information about transport path. Thus, through collaboration meta-agents make decisions to achieve common desired goals without revealing secret information.

The supply chain depicted in Figure 2 is simplified for the sake of illustration. Software agents are noted as SA and meta-agents are noted as MA. We assume that buyer enterprise has three primary suppliers. Each of them also has three primary suppliers, where each of these has three primary suppliers.

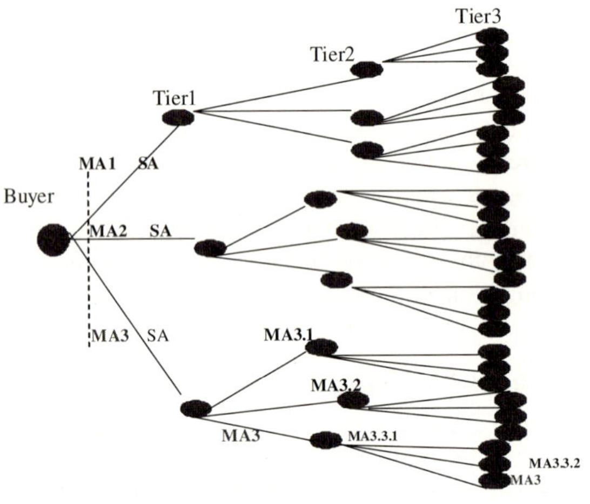

Fig. 2. Intelligent agents in multiple tire supply chain

In our example the process starts when particular request/order is made by buyer organisation. We assume that stock is empty. First a test message is send. The software program starts to execute and three software agents start travelling between buyer and suppliers on Tier1 through transport level where they check for availability of ordered product/service, price and amount. Three meta-agents are created and gather information from software agents, i.e., a meta-agent for each branch. These meta-agents collaborate with each other and make a decision how to proceed (by which branch) depending on information about time, amount and price. According to made decision one meta-agent (in our example MA3) continues its travel forward. The software agents between suppliers on Tier1 and Tier2 start execute and meta-agent is following along to the next node. For the first executed branch, the meta-agent is expanded (M3). For the rest of branches, a copy of the meta-agent is created for each branch (M3.1), (M3.2) and expanded with the branch. The procedure continues until agents reach the last node in supply chain. Meta-agents cooperate in decision making on every tier. Meta-agents have recorded all information about the nodes visited, which is sent back to the start node and becomes the path for the actual transaction. Second time real business message is send, which follows exactly the same way as the test message. If meta-agent discovers deviation in the path it can mean that an attack is on the way. Software agent must stop execution, recognize what the problem is and correctly respond to the problem.

The meta-agent can reroute the last SOAP message to another already tried path that is considered to be safe, or retry the complete path, when it operates on transport level. Meta-agents create reports with necessary information. In case of successful execution of all involved agents response information about message delivery can be returned as well as cheapest price and shortest time. Successful execution means that only reliable business partners' nodes were visited and that the order/business message was not altered by unauthorized parties, while moving at application level and transport level. In case of unsuccessful execution, such as, disclosure of business information by unauthorized parties or downtime of web server, the failure report will be generated by meta-agent with information about fault type/reason and time and place. Thus, using meta-agents upon software agents improve message security and eliminate or reduce complexity of supply chains.

In some senses, software agents can be compared with web crawlers, also known as web spiders or software robots, because of using the same technology. Different types of intelligent agents/software robots are used in e-business today. Search engines, for example, employ web crawlers/web spiders that actually are intelligent agents, which search the Web for different reasons [16]. In our system, intelligent agents perform focused crawling; i. e., agents search domains/IP addresses specified in the message. The starting point is determined by buyer. Meta-agents use the same technology as Web spy and packet sniffers, also known as a network analyzer or protocol analyzer, which are used in a multi platform environment for monitoring purposes [17, 19]. Our meta-agent monitors the nodes and coordinates software agents on application and transport level for message monitoring purposes, intrusion detection purposes, as well as, for web server availability purposes and, then, analyzes the results. To secure message on transport level meta-agent will analyze network traffic and detect intrusion attempts, monitor network usage, gather and report network statistics and filter suspect content from network traffic.

7 Conclusions and Further Work

We have presented some important issues within e-supply chain emphasized by the aim of solving security problems concerning sensitive business messages in supply chain networks. The result is the integration of software agents and meta-agents together with web services into e-supply chain, which improves efficiency of decision making and message security. In this paper, we have presented an approach to use meta-agents on top of software agents in multi-agent systems to decrease security problems. Integration of intelligent meta-agents with other web services technologies, protocols, and encryption technologies can be considered as a building block that will provide intelligent and secure information management throughout supply-chain. Current solution is still in its infancy: the software program is under development and the results are not available yet. We still belief that use of web services in conjunction with intelligent agents can improve security of e-supply chain management. The next step is to implement the current solution. We consider using multi-agent development platform, namely, iJADE (intelligent Java agent development environment), which supports various e-commerce applications.

References

1. Atallah, M.J., Elmongui, H.G., Deshpande, V., Schwarz, L.B.: Secure Supply chain protocols. In: IEEE International Conference on E-Commerce (CEC 2003) (2003)
2. Galbraith, B., Hankinsson, W., Hiotis, A., Janakiraman, M.D.V., Triverdi, R., Whitney, D.: Professional Web Services Security. Wrox Press (2002)
3. Fasli, M.: On agent technology for e-commerce: trust, security and legal issues. The Knowledge Engineering Review 22(1), 3–35 (2007)
4. Awad, E.M.: Electronic Commerce: From Vision to Fulfillment, 3rd edn. Prentice Hall, Englewood Cliffs (2006)
5. Papazoglou, M.P.: Agent-oriented technology in support of e-business. Communication of ACM 44(4), 71–77 (2001)
6. Rosenberg, J., Remy, D.: Securing Web Services with WS-security (2004)
7. Web services architecture. W3C working group note (2004), http://www.w3.org/TR/ws-arch/
8. Russell, S., Norvig, P.: Artificial Intelligence: A Modern Approach. Prentice-Hall, Englewood Cliffs (1995)
9. Apelkrans, M., Håkansson, A.: Information coordination using meta agents in information ligistics processes. In: KES 2008 (accepted, 2008)
10. Håkansson, A., Hartung, R.: Calculating optimal decision using Meta-level agents for Multi-Agents in Networks. In: Apolloni, B., Howlett, R.J., Jain, L. (eds.) KES 2007, Part I. LNCS (LNAI), vol. 4692, pp. 180–188. Springer, Heidelberg (2007)
11. Håkansson, A., Hartung, R.: Using Meta-Agents for Multi-Agents in Networks. In: Arabnia, H., et al. (eds.) Proceedings of The 2007 International Conference on Artificial Intelligence, ICAI 2007, vol. II, pp. 561–567. CSREA Press, U.S.A (2007), http://www.world-academy-of-science.org
12. Wooldridge, M.J. (ed.): Introduction To Multi-Agent Systems. John Wiley and Sons Ltd, Chichester (2002)
13. Moradian, E.: Web applications and Web services – similar and different attacks. MSc. Thesis. Uppsala University (2005)
14. Hartman, B., Flinn, D.J., Beznosov, K., Kawamoto, S.: Mastering Web Services Security. Wiley, Chichester (2003)
15. Papazoglou, M.: Web Services: Principles and Technology. Pearson Education, Essex England (2008)
16. Laudon, K.C., Traver, C.G.: E-Commerce business technology society. Pearson education, London (2008)
17. Thirukonda, M.M., Becker, S.A.: WebSpy: An Architecture for Monitoring Web Server Availability In a Multi-Platform Environment. Technical Report CS-2002-07
18. Moradian, E., Håkansson, A.: Approach to solving security problems using meta-agents in multi-agent system. In: Nguyen, N.T., Jo, G.S., Howlett, R.J., Jain, L.C. (eds.) KES-AMSTA 2008. LNCS (LNAI), vol. 4953, pp. 122–131. Springer, Heidelberg (2008)
19. Meehan, A., Manes, G., Davis, L., Hale, J., Shenoi, S.: Packet sniffing for automated chat room monitoring and evidence preservation. In: Proceedings of the 2001 IEEE Workshop of Information Assurance and Security (2001)

The User Centred Knowledge Model - t-UCK

Anne Håkansson

Department of Information Science, Computer Science,
Uppsala University, Uppsala, Sweden
`Anne.Hakansson@dis.uu.se`

Abstract. In knowledge engineering, modelling knowledge is the process of structuring knowledge before implementation. A crucial part of system development depends on the acquiring and structuring, since the quality of system's contents is of decisive importance for making good decisions. Models are needed to assure that all the required knowledge is present. However, the current models tend to be large and this makes it hard to get a grip on the knowledge presented by the model. Also, many models are difficult to use and the users have to be experts on the models before using them. To avoid these problems, we introduce the User-Centred Knowledge Model (t-UCK) for modelling knowledge. The model supports different users, i.e., domain experts, knowledge engineers and end-users, to model, implement, test, consult, and educate through the use of graphic representation and visualisation.

Keywords: Rule-based systems, knowledge modelling, visualisation, graphic modelling, user-centred modelling.

1 Introduction

A purpose with knowledge-intensive systems is to transfer different kinds of knowledge from the organization to end-users [15]. Within this transfer process, domain specific knowledge is transferred from a source through a system to eventually reach a receiver [14]. The system needs to communicate the knowledge to an end user, who to a greater extent can understand and utilize the knowledge. Developing these systems is known to be time-consuming and tedious, which would benefit from graphical modelling supporting the users to easily grasp and comprehend the contents of the system.

Some problems found in the knowledge engineering process are communication problems, language confusion and cultural differences. There are also problems with correct modelling, i.e., develop valid models and adjust the models to the knowledge representation. According to Alonso et al., [1] the hardest part of knowledge acquisition is developing models of the domain knowledge and reasoning strategy, since these are the main components in knowledge-based systems, and creating correct models of these is the key factor for success. Furthermore, not all knowledge can be captured or expressed in models. A problem with model adjustment is that each part of the contents of the model must fit the knowledge representation using a refinement process where each step of the refinement can be troublesome and lead to both logical and physical errors.

I. Lovrek, R.J. Howlett, and L.C. Jain (Eds.): KES 2008, Part III, LNAI 5179, pp. 779–787, 2008.

Several graphical maps, charts and tables have been created to avoid some of the problems with analysing and modelling. Most common are conceptual maps, inference networks, flowcharts and decision trees [4], which use concepts or objects and relationships connecting the concepts to present knowledge. Others are decision tables [7] and repertory grids [15], which present knowledge in tables. These tools are expressive but have difficulties with handling large knowledge bases. The models tend to grow fast and become large and complex, which makes it hard for the different actors to get a grip on the knowledge presented in the model. Moreover, some of the models are difficult to apply and they require skilled knowledge engineers to work with the models.

To minimize the problems with modelling and adjusting models, we have designed a model called the User-Centred Knowledge model, so called t-UCK. The model uses graphic representation and visualization to illustrate the domain knowledge, the reasoning strategy and other functionality. The purpose is to support the different users to model the relevant knowledge without having skills in representing the knowledge. As graphic representation support, we use a modified form of the Unified Modeling Language (UML) [2; 8].

2 Related Work

CommonKADS is a methodology for model-based knowledge engineering and is a well-known methodology for knowledge engineering in Europe. CommonKADS is used to build knowledge-based systems in an object-oriented fashion where KADS splits the overall model into other more specific models: organizational model, application model, task model, conceptual model and design model. For the models, CommonKADS uses different kinds of diagrams where some of them are UML diagrams [15]. However, one of problems is that CommonKADS requires expert skills in the method and this is the problem we try to tackle.

In our approach, we minimize interference of the knowledge engineers by supporting the different users while modelling the system. The model can be seen as a merge of the KADS conceptual model and design model because the transformation between the model and the code will be made automatically. To support different users, we use two different views of the system, a modelling view and a consulting view of the model. The modelling view supports developing the system whereas the consulting view is only for sessions. These views, share some of diagrams containing the relevant knowledge for both the developers and the end-users.

3 The User-Centred Knowledge Model

The user-centred knowledge model (t-UCK) uses knowledge transfer to pass on domain specific knowledge from a sender through a system to eventually reach a receiver. Graphic modelling of the knowledge is achieved by incorporating a conceptual model. The domain expert's expertise forms an integrated part of the system when the conceptual model is visualised at the user interface. This conceptual model must, therefore, be beneficial for the domain experts and the knowledge engineer and represent the domain experts' view of the domain and relevant domain knowledge. The model must also be beneficial for the end users using the domain knowledge and represent views of the domain supporting the end users. Since the different users' interests in the system are likely to differ, they have different views of the conceptual model, i.e., we have to cater for modelling and consulting. First and foremost, the modelling view supports the domain expert and the knowledge engineer whereas the consultation view supports the end users.

In t-UCK model, the conceptual design uses the notions of design model, conceptual model, and user model. The design model is held by the designer and is used to bridge the gap between the conceptual model and the users' models. The user model is the end user's model of the system, which is formed through interaction with the system and is usually built on the functionality of the system, the on-line help and documentation.

To conceptual design is incorporated into t-UCK to minimize the distinction between the users' intentions and the execution of these intentions in the system, realised by taking users' previous knowledge and experience into account. The essence of conceptual design is to take into account the users' mental models and the correspondence of these mental models to the actual artefact. The closer the users' models match the way in which the application actually works, the more successful the operations of the application for the user will be. The resulting user-centred knowledge model is presented in Figure 1.

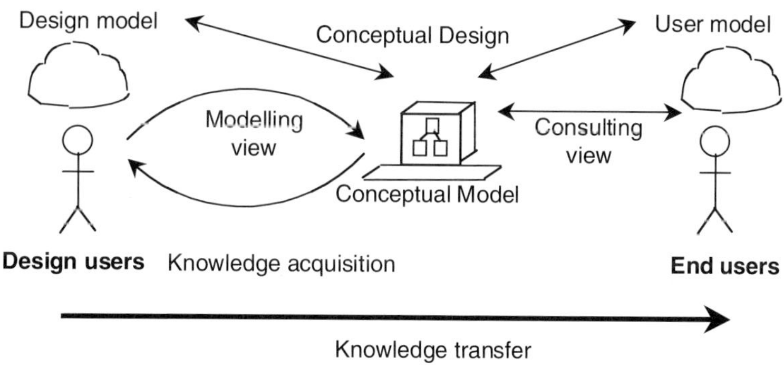

Fig. 1. The User-Centred Knowledge Model, t-UCK

The model is developed to support *knowledge transfer* [14] from design users to end users. The transfer is to structure the *design model* of the design users, to the *user model* of the end users via a *conceptual design* [12]. Introducing the term design developer and saying that this particular user is a middleman acting both as a user and designer should not be confused with the end user who only uses the final application. Thus, when developing a knowledge-based system using a tool, the more appropriate term for this user (or developer) is design developer or design user. The *conceptual model* [9; 1] on the other hand, is a framework for developing the system that will support clarifying the different terms but also correctly applying these terms. The design users have a *modelling view* of the conceptual model helping to develop the system. The end users have a *consulting view*, which support using the system for different sessions.

The *Knowledge acquisition* includes eliciting, modelling and representing the domain knowledge as a formally organized knowledge representation of the knowledge base. The acquisition is supported by the system without intermediary of a knowledge engineer. Instead, the domain experts should be able to work directly with the system, with some help from the engineer. Since these two different experts have different experiences and skills, their competence can be an asset during the implementation. Thus, ideally, there should be a co-operation between the knowledge engineer and the domain expert during the development.

3.1 Conceptual Model of the System

Conceptual modelling addresses the manner of inferring and controlling knowledge. The modelling is one of two activities for knowledge structuring, also called conceptualisation [1]. The other activity is the formal modelling of the system to address the domain knowledge to be implemented in the system and its implementation. A conceptual model can be regarded as "a general conceptual framework through which the functionality is presented" [11]. A good conceptual model should be effectively presented through the user interface.

In our approach, the conceptual model is used as a medium for representing the contents of the system to the design user and the end users. The design user uses the conceptual model as a framework to develop a system by modelling the domain knowledge, reasoning strategy, and other functionality. To support these tasks, the conceptual model is transparent and reflects the contents of the system. It supports the design users work with the system during development. The end users utilise the conceptual model during the consultation with the system. In this case, the model supports the end users' understanding of the strategy behind a conclusion. To support both the design users and the end users, the conceptual model must assist the users in building a mental model of the system's contents and of the domain.

The conceptual model is build to support a range of design users by clarifying the different terms in the system and helping the users in applying these correctly. Thus, terms such as questions are supported with attendant questions, production rules, consultation, inference mechanism and conclusions. The model's supportiveness is intended to show the manner in which these terms are related to each other and give an understanding of how they are related internally. For example, the contents of one rule and relations between rules form the knowledge in the knowledge base.

For knowledge modelling and consultation, the model permits entering new knowledge, but also browsing and editing the knowledge implemented, i.e., the contents of the system. The model also allows the searching and retrieving of specific parts of the knowledge base and sequencing the knowledge, which is entered by controlled and dialog-driven communication. Moreover, the model permits different users to control the behaviour of the system and informs them about the manner in which the system is utilising the domain knowledge. This knowledge is the reasoning strategy of the system and it has to correspond to domain problem solving. Thus, the system's interpretation of the domain knowledge follows the domain expert's problem solving to be advisory and educational within the domain. Furthermore, the model allows the design user to change the reasoning strategy of the system, i.e., it allows the execution order of the strategy during the knowledge modelling to be changed. The system usually uses one strategy, but sometimes it will use several strategies, e.g., deductive reasoning together with inductive reasoning. The system allows a switch from deductive to inductive reasoning and a combination of the two.

The other functionality defines what tasks can be performed in the system and determines how these tasks are performed. This functionality sometimes needs to be tailored for a specific problem. However, the already implemented functionality of the system may not be adequate to accomplish a specific task, so the user might have to insert new or change the existing functionality. Presenting programming language code textually is probably useless to programming novices and would be too time-consuming for the experts. Therefore, the model lets the design user utilise and change the functionality without having to learn the syntax.

The domain knowledge in the system must be of adequate quality. To check the adequacy, relations between parts in different sets have to be controlled for quality assurance, i.e., verification and validation must be conducted. One of the challenges is to illustrate, through the model, how different rules are related to each other, thereby enabling the design user to decide the quality, which is done by visualisation.

3.2 Modelling View and Consultation View

A knowledge-based system is interactive in nature because it must be able to form a dialogue between the users and itself [3]. The inputs and the outputs constitute the communication with the system; input is the knowledge required for modelling and for consultation; output reflects the contents of the system and presents it to the users. To support the communication in the interface, the system should provide high-quality graphical user interfaces with views and dynamic interactions.

To support many different users, several views of the contents of the system can be supplied and, in so doing, the system will constitute an intermediate link between different users. The most significant parts are the different users mental models. These users' models are the foundation for developing the system by using several different visual diagrams versus for utilising previously built systems. The former provides access to t-UCK for building the system and the latter provides access to operations of the system, which requires a clear distinction to be made between the knowledge engineer's interface and the end-user's interface.

The design user's interface is used for learning and designing the system under development. The end users' interface is used for learning and operating with the developed system. Since the design users and end users will use the system for different purposes, most of their views of the system will differ. The design user is interested in using the framework to insert domain knowledge, but also in the knowledge already inserted. This knowledge makes it possible for the design user to evaluate the contents in the system and judge whether the system draws the appropriate conclusions. Moreover, the design user is interested in the lack of knowledge to understand what additional knowledge needs to be inserted. The end user is interested in the conclusions presented by the system and the reasoning strategy adopted. Hence, this user is interested in the values and to understand the reasoning strategy used, but also in the rules that have been used to reach a specific conclusion. Moreover, the end users may be interested in the different sets of rules implemented in the system and the relation between them because this will enable them to elaborate on the reasoning and identify alternative conclusions. Because of the different tasks relevant for design user and end user, the interface is divided into two different parts, each dealing with a different perspective; called the modelling view and the consultation view.

The modelling view is the design user's user interface corresponding to the knowledge acquisition for the engineering and involves the interaction for modelling domain knowledge, as well as, changing reasoning strategies and other functionality. Knowledge modelling involves insertion, editing, and testing of the domain knowledge through different types of knowledge specifying what is generally true about the world; it is what the system knows or thinks about a domain. Procedural knowledge can be notified through procedures or simulations in a domain and is illustrated as objects in sequences simulating the inference mechanism [6]. Declarative knowledge can be announced as facts and semantic concepts, which are illustrated as templates for facts and rules [5] with concepts applied to the rules. Meta-knowledge is illustrated at different levels of abstraction in hierarchies and is represented as classifications of knowledge by utilising clusters of concepts to express statements about a domain [6]. Heuristic knowledge is also expressed as sequences of strategies and incorporates leaps when details are lacking while conducting the reasoning with heuristics. Moreover, structural knowledge is expressed as sequences by using concepts and relationships with a folding and unfolding effect for hiding currently irrelevant knowledge [5].

The design user must be able to handle the reasoning strategy of the system. This strategy is illustrated as a stepwise exposition of the parts involved in an interpretation used to draw a particular conclusion. If the view facilitates the changing of rules and facts during the exposition, it is possible to check the impact of using different sets of data. Moreover, the view expresses the current functionality and allows the user to make changes. This usually demands a considerable amount of knowledge about the programming language and the architecture of the system, but this can be supported by visual programming techniques. Visualisation of programming language code is illustrated in commonly encountered diagrams, e.g., state chart and activity diagrams.

Modelling also means providing and utilising explanation facilities. The explanation mechanism can be utilised by the design user during the development of the system to see a rule's involvement in a context. In addition, the design user should be able to control the design of the user interface of the system. Developing a user interface can be a difficult task, but it is an important one since the end users will form their decisions whether or not they like the system on the basis of the interface.

The "consultation view" is the end user's interface, which supports consultation, explanation facilities for questions, conclusions and reasoning strategies and on-line help. The end users' interaction consists of utilising the system to perform the tasks desired from the end users' point of view. During the consultation, each part involved in the consultation and interpretation is presented to the user. A specific part can be studied in the system to get information about when and in what content it is used, which can serve to clarify the system's reasoning process. For example, the system can give the end user information about the questions in advance, i.e., before answering the questions. Sometimes one question is narrower than another is and if the end user is aware of the subsequent questions, the user might have answered the earlier questions differently. Beside consultation, the interaction involves other options, such as consultation and fetching knowledge from an external source.

The explanation mechanism is important because it makes the system transparent, i.e., the end users will be aware of what kind of knowledge the system contains and the reasoning behind a conclusion. This can support the users to understand why they have to supply certain knowledge. The end user may be interested in checking a question's role in the consultation to find out why it is asked. Understanding the role of this question may simplify the comprehension of the reasoning process of the system by putting the question in an inference mechanism order. The end user may also be interested in checking a conclusion, i.e., in determining how the system reached a conclusion. The how explanation may also simplify the comprehension of the reasoning process by checking the facts and rules involved in the task. How-explanations can also simplify the control of the knowledge used by the inference mechanism. This is interesting if the end user is not satisfied with the solution the system has reached. The end user can check how knowledge has been used and decide what can be changed in the next consultation. The consultation view must also provide on-line help for the questions involved in a consultation [5]. Sometimes the alternative answers are not clear and, therefore, need to be explained before an answer can be given. Moreover, other options utilised by system are explained.

To support the end user while providing complementary domain knowledge, the system needs to be both useable and give manageable domain knowledge through an interactive end user interface. This is especially important if the domain expert is not at hand. If the end user is forced to use the system without any help from the domain expert or knowledge engineer, the questions and the alternative answers must be clear enough. It is important to phrase a question correctly if one is to get the proper information. Even the conclusions presented to the end user, must be comprehensible and be relevant to the users' tasks. Terms may have to be translated or be explained. Moreover, the system may explain how a conclusion is reached to clarify the domain expert's expertise and approach to problem solving to make a conclusion useable. It can also be used for learning matters.

3.3 Utilising Graphics for Modelling

A user interface of a system should be built on knowledge, habits, routines and the expectations people already have, as well as familiar objects, relationships, object attributes and actions [15]. Standard notions and well-known illustrative diagrams such as modified Unified Modeling Language (UML) are used. Moreover, the use of several graphical diagrams can support different users' mental models. T-UCK is providing the same domain knowledge, but allows users to view the system from the perspective relevant to them. Hence, the framework includes several complementary views of the domain knowledge.

Experienced users of graphical user interfaces tend to make fewer mistakes and feel less frustrated than novices [10]. In graphical interfaces, direct manipulation of visual objects and actions can replace a complex command-language syntax [15] even making it possible to visualize programming language code. The graphical objects can correspond to compositions of programming language components, which can simplify adding or changing the functionality in large systems. In conclusion, the conceptual model needs to be visualized by using different diagrams for different purposes. For this, we use UML's class diagrams, object diagrams, interaction diagrams, state-chart diagrams, activity diagrams, and packages.

In user-centred knowledge modelling activity, the conceptual model constitutes the graphical user interface of the knowledge-modelling tool and the system. This model is relevant to both the modelling view and the consulting view. The contents in these views are different, but at least both users, i.e., design user and end user, share one of the graphical diagrams. Hence, different users share the same view of the domain, but their purposes are different. Using several different views may bring the user closer to a system and vice versa and, thus, may provide a more user-centred approach. However, the users also utilise more "user-specific" diagrams. Using several different graphical views of the domain knowledge can provide a more user-centred modelling of the domain. Some of these different views may also support a more user-centred consultation with the system. The combination of views provides a means for making the system more transparent by viewing the domain knowledge and the reasoning strategies.

The use of multiple aspects allows the same model to be viewed in several ways to support a number of tasks and people [13]. Each view of the system presents the object in a different context where each object shows an operation in many ways. Objects require dividing complex views into a series of smaller less complex problems and separating the concerns into individual issues. Hereby, the complexity can be decomposed in a hierarchical way where components can be decomposed into various sub-components [13]. Strictly, hierarchical decomposition is well suited for structuring large amounts of complex information in one way. Concerns are likely to be addressed differently by various kinds of people when they are performing various tasks.

4 Conclusions and Further Work

In this paper, we presented the User-Centred Knowledge model, t-UCK. The model is a tool for developing all kinds of knowledge-intensive systems. It concentrates on supporting the design users and the end users to be involved in the development process and capture the relevant knowledge without having insight in knowledge representation. The model has been used for three years for several different knowledge-management system projects and has proved to be useful. Some examples of projects are modelling knowledge about museum objects, decision-making of assistance policies, acidification in lakes but also classification of mushrooms and water-weeds.

However, the model has some drawbacks. The views become very big very fast and need to be fully automated. The automation must better support hiding parts of the views that are not relevant for the moment. Thus, the parts that are not affected by the knowledge implemented for the moment. Moreover, the transformation from the views to the knowledge representation needs to be automated. To assure valid code, it needs to be a one-to-one translation.

References

1. Alonso, F., Fuertes, J.L., Martinez, L., Montes, C.: An incremental solution for developing knowledge-based software: its application to an expert system for isokinetics interpretation. Expert Systems with Applications 18, 165–184
2. Booch, G., Rumbaugh, J., Jacobson, I.: The Unified Modeling Language User Guide. Addison Wesley Longman, Inc., Amsterdam (1999)
3. Darlington, K.: The Essence of Expert Systems. Prentice Hall, England (2000)
4. Durkin, J.: Expert System Design and Development. Prentice Hall International Editions. Macmillian Publishing Company, New Jersey (1994)
5. Håkansson, A.: Graphic Representation and Visualisation as Modelling Support for the Knowledge Acquisition Process. Ph. D. Thesis Computer Sciences, Department of Information Science, University of Uppsala, Sweden (2004) ISBN 91-506-1727-3
6. Håkansson, A.: Transferring Problem Solving Strategies from the Expert to the End Users - Supporting understanding. In: Proceedings of 7th International Conference on Enterprise Information Systems, ICEIS 2005, vol. II, pp. 3–10. INSTICC, Portugal (2005) ISBN: 972-8865-19-8
7. Jackson, P.: Introduction to Expert Systems. Addison Wesley, Reading (1999)
8. Larman, C.: Applying UML and Patterns: An Introduction to Object-Oriented Analysis and Design and Iterative Development, 3rd edn. Prentice Hall, Englewood Cliffs (2004)
9. Luger, G., Stubblefield, W., Cummings, B.: Artificial Intelligence: Structures and Strategies for Complex Problem Solving. The Benjamin/Cummings Pub. Company, Inc. (1993)
10. Macaulay, L.: Human-Computer Interaction for Software Designers. International Thomson Publishing, UK (1995)
11. Mayhew, D.: Principles and Guidelines in Software User Interface Design. Prentice Hall Inc., New Jersey (1992)
12. Norman, D.A.: Cognitive engineering. In: Norman, D.A., Draper, S.W. (eds.) User Centred System Design: New Perspectives on Human-Computer Interaction, Hillsdale, Lawrence Erlbaum Associates, New Jersey (1986)
13. Robbins, J., Morley, D., Redmiles, D., Filatov, V., Kononov, D.: Visual Language Features Supporting Human-Human and Human-Computer Communication. In: IEEE Symposium on Visual Language, Bolder, Colorado (1996)
14. Sandahl, K.: Developing Knowledge Management Systems with an Active Expert Methodology. Dissertation No. 277, Linköping University, Sweden (1992)
15. Schreiber, G., Akkermans, H., Anjewierden, A., de Hoog, R., Shadbolt, N., Van de Velde, W., Wielinga, B.: Knowledge Engineering and Management – the CommonKADS Methodology. The MIT Press, Cambridge (2001)

Information Coordination Using Meta-agents in Information Logistics Processes

Mats Apelkrans and Anne Håkansson

Jönköping International Business School, Department of Informatics,
Jönköping, Sweden
`Mats.Apelkrans@jibs.hj.se`
Department of Information Science, Computer Science,
Uppsala University, Uppsala, Sweden
`Anne.Hakansson@dis.uu.se`

Abstract. In order to coordinate and deliver information in the right time and to the right place, theories from multi-agent systems and information logistics are combined. We use agents to support supply chain by searching for company specific information. Hence, there are a vast number of agents working at the Internet, simultaneously, which requires supervising agents. In this paper, we suggest using meta-agents to control the behaviour of a number of intelligent agents, where the meta-agents are working with coordination of the communication that takes place in a supply chain system. As an example, we look at a manufacturing company receiving orders on items from customers, which need to be produced. The handling of this distributed information flow can be thought of as an Information Logistics Processes and the similarities of the functioning of processes and intelligent agents' behaviour are illuminated.

Keywords: Intelligent agents, Meta-agents, Multi-agent systems, Information Logistics, Supply chain.

1 Introduction

The goal for manufacturing companies is to deliver the right products in the right time with good quality to be competitive. This requires good planning, optimized purchases and well functioning distribution channels. It also requires an efficient handling of the information flow in the company. One of the problems in Business Informatics is the coordination of the information flow between the manufacturing company, its customers, and its suppliers. The goal of this coordination is to find a more efficient product manufacturing process. Information coordination between the actors should speed up the information exchange and, hence, optimize the production cost. Our approach is to combine theories from Multi Agent Systems (MAS), meta-agents and Information Logistics in order to coordinate and deliver information at the right time and to the right place at an acceptable cost.

To enforce qualified information management and provide the right information at the right time, we apply a multi-task system using multi-agents performing these tasks in parallel, which can handle a large number of tasks faster. In this paper, we study

I. Lovrek, R.J. Howlett, and L.C. Jain (Eds.): KES 2008, Part III, LNAI 5179, pp. 788–798, 2008.

the information flow needed in a manufacturing company to fulfill their order stock. In the production process, a major issue is the double directed information flows, one from customers to company and another from company to its suppliers. This information can concern orders, requirements and production plans.

In our work, we use intelligent agents to take care of different tasks in the information flow. Each agent's behaviour can be thought of as an Information Logistics Process (ILP) handling information between several sources with given input and output. Still there is a need for coordination of the intelligent agents' performances in order to fulfill the global task of controlling a successful company. The solution is using meta-agents. The use of meta-agents on top of the information flow chart can monitor the action of the intelligent agents and be used to control the information passing between the agents. The meta-agents can provide the users with requested information. An examination of the behaviour of an intelligent agent shows that it can perform as an ILP. Hence, there will be a number of ILP calls. The ILP is managed in overall strategy by the use of meta-agents. Each ILP has to meet the demands on e.g. time, content, and presentation.

2   Related Work

The ILP approach has been used in a number of test cases. One example is handling of content management problems [1]. These ideas were further developed using visual modelling of the e-invoice process [2]. Moreover, the work concerning ERP configuration was of a more surveying perspective i.e. to choose the right ERP implementation for a given company, hopefully, lasting for a number of years [3]. In this paper we like to make the right decision in a short time (dynamical) perspective. Looking at the business process perspective many papers are written in order to optimize the work-flow in a manufacturing situation [14]. However, these contributions do not include ILP. Our test case is more on the decision side, i.e., choosing the right solution to supplier problems to the best cost and at right time.

3   Information Logistics

The concept of Information Logistics (IL) has been discussed for a number of years and a number of definitions have been established. A recent definition is [16]:

"The main objective of Information Logistics is optimized information provision and information flow. This is based on demands with respect to the content, the time of delivery, the location, the presentation and the quality of information. The scope can be a single person, a target group, a machine/facility or any kind of networked organization. The research field Information Logistics explores, develops and implements concepts, methods, technologies and solutions for the above mentioned purpose"

It is possible to consider IL as a process that manufactures an information product and, hence, to introduce thinking from the production area, like Just-In-Time (JIT). Information Logistics could be defined as the discipline [2]:

- that will supply the right information at the right time, in the right shape and at the right place
    - in a user-friendly way
    - with desired quality
    - to the lowest possible cost in where the final product is distributed by some kind of information carrier like paper, card, CD, Smart Card, Internet.

It is easy to pinpoint a number of differences between ordinary logistics and IL. The information supply in ordinary logistics is essential and often time consuming and expensive. Much effort must be taken in order to handle inbound logistics, checking deliveries or out-of-stock problems or damaged materials. In IL stock replenishment, there is no problem. It is just make a new copy of the information product.

Looking at Information Logistics from a process perspective, we can see that an information logistics process transforms a given input into some form of output. The input is some kind of fragmented information or knowledge description, which is derived from a so-called information supplier. This input information can be handled, either manually or automatically, by the system. The process output is an information product that becomes accessible and delivered to the information receiver who can make use of the information. This workflow is called the Information Logistics Process (ILP). The input to the ILP comes from the application and the receiver can be either a new ILP or a database or system user. From the input the ILP communicates with the knowledge base to produce a desired output, see Figure 1. We can request the system to send output information as messages.

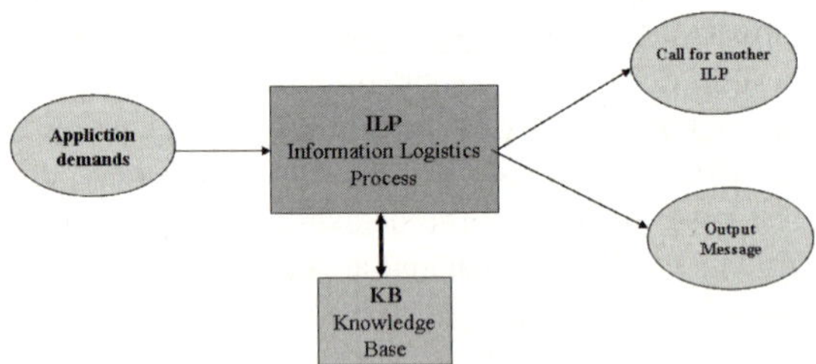

Fig. 1. The Information Logistic Process (ILP)

There are methods and discussions of ILP in the literature, where knowledge is applied in the ILP to support sending e-invoices between companies using external network [2] but also to automate configuration of enterprise systems [3].

The ILP processes are implemented with different methods found in the computer science area. Simple ILP processes are just straightforward database solutions; others need real-time machinery in order to deliver the information product at right time to the right place. Real-time components can, e. g., handle time management and communication details to facilitate the distribution of information to right place in right time [7]. Still others need knowledge management. A question is what can possibly be automated and what will still be contained in the dialogue between users of ILP and the ILP process. The ILP processes have to handle knowledge, or more properly, expressions of knowledge descriptions.

There are similarities between an ILP and the behaviour of an intelligent agent. Both concepts have the ability of communicate with others, produce messages in an intelligent way, and to send a message at right time. Hence, the use of intelligent agents offers a solution to the knowledge management tasks of an ILP.

4 Intelligent Agents and Meta Agents in Multi Agent Systems

Intelligent agents in multi-agent systems can be used to perform tasks in complex environments such as searching and retrieving information from sources and accessing services at the web. Some of the agents can work efficiently with commodity products, usually offering a list of products based on some criteria. Some of these agents can take actions on production process requirements and support assembling important information from several sources.

In logistics area, a multi-agent environment is needed since the tasks are too complex to be accomplished with one single agent. To our best knowledge, not many of the logistic agents work at the intranet and extranet, simultaneously, supporting information logistics process by supplying the process with external information. The extranet is extended to users outside the company and can support doing business with a pre-approved set of other companies over the network.

The agents must be capable of autonomous actions, situated in different environments [17] where some are intelligent and, thus, adaptable to the environment. The environment, i.e., the web, has some characteristics that the agents observe and act upon. The web is a vast network with over 100 million web sites where 74% are commercial sites or other sites operating at the .com. Hence, the environment is partially observable, stochastic, dynamic, continuous, and episodic.

In a partially observable environment only some of the information is known and therefore, can be a limitation. Environments such as the intranet and extranet are considered to be partially observable [17] because of its nature of continuously growth. However, the agents do not have to maintain their internal state to find satisfactory information. The agents, in our work, are working in the Information Logistics Process system to solve tasks, but also use the intranet and the extranet finding information. Either though we consider the environment to be partially observable, the agents must still able to find the significant information needed for finding solutions. Finding the necessary information corresponds to fully observable environment, where the agents can obtain complete, accurate and up-to-date information about the environment state [17]. The agents do not need to maintain any internal state to track the world and can easily achieve the task based on the information in the environment.

In a partially observable environment, the environment can appear to be stochastic [15]. The agent cannot predict the behaviour of the environment, as in real-world cases, since the state, which will result from performing an action, is unclear. Even though the environment is deterministic, there can be stochastic elements that randomly appear. A partially observable environment with stochastic elements is what will be expected for our agents. The agents have a task of searching and combining data but the information and combinations will vary with time, i.e., when actions these are performed.

A dynamic environment [17], the environment can change while agents are deliberating its contents [15], which is beyond the agents' control. In these dynamic environments, the agents need to interact with the environment and continuously check the surroundings to act properly. This characterises the intranet and the extranet since these change on a daily bases. However, the environment might be static over smaller time intervals, thus, remaining unchanged while the agents consider their course of action between states. Nonetheless, we will consider continuous changes, which will affect the agents and make them monitor changes during each task.

In continuous environments, there might be uncountable many states, arising from the continuous time problem [17]. Continuous time can be a problem for the intranet and the extranet agents because of the number of states and actions. This requires special treatment of the agents using an execution suspension to control the agents' performance. A common suspension of execution occurs when the agents have found information and returned with result. This limits the possibility of finding several solutions. A better solution in our system is an execution suspension for a short time interval followed by a resumption of the search.

In an episodic task environment [15] the choice of action in each episode depends on the episode itself. The agents, in our work, perform one task at the time while moving between commercial sites finding information. The task itself can be more complex, but before it is assigned to agents, the task is divided into smaller, single tasks, where each task is applied to an agent. In multi-agent systems for information logistics, a lot of intelligent agents will be used to perform a task, which in complex task environments will require a vast number of the agents. In these cases, meta-agents can be used to keep track of all the agents and their results produced from accomplishing tasks within acceptable time-range.

The concept of meta-agents is based on the idea of meta-reasoning. Meta-reasoning is a technique that supports the system to reason about its own operation. This reasoning can be used to reconstruct the agents' behaviour [13], but also to help in the interaction among agents. The meta-reasoning can be applied in the implementation strategies or plans to respond to requests [6].

Besides reasoning the meta-level agents (so called meta-agents) can plan actions and maintain individual agents' state information but also attempt to control future behaviour, classify conflicts and resolve these conflicts [5]. Another benefit with meta-agents is that they can work with intelligent agents in network to control and guide the performance [11] but also calculate the optimal decisions based on the intelligent agents performances in the time-dependent characteristics in the environment [10]. For our work, it is interesting to use meta-agents to combine different information pieces and calculate and select the best option for company on the basis of predefined features. Thus, evaluation is important since the results from the intelligent agents must be compared to find the best result that is most useful for the company.

Moreover, using strategies and plans to either look for information at the intranet or the extranet is important. There might be a mix of the searching, which arises when the information is not found at the intranet. Then the agents need to redirect the work to the extranet. The features of the meta-agents are applied in our system to support information flow in the information logistic process.

5   Information Flow Around a Manufacturing Company

The example company we like to study in this paper produces goods that are assembled from a number of items, some of them manufactured at place and some of them bought from different suppliers see Figure 2.

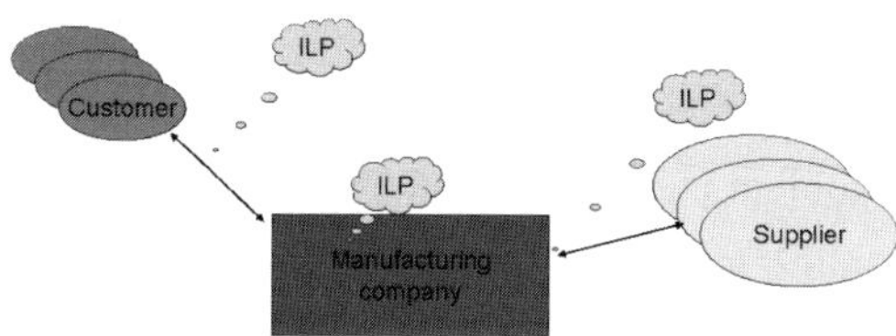

Fig. 2. The example company and its surroundings; ILP is Information Logistics Processes

As mentioned above, a lot of the information flow in the system consists of messages, some of them are input(s) to an ILP and others output from ILP directed towards a final goal or another ILP. Some examples of messages are:

- "Buy 25 pieces of item #1234 to best possible price with delivery not later then xx-xx-xx"
- "Take care of a new order of 30 pieces of item #2345 which should be delivered yy-yy-yy"
- "Check inventory for pieces of item#3456"
- "Tell customer ABC that his order is delayed 4 days"

Specialized intelligent agents that fulfill special tasks at right time can cover the behaviour of the ILPs. Thus, we can look at them as a simple information searching engines traveling between parts in the information flow systems, where the intelligent agents (IA) are solving the ILP information problems. This is, in fact, a multi-agent system and, in order to get useful solutions, their global effects have to be controlled by a number of meta-agents, which have the over all strategy for a controlled solution. Much of a manufacturing process is conducted by information stored in a database, hence the ILP have to communicate with a number of registers like:

Customer records
Supplier records
Load Planning
Returns Management
Product description with drawings
Price lists

The messages examples given above are not quite realistic since they are simplified in their task nature. In fact the intelligent agents have to penetrate much harder tasks like:

- Find out how to manufacture a product ordered from a customer
- Plan the master schedule
- Calculate Bills of material
- Produce Buying lists

For these different tasks both intelligent agents and meta-agents are applied. The intelligent agents work with information between customers, the manufacturing company and suppliers. The meta-agents work with monitoring and controlling the intelligent agents, and also accepting and combing information for optimal purchases for the manufacturing company.

6 The Multi-agent System

In the multi-agent system, the intelligent agents and the meta-agents have several different tasks. The intelligent agents' tasks are searching for suppliers of material and finding material according to pre-decided attributes for purchasing the raw material suiting optimal conditions. Additionally, the intelligent agents are keeping track of the material and the attributes. The meta-agents' tasks are combining information based on the results of the intelligent agents. Moreover, the meta-agents are planning manufacture, calculating the total cost of the manufactured product but also producing lists and plans. Furthermore, the meta-agents keep information about time delivery and records about the customers.

The process starts with customers ordering manufactured products. The orders are handled by the system, where each order becomes a planned action. The action launches the production line, which will begin to look for the parts and amount of pieces used in manufacturing. Each time a part is needed, the multi-agent system is invoked and the intelligent agents and meta-agents are activated.

As an example, let´s look at an order from customer ABC on 500 pieces of end-item #1. An intelligent agent finds the corresponding Bill of materials (BOM) which describe the "parts list" of components needed to complete a saleable end-item. The result can be 1500 pieces of item #1234, 1000 pieces of #2345 and finally 500 pieces of items #3456 and #4567 each.

In the case of suppliers, the intelligent agents search for information at the intranet or the extranet depending on the requested part. The intelligent agents move between the manufacturing company and the different suppliers, offering the parts for sale, and ask for information about the particular part. If these suppliers are subsidiaries, the agents are looking for information at the intranet, otherwise, at the extranet. Intranet is used when the agents need information from the warehouse of the manufacturing company.

When the intelligent agents identified all the suppliers offering the raw material, they need to find out the price for the pieces. Moreover, the agents need to know the amount of parts to order and find out if the suppliers have the parts in their stock or need to order these, in their turn. Also the price can change with the quantity and the agents must derive the price for the entire stock.

Finally, the delivery date is important. It is the supplier that sets this date but the agents need to consider the date and keep it stored. Moreover, if several customers want to buy the same product from the manufacturing company, the intelligent agents need to find enough raw materials for the product to cover the orders from all customers. This requires buying extra material from the suppliers and, moreover, might require buying the material from several different suppliers.

The results from the intelligent agents are compared to find the best options for the products. Moreover, commonly several raw material parts are needed to produce a product and all these pieces are important to assemble the final product and to calculate the delivery time to the customer. Since many intelligent agents are involved in the process, it is too complex to let them compare all the information about the raw material collected at the intranet and the extranet. Therefore, the meta-agents are applied to the intelligent agents.

The meta-agents correspond to goals of the intelligent agents. The intelligent agents work to find information on the parts and report to the meta-agents. Since many agents are involved, a strong part of the meta-agent needs to be a "reasoner" that evaluated alternative strategies.

As the information reaches the meta-agents, they need to select the best solutions based on the user-given attributes such as price, time, quantity, and quality. Hence, the user describes the attributes that the raw material needs to meet and the meta-agents have to combine these attributes to select the best option.

The calculation for the combination is complex. To evaluate the different attributes, they are given values within a range between 1-5, with the linguistic from of extremely significant to total irrelevant. The combination is then used together with the cost of the raw material to meet the conditions to the best possible price.

It has to be a balance of the cost, the material and the sacrifice of other aspects. For example, a cheap product may not be the best in quality and vice versa. Quality versus price raises questions, such as, is the quality of more important than the price and if so, what is the acceptable price for the material. Also time can be a cost for the company, which is why a slightly more expensive raw material can be acceptable for a particular production. Another aspect is quantity. If the company can buy a larger quantity, which might be more than needed for the manufacturing, they may get an offer that is price-worthy.

Except for keeping track of the amount of raw material, the meta-agents also need to keep track of the inventory of the pieces in the warehouses. The parts in the warehouse together with new material should barely cover the orders from the customers. Especially, for the companies that consider the importance of keeping the efficiency of lean production and reduce the stock in warehouses.

So far, we have only presented the process of finding information about suppliers for one part of raw material. However, in manufacturing, usually several parts are involved. The meta-agent needs to be the controlling agent, which holds information about all the parts that are being assembled. Thus, they need to direct the tasks of finding different suppliers for these parts to the intelligent agents. Then, from the results of intelligent agents, the meta-agents need to combine all parts to assemble the product and calculate the total cost and delivery time for the manufacturing. This combination is also used for planning the manufacturing.

To accomplish the search of all parts needed for manufacturing the product, information about the parts is essential. There are many ways to represent this information, such as, bills of material, frames and rules. For example: when the goal is to build a widget, we need to have a part-whole relation to describe the parts of the widget. Note, that a widget can be either information or a physical thing and that some widgets may have multiple possible parts. Either, it is possible to buy a set of parts to build a subassembly or buy the finished assembly.

Another important roll of meta-agents is to keep record of the customers. The meta-agents keep track of the information about the quantity of the manufactured products, the customer ordered. They also need to the information about requested delivery time of each product. Finding the raw material for all products, planning the manufacturing and calculate the delivery is a scheduling task for the meta-agents.

In the case of changes, such as the manufacturing, e.g., a broken machine that needs repairing, or the customer made changes in the order, the meta-agents need to react. The meta-agents have to take necessary steps to handle the situation, and depending on the problem, there will be different options. If they are machine problems, the meta-agents need to recalculate the delivery time. In the case of changed orders, they may redo the complete process from searching for the parts needed to planning manufacturing and scheduling the delivery. Another task for the meta-agents and Multi-agent system is to check the producing capacity. If a customer, for example, extends the order with 50 more pieces, the meta-agents have to check if the producer has sufficient capacity for the additional quantity. Capacity concerns both personnel and machines, and a suggested solution may be to delay delivery time a number of days.

6.1 Presenting the Information

An important part of the multi-agent system is to present the result to the users. For this we use similar diagrams of UML, more precisely sequence diagrams and collaboration diagrams [4; 12].

The purpose of using sequence diagrams is that it suits well when illustrating static information of knowledge bases [8], which can also be used for production systems and show strategies [9]. Collaboration diagrams on the other hand are good at illustrating dynamic phenomena, as inserted data can affect executing order [8].

The sequence diagrams are used to show the parts found in the manufactured product. These diagrams are static in the sense that it is always the same parts assembled to constitute the product.

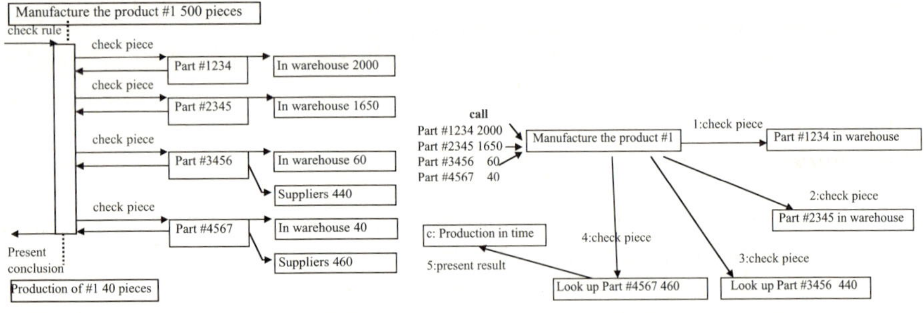

Fig. 3. Sequence diagram and collaboration diagrams illustrate the manufacturing chain

The sequence diagram and collaboration diagrams are used illustrate the manufacturing chain by showing the raw material and time aspect. The sequence diagram in Figure 5, illustrates the parts that are needed to manufacture the product #1. Moreover, the diagram shows where the parts can be found, that is either in the warehouse or at suppliers. In this case, the part #1234 and the part #2345 are in the stock already. The part #3456 and the part #4567 are not in the warehouse and need to be supplied for. The amount of pieces to produce is presented in the diagram. The order is 500 pieces, "Manufacture the product #1 500 pieces. Then to the left in the figure, the pieces found in the warehouse is presented but also how many must be purchased. Finally, the conclusion is presented, which is the amount of parts that can be produced directly from the pieces found in the warehouse.

The collaboration diagram illustrates how the manufacturing chain looks for parts to be used for manufacturing the product. If all the pieces are present, the product can be delivered in time. This is also the result presented in the diagram. This is a simplified diagram but yet it illustrates what is needed to assembly the product and what is found in the warehouses. The call to the diagram is two different parts provided by external suppliers. The square with look up pieces is to search for the pieces in the warehouse.

7 Conclusions and Further Work

In this paper, we have presented an approach using meta-agents to control the behaviour of a number of intelligent agents. The meta-agents work with the coordination of the communication that takes place in a supply chain system. The intelligent agents perform tasks between the supplier companies, searching for information about the products, price and quantity. The meta-agents combine information from the different agents to find the optimal raw material for manufacturing the product. The meta-agents can also schedule the time for production.

The current multi-agent systems search for environment characteristics while moving in the network and build meta-agents from the agents' result. Next step is to compute the combination of the result and schedule production time.

Further research has do be done on efficiency measures, and quality measures as well. Logistics should be a value adding and quality raising process. Hopefully this can be true also for Information Logistics. Our current work holds limitations that we will be addressed in the future. Some shortcomings have been identified: (*i*) problems with representation of complicated business situations, (*ii*) a tool for automating our approach with a highly interactive design.

References

1. Apelkrans, M., Braf, E.: Content Management: An Integrated Approach of Information Logistics and Knowledge Management. In: Proceedings of the Is One World conference, Las Vegas (2004)
2. Apelkrans, M., Håkansson, A.: Visual knowledge modeling of an Information Logistics Process - A case study. In: ICICKM 2005, 2nd International Conference on Intellectual Capital, Knowledge Management and Organisational Learning, Dubai, Förenta Arab Emiraten (2005)
3. Apelkrans, M., Håkansson, A.: Enterprise systems Configuration as an Information Logistics Process - A Case Study. In: Cardoso, J., et al. (eds.) Proceedings of 9th International Conference on Enterprise Information Systems, ICEIS 2007, pp. 212–220. INSTICC, Portugal (2007)

4. Booch, G., Rumbaugh, J., Jaconson, I.: The Unified Modeling Language User Guide. Addison Wesley Longman, Inc., Amsterdam (1999)
5. Chelberg, D., Welch, L., Lakshmikumar, A., Gillen, M., Zhou, Q.: Meta-Reasoning For a Distributed Agent Architecture (2000), `http://zen.ece.ohiou.edu/~robocup/papers/HTML/SSST/SSST.html`
6. Costantini, S.: Meta-reasoning: a survey. In: Kakas, A., Sadri, F. (eds.) Computational Logic: From Logic Programming into the Future: Special volume in honour of Bob Kowalski, Springer, Berlin (2002)
7. Frauenhofer Institute (December 10, 2006), `http://www.isst.fhg.de/englisch/download/34868_I-Log-4-Seiter-engl-2.pdf`
8. Håkansson, A.: UML as an approach to Modelling Knowledge in Rule-based Systems. In: ES 2001 The Twenty-first SGES International Conference on Knowledge Based Systems and Applied Artificial Intelligence, December 10–12, 2001. Peterhouse College, Cambridge (2001)
9. Håkansson, A.: Transferring Problem Solving Strategies from the Expert to the End Users - Supporting understanding. In: Proceedings of 7th International Conference on Enterprise Information Systems, ICEIS 2005, vol. II, pp. 3–10. INSTICC, Portugal (2005)
10. Håkansson, A., Hartung, R.: Using Meta-Agents for Multi-Agents in Networks. In: Arabnia, H., et al. (eds.) Proceedings of The 2007 International Conference on Artificial Intelligence, ICAI 2007. U.S.A., vol. II, pp. 561–567. CSREA Press, USA (2007)
11. Håkansson, A., Hartung, R.: Calculating optimal decision using Meta-level agents for Multi-Agents in Networks. In: Apolloni, B., Howlett, R.J., Jain, L. (eds.) KES 2007, Part I. LNCS (LNAI), vol. 4692, pp. 180–188. Springer, Heidelberg (2007)
12. Jacobson, I., Booch, G., Rumbaugh, J.: The Unified Software Development Process. Addison Wesley, USA (1999)
13. Pechoucek, M., Stepánková, O., Marík, V., Bárta, J.: Abstract Architecture for Meta-reasoning in Multi-agent Systems. In: Mařík, V., Müller, J.P., Pěchouček, M. (eds.) CEEMAS 2003. LNCS (LNAI), vol. 2691, p. 84. Springer, Heidelberg (2003)
14. Reijers, H., Limam Mansar, S.: Best practices in business process redesign: An overview and qualitative evaluation of successful redesign heuristics. Omega: The International Journal of Management Science 33(4), 283–306 (2005)
15. Russell, S., Norvig, P.: Artificial Intelligence: A Modern Approach. Prentice Hall, Upper Saddle River (2003)
16. Sandkuhl, K.: Information logistics in networked Organisations Issues, Concepts and Applications. In: Cardoso, J., et al. (eds.) Proceedings of 9th International Conference on Enterprise Information Systems, ICEIS 2007, pp. 23–30. INSTICC, Portugal (2007) ISBN: 978-972-8865-90-0
17. Wooldridge, M.: An Introduction to MultiAgent Systems. John Wiley & Sons Ltd, Chichester (2002)

Ontology for Enterprise Modeling

Ronald L. Hartung, Jay Ramanathan, and Joe Bolinger

Chair of Computer Science,
Franklin University, Columbus, Ohio, USA
Hartung@franklin.edu
CETI, Dept. of Computer Science and Engineering, The Ohio State University,
Columbus Ohio USA
jayram@cse.ohio-state.edu

Abstract. Enterprise modeling has a history of related research that has advanced high-level concepts that do not map well to implementation-oriented models like UML. Here acknowledging that the enterprise and business context impacts implementation requirements for adaptive services, we introduce the notation for representing the Adaptive Complex Enterprises through the Requirements-Execution-Delivery interaction or primitive. This simple primitive allows complex models to be developed that treats business processes at all levels and underlying data processing systems as well as human processing systems in a holistic fashion. In this paper we introduce the ontology underlying the Requirements-Execution-Delivery interactions, thus allowing Requirements-Execution-Delivery to be precisely specified as a single primitive for modeling, execution, and monitoring the behavior of Adaptive Complex Enterprises. We show how this takes the concepts developed in artificial intelligence and applies it into business activity monitoring.

1 Introduction

We begin by introducing and illustrating the underlying enterprise modeling concepts, To do this we use an easy-to-understand illustrative example based on racing. Winning the famous Indy 500 today involves several team cars that race together ensuring the lead car wins for the whole team. In addition numerous suppliers and experts collaborate to address the most challenging tasks. Some of the resources are shared – for example the fuel trucks. The enterprise modeling challenge is understood by contemplating the underlying many-to-many *interactions*, many of which are determined dynamically. Similarly the teams in a complex enterprise must execute agile strategies that require minimal coordination to achieve overall goals.

For modeling the dynamic interactions between inputs, outputs, goals and the stakeholder value of achieved over a period of time, we introduce a very basic and widely applicable ontology for customer–provider service interactions. This *interaction* abstraction can be assembled into a network structure. The resulting Adaptive Complex Enterprise (or ACE) structure captures metrics critical to the various stakeholders. This representation and method have been applied for sense-respond strategic planning [6, 10, 11] and process improvement for mixed-mode (i.e. human and IT) processes in a health-care environment [11]. We have also extended it to include linking business goals to operations [12].

I. Lovrek, R.J. Howlett, and L.C. Jain (Eds.): KES 2008, Part III, LNAI 5179, pp. 799–807, 2008.

In this paper we introduce the underlying ontology and show that this dynamic and precise representation can also become the basis for a variety of monitoring and intelligence gathering methods.

2 Related Work

There isa range of interdisciplinary work that is currently not integrated for holistic enterprise modeling. The rationale behind any enterprise modeling activity is the achievement of enterprise goals [17]. To this end, several goal modeling and specification techniques such as KAOS [5], Goal-Based Requirements Analysis Method (GBRAM) [1], [16], and the NFR Framework [4], have been proposed in support of requirements engineering and related activities such as representation of motivations, elaboration, consistency and completeness checking, evaluation of alternatives, and evolution. They provide a systematic, comprehensive and pragmatic notation and approach to building quality into enterprise models and software systems. In these frameworks, non-functional requirements (design requirements such as scalability, security) that describe how a system should deliver its function are captured in softgoals that may be decomposed or refined and then operationalized by design elements. In paticular, [8] which proposes an integrated top-down and bottom-up approach to identifying misaligned goals and planning organizational change, [7] which describes a method for associating goals with justifications and [9] which uses goal-models to guide the adaptation of pervasive systems to changes in the environment. The NFR Framework initially focused on system requirements, but has been extended [14, 19] to connect enterprise architecture goals to system architecture goals, and to connect to enterprise security goals. The customer-provider interaction model itself was introduced early on by Flores and Winograd [15].

The interaction (RED) model presented here defines structural and control relationships between the elements of the complex system.. The interfaces define the objects that flow between the transactions. In order to extend this model, ontology is an appropriate representation system. Ontology is in wide use with a variety of applications areas and for a number of reasons documented in [3] [13]. An ontology also provides a useful reasoning basis using frame and description logics[2].

3 Enterprise Modeling

The interaction model is used for enterprise modeling.The interaction occurs between the inputs (and roles) of the business that are engaged in producing the outputs based on targets to meet goals(s), see figure 1A. We use the term *role* to reflect an input that *provides a typical service*. We use the term *goal* to mean a statement of intent whose satisfaction requires performance contributions due to interactions. The inputs might include all types of physical and electronic services provided within the underlying infrastructure of the organization. Next, the interaction in A can be used to represent Pit Stop interactions, as illustrated in figure 1B. Firstly, the Pit Stop representation sequences two interactions called 'Refuel car' and 'Systems test'. These associate all the inputs (the car etc.), the roles (the driver, the pit crew, the equipment)

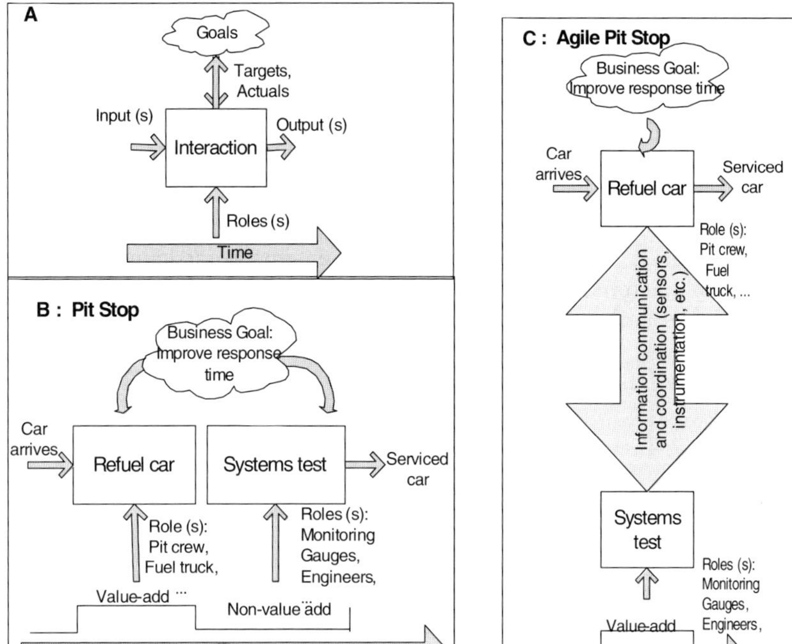

Fig. 1. The interaction concept between inputs controls and outputs for a Pit Stop

to the business goals. Secondly, note that the Pit Stop representation has a notion of time and *value-add*. Here we can quickly see that the from the customer's (driver in this case) perspective, the systems test does not add value.

The improved Agile Pit Stop process made up of *parallel* activities, as seen in figure 1C. Agility is achieved through improved communication and coordination between the different roles (refueling team, test team etc.). The underlying technology includes sensors, gauges, and other sources information that allow the engineers to monitor and test as the car is being refueled. In other words, the communication is based on sophisticated instrumentation providing instantaneous visibility into the most current conditions of the car. Also, then non-value-add time is eliminated.

3.1 RED Interaction Perspectives

RED expands the conceptualization of an interaction to include the following perspectives:

- Customer organization that requests the interaction and
- Provider organization that services the request using the input services in the *infra-structure[i]* needed for the execution of the interaction.

Thus, as illustrated in figure 2,an interaction executes due to a *Request* from a customer - an external customer, or an internal customer on behalf of an external one.

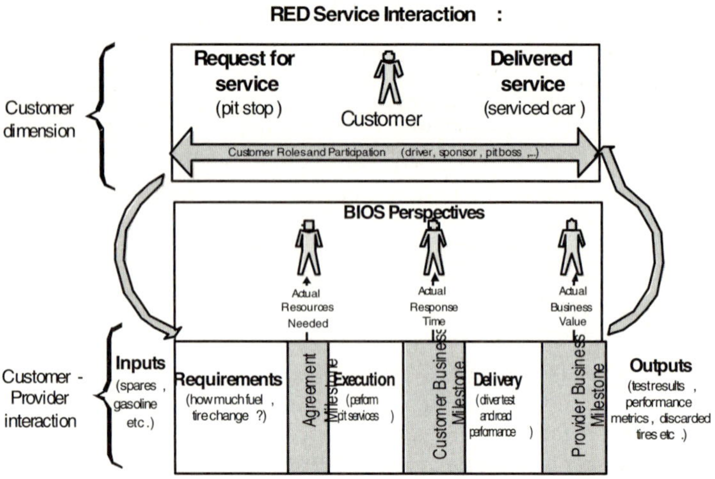

Fig. 2. Customer-provider service interaction

The objective of the RED interaction structure is to make explicit all the perspectives and points of metrics collection. This is achieved by the layers in the RED interaction:

- *Customer perspective:* A *primary* interaction begins with the *request event* (of a particular type) within the end-customer's environment and ends with the related *service deliverable*(s) that adds value to that environment. Many customer roles might be involved during the interactions (e.g. race car driver, sponsor, etc.). During the interaction, there maybe many customer roles that review and participate as the service progresses.

- *RED Conceptual perspective:* During the course of the interaction explicit *transitions* (represented as ⤶) occur between the *milestones* - Requirements, Execution, and Delivery (also called R, E, and D steps or just RED). The request is examined and the provider negotiates and understands the service requirements. Then both the customer and provider agree to proceed based on a formal/informal proposal or 'agreement milestone'. For example, here the fuel provider would agree on the types of fuel combinations that could potentially be provided, based on future conditions. Next, transitioning from Requirements to Execution step, the provider assigns inputs needed to service the request. During the interaction execution, the service[ii]completed and *create value* that is delivered in the customer's environment, thus completing the next milestone. The final milestone is when the provider delivers and is compensated. In this Pit Stop, compensating for the fueling and system check services is included.

A service interaction can be simple as in the example. Or it can be end-to-end like order fulfillment or the incident resolution. Or it can be a complex 'project'. In the latter case, interactions often use sub-interactions, also initiated with a request.

The RED transitions are important because they are achieved using provider inputs and services. For each service *Request type* (e.g. complaints, standard service, non-routine service) handled by the organization, there is a RED service interaction template with required inputs needed to execute the interaction.

Note: It is useful to compare an activity and an interaction. In the enhanced interaction semantics, we take a structured view of the interaction between the customer-provider perspectives, the Request and Deliverables perspective, infrastructure roles and resources perspective, and milestones that occur over time. The metrics is measured at milestones.

4 Ontology for RED

In order to implement and operationalize the RED interaction within an IT infrastructure, an ontology is required to precisely define the operations and the data within. The ontology representation can be almost any of the currently available forms. In this paper we use a simple text form for easier readability.The RED interaction itself can be described in the ontology or separate from the ontology. RED is described in three main classes of entities - operation, data object and RED interactions.

Data objects can take a number of forms, depending on the level of modeling attempted. There are some data objects that are given by user input or requests. There are also data objects that represent the delivered entities, whatever they may be. It is likely that a system will also have some form of internal data objects. These can be information passed inside the system or they could be parts to be assembled into the final delivered object. It should be noted, RED can be used for modeling physical objects and equally well RED may model software or information objects. The separation of these types can useful.

There are three ways to deal with the data objects in the *recursive invocation* of RED interactions. One is to allow internal input objects to be used in any role. A second approach is to allow multiple inheritance so that an internal object can also be a requirement or delivery object. The third would be to provide transformations between types. In this model, we allow multiple inheritances. Many ontologists disapprove of multiple inheritance, and the authors tend to agree. However, limiting multiple inheritance to this one case does not produce harm to the ontology. When building a model, the ontology will be populated with data objects under these main types. The modeler can build a hierarchy of sub types as seems most useful to describe the domain being modeled.

The data objects can have properties, treated as name-value pairs. A single data object can have a set of properties that describe options or detailed specifications for the data object. One can also choose to use separate types without properties. The reason to choose is a question of what will best represent the domain. Using a very large number of types can be difficult to understand and may be easier to work with if the number of types are reduced and differentiated by properties. Similarly too many properties can make the use of the object more difficult. The modeler will also find that two kinds of properties are useful. The properties can represent requirements, resources and outputs (delievery objects) used in the transactions. The other kind will be properties that describe the value aspects of the transactions.

This paper will use the following formal notation. This is used as s simpler alternative to the XML format of OWL. The data object definition can be expressed as follows:

TypeName :: parentTypeName {

    propertyName: propertyype

    ….

}

Properties can be given a default value. The syntax for that is to add an assignment operation with the default value to the end of the declaration, e.g. "propertyName: propertyType := value".

Beside the properties specific to the domain, the metrics, or value aspects, for the RED model are included. For a requirements object, these include lead time, reliability, price and risk. These are not mandatory, but when they are present, we can apply axioms as constraints.Delivery objects can have properties for the value of the deliverable (measured from the customers/requestors view), value of the overall interaction to the business, and defects detected.

Internal objects collect a number of properties. These include response time of the entire operation, span and wait time of underlying agents, capacity of resources used cost of resources, and early alert conditions.

As the RED interaction has three parts, there are three types of operations. There are requirements, execution and delivery operations, or R, E, and D. In the ontology these are sub types of the part object. The basic operation has a set of properties to define the operation. The description can be very complex or simple as best supports the model. In the simplest case, the operation will have a set of input data objects and a set of output data objects. In addition, the operation has a time value, the time that the real world equivalent will require in execution.

A more complex option is to define functions to help define the operation. These will be used to express the production of an output type by allowing the properties of the output objects to be set based on computations on the property values of the input objects.

While it is possible to extend this definition, this is enough for most systems and simplicity argues to keep the computations simple, substituting multiple operations rather than a few very complex operations. The functions act more as property value constraints and should be used to clarify model.

An operation is is an entity with for groups of properties. Input and output specify data objects required for the operation. A set of functions define the operation semantics and a separate set of functions define the execution time.

The specification of a function is a set of expressions. The expressions are in the form of assignments to properties of the output data objects. Property values are referenced by dataTypeName.propertyName. For aggregated types, apply the standard indexing operation, dataTypeName.propertyName[i]. We notate assignment with the usual ":=" operator. The operations used in the expression can be largely whatever the modeler desires. In addition, it is useful to allow iteration and if operations as part of the specification. The iteration takes the form of "forall" to iterate over a set of aggregated property values. If operations take the form of "if" condition "then" expression "else" expression.

The execution time can be a simple constant. However, the expression syntax can be used to compute a time value based on data object properties. If an expression is used, no assignment operator is required.

As mentioned above, there are three parts, R, E, and D. Each part has a definition of the operations required to complete its actions, a specification of the required input data objects, and the metrics to be collected. The input data to each part is specified as a set of data objects. The input data is considered globally available to any action contained in that part of the RED interaction.

The actions are composed of two basic types, operations or other RED interactions. These are organized as a set of sequences, each sequence is considered to operate in parallel with the other sequences. A sequence is a series of either operations or interactions. The actions in a sequence are executed sequentially.

The binding of arguments to the operations is not explicitly specified in the sequence. RED interactions get treated a little differently.When an action is executed, it can use any data object in the global pool, as defined by the data types in the operation's input. Any output data object produced is placed in a set of data objects local for the sequence, a local set. That set of objects is available for any subsequent operation or interaction in the sequence.

Binding data objects to a interaction is done by building a set of requirement data objects from the available global and sequence local pools. The required set of data objects will be found in the definition of the RED interaction.

Notation is simply: RED TranactionName { dataTypeName,....}
In order to validate a model, a set of axioms is used to specify constraints.

1. Top-level RED interactions, those invoked by external requests, can only produce delivery data objects.
2. Top-level RED interactions can only accept requirements data objects as inputs.
3. All data objects passed into any part of a RED interaction must be used as arguments to the operations or passed as arguments to another RED interaction.Properties of data objects produced by operations or RED interactions must have their values set, either by a default value or by the operations.
4. Lead time is the total of all time results for each operation. This requires the detection of parallel paths and applying a maximum function be used to compute to the results of the parallel paths.
5. Price is a simple accumulation of the price of each sub interaction.
6. Wait time is like the lead-time, accumulated over the interactions. But there are two possible axioms to choose. One is to look at total wait time over all the interactions. This can be a measure of idle time in the system. Second, one can compute it with the same axiom as lead-time. This gives a view of wait time delays in the process.
7. Reliability requires a more complex set of axioms. There is a large body of work regarding reliability calculations.

Other axioms can be added to the above seven axioms, adding specific axioms for the problem domain under consideration. The axioms are central to value chain computations. The axioms will be used to express relationships on the values within the domain and can drive computation of these values for the stakeholders.

5 Applications and Conclusions

We have shown that the REDs allow us to trace the internal value-chain commitments and value-add contributions from the perspective of Business, IT Infrastructure, Operations, Strategy (BIOS) stakeholders. The typical commitments are:

$R \rightleftharpoons E$:Customer-provider agreement (service level agreements, proposals, quotes, contracts, verbal commitments etc.) on the requirements of the deliverable. The provider agrees to provide services and invest in and maintain the infrastructure. The customer agrees to compensate for the services. Metrics from the perspective *of the customer* include 'Lead time, Reliability, Price, Risk'.

$\tilde{E} \rightleftharpoons D$:Provider used service resources (shown as $\tilde{\ }$) to execute and created customer value and contributed to strategyas agreed to. Here, in order to meet the SLA agreements, secondary agreements with *provider agents* that perform the services during execution are put into place.   These take the form of operational level agreements, procurement contracts, verbal instructions, work instructions, job descriptions, and prescribed processes etc. The metrics from the provider perspective include 'Response time of the entire execution, Span time and wait time for each of the underlying agents used, Capacity of each resource used, Which resources were used and for how long, Cost of each resource used, Delivery such as when, where (location), Early alert conditions, such as long wait times'.

$D \triangle$Closure:Here we identify metrics both for continuous Improvement and the business. Note that at this point we are closing the interaction by delivering within the customer's environment. We can identify the following metrics 'Value of the deliverable and interaction closure within the customer's environment in monetary or relative terms; Value of the interaction to the business in monetary or relative terms; Defects detected (time, resources, etc)'.

Thus, the RED structure goes beyond typical project and task management to become an interaction structure with precise semantics that can actually be executed (manually or electronically) to provide a conceptual layer integrating business and infrastructure use. The result is a conceptual performance structure for real-time monitoring and predictive capabilities.By monitoring the RED execution we look for a variety of symptoms. For example, if an interaction is queued at execution too long, we can infer that the committed resources are not available. Another example is that if an interaction has moved into execution, then the business value being achieved is less at risk. If an interaction never moves to execution, like a no-bid decision, the cause has to be examined. And so on.

Our future work is to apply this framework in real examples, and present implementations resulting from these applications, along with performance data – such as reductions in interaction costs.

References

1. Anton, A.I.: Goal-based requirements analysis. In: Proceedings of the 2nd International Conference on Requirements Engineering (ICRE 1996) (1996)
2. Brachman, R.J., Levesque, H.J.: Structured Representations. In: Knowledge Representation and Reasoning, ch. 9. Elsevier, Amsterdam (2004)

3. Chandrasekaran, B., Jsephson, J.R., Benjamins, V.R.: What Are Ontologies and Why Do We Need Them. IEEE Expert Systems, 1094–7167 (1999)
4. Chung, L., Nixon, B.A., Yu, E., Mylopoulos, J.: Non-Functional Requirements in Software Engineering. Kluwer Academic Publishers, Boston (2000)
5. Dardenne, A., van Lamsweerde, A., Fickas, S.: Goal-directed requirements acquisition. Science of Computer Programming 20(1-2), 350 (1993)
6. Haeckel, S.H.: The Adaptive Enterprise: Creating and Leading Sense-And-Respond Organizations. Harvard Business School Press (1999)
7. Jureta, I.J., Faulkner, S., Schobbens, P.: Justifying Goal Models. In: 14th IEEE International Requirements Engineering Conference (RE 2006) (2006)
8. Kavakli, E., Loucopoulos, P.: Experiences With Goal-Oriented Modeling of Organizational Change. IEEE Interactions On Systems, Man, And Cybernetics Part C: Applications and Reviews 36(2) (March 2006)
9. Mei, L., Easterbrook, S.: Evaluating User-centric Adaptation with Goal Models. In: First International Workshop on Software Engineering for Pervasive Computing Applications, Systems, and Environments (May 2007)
10. Ramanathan, J.: Fractal Architecture for the Adaptive Complex Enterprise. Communications of the ACM 48(5) (May 2005); Special issue: Adaptive Complex Enterprises
11. Ramanathan, J., Ramnath, R., Glassgow, R.: An Adaptive Complex Enterprise Framework for Lean Information Technology-Enabled Services Delivery, OSU-CISRC-5/07-TR37. Computers in Industrial Engineering Journal (in submission)
12. Ramnath, R., Ramanathan, J.: Integrating Goal Modeling and Execution in Adaptive Complex Enterprises. In: Proceedings of the Symposium for Applied Computing, Organizational Engineering Track, Fortaleza, Ceara, Brazil (2008)
13. Sowa, J.F.: Knowledge Representation Logical. In: Philosophical and Computational Foundations. Brooks/Cole (2000)
14. Subramanian, N., Chung, L.: An NFR-Based Framework for Establishing Traceability between Enterprise Architectures and System Architectures. In: Seventh ACIS International Conference on Software Engineering, Artificial Intelligence, Networking, and Parallel/Distributed Computing (2006)
15. Winograd, T., Flores, F.: Understanding Computers and Cognition - A New Foundation for Design. Addison Wesley Publishing Inc., Reading (1987)
16. Yu, E.S.K., Mylopoulos, J.: An actor dependency model of organizational work: with application to business process reengineering. In: Proceedings of the Conference on Organizational Computing systems (December 1993)
17. Zachman, J.A.: A Framework for Information System Architecture. IBM Systems Journal 26(3), 276–292 (1987)

Author Index

Printing: Mercedes-Druck, Berlin
Binding: Stein+Lehmann, Berlin

Lecture Notes in Artificial Intelligence (LNAI)

Vol. 5221: B. Nordström, A. Ranta (Eds.), Advances in Natural Language Processing. XII, 512 pages. 2008.

Vol. 5208: H. Prendinger, J. Lester, M. Ishizuka (Eds.), Intelligent Virtual Agents. XVII, 557 pages. 2008.

Vol. 5195: A. Armando, P. Baumgartner, G. Dowek (Eds.), Automated Reasoning. XII, 556 pages. 2008.

Vol. 5179: I. Lovrek, R.J. Howlett, L.C. Jain (Eds.), Knowledge-Based Intelligent Information and Engineering Systems, Part III. XXXVI, 817 pages. 2008.

Vol. 5178: I. Lovrek, R.J. Howlett, L.C. Jain (Eds.), Knowledge-Based Intelligent Information and Engineering Systems, Part II. XXXVIII, 1043 pages. 2008.

Vol. 5177: I. Lovrek, R.J. Howlett, L.C. Jain (Eds.), Knowledge-Based Intelligent Information and Engineering Systems, Part I. LVI, 781 pages. 2008.

Vol. 5144: S. Autexier, J. Campbell, J. Rubio, V. Sorge, M. Suzuki, F. Wiedijk (Eds.), Intelligent Computer Mathematics. XIV, 600 pages. 2008.

Vol. 5118: M. Dastani, A. El Fallah Seghrouchni, J. Leite, P. Torroni (Eds.), Languages, Methodologies and Development Tools for Multi-Agent Systems. X, 279 pages. 2008.

Vol. 5113: P. Eklund, O. Haemmerlé (Eds.), Conceptual Structures: Knowledge Visualization and Reasoning. X, 311 pages. 2008.

Vol. 5110: W. Hodges, R. de Queiroz (Eds.), Logic, Language, Information and Computation. VIII, 313 pages. 2008.

Vol. 5108: P. Perner, O. Salvetti (Eds.), Advances in Mass Data Analysis of Images and Signals in Medicine, Biotechnology, Chemistry and Food Industry. X, 173 pages. 2008.

Vol. 5097: L. Rutkowski, R. Tadeusiewicz, L.A. Zadeh, J.M. Zurada (Eds.), Artificial Intelligence and Soft Computing – ICAISC 2008. XVI, 1269 pages. 2008.

Vol. 5078: E. André, L. Dybkjær, W. Minker, H. Neumann, R. Pieraccini, M. Weber (Eds.), Perception in Multimodal Dialogue Systems. X, 311 pages. 2008.

Vol. 5077: P. Perner (Ed.), Advances in Data Mining. XI, 428 pages. 2008.

Vol. 5076: R. van der Meyden, L. van der Torre (Eds.), Deontic Logic in Computer Science. X, 279 pages. 2008.

Vol. 5064: L. Prevost, S. Marinai, F. Schwenker (Eds.), Artificial Neural Networks in Pattern Recognition. IX, 318 pages. 2008.

Vol. 5049: D. Weyns, S.A. Brueckner, Y. Demazeau (Eds.), Engineering Environment-Mediated Multi-Agent Systems. X, 297 pages. 2008.

Vol. 5043: N. Jamali, P. Scerri, T. Sugawara (Eds.), Massively Multi-Agent Technology. XII, 191 pages. 2008.

Vol. 5040: M. Asada, J.C.T. Hallam, J.-A. Meyer, J. Tani (Eds.), From Animals to Animats 10. XIII, 530 pages. 2008.

Vol. 5032: S. Bergler (Ed.), Advances in Artificial Intelligence. XI, 382 pages. 2008.

Vol. 5027: N.T. Nguyen, L. Borzemski, A. Grzech, M. Ali (Eds.), New Frontiers in Applied Artificial Intelligence. XVIII, 879 pages. 2008.

Vol. 5012: T. Washio, E. Suzuki, K.M. Ting, A. Inokuchi (Eds.), Advances in Knowledge Discovery and Data Mining. XXIV, 1102 pages. 2008.

Vol. 5009: G. Wang, T. Li, J.W. Grzymala-Busse, D. Miao, A. Skowron, Y. Yao (Eds.), Rough Sets and Knowledge Technology. XVIII, 765 pages. 2008.

Vol. 5003: L. Antunes, M. Paolucci, E. Norling (Eds.), Multi-Agent-Based Simulation VIII. IX, 141 pages. 2008.

Vol. 5001: U. Visser, F. Ribeiro, T. Ohashi, F. Dellaert (Eds.), RoboCup 2007: Robot Soccer World Cup XI. XIV, 566 pages. 2008.

Vol. 4999: L. Maicher, L.M. Garshol (Eds.), Scaling Topic Maps. XI, 253 pages. 2008.

Vol. 4994: A. An, S. Matwin, Z.W. Raś, D. Ślęzak (Eds.), Foundations of Intelligent Systems. XVII, 653 pages. 2008.

Vol. 4953: N.T. Nguyen, G.S. Jo, R.J. Howlett, L.C. Jain (Eds.), Agent and Multi-Agent Systems: Technologies and Applications. XX, 909 pages. 2008

Vol. 4946: I. Rahwan, S. Parsons, C. Reed (Eds.), Argumentation in Multi-Agent Systems. X, 235 pages. 2008.

Vol. 4944: Z.W. Raś, S. Tsumoto, D.A. Zighed (Eds.), Mining Complex Data. X, 265 pages. 2008.

Vol. 4938: T. Tokunaga, A. Ortega (Eds.), Large-Scale Knowledge Resources. IX, 367 pages. 2008.

Vol. 4933: R. Medina, S. Obiedkov (Eds.), Formal Concept Analysis. XII, 325 pages. 2008.

Vol. 4930: I. Wachsmuth, G. Knoblich (Eds.), Modeling Communication with Robots and Virtual Humans. X, 337 pages. 2008.

Vol. 4929: M. Helmert, Understanding Planning Tasks. XIV, 270 pages. 2008.

Vol. 4924: D. Riaño (Ed.), Knowledge Management for Health Care Procedures. X, 161 pages. 2008.

Vol. 4923: S.B. Yahia, E.M. Nguifo, R. Belohlavek (Eds.), Concept Lattices and Their Applications. XII, 283 pages. 2008.

Vol. 4914: K. Satoh, A. Inokuchi, K. Nagao, T. Kawamura (Eds.), New Frontiers in Artificial Intelligence. X, 404 pages. 2008.

Vol. 4911: L. De Raedt, P. Frasconi, K. Kersting, S. Muggleton (Eds.), Probabilistic Inductive Logic Programming. VIII, 341 pages. 2008.

Vol. 4908: M. Dastani, A. El Fallah Seghrouchni, A. Ricci, M. Winikoff (Eds.), Programming Multi-Agent Systems. XII, 267 pages. 2008.

Vol. 4898: M. Kolp, B. Henderson-Sellers, H. Mouratidis, A. Garcia, A.K. Ghose, P. Bresciani (Eds.), Agent-Oriented Information Systems IV. X, 292 pages. 2008.

Vol. 4897: M. Baldoni, T.C. Son, M.B. van Riemsdijk, M. Winikoff (Eds.), Declarative Agent Languages and Technologies V. X, 245 pages. 2008.

Vol. 4894: H. Blockeel, J. Ramon, J. Shavlik, P. Tadepalli (Eds.), Inductive Logic Programming. XI, 307 pages. 2008.

Vol. 4885: M. Chetouani, A. Hussain, B. Gas, M. Milgram, J.-L. Zarader (Eds.), Advances in Nonlinear Speech Processing. XI, 284 pages. 2007.

Vol. 4874: J. Neves, M.F. Santos, J.M. Machado (Eds.), Progress in Artificial Intelligence. XVIII, 704 pages. 2007.

Vol. 4870: J.S. Sichman, J. Padget, S. Ossowski, P. Noriega (Eds.), Coordination, Organizations, Institutions, and Norms in Agent Systems III. XII, 331 pages. 2008.

Vol. 4869: F. Botana, T. Recio (Eds.), Automated Deduction in Geometry. X, 213 pages. 2007.

Vol. 4865: K. Tuyls, A. Nowe, Z. Guessoum, D. Kudenko (Eds.), Adaptive Agents and Multi-Agent Systems III. VIII, 255 pages. 2008.

Vol. 4850: M. Lungarella, F. Iida, J.C. Bongard, R. Pfeifer (Eds.), 50 Years of Artificial Intelligence. X, 399 pages. 2007.

Vol. 4845: N. Zhong, J. Liu, Y. Yao, J. Wu, S. Lu, K. Li (Eds.), Web Intelligence Meets Brain Informatics. XI, 516 pages. 2007.

Vol. 4840: L. Paletta, E. Rome (Eds.), Attention in Cognitive Systems. XI, 497 pages. 2007.

Vol. 4830: M.A. Orgun, J. Thornton (Eds.), AI 2007: Advances in Artificial Intelligence. XIX, 841 pages. 2007.

Vol. 4828: M. Randall, H.A. Abbass, J. Wiles (Eds.), Progress in Artificial Life. XII, 402 pages. 2007.

Vol. 4827: A. Gelbukh, Á.F. Kuri Morales (Eds.), MICAI 2007: Advances in Artificial Intelligence. XXIV, 1234 pages. 2007.

Vol. 4826: P. Perner, O. Salvetti (Eds.), Advances in Mass Data Analysis of Signals and Images in Medicine, Biotechnology and Chemistry. X, 183 pages. 2007.

Vol. 4819: T. Washio, Z.-H. Zhou, J.Z. Huang, X. Hu, J. Li, C. Xie, J. He, D. Zou, K.-C. Li, M.M. Freire (Eds.), Emerging Technologies in Knowledge Discovery and Data Mining. XIV, 675 pages. 2007.

Vol. 4811: O. Nasraoui, M. Spiliopoulou, J. Srivastava, B. Mobasher, B. Masand (Eds.), Advances in Web Mining and Web Usage Analysis. XII, 247 pages. 2007.

Vol. 4798: Z. Zhang, J.H. Siekmann (Eds.), Knowledge Science, Engineering and Management. XVI, 669 pages. 2007.

Vol. 4795: F. Schilder, G. Katz, J. Pustejovsky (Eds.), Annotating, Extracting and Reasoning about Time and Events. VII, 141 pages. 2007.

Vol. 4790: N. Dershowitz, A. Voronkov (Eds.), Logic for Programming, Artificial Intelligence, and Reasoning. XIII, 562 pages. 2007.

Vol. 4788: D. Borrajo, L. Castillo, J.M. Corchado (Eds.), Current Topics in Artificial Intelligence. XI, 280 pages. 2007.

Vol. 4775: A. Esposito, M. Faundez-Zanuy, E. Keller, M. Marinaro (Eds.), Verbal and Nonverbal Communication Behaviours. XII, 325 pages. 2007.

Vol. 4772: H. Prade, V.S. Subrahmanian (Eds.), Scalable Uncertainty Management. X, 277 pages. 2007.

Vol. 4766: N. Maudet, S. Parsons, I. Rahwan (Eds.), Argumentation in Multi-Agent Systems. XII, 211 pages. 2007.

Vol. 4760: E. Rome, J. Hertzberg, G. Dorffner (Eds.), Towards Affordance-Based Robot Control. IX, 211 pages. 2008.

Vol. 4755: V. Corruble, M. Takeda, E. Suzuki (Eds.), Discovery Science. XI, 298 pages. 2007.

Vol. 4754: M. Hutter, R.A. Servedio, E. Takimoto (Eds.), Algorithmic Learning Theory. XI, 403 pages. 2007.

Vol. 4737: B. Berendt, A. Hotho, D. Mladenic, G. Semeraro (Eds.), From Web to Social Web: Discovering and Deploying User and Content Profiles. XI, 161 pages. 2007.

Vol. 4733: R. Basili, M.T. Pazienza (Eds.), AI*IA 2007: Artificial Intelligence and Human-Oriented Computing. XVII, 858 pages. 2007.

Vol. 4724: K. Mellouli (Ed.), Symbolic and Quantitative Approaches to Reasoning with Uncertainty. XV, 914 pages. 2007.

Vol. 4722: C. Pelachaud, J.-C. Martin, E. André, G. Chollet, K. Karpouzis, D. Pelé (Eds.), Intelligent Virtual Agents. XV, 425 pages. 2007.

Vol. 4720: B. Konev, F. Wolter (Eds.), Frontiers of Combining Systems. X, 283 pages. 2007.

Vol. 4702: J.N. Kok, J. Koronacki, R. Lopez de Mantaras, S. Matwin, D. Mladenič, A. Skowron (Eds.), Knowledge Discovery in Databases: PKDD 2007. XXIV, 640 pages. 2007.

Vol. 4701: J.N. Kok, J. Koronacki, R. Lopez de Mantaras, S. Matwin, D. Mladenič, A. Skowron (Eds.), Machine Learning: ECML 2007. XXII, 809 pages. 2007.

Vol. 4696: H.-D. Burkhard, G. Lindemann, R. Verbrugge, L.Z. Varga (Eds.), Multi-Agent Systems and Applications V. XIII, 350 pages. 2007.

Vol. 4694: B. Apolloni, R.J. Howlett, L. Jain (Eds.), Knowledge-Based Intelligent Information and Engineering Systems, Part III. XXIX, 1126 pages. 2007.

Vol. 4693: B. Apolloni, R.J. Howlett, L. Jain (Eds.), Knowledge-Based Intelligent Information and Engineering Systems, Part II. XXXII, 1380 pages. 2007.